Bayesian Statistics and
Its Applications

Bayesian Statistics and Its Applications

Edited by

Satyanshu K. Upadhyay

Umesh Singh

Department of Statistics, Banaras Hindu University
Varanasi-221 005, India

Dipak K. Dey

Department of Statistics, University of Connecticut
Storrs, CT 06269-4120, USA

Tunbridge Wells, UK

Anamaya Publishers
New Delhi

A catalogue record for the book is available from the British Library

ISBN 10 1-905740-00-X
ISBN 13 978-1-905740-00-0

Copublished by
Anshan Limited
The Control Centre, 11 Little Mount Sion
Tunbridge Wells, Kent TN1 1YS
United Kingdom
Tel.: +44 (0) 1892 557767
Fax: +44 (0) 1892 530358
Email: info@anshan.co.uk
Website: www.anshan.co.uk

Sold and distributed in all the countries, except
India, Pakistan, Sri Lanka, Bangladesh and Nepal by
Anshan Limited, The Control Centre, 11 Little Mount Sion
Tunbridge Wells, Kent TN1 1YS, United Kingdom

In India, Pakistan, Sri Lanka, Bangladesh and Nepal by
Anamaya Publishers, F-154/2, Lado Sarai, New Delhi-110 030, India
Email: anamayapub@vsnl.net

Printed in India.

To

My Parents

–Satyanshu K. Upadhyay

My Mother

–Umesh Singh

My Mother

–Dipak K. Dey

Preface

Bayesian world is fast changing. In the last two to three decades, the subject has acquired immense height and it has penetrated almost every area including those where the application of statistics appeared to be a remote possibility. Now whether we talk of communication technology, biotechnology, official statistics, sample surveys, data mining, pattern recognition, reliability, image processing, legal proceeding or criminology, the subject has emerged as an important paradigm to deal successfully with all sorts of possibilities and that too with often astonishing results.

This volume provides both theoretical and practical insights of the subject with a detailed up-to-date material on various aspects. It primarily aims at two important objectives. First, it is expected to provide a thorough background material for theoreticians and second, it is aimed to give a variety of application areas where the subject is successfully serving the needs of applied statisticians and practitioners.

The volume consists of 33 chapters. The main topics covered include biostatistics, econometrics, reliability, image analysis, Bayesian computation, neural networking, prior elicitation, objective Bayesian methodologies, role of randomization in Bayes analysis, spatial data anlaysis, Bayesian nonparametrics, etc.

The contributions given in the volume have been refereed and accordingly modified so as to provide the most up-to-date material on the concerned topics. A few of the chapters are exhaustive in the sense that they have been written with a proper introduction of background material so that the readers may understand the insights of the subject even if they are not much familiar with the concerned topic. We are thankful to our contributors who have taken care of this important aspect while writing and finalizing their contributions.

All the chapters included in the volume focus on Bayesian methodologies but each is self-contained and independent. There may be topics closely related to each other, but we feel that there is no need to maintain any kind of sequencing for reading the chapters. The chapters have been arranged purely in alphabetical order according to the last name of the first author and, in no way, it should be taken to mean any preferential order.

The graphs and pictures are the important components for understanding a few of the chapters. We feel important to mention that every attempt has been made to reproduce the graphs and pictures exactly in the format that have been provided to us, however, minute and little variations cannot be denied.

We express our thankfulness to Ashok Bansal, James Berger, José Bernardo, David Draper, Alan Gelfand, Edward George, John Geweke, Jayanta Ghosh, Malay Ghosh, Prem Goel, Kanti Mardia, Peter Müller, Anthony O'Hagan, Adrian Smith and Arnold Zellner, to name a few. It is the constant encouragement and guidance provided by them that enabled us to take up and finalize this important task. The present volume is a physical insignia of their inspiration, which will be quite helpful for the future researches using Bayesian paradigm.

We fail in our duties if we do not express our sincere indebtedness to our referees, who were quite critical and unbiased in giving their opinion. We do realize that in spite of their busy schedule they offered us every support in timely commenting on various manuscripts. Undoubtedly, it is the joint endeavor of the contributors and the referees that emerged in the form of such an important and significant volume for the Bayesian world.

We thankfully acknowledge the Rights Manager, Elsevier, for his permission to reproduce the article "Theoretical and Applied Bayesian Information Processing" by Arnold Zellner in a slightly modified form. This article entitled as "Some Aspects of the History of Bayesian Information Processing" was due to appear in the 'Journal of Econometrics' in 2005.

We express our sincere thanks to each and everyone who was associated with us directly or indirectly while the work was in progress. The list is certainly too lengthy to be exhaustive but we would like to give special mention to Anuradha, Asha, Rita, Geetika, Vertika, Abhishek, Tanaya, Debosri, Shashi Kant and Satya Prakash among others. Our students at BHU, especially Meena, Bhaswati, Puja, Shruti, Archana and Ashutosh, deserve special mention, who were not only critical, but always provided concrete suggestions while the editing work was in progress. Undoubtedly, they helped us improve the final form of the volume. And last, but not the least, it is our foremost duty to express our thankfulness to M/s Anamaya Publishers, New Delhi and M/s Anshan Limited, UK, for making the volume available to world audience.

SATYANSHU K. UPADHYAY
UMESH SINGH
DIPAK K. DEY

Contents

Bayesian Statistics and Its Applications
Edited by S.K. Upadhyay, U. Singh and D.K. Dey
Anamaya Publishers, New Delhi, India

Two-Fold Spatial Zero-Inflated Models for Analysing Isopod Settlement Patterns

Deepak K. Agarwal

Member, Technical Staff, AT&T Shannon Labs, Florham Park, NJ 07932-0791, USA

Abstract

We consider the problem of modeling association in count data that have excess number of zeroes and are spatially correlated after adjusting for all known covariate effects. Such problems are commonplace in ecology where the goal is to understand species settlement patterns. A discrete random variable is zero inflated if with probability p it is exactly zero and with probability $1 - p$ it follows a standard discrete distribution $\pi(.|\theta)$ like Poisson, binomial etc. Regression to explain variation in p and θ as functions of covariates was introduced in Lambert (1992). In an earlier paper Agarwal, Gelfand and Citron-Pousty (2002) model spatial association in a hierarchical framework by introducing random effects (modeled using an Intrinsic Autoregressive Prior (IAR)) only in $\pi(.,|\theta)$ and remarked that introducing random effects in both p and $\pi(.,|\theta)$ is computationally expensive and could lead to problems of identifiability. In this paper, we investigate the possibility of introducing random effects at both stages of regression resulting in a two-fold random effects hierarchical spatial model. Model fitting is illustrated with a data set involving counts of isopod nest burrows for 1649 pixels over a portion of the Negev desert in Israel. In our example, we do not find problems fitting the model using MCMC, nor do we find problems with identifiability. However, the spatial association as well as the variance component for random effects associated with p are quite weak relative to the ones associated with $\pi(.|\theta)$.

1. Introduction

Count data with excess number of zeroes arise in many applications and have been well studied in the literature. An attractive approach to model extra heterogeneity not captured by the usual parametric models in such data is through a zero-inflated model. A discrete random variable is said to have a zero-inflated distribution if, with probability p it is degenerate at 0 and with probability $1 - p$ it follows a law given by some commonly used parametric probability mass function $\pi(.|\theta)$ like Poisson, binomial, negative binomial, etc. Henceforth, we refer to the degenerate component of the model as the p-part and the non-degenerate component as the q-part. For instance, Lambert (1992) considers an industrial setting where a reliable manufacturing process moves back and forth between a perfect state with probability p in which defects are extremely rare and an imperfect state with probability $1 - p$ in which number of defects follow a Poisson distribution with mean λ. Both p and λ are allowed to depend on covariates. However, there are situations where excess heterogeneity persists after accounting for the effect of covariates in a zero-inflated model. In such cases, modeling extra heterogeneity by indroducing random effects in a generalized linear model framework becomes attractive. For instance, Hall (2000) models overdispersion in zero-inflated Poisson and binomial models using a random effects model; Lee, Wang and Yau (2001) present an interesting study using a zero-inflated Poisson model with random effects introduced to model the Poisson intensity parameter. For spatial data which is our focus in this paper, Agarwal, Gelfand and Citron-Pousty (2002) model species settlement patterns using zero-inflated Poisson model introducing random effects in the Poisson part. The random effects are modeled using the intrinsic autoregressive prior (IAR) introduced by Besag, York and Mollié (1991). In this article, we extend our earlier work to allow for spatial random effects both in the p and q parts. The random effects are modeled using a modification to Conditionally Autoregressive (CAR) prior introduced by Sun, Tsutakawa, Kim and He (2000) which allows an additional hyperparameter to measure the strength (or degree) of spatial association. To the best of our knowledge, the only other work with a spatial flavor in this area appeared recently in the context of disease mapping by Ugarte, Ibanez and Militino (2004) where they introduce a bootstrap technique to test for zero-inflation in the Poisson model. However, all the above mentioned papers (except our earlier work) use maximum likelihood based methods for model fitting. We take a fully Bayesian approach to fit models and use the Deviance Information Criteria (DIC) introduced by Spiegelhalter et al. (2002) to address model selection issues. A fully Bayesian approach can have several advantages in this scenario. First, it provides more realistic estimates of uncertainty in the parameters, especially those involving variance components and spatial association parameters. This is especially useful when the MLE of variance components is zero or close to zero. The Bayesian approach provides exact inference and does not depend on asymptotic approximations although for large data sets the answers are close (true for the regression parameters in our application). Finally, the Bayesian method typically implemented using MCMC provides the entire posterior distribution of parameters and any arbitrary functional of the parameters.

An important problem in ecology is to model species abundance or understand species settlement patterns. In such situations, abundance data is often in the form of species counts where a large fraction of counts are zeroes and a small fraction are large, i.e. species settlements tend to exhibit a clustering effect. One could attribute part of this clustering effect to covariates like habitat conditions and topographical features. Measurements on these covariates are often available from a geographic information system. However, often these covariates do not explain the clustering effect completely which can be seen by examining the map of residuals obtained from the covariate only zero-inflated model. This is also intuitive from an ecological perspective if one believes that other unobservable or hard to quantify spatial processes like reproduction, dispersal etc. are instrumental in determining the settlement patterns. Prompted by remarks from a referee, we investigate the possibility of simultaneously estimating spatial association in the p and q parts in such scenarios. The possibility of introducing random effects in both the p-part and q-part raises issues of model choice which we resolve by using the Deviance Information Criteria (DIC). Model fitting is done in a fully Bayesian framework using an MCMC approach. We illustrate our method by fitting the proposed model to a dataset which records isopod nest burrow counts for pixel areal units. We note that in general our techniques can be used to study species settlement patterns in ecology.

Section 2 gives modeling details including issues of posterior propriety, Section 3 briefly summarises the implementation details of the algorithm, Section 4 describes our data and present results from non-spatial models, which motivate the need to consider spatial models. Section 5 present results of our Bayesian analyses to the isopod data including issues related to model choice and Section 6 concludes with summary and scope for future work.

2. Modeling Details

A portion of this section overlaps with the theoretical development described in Agarwal et al. (2002) but is repeated here to set up notations and for the sake of completeness. Given a parametric distribution $\pi(y|\theta)$ on the integers $y = 0, 1, 2, \cdots$, its corresponding zero-inflated distribution is a mixture of $\pi(y|\theta)$ and a degenerate distribution at zero. Formally, a discrete random variable is zero inflated relative to $\pi(y|\theta)$ if it has distribution given by

$$P(Y = 0|p, \theta) = p + (1 - p)\pi(0|\theta) \tag{1}$$

$$P(Y = y|p, \theta) = (1 - p)\pi(y|\theta), \ y > 0.$$

Note that the idea in Eq. (1) can be used to inflate the probability at any integer but, in practice, 0 is usually the value of interest. Candidates for $\pi(y|\theta)$ include the Poisson, negative binomial, binomial, beta binomial and hypergeometric distributions. For future use we note that $P(Y = y > 0|Y > 0, p, \theta) = \pi(y|\theta)/(1 - \pi(0|\theta))$, free of p. Also, $E(Y|p, \theta) = (1 - p)E_\pi(Y|\theta)$, and $\operatorname{var}(Y|p, \theta) = p(1 - p)(E_\pi(Y|\theta))^2 + (1 - p)\operatorname{var}_\pi(Y|\theta)$, where $E_\pi(Y|\theta)$ and $\operatorname{var}_\pi(Y|\theta)$ denote, respectively, the expectation and variance of a random variable Y with probability mass function π.

Introduction of a latent indicator variable facilitates computation with (1). Introduce a latent variable Z such that the joint distribution for Y and Z is as follows:

$$P(Y = 0, Z = 1|p, \theta) = p, P(Y = y, Z = 0|p, \theta) = (1 - p)\pi(y|\theta).$$

Then, marginally Z is a Bernoulli random variable with success probability p. Conditionally, $P(Y = 0|Z = 1) = 1, P(Y = y|Z = 0, p, \theta) = \pi(y|\theta), P(Z = 1|Y = y > 0) = 0, P(Z = 1|Y = 0, p, \theta) = p/(p + (1 - p) \ \pi(0|\theta))$.

For a sample of size n with Y_i given p_i and θ_i distributed as in (1), the full data likelihood arises from $\prod_i P(Y_i = y_i|Z_i = z_i)P(Z_i = z_i)$ and takes the form

$$L(\boldsymbol{p}, \boldsymbol{\theta}; \boldsymbol{Y}, \boldsymbol{Z}) = \prod_{i=1}^{n} p_i^{z_i}((1 - p_i)\pi(y_i|\theta_i))^{1-z_i}. \tag{2}$$

In fact, because Z_i is degenerate at 0 if $Y_i > 0$, the only nondegenerate Z_i (and thus the only ones we need to introduce) are those associated with the $Y_i = 0$. Therefore, (2) can be rewritten as

$$\prod_{y_i > 0} (1 - p_i)\pi(y_i|\theta_i) \prod_{y_i = 0} p_i^{z_i}((1 - p_i)\pi(0|\theta_i)^{1-z_i}. \tag{3}$$

The marginal or observed data likelihood takes the form

$$L(\boldsymbol{p}, \boldsymbol{\theta}; \boldsymbol{Y}) = \prod_{i=1}^{n}\{p_i 1(y_i = 0) + (1 - p_i)\pi(y_i|\theta_i)\}. \tag{4}$$

In the i.i.d. case

$$\log L(p, \theta; \boldsymbol{Y}) = V_0 \log(p(1 - \pi(0|\theta)) + \pi(0|\theta)) + (n - V_0)\log(1 - p) + \sum_i \log \pi(y_i|\theta)$$

where $V_0 = \#(Y_i = 0)$, which is immediately unimodal in p for each θ. In this article we choose $\pi(y|\theta)$ to be binomial (K, θ) (K known) leading to the ZIB (p, θ)(zero inflated binomial) model. For ZIB, $E(Y|p, \theta) = (1 - p)K\theta$ and $\operatorname{Var}(Y|p, \theta) = p(1 - p)K^2\theta^2 + (1 - p)K\theta(1 - \theta)$. In fact, the square of the coefficient of variation for ZIB equals $p/(1 - p) + CV_\pi^2/(1 - p)$ (CV_π is coefficient of variation for binomial distribution) which is larger than CV_π^2, indicating overdispersion relative to the binomial. Reasons for choosing binomial in our application will be explained in section 4.

Incorporating covariates into the ZIB model is straightforward explained as follows: Let $Y_i \sim$ ZIB (p_i, θ_i), where Y_i's are independently distributed. Using canonical links, we assume the parameters $\boldsymbol{\theta} = (\theta_1, \cdots, \theta_n)'$ and $\boldsymbol{p} = (p_1, \cdots, p_n)'$ satisfy

$$\text{logit}\,(\boldsymbol{\theta}) = \boldsymbol{B\beta}, \text{logit}\,(\boldsymbol{p}) = \boldsymbol{G\alpha} \tag{5}$$

where $\boldsymbol{B} = \begin{pmatrix} b_1' \\ \cdot \\ \cdot \\ \cdot \\ b_n' \end{pmatrix}$ and $\boldsymbol{G} = \begin{pmatrix} g_1' \\ \cdot \\ \cdot \\ \cdot \\ g_n' \end{pmatrix}$ are specified design matrices with $\beta = (\beta_0, \beta_1, ..., \beta_{(q-1)})'$ and

$\alpha = (\alpha_0, \alpha_1, ..., \alpha_{(m-1)})'$, the associated parameter vectors. $\boldsymbol{B}$ and $\boldsymbol{G}$ may share some common covariates but need not be the same. Inference using a maximum likelihood approach (Lambert, 1992) or a Bayesian approach (Agarwal et al., 2002) can be implemented. To capture additional heterogeneity in the Y_i's we can assume that (5) takes the form

$$\text{logit}\,(\boldsymbol{\theta}) = \boldsymbol{B\beta} + \boldsymbol{W}_1 \boldsymbol{u}; \ \text{logit}\,(\boldsymbol{p}) = \boldsymbol{G\alpha} + \boldsymbol{W}_2 \boldsymbol{v} \tag{6}$$

where $\boldsymbol{u}$, $\boldsymbol{v}$ are random effects and $\boldsymbol{W}_1$, $\boldsymbol{W}_2$ are appropriate incidence matrices. If $\boldsymbol{p}$ and $\boldsymbol{\lambda}$ are not related, it is reasonable to assume that $\boldsymbol{u}$ and $\boldsymbol{v}$ are independent of each other. Conditionally on $\boldsymbol{u}$ and $\boldsymbol{v}$ (as well as $\boldsymbol{\beta}$ and $\boldsymbol{\alpha}$) the Y_i's are independent. Marginalizing over $\boldsymbol{u}$ and $\boldsymbol{v}$ they are not. Hence (6) provides a specification for modeling dependencies in zero-inflated counts. Inference using sampling based methods (in particular Gibbs sampling) under a fully Bayesian formulation provides an attractive approach for fitting (6). Samples obtained from the posterior distribution of $(\boldsymbol{\alpha}, \boldsymbol{\beta}, \boldsymbol{u}, \boldsymbol{v})$ provide exact inference for the parameters (or functions of the parameters) without relying on numerical or asymptotic approximations.

We now discuss modeling spatial association between counts. This is accomplished by assuming $(\boldsymbol{u}, \boldsymbol{v})$ to be spatial random effects. Since our focus is on lattice data, we model $(\boldsymbol{u}, \boldsymbol{v})$ using Conditionally Autoregressive Prior (CAR). For instance, one could model $(\boldsymbol{u}, \boldsymbol{v})$ using a bivariate CAR (Hoon, Sun and Tsutakawa, 2001) which, apart from modeling spatial association in the p and q parts also models correlation between the two spatial processes. However, in this article we assume $(\boldsymbol{u}, \boldsymbol{v})$ to be independent spatial processes. In fact, we assume $\boldsymbol{v}$ follows CAR with spatial association parameter ρ_1 and variance component δ_1 while $\boldsymbol{u}$ follows CAR with the corresponding parameters being ρ_2 and δ_2.

2.1 Posterior Propriety

We briefly discuss issues of posterior propriety associated with the model and refer the reader to Agarwal et al. (2002) for further details (including methods to elicit informative but vague priors).

The full Bayesian model arising from (2) and (4) may be expressed in the form

$$\prod_i f_{Y|Z,\beta}(Y_i|Z_i, \beta) \prod_i f_{Z|\alpha}(Z_i|\alpha) f_{\alpha}(\alpha) f_{\beta}(\beta) \tag{7}$$

with resulting posterior for $\boldsymbol{\alpha}, \boldsymbol{\beta}, \boldsymbol{Z}$ given $\boldsymbol{Y}$ proportional to (7). Here $\boldsymbol{Z}$ denotes the vector of all unobserved Z's. Since prior information regarding $\boldsymbol{\beta}$ and especially $\boldsymbol{\alpha}$ is likely weak, improper priors are often proposed for $\boldsymbol{\alpha}$ and/or $\boldsymbol{\beta}$. Thus, the question of posterior propriety arises under (7).

We note that given $\boldsymbol{Z}$ and $\boldsymbol{Y}$, $\boldsymbol{\alpha}$ and $\boldsymbol{\beta}$ are independent. Therefore,

$$f_{\alpha,\beta|Y}(\alpha, \beta|Y) = \sum_z f_{\alpha|Z}(\alpha|Z) f_{\beta|Z,Y}(\beta|Z, Y) \tag{8}$$

where the sum in (8) is overall possible configurations of the unobserved z. Hence propriety of (8) requires propriety of $f_{\alpha|Z}(.|Z)$ and $f_{\beta|Z,Y}(.|Z,Y)$ for *each* possible z. In the ZIB case, both the posteriors are associated with a logistic regression.

Work dating to Diaconis and Ylvisaker (1979) has addressed the propriety of posteriors associated with generalized linear models (GLM's) under both flat and Jefferys' prior specifications. The paper of Ibrahim and Laud (1991) is noteworthy here, as is a survey paper of Gelfand and Ghosh (2000). For instance, Diaconis and Ylvisaker (1979) note that a flat prior on α need not give a proper posterior. (The case where α is just a common intercept and all the Z's are 0 or all of the Z's are 1 makes this clear.) Under Jeffreys' prior, Ibrahim and Laud (1991) reduce the problem of posterior propriety to checking the existence of a set of one dimensional integrals. But, in general, propriety is not assured.

Hence, to avoid propriety concerns we take $f_{\alpha}(.)$ and $f_{\beta}(.)$ proper in the sequel. In fact, employing simulation to fit (7), fairly informative priors are required to achieve a well behaved Gibbs sampler.

Next, we briefly discuss the CAR model and refer the reader to Besag (1974) and Cressie (1993, pp. 430-441) for complete details. The central idea in a CAR model is specifying a multivariate normal distribution exclusively in terms of its full conditional distributions. It is well known that all conditional distributions of a multivariate normal are normal; the converse of specifying the joint distribution through full conditional distributions is true as well under certain constraints on the conditionals. Fortunately, the constraints imposed are not severe and allow a flexible class to model spatial associations. Specifically, let v be a random vector such that the conditional distribution of $v_i|v_{-i} \sim N(\sum_{j:j\neq i} \beta_{ij}v_j, \sigma_i^2)$ (v_{-i} denotes the vector v with the ith element missing). Then, under the constraints $\beta_{ij}\sigma_j^2 = \beta_{ji}\sigma_i^2$ the vector v has a multivariate normal distribution with mean $\mathbf{0}$ and precision matrix $D - B$, where $D = \text{diag}(1/\sigma_1^2, \cdots, 1/\sigma_n^2)$ and $B = ((\beta_{ij}))$. In spatial statistics, it is customary to define a symmetric neighborhood structure (for instance, this could be based on adjacency or geographic proximity) $\{N(i)\}$ ($N(i)$ is the neighbor set of location i) and then assume $\beta_{ij} = \rho w_{ij}/w_{i+}$, where w_{ij} is the weight assigned to the jth neighbor of location i ($w_{ij} = w_{ji}$ and $w_{ik} = 0$ if k is not a neighbor of i), $w_{i+} = \sum_{j\in N(i)} w_{ij}$ and ρ is a constant between 0 and 1. To appropriately scale the local variance of each location based on the neighbor weights, one assumes $\sigma_i^2 = \delta/w_{i+}$ for some positive variance component δ. Note that this parametrization *shrinks* each v_i towards its neighborhood mean with the amount of shrinkage determined by ρ and δ. A value of ρ close to zero would mean little shrinkage towards spatial neighbors and signifies little spatial association; values closer to 1 indicate stronger spatial association. Thus, ρ serves as a measure of spatial association and gets estimated from the data. The variance component δ is informative of the role played by the random effects in the model. A value close to zero indicates the random effects do not contribute much; on the other hand, a very high value is indicative of a severe conflict between data and the model, i.e. the random effects almost mimic the residuals and overfit the data. Hence, the scale of δ should ideally be somewhere between these two extremes where it enhances model performance by achieving the right amount of Bayesian shrinkage and borrowing of strength. With the parametrization above, the joint distribution of v is multivariate normal with mean $\mathbf{0}$ and precision matrix $(\text{diag}(w_{1+}, \cdots, w_{n+}) - \rho W)/\delta$, where $W = ((w_{ij}))$. Note that for $\rho = 1$ the precision matrix is singular and we have a degenerate normal which puts all its mass on a lower dimensional subspace. In fact, the IAR model is a special case with ρ fixed at 1 instead of being estimated from the data. To resolve the impropriety in IAR, one normalizes the random effects vector v to satisfy the sum-to-zero constraint after each iteration of the Gibbs sampler. The weights w_{ij} are normally chosen to be binary (1 if j is neighbor, 0 otherwise) or some monotonically decreasing function of distance between sites. In principle, one would like to estimate the appropriate weight structure from the data itself but this imposes substantial computational overhead and has not been pursued in the literature. In this paper, we try a few neighborhood structures by varying the size of the neighborhood and choose the one that is best in terms of DIC.

3. Implementation Issues

In order to carry out the full Bayesian analysis, we need to specify priors for our unknown parameters $(\alpha, \beta, \delta_1, \rho_1, \delta_2, \rho_2)$. We assume the parameters are independent *a priori*, i.e.

$$f(\alpha, \beta, \delta_1, \rho_1, \delta_2, \rho_2) = f(\alpha)f(\beta)f(\delta_1)f(\rho_1)f(\delta_2)f(\rho_2)$$

where $\alpha \sim N(\alpha_0, \Sigma_\alpha)$, $\beta \sim N(\beta_0, \Sigma_\beta)$ with the prior parameters derived as discussed earlier. We assume δ_1 and δ_2 follow an inverse gamma with shape=scale=.0001. This is proper but rather vague. We assume ρ_1 and ρ_2 have a uniform prior in the interval $(0,1)$. The full conditionals for the regression parameters and the spatial random effects are all log-concave and easily sampled using adaptive rejection sampling (Gilks and Wild, 1992). The full conditionals for the variance components δ_1 and δ_2 are inverse gamma and sampled directly. The full conditionals of the spatial association parameters ρ_1 and ρ_2 are sampled using slice sampling (Agarwal and Gelfand, 2005). Full conditionals of Z_i's are Bernoulli and sampled directly.

4. Data Description and Preliminary Analysis

The main goal here is to enhance our understanding of why a particular animal chooses to live in one location rather than another. The particular organism used in this study is a terrestrial isopod, *Hemilepistus reaumuri*. Its life history is well studied. See, e.g., Baker et al. (1998), Citron-Pousty and Shachak (1998) and further references therein. These isopods live in family burrows which provide shelter and humidity in the arid environment. Previous studies have shown that the isopods respond to rainfall runoff redistribution (Yair and Shachak, 1982; Shachak and Yair, 1984; Shachak and Brand, 1991). Areas with increased runoff tend to have increased probability of settling (Citron-Pousty and Shachak, 1998). Because of the dry conditions, water is a limiting factor in the environment.

Small first order watersheds were surveyed in the Haluqim ridge of the Negev desert, Israel and rasterized to a 1 m × 1 m binary presence/absence map. In this desert, dewfall can contribute relatively large amounts of water per year and has less annual variability when compared to rainfall (Zagvil, 1996; Kidron, 1998). Our watershed opened in the west direction. The watershed was exhaustively surveyed for burrows in the summer of 1995. All burrow locations were mapped in the Israel coordinate system. For this watershed we aggregate isopod burrow counts to a resolution of 5 m × 5 m from the presence/absence map at 1 m × 1 m putting an upper bound of 25 burrows for each of the 5 m × 5 m pixel. This explains why a binomial sampling model (or its zero-inflated counterpart) is appropriate. The watershed covers an area of approximately 40,000 m² (approximately 200 m × 200 m). We note that the analyses could have been done on the finer scale of 1 m × 1 m but would have led to enormous run times for the MCMC sampler. For ease of illustration and due to the clustering effect seen in the presence/absence map at 1 m × 1 m, we aggregate counts to the 5 m × 5 m resolution. Fig. 1 shows the region under study with the burrow counts overlaid. We have a total of 1649 pixels of which 82.1% have zero counts.

The explanatory variable dew duration measures time (in minutes from 8 am) to evaporation of dew. Fig. 2 shows the distribution of dew duration over the region. The explanatory variables shrub density and rock density measure the abundance of shrub and rock, respectively, at a given pixel.

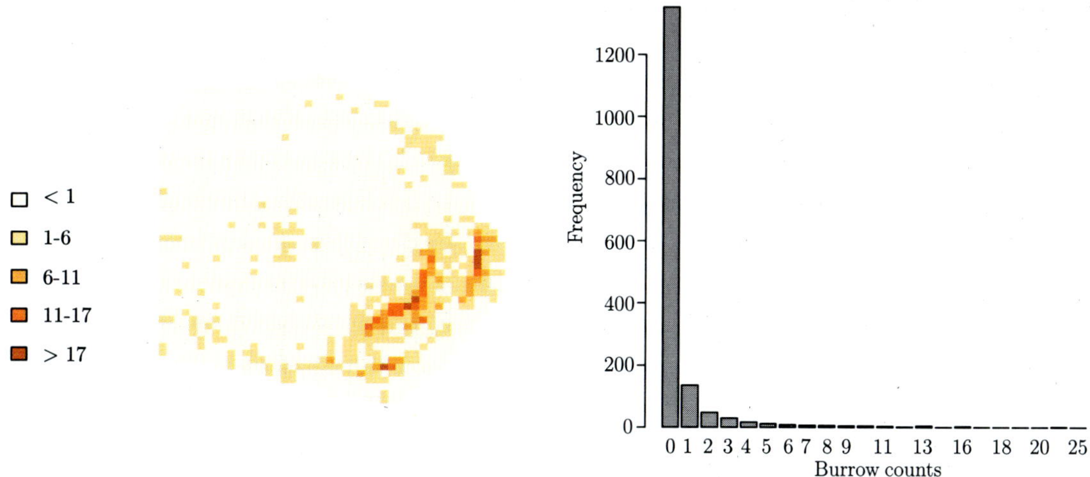

Fig. 1.   Distribution of burrow counts at 5 m × 5 m resolution. 82.1% of pixel counts are zero, there is preponderance of isopods in the west direction where the watershed opens.

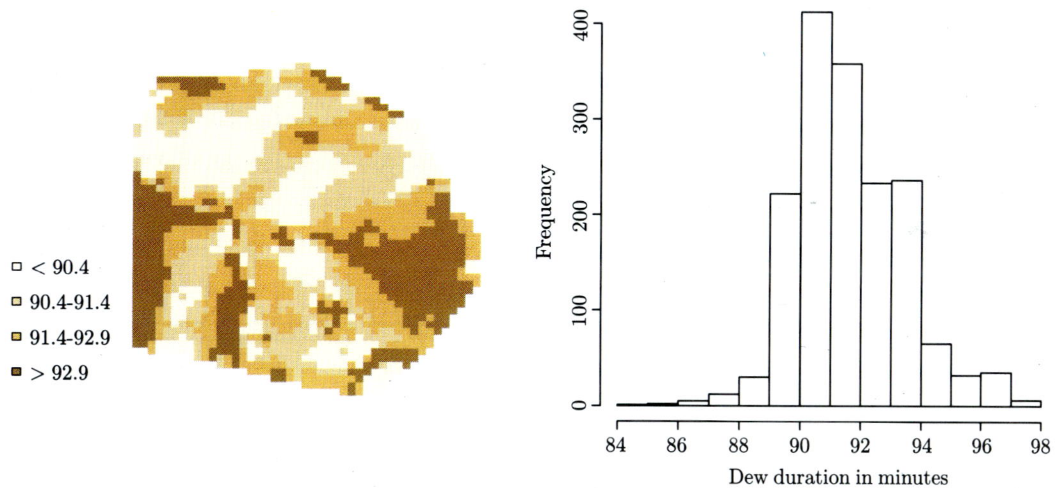

Fig. 2.   Distribution of dew duration in minutes from 8 am to evaporation of dew.

At a resolution of .1 m × .1 m, we have binary maps showing shrub and rock incidences. These were aggregated to a resolution of 5 m × 5 m giving us the shrub and rock densities (as a proportion). Figs. 3 and 4 show the distribution of shrub and rock density over the region. Based upon previous research we expect a positive relationship with regard to burrow survival (hence desirability of the site) for each of dew availability, rock density, and shrub density.

We fitted binomial and zero-inflated binomial (ZIB) models to the data with dew duration, shrub and rock densities as our covariates. Due to the preponderance of burrow counts from east to west, a linear trend surface was included as well (in ZIB, the trend was used only in the q-part).

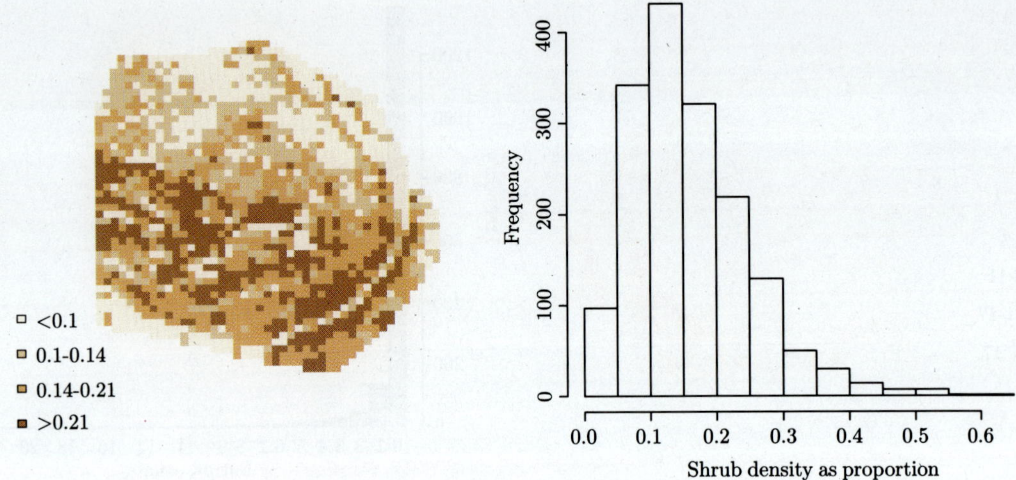

Fig. 3.   Distribution of shrub density as a proportion.

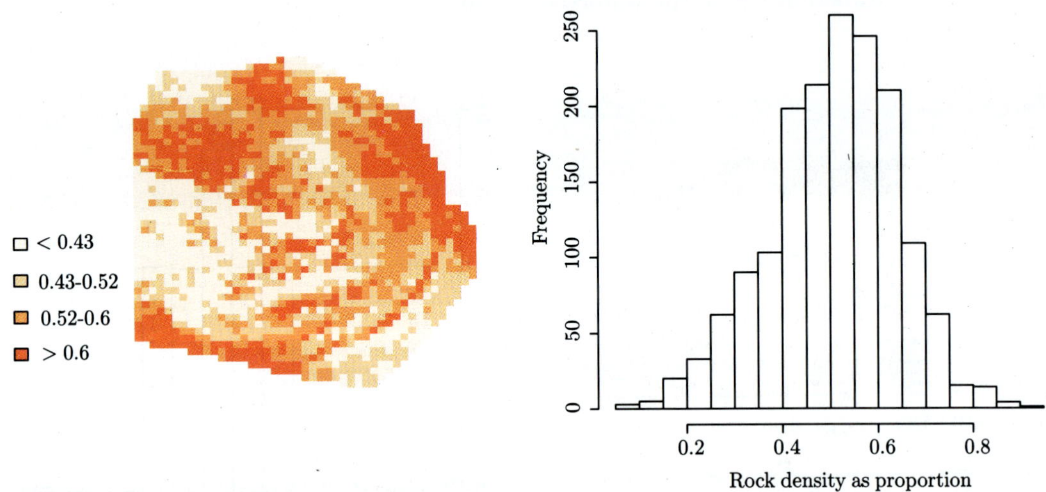

Fig. 4.   Distribution of rock density as a proportion.

Table 1 compares the models in terms of AIC, BIC and DIC. The ZIB model shows a substantial improvement relative to the binomial model indicating that accounting for overdispersion due to excess zeros is advantageous in our application. We note that AIC and BIC were obtained using maximum likelihood while DIC was obtained using our MCMC implementation. Note that DIC is

Table 1.   Comparing binomial and zero inflated binomial in terms of AIC, BIC and DIC. First two measures were computed using maximum likelihood and the last measure was obtained from MCMC algorithm

Model	AIC	BIC	DIC
Binomial	3386.0	3462.9	3385.7
ZIB	2814.6	2942.8	2813.1

very close to AIC in this case as expected. Also, the coefficient estimates (including standard errors) obtained from MCMC are in agreement with the ones obtained from maximum likelihood showing that the data overwhelms the information contained in the somewhat informative priors (for the regression coefficients) we begin with. It also shows that the asymptotic approximations used in MLE are appropriate for the regression parameters. Finally, it also provides a sanity check for our MCMC code. Fig. 5 shows the distribution of Pearson residuals obtained from the ZIB model fitted to the data above.

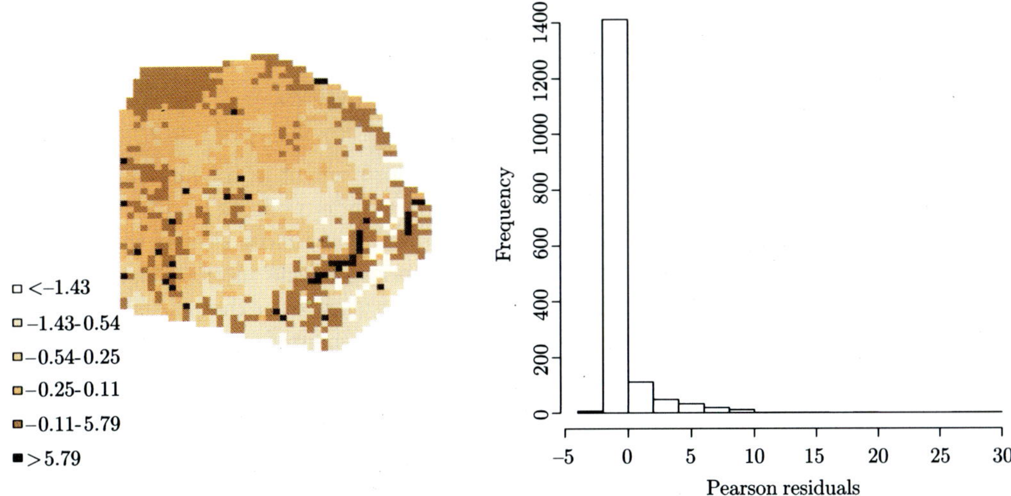

Fig. 5.   Distribution of Pearson residuals from ZIB, the clustering effect indicates spatial association.

The distribution of the residuals indicate a clustering effect providing a hint that a spatial model might be appropriate here. To quantify the spatial association in the residuals, we estimated the spatial association parameter ρ by fitting a CAR model to the residuals using adjacency to define neighbors and using 4, 8 and 12 as the neighborhood size. The estimates of ρ were .73, .81 and .9 for 4, 8 and 12 neighbors, respectively, indicating strong spatial association.

5.   Fitting Spatial Models

We describe the results of fitting spatial models to our data. All DIC computations are done using canonical parametrization for the binomial model since it is close to approximate normality as suggested in Spiegelhalter et al. (2002). We compute DIC for the full model (spatial random effects in both parts of ZIB) using 4, 8, 12, 16, 20 and 24 nearest neighbors and observe that performance gets slightly better with increasing neighborhood size until 12 neighbors beyond which it starts deteriorating slowly. All subsequent analyses are presented with 12 nearest neighbors for both parts of ZIB.

Since our aim is to compare models, the deviance function in DIC is chosen to be negative of twice the log-likelihood. Table 2 shows the results of model fit for three models M_1 (spatial effect only in the p part), M_2 (spatial effect only in the q part) and M_3 (spatial effect in both p and q parts).

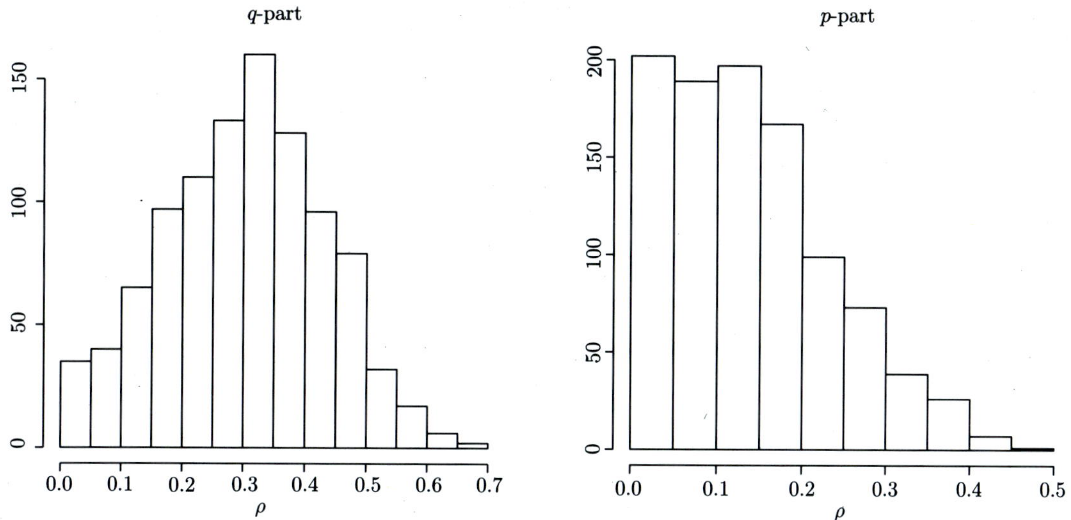

Fig. 7.   Posterior distributions of spatial association components for model M_3.

Figure 8 shows the posterior means of the spatial random effects both for the p and q-parts. Again, we see the spatial adjustments in the p part are small. Spatial adjustments in the q-part are high at locations where we have a preponderance of isopod counts. The two random effects are negatively correlated. In fact, the correlation between the posterior means is $-.65$. Note that for a ZIB model, $E(Y) = (1-p)K\theta$ and hence it tends to adjust for underestimating (overestimating) the mean by simultaneously adjusting random effects (in opposite directions) in the p and q-parts.

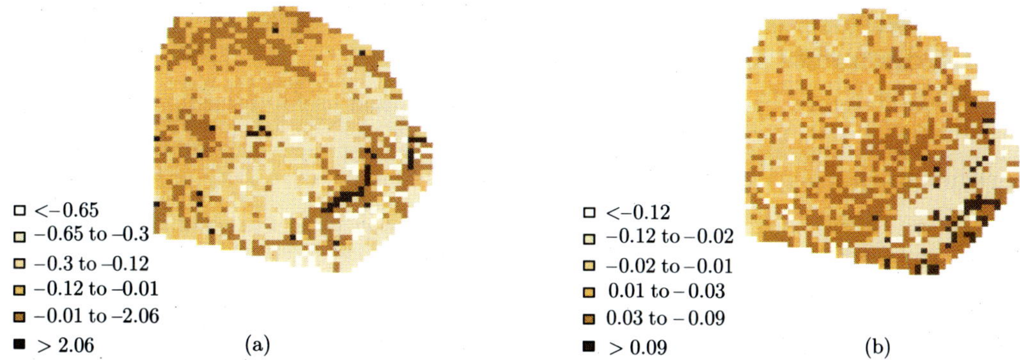

Fig. 8.   Posterior means of spatial random effects model M_3. Random effects for: (a) q-part and (b) p-part.

Finally, we present inference for the regression parameters obtained from model M_3 in Table 3. As expected from previous research, increased shrub density, dew duration and rock density tend to increase the probability of isopod presence at a pixel. The regression coefficients in the p-part are statistically insignificant although the posterior for shrub density is suggestive of a negative effect.

Table 3. **The 95% credible intervals for the posterior of regression parameters obtained from model M_3. All covariates were centered around 0 and scaled to have a standard deviation of unity**

Quantiles	2.5%	50%	97.5%
Parameter estimates for q-part			
Shrub density	.73	1.02	1.33
Dew duration	.10	.29	.48
Rock density	.63	.97	1.39
Latitude	.93	1.11	1.31
Longitude	−1.11	−.88	−.64
Parameter estimates for p-part			
Shrub density	−.67	−.31	.09
Dew duration	−.58	−.23	.13
Rock density	−.68	−4.21	.19
Intercept	−.99	−.60	−.28

6. Discussion

We extended our earlier work incorporating spatial random effects in both p and q parts. We show that this leads to well behaved simulation based fitting in a fully Bayesian framework and does not pose issues with identifiability. We replaced the intrinsic autoregressive prior with the one that allows estimation of spatial strength from data. Working with two-fold random effects raises important model selection issues which we addressed using the DIC criterion. The DIC criterion was also used to select a sensible spatial neighborhood structure. Future work might consider a multivariate spatial model for $(\boldsymbol{u}, \boldsymbol{v})$ which, apart from spatially smoothing, would also model correlation between u's and v's.

Acknowledgements

I thank Citron-Steven Pousty for providing the isopod data and useful discussions.

References

Agarwal, D. and Gelfand, A.E. (2005). Slice Sampling for Simulation based Fitting of Spatial Data Models *Statistics and Computing*, **15**, 61-70

Agarwal, D., Gelfand, A.E. and Citron-Pousty, S (2002). Zero-Inflated Models with Application to Spatial Count Data, *Environmental and Ecological Statistics*, **9**, 341-355.

Baker, M.B., Shachak, M., Brand, S. and Yair, A. (1998). Settling behavior of the desert isopod *Hemilepistus reaumuri* in response to variation in soil moisture and other environmental cues, *Israel Journal of Zoology*, **44**, 345-354.

Besag, J. (1974). Spatial interaction and the analysis of lattice systems (with discussion), *Journal of the Royal Statistical Society*, Ser. B, **36**, 192-236.

Besag, J., York, J. and Mollié, A. (1991). Bayesian Image Restoration With Two Applications in Spatial Statistics, (with discussion) *Annals of the Institute of Statistical Mathematics*, **43**, 1-59.

Citron-Pousty, S. and Shachak, M. (1998). Comparative spatial patterns of the terrestrial isopod *Hemilepistus reaumuri* at multiple scales. *Israel Journal of Zoology*, **44**, 355-368.

Cressie, N.A.C. (1993). *Statistics for Spatial Data*(rev.ed.). New York: John Wiley.

Diaconis, P. and Ylvisaker, D. (1979). Conjugate Priors for Exponential Families, *Annals of Statistics*, **7**, 269-281.

Gelfand, A.E. and Ghosh, M. (2000). *Generalized Linear Models: A Bayesian Perspective*, (eds. D.K. Dey, S.K. Ghosh and B.K. Mallick), New York: Marcel Dekker, Inc, 3-14.

Gilks, W.R. and Wild, P. (1992). Adaptive rejection sampling for Gibbs sampling. *Applied Statistics*, **41**, 337-348.

Hall, D. B. (2000). Zero-inflated Poisson and binomial regression with random effects: A case study, *Biometrics*, **56**, 1030-1039.

Hoon, K., Sun, D. and Tsutakawa, R. K. (2001). A Bivariate Bayes Method for Improving the Estimates of Mortality Rates With a Twofold Conditionally Autoregressive Model. *Journal of the American Statistical Association*, **96**, 1506-1521.

Ibrahim, J.G. and Laud, P.W.(1991). On Bayesian Analysis of Generalized Linear Models Using Jeffreys's Prior, *Journal of the American Statistical Association*, **86**, 981-986.

Kidron, G.(1998). Dew variability, lichen and cyanobacteria distribution along slopes at Sde Boqer-northern Negev. *Physical Geography*, Jerusalem, The Hebrew University of Jerusalem: 92.

Lambert, D. (1992). Zero-Inflated Poisson Regression, with an Application to Defects in Manufacturing, *Technometrics*, **34**, 1-14.

Lee, A. H., Wang, K. and Yau, K. K. W. (2001). Analysis of zero-inflated Poisson data incorporating extent of exposure. *Biometrics Journal*, **43**, 963-975.

Shachak, M., and Brand, S. (1991). Relations among spatiotemporal heterogeneity, population abundance, and variability in a desert. *Ecological Heterogeneity*, eds. J. Kolasa and S.T.A. Pickett, New York: Springer-Verlag, 202-223.

Shachak, M. and Yair, A. (1984). Population dynamics and role of Hemilepistus reaumuri (Audouin and Savigny) in a desert ecosystem. *Symposium of Zoological Society of London*, **53**, 295-314.

Spiegelhalter, D.J., Best, N.G., Carlin, B.P. and van der Linde, A. (2002). Bayesian measures of modelcomplexity and fit. *J. Royal Statist. Society Series B*, **64**, 583-639.

Sun, D., Tsutakawa, R. K., Kim, H. and He, Z. (2000). Spatio-Temporal Interaction with Disease Mapping, *Statistics in Medicine*, **19**, 2015-2035.

Ugarte, M. D., Ibanez, B. and Militino, A.E. (2004). Testing for Poisson Zero Inflation in Disease Mapping, *Biometrical Journal*, 46, 526-539.

Yair, A. and Shachak, M. (1982). A case study of energy, water and soil flow chains in an arid ecosystem, *Oecologia*, **54**, 389-397.

Zagvil, A.(1996). Six years of dew observation in the Negev Desert, Israel. *Journal of Arid Environments*, **32**, 361-371.

Bayesian Statistics and Its Applications
Edited by S.K. Upadhyay, U. Singh and D.K. Dey
Anamaya Publishers, New Delhi, India

Prior Model for Reconstruction of Contingency Table

J.-F. Angers[1,2]**, C. Laberge-Nadeau**[2,3]**, F. Bellavance**[2,4]**,
S. Courchesne**[1]**, L.-F. Poirier**[1] **and V.C. Allaire**[5]

[1]DMS, [2]CRT, [3]DMSP Université de Montréal
[4]HEC, Montréal
[5]Collège du Bois de Boulogne, Montréal

Abstract

The goal of this paper is to assess what influence the use of cell phone while driving may have on the risk of having a car accident. When a driver has an accident, we can tell, by combining various sources of data, if he was using his cell phone during a given time interval prior to the time of the accident. However, if the driver has not had an accident, we cannot tell whether he was using his cell phone or not while driving. Hence, we end up with an incomplete 2×2 contingency table. Consequently, we cannot compute the instantaneous risk directly, which is used to measure the association between the two variables "use of a mobile phone while driving" and "having a car accident." This paper introduces a method for reconstructing the contingency table. In this method the marginal densities are first modelled, using auxiliary source of information. Then, conditioned on these marginals, a Monte Carlo simulation study can be run to complete the 2×2 table and to estimate the instantaneous risk. Using the proposed approach, we have estimated the instantaneous risk and found an overall instantaneous risk of 1.735.

---

1. Introduction

Is there an association between the use of cell phone while driving and the risk of having a car accident?

In the past few years, the use of cell phones has increased substantially and many people and legislators are now concerned about the safety of using this device while driving. Many drivers will often take their eyes off the road while dialing and then become so absorbed in their conversations that they may be distracted from the act of driving. This can increase the overall risk of accident for all road users, including collisions with pedestrians and cyclists. Some researchers have already assessed the risk of traffic accidents linked to cell phone use. Most of the earlier studies on the link between cell phone and accidents have used data derived from simulations on instrumented vehicles or from test trials conducted off-road sites where risks were assessed for a relatively small number of subjects in conditions disconnected from real life. But very few epidemiological studies have investigated the relation between the cell phone use and driving, based on actual accidents.

Such an epidemiological study has been carried out by Violanti and Marshall (1996). They used a case-control design and logistic regression techniques to examine the association between cell phone use in motor vehicles and the risk of traffic accidents. They reported that talking for more than 50 min per month on the cell phone while driving was associated with a 5.59-fold increase in the risk of accident. A case-crossover study conducted by Redelmeier and Tibshirani (1997) concluded that the risk of collision is more than four times higher while using a cell phone. These results were later challenged in Laberge-Nadeau et al. (2001). Laberge-Nadeau et al. (2001) found that the risk is 38% higher for cell phone users. However, none of these studies, except the one by Redelmeier and Tibshirani (1997) investigate whether or not drivers were actually using their cell phones while driving. Their association was obtained by means of a case-crossover design. However, it is known that this particular design overestimates the instantaneous risk when the data are subject to error (see Bourhattas, 2002).

This article proposes a new method to assess more accurately the instantaneous risk of having a car accident occurring when the driver is using his cell phone. Section 2 discusses the motivation for this research and introduces the database. In Section 3, the auxiliary information is modeled. These models are used in Section 4 to obtain the density on the margin of the 2×2 contingency table. Section 5 describes the instantaneous risk estimated under several hypotheses while some conclusions are drawn in Section 6.

2. Preliminary

This research was motivated by an epidemiological study which was undertaken by the Laboratory on Transportation Safety at the Center for Research on Transportation (CRT) of the Université de Montréal in 2000 (see Laberge-Nadeau et al., 2001). The study produced a data-set containing information on the subject's past accidents, on their record of violations from January 1, 1996 to August 3, 2000 and on several other variables. The power of this database is that it gives the possibility of combining information from a questionnaire, 4 years of past accidents and 2 years of cell phone use in order to generate a greater amount of information for analysis.

Since we wanted to model the relation between cell phone use while driving and the risk of collision, only individuals who were the sole user of their cell phone were retained. This decision was necessary to insure that, at the time of accident, it was actually the driver who was using his phone. The study contains information about 36,079 drivers but only 6,391 of them were sole users of cell phones.

2.1 Set-up

From the data at hand, we can build a contingency table like Table 1, where X corresponds to the event "using the cell phone while driving" and Y to the event "having a car accident". However the observations n_{12} and n_{22} are unknown. To retrieve this information, we have to model the probability of having an accident and the probability of making a phone call at the time of accident. Let $\underset{\sim}{n} = (n_{11}, n_{12}, n_{21}, n_{22})$. Hence, $\underset{\sim}{n}$ can be thought of as a multinomial random vector with parameters n_{++} and $\underset{\sim}{\nu} = (\nu_{11}, \nu_{12}, \nu_{21}, \nu_{22})$. However, only n_{11} and n_{21} are observed.

Let t_i, $i = 1, 2, \ldots, n_{\text{acc}}$ be the time of the ith accident and consider a time window of 15 minutes prior to t_i (see N'Zué, 2002 for the motivation of this time window. Other time windows are also considered in Section 5). We can defined the following quantities:

$n_{11i} : =$ the length (in min) of the last phone call made by the driver involved in the ith accident in the time interval $(t_i - 15, t_i]$

$n_{21i} :=$ the rest of time in the interval $(t_i - 15, t_i]$ prior (and after) the last phone call made by the driver who had the ith accident $= 15 - n_{11i}$.

Hence, $n_{11} = \sum_{i=1}^{n_{acc}} n_{11i}$ and $n_{21} = \sum_{i=1}^{n_{acc}} n_{21i} = \sum_{i=1}^{n_{acc}} (15 - n_{11i})$. The unobserved variables n_{12} and n_{22} can be defined in a similar way, that is

$n_{12ij} :=$ is the length (in minute) of the last phone call made by the jth driver who was not involved in an accident, in the time interval $(t_i - 15, t_i]$

$n_{22ij} := 15 - n_{12ij}$

Doing so, we have that $n_{12} = \sum_{i=1}^{n_{acc}} \sum_{j \in J^*} n_{12ij}$ and $n_{22} = \sum_{i=1}^{n_{acc}} \sum_{j \in J^*} n_{22ij} = \sum_{i=1}^{n_{acc}} \sum_{j \in J^*} (15 - n_{12ij})$, where J^* denotes the set of indices corresponding to the drivers without accident (other hypotheses on the cell phone call used to compute the n_{ij} are considered in Section 5). This is similar to the case-crossover design. However, instead of using the same subject as control and case, we shall use a theoretical population constructed from the auxiliary information.

Our main goal is to develop a method to reconstruct the contingency table and to estimate the instantaneous risk of having a car accident while using a cell phone. The instantaneous risk is defined as

$$\text{IR} = \frac{\nu_{1|1}}{\nu_{2|1}} = \frac{\nu_{11}/(\nu_{11} + \nu_{12})}{\nu_{21}/(\nu_{21} + \nu_{22})}$$

where $\nu_{i|j} = \Pr(Y = i \mid X = j)$.

It is usually estimated by

$$\widehat{\text{IR}} = \frac{n_{11}(n_{21} + n_{22})}{n_{21}(n_{11} + n_{12})}$$

where n_{ij} are the number of outcomes in the cell (i, j), $i, j = 1, 2$ of a 2×2 contingency table like Table 1 for the categorical variables X and Y. The method proposed in Sections 3 and 4 is a novel approach to the contingency table reconstruction problem.

Table 1. The 2×2 contingency table

$X \backslash Y$	Yes	No	Total
Yes	n_{11}	n_{12}	n_{1+}
No	n_{21}	n_{22}	n_{2+}
Total	n_{+1}	n_{+2}	n_{++}

3. Statistical Models Obtained from the Auxiliary Information

Let us first consider the phone calls data. We are interested in modeling the distribution of the phone calls made and received. For each individual, we have the time when the phone was in use (starting and ending times for incoming and outgoing calls) from August 1, 1998 to August 31, 2000 (the information given by cell phone companies is accurate within one second). We have transformed those times into the number of single minutes after midnight. For example, if a call started at 00:00 am and ended at 1:35 am, then the two measures associated with this phone call would be 0 and 95 which is equivalent to 95 one-min calls because we count calls by minute. Since the data correspond to a time in a 24-hour period, the circular data theory is used to model the phone call distribution.

The equivalent of the normal distribution for circular data is the von Mises distribution (see Fisher, 1993, Sec. 3.3.6) whose density function is given by

$$f(y \mid \mu, \kappa) = \frac{1}{2\pi I_0(\kappa)} \exp[\kappa \cos(y - \mu)] \tag{1}$$

where $0 \leq y < 2\pi$, $0 \leq \mu < 2\pi$, $0 \leq \kappa < \infty$ and $I_0(\kappa)$ is the modified Bessel function of order 0 given by

$$I_0(\kappa) = \sum_{r=0}^{\infty} (r!)^{-2} \left(\frac{\kappa}{2}\right)^{2r}.$$

We say that $Y \sim \text{VM}(\mu, \kappa)$ if its density is given by Eq. (1). Since the histogram of the phone calls is multimodal (Fig. 1), a mixture of von Mises densities is considered, that is

$$f(y|\underset{\sim}{\mu}, \kappa) = \sum_{j=1}^{g} p_j f_j(y \mid \mu_j, \kappa)$$

with $\underset{\sim}{\mu} = (\mu_1, \mu_2, \ldots, \mu_g)$, $0 \leq p_j \leq 1$, $\sum_{j=1}^{g} p_j = 1$ (see Titterington, Smith, Makov, 1985 for details). Note that the three modes correspond to 11:45 am (just before lunch time), 4:21 pm (beginning of afternoon rush hour) and 8:07 pm.

It can be shown that if $\mu_j \sim \text{VM}(\mu_{(0)j}, \omega_j \kappa)$, then the marginal of y is given by

$$m(y \mid \underset{\sim}{\psi}_g) = \sum_{j=1}^{g} p_j m_j(y \mid \underset{\sim}{\theta}_j)$$

where $\underset{\sim}{\psi}_g = (p_1, \underset{\sim}{\theta}_1, \ldots, p_g, \underset{\sim}{\theta}_g)$, $\underset{\sim}{\theta}_j = (\mu_{(0)j}, \omega_j, \kappa)$ the vector of unknown parameters and

$$m_j(y \mid \underset{\sim}{\theta}_j) = \frac{I_0(\kappa R_j)}{2\pi I_0(\kappa) I_0(\omega_j \kappa)}$$

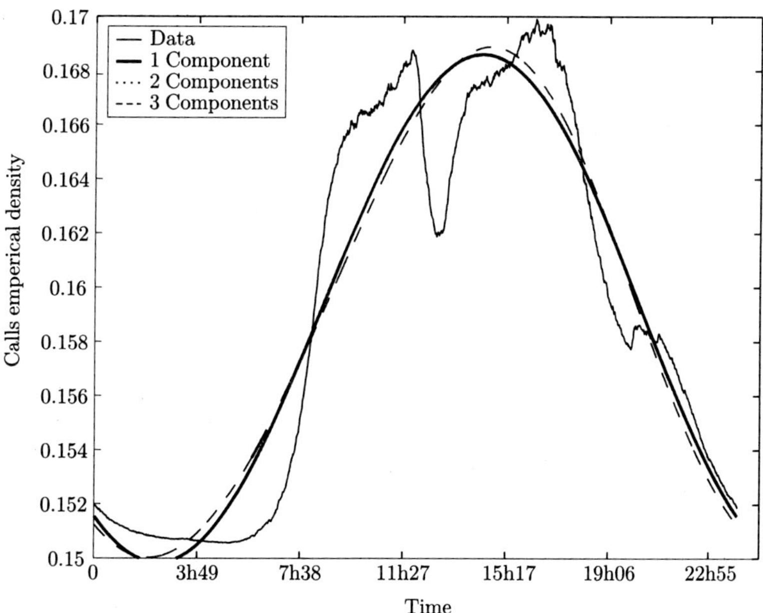

Fig. 1. **Comparison between the data and the different approximations depending on the number of components for the calls.**

with $R_j = \sqrt{C_j^2 + S_j^2}$, $C_j = \cos(y) + \cos(\mu_{(0)j})$, $S_j = \sin(y) + \sin(\mu_{(0)j})$. Several values for g have been tested but $g = 1$ seems to be more appropriate (Fig. 1) and the ML-II estimates for the hyper-parameters are given in Table 2.

Table 2.   Estimation of the parameters with ML-II method for the calls

		Final estimate
$\mu_{(0)1}$		3.764
		(2:23PM)
	ω_1	0.132
	κ	0.999

Similarly, we apply the same modeling technique to the accident data. The only measure we have on the accidents is the time indicated on the police report. Since this time is not precise (see N'Zué, 2002), it has to be considered as an upper bound of the "real" time of the accident. Based on discussion with experts, we considered a 15 min. interval before the time indicated on the police report to be a suitable length for the real time of the accident (other lengths for the time window are considered in Section 5). Using the same method, a model with 4 components ($g = 4$) is selected (Fig. 2) with the estimated hyperparameters given in Table 3. The histogram in Fig. 2 is clearly multimodal. The major modes are 7:58 am (morning rush hour) and 4:08 pm (early afternoon rush hour). The mode around midnight is an artifact of the reporting system used by the road police, midnight being the default value.

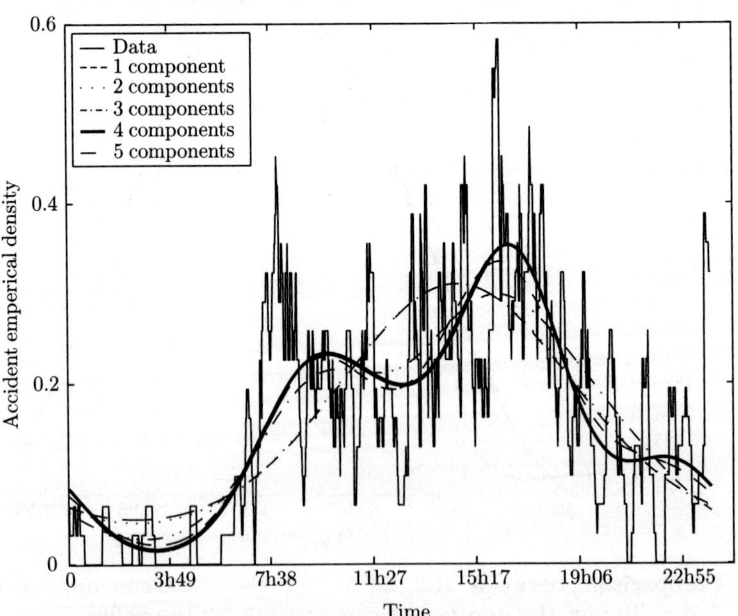

Fig. 2.   Comparison between the data and the different approximations depending on the number of components for the accidents.

Table 3. Parameter estimates for the 4 components model for the accidents

4 components			
Parameter	Final estimate	Parameter	Final estimate
p_1	0.145	ω_1	133.350
p_2	0.286	ω_2	0.950
p_3	0.418	ω_3	90.000
p_4	0.151	ω_4	136.500
$\mu_{(0)1}$	2.312 (8:50AM)	$\mu_{(0)3}$	4.382 (4:44PM)
$\mu_{(0)2}$	3.026 (11:34AM)	$\mu_{(0)4}$	5.928 (10:39PM)
κ	3.763		

4. The Margin Probabilities

In Table 1, the quantities n_{11}, n_{12} and n_{++} are known. However, if n_{21} is specified, the contingency table is fully specified. One method used to specify n_{21} in this context is the case-crossover design (see Redelmeier and Tibshirani, 1997). The case-crossover design was introduced by Maclure (1991) to study the transient effect of brief exposure on the occurrence of a rare acute-onset disease. As in most epidemiological studies control data are needed to determine what is "unusual." However, since everyone is different, it is not possible to make comparison among individuals. The comparison must instead be made on an intra-individual basis. In other words, the individual's activity before the car accident should be compared to his/her usual activity. This gives us a "case" component as well as a "control" component, but the information for both components will be derived from the same individual. In our context, the "case component" can be defined as the time period right before the car accident whereas the "control component" would be the same time period but on a previous day. The information of the individual's cell phone use during the critical pre-accident period and the control period would be compared. If such information is collected from many subjects who use cell phones while driving, we can test for a consistent relationship between cell phone use and the risk of car accident. Since we consider both, a hazard period and a control period, and each individual provides the exposure information for both the hazard and control periods, the case-crossover design can be viewed as a matched case-control study design involving cases where each individual serves as his/her own control (see Chang, Y-F, www.pitt.edu/~super1/lecture/lec0821 for more details).

In this article, instead of using the case-crossover design, we shall obtain the "control component" from the fitted models specified in Tables 2 and 3. In order to do so, let t_i, for $i = 1, 2, \ldots, n_{\mathrm{acc}}$ be the time of the n_{acc} accidents. Integrating the calls curve (Fig. 1) and the accidents curve (Fig. 2) over $(t_i - 15, t_i]$ for $i = 1, 2, \ldots, n_{\mathrm{acc}}$, we obtain n_{acc} probabilities of using the cell phone and of having an accident, denoted by $\nu_{1+,i}$ and $\nu_{+1,i}$ for $i = 1, 2, \ldots, n_{\mathrm{acc}}$. Histograms of these probabilities are given in Figs. 3 and 4 for the phone calls and the accidents, respectively.

The next step is to model these histograms using a mixture of beta densities. Since the probabilities on "using the cell phone" or "having an accident" are very small, the value of the second parameter in the beta densities has to be very large. To avoid this problem, we considered

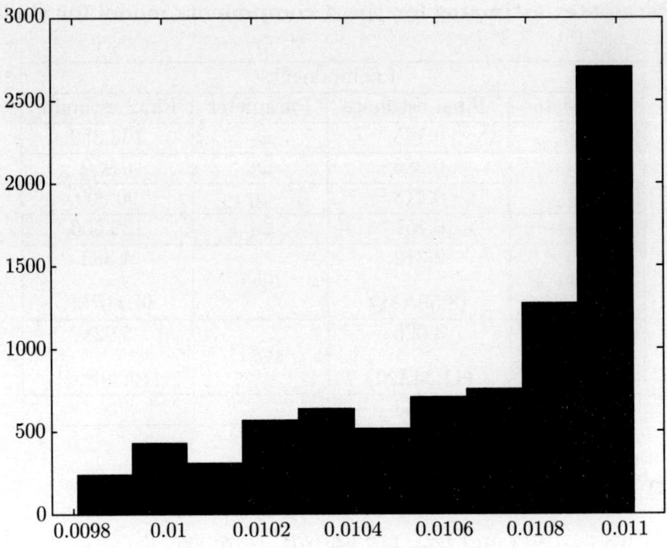

Fig. 3.   Call probabilities.

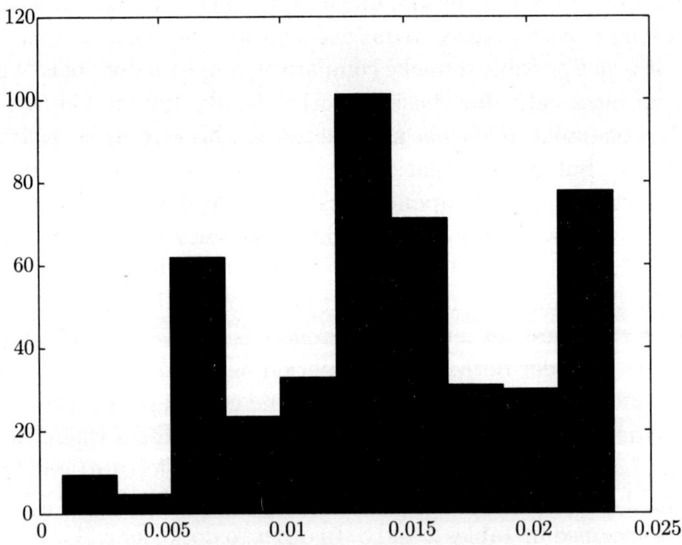

Fig. 4.   Accident probabilities.

the transformations

$$v_{1+,i} = \frac{\nu_{1+,i} - \min_{1 \le i \le n_{\mathrm{acc}}} \{\nu_{1+,i}\}}{\max_{1 \le i \le n_{\mathrm{acc}}} \{\nu_{1+,i}\} - \min_{1 \le i \le n_{\mathrm{acc}}} \{\nu_{1+,i}\}} = \frac{\nu_{1+,i} - 0.0098}{0.0012}$$

$$v_{+1,i} = \frac{\nu_{+1,i} - \min_{1 \le i \le n_{\mathrm{acc}}} \{\nu_{+1,i}\}}{\max_{1 \le i \le n_{\mathrm{acc}}} \{\nu_{+1,i}\} - \min_{1 \le i \le n_{\mathrm{acc}}} \{\nu_{+1,i}\}} = \frac{\nu_{+1,i} - 9.2 \times 10^{-4}}{0.0023}.$$

Since we can suspect the presence of several modes, we will also use the EM algorithm to find the maximum likelihood estimates for different models. In order to select the appropriate model, we computed the Bayesian information criterion (BIC) for $g = 1$, 2 and 3. The BIC values are given in

Table 4. Based on these results, we decided to use 3 components to model the probability curves for the call and for the accidents.

Table 4.   BIC values for the choice of the number of components for the calls and the accidents

g	BIC	
	Calls	Accidents
1	−2.795	−3.006
2	−7.810	−7.839
3	−11.005	−7.714

The results are shown in Table 5 and displayed in Fig. 5 for the cell phone calls and Fig. 6 for the accidents.

Table 5.   Parameter estimates for the 3 components model for the calls and the accidents

	Calls	Accidents
	Final estimate	Final estimate
p_1	0.097	0.410
p_2	0.245	0.283
p_3	0.608	0.307
α_1	2.420	2.402
α_2	5.340	61.317
α_3	9.707	9.854
β_1	13.865	3.744
β_2	5.828	42.933
β_3	1.461	1.456

The distribution of the n_{ij}, $i, j = 1, 2$, in a 2×2 contingency table is a multinomial with parameters $n_{++} = n_{11} + n_{12} + n_{21} + n_{22}$ and $\nu = (\nu_{11}, \nu_{12}, \nu_{21}, \nu_{22})$, where $0 \leq \nu_{ij} \leq 1$, $i, j = 1, 2$, and $\sum_{i=1}^{2} \sum_{j=1}^{2} \nu_{ij} = 1$. In order to obtain the Bayes estimator of the instantaneous risk, we then have to integrate

$$g(\underset{\sim}{\nu}) = \frac{\nu_{11}(\nu_{21} + \nu_{22})}{\nu_{21}(\nu_{11} + \nu_{12})} = \frac{\nu_{11}(\nu_{21} + [1 - \nu_{11} - \nu_{12} - \nu_{21}])}{\nu_{21}(\nu_{11} + \nu_{12})}$$

$$= \frac{\nu_{11}(1 - [\nu_{11} + \nu_{12}])}{\nu_{21}(\nu_{11} + \nu_{12})} = \frac{\nu_{11}(1 - \nu_{1+})}{(\nu_{+1} - \nu_{11})\nu_{1+}} \tag{2}$$

where $\nu_{1+} = \nu_{11} + \nu_{12}$, $\nu_{+1} = \nu_{11} + \nu_{21}$, with respect to the posterior density of $\underset{\sim}{\nu}$. Consequently, to apply this method, we need to know the joint distribution of the observations as well as the prior density of each involved parameter. We know the distribution of the n_{ij}, $i, j = 1, 2$, is multinomial with parameters n_{++}, $\underset{\sim}{\nu}$. Since n_{12} is unobserved, we write the likelihood of $\underset{\sim}{\nu}$ as

$$L(\nu_{11}, \nu_{12}, \nu_{21} | \underset{\sim}{n}) = L(\nu_{12} | \nu_{11}, \nu_{21}, \underset{\sim}{n}) L(\nu_{11}, \nu_{21} | \underset{\sim}{n})$$

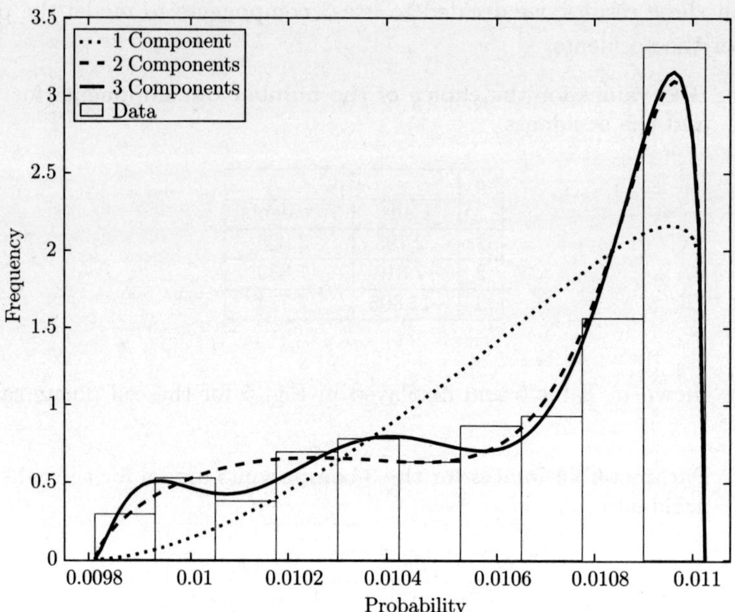

Fig. 5. Probability curves for the calls.

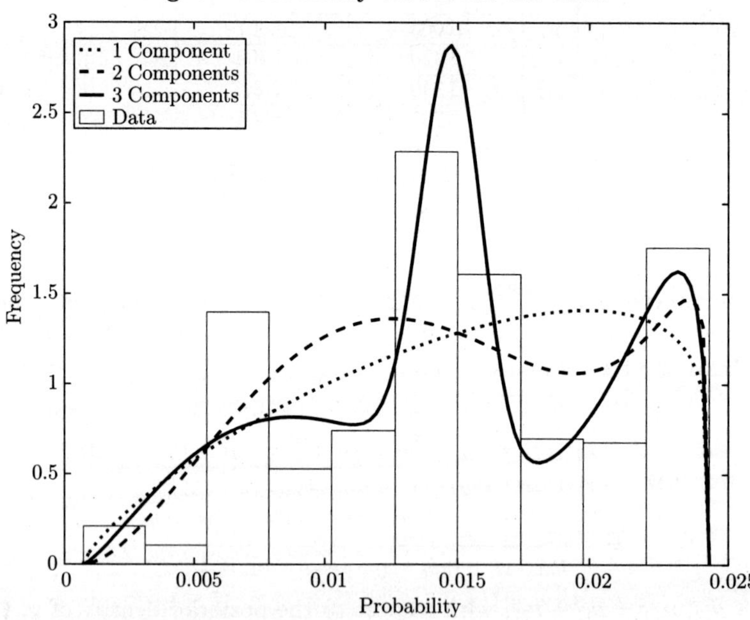

Fig. 6. Probability curves for the accidents.

where

$$L(\nu_{11}, \nu_{21} \mid \underset{\sim}{n}) = \binom{n_{++}}{n_{11}, n_{21}} \nu_{11}^{n_{11}} \nu_{21}^{n_{21}} (1 - \nu_{11} - \nu_{21})^{(n_{++} - (n_{11} + n_{21}))}$$

$$L(\nu_{12} \mid \nu_{11}, \nu_{21}, \underset{\sim}{n}) = \binom{n^*}{n_{12}} (\nu_{12}^*)^{n_{12}} (1 - \nu_{12}^*)^{n^* - n_{12}}$$

Table 2. **Comparing three spatial models in terms of DIC, variance components and spatial association. Numbers in parantheses indicate standard deviation, see text for further details**

Model	M_1	M_2	M_3
Deviance	2720.2	1397.4	1371.3
df	46.9	319.3	327.0
DIC	2814.0	2036.1	2025.3
$\sqrt{(\delta_1/w_{i+})}$	.54(.1)	NA	.53(.2)
ρ_1	.15(.1)	NA	.13(.1)
$\sqrt{(\delta_2/w_{i+})}$	NA	1.5(.3)	1.4 (.3)
ρ_2	NA	.29(.1)	.31(.1)

Spatial random effects in the q part lead to substantial improvement in the model fit. The spatial story in the p part is weak in this application. The two stage random effects lead to a better fit but increased degrees of freedom and overall the DIC for model M_3 is almost similar to that of model M_2 indicating we do not gain much by putting spatial random effects in the p part. The spatial component ρ_1 is close to zero indicating negligible spatial association in the p part compared to a moderate value of .31 for ρ_2 in the q part. The variance component δ_1 is almost 9 times smaller than δ_2. However, the 95% credible interval for $\sqrt{(\delta_1/w_{i+})}$ is (.21, .96) and does not include zero indicating that excess heterogeneity, though small is present in the p-part and gets estimated by our Bayesian model. We did not encounter convergence problems with our MCMC algorithm.

One of the advantages of working in a Bayesian framework is the ability to obtain the entire posterior distribution of parameters without relying upon asymptotic approximations. Figs. 6 and 7 show the posterior distribution of the variance components and the spatial association parameters, respectively, for model M_3.

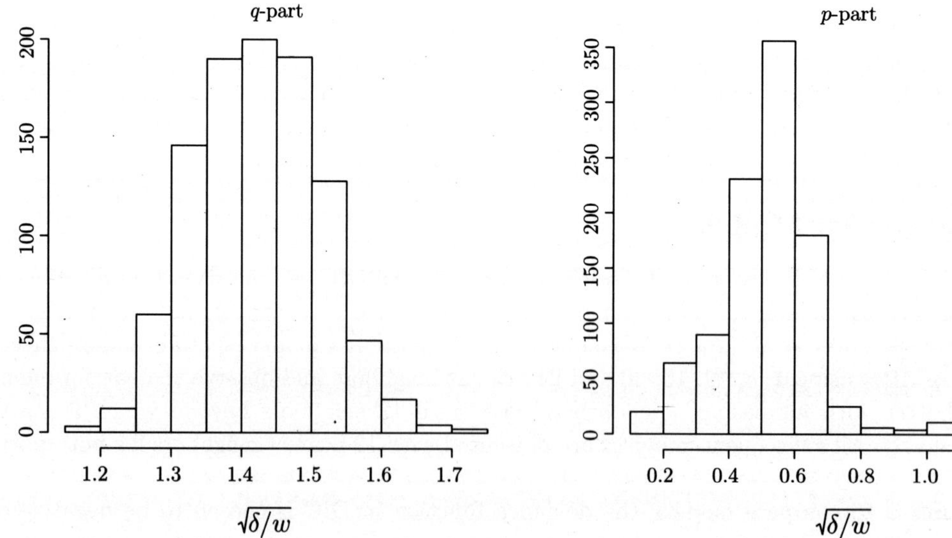

Fig. 6. **Posterior distributions of variance components for model M_3.**

$n^* = n_{++} - n_{11} - n_{21}$ and $\nu_{12}^* = \dfrac{\nu_{12}}{1 - \nu_{+1}} = \dfrac{\nu_{1+} - \nu_{11}}{1 - \nu_{+1}}$. Hence, using the prior densities for ν_{1+} and ν_{+1} given in Table 5, next putting a uniform prior for ν_{11} on the interval $(\max(0; \nu_{1+} + \nu_{+1} - 1); \min(\nu_{1+}, \nu_{+1}))$ and then using Monte Carlo integration with importance sampling, one can compute the expectation of Eq. (2) with respect to ν_{11}, ν_{1+} and ν_{+1}.

5. Results

Using our data set, we obtain $n_{11} = 140$ and we are able to find out that 442 drivers were involved in 473 accidents. Thus, we deduce $n_{21} = (473 \times 15) - 140 = 6955$. Our study involves 6360 individuals but 31 have to count for more than one because they had more than one accident. Each individual in the study (plus the 31 additional cases) counts for 15 one-min. individuals, which gives $n_{++} = (6360 + 31) \times 15 = 95865$. We are now able to build an incomplete 2×2 contingency table given in Table 6.

Table 6. Incomplete 2×2 contingency table

$X \setminus Y$	Yes	No	Total
Yes	140		
No	6955		
Total	7095	88870	95865

Since, in some cases more than one call was made in the 15 min. period observed before the time of the accident as noted on the police report, we need hypotheses to test whether the driver was on the phone in that interval. Hence, we selected the calls in three ways:

(a) by considering the last call made in that 15 min frame ($n_{11} = 140$ and $n_{21} = 6955$ as in Table 6)

(b) taking the shortest call in that period of time ($n_{11} = 111$ and $n_{21} = 6984$)

(c) taking the longest call within that period of time ($n_{11} = 184$ and $n_{21} = 6911$).

With the method proposed in Sections 3 and 4, we get an estimate of the instantaneous risk of 1.735 for the first hypothesis. For the second hypothesis the estimate is of 1.407, and for the last it is 2.211 (Table 7). (The precision of the Monte Carlo estimate is 0.001.) This means that anyone using a cell phone while driving is approximately 74% more likely to have an accident. We were also able to find an α-credible set for the instantaneous risk (IR) using Monte Carlo of the form

$$\mathbb{P}(a \leq \text{IR} \leq b) = \int_{\Theta} \mathbb{I}_{(a,b)}(\text{IR})\pi(\underset{\sim}{\nu} \mid y)d\underset{\sim}{\nu} = 1 - \alpha.$$

These credibility intervals are given in Table 7. Hence, the choice of the phone call considered does not have much effect on the size of the instantaneous risk.

Table 7. Credible intervals for the instantaneous risk

Hypothesis on n_{11}	Instantaneous risk	Credible interval
Last call	1.735	(1.632; 1.841)
Shortest call	1.407	(1.323; 1.491)
Longest call	2.211	(2.090; 2.332)

The other hypothesis needed to compute this risk was the size of the time window considered before the accident. As in Redelmeier and Tibshirani (1997), three time windows, that is, 5, 10 and 15 min. prior to the accident, have been considered. We have also found that the risk increases as the length of the time window decreases (Table 8).

Table 8. Instantaneous risk and length of the time window

Length	Instantaneous risk	R&T risk
5 min	6.5	4.8
10 min	2.9	4.3
15 min	1.7	1.3

6. Conclusion

We adopted a Bayesian approach to model the probability of having a car accident and the probability of cell phone use during the time interval in which a car accident happened. Using auxiliary information, the densities for the car accidents and cell phone use are obtained. With an approach similar to the case-crossover design, the densities on the marginal of a 2×2 contingency table can then be obtained. We then used those densities to perform a Monte Carlo study to get a point estimate and an α-credible set for the instantaneous risk. Finally, we concluded that using a cell phone while driving increases the risk by 74% of having a car accident.

References

Bourhattas, M. (2002). *Estimation du risque relatif avec le devis cas-chassé-croisé : étude de simulations pour le cas du téléphone mobile au volant et le risque d'accidents.* Mémoire de maîtrise, HEC-Montréal.

Fisher, N.I. (1993). *Statistical Analysis of Circular Data*, Cambridge University Press, Cambridge.

Laberge-Nadeau,C., Maag, U., Bellavance, F., Lapierre, S.D., Desjardins, D., Messier, S., Sadi, A. (2001). Wireless telephones and the risk of road crashes. *Accident Analysis and Prevention*, **35**, 649–660.

Maclure, M. (1991). The case-crossover design: a method for studying transient effects on the risk of acute events. *American Journal of Epidemiology*, **133**, 144–153.

N'Zué, A.K. (2002). *Évaluation du système de gestion autoroutière de Montréal sur les accidents et sur les délais d'intervention des services d'urgence.* PhD thesis, Faculté de médecine, Université de Montréal.

Redelmeier, D.A. and Tibshirani, R.J. (1997). Association between cellular-telephone calls and motor vehicle collisions. *The New England Journal of Medicine*, **336**, 453–458.

Titterington, D.M., Smith, A.F.M and Makov, U.E. (1985). *Statistical analysis of finite mixture distributions.* Wiley, New York.

Violanti, J.M. and Marshall, J.R. (1996). Cellular phones and traffic accidents: an epidemiological approach. *Accident Analysis and Prevention*, **28**, 265–270.

Bayesian Statistics and Its Applications
Edited by S.K. Upadhyay, U. Singh and D.K. Dey
Anamaya Publishers, New Delhi, India

Bayesian Estimation of the Key Parameters of a Fish Population

Manuela Azevedo

Institute for Fisheries and Sea Research, IPIMAR, Avenida de Brasília, 1449-006 Lisbon, Portugal

Abstract

An age-structured Bayesian assessment is illustrated with an application to the Iberian hake stock using 22 years of reported commercial catch and abundance indices at age from surveys. Aspects related to setting up the probability model, interpreting the posterior distribution and evaluating the fit of the model are presented. Uncertainty estimates of the population size, spawning biomass and fishing mortality are provided for the first time for this stock. The posterior estimate of the precision of the reported catches was higher than the surveys. The population size and spawning biomass were estimated to be smalller and the fishing mortality to be higher than those currently used for management advice. The re-evaluation of the future trajectory of the stock is recommended using the Bayesian estimates and uncertainty of the parameters provided for the last year.

1. Introduction

The age-structured stock assessment usually involves estimating the parameters of a non-linear population dynamics model by fitting it to commercial catch at age and mean abundance observed in the fishery or scientific surveys. The major concern about current stock assessment estimates relates to the general lack of allowance made for the uncertainty. In fact, fisheries management decisions should be founded on confidence statements for population key parameters. Bayesian analysis provides an appropriate means of providing quantitative support to fish stock management in the presence of uncertainty (e.g. Patterson, 1999; Millar and Meyer, 2000; Punt and Hilborn, 2001; McAllister et al., 2001; Myers et al., 2002; Lewy and Nielsen, 2003; Michielsens et al. 2004; Monteiro et al., 2004 and references therein).

In this paper, a Bayesian age-structured fish stock assessment is illustrated with an application to the hake (*Merluccius merluccius*) stock distributed in Iberian waters. This stock is exploited by Spanish and Portuguese trawlers and artisanal gears (mainly gillnets) and is of great commercial importance for both countries.

It is considered that the Iberian hake stock is overfished, suffering reduced reproductive capacity and at risk of being harvested unsustainably (ICES 2005a). Given this perception of the stock abundance and exploitation, it is additional cause for concern that the assessment method currently used (XSA, Shepherd, 1999) does not provide uncertainty of the fish stock key parameter estimates (ICES, 2005b) and results in annual overestimation of the stock spawning biomass and underestimation of the fishing mortality. The assessment method XSA uses backward-recursion population equations and an iterative process to minimize the discrepancies between the virtual population analysis estimates of fish abundance and those determined by abundance indices. Here, the most important characteristics for management purposes of an exploited fish stock, that is, the population size, the biomass of adult fish (spawning biomass) and the mortality due to fishing, are estimated for the last 22 years using a Bayesian age-structured model. The population survivor model currently used for assessment purposes is adopted by assuming separable fishing mortality. Prior information is available for the errors in observables due to measurement and sampling variability and for some characteristics of the exploitation of the stock, incorporated via the Bayesian framework. Discussion focuses on the adopted priors, modelling issues, perception of the stock abundance and on the uncertainty of the parameter estimates.

2. Methods

2.1 Data

Bayes' rule combine the information in a set of data or observables (vector $\mathbf{X}$), through the likelihood $p(\mathbf{X} \mid \theta)$, with the *prior* probabilities for the unknown parameters or unobservables $p(\theta)$, to calculate the *posterior* distribution of the model parameters, given the data $p(\theta \mid \mathbf{X})$:

$$p(\theta \mid \mathbf{X}) = \frac{p(\theta)p(\mathbf{X} \mid \theta)}{\int p(\theta)p(\mathbf{X} \mid \theta)d\theta}. \tag{1}$$

Since the denominator of Eq. (1) is only a rescaling constant (does not depend on θ), $p(\theta \mid \mathbf{X})$ is often expressed as $p(\theta \mid \mathbf{X}) \propto p(\theta)p(\mathbf{X} \mid \theta)$.

For the present analysis the vector $\mathbf{X}$ includes the reported commercial catches of hake (number at age; years: 1982-2003; age range: 1-8+) and mean abundance of hake from three scientific surveys: the Portuguese bottom trawl survey carried out in summer time (number per hour; years: 1989-1993, 1995, 1997-2001; ages: 0-8+) and two bottom trawl surveys carried out in the fall, the Portuguese survey (number per hour; years: 1989-2003; ages: 0-8+) and the Spanish survey (number per half an hour; years: 1983-1986, 1988-2003; ages: 0-8+) (ICES, 2005b).

Due to low ageing reliability (Godinho et al., 2001) the last age group represents the accumulated catches or the mean abundance of hake of age 8 and older and is, therefore, denoted by 8+. The zero values, reported for the oldest ages in the abundance indices in the fall (age 7 in 2002 and age 8+ in 1997, 2002 and 2003 in the Portuguese survey; age 7 in 1998 and age 8+ from 1997 to 1999 in the Spanish survey) and likely due to sampling problems were replaced by the minimum values observed in the respective age group and survey.

2.2 Survivor Model

The basic assumption of the evolution of the number of fish of a cohort, in a time interval $T = 1$, is that it follows an exponential decay model. Let $N_{a,y}$ denote the true number of age a fish in the population in the beginning of year y. With the assumption of separable fishing mortality (e.g. Pope and Shepherd, 1982), the number of survivors of age a fish in the end of year y, $N_{a+1,y+1}$ is expressed as:

$$N_{a+1,y+1} = N_{a,y} \exp(-f_y r_a - M_{a,y}) \tag{2}$$

where f_y is the level of intensity of mortality by fishing during the year y and is associated with the quantity of fishing effort (number of fishing vessels, number of hours fishing, etc.) and with the efficiency or fishing power of the vessels or gear; r_a is the relative exploitation pattern at age a and is associated with the selective properties of the fishing gears relative to the ages of the fish available to be captured during the year y; and $M_{a,y}$ the coefficient of natural mortality (all other causes of death) by age and year.

The separable fishing mortality model relies on the assumption that the multiplicative combined set of the fishing level (usually a unique value for all ages) and the relative exploitation pattern (different values according to the age of the fish) represents the fishing mortality coefficient applied to fish age a in year y, expressed as

$$F_{a,y} = f_y r_a \tag{3}$$

For most fish stocks prior information on $M_{a,y}$ is not available hence natural mortality is usually assigned a fixed value based on demographic or life history information. For the Iberian hake stock, a fish species with mean longevity of around 20 years, M is assumed to be 0.2 year^{-1} (ICES, 2005b). Note also that information encoded in observables does not allow determining M.

2.3 Observation Model

Observation models are needed to relate the stock size to a variable that is directly related to stock size and can be observed. Let $C_{a,y}$ denote the catch, in number, of fish age a in year y. Although the catch in number can be expressed in several ways (e.g. Cadima, 2003), with the assumed constant natural mortality, the relationship between catch and stock size used is

$$C_{a,y} = \frac{f_y r_a}{f_y r_a + M} N_{a,y} \left(1 - \exp\left(-f_y r_a - M\right)\right). \tag{4}$$

Let $I_{a,y}^s$ denote the mean fish abundance in number observed in the scientific survey s. A linear relationship between indices of abundance and true stock size is assumed:

$$I_{a,y}^s = q_a^s N_{a,y} \exp\left(t^s \left(-f_y r_a - M\right)\right) \tag{5}$$

where q_a^s (catchability coefficient) is the fraction of the stock size at age a that is caught by unit of effort of survey s and t^s is the proportion of the year that has been completed when the survey s took place (0.58 and 0.83 for the Portuguese surveys in summer and fall, respectively, and 0.75 for the Spanish survey in the fall).

Because catch and abundance indices are positive numbers and observation errors are usually assumed to be multiplicative with a constant coefficient of variance (e.g. Gavaris and Ianelli, 2002; Maunder and Punt, 2004) a log-normal likelihood is adopted for the analysis. It might be argued that once we are dealing with counting random variables it would be more appropriate to adopt a discrete distribution, such as the binomial distribution. However, since the population size and catch are large counts and the mortality coefficients are small values this would result in a very small spread of the binomial distributions and, hence, a model nearly deterministic.

Therefore, it is assumed that the log-transformed catch (Eq. (6)) and abundance indices (Eq. (7)) are independently and identically normally distributed:

$$\log\left(C_{a,y}\right) \sim N\left(\mu_{a,y}^c,\ \sigma_c^2\right) \tag{6}$$

$$\log\left(I_{a,y}^s\right) \sim N\left(\mu_{a,y}^s,\ \sigma_s^2\right). \tag{7}$$

The parameters μ and σ^2 in Eqs. (6) and (7) are the mean and variance of the log of the variables, which from Eqs. (4) and (5) can be expressed as

$$\mu_{a,y}^c = \log(f_y) + \log(r_a) - \log(f_y r_a + M) + \log(N_{a,y}) + \log\left(1 - \exp\left(-f_y r_a - M\right)\right) \tag{8}$$

and

$$\mu^s_{a,y} = \log(q^s_a) + \log(N_{a,y}) + t^s \left(-f_y r_a - M\right).\tag{9}$$

Thus, the likelihood function for a single catch (Eq. (10)) or single survey abundance (Eq. (11)) observation (age a in year y) is the normal probability density function of the difference between the log-transformed predicted and observed catch or survey abundance:

$$p\left(C_{a,y}|\theta\right) = \left(\sqrt{2\pi}\sigma_c\right)^{-1} \exp\left(-\frac{\left(\log\left(C_{a,y}\right) - \mu^c_{a,y}\right)^2}{2\sigma^2_c}\right)\tag{10}$$

and

$$p\left(I^s_{a,y}|\theta\right) = \left(\sqrt{2\pi}\sigma_s\right)^{-1} \exp\left(-\frac{\left(\log\left(I_{a,y}\right) - \mu^s_{a,y}\right)^2}{2\sigma^2_s}\right).\tag{11}$$

2.4   Parameter Specification and Priors

For the Bayesian implementation of the Iberian hake age structured model presented here, priors are assigned to the population size in the first age group by year, denoted $N_{0,y}$ (commonly designated by recruits), on the population size at age in the first year, denoted $N_{a,1982}$, on the annual fishing level f_y and on the relative exploitation pattern r_a. By assigning prior distributions to the parameters referred above, the remaining population size at age is projected forward using recursively the survivor model (Eq. (2)).

The remaining unobservables are the age dependent catchability coefficients for each survey, denoted q^{PS}_a, q^{PF}_a and q^{SF}_a (superscripts PS and PF refer to the Portuguese surveys in summer and in the fall, respectively, and SF to the Spanish survey in the fall) and the variances of the log of reported catch σ^2_c, and of the log of survey abundance indices, denoted σ^2_{PS}, σ^2_{PF} and σ^2_{SF} (using under scripts as survey indexes).

Table 1 presents the priors assigned to the parameters. The upper bound on $N_{0,y}$ and $N_{a,1982}$ are set such that encompasses the magnitude of the population size as perceived from previous analysis (e.g. ICES, 2005b). The level of fishing intensity in the first year f_{1982} was fixed at 0.8 year^{-1}, this value being estimated by a catch curve analysis of the hake abundance by length from the Portuguese bottom trawl surveys carried out in September and October of 1982 and available in Cardador (1988). The upper bound of the uniform prior for f_y, set at 2.0, represents a rather low survival rate due to fishing (around 14%) for the fully exploited ages and encompasses values estimated for other long-lived demersal stocks that have been subject to high exploitation rates (e.g. cod stock in the North Sea, northern hake stock (ICES, 2005a)).

The relative exploitation pattern for the Iberian hake stock is a combination of the selectivity properties of several gears and the relative contribution to the catches of each of these gears. Fig. 1 shows the selectivity properties of the bottom trawl (Fonseca et al., 2000) and gillnets (Fonseca et al., 2005), the gears contributing to more than 90% of the reported catches of hake. In the present analysis the parameter vector was reduced by fixing the relative exploitation pattern at age 6 at 1.0 ($r_6 = 1.0$), supported by the selection curves for combined gears (Fig. 1).

According to experts there was a change in the relative exploitation pattern around 1995 (Cardador et al., 2002; ICES, 2005b) and, accordingly, two periods were assumed for r_a, one between 1982 and 1994 and the other from 1995 to 2003.

Gamma prior distributions with hyperparameters were specified for the inverse of variance (precision) of the log of the reported catches and survey abundance indices (Table 1). The hyperparameters were set such that the 2.5% and 97.5% percentiles of the coefficients of variance of the

Table 1. Prior distributions assigned to the parameters

Parameter	Prior
$N_{0,1982}, N_{0,1983}, \ldots, N_{0,2003}$	$U(0, 10^6)$
$N_{1,1982}, N_{2,1982}, \ldots, N_{8+,1982}$	$U(0, 10^6)$
$f_{1983}, f_{1984}, \ldots, f_{2003}$	$U(0, 2)$
$r_1, \ldots, r_5, r_7, r_{8+}\,(1982\text{-}1994)$	$U(0, 1)$
$r_1, \ldots, r_5, r_7, r_{8+}\ (1995\text{-}2003)$	
$q_0^{PS}, q_1^{PS}, \ldots, q_{8+}^{PS}$	$U(0, 1)$
$q_0^{PF}, q_1^{PF}, \ldots, q_{8+}^{PF}$	
$q_0^{SF}, q_1^{SF}, \ldots, q_{8+}^{SF}$	$G(\alpha_c, \beta)$
$1/\sigma_c^2$	
α_c	$N(4.51, 1.28)$
$1/\sigma_{PS}^2 \ ; \ 1/\sigma_{PF}^2$	$G(\alpha_P, \beta)$
α_P	$N(62.5, 0.003)$
$1/\sigma_{SF}^2$	$G(\alpha_S, \beta)$
α_S	$N(72.22, 0.005)$
β	$G(0.1, 0.1)$

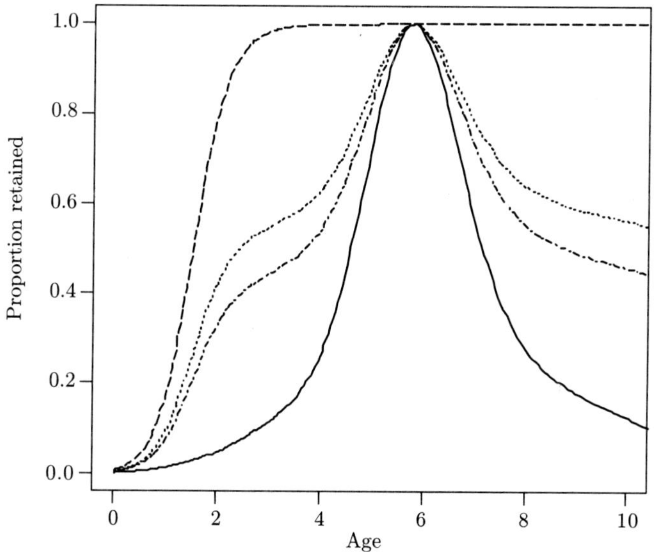

Fig. 1. Selection curves by age for the trawl (65 mm mesh size; dashed line) and gillnet (80 mm mesh size; solid line) gears exploiting Iberian hake. Superimposed lines illustrate the expected proportion of fish retained corresponding to higher (dotted line) and lower (dot-dashed line) contribution of the trawl gear to the total hake catches.

catches were 40 and 60% (Jardim et al., 2004), were 10 and 20% for the Portuguese surveys and were 10 and 15% for the Spanish surveys (ICES, 2005b; IPIMAR, 2004). Therefore, for example, the shape parameter of the gamma distribution for the reported catch α_c is modelled as a normal distribution with $\mu = 4.25$ and $1/\sigma^2 = 1.28$. To avoid negative values for the hyperparameter, the normal distributions were left truncated.

2.5 Posterior Distributions for the Parameters

The vector θ_i, $\theta = (\theta_1, \ldots, \theta_n)$ in the model presented for the Iberian hake stock has 95 parameters $(i = 1, \ldots, n)$ as follows:

$$
\begin{aligned}
\theta = (&N_{0,1982}, N_{0,1983}, \ldots, N_{0,2003}, N_{1,1982}, N_{2,1982}, \ldots, N_{8+,1982}, \\
&f_{1983}, f_{1984}, \ldots, f_{2003}, r_1^{1982\text{-}1994} \ldots, r_5^{1982\text{-}1994}, r_7^{1982\text{-}1994}, r_{8+}^{1982\text{-}1994}, \\
&r_1^{1995\text{-}2003} \ldots, r_5^{1995\text{-}2003}, r_7^{1995\text{-}2003}, r_{8+}^{1995\text{-}2003}, \\
&q_0^{PS}, q_1^{PS}, \ldots, q_{8+}^{PS}, q_0^{PF}, q_1^{PF}, \ldots, q_{8+}^{PF}, q_0^{SF}, q_1^{SF}, \ldots, q_{8+}^{SF}, \sigma_c^2, \sigma_{PS;PF}^2, \sigma_{SF}^2).
\end{aligned}
\tag{12}
$$

The joint prior $p(\theta_i)$, is the product of the priors for each of the 95 parameters (all parameters were assumed to be independent of each other). For a model with so many parameters, the application of Bayesian inference requires a high dimensional integration to find the posterior distributions of the model parameters, given the data, since involves the computation of the multiple full conditional densities, $P(\theta_i|\theta_{-i})$. This was solved simulating the posterior distributions of the parameters using the Markov Chain Monte Carlo (MCMC) methodology with Gibbs sampling (Gilks et al., 2000; Gelman et al., 2000). Computations were performed within WinBUGS (BUGS Project, 2005).

There are some parameters derived quantities that are used for stock assessment and management purposes. Routinely computed quantities are the spawning stock biomass SSB_y, and the average fishing mortality over some selected ages $Fbar_y$. SSB_y is used to measure the annual reproductive capacity of the stock, being a function of the population size at age and the biological parameters mean weight and proportion at age of mature fish (the biological information necessary to compute SSB_y was adopted from ICES, 2005b). $Fbar_y$ summarises the annual fishing mortality and, for the hake stock, can be expressed as (using Eq. (3)):

$$
\text{Fbar}_y = \sum_{a=2}^{5} \frac{F_{a,y}}{4}.
\tag{13}
$$

The main diagnostics related to the required properties of MCMC chains were performed aiming to ensure numerical convergence of the chain, removal of parameter's autocorrelation and adequate precision. Standardised residuals and predictive p-values, computed using posterior predictive simulations of $(\theta, \mathbf{X}^{\text{rep}})$ were used for the Bayesian test of model fit (Gelman and Meng, 2000; Paulino et al., 2003).

3. Results

3.1 Convergence of the Chain and Parameter Autocorrelation

A first run with two chains, started in over-dispersed values, showed that they overlapped and exhibited similar pattern of regularity after few iterations. Therefore, one chain of length 51000 iterations was run. The visual diagnosis of the parameters' trace plots and kernel densities showed that it was sufficient a burn-in of 1000 iterates to remove the influence of initial values. Stopping the chain at 51000 iterations was considered to provide adequate precision: the Monte Carlo standard errors were less than 5% and the summary statistics of the iterations in the beginning of the chain and those of the last iterations were very similar for all the parameters (not presented). The parameter's autocorrelation was removed by adopting a thinning interval of 10 iterations. Fig. 2 presents, for some of the parameters and for illustrative purposes, the trace, autocorrelation plots and kernel estimates for the marginal posterior densities from the 5000 samples.

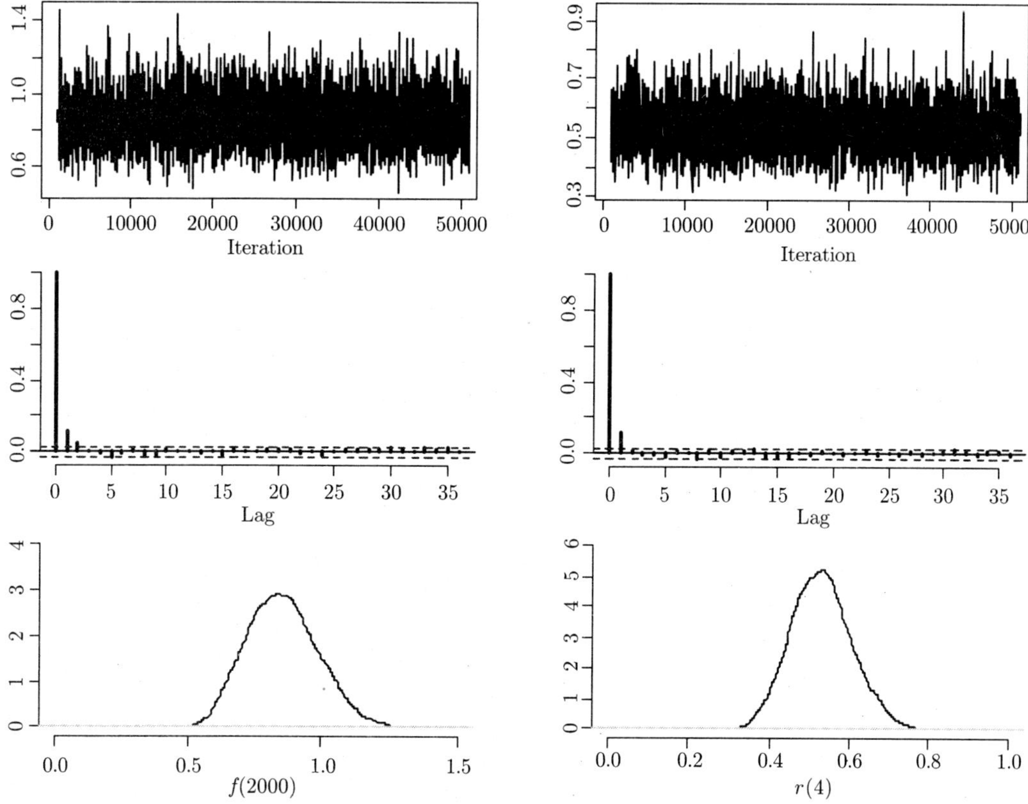

Fig. 2. Trace plots (upper panel), autocorrelation (middle panel) and kernel densities (lower panel) for the annual fishing level in 2000 (f_{2000}, left side) and the relative exploitation pattern at age 4 in the period 1982-1994 ($r_4^{1982-1994}$, right side).

The parameter estimates were summarised by the posterior medians and the 95% credible intervals (95% CI) of the 5000 samples. The posterior medians were, however, very close to the means for most parameters.

3.2 Model Adequacy

Inspection of standardised residuals revealed no major discrepancies between observed and predicted values and, despite some few exceptions, standardised residuals were between acceptable values (Fig. 3). The normal probability plots (Fig. 3, qq-plots) show moderate skewness for the catch (positive and negative) and very moderate skewness for the Portuguese survey in the summer (positive) and the Spanish survey in the fall (negative). The high negative residual in the catch, corresponds to the lowest reported catch at age 1 in 1993. It is noted that age 1 is usually poorly sampled, due to discarding practices of small and young hakes. The high negative residual in the Portuguese survey in the fall, corresponds to an abnormal low abundance at age 0 in 1995. Taking into account that the survey sampled area in 1995 covered the main recruitment areas (Cardador, pers. comm.), the low recruitment abundance observed in that survey is likely to be due to a change in the main recruitment season in 1995 once this yearclass was of average strength at age 1 in 1996 and the most abundant at age 2 in 1997.

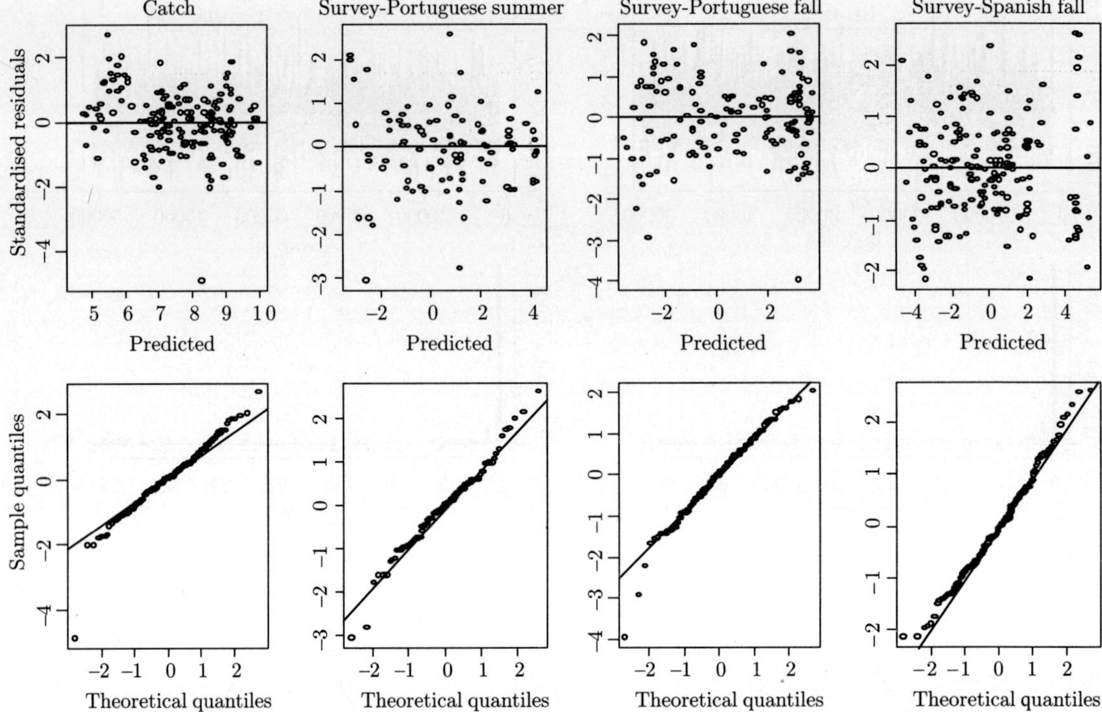

Fig. 3. **Scatterplots of standardised residuals versus predicted values (upper panel) and *qq*-plot of standardised residuals (lower panel) for each observable.**

Data was replicated for all observables (catch and abundance indices) by generating new sequences of data for each of the 5000 values sampled from the joint posterior distribution of θ. The posterior predictive model check indicates that the model is reasonably accurate once the points in the scatterplot of observed versus predicted data are evenly distributed above and below the diagonal line (45° line). This is shown in Fig. 4 for the catch and survey abundance indices in the fall. The posterior predictive *p*-values, given by the proportion of points lying above the 45° line, indicated that the probability that the replicated data were more extreme than the observed data was around 0.5.

3.3 Parameter Estimates

The posterior estimates of the variance of the log-transformed observations indicated that the catches were more precise than the survey abundance indices, once posterior estimates were 0.23 for the catch, 0.93 for the Portuguese surveys (summer and fall) and 0.81 for the Spanish survey. These values correspond to CVs (computed as CV $=\sqrt{\exp(\sigma^2)-1}$) of 51%, 124% and 112%, respectively.

Survey Catchability The catchability at age estimates is presented for each survey (log scale) in Fig. 5. The results indicate that the time of the year (summer and fall) has no major effect on the catchability at ages 1-8+ for the Portuguese surveys. The lower catchability estimated at age 0 in the Portuguese summer survey reflects the lower availability of recruits in the summer time. A decreasing catchability at older hakes is estimated for all surveys. The Spanish survey is, from all

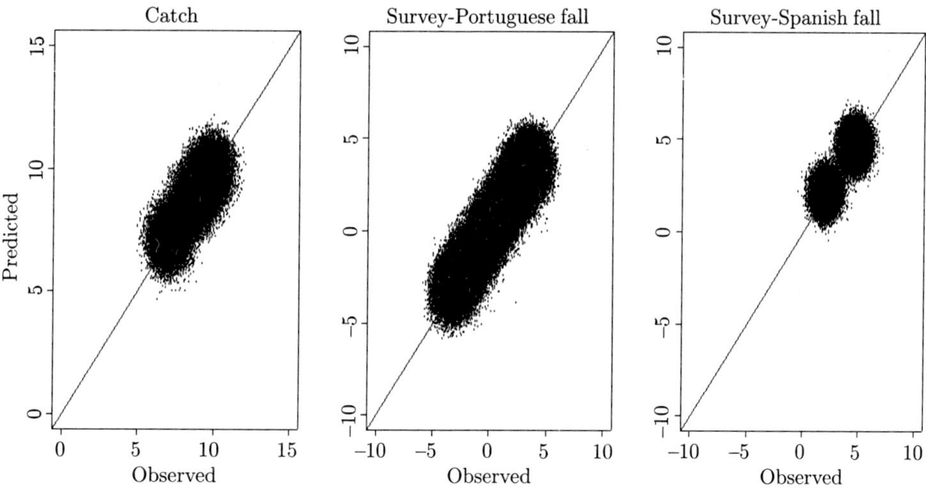

Fig. 4.   Scatterplots of observed versus predictive values with the 45° line superimposed.

three surveys, the one with the highest precision on the catchability at age estimates. The estimates also indicate for all surveys higher estimates uncertainty for the oldest age group (8+), with CV's between 26 and 31%, while in the ages 1 to 7, the CV's ranged between 15 and 21%.

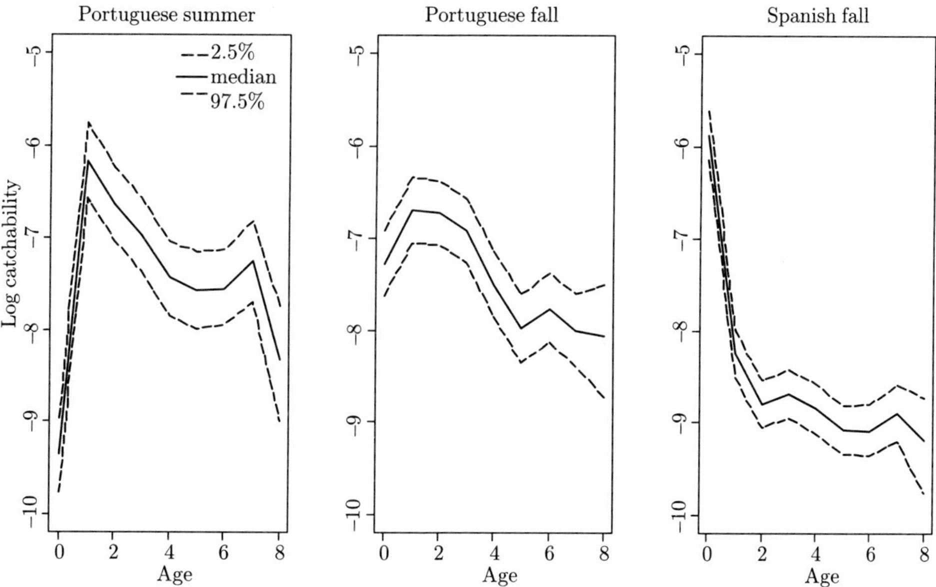

Fig. 5.   Estimated catchability (log scale) at age and 95% CI by survey.

Fishing Level and Relative Exploitation Pattern   The Bayesian estimates of the fishing level f_y, indicated quite high fishing intensity from 1983 to 2003, from 0.7 to 1.2 year^{-1} and moderate CV's (with the exception of the year 2003, CV's were between 14 and 17%). The highest estimated fishing level was in 1988 and 2001. In the last year of the analysis the Bayesian estimate was 0.92 year^{-1}, although with 95% CI of (0.60, 1.41).

The estimated relative exploitation pattern in each period (Fig. 6) exhibited an acceptable age-specific selectivity given the selectivity characteristics of the main gears operating in the fishery (Fig. 1).

After 1995 the relative exploitation pattern was smaller for the younger hakes (ages 1 and 2) and higher in the oldest age group. In 1995 the proportion of hake total landings of the fishing trawl component increased from an average of 0.38 (1982-1994) to 0.54 (1995-2003). The expected effect of the higher contribution of the trawl fleet landings on the relative exploitation pattern is opposite to the one estimated. Nevertheless, the estimated decrease in the exploitation pattern in younger ages is likely to be due to the recent enforcement of the minimum landing size for hake (27 cm; mostly ages 0 and 1) resulting in a decrease in the reported catches for those ages.

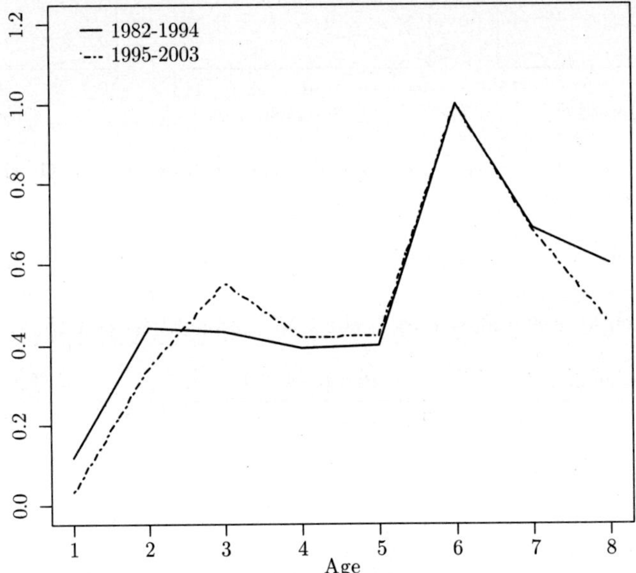

Fig. 6. Estimated relative exploitation pattern (r_a at age 6 fixed at 1.0) by period.

Recruitment, Spawning Biomass and Fishing Mortality Fig. 7 presents for the entire assessment period, the Bayesian estimates for the recruitment, spawning biomass and Fbar (computed using Eq. (13)).

The recruitment sharply decreased from 1982 to 1988 from 110500 to 51000 thousand fish, fluctuated during the nineties around 46000 thousand and decreased again in 2000 and 2001 to 31000 thousand. In the most recent years (2002-2003) it seems that recruitment has slightly increased again, although high uncertainty is associated with these estimates. In fact the CV's from 1982 to 2001 were in the range 12-19% but in 2002-2003 were between 25-35%. There is a high probability that the spawning biomass has continuously declined from 55000 to 11000 tonnes in the period 1982-1995, has remained at around 10500 tonnes from 1996 to 2001 and has decreased again in 2002 and 2003. The lowest spawning biomass, estimated to be 7400 tonnes in 2003 (CV of 9%) represents only 13% of the 1982 level. The Bayesian estimates of Fbar_y (from 0.40 to 0.73 year^{-1}) show a time trend similar to the fishing intensity. With the exception of 2003, the uncertainty around the estimate of Fbar is quite small (CV's between 9 and 16%).

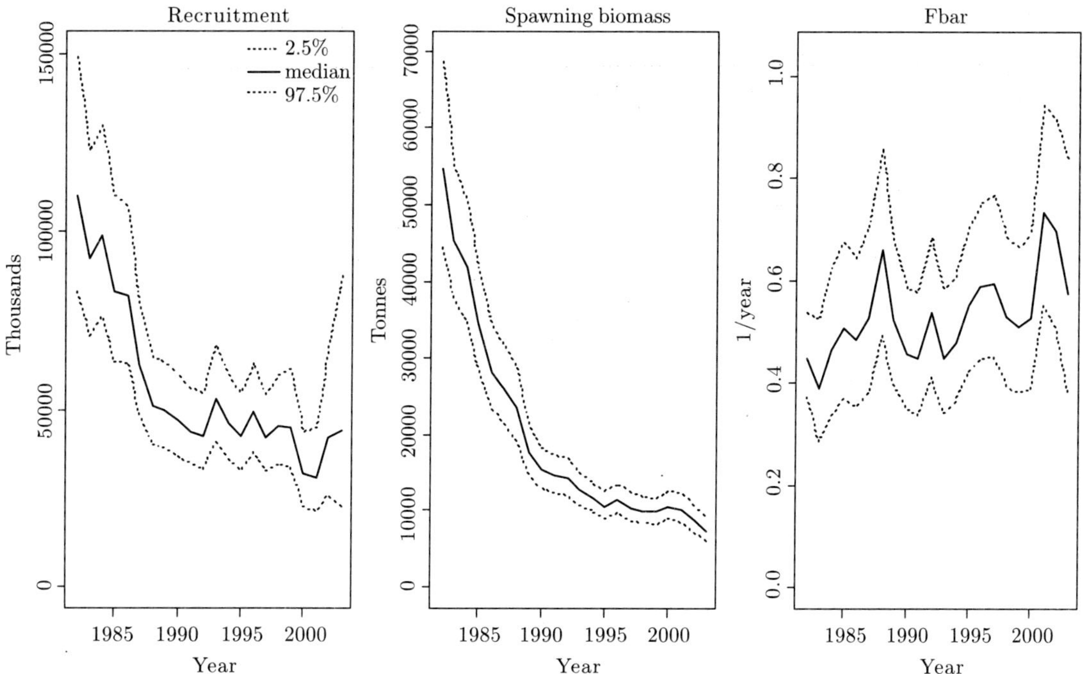

Fig. 7. Recruitment, spawning biomass and average fishing mortality (Fbar: mean fishing mortality for ages 2 to 5) estimated by the Bayesian age based model.

4. Discussion

4.1 Priors and Modelling

In the Bayesian analysis of the hake stock, with the exception of the variances of the log-transformed data, uniform priors were assigned to the parameters. The background information on these parameters was used to set bounds that would encompass most of the exploitation and biologically reasonable parameter combinations that were likely given the data. When the data contains much more information than the prior, the likelihood dominates. So the posterior, the product of prior and likelihood, looks more like the likelihood (e.g. Gelman et al., 2000). Also, it has been shown that when priors are diffuse and the observation error is large compared to the process error (a deterministic component of a population model inadequately describing population processes), the posterior modes of the parameters may be similar to the maximum likelihood estimates under observation error model (Millar and Meyer, 2000). However, Nielsen and Lewy (2002), using simulated data and flat priors, compared the frequentist properties of Bayes and the maximum likelihood estimators in an age-structured fish stock assessment model and showed that Bayesian analysis can produce less biased estimates and with lesser variance. This is a desirable feature given that decision analysis on fisheries management actions depend on the fish stock abundance estimates and, therefore, any increase in the precision of these estimates is most desirable. Therefore, albeit that the parameter estimates for the Iberian hake stock presented here were mainly dominated by the assumed likelihood function, they should be less biased than estimates from other assessment methods. Moreover, with the Bayesian age based assessment, parameter estimates uncertainty is provided for the first time for this stock.

Different coefficients of variance were assumed for the observables, with bounds on the priors set according to scientific background information. However, assuming the same bounds for all ages and years may not be the most appropriate as catch data, for the young ages is more noisy because of discarding practices (not yet quantified) and mostly in recent years, due to the enforcement of the minimum landing size.

Time invariant catchability was assumed. This assumption is considered reasonable for abundance indices from surveys that us the same vessel/gear every year and follow a random sampling plan. Note that the Portuguese bottom trawl surveys in the fall of 1996, 1999 and 2003, used in the analysis, were carried out with a different gear but according to the results from Cardador and Azevedo (1996) the hake mean abundance between the two gears is not significantly different. The estimated decreasing catchability at older hakes (Portuguese and Spanish surveys) indicates that modelling survey catchability with a logistic curve or assuming a constant catchability above a certain age is not appropriate for the hake stock.

Also, note that once the Bayesian estimates of the relative exploitation pattern indicated a decrease at older ages in the period 1995-2003 (Fig. 6), any assessment method that assumes fishing mortality at age 8+ equal to age 7, as is the case of the assessment method currently used for the Iberian hake stock, is likely to underestimate the survivors at the oldest age group.

4.2   Perception of the Stock Abundance and Exploitation

The Bayesian assessment indicates a declining stock, subject to high fishing mortality levels. The last assessments of the hake stock, carried out with XSA, show a tendency to overestimate the spawning biomass and to underestimate the fishing mortality (ICES, 2005a).

A comparison between the posterior predictive estimates of spawning biomass and average fishing mortality from the Bayesian model and the estimates from the assessment carried out in 2004 was made computing their ratio (Bayes/XSA). The Bayesian estimates indicated, for most of the years, lower spawning biomass and higher fishing mortality (Fig. 8a). It is emphasised that the discrepancies were not due to the fact that abundance indices at age of some commercial fleets were used by XSA because the catch data from these fleets were already part of the commercial catch and, within the Bayesian framework; the same information should not be used twice.

The assessment with XSA excluded from the analysis the survey abundance indices observed for ages 6 to 8+ due to the low catches of large hakes, that is, only data for ages 0 to 5 were used in the analysis (ICES, 2005b). To investigate whether the data points corresponding to the excluded ages could explain the discrepancies a second Bayesian analysis was performed. Despite that some approximation between estimates has occurred, the Bayesian estimates still indicate lower spawning biomass and higher average fishing mortality (Fig. 8b). The survey abundance indices are the only independent source of information on the changes in the abundance of the older fish and since there is no *a priori* reason to exclude these data points in the Bayesian model, the remaining discussion is based on the results from the first analysis (including all observations from the surveys).

4.3   Uncertainty in the Parameter Estimates

For management advice it is necessary to project the stock forward for different levels of fishing mortality to anticipate the stock abundance trajectory in future years. It is, therefore, important to take account of the uncertainties in the parameter estimates and, in particular, in the population size and fishing mortality at age in the last year of the assessment. Table 2 presents the posterior distributions of these parameters summarised in terms of their median estimate, CV and 95% CI. As

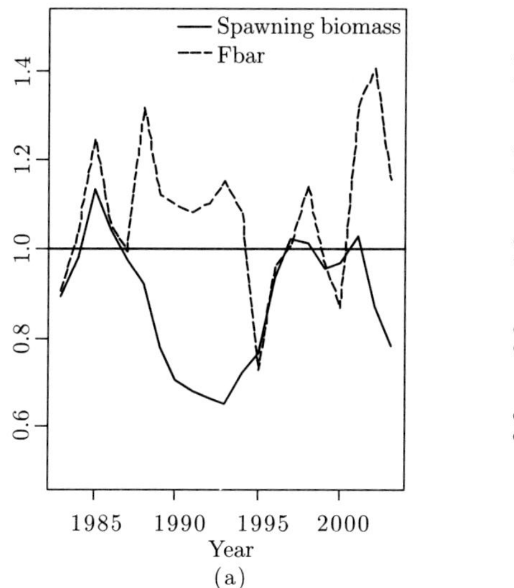

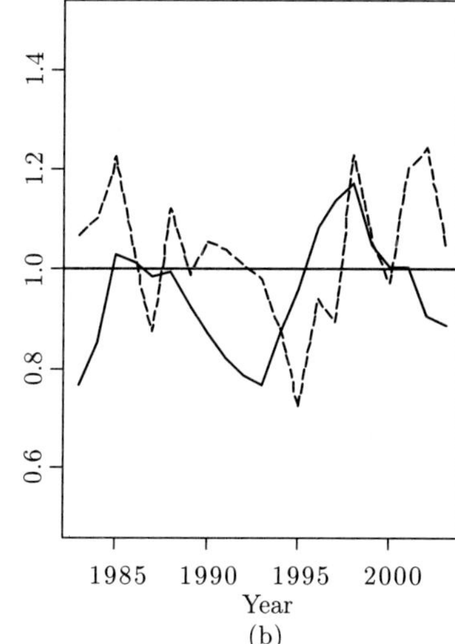

Fig. 8. **Ratio between the Bayesian estimates and XSA estimates of spawning biomass and average fishing mortality (Fbar): (a) using the observed survey abundance indices for ages 0 to 8+ and (b) only abundance indices for ages 0 to 5.**

already stressed, higher uncertainty was estimated for the recruitment in 2003, CV of 35%, mainly because only a limited number of observations were available.

Table 2. **Summary of the posterior distributions of the population size at age $N_{a,y}$ (start of the year) and fishing mortality at age $F_{a,y}$ in 2003: median estimates, 95% credible interval and coefficient of variance (CV)**

Age	Population size				Fishing mortality			
	Median	2.5%	97.5%	CV	Median	2.5%	97.5%	CV
0	44360	23030	87760	35%				
1	34190	21070	54800	25%	0.04	0.02	0.07	27%
2	19650	13440	28600	20%	0.45	0.28	0.73	25%
3	9509	6535	13590	19%	0.70	0.44	1.08	23%
4	4578	3225	6591	19%	0.55	0.34	0.86	24%
5	2188	1581	3052	17%	0.58	0.36	0.90	24%
6	1023	722	1426	18%	0.92	0.59	1.41	22%
7	370	250	535	19%	0.82	0.52	1.24	23%
8+	196	124	322	25%	0.69	0.38	1.13	27%

Because the Iberian hake stock was considered to be in a severely depleted state, a recovery plan was proposed in 2003 (STECF, 2003), though not yet implemented. The stock abundance projections then performed with annual reductions in the fishing mortality were based on assumed uncertainty for the population size and fishing mortality at age. In light of the updated information on the parameters point estimates and uncertainty from the Bayesian analysis, it is suggested to have a re-evaluation of the trajectory of the stock. The Bayesian estimates indicate lower population size at age, higher fishing mortality at age and uncertainty above that used.

Prediction of the recruitment in coming years is usually associated with high uncertainty. Often the procedure is to adopt the geometric mean of the recruitment time series or, alternatively, to fit a theoretical relationship (stock and recruitment function) that associates the spawning biomass in each year with the recruitment resulting from that spawning biomass (e.g. Cadima, 2003). Plotting the 5000 estimates from the posterior predictive distributions of the hake recruits against the spawning biomass for the 22 years gives the cloud presented in Fig. 9. The median estimates are also displayed indicating a decreasing trend of the recruitment with decreasing spawning biomass.

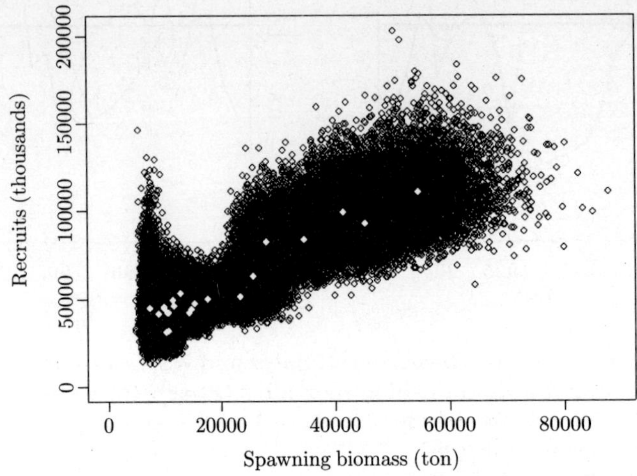

Fig. 9.   Scatterplot of recruitment and spawning biomass for the entire assessment period: using all Bayesian estimates (black circles: 5000 samples × 22 years for each parameter) and only the annual median estimates (white dots).

Since the recent levels of spawning biomass were the lowest of the 22 years of data, the recruitment in 2004 was not expected to increase. Notwithstanding, the cloud of points also indicates that at low spawning biomasses the recruitment range can be quite high. Knowing that both the Portuguese (Fig. 10) and Spanish surveys carried out in the fall of 2004 indicated higher recruitment levels compared to previous years (NeoMAv project, Sanchez, pers. comm.) it is tempting to associate an increase in the recruitment to other causes (e.g. ecological, environmental) rather than to a direct relationship with the spawning biomass. Indeed, previous fits of a stock-recruitment function for the Iberian hake stock were poor (ICES, 2003).

The analysis illustrated the application of an age based Bayesian assessment. Considering the Iberian hake stock, used as an example here, there are several aspects that should be improved both in data and modelling, namely extending the model to include prior information on hake discards, known to occur mainly for the young fish. This is by far the most important aspect once, at present, the survival of the recruits is only determined by the assumed natural mortality.

Acknowledgements

The author would like to thank Fátima Cardador and Ernesto Jardim (IPIMAR) for their input knowledge on the Iberian hake stock and to Professor Maria Antónia Turkman, University of Lisbon, who introduced me to the Bayesian Inference and methods. This work was carried out within the IPIMAR project NeoMAv (UE-cofinanced, QCA III-FEDER).

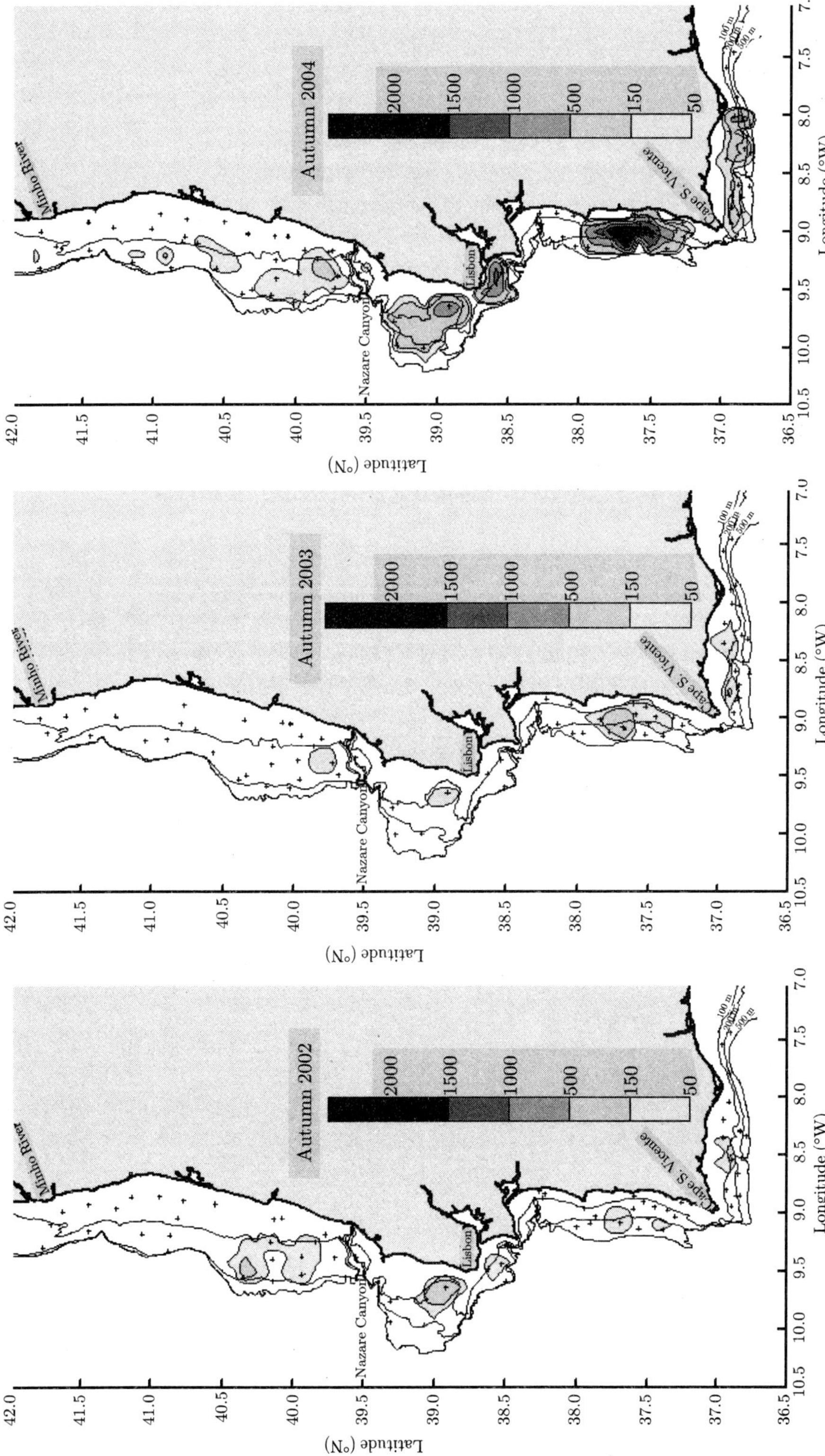

Fig. 10. Mean abundance of hake recruits (age 0) along the Portuguese coast from the survey carried out in the fall of 2002, 2003 and 2004 (redrawn from Cardador et al., 2005).

References

BUGS Project, (2005). Bayesian Inference Using Gibbs Sampling. URL www.mrc-bsu.cam.ac.uk/bugs.

Cadima, E.L. (2003). Fish stock assessment manual. *FAO Fisheries Technical Paper*, 393, Rome, FAO, 161p.

Cardador, F. (1988). Estratégias de exploração do stock de pescada (*Merluccius merluccius* L.) das águas Ibero-Atlânticas. Efeitos em stocks associados. Dissertação original apresentada para Provas Públicas de acesso à categoria de Investigador Auxiliar. IPIMAR, 98pp (Thesis in Portuguese).

Cardador, F. and Azevedo, M. (1996). A first attempt to compare efficiency of three bottom trawl nets used in research surveys along the Portuguese coast. *Bol. Inst. Port. Invest. Marít.*, Lisbon, **2**, 23-39.

Cardador, F., Jardim, E., Morgado, C., Chaves, C. and Velasco, F. (2002). Exploring the exploitation pattern of fleets catching southern hake. 18pp. *In* Working Group on the assessment of hake, monkfish and megrim, Lisbon, 21-30 May 2002. ICES, CM 2003/ACFM:02.

Cardador, F., Borges, M.F., Chaves, C. and Groom, S. (2005). Hake (*Merluccius merluccius* L.) spatial and temporal recruitment variability: a key species in the Portuguese Upwelling Ecosystem. ICES CM 2005/L: 37 (poster).

Fonseca, P., Campos, A., Garcia, A. and Cardador, F. (2000). Trawl selectivity studies in Region 3. Final Report of the Study Contract $n°96/61$ (UE–FAIR), 178 pp.

Fonseca, P., Martins, R., Campos, A. and Sobral, P. (2005). Gill-nets selectivity off the Portuguese western coast. *Fisheries Research*, **73**, 323-339.

Gavaris, S. and Ianelli, J.N. (2002). Statistical issues in fisheries' stock assessments. *Scan. J. Statist.*, **29**, 245-271.

Gelman, A. and Meng, X.L. (2000). Model checking and model improvement. In: Gilks, W.R., Richardson, S. and Spiegelhalter, D.J. *Markov Chain Monte Carlo in Practice.* Chapman & Hall/CRC, USA: 189-201.

Gelman, A., Carlin, J.B., Stern, H.S. and Rubin, D.B. (2000). *Bayesian Data Analysis.* Chapman & Hall, N.Y.

Gilks, W.R., Richardson, S. and Spiegelhalter, D.J. (2000). *Markov Chain Monte Carlo in Practice.* Chapman & Hall/CRC, USA.

Godinho, M.L., Afonso, M.H. and Morgado, C. (2001). Age and growth of hake *Merluccius merluccius* Linnaeus, 1758 from the Northeast Atlantic (ICES division IXa). *Bol. Inst. Esp. Oceanogr.* **17**(3 y 4), 255-262.

ICES (2003). Report of the Study Group on Precautionary Reference Points for Advice on Fisheries Management. Copenhagen. 24-26 February 2003. ICES CM 2003/ACFM:15.

ICES (2005a). Report of the ICES Advisory Committee on Fishery Management and Advisory Committee on Ecosystems, 2004. ICES Advice. 1, 2, 1544.

ICES (2005b). Report of the Working Group on the Assessment of Southern Shelf Stocks of Hake, Monk and Megrim, May 2003. ICES CM 2005/ACFM:02.

IPIMAR (2004). Report of the Survey Workshop under NeoMAv project. IPIMAR, Lisbon, 6-10 December 2004 (available through mazevedo@ipimar.pt).

Jardim, E., Trujillo, V. and Sampedro, P. (2004). Uncertainties in sampling procedures for age composition of hake and sardine in Iberian Atlantic waters. *Scientia Marina*: **68** (4), 561-569.

Lewy, P. and Nielsen, A. (2003). Modelling stochastic fish stock dynamics using Markov Chain Monte Carlo. *ICES Journal of Marine Science*, **60**, 743-752.

Maunder, M.N. and Punt, A.E. (2004). Standardizing catch and effort data: a review of recent approaches. *Fisheries Research*, **70**, 141-159.

McAllister, M.K., Pikitch, E.K. and Babcock, E.A. (2001). Using demographic methods to construct Bayesian priors for the intrinsic rate of increase in the Schaefer model and implications for stock rebuilding. *Can. J. Fish. Aquat. Sci.*, **58**, 1871-1890.

Michielsens, C., McAllister, M.K., Pakarinen, P., Karlsson, L., Romakkaniemi, A., Kuikka, S., Perä, I. and Mäntyniemi, S. (2004). Bayesian state-space mark-recapture model to estimate exploitation and abundance of salmon in different assessment areas of the Baltic Sea. *In* Report of the Working Group on the Assessment of Baltic Salmon and Trout. Estonia, 21-30 April 2004. ICES, CM 2004/ACFM:23, 164-192.

Millar, R.B. and Meyer, R. (2000). Bayesian state-space modeling of age-structured data: fitting a model is just the beginning. *Can. J. Fish. Aquat. Sci.*, **57**, 43-50.

Monteiro, J.F., Azevedo, M. and Alpizar-Jara, R. (2004). Análise de coortes do stock Ibérico da sardinha (*Sardina pilchardus*) utilizando a Inferência Bayesiana. *Actas do XII Congresso Anual da SPE*. Évora, 27 Setembro – 02 Outubro 2004, 467-476.

Myers, R.A., Barrowman, N.J., Hilborn, R. and Kehler, D.G. (2002). Inferring bayesian priors with limited direct data: applications to risk analysis. *North American Journal of Fisheries Management* **22**, 351-364.

Nielsen, A. and Lewy, P. (2002). Comparison of the frequentist properties of Bayes and the maximum likelihood estimators in age-structured fish stock assessment model. *Can. J. Fish. Aquat. Sci*, **59**, 136-143.

Patterson, K.R. (1999). Evaluating uncertainty in harvest control law catches using Bayesian Markov chain Monte Carlo virtual population analysis with adaptive rejection and including structural uncertainty. *Can. J. Fish. Aquat. Sci*, **56**, 208-221.

Paulino, C.D., Turkman, M.A.A. and Murteira, B. (2003). *Estatística Bayesiana*. Fundação Calouste Gulbenkian, Lisboa, 446 p.

Pope, J.G. and Shepherd, J.G. (1982). A simple method for the consistent interpretation of catch-at-age data. *J. Cons. Int. Explor. Mer*, **40**, 176-184.

Punt, A.E. and Hilborn, R. (2001). BAYES-SA. Bayesian stock assessment methods in fisheries. User's manual. *FAO Computerized Information Series (Fisheries)*, 12, Rome, FAO, 56 p.

Shepherd, J.G. (1999). Extended survivors analysis: An improved method for the analysis of catch-at-age data and abundance indices. *ICES Journal of Marine Science*, **56**, 584-591.

STECF (2003). Report of the recovery plans of southern hake and Iberian norway lobster stocks. Subgroup on Management Objectives (SGMOS) of the Scientific, Technical and Economic Committee for Fisheries (STECF). IPIMAR, Lisbon, 9-13 June 2003.

Bayesian Statistics and Its Applications
Edited by S.K. Upadhyay, U. Singh and D.K. Dey
Anamaya Publishers, New Delhi, India

Why Bayesianism?[1]
A Primer on a Probabilistic
Philosophy of Science

Prasanta S. Bandyopadhyay[1]

Department of History & Philosophy, Montana State University, Bozeman, MT 59717, USA

Abstract

Several attempts have been made both in the present and past to impose some *a priori* desiderata on statistical/inductive inference (Fitleson, 1999, Jeffreys, 1961, Zellner, 1996, Jaynes, 2003, Lele, 2004). Bringing this literature on desiderata to the fore, I argue that these attempts to understand inference could be controversial. This is why Royall's (1997) views on the foundations of statistics are more fruitful. Royall distinguished among three types of questions, (i) *the belief question*, (ii) the *evidence question* and finally (iii) the *acceptance question* (van Fraassen 1991). He thought that Bayesians could only handle the first question, whereas classical statistics (error-statistics), can address the third question. He contended why the Likelihood framework alone is able to answer the second question.

Royall's work makes it clear that statistical inference has multiple goals. As a result, it is unlikely that one measure is able to address all problems in statistical inference at the core of a probabilistic philosophy of science. The purpose of the paper is to evaluate Royall's work from a Bayesian perspective. Contra him, I contend that Bayesianism and Bayesianism alone is able to address all three questions in a manner that is at least as satisfactory as classical statistics (error-statistics) or likelihood approach. So the answer to the questions, "Why Bayesianism?" is that Bayesian School alone provides a unified approach to probabilistic philosophy of science.

Overview

The answer to the question, "Why Bayesianism?" is that the Bayesian School alone provides a unified approach to the statistical inference at the core of scientific methodology. Deductive and inductive arguments are essential aspects of scientific methodology. An argument is deductively valid if and only if it is logically impossible for its conclusion to be false whenever its premises are

---

[1] The author wishes to thank Jose M. Bernardo, Robert J. Boik, Gordon Brittan Jr., Colin Howson, Jayanta Ghosh, Mark Greenwood, Subhash Lele, Sue Monahan, James Robison-Cox, Tasneem Sattar, Elliott Sober, C. Andy Tsao and an anonymous referee for suggesting several improvements in the contents of the paper by their direct or indirect comments regarding the issues raised here. The author is especially indebted to both John G. Bennett and Mark L. Taper for numerous discussions/comments regarding the various aspects of the paper. He also thanks the participants of the Philosophy of Science Association meetings in the workshop on Royall's work held in Milwaukee 2002 and the International Workshop/Conference on Bayesian Statistics held in Varanasi 2005 for their comments and encouragement.

true. A deductively valid argument is distinguished from its inductive counter-part by the property of "monotonicity." The monotonicity property states that one can't undermine the conclusion of a deductively valid argument by adding a new premise provided all the original premises remain intact. Inductive inference does not satisfy the monotonicity property (Adams, 1998; and Bandyopadhyay and Bennett, 2004). Nonetheless, it is inductive inference that underlies much of statistical inference.

Unlike the conclusion of a deductive argument, the conclusion of an inductive argument does not follow necessarily from its premises. As a result, several attempts have been made both in the present and past to understand and ultimately, if possible, to develop a "logic" analogous to deductive logic for inductive arguments. Rudolf Carnap (1952), for example, searched for general inductive rules by which, given information about past results, one could make the best possible prediction of future events. In spite of Carnap's impressive work, his program was universally judged as failing to deliver such rules. Although statisticians and philosophers have abandoned the Carnapian research program, many still believe in the need for imposing some *a priori* desiderata on statistical/inductive inference (Jeffreys, 1961; Zellner, 1996; Zellner, 1997; Jaynes, 2003; Lele, 2004; Christensen, 1999; and Fitleson, 1999). I argue that these attempts to understand inductive inference fail to do justice to various dimensions of inductive inference. The justification for satisfying specific desiderata is contingent on which goals have been adopted, and desiderata could vary from one agent to another. This is why Richard Royall's (1997) views on the foundations of statistics, which take account of this fact, are more fruitful.

Royall distinguished three types of questions. He calls the first two, (i) the *belief question*, and (ii) the *evidence question*. Lastly (iii), borrowing an expression from philosophical literature, I will call Royall's third question the *acceptance question* (van Fraassen 1991). Royall thought that Bayesians could only handle the first question, called the belief question, whereas classical statistics (error-statistics) can address the acceptance question. Royall contended that the likelihood framework alone is able to answer the evidence question. The notion of likelihood is used to answer the question, "how likely are the data given the hypothesis?" Those who believe that the notion of likelihood alone is adequate to capture statistical inference belong to the likelihood framework.

Royall's work makes it clear that statistical inference has multiple goals. As a result, it would seem unlikely that one account is able to address all problems in statistical inference. In response to Royall, I have developed three accounts to satisfy these multiple goals within Bayesianism. They are (i) the confirmation account, (ii) the evidence account, and (iii) the acceptance account. Contrary to Royall, I contend that Bayesianism and Bayesianism alone is able to address all three questions in a manner that is at least as satisfactory as classical statistics (error-statistics) or the likelihood approach.

The central theme of this paper is to bring Royall's three questions to a larger audience and then provide a Bayesian evaluation of these questions. In light of this theme, the paper is divided into nine sections. In the first section, I will broach Royall's three questions. In the second section, I will discuss why Bayesians could be taken to address solely the belief question by drawing on the works of Harold Jeffreys and Arnold Zellner, two kindred-spirited Bayesians who hold the same kind of views on inductive/statistical inference. In the third, I will discuss Subhash Lele's recent non-Bayesian approach to statistical evidence and point out that his work is exclusively tied to the evidence question. In the fourth section, I will address why classical statistics deals only with the acceptance question. The next section will be devoted to evaluating both Bayesian and non-Bayesian approaches to inductive/statistical inference. Here, I will argue that Royall's emphasis on multiple goals is crucial in understanding foundational questions in statistics. In the sixth section, I will propose two accounts, the confirmation account and the evidence account as part of a Bayesian response to Royall's belief and evidence questions. Furthermore, I will detail some crucial unnoticed differences between an account of confirmation and an account of evidence. In the seventh, I will revisit the belief and evidence questions with diagnostic examples involving simple hypotheses. Here, I will also discuss, among other things, Deborah Mayo's error-statistical account (Mayo, 2004). The eighth section will address one model selection problem involving cases that are not liable to the same treatments that simple hypotheses will receive in section seven. I will propose a criterion called the Bayes' Theorem Criterion (BTC) to address the problem in model selection. In section nine, I will develop a decision theoretic account of acceptance. Finally putting the pieces together, I will argue the merit of having a unified account and why Bayesianism alone is able to provide that unification adequate for a probabilistic philosophy of science.

1. Royall's Three Questions

Royall argues that one needs to distinguish three types of questions pertaining to three problem areas in statistics. Consider two hypotheses, H, representing that the patient suffers from cervical cancer, and $\sim$ H, its denial. Assume that a PAP smear test, which is administered as a routine test for screening for cancer, comes out positive for the patient. Given the datum that the test comes out positive, one could ask three questions:

(i) Given the datum, what should we *believe* and to what degree?
   Or,
   Given the datum, how should we *change* our degree of belief?

(ii) What does the datum say regarding *evidence* for H against its counter-parts?

(iii) What should we *do*?

The first question Royall calls the *belief question*, the second one the *evidence question* and the third one is what I call the *acceptance question*. He further argues that, depending on which school of statistics one belongs to, one asks the particular question appropriate for that school. He thinks that Bayesians ask the belief question, likelihoodists like him ask the evidence question, and classical statisticians ask the acceptance question. Within the likelihood framework, Royall has done three things. (1) He has clearly spelled out what is really the evidence question. (2) He argues that the likelihood ratio (LR) is the correct measure of evidence. Finally (3), he contends that the law of likelihood (LL) justifies the use of likelihood ratio as a measure of evidence. According to the LR, D is evidence for H over $\sim$H2 if the ratio Prob(D|H)/Prob(D| $\sim$H) is > 1, whereas the LL says that if Prob(D|H) $>$ Prob(D| $\sim$H) then D is evidence for H over $\sim$H.

2. The Belief Question

I will address Jeffreys and Zellner's Bayesian approach to imposing desiderata on a statistical/ inductive theory and conclude that they are concerned with the belief question.

Although Jeffreys wrote before Zellner, they share a great many common convictions about how and why a Bayesian theory needs to be developed. Both believe that "Bayesian statistics is the technology of inductive inference." They know the fundamental difference between inductive and deductive inferences. They share similar views about imposing *a priori* constraints on inductive/statistical inference to make it more effective, especially in scientific practice. For the sake of brevity, I address those aspects in which their views are similar. I call their views the Jeffreys-Zellner approach (JZA) to inductive/statistical inference.

The JZA consists of eight *a priori* desiderata. The first five Jeffreys consider "essential" and the rest of them he thinks are "useful guides." Here are Jeffreys-Zellner's eight rules.

(i) All hypotheses should be clearly stated and the conclusions should follow deductively from these hypotheses.

(ii) A theory of induction should be self-consistent implying that it is logically impossible to derive contradictory claims from that theory along with a given set of observations.

(iii) Any rule should be applicable in practice. Any definition becomes useless unless the object of a definition can be recognized in terms of the definition when it occurs. Any reference to an existing thing or the estimate of a quantity should involve an empirically possible experiment.

(iv) A theory of induction should make room for the possibility that a particular inference using the theory might turn out to be incorrect.

(v) A theory of induction should not eliminate any empirical proposition *a priori*.

(vi) The number of postulates in a theory should be as few as possible.

(vii) Although a theory of induction need not regard a human mind as a perfect reasoner, it should at least agree with actual thought process in outline.

(viii) Because of the vast complexity of inductive inferences, one should not hope to develop a theory of inductive inference as thoroughly as a theory of deductive inferences.

Consider now how these *a priori* constraints help both Jeffreys and Zellner to develop a theory of inductive inference, especially of a Bayesian kind. Zellner concurs with Jeffreys when the latter writes,

They [these eight rules] rule ... out any definition of probability that attempts to define probability in terms of infinite sets of possible observations, for we cannot in practice make an infinite number of observations. The Venn limit, the hypothetical infinite population of Fisher and the ensemble of Willard Gibbs are useless to us by [rule iii].

To explain why Jeffreys thinks that the rule eliminates the Venn limit, I will discuss how he looks at it. According to the Venn limit, if the number of trials tends to infinity and an event in these trials occurs a large number of times, then the probability p is the limit of the ratio of the number of times when p will be true to the whole number of trials. Jeffreys thinks that the defect of this definition is that it is inapplicable in nature because the presupposition of the limit eliminates total randomness in nature. In addition, he argues that the Venn limit goes against the rule (iv) because the former restricts *a priori* how nature should behave.

He goes on to explain why he thinks that the Venn limit is inapplicable in scientific practice. He writes,

"The form of this definition restricts the field of probability very seriously. It makes it impossible to speak of probability unless there is a possibility of an infinite series of trials. When we speak of the probability that the Solar System was formed by the close approach of two stars, or that the stellar Universe is symmetrical, the idea of even one repetition is out of question; but it is just in these cases that the epistemological problem is most acute. But this is not all, for the definition has no practical application whatever." (Jeffreys, 1957; p 182).

In contrast, both Jeffreys and Zellner attribute probability statements to a parameter. Zellner writes, "The operations of Bayesian statistics enable us to make probability statements about parameters' values and future values of variables." (Zellner, 1984; p.6). Jeffreys thinks that the Venn

(iv) *The invariance under transformation of the data:* The evidence function should remain unaffected insofar as the quantification of the strength is concerned if one uses different measuring units.

Lele has also added two more conditions on the evidence function, which he calls "regularity conditions," so that the probability of strong evidence for the true hypothesis should converge to 1 as the sample size increases. He shows that the likelihood ratio becomes an optimal measure of evidence under those epistemological and regularity conditions, providing a justification for the use of likelihood ratio as a measure of evidence. He believes that showing the optimality of the likelihood ratio amounts to providing necessary and sufficient conditions for the optimality of the law of likelihood.

One could find a flaw in Lele's approach, because the law of likelihood is faulty. Consider an example. A fair die has been thrown, but we do not know which face is up. The investigator has proposed H1 and H2 separately as two likely hypotheses to explain the true nature of the die. H1 = 1 is up; H2 = 6 is not up; and the data (D) is the number of up which is not greater than 2. Based on the law of likelihood, the agent is forced to conclude that D provides more support for H1 over H2 because the probability of the data given H1 is 1, whereas that of the latter is 2/5. However, if D is true then it follows deductively that H2 must be true. The relationship between D and H2 is a logical relation, therefore, Prob(D|H2) must be 1. No legitimate inference could be made from H1 to H2 insofar as their comparative evidential support relative to the data is concerned (see section 6 for more.) The law fails to respect this deductive relationship between D and H2 yielding an erroneous conclusion (See Levine and Schervish, 1999; Sober, 2005. However, my example is due to John Bennett.) This counter-example to the law raises a problem for Lele's argument that the likelihood ratio is an optimal evidence function, because, in the evidentialist framework, the law justifies the use of likelihood ratio in the first place.

One could, however, appreciate the theme of the section while overlooking the defect of the law of likelihood. Consider how Lele has chosen regularity conditions that would yield the likelihood ratio as an optimal evidence function. Lele is aware of this and writes, "Toward this goal, some additional regularity conditions are imposed." It is interesting to note that those conditions on a class of evidence functions are quite general because they could yield the Kullback-Leibler disparity measure (see Burnham and Anderson, 1998) as an instance of that evidence function. However, they eliminate the posterior distribution as an instance of this evidence function. In fact, the parameter transformation invariance is not satisfied by the posterior distribution. That is, the latter violates condition (iii) on this class of evidence functions.

In our modeling of nature, the same phenomena may be represented equally well by different formalisms. It is arbitrary whether we quantify spread by the variance, v, or by the standard deviation, s. Exactly the same information is conveyed by these two parameterizations because there is an exact transformation between them. Bayesian inference, however, is not invariant under parameter transformation. At the core of Bayesian inference lies posterior probability distribution. A Bayesian takes an estimate to be of some function of a mean or mode of a posterior distribution. For discussion, let us say that the researcher chooses to use the mean of the posterior distribution of a parameter as a point estimate. One can transform a probability distribution, f, for the parameter v into a probability distribution, h, for the parameter s using the idea of change of variable technique (e.g. Casella and Berger, 1990). The mean of $f(v)$ need not be equal to the square of the mean of $h(s)$ within a Bayesian framework. As a result, the Bayesian point estimates for the two parameterizations need not match.

Consider an example to see why the Bayesian point estimates for the two parameters may not yield the same conclusion after transforming one parameter into another via change of variable technique. Suppose a set of data is distributed with log-normal *(m, sd)* in which $sd = 2$ and $m = 2$. Based on this information about the distribution of the data, we would like to make an inference about the parameter v. As a Bayesian, we need to have a posterior distribution of v, which is as follows:

$$f(v) = \frac{\exp\left[\frac{-(\ln(v) - m)^2}{2sd^2}\right]}{vsd\sqrt{2\pi}}.$$

Because the above function along with other functions applied in this specific example are not defined at zero, I have used numbers different from zero as their lower limits in all my integrations.

The value of the integration, $\int_{.000001}^{\infty} f(v)dv = 1$ proves that it is a probability distribution. Now by using the same distribution of the data, we would like to make an inference about the parameter s, which is a posterior distribution of s derived via change of variables

$$h(s) = \frac{|2s|}{s^2\sqrt{2\pi sd^2}} \exp\left[-\left[\frac{(\ln(s^2) - m)^2}{2sd^2}\right]\right]$$

$\int_{.001}^{\infty} h(s)ds = 1$. This turns out to be a probability distribution too. The mean of the posterior distribution for s is

$$mnh = \int_{.00001}^{\infty} sh(s)ds, \qquad mnh = 4.482.$$

In contrast, the mean of the posterior distribution for v is

$$mnv = \exp\left(m + \frac{sd^2}{2}\right); \quad \int_{.000000001}^{\infty} vf(v)dv = 54.598; \quad mnv = 54.598; \quad \sqrt{mnv} = 7.389.$$

These two values, 4.482, which is the mean of the posterior distribution for the parameter "s" via the probability distribution h, and 7.389, which is the mean of the posterior distribution for the parameter "v" via the probability distribution f, are not equal.

What this example shows is that if one imposes a desideratum on the evidence function that the latter must satisfy the reparameterization invariance property, then one won't be able to use posterior distribution, a pillar of Bayesianism, as a suitable evidence function.[3] I will argue

---

[3] One might worry how effective this objection to Bayesianism is when one argues that posterior distribution does not remain invariant under parameterization. The reason for this worry might be that even within a frequentist paradigm, some properties of an estimate are also *not* invariant under transformation. Even though T is an unbiased estimator of θ, T^2 might not be an unbiased estimator of θ^2. If within a frequentist's framework an unbiased estimator of θ could become a biased estimator of θ^2 after transformation, then, the objector wonders, why does the failure of posterior distribution to satisfy parameter invariance should be touted as a defect for Bayesians? One response is that although the argument that the failure for posterior distribution to satisfy parameter invariance is related to the above point regarding the failure to satisfy unbiasedness of a parameter after transformation, the former is a different objection than the latter one. For example, although MLE is not necessarily an unbiased estimator, the information contained in MLE point estimate remains unchanged under reparameterization. To put the latter point formally, MLE of $V = (MLE\sigma)^2$. The point is that whether two estimates yield equivalent value under parameterization is different from whether that estimate is an unbiased estimator. I thank both an anonymous referee and Mark Taper for clarifying two related issues here.

(section VI) that posterior distribution provides a framework for answering the belief question, which is different from the evidence question. So Royall is right that the likelihood framework or the evidentialist framework, which is a generalized version of the likelihood framework, is able to answer the evidence question, and not the belief question.

4.   The Acceptance Question

According to Royall, classical (error-statistics) is decision-theoretic in character; as a result it is interested in the acceptance question. Classical statistics has to do with computing error probabilities that are either tied to the data and particular experimental set-ups or generated by human intervention in these set-ups. Classical Neyman-Pearson statistics neatly summarizes these two types of errors as Type I and Type II. Suppose the clinician is interested in knowing the effect of the drug AZT among patients suffering from HIV after it has been administered (Neyman, 1967). To learn the effect of the drug, and to understand how the situation is related to the two effects of errors, consider two hypotheses: H_a and H_0. H_0 is the null hypothesis; it is that there is no difference in patients before and after treatment. H_a is the alternative hypothesis that there is a difference. The Type I error occurs when the decision is to reject the null hypothesis when it is actually true. The Type II error occurs when the decision is not to reject the null hypothesis when it is actually false. The *goal* of error-statistics is thus twofold: to reject the null hypothesis as a function of some pre-assigned significance and to minimize these two errors. Now whether Mayo would accept this characterization of error-statistics as decision-theoretic is another story. However, it seems clear that Royall is right in pointing out this decision-theoretical character of the notion of acceptance in the case of classical statistics.

From this discussion, one pattern of reasoning emerges insofar as (especially) both the Bayesian and likelihood/evidentialist approaches to desiderata are considered. Whether one is a Bayesian or non-Bayesian, one would like to impose those *a priori* constraints on one's methodology that would yield only one's favorite theory, Bayesian or otherwise. I have only provided two cases in which this pattern is realized. However, this pattern is typical among both philosophers and some working statisticians.

5.   Three Questions, Different Desiderata, and Various Goals

There could be several desiderata for a model/theory to be good. However, working scientists are well aware that one might not be able to capture all desiderata of a model when confronted with a hard problem. The curve-fitting problem, which is an instance of the model selection problem, provides an example of this kind. In the curve-fitting problem, the investigator is confronted with how to fit a line to a set of data. If she would like to capture the data well, then she would have a high goodness-of fit value, which is one desideratum for a model, but the mathematics of that line could get complicated. In contrast, if she would like to make the model simple, then she would prefer a straight line, thus failing to capture the data well. So the curve-fitting problem arises because one can't simultaneously optimize disparate desiderata such as "goodness-of-fit" and "simplicity." Even knowing this conundrum of model selection well, why do working scientists, Bayesian and non-Bayesian, search for some *a priori* constraints on their theories? The rationale is that they have different goals, which they would like to achieve with the help of those desiderata that would not generate a tension in theory selection. To escape this possible tension stemming from conflicting desiderata, Jeffreys and Zellner have adopted (ii), that is, a theory of induction should be self-consistent where it is logically impossible to derive conflicting results from that theory along with a given set of data.

This Bayesian/non-Bayesian attitude to desiderata is not new in scientific methodology. It dates back to Descartes who recommends rules for how one should conduct a correct discourse in epistemology. Descartes in his *Discourse on the Method* (VI 18-19, HR 1 92) lays down four rules that should precede any methodology.

(i) "Accept nothing as true which I didn't clearly recognize to be so: that is to say, to avoid carefully precipitation and prejudices, and to accept nothing in my judgment beyond what presented itself no clearly and distinctly to my mind, that I should have no occasion to doubt it.

(ii) Divide each of the difficulties which I examined into as many parts as possible, and as might be necessary in order best to resolve them.

(iii) Carry on my reflections in order, starting with those objects that were most simple and easy to understand, so as to rise little by little, by degrees, to the knowledge of the most complex: assuming an order among those that did not naturally fall into a series.

(iv) Last, in all cases make enumerations so complete and reviews so general so I should be sure of leaving nothing out."

It is no wonder that Descartes himself has followed these rules in building a foundation for his epistemology. Commenting on Descartes' rules of method, Leibniz gibes that they were 'like the precepts of some chemist; take what you need and do what you should, and you will get what you want.' (*Philosophical Schriften von G.W. Leibniz*, ed. C.I. Gerhardt, Berlin, 1875-90, vol. IV, p. 328). My own position on imposing *a priori* constraints is like the position Leibniz has adopted toward evaluating Descartes. Leibniz criticized Descartes for recommending that methodologists follow those rules. He argued that since Descartes had a goal to build a foundation of knowledge, Descartes needed those rules badly. However, if somebody has a different goal, she might ask for different rules to follow. Leibniz concludes that satisfying those desiderata are solely contingent on what one would like to achieve. As a result, those desiderata could vary from one agent to another. This variation in desiderata is likely to be due to some specific questions/problems investigators are confronted with. This is why I believe that the way Royall has addressed the foundations of statistics is much more fruitful.

The significance of these questions is that, unlike Jeffreys, Zellner and Lele, Royall makes it explicit that statistical inference has multiple goals. Depending on what goal the investigator has, it is perfectly legitimate for her to develop one's statistical machinery to achieve that goal. However, there is a further implication of Royall's three questions, which Royall thinks, is that only the likelihoodist could handle the evidence question and no other school can.

6. Bayesian Accounts of Belief and Evidence Questions

In this section, I will develop two distinct Bayesian accounts to address both belief and evidence questions and argue that Royall is mistaken about his claim that Bayesians can't handle the evidence question. Although this paper disagrees with Royall, it draws its inspiration from his work regarding the nature and importance of three questions.

My two accounts to address the belief and evidence questions are an account of confirmation and an account of evidence. For Bayesians, an account of confirmation provides a confirmation relation,

C(D, H,B) among data, D, hypothesis, H, and the agents' background knowledge, B. Because a confirmation relation is a belief relation, it must satisfy the probability calculus, including the rule of conditional probability together with some reasonable constraints on one's *a priori* degree of belief in an empirical proposition. Learning from experience is, of course, part and parcel of Bayesianism. The rule of conditional probability ensures this inductive basis of our learning. As a result, the agent should not have an *a priori* belief about an empirical proposition with full certainty (probability 1 or probability 0) if she would like to learn from experience, but rather something in between these two extremes.[4]

For Bayesians, then, belief is fine-grained. They allow belief to admit of any degree of strength between 0 to 1. A satisfactory Bayesian account of confirmation should be capable of capturing this notion of degree of belief. According to the confirmation condition (CC), D confirms H if and only if Prob(H|D) >0 but < 1.[5] The posterior probability of H could vary between 0 and 1 exclusive. Confirmation becomes strong or weak depending on how high or low this posterior probability is.

The posterior and prior probabilities are essential parts of Bayes' theorem, a foundation of any Bayesian approach. According to the theorem, the posterior probability of a hypothesis H equals its prior probability multiplied by the likelihood of H (Prob(D|H)) then divided by the marginal probability of D (Prob(D)):

$$\text{Prob}(H|D) = \frac{\text{Prob}(H) \times \text{Prob}(D|H)}{\text{Prob}(D)}. \tag{1}$$

Prob(H|D) is also called the conditional probability of H given D. Prior probability of a hypothesis depends on the agent's degree of belief in that hypothesis before the data for the hypothesis have been gathered. The likelihood function provides an answer to the question, "How likely are the data given the hypothesis?" The marginal probability represents the probability that D would obtain, averaged over the hypotheses being true and false.

My account of evidence lays down conditions for the evidence relation to hold among D, H, and B. However, because the evidence relation is not a belief relation it need not satisfy the probability calculus. While the account of confirmation is concerned with a single hypothesis embodied in equation 1, an account of evidence must compare the merits of two competing hypotheses, H1, and H2 (or ~H1) using data D. The Bayes factor (BF) captures the bare essentials of a Bayesian account of evidence (see Raftery, 1995). For simple hypotheses, the Bayes Factor and the likelihood ratio are identical (Berger, 1985, p. 146). The BF can either be represented by (a) or (b):

(a) D becomes E (evidence) for H1&B against H2&B if and only if

$$\left[\frac{\text{Prob}(D|H1\&B)}{\text{Prob}(D|H2\&B)} \right] > k > 1.$$

---

[4] However, I take the data in light of which the agent updates her belief function to have probability 1.

[5] One might object that this might not be a representative probabilistic measure for Bayesian confirmation. One suggestion might be to use the measure Prob(H|D) > Prob(H) instead. To respond to this objection, we would like to point out first that there are also other measures, for example, the difference between Prob(H|D and Prob(H| ~D), and many more. Second, we need to distinguish between two concepts of confirmation; the *absolute* confirmation and the *incremental* confirmation (see Carnap, 1952). This notion of confirmation is intended to capture the absolute confirmation, while at the same time, making Bayesianism to be both minimally constrictive and maximally tolerant. In addition, Gill, a Bayesian sociologist, also endorses this absolute confirmation measure because, according to him, to judge the quality of a single model, there is no better measure than referring to their posterior summaries (Gill, 2002, p. 271).

(b) D provides the same evidential support (E) for both H1&B and H2&B if and only if

$$\left[\frac{\text{Prob(D|H1\&B)}}{\text{Prob(D|H2\&B)}}\right] = 1.$$

Call these two conditions, Eq. (2(a)) and Eq. (2b)) respectively.[6] Note that in Eq. (2(a)) if $1 < K \leq 8$, then D is often said to provide weak evidence for H1 against H2, while when $K > 8$, D provides strong evidence (Royall, 1997). This cut-off point is determined contextually and may vary depending on the nature of the problem that confronts the investigator. The range of values for BF varies from 0 to ∞ inclusive. Note that in both (a) and (b) D represents "actual" observed data, and not "possible" unobserved data. This is the backbone of the likelihood principle. To know whether a theory is supported by data, the likelihood principle says that the only part of data that is relevant is the likelihood of the actual data given the theory. For my Bayesian account of evidence, it is the likelihood principle (LP) and not the law of likelihood that justifies the use of the Bayes Factor as a measure of evidence (Birnbaum, 1962; Berger and Wolpert, 1988, Berger, 1985, Berger and Pericchi, 1996, Good, 1983 and Rosenkrantz, 1977).[7]

The LP is derivable from two principles, the first one is called the weak sufficiency principle (WSP) and the other one is called the weak conditionality principle (WCP), which philosophers like Elliott Sober calls the principle of actualism (Sober, 1993). The WSP says if T is sufficient and if $T(x_1) = T(x_2)$, then both x_1 and x_2 will provide equal evidential support. A statistic is sufficient, if given it, the distribution of any other statistic does not involve the unknown parameter θ. This statistic is considered to be a desirable feature for a good estimator, and as such is embraced by both classical and Bayesian statisticians. Hence, one is ill-advised simply to reject the latter. As a result, a defense of the LP could rest on a defense of the WCP, or the principle of actualism. The WCP says that experiments not actually performed are irrelevant. According to the principle of actualism, one should make judgments about the tenability of a hypothesis based on data we actually have, rather than on unobserved, possible data. Borrowing an insight from Sober, I will justify the principle of actualism, which, indeed, will justify the LP in the end.

---

[6] There could be several objections to using the Bayes Factor as a measure of evidence. I will consider two related objections. The first objection to the Bayes Factor is that the Bayes factor could be arbitrarily large when Prob(H1) = 0. But, the objection continues, Prob(H1) = 0 means that one didn't believe that there can be any evidence for H1. In response to this objection, I could say that if Prob(H1) = 0, then the numerator in the Bayes Factor is not defined because it is of the form of the ratio 0/0, as a result, the entire Bayes factor ratio can't be evaluated. Nonetheless, if 0 < Prob(H1) < 1, and there are only two hypotheses, then BF = LR, and LR can be arbitrarily large. Large LR means strong evidence in favor of H1 over H2, but large LR does not imply large posterior probability of H1. The second objection which is a corollary to the first objection is that no account of evidence can succeed when it ignores prior probability. I have already evaluated this charge in sections 6 and 7, implying that this charge is not generally correct, because in comparing two simple hypotheses, one is not necessarily committed to invoking prior probabilities (See also Berger, 1985, p.146). However, when one is involved with a complex model, then one must appeal to the prior probability of that model. I have addressed this point in section 8. On these two points, I owe both to Robert Boik and Colin Howson for comments and suggestions. (See, Bernardo and Smith, 2000, and also Bernardo, forthcoming for their objections to the use of Bayes Factor as a measure of evidence.)

[7] Two remarks are in order. The first remark pertains to an anonymous referee's comment and its subsequent follow-up. S/he thinks that the likelihood principle does not justify the use of Bayes factor. The referee adds that there are Bayesians who criticize Bayes Factor, because they don't meet the coherence condition. What the referee says is correct. However, there are several notions of coherence in Bayesian literature. One needs to be careful about it. (For various senses of coherence, see Bandyopadhyay forthcoming). The second remark has to do with a possibility that whether one could support the LP without supporting the LL. I contend that they are different concepts, hence supporting or denying one does not necessarily lead to supporting the other and conversely. The LP only says where the information about the parameter θ is and suggests that the data summarization can be done through the likelihood function. It does not clearly indicate how to compare the evidential strength of two competing hypotheses. In contrast, the LL compares the evidential strength of two contending hypotheses. I thank both C. Andy Tsao and Jayanta Ghosh for some clarifications on the relationship between the LP and the LL.

Consider a doctor making a diagnosis about whether a patient has diabetes (H1) or small pox (H2). There are two laboratory tests conducted, one for diabetes and the other for small pox. Their results contain both "positive" and "negative" outcomes with errors associated with these tests. Consider another test to be conducted to examine whether the patient has diabetes and this test is, in fact, infallible in the sense that it has probability one of detecting correctly if the patient suffers from diabetes. It is crucial for this defense to assume that there is no such infallible test for small pox. Suppose that the doctor runs the first two tests and not the infallible third test. Suppose each test yields a positive result with Prob(D|H1) = Prob(D|H2). Consequently, each has equal epistemological footing given actual evidence. Should we be more confident that the patient has diabetes than we should be that the patient has small pox based on our reasoning that we have an infallible test for the diabetes, but not for small pox? Our intuition about this question is "no". The principle of actualism exploits this intuition about evidence. We should base our decision relying on these *actual* results, rather than on the *possible* test results. The mere possibility of running the third test is irrelevant for the study under consideration. So far, I have defended the principle of actualism at the core of the LP, which underlies any Bayesian account of evidence.

The crucial feature for understanding the distinction between an account of evidence and that of belief confirmation is the role of the "coherence condition." Confirmation has a coherence condition (CCC) that says if H1 entails H2, then Prob(H2) $\geq$ Prob(H1). The coherence condition is a consequence of the axioms of the probability theory. As a result, any account of confirmation must satisfy it. However, an account of evidence does not need to have a corresponding coherence condition (ECC) which says: If H1 implies H2, then whatever is evidence for H1 must also be (as good) evidence for H2.

A simple dice example will illustrate why the account of evidence does not need to satisfy this condition. Consider a pair of standard, fair dice, consisting of one red and one white die that have been thrown, but the result of the throw is unknown. The two hypotheses, H1 and H2, regarding the value of their faces would be:

(H1): the value on the faces of the two dice is equal to 3.

(H2): the value on the faces of the two dice is $\geq$ 3.

Note that H1 entails H2. Then an "evidence coherence condition" would imply that whatever is evidence for H1 must also be evidence for H2. Suppose our datum (D) is that we have learned that the red die showed 1. We have no information about the other die. Here, D favors H1 over H2, although H1 entails H2 (See Levine and Schervish, 1999 on related issues). The Bayes's factor for H1 vs. H2 is $\frac{1/2}{5/35} = \frac{1/2}{1/7} = 3.5$. So although H1 entails H2, given D, whatever is evidence for H1 need not be as strong evidence for H2, thus violating the "evidence coherence condition (ECC)."[8]

To explain the above scenario, one could say that D makes H2 makes D less likely than H1, while H1 makes D more likely. There were more ways H2 could be true than H1 before the occurrence of D. However, given H1, the occurrence of D is a bit more likely than given H2. It is important to realize that this dice example showing the violation of ECC does not rely on any prior probability or belief for either of the hypotheses.

---

[8] Bernardo informs me (in a private conversation) that this should be an adequate reason for rejecting the Bayes Factor as a measure of evidence, because the Bayes Factor is unable to handle evidential support involving two nested hypotheses (see also footnote 4). There is a saying among philosophers that one man's *modus ponens* is another man's *modus tollens*. As a result, contrary to Bernardo, I think that this example shows why an account of evidence should be different from that of confirmation, and in fact, I think, the Bayes Factor is able to capture this crucial feature of evidence.

7. Revisiting the Belief/Evidence Distinction and Four Accounts of Evidence

Continue with the PAP smear example mentioned in section 1 (Pagano and Gauvreau, 2000). An on-site proficiency test conducted in 1972, 1973, and 1978 evaluates the competency of technicians who examine PAP smear slides for abnormalities. I will assume from a large number of data that we are nearly certain that the propensity for having a positive PAP smear for members with cancer of the cervix is Prob(D|H) = 0.8375, and the propensity for having a positive PAP smear for population members without cervical cancer is Prob(D| $\sim$ H) = 0.186. Call this background theory of the propensities for PAP smear test outcomes, "B".

Let H represent the hypothesis that an individual is suffering from cervical cancer and $\sim$ H the hypothesis that she is not. These two hypotheses are mutually exclusive and jointly exhaustive. In addition, assume D presents a positive PAP smear test result. We would like to know Prob(H|D), i.e., the posterior probability that a person with a positive test result actually does have the disease. To apply Bayes' theorem, we need to know the prior probability for H. Prob(H) is the probability that a woman suffers from cervical cancer when randomly selected from the population. One source reports that the rate of cases of cervical cancer among women studied in 1983-1984 was 8.3 per 100,000 (Pagano and Gauvreau, 2000). The data yield Prob(H) = 0.000083. Then, Prob ($\sim$H) = 0.999917. By Bayes' theorem, we get Prob(H|D) = 0.000373. Here, it tells us that for every 1,000,000 positive PAP smears, only 373 represent true cases of cancer, So, Prob(H|D) is very low; there is weak confirmation for the hypothesis. Does D provide evidence for H against $\sim$H? The Bayes Factor-based account of evidence yields Prob(D|H)/Prob(D| $\sim$ H), which equals 4.49 times. What does the value 4.49 times mean? It means that after we know that the individual has a positive result, she is five times more likely to have cervical cancer. D provides weak evidence for H against $\sim$H. So the PAP case illustrates that the agent could have both weak belief and weak evidence for the hypothesis.

Consider another example, which I call the tuberculosis case (TB). The X-ray is administered to examine whether someone is suffering from the disease. I will suppose from a large number of data that we are nearly certain that the propensity for having a positive X-ray for members with TB is 0.07333, and the propensity for a positive X-ray for population members without TB is 0.0285. Call this background theory of the propensities for X-ray outcomes "B". Let H represent the hypothesis that an individual is suffering from tuberculosis and $\sim$H the hypothesis that she is not. These two hypotheses are mutually exclusive and jointly exhaustive. In addition, assume D represents a positive X-ray test result. We would like to find Prob(H|D), the posterior probability that an individual who tests positive for tuberculosis actually has the diseases. Bayes' theorem helps to obtain that probability. However, to use the theorem, we need to know first Prob(H), Prob($\sim$H), Prob (D|H), and Prob(D| $\sim$H).

Prob(H) is the prior probability that an individual in the general population has tuberculosis. Because the individuals in different studies need not be chosen form the population at random, the correct frequency based prior probability of the hypothesis could not be obtained from them. Yet in a 1987 survey, there were 9.3 cases of tuberculosis per 100,000 population (Pagano and Gauvreau, 2000). Consequently, Prob(H) =0.000093. Hence, Prob($\sim$H) = 0.999907. Based on a large dataset kept as medical records, we are certain about these following probabilities: Prob(D|H) is the probability of a positive X-ray given that an individual has tuberculosis. Prob(D|H) = 0.7333. Prob(D| $\sim$H), the probability of a positive X-ray given that a person does not have tuberculosis, is 1$-$ Prob($\sim$D| $\sim$H) = 1 $-$ 0.9715 = 0.0285.

Using all this information, I compute Prob(H|D) = 0.00239. For every 100,000 positive X-rays, only 239 signal true cases of tuberculosis. The posterior probability is very low, although it is slightly higher than the prior probability. Although CC is satisfied, the hypothesis is not very well confirmed. Yet at the same time, the BF, i.e., 0.7333/0.0285 (i.e., Prob (D|H)/Prob(D| $\sim$ H)) = 25.7, is very high. Therefore, the test for tuberculosis has a great deal of evidential significance.

There is little point in denying that the meanings of "evidence" and "confirmation" (or its equivalents like "belief") often overlap in ordinary English as well as among epistemologists. There is a theorem, BF > 1 if and only if Prob(H|D) > Prob(H), which shows this connection. However, strong belief does not imply strong evidence and the latter also does not imply the former, as illustrated by the TB example. My case for distinguishing them rests not on usage, but on the clarification in our thinking that is thus achieved and supported by inferences frequently made in diagnostic studies (see Bandyopadhyay and Brittan, forthcoming).

How good is a Bayesian account of evidence compared to Royall's likelihood-based account of evidence and Mayo's error-statistical account of evidence? Like me, Royall himself uses the likelihood ratio as a measure of evidence. Since the former and the latter is the same measure, there is no difference between these two accounts insofar as quantification of the strength of evidence of competing *simple hypotheses* is concerned. Hence, there is no difference between a Bayesian account of evidence and a likelihood account of evidence with regard to the PAP and TB cases. However, Royall and my account tend to diverge when we are confronted with a complicated problem in model selection involving a complex hypothesis (see section 8 on this point).

Royall thinks that classical statistics or error-statistics is only able to handle the acceptance question, which is, according to him, decision theoretic. Contrary to Royall, Mayo has argued that it is a mistake to think that error-statistical account can't provide an account of evidence. I will briefly discuss Mayo's account and argue that a Bayesian account of evidence is at least as good as her error-statistical account. Mayo's error-statistical account rests heavily on the notion of severity, which she has borrowed from Karl Popper (Popper, 1959) with two major differences from the latter. First, she disagrees with Popper on the appropriate scope of falsification that underlies Popper's use of the severe test. A necessary condition for a theory to be scientific, according to Popper, is that the theory in question is in principle falsifiable. Mayo thinks that Popper is misguided in attempting to falsify a whole theory. According to her, theory testing should begin with small scale testing, such as testing a particular hypothesis with a particular outcome. Her notion of good evidence or what she calls a "severe test" is "always attached to a particular hypothesis passed or a particular inference reached" (Mayo, 1996, p. 184). The second difference between them is that while her notion of severity is probabilistic, his is deductivistic.

Consider H, and its denial, $\sim$ H, to be mutually exclusive and jointly exhaustive of all possible hypotheses in a domain. Assume that H is a simple hypothesis, and D is the datum for H. For Mayo, H passes severely with D just in case, (i) Prob(D|H) is very high, and (ii) Prob (D| $\sim$H) is very low. One distinct part of her account is that the notion of probability invoked for making statistical inference is frequentist/objectivist in spirit. It is qualities of experimental procedures that supply error probabilities. In her account, error-statistics rest on these error probabilities, which provide adequate information just to the extent that one should be able to make reliable statistical inferences based solely on them.

If we apply her account to the PAP and TB cases, then we find that her account considers both tests to be severe.[9] Both of her conditions for the severe test are satisfied by these two cases. In the PAP case, $\text{Prob}(D|H) = 0.8375$ and $\text{Prob}(D| \sim H) = 0.186$, whereas in the TB case, $\text{Prob}(D|H) = 0.7333$, and $\text{Prob}(D| \sim H) = 0.0285$. In contrast, my Bayesian account distinguishes the strength of evidence in those two cases. In the PAP case, my account says that evidence is weak, whereas in the TB case, it yields strong evidence.

Often, the Akaikean Information Criterion (AIC) has been proposed as a measure of evidence. (For Akaike's own works, see Akaike, 1973; for applications of the AIC framework in philosophy of science, see Forster and Sober 1994, Forster and Sober, 2004. See also Taper, 2004, to know more about a different slant on the possibility of arriving at better information criteria than the AIC.) The goal in using AIC is to maximize predictive accuracy and AIC provides a consistent estimate of an index of predictive accuracy. Since AIC is an estimate, the latter is computed based on both actual and possible data. Forster and Sober's recommendation is that one should choose the model with maximal AIC. The measure of evidence they recommended based on the AIC framework says that D is evidence for H1 over H2 if and only if the AIC(H1) > AIC(H2). If one uses the AIC as a measure of evidence for the PAP and TB cases, then one finds that the AIC-based measure will yield exactly the same results as given by the likelihood-based measure in those cases, because these two cases don't involve any adjustable parameters. Since the AIC violates the likelihood principle, like the error-statistical account it fails to be a viable account of evidence (see Boik, 2004 for more detail).

In short, my Bayesian account of evidence is at least as good as any account of evidence when the former is compared with Mayo's error-statistical account or Royall's account or for that matter with the AIC based account of evidence.

8.   The Curve-Fitting Problem: Accommodating Two Bayesian Accounts

So far, I have addressed only simple diagnostic examples involving a statistical hypothesis to explain the belief/evidence distinction. However, what happens when one is confronted with a complicated scenario and how do the two previous Bayesian accounts handle that case? Consider a hypothetical example to get a handle on how a complex scenario could be addressed within a Bayesian framework. Salam is a fishmonger in a small village, Shrimongal, in the northern part of Bangladesh. Two elements in Shrimongal fish-mongering are (i) torrential rains during the rainy season and (ii) lack of electrification. Salam supports his family by selling fish that he keeps under chunks of ice so that the fish will stay fresh. Now assume we have data that reflect an average monthly temperature in centigrade in Shrimongal from April to December with corresponding consumption of ice measured in cubic feet. Salam's goal is to predict how much ice he will need at different temperatures based on a given data set. To do this he needs to find a relationship, if any, between average monthly temperature and ice consumption per day. Table 1 represents historical data for one season.

In Table 1, the explanatory variable, x, is average monthly temperature, and the response variable, Y, is ice consumption per day. One can see that as soon as the monsoon season in June sets in, gradually temperature decreases; as a result, Salam's ice consumption goes down, hitting rock

---

[9] Since Mayo is not a friend of the likelihood framework, she might object to my reducing her account to two likelihoods. There could be two responses to it. First, Mayo herself has formulated her account in terms of the likelihood (Mayo, 1996, pp.179-81). Second, if she would raise this objection to my reconstruction of her position, then her account would turn out be incomplete because it then fails to explain even simple diagnostic cases like the ones discussed above.

Table 1. Shrimongal data

Variable	Month								
	April	May	June	July	Aug.	Sept.	Oct.	Nov.	Dec.
x	Average monthly temperature								
	15.6	26.8	37.4	36.4	35.5	18.6	15.3	7.9	0.0
Y	Average daily ice consumption in (cu ft)								
	5.2	6.1	8.7	8.5	8.8	4.9	4.5	2.5	1.1

bottom in December, which is winter in Bangladesh. (This example assumes, by the way that Salam buys same amount of fish each month.)

The relationship of x to Y may be represented through the following equation:

$$Y_i = \alpha_0 + \sum_{j=1}^{k} \alpha_j x_i^j + \varepsilon_i, \text{ for } i = 1, ..., n$$

where n is the sample size; $\alpha_j, j = 0, ..., k$ are unknown regression coefficients, k is the order of the polynomial model, and ε_i, is random error. The error terms, $\varepsilon_i, i = 1, ..., n$ are assumed to be independently distributed with mean zero and variance σ^2.

Salam wants to know what his ice consumption will be for a month this year that has $x = 15.3\,°C$. In October of the historical data set, 15.3°C corresponded to 4.5 cubic feet of ice. I shall forecast from three different regression line equations how much ice Salam will require at $15.3\,°C$. I consider three regression equations corresponding to three mutually exclusive hypotheses, H1, H2, or H3 in a domain. The hypotheses are,

H1: $E(Y|x) = \alpha_0 + \alpha_1 x$
H2: $E(Y|x) = \alpha_0 + \alpha_1 x + \alpha_2 x_2$ and
H3: $E(Y|x) = \alpha_0 + \alpha_1 x + \alpha_2 x_2 + \alpha_3 x_3$.

Here, $E(Y|x)$ is the conditional expectation of Y given x. To say that these hypotheses are mutually exclusive is to say that the coefficients of x_k under H_k are non-zero.

The least squares forecast under H1 is $\widehat{Y} = 1.22 + 0.20 \times 15.3 = 4.33$ hundred cubit feet. Under H2, the prediction is $\widehat{Y} = 1.09 + 0.22 \times 15.3 - 0.0005 \times 15.3^2 = 4.39$ hundred cubit feet per day. This is closer to the historical value of 4.5 than that based on H1. If we use H3, then we will find that the prediction, $\widehat{Y} = 4.45$ is even closer to the historical value. In general, as the order of the polynomial regression model increases from H1 to H3, the goodness of fit of the model to the observed data increases. This is measured by maximizing the likelihood function under H_k denoted by $\widehat{L}_k$, that is,

$$\widehat{L}_k = \max_{\sigma^2;\ \alpha_0, ..., \alpha_k} \text{Likelihood}(\sigma^2; \alpha_0, ..., \alpha_k | H_k; Y_1, ..., Y_n). \tag{3}$$

A model having too large an order will over-fit the data. Predictions of future data from such a model will, in general, have larger errors than will predictions from a model with a smaller, but sufficient, number of parameters. So to both maximize better prediction, and minimize over-fitting error, I propose that Salam use Bayes' Theorem Criterion (BTC) to calculate how much ice he needs. The BTC implies that if one adopts certain non-informative priors on σ_2 and α_j for $j = 1, ..., k$, then the posterior probability, Prob (H_k|data), of a hypothesis is proportional to its prior probability, Prob (H_k), multiplied by the maximum likelihood function, $\widehat{L}_k$ where $\widehat{L}_k$ is given by Eq. (3). (For more details, see, Bandyopadhyay, Boik and Basu, 1996; and Bandyopadhyay and Boik, 1999.)

I consider H1 as the simplest hypothesis because it contains the fewest number of parameters (α_0, α_1) while in light of epistemological|pragmatic considerations, it seems to be the best, and therefore I assign the highest prior probability, 1/2, to it. (Our work was here influenced by Jeffreys's. However, our account of simplicity is different from his. For our work, see two immediately cited papers. For more on an account of simplicity that rests on Bayesianism, see Bandyopadhyay, 2002.) Because of epistemological|pragmatic considerations, H2 gets assigned 1/4 and H3 1/8. Recall that I considered only three hypotheses. The other hypotheses are lumped together. This is called the catch-all hypothesis, denoted by H_c and gets assigned 1/8. In short, BTC says, choose the one that maximizes posterior probability, that is, maximizes.

$$\text{Prob } (H_k|\text{Data}) \propto \hat{L}_k \times \text{ Prob}(H_k). \tag{4}$$

The result of applying Eq. (4) to the Shrimongal data yields the following table. So Salam chooses H1 as the best model for his prediction.

Table 2. Applications of BTC (Eq. (4)) to the shrimongal data

H1	H2	H3
-4.75	-5.30	-5.55

Consider how this account based on the BTC is able to provide a unified account of both belief and evidence while being careful about the distinction between the two questions (Schwarz's Bayesian Information Criterion called BIC is, however, different from BTC, although we proved in the already cited papers that the former is logically equivalent to the latter with a choice of priors|substitutions. For more on the BIC see, Schwarz, 1978). Given the data about ice consumption and average monthly temperatures for a specific season, if Salam's interest can be shown to be in the evidence question, then the BF based account provides that answer. To compute the likelihood of a family under H1, and then under H2, we assign non-informative priors over regresssion co-efficients, variance and error terms. After doing this computation, what we get is that a hypothesis with higher parameters always fits the data better than a hypothesis with fewer parameters. Here, our claim just made rests on the assumption that we work within a nested model. However, if the agent opts for a hypothesis with higher dimension, then she might be involved both in an over-fitting error and the penalty due to introductions of more parameters. We need to take into account the role of simplicity in theory choice. This leads to the curve fitting problem, which arises when one tries to optimize two conflicting desiderata, simplicity and goodness of fit. The Bayes' theorem Criterion (BTC) has been proposed to resolve the problem. An account that rests on BTC accommodates both the belief and evidence questions in a natural way and satisfies both the probability calculus and the likelihood principle that rests on using actual observed data. It could be shown that BTC satisfies the probability calculus. The former also automatically satisfies the likelihood principle which is the foundation for a Bayesian measure of evidence because the data's support for the theory stems only from the likelihood function in any Bayesian analysis (For connections between BIC and the Bayes factor see, Raftery, 1995).

9. A Decision-Theoretic Account of Acceptance

In response to Royall's acceptance question, I will provide a decision-theoretic account of acceptance. The notion of acceptance is crucial both for statistical theories as well as for theories in physical and biological sciences. Two questions arise in connection with acceptance of a theory: (i) *what* is acceptance of a theory and (ii) if we sometimes accept a theory, then what *justifies* this acceptance?

To answer these questions, I provide a Bayesian theory of acceptance. Building on van Fraassen's theory (van Fraassen, 1991), I defend a double aspect theory of acceptance: a) the belief aspect that states my degrees of belief in a theory and b) the pragmatic aspect that states my non-epistemic reasons for pursuing a theory, such as getting an NSF grant. In my account, I also have a justification for my double- aspect account of acceptance. A) Like Bayesians, on my view, an agent's degrees of belief must obey the probability calculus and any change in her degrees of belief must be done in accordance with the rule of conditionalization. B) As a Bayesian, I justify the agent's pragmatic reasons for pursuing a theory by invoking the principle of maximizing expected utility (hereafter, MEU).

Bayesians hold that our degrees of belief admit of a numerical representation that obeys the rules of the probability calculus. If our degrees of belief disobey the rules of probability calculus, then *some* Bayesians argue, a clever bettor can make a book against us so that we are bound to lose no matter how the world turns out to be. This argument for justifying why the agent's degree of belief should satisfy the probability calculus is known as *the Dutch-Book argument* (see, de Finetti, 1962; Skyrms, 1995. For a different perspective on the Dutch-Book argument, see Howson and Urbach, 2005). If a rational agent updates her belief in light of new data according to rules of belief change, then the rational agent's belief must satisfy the principle of conditionalization.

My Bayesian account of acceptance rests on MEU. According to MEU, in a given decision situation, the decision maker should choose the alternative with maximal expected utility. It is commonly assumed in Bayesianism that the decision maker can assign numerical values to the utilities of various outcomes in the decision situation. In Bayesianism, it is further assumed that the decision maker can assign probabilities to the states of the world. This is why Bayesian decision theory is based on the subjective probability of the event in question. For any state of the world, the decision maker can assign any probability value between 0 and 1. Although the rational agent can pick any probability value between 0 and 1, new information gathered from experience, and subject to some conditions, can change the subjective probability of the agent. Given the utility of the outcomes and the probabilities of the states, the decision maker can compute the expected utility of the various alternatives. Here, utilities are taken to be linearly related to money.

As a Bayesian I contend that in a decision situation one ought to accept the theory which has a higher expected utility than any other. Sometimes, it may happen that we accept a theory that has a lower probability than the rest of the theories in a domain, though the former has a significantly higher utility than the rest of them. As a result, based on our expected utility calculation, we end up getting a higher expected utility if we accept the theory. In contrast, we may embrace a theory that has a lower utility, but which has an appreciably higher probability than the rest of them. In the end, we embrace the former, because we obtain a higher expected utility if we embrace it.

10. Summing Up: Dreams of a Final Theory

The purpose of this paper was to bring Royall's three questions to a wider audience and assess them from a Bayesian perspective. In addition, the paper also provided analyses for other topics. I discussed both Bayesian and non-Bayesian considerations for imposing *a priori* desiderata on inductive/statistical inference. I diagnosed that the choice of desiderata depends on one's goals, which could be multiple. I discussed why Royall's emphasis of multiple goals in terms of three questions is rewarding, although I pointed out a defect in the law of likelihood on which his likelihood framework rests. I also defended the likelihood principle. I developed two distinct accounts within Bayesianism to respond to the belief and evidence questions and discussed how my Bayesian account could tackle complicated problems that often arise in model selection. Based on the idea of maximizing expected utility, I have provided a response to Royall's acceptance question. My Bayesian answer to the acceptance question is that we should choose the theory that has a higher expected utility than its rival.

In the literature on scientific explanation, an explanation is considered to be a good scientific explanation if it is able to unify diverse phenomena. A theory that has the ability to provide a unifying explanation for several apparently unrelated phenomena is hailed as a good theory. For example, we prefer Einstein's theory of relativity to Newton's theory because of this reason. Using Einstein's theory, we could explain (i) the occurrence of a red-shift, (ii) the bending of light in front of massive nearby objects and finally (iii) the precession rate of Mercury's perihelion with sufficiently precise details. Although the ability to unify diverse items under one banner is canvassed as a plus point for scientific explanation, I think that this ability should also be counted as an added advantage for a statistical account that provides a unifying view of several apparently disjoint questions. I argued that Bayesianism and Bayesianism alone is able to provide an unified account to all three questions thus providing a primer for philosophers of science interested in a unified approach to issues like belief, evidence, statistical inference and decisions about which theories they should accept.

References

Adams, E. (1998). *A Primer of Probability Logic*. Stanford, CA:CSLI.

Akaike, H. (1973). Information Theory as an Extension of the Maximum Likelihood Principle. In Petrov, B.N., and Csaki, F. ed., *Second International Symposium on Information Theory*. Budapest: Akademia Kiado.

Bandyopadhyay, P.S. Types of Coherence and Coherence among Types. In Pereira L. M., and Wheeler, G. (eds.). In *Computational Models of Scientific Reasoning and Applications*: CMSRA-IV, Lisbon, 2005, 129-143.

Bandyopadhyay, P.S. and Brittan. G. Jr. Acceptance, Severity and Evidence. *Synthese* (forthcoming): 148 (2). 259-293.

Bandyopadhyay, P.S. and Bennett, J.G. (2004). Commentary on Mauer. In M.L. Taper and S.R. Lele, eds., *The Nature of Scientific Evidence*. Chicago: University of Chicago Press, 32-39.

Bandyopadhyay, P.S. (2002). Simplicity: Our View, Their View. Presented at the *American Philosophical Association,* Chicago: Central Divisional meetings.

Bandyopadhyay, P.S. and Boik, R.J. (1999). The Curve Fitting Problem: A Bayesian Rejoinder. *Philosophy of Science*, 66 (supplement): 391-402.

Bandyopadhyay, P.S., Boik, R.J. and Basu, P. (1996). The Curve Fitting Problem: A Bayesian Approach. *Philosophy of Science*, 63 (supplement): 264-272.

Berger, J.O. (2000). Bayesian Analysis: A Look at Today and Thoughts of Tomorrow. *Journal of the American Statistical Association,* **95**, 1269-1276.

Berger, J.O. (1985). *Statistical Decision Theory and Bayesian Analysis*. Second edition, New York: Springer.

Berger, J.O. and Wolpert, R.L. (1988). *The Likelihood Principle*. Hayward, CA: Institute of Mathematical Statistics.

Berger, J.O. and Pericchi, L.R. (1996). The Intrinsic Bayes Factor for Model Selection and Prediction. *Journal of the American Statistical Association,* 91: 109-122.

Bernardo, J.M. and Smith, A.F.M. (2000). *Bayesian Theory*. Weinheim: John Wiley.

Bernardo, J. Reference Analysis, 1-86. In *Handbook of Bayesian Statistics*. D. Dey and C.R. Rao (eds.), Elsevier (forthcoming).

Birnbaum, A. (1962). On the Foundations of Statistical Inference (with discussion). *Journal of the American Statistical Association,* **57**: 269-326.

Boik, R.J. (2004). Commentary on Forster and Sober. In M.L. Taper and S.R. Lele, eds., *The Nature of Scientific Evidence*. Chicago: University of Chicago Press.

Burnham, K.P. and D.R. Anderson, (1998). *Model Selection and Inference*, New York: Springer.

Carnap, R. (1952). *The Continuum of Inductive Methods*. Macmillan, New York.

Cassella, G. and Berger., R.L. (1990). *Statistical Inference*. Belmont, Ca: Duxbury.

Christensen, D. 1999. Measuring Confirmation. In *Journal of Philosophy*, **99**, 9, Sept, 437-461.

Dawid, A.P. and Morera, J. (1996). Coherent Analysis of Forensic Identification of Evidence. In *Journal of the Royal Statistical Society*. Series B (Methodological), **58**, 2, 425-443.

de Finetti, B. 1937. La prevision: ses lois logiques, ses sources subjectives. English trans., "Foresight: Its Logical Laws, Its Subjective Sources. In H.E. Kyburg, Jr. and Smokler, H.E. eds., *Studies in Subjective Probability*. Huntington, NY: Kreiger, 1962.

Descartes, R. 1911. *The Philosophical Works of Descartes*, eds. Haldane, E. and G.R.T. Ross, Cambridge: Cambridge University Press.

Fitelson, B. 1999. The Plurality of Bayesian Measures of Confirmation and the Problem of Measure Sensitivity. *Philosophy of Science*, 66 (supplement), 362-378.

Forster, M. and Sober, E. (2004). Why Likelihood? Chapter 6. In M.L. Taper and S.R. Lele, eds., *The Nature of Scientific Evidence*. Chicago: University of Chicago Press.

Forster, M. and Sober, E. (1994). How to Tell When Simpler, More Unified, or Less *Ad Hoc* Theories Will Provide More Accurate Predictions. *British Journal for the Philosophy of Science*, **45**:1-35.

Ghosh, J.K. (ed.) (1988). *Statistical Information and Likelihood*. New York: Springer.

Gill, J. (2002). *Bayesian Methods*. London. Chapman & Hall.

Good, I.J. (1983). *Good Thinking*. Minneapolis: University of Minnesota Press.

Howson, C. and Urbach, P. (2005). *Scientific Reasoning: The Bayesian Approach*. Third edition, La Salle, IL: Open Court.

Jaynes, E.T. (2003). *Probability Theory: The Logic of Science*. Cambridge: Cambridge University Press.

Jeffreys, H. (1957). *Scientific Inference*. Second edition. Cambridge: Cambridge University Press.

Jeffreys, H. (1961). *Theory of Probability*. Third edition. Oxford: Clarendon Press.

Leibniz, G. *Philosophical Schriften von G.W. Leibniz*, ed. Gerhardt, C.I., Berlin, 1875-90, IV.

Lele, S.R. (2004). Evidence Functions and the Optimality of the Law of Likelihood. In M.L. Taper, and S.R. Lele, eds., *The Nature of Scientific Evidence*. Chicago: University of Chicago Press.

Levine, M. and Schervish, M.J. (1999). Bayes Factor: What They Are and What They Are Not. *The American Statistician*, **53**, 2, 119-122.

Mayo, D. 2004. An Error-Statistical Philosophy of Evidence. In M.L. Taper and Lele, S.R. eds. *The Nature of Scientific Evidence*. Chicago: University of Chicago Press.

Mayo, D. 1996. *Error and the Growth of Experimental Knowledge*, University of Chicago Press: Chicago.

Neyman, J. 1967. *A Selection of Early Statistical Papers of J. Neyman*. Berkeley: University of California Press.

Pagano, M. and Gauvreau K. (2000). *Principles of Biostatistics*. Duxbury, Australia.

Popper, K.R. (1959). *The Logic of Scientific Discovery*. New York: Basic Books.

Raftery, A.E. (1995). Bayesian Model Selection in Social Research (with discussion). In *Sociological Methodology 1995*, Marsden, P.V. ed.

Rosenkrantz, R.D. (1977). *Inference, Method, and Decision*. Dordrecht: Reidel.

Royall, R. (1997). *Statistical Evidence: A Likelihood Paradigm*. New York, Chapman Hall.

Schwarz, G. (1978). Estimating the Dimension of a Model. *Annals of Statistics*, 6, 461-464.

Sober, E. (1993). Epistemology for Empiricists. In *Midwest Studies in Philosophy*, XVIII (eds.) French, P., Uehling, T. and Weinstein, H.

Sober, E. (2005). Is Drift a Serious Alternative to Natural Selection as Explanation of Complex Adaptive Traits. In O'Hear, A., ed., *Philosophy, Biology and Life*. Cambridge, Cambridge University Press.

Skyrms, B. (1995). Strict Coherence, Sigma Coherence and the Metaphysics of Quantity. *Philosophical Studies*. January issue.

Taper, M.L. (2004). Model Identification from Many Candidates. In M.L. Taper and S.R. Lele (eds), *The Nature of Scientific Evidence*. Chicago, University of Chicago Press.

Taper, M.L and Lele, S.R., eds. (2004). *The Nature of Scientific Evidence*, Chicago, University of Chicago Press.

Van Fraassen, B.C. (1991). *Quantum Mechanics: An Empiricist View*. Oxford: Clarendon Press.

Weinberg, S. (1992). *Dreams of a Final Theory*. New York: Pantheon Books.

Zellner, A. (1997). *Bayesian Analysis in Econometrics and Statistics*. Cheltenham, UK: Edward Elgar Publishing Limited.

Zellner, A. (1996). *An Introduction to Bayesian Inference in Econometrics*. New York, John Wiley.

Zellner, A. (1984). *Basic Issues in Econometrics*. Chicago: University of Chicago Press.

Zellner, A. (2001). Keep it Sophistically Simple. In A. Zellner, H. Keuzenkamp and McAleer, M., eds., *Simplicity, Inference and Modelling*, Cambridge: Cambridge University Press.

Bayesian Statistics and Its Applications
Edited by S.K. Upadhyay, U. Singh and D.K. Dey
Anamaya Publishers, New Delhi, India

On Coregionalized Models for Spatially Replicated Experiments in Weed Proliferation Studies

Sudipto Banerjee[1] and Gregg A. Johnson[2]

[1]Division of Biostatistics, [2]Department of Agronomy and Plant Genetics, University of Minnesota, Minneapolis, Minnesota 55455

Abstract

Agricultural experiments are often conducted in spatially replicated fields, where each plot comprises an array or "lattice" of subplots. In devising statistical models to assess spatial variation, one must account for this design. This article looks at such a class of spatiotemporal models which model weed proliferation in agricultural fields. Weeds significantly hamper crop productivity causing huge losses for the farming community. Recent advances in Geographical Information Systems (GIS) now allow geocoding of agricultural data, such as ours, enabling more sophisticated spatial analysis. Our data come from experiments conducted in Waseca, Minnesota, that recorded density of the weed called *Setaria* spp. We develop a Bayesian hierarchical framework to deal with this spatially replicated experiment. Flexible classes of models arise which are fitted using simulation-based methods.

1. Introduction

Weed presence in agricultural fields causes significant damage to crop productivity often leading to deeper economic repercussions. Better modelling of weed proliferation assists farmers in making critical decisions regarding the location and type of control measures to implement. Traditionally, weed control strategies have been applied homogeneously across the field largely ignoring spatial variation (Colbach et al., 2000; Johnson et al., 1995; Johnson, et al. 1996). New technologies are available that now enable farmers and other land managers to apply weed control tactics site-specifically across a field. Recent studies have attempted to understand the spatial arrangement of weeds using multivariate statistical methods such as canonical correlation analysis (Dieleman et al., 2000). Alternatively, crop growth models have been used to investigate the temporal aspects of plant growth, but these mechanistic models are based upon data with relatively large time intervals (e.g., between years) and with homogeneous soil units (Deen et al., 2001; Batchelor et al., 2002).

Our current work focuses primarily upon the spatial modelling of replicated data. We restrict ourselves to a single time point while recognizing that several spatiotemporal approaches could be implemented in our setting. Agricultural experiments are often spatially replicated, where each plot consists of subplots where weed densities are recorded, producing replicated measurements with enhanced precision. Ideally, the subplots should be homogeneous, but some variation between them is inevitable. Hence, spatial models designed for simply capturing variation between plots are inadequate and one needs to accommodate variation between the subplots as well. We refer to the variation between the main plots as *macro*-level spatial variation and that between the subplots as the *micro*-level variation nested within each site.

Recently, Banerjee and Johnson (2006) constructed a class of multi-resolution coregionalized models with spatially varying coefficients using parametric linear growth functions whose coefficients were spatial processes. Using cross-covariance matrix specifications, they captured spatial association using geostatistical processes at both the micro and macro resolutions as well as association between the regression coefficients. In particular, they demonstrated that simpler univariate geostatistical models (e.g. Banerjee et al., 2004; Chapter 5) may be inadequate for capturing all inherent associations. While their framework encompasses rich modelling, parametric growth curves may only have a limited ability to capture spatial variation. Here we address modelling of such replicated data directly using spatial processes. In that sense our approach has a more "nonparametric" flavour than those adopting parametric growth curves. Another difference from the aforementioned work is our modelling the micro variation using array *adjacencies*, while using geostatistical processes to model the macro variation between the main plots.

Section 2 outlines the design and scheme of the experiment. Section 3 develops different models for the data, while Section 4 discusses model implementation. Section 5 focuses upon the analysis of the weed density data and Section 6 concludes with a discussion.

2. Data and Design

Our data was collected from field research conducted in the year 2001 at the University of Minnesota's Agricultural Ecology Research Farm (AERF) located at the Southern Research and Outreach Center in Waseca, Minnesota. The field was naturally drained and chisel ploughed in the fall, followed by a field cultivator to prepare the seed bed in the spring. The experimental design comprised six replicates (subplots) nested as a 3×2 lattice within each main plot. Fig. 1 shows the spatial orientation of the main plots, where measurements were taken. Main plots, often called sites or locations, are ten research sites that were established within a sixteen hectare area of the AERF based upon detailed terrain analysis and soil taxonomy information. The maximum distance between the main plots was approximately 186 m. Uniform geometric dimensions of the lattices yield a common inter-subplot distance matrix for each main plot. The maximum distance between locations of the subplots within a main plot is approximately 7 m, and this is the same for all main plots.

Several terrain attributes that define a treatment, e.g., aspect, hill-slope position, and soil characteristics, are identified. However, we expect the soil environment within these treatments or sites to have a certain degree of variability across the plot area. Thus, we want to be able to capture the subtle spatial variability within each site. This subtle variability is the *micro*-level variability, which is nested within the *macro*-level variability between the main plots.

In order to model variation in growth patterns within each subplot, weed density data was collected using non-destructive methods and development data were collected weekly for five consecutive weeks. The field was tilled a day ahead of planting to prepare a level seedbed. The tillage operation (and to some extent planting operation) essentially eliminated all existing plants. Therefore, plants that emerged after planting would have come from seeds in the soil. Seed germination and emergence from the soil takes anywhere from 7 to 14 days. Thereafter, we started recording density measurements after such time that there was enough vegetation above the ground to get a good measurement, i.e. 14 days after planting (or elimination of all weeds).

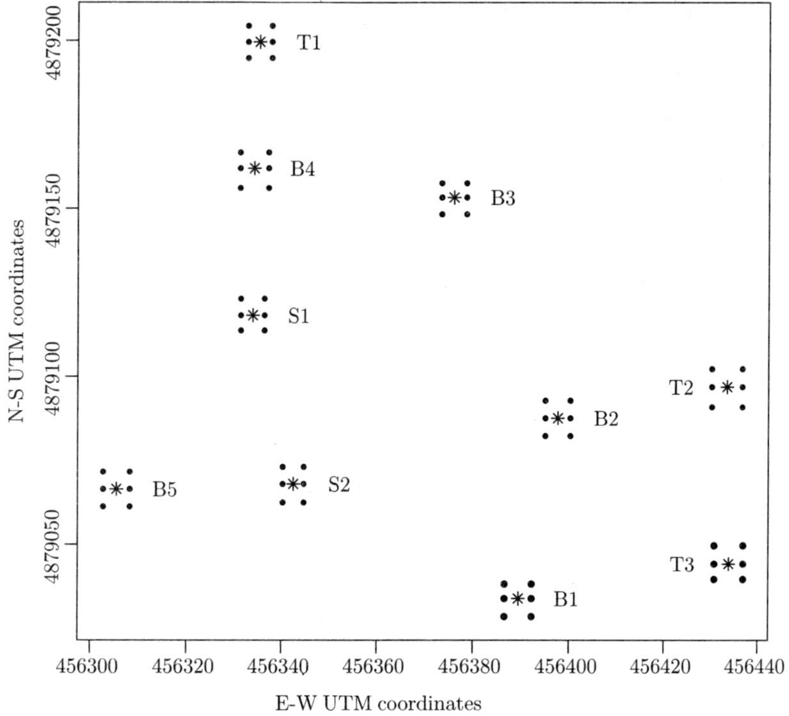

Fig. 1. Location of the sites in Waseca. Each site, when zoomed in, is a rectangular 3×2 lattice of spatially arranged subplots (replicates). Labels indicate back-slopes (B's), toe-slopes (T's) and summits (S's).

The weed of interest in the current experiment is *Setaria* spp. an especially infamous species of grasses in highly disturbed agricultural systems. They occur throughout North America and are very competitive with crop plants. This species of weeds germinates early in the season, has a high growth rate, and is a very prolific seed producer. Wilson and Tilman (1995) studied the competitive responses of eight weed species, including *Setaria* spp., concluding that competitive effects shift from roots to shoots as fertility increases and the competitive effects decline with disturbance.

Knowledge of soil hydrology is also essential to evaluating the processes related to weed growth. Thompson and Bell (1996) introduced a soil color index, the Profile Darkness Index (PDI), which accounts for the observed morphological difference from shoulder to depressional areas in Mollisol hydrosequences. In these landscapes, PDI is well correlated with duration of saturated conditions at various soil depths. Successful weed seed germination, establishment, and growth is partly a function of soil moisture and nutrient availability. High PDI values identify areas that are saturated and high in nutrient availability. Conversely, low PDI values indicate lower organic matter, higher sand content, and relatively low nutrient availability. Therefore, it is reasonable to expect this index to be correlated to weed growth along a hill-slope and forms an important covariate.

3. Models for Spatially Replicated Growth Data

Here we discuss model development for our application. We begin with simpler models and graduate to the more complex models by incorporating richer structure to capture association at different resolutions. Let N_s denote the number of main plots (indicated as the array centroids in Fig. 1);

within each site there are N_r replicates (subplots denoted by solid circles in Fig. 1), so the total number of observed locations $N = N_s N_r$, each yielding a single observation (we look at a single time point) for weed density data points. Thus, in our experimental setting we have $N_s = 10$ and $N_r = 6$ yielding a total of $N = 60$ locations.

Momentarily ignoring the nested structure of the sites, and treating each subplot as a location providing observations, we can generate a class of spatial models as

$$Y(s) = \boldsymbol{x}^T(s)\boldsymbol{\beta} + W(s) + \varepsilon(s) \tag{1}$$

where $Y(s)$ is the natural log of the weed density in subplot $s \in \mathcal{S}$ at time t ($\mathcal{S}$ is our spatial domain), and $\boldsymbol{x}(s)$ is the associated vector of covariates. Also, $\varepsilon(s) \overset{\text{i.i.d.}}{\sim} N(0, \tau^2)$ captures measurement error (also called the *nugget effect*) and $W(s)$ is a spatial process capturing the effect of site s at time t.

A simple linear regression model is obtained from setting

$$\text{Model 1}: W(s) = 0 \text{ a.s.}$$

in (1) and would suffice with negligible extraneous variation—beyond what is explained by the covariates. A simple extension of this model would include site-specific random effects

$$\text{Model 2}: W_i \overset{\text{i.i.d.}}{\sim} N(0, \sigma_i^2), \ i = 1, \ldots N_s$$

Note that the variability is captured specific to each main-plot i, yielding N_s parameters for the σ_i^2's, which are identifiable (with weak prior specifications) from the subplot level data.

Model 2 does not attempt to model *spatial associations*. However, we expect similar weed densities in proximate locations, possibly resulting from similar topographic and environmental conditions. This suggests modelling association as a function of distance between locations: association diminishes as distance increases. Scientific interest resides in capturing this phenomenon, hypothesized through spatial models. A popular specification is the Gaussian Process, denoted by $W(s) \sim GP(\mu(s), \ K(\cdot))$, where $\mu(s)$ is the process mean (or trend surface) and $K(\cdot)$ is a positive definite covariance function. Gaussian processes are extremely popular in modelling spatial variation for their ability to directly model spatial correlation. More extensive details about Gaussian Processes can be found, for example, in Cressie (1993) and Banerjee et al. (2004). The rich class of space-time models in this framework has been mentioned in Section 1. Following Banerjee and Johnson (2006), we further classify these spatial models as *single-resolution* or *multi-resolution*. The former ignores the nested structure of the spatial domain and accounts for spatial association based upon all the $N = N_s N_r$ subplot locations, assuming

$$\text{Model 3}: W(s) \sim GP(0, K(\cdot); K(s - s'; t)) = \sigma^2 \rho(||s - s'||; .)$$

Here $K(s - s')$ does not depend upon t and the strength of spatial association is captured through a purely spatial correlation function $\rho(\cdot)$. A versatile class of spatial correlation functions, allowing control of spatial association as well as smoothness, is the Matérn class (see, e.g., Stein, 1999, p. 51) given by

$$\rho(d, \phi, \nu) = \frac{1}{2^{\nu-1}\Gamma(\nu)}(2\sqrt{\nu}d\phi)^\nu K_\nu(2\sqrt{\nu}d\phi)$$

where K_ν, with order ν, is the Bessel function of the second kind, ϕ controls the decay in spatial correlation and ν is a smoothness parameter with higher values yielding smoother process realizations (see Banerjee, Gelfand and Sirmans, 2003). We resort to this class for our subsequent analysis.

We next expand the above model to a multi-resolution setting that recognizes the nested nature of our plots. Hence, we refer to the jth subplot within s as "location" $s(j)$. Let $Y(s(j))$ be the response (logarithm of *Setaria* spp. density) from the jth subplot (replicate), $j = 1, \ldots, N_r$, residing within main plot s. Let $\mathbf{x}(s(j))$ be a $p \times 1$ vector of covariates associated with $Y(s(j))$. Then, we have

$$Y(s(j)) = \mathbf{x}^T(s(j))\boldsymbol{\beta} + W(s(j)) + \varepsilon(s(j)) \tag{2}$$

where $\epsilon(s(j)) \overset{\text{i.i.d.}}{\sim} N(0, \tau^2)$, captures measurement error (also called the *nugget effect*) and $W(s(j))$ is a spatial process capturing the effect of site $s(j)$. Collecting over the subplots, we now have a multivariate process, $\mathbf{W}(s) = (W(s))_{j=1}^{N_r}$ associated with each main plot s.

The multivariate spatial process $\mathbf{W}(s)$ is characterized by its *cross-covariance* matrix function, $\Gamma_W(s, s')$, which is defined for any two sites s and s' as the $N_r \times N_r$ matrix whose jj'-th element is given by $[\Gamma_W(s, s')]_{jj'} = \text{Cov}(W(s(j)), W(s'(j')))$. In particular, note that with $s = s'$, $\Gamma_W(s, s)$ is a covariance matrix for the elements of $\mathbf{W}(s)$ within site s. Given the process realization over a set of sites $(s_i)_{i=1}^{N_s}$, the joint dispersion matrix of $\mathbf{W} = (\mathbf{W}(s_i))_{i=1}^{N_s}$ is determined by the cross-covariance matrix as $\Sigma_W = [\Gamma_W(s_i, s_j)]_{i,j=1}^{N_s}$, that is, Σ_W is a $N_r N_s \times N_r N_s$ block matrix of $N_r \times N_r$ matrices with the $(i, j)^{\text{th}}$ block being $\Gamma_W(s_i, s_j)$. A valid $\Gamma_W(\cdot)$ is one that ensures a symmetric and positive definite Σ_W. Note, however, that $\Gamma_W(s, s')$ for $s \neq s'$ need not be positive definite or even symmetric, except in the limiting sense as $s \to s'$, but must satisfy $\Gamma_W(s, s') = \Gamma_W^T(s', s)$ to ensure the symmetry of Σ_W. These multivariate processes are *stationary* when the cross-covariances are functions of the separation between the sites, in which case we write them as $\Gamma_W(s - s')$. Further details on cross-covariance matrices can be found in Cressie (1993) and, specific to multi-resolution contexts, in Banerjee and Johnson (2006).

A valid cross-covariance matrix must ensure the positive-definiteness of Σ_W; this offers a greater challenge than finding correlation functions in univariate spatial models. Here, we require that for an arbitrary number and choice of locations, the resulting Σ_W be positive definite. A theorem by Cramér (see, e.g., Chilés and Delfiner, 1999) provides a characterization of cross-covariance functions, akin to what Bochner's theorem says for correlation functions. However, using Cramér's result is less trivial from a computational perspective, especially so in multi-resolution settings. Following Banerjee and Johnson (2006), therefore, we pursue a constructive approach to obtain a flexible class of models known as coregionalization (Wackernagel, 2003).

This approach proceeds by letting $\mathbf{W}(s)$ as an affine transformation of a *latent* independent, zero-centered stationary processes with unit variances $(v_1(s), \ldots, v_{N_r}(s))$ such that $\mathbf{W}(s) = \boldsymbol{\mu}(s) + A(s)\mathbf{v}(s)$, where $\mathbf{v}(s) = (v_j(s))_{j=1}^{N_r}$ is a multivariate $GP(\mathbf{0}, \Gamma_{\mathbf{v}}(\cdot))$ and $A(s)$ is a non-singular (possibly space-varying) transformation matrix. This yields $\mathbf{W}(s)$ as a multivariate $GP(\mu(s), \Gamma_W(\cdot))$ with $\Gamma_W(s, s') = A(s)\Gamma_{\mathbf{v}}(s - s')A(s')^T$. Here, the components of $\mathbf{v}(s)$ model the macro-level correlation structure, so that $\Gamma_{\mathbf{v}}(s - s') = \oplus_{j=1}^{N_r}\rho_j^{\text{mac}}(s - s')$, where the $\oplus$ is the "block-diagonal" operator. Thus, $\Gamma_{\mathbf{v}}(s - s')$ is a diagonal matrix with $\rho_j^{\text{mac}}(s - s')$ as the diagonal elements. This induces the joint dispersion matrix of $\mathbf{W}$ as $\Sigma_W = \mathcal{A}\Sigma_{\mathbf{v}}\mathcal{A}^T$, where $\mathcal{A} = \oplus_{i=1}^{N_s}A(s_i)$ is the block-diagonal matrix with $A(s_i)$'s as the blocks. In particular, note that when $s = s'$, $\Gamma_W(s, s) = A(s)A^T(s)$ since $\Gamma_{\mathbf{v}}(0) = I$. This *identifies* $A(s)$ as a square-root (e.g. a lower-triangular Cholesky) of the within-site micro-level dispersion matrix $\Gamma_W(s, s)$.

This framework includes some interesting special cases. Suppose that the micro-level dispersion matrix remains invariant over the main plots, i.e., $\Gamma_W(s, s) = \Lambda$, where Λ is a single positive definite matrix modelling covariances between the subplots in each array s. In addition, suppose we specify $\rho_j^{\text{mac}}(\cdot) = \rho^{\text{mac}}(\cdot)$—a common correlation function. This means, from the above discussion, that

$A(s) = A$ is independent of s, which leads to a *separable model* (Smith et al., 2001a, 2001b):

$$\text{Model 4}: \Gamma_W(s - s') = \rho^{\text{mac}}(s - s')\Lambda.$$

This translates to $\Sigma_W = R^{\text{mac}}(\phi) \otimes \Lambda$, where $\otimes$ is the Kronecker product (see, e.g., Harville, 1997) and R^{mac} a correlation matrix with $[R^{\text{mac}}]_{ii'} = \rho^{\text{mac}}(s_i - s_{i'})$. Here Λ is a completely unspecified positive definite matrix that is treated as a parameter (assigned, perhaps, an Inverse-Wishart prior).

However, we may want to use the adjacency information (they form a lattice; see Fig. 1) of the subplots to model the correlations using a Conditionally Autoregressive (CAR) prior (see, e.g., Besag and Higdon, 1999). Then, $\Lambda^{-1} = \lambda(\text{Diag}(m_j) - \psi B)$, where m_j are the number of neighbors of the jth replicate, ψ is a parameter controlling the extent of spatial association, B is the adjacency matrix (a binary matrix with $[B]_{jj'} = 1$ if and only if j and j' are neighbors), and λ is a spatial precision parameter. This gives

$$\text{Model 5}: \Gamma_W(s - s') = \rho^{\text{mac}}(s - s')(\text{Diag}(m_j) - \psi B)^{-1}.$$

This is still a separable model, like model 4, but with additional parametrization of B. Restricting ψ to lie between $(0, 1)$ is sufficient to ensure positive definiteness of the above matrices. A spatially strong prior is obtained by setting these parameters close to 1. The extreme case results in a singular specification for Λ, but can still be employed in Bayesian settings since they result in proper posteriors (e.g. Banerjee et al., 2004, Ch. 6).

Separable models, while being simple and easily interpretable, forces a common micro-level correlation structure for all the sites. Separability also collapses if the subplots have different neighborhood relations between main plots. This latter point is not an issue in the present application, but richer models allowing each site its own micro-level modelling parameters are still desirable. For instance, the assumption that every array or main plot have the same ψ and λ is perhaps inappropriate. Then, we can incorporate array-specific parameters ψ_s and λ_s resulting in a non-separable and non-stationary model

$$\text{Model 6}: \Gamma_W(s, s') = \rho^{\text{mac}}(s - s')A(s)A^T(s'); A^T(s) = \lambda_s^{-1/2}(\text{Diag}(m_j) - \psi_s B)^{-1/2}.$$

This model is quite flexible, yet still retains easy interpretation. The non-stationary behaviour here is caused precisely because of the variation of the smoothness parameters across the sites.

In our analysis here, we will consider the performance of these six models. We remark, however, that these techniques can be extended to a number of other settings. The aforementioned work of Banerjee and Johnson (2006) does so for spatially varying growth curves; other situations with multiple observations, misaligned variables and spatiotemporal processes can also be addressed.

4.   Bayesian Implementation and Model Comparisons

We adopt a Bayesian approach specifying prior distributions on the parameters to build hierarchical models that are estimated using a Gibbs sampler, with Metropolis updates when required, for fitting our models (see, e.g., Gelman et al., 2004; Chapter 11). Since such algorithms often demand intensive coding, casting the problem in a general template enables several models to be fit without rewriting vast amounts of code.

Generally, we can cast our models into a data equation resembling a first-stage mixed model

$$\mathbf{Y} = X\boldsymbol{\beta} + \mathbf{W} + \boldsymbol{\epsilon}; \boldsymbol{\epsilon} \sim N(\mathbf{0}, \tau^2 I) \tag{3}$$

where $\mathbf{Y}$ is the response vector, X the covariate design matrix, $\boldsymbol{\beta}$ the corresponding vector of regression coefficients, $\mathbf{W}$ are the multivariate process realizations and $\boldsymbol{\epsilon}$ is a vector of uncorrelated random errors. Markov Chain Monte Carlo model fitting proceeds with a Gibbs sampler with Metropolis steps (see, e.g., Gelman et al., 2004) on the marginalized scale, after integrating out $\mathbf{W}$, to reduce the parameter space. The marginalized likelihood becomes $N(X\boldsymbol{\beta} + \boldsymbol{\mu}, \Sigma_W + \tau^2 I)$, where $\boldsymbol{\mu} = E[\mathbf{W}]$ and Σ_W is the spatial covariance matrix for the respective model. Often, we set the GP mean as a constant, $\boldsymbol{\mu} = \mu\mathbf{I}$, which acts like an "intercept" and precludes inclusion of a separate intercept term in the covariate design matrix X for preserving model identifiability. Also, τ^2 acts as a residual variance after the spatial variation has been accounted. In principle, one may have different residual variances for each array (say, τ_s^2) which would result in a diagonal dispersion matrix for $\boldsymbol{\epsilon}$, but identifiability of these additional parameters is practically impossible. To be specific, for the multi-resolution models (Models 4, 5 and 6) we let $\mathbf{Y}(s) = (Y(s_i(j)))_{j=1}^{N_r}$ be the observations across subplots from main plot s_i and form $\mathbf{Y} = (\mathbf{Y}(s_i))_{i=1}^{N_s}$. Analogously, we have $X = (\mathbf{x}(s_i(j)))_{i,j=1}^{N_s,N_r}$ as the $N_r N_s \times p$ dimensional covariate matrix.

Bayesian hierarchical specifications assign prior distributions to the parameters. Choice of priors can play an important role in the efficiency of the algorithm. A flat prior is assigned for the covariate slope vector $\boldsymbol{\beta}$, while the error variance τ^2 is assigned an inverse-Gamma prior. Prior distribution for $\boldsymbol{\mu}$ is again taken as flat, although typically some simplification is made. We assume that the mean level of the process vary by main plots, but remain constant across the subplots within a main plot. For instance, in the multi-resolution setting this implies $E[\mathbf{W}(s_i)] = \boldsymbol{\mu}(s_i) = \mu_i\mathbf{1}$. The variance covariance parameters in Σ_W depend upon the specific model and how the associated cross-covariance(s) is (are) defined. When random effect (or spatial) variances are characterized by scalars, say σ_i^2 (as in Model 2), we assign inverse-Gamma distributions to them. More generally, when the dispersion is captured through a matrix Λ (e.g. in Model 4), an inverse-Wishart prior is assigned to Λ, while with the CAR specification (Model 5), a Beta $(18, 2)$ (mean $18/20 = 0.90$) is assigned to ψ and a Gamma prior is assigned to λ. In the coregionalized setting (Model 6), we use the same families for priors on ψ_s and λ_s. It is important to note the use of fairly informative proper priors for the parameters embedded in the spatial correlation structure that ensures proper and well-identified posteriors resulting in stabler convergence of algorithms (e.g. Berger et al., 2001).

Note that in updating the parameters using the marginal model as outlined above, we do not directly sample the spatial coefficients $\mathbf{W}$. This diminishes the dimension of the parameter space resulting in a more efficient MCMC algorithm. Nevertheless, the posterior distribution of $\mathbf{W}$ can be recovered in a posterior predictive fashion by sampling from

$$P(\mathbf{W}|\text{Data}) \propto \int P(\mathbf{W}|\Omega, \text{ Data})P(\Omega|\text{Data})d\Omega \qquad (4)$$

where Ω generically denotes the set of all parameters being updated. Once the posterior samples from $P(\Omega|$, say $\{\Omega^{(g)}\}_{g=1}^{G})$, have been obtained, posterior samples from $P(\mathbf{W}|\Omega^{(g)}, \text{Data})$ are drawn by sampling $W^{(g)}$ for each $\Omega^{(g)}$ from $P(\mathbf{W}|\Omega^{(g)}, \text{Data})$. This *composition sampling* is routine because $P(\mathbf{W}|\Omega, \text{Data})$ in (4) is Gaussian. The posterior estimates of these realizations can subsequently be mapped with contours to produce the image and contour plots, revealing trends (or hot-spots) in the spatial distribution. Spatial interpolation and prediction can also be carried out at any arbitrary site, even where no data monitoring has been performed, by computing

$$P(\mathbf{W}(s_0)|\text{Data}) \propto \int P(\mathbf{W}(s_0)|\mathbf{W}, \Omega, \text{Data})P(\mathbf{W}|\Omega\text{Data})P(\Omega|\text{Data})d\Omega d\mathbf{W} \qquad (5)$$

where s_0 is any location. In particular, for the multi-resolution models, we interpolate $\mathbf{W}(s_0)$. One added complication here is that we need to marginalize over the process parameters that depend upon the main plot s_0. Rather than a posterior predictive approach, Eq. (5) is more conveniently implemented using a full Gibbs sampler that updates all the unknown parameters from their full conditional distributions (using appropriate Metropolis steps when necessary).

Turning to issues of model comparison, we use the Deviance Information Criteria (DIC) (Spiegelhalter et al., 2002) as a measure of model choice. This criteria is the sum of the Bayesian deviance (a measure of model fit) and the (effective) number of parameters (a penalty for model complexity). It rewards better fitting models through the first term and penalizes more complex models through the second term, with lower values indicating favorable models for the data. While more formal decision theoretic measures of model fit (see, e.g., Gelfand and Ghosh, 1998) are available, the DIC is preferred here for its computational ease.

5. Analysis of Data

We now fit several competing models using our template to our spatially replicated data described in Section 2. The comparisons using the DIC are presented in Table 1. All the models contain the same set of covariates and differ only in the specification of the spatial process as described in Section 3. For each of these models three parallel MCMC chains were run for 5000 iterations. The CODA package in R (www.r-project.org) was used to diagnose convergence by monitoring mixing, Gelman-Rubin diagnostics, autocorrelations and cross-correlations. For each of the models, 1000 iterations revealed sufficient mixing of the chains, so the remaining 12000 samples (4000×3) were retained for posterior analysis.

Table 1. Model comparisons using the DIC criteria

	Model	pD	DIC
Model 1	Fixed growth	5.6	985
Model 2	i.i.d. varying growth	23.2	810
Model 3	Single-resolution spatial	23.6	773
Model 4	Separable: unstructured Λ	29.8	768
Model 5	Separable: CAR specification	25.1	758
Model 6	Coregionalized CAR	24.6	747

The DIC scores in Table 1 clearly indicate the advantages of spatial modelling. In fact, the incorporation of non-spatial but i.i.d. varying growth curves (Model 2) brings about a sharp decrease in the score from Model 1. Single resolution spatial association between the sites (Model 3) further reduces the scores. The next three models are the multi-resolution models, which perform much better than the ones above, highlighting the need for the multi-resolution, and particularly, coregionalized models. Note the further reduction in the DIC score of the coregionalized model, in spite of the increased number of parameters. For purpose of illustration, we present detailed analysis of the coregionalized Model 6.

Table 2 presents the parameter estimates for Model 6. The effects of the covariates are expected. Percentages of sand and potassium negatively impact the growth of *Setaria* spp., while phosphorus seems to be favorable. Higher PDI, which is related to higher organic matter and saturated soils, seems to be conducive to the growth of these weeds. The macro-level spatial parameters are presented in the next segment of Table 2. These parameters correspond to the macro-level process. The spatial decay parameter suggests a spatial range of approximately 98.5.

Table 2.   Parameter estimates for the coregionalized Model 6

Parameter	Estimate: 50%	(2.5%,	97.5%)
Sand	−0.097	(−0.166,	−0.057)
Phosphorus	−0.033	(−0.051,	−0.015)
Potassium	−0.013	(−0.017,	−0.011)
PDI	0.046	(0.008,	0.088)
ϕ	0.081	(0.022,	0.387)
ν	1.077	(0.662,	1.588)
τ^2	0.002	(0.001,	0.004)

As a result of differences in topography, soils developed from similar parent material may vary greatly within a small area (Foth, 1990). A sequence of soils along a transect that differs because of topography (e.g. summit S to back-slope B to toe-slope T) is called a soil catena or topose-quence. Some variation in growth patterns is attributed to type of site. Of our ten main plots, we had five sites in the back-slope, three in the toe-slope and two in the summit. The site-varying process means (μ_i's), which play the role of spatially-varying intercepts are presented in Fig. 2.

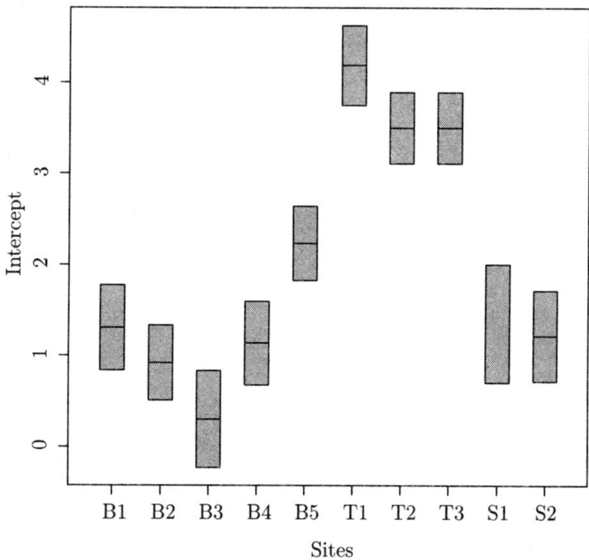

Fig. 2.   The central 95% posterior credible intervals for the site-wise process means (the μ_i's), which are the site-specific "intercepts" of the spatial process.

Figure 3 is the image plot (with overlaid contours) for the posterior mean surface of $\mathbf{W}(s)$, obtained by slicing the multivariate process by each replicate. These show the variation over the spatial domain, with respect to each replicate, and would have been inaccessible without our multi-resolution spatial process formulation. The lighter shades correspond to higher values of the process, while the darker shades correspond to the lower values. These plots offer a visual perspective

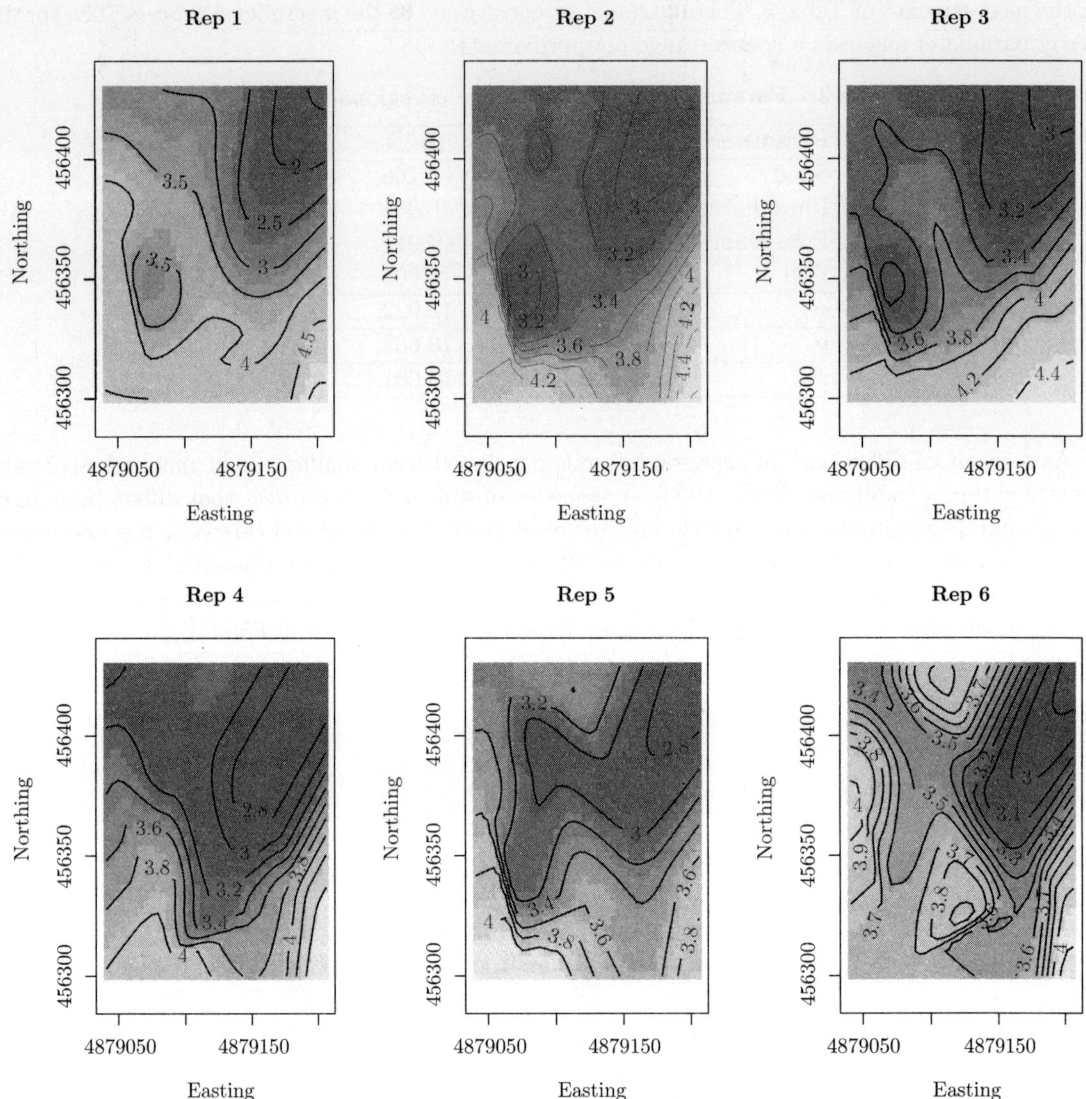

Fig. 3.   Replicate-sliced image plots of $\mathbf{W}(s)$ over the spatial domain.

to departure from separability; if these sliced plots revealed identical patterns, a separable model would suffice. While some similarities across the replicates are seen, like increasing gradients towards the south for the intercept process, the differences among the replicates are also apparent, somewhat justifying our adoption of the coregionalized models. Since each replicate is often designed as surrogates for local terrain properties and other soil-landscape attributes, these maps provide insight into the nature of spatial variation, with regard to those attributes. Finally, we slice the multivariate processes by sites, and show the variation of the posterior means $E[Y(r, s)\mid \text{Data}]$ for the six replicates within each site s. These are presented in the box-plots in Fig. 4. The site-type (B's or S's or T's) does not seem to affect the pattern of variation in the sub-plots, although modest variation is seen across all the sites.

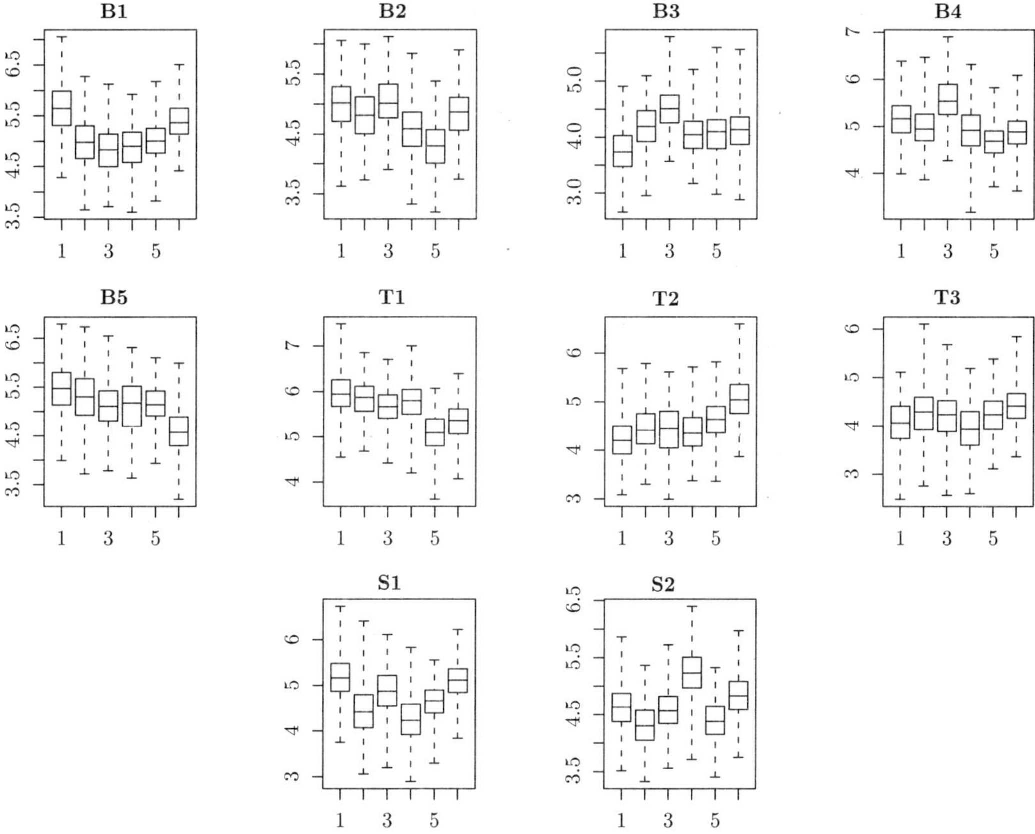

Fig. 4. Site-sliced box-plots showing the variation in the posterior mean $E[Y(r, s)|\text{Data}]$ for the six replicates within each site, bringing out the variation in the process between replicates within each site.

6. Discussion

The long-term scientific objective of our study is to characterize the extent of spatial variability in the patterns of weed densities across an agricultural landscape and enhance our understanding of factors contributing to differences in weed growth across this field landscape. Our long-term goal is to develop a decision support model that can be used by agricultural professionals to target application of integrated weed control strategies in a way that will reduce the risk of yield loss and address more long-term risk management issues. This type of analysis will also be used as a basis for optimizing the placement of alternative crops across the landscape with the goal of increasing profit while providing ecosystem services.

While the above goal is perhaps unattainable with single field experiments, our current work should contribute towards that goal. Spatial analysis of weed density patterns will help to identify unsuitable placement zones, and improve our understanding of the phenomena of weed growth. We propose to extend our models to accommodate for terrain attributes. A substantive part of the spatial variation is often attributed to local and global terrain attributes, although methods for testing such hypothesis have remained unaddressed. Recent work by Banerjee, Gelfand and Sirmans

(2003) develops statistical theory for digital terrain models, enabling joint spatial modelling of weed-growth and digital terrain features such as slopes and aspects.

We conclude with some comments about model implementation and software accessibility. For the single-resolution models, where the between-subplot spatial variation can be ignored, the Win-BUGS (**www.mrc-bsu.cam.ac.uk/bugs**) package can be used to fit the models. However, when the multivariate spatial processes are used, WinBUGS is much less efficient in fitting these models. Furthermore, the Matérn correlation family is not accessible in WinBUGS. In these, more general, settings our representation of the data equation and the subsequent hierarchical structure, leads to efficient coding with object oriented languages (C++, Java or S).

References

Banerjee, S., Carlin, B.P. and Gelfand, A.E. (2004). *Hierarchical Modeling and Analysis for Spatial Data.* Boca Raton, FL: Chapman and Hall/CRC Press.

Banerjee, S. and Johnson, G.A. (2006). Coregionalized single- and multi-resolution spatially-varying growth curve modelling with applications to weed growth. *Biometrics*, to appear.

Banerjee, S., Gelfand, A.E. and Sirmans, C.F. (2003). Directional rates of change under spatial process models. *Journal of the American Statistical Association*, **98**, 946–954.

Batchelor, W.D., Basso, B. and Paz, J.O. (2002). Examples of Strategies to Analyze Spatial and Temporal Yield Variability Using Crop models. *European Journal of Agronomy*, **18**, 141–158.

Berger, J.O., De Oliveira, V. and Sansó, B. (2001). Objective Bayesian analysis of spatially correlated data. *Journal of the American Statistical Association*, **96**, 1361–1374.

Besag, J. and Higdon, D. (1999). Bayesian analysis of agricultural field experiments. *Journal of the Royal Statistical Society B*, **61**, 691–746.

Chilés, J.P. and Delfiner, P. (1999), *Geostatistics: Modelling Spatial Uncertainty.* New York: John Wiley & Sons.

Colbach, N., Forcella, F. and Johnson, G.A. (2000). Spatial and temporal stability of weed populations over five years. *Weed Science*, **48**, 366–377.

Cressie, N.A.C. (1993). *Statistics for Spatial Data*, 2nd edition. New York: Wiley.

Deen, W., Swanton, C.J. and Hunt, L.A. (2001). A Mechanistic Growth and Development Model of Common Ragweed. *Weed Science*, **49**, 723–731.

Dieleman, J.A., Mortensen, D.A., Buhler, D.D., Cambardella, C.A., and Moorman, T.B. (2000). Identifying Association Among Site Properties and Weed Species Abundance. *Weed Science*, **48**, 567–575.

Foth, H.D. (1990). Fundamentals of Soil Science. New York: John Wiley & Sons.

Gelfand, A.E. and Ghosh, S.K. (1998). Model choice: A minimum posterior predictive loss approach. *Biometrika*, **85**, 1-11.

Gelfand A.E., Schmidt, A., Banerjee S. and Sirmans C.F. (2004). Nonstationary multivariate process modelling through spatially varying coregionalization. *Test*, **13**, 263–312.

Gelman, A., Carlin, J.B., Stern, H.S. and Rubin, D.B. (2004). *Bayesian Data Analysis.* Second Edition. Boca Raton, FL: Chapman and Hall/CRC Press.

Harville, D.A. (1997). *Matrix Algebra from a Statistician's Perspective.* New York: Springer.

Johnson, G.A., Mortensen, D.A and Martin, A.R. (1995). A Simulation of Herbicide Use Based on Weed Spatial Distribution. *Weed Research*, **35**, 197–205.

Johnson, G.A., Mortensen, D.A. and Gotway, C.A. (1996). Spatial and Temporal Analysis of Weed Seedling Populations Using Geostatistics. *Weed Science*, **44**, 704–710.

Smith, A.B., Cullis, B.R. and Thompson, R. (2001a). Analysing variety by environment data using multiplicative mixed models. *Biometrics*, **57**, 1138–1147.

Smith, A.B., Cullis, B.R., Appels, R., Campbell, A.W., Cornish, G.B., Martin, D. and Allen, H.M. (2001b). The statistical analysis of quality traits in plant improvement programs with applications to the mapping of milling yield in wheat. *Australian Journal of Agricultural Research*, **52**, 1207–1219.

Stein, M. L. (1999). *Interpolation of Spatial Data: some theory for kriging*, New York: Springer.

Spiegelhalter, D.J., Best, N.G., Carlin, B.P., and van der Linde, A. (2002). Bayesian measures of model complexity and fit (with discussion and rejoinder). *Journal of the Royal Statistical Society, Series B*, **64**, 583–639.

Thompson, J.A. and Bell, J.C. (1996). Color Index for Identifying Hydric Conditions for Seasonally Saturated Mollisols in Minnesota. *Journal of the Soil Science Society of America*, **60** 1979–1988.

Wackernagel, H. (2003). *Multivariate Geostatistics*. Third Edition. New York: Springer.

Wilson, S.D. and Tilman, D. (1995). Competitive Responses of Eight Old-field Plant Species in Four Environments. *Ecology*, **76**, 1169–1180.

Bayesian Statistics and Its Applications
Edited by S.K. Upadhyay, U. Singh and D.K. Dey
Anamaya Publishers, New Delhi, India

Bayes Factors for One Sided Hypothesis Testing in Linear Calibration

Maria Maddalena Barbieri[1] and Brunero Liseo[2]

[1]Dipartimento di Economia, Università Roma Tre, via Ostiense 139, 00154 Roma, Italy
[2]Dipartimento di Studi Geoeconomici, Linguistici, Statistici, e storici per L'analisi Regionale,
Università "La Sapienza", via del castro Laurenziano, 9, 00161 Roma, Italy

Abstract

This article considers one side hypothesis testing on the unknown value of the explanatory variable in univariate linear calibration. The problem is formulated in a general form and the solution is suited for a large set of models, namely when the sampling distribution belongs to a particular class, defined in Gleser and Hwang (1987). We discuss the drawbacks of frequentist solutions and we show how a proper Bayesian analysis encounters relatively similar difficulties. We explore the performances of some noninformative Bayesian approaches to testing, namely, default Bayes factors. Default Bayes factors based on Jeffreys' priors seem to provide sensible results although not all the problems seem to be solved.

1. Introduction

Suppose we observe n independent replications of two independent normal random variables

$$X \sim N\left(\alpha, \sigma^2\right) \text{ and } Y \sim N\left(\beta, \sigma^2\right) \tag{1}$$

with variance σ^2 known and $\xi = \alpha/\beta$. The goal of the analysis is to test the null hypothesis

$$H_1 : \xi \leq \xi_0, \text{ versus } H_2 : \xi > \xi_0 \tag{2}$$

or equivalently to choose between model M_1 with $\xi \leq \xi_0$, and M_2 with $\xi > \xi_0$.

This is a quite general scenario, since numerous statistical models may be reduced in the form (1). Examples include calibration and the linear errors-in-variables model (Lindley and El-Sayyad, 1968) and principal component analysis. This testing problem is standard because all possible approaches to model selection, either frequentist or Bayesian, can be used. However, it is also challenging because none of these methods work satisfactorily. Loosely speaking, there is a problem of local nonidentifiability in the model: as β tends to zero, ξ cannot be estimated.

The hypothesis testing problem outlined above, known as Fieller's problem, belongs to a category of problems which we refer to as the Gleser-Hwang class of problems (Gleser and Hwang, 1987). Cheng and Van Ness (1994) provide a complete review of the difficulties that a frequentist approach to estimation encounters in this class of models. Berger, Liseo and Wolpert (1999) consider various types of conditional inference in related problems and discuss the difficulties of non-Bayesian approaches. A Bayesian analysis with partially informative priors is developed in Buonaccorsi and Gatsonis (1988), while Kubokawa and Robert (1994) propose a reference prior approach. All these papers are focused on point and set estimation of the parameter of interest. Barbieri, Liseo and Petrella (2000) provide a solution for the point null hypothesis $H_0 : \xi = \xi_0$ versus a two-side alternative based on default Bayes factors and reference priors. Liseo (2003) gives a general review of estimation issues with Fieller's problem.

In this paper we analyze the general behavior of the standard Bayesian tool for model comparison, i.e. the Bayes factor, in the case where the two competing hypotheses are not nested. It is well known that this problem is quite more complicated than the one involving point null hypothesis and few references are available, basically Berger and Mortera (1999) and Moreno (2005).

In what follows we show that a proper Bayesian analysis of Gleser-Hwang problems is plagued with difficulties similar to those of the frequentist approach. In particular, Bayesian answers are very sensitive to the prior distribution, both on the parameter of interest and, less typically, on the nuisance parameters.

We examine the results obtained using the 'default' testing approach, namely the 'default' Bayes factors, such as the intrinsic Bayes factor (Berger and Pericchi, 1996) and the fractional Bayes factor (O'Hagan, 1995).

In one sided testing a default Bayesian analysis is often performed using directly the standard noninformative prior distributions. There are some problems related to this approach, suggesting the need to explore the applicability of new 'default' methods. First, it does not seem to be suited for small and moderate sample sizes. Besides the use of a conventional prior in this setting results in answers that often coincide with the classical p-value (Casella and Berger, 1987). In particular for location parameter problems, the resulting posterior probability of M_1 (and the p-value) is the lower bound of the posterior probability of M_1 over reasonable classes of prior densities. It is questionable whether the lower bound is the best evidential summary to provide. While the fractional and the intrinsic Bayes factor usually produce answers which are not as extreme as the standard classical or Bayesian answers (Berger and Mortera, 1999).

The main contribution of this paper is to show that default Bayes methodology in testing and model selection is not only useful when prior information is vague; it also provides sensible results in situations, like the Gleser-Hwang class, where a frequentist or likelihood analysis may be very difficult or even impossible and a subjective Bayesian answer would be rather sensitive to prior inputs.

Section 2 briefly recalls the Gleser-Hwang family of models. Section 3 explains how the calibration testing problems can be reduced to a particular case of the Gleser-Hwang class of problems. The behavior and the performances of the various default Bayes factors are analyzed in Section 4. Some discussions are given in the final section.

2.   The Gleser-Hwang Class

Gleser and Hwang (1987) consider the following situation. Let $\Phi \times \Psi$ be the parameter space and $\mu(\phi)$ a scalar function of $\phi \in \Phi$. Suppose there exists a subset $\Phi^* \subset \Phi$ such that $\mu(\phi)$ has an unbounded range over Φ^*. Also, suppose there exists a point $\psi^* \in \overline{\Psi}$, the closure of Ψ, such that, for each $\phi \in \Phi^*$ and for any sample $z = (z_1, \ldots , z_n)$ from the sampling model $p(z; \phi, \psi)$, the limit

$$\lim_{\psi \to \psi^*} p(z; \phi, \psi) = p(z; \psi^*) \tag{3}$$

exists and is a sampling distribution independent of ϕ. Gleser and Hwang (1987) proved that every confidence procedure $C_\gamma(z)$ for $\mu(\phi)$ with a positive confidence level γ satisfies $P_{(\phi,\psi)}\left[d\{C_\gamma(z)\} = \infty\right] > 0$, where $d\{C_\gamma(z)\}$ is the diameter of $C_\gamma(z)$.

We can see that Fieller's problem falls into this setting by taking $\phi = \mu(\phi) = \xi$, $\psi = \beta$, $\Phi^* = \Phi = R$ and $\beta^* = 0$.

Barbieri, Liseo and Petrella (2000) investigate the Bayesian implications of condition (3). Let $\pi(\phi,\psi)$ be an absolutely continuous proper prior distribution; in terms of the likelihood surface, condition (3) implies that the likelihood function is constant along the line $\psi = \psi^*$, when ϕ varies in Φ^*. In terms of the posterior distribution, if we switch the role of ϕ and z, it is easy to see that (Barbieri, Liseo and Petrella, 2000)

LEMMA 1. *If the sampling model $p(z;\phi,\psi)$ satisfies condition (3), then*

$$\lim_{\psi \to \psi^*} \pi(\phi|\psi, z) = \pi(\phi|\psi^*).$$

Conditionally on ψ^*, data are not informative about the parameter of interest ϕ. Also, assume that the likelihood function is constant over the cylinder $L_\delta(\psi^*) = \Phi \times I_\delta(\psi^*)$, where $I_\delta(\psi^*)$ is a neighborhood of ψ^* of radius δ; this approximation can be justified by (3). Then it is easy to see that

$$\pi(\phi|z) \approx \pi\{\phi|I_\delta(\psi^*)\}\pi\{I_\delta(\psi^*)|z\} + \pi\{\phi|z, I_\delta^c(\psi^*)\}\pi\{I_\delta^c(\psi^*)|z\}. \tag{4}$$

Note that the first factor depends on the data only through the nuisance parameter in a neighborhood of the critical value. In the one side testing scenario the above result is important. Suppose, in fact, that the prior probability of the two competing hypotheses is the same; then the Bayes factor is equal to the ratio of the posterior probabilities of the two hypotheses which, from (4), both suffer from the Gleser and Hwang pathology. This implies that the resulting Bayes factor is inconclusive, since it heavily depends on the information about the nuisance parameter, which is hardly available.

3.   The Calibration Problem

The well known linear calibration problem arises in measurement settings. It may be briefly described referring to a classical situation where we have two variables V and Z. The variable Z comes from the measurement of a characteristic of interest determined by some difficult and expensive method, while V is the measurement related to the same characteristic obtained by an easy and cheap method. The calibration problem is concerned with making inferences about an unknown value z^* from a set of *new* random observed responses v_j^* $(j = 1, 2, \ldots, k)$ and the available calibrated data (z_i, v_i) $(i = 1, 2, \ldots, n)$ from the past, where the v_i's are known values of the precise measurements. In linear calibration it is usually assumed that

$$v_i = a + b\,z_i + \epsilon_i \qquad\qquad i = 1, 2, \ldots, n$$

and

$$s\,v_j^* = a + b\,z^* + \epsilon_j^* \qquad\qquad j = 1, 2, \ldots, k$$

where ϵ_i and ϵ_j^* are independent $N(0, \sigma^2)$.

We consider a situation where we are interested in the following testing problem:

$$H_1 : z^* \leq z_0 \text{ versus } H_2 : z^* > z_0. \tag{5}$$

The solution is based upon v_j^* $(j = 1, 2, \ldots, k)$ given that the regression parameters a and b have been estimated using the calibration data.

The problem frequently arises in the context of decision and detection limit problems in bio-assays and chemical assays. To illustrate the procedure we will refer to example (Krishnamoorthy, Kulkarni and Mathew, 2001) dealing with the problem to test if the unknown blood alcohol concentration of an individual z^* exceeds a threshold z_0 (concentrations below z_0 are considered *safe* for driving), based on breath measurement of the alcohol concentration v^*.

We denote with

$$\hat{b} = \frac{\sum_{i=1}^{n}(z_i - \overline{z})(v_i - \overline{v})}{\sum_{i=1}^{n}(z_i - \overline{z})^2}$$

where $\overline{z} = n^{-1}\sum_{i=1}^{n} z_i$ and $\overline{v} = n^{-1}\sum_{i=1}^{n} v_i$, the least square estimate of b and with s_1 and s_0 the two sums of residual squared errors based on the calibration and the prediction experiments, respectively, while $\overline{v}^* = k^{-1}\sum_{j=1}^{k} v_j^*$.

As suggested by Fieller (1954) and following Kubokawa and Robert (1994), a linear calibration problem may be reduced, without loss of generality, to the problem of making inference on a ratio of two normal means. In fact the statistics $\hat{b}$, $\overline{v}$, $\overline{v}^*$ and $s = s_1 + s_0$ are minimal sufficient and mutually independent with the following distributions:

$$\hat{b} \sim N\left(b, \frac{\sigma^2}{\sum_{i=1}^{n}(z_i - \overline{z})^2}\right), \qquad \overline{v} = N\left(a + b\overline{z}, \frac{1}{n}\sigma^2\right).$$

$$\overline{v}^* = N\left(a + z^*, \frac{1}{k}\sigma^2\right), \qquad s \sim \sigma^2 \chi^2_{(n-2)+(k-1)}.$$

Now if we let

$$\beta = \left[\sum_{i=1}^{n}(z_i - \overline{z})^2\right]^{1/2} b, \quad y = \left[\sum_{i=1}^{n}(z_i - \overline{z})^2\right]^{1/2} \hat{b}$$

$$x = \frac{(\overline{v}^* - \overline{v})}{(n^{-1} + k^{-1})^{1/2}}, \quad \xi = \frac{\left[\sum_{i=1}^{n}(z_i - \overline{z})^2\right]^{-1/2}(z^* - \overline{z})}{(n^{-1} + k^{-1})^{1/2}}$$

the model may be rewritten as

$$x \sim N(\beta\xi, \sigma^2) \qquad y \sim N(\beta, \sigma^2).$$

Due to translation invariance, the testing problem (5) is equivalent to (2).

For simplicity, we consider the unidimensional and univariate linear calibration, meaning that v and z and also x and y are scalar. However, the proposal may be easily generalized to the case when we deal with vectors.

4. Default Bayes Factors

4.1 Fractional Bayes Factor, Encompassing Model and Intrinsic Priors

The main tool used for hypotheses testing and model selection in the Bayesian setting is the Bayes factor. For an exhaustive review and discussion on this topic see Kass and Raftery (1995).

Suppose data z have density $p(z; \theta)$, with θ being a real k-dimensional unknown parameter vector. The models $M_1 : \theta \in \Theta_1$ and $M_2 : \theta \in \Theta_2$ are under consideration, where Θ_i are non-nested, such as in the one-sided testing problem (2). If $\pi_i(\theta)$ denote the prior distribution for the parameters under model i ($i = 1, 2$), the Bayes factor B_{ij} for comparing M_i with M_j is defined by

$$B_{ij} = \frac{m_i(z)}{m_j(z)} = \frac{\int p(z; \theta)\pi_i(\theta)d\theta}{\int p(z; \theta)\pi_j(\theta)d\theta}$$

where $m_i(z)$ is the marginal density of z under model M_i.

Often in Bayesian analysis conventional noninformative or default priors are used to compute B_{ij}. They are typically improper distributions and they depend on some constants k_i. Then, the resulting Bayes factor

$$B_{ij}^N = \frac{m_i^N(z)}{m_j^N(z)} = \frac{\int p(z; \theta)\pi_i^N(\theta)d\theta}{\int p(z; \theta)\pi_j^N(\theta)d\theta} \tag{6}$$

has an arbitrary value, depending on the ratio of constants k_i and k_j.

A solution to the problem is to use the so-called *intrinsic Bayes factor* (IBF) proposed by Berger and Pericchi (1996). The idea is to divide the data into two parts: the first part $z(l)$ is used as a training sample to obtain a proper posterior distribution for the parameters

$$\pi_i(\theta \,|\, z(l)) = \frac{p(z(l); \theta)\pi_i^N(\theta)}{m_i^N(z(l))}$$

where $p(z(l); \theta)$ is the marginal density of $z(l)$ under M_i and $m_i^N(z(l)) = \int p(z(l); \theta)\pi_i^N(\theta)d\theta$.

Berger and Pericchi (1996) suggest to use training samples of minimum size. Now adopting $\pi_i(\theta \mid z(l))$ as the prior, the remainder of the data $z(-l)$ are used to compute a Bayes factor

$$B_{ij}(l) = \frac{\int p(z(-l); \theta, z(l))\pi_i(\theta \mid z(l))d\theta}{\int p(z(-l); \theta, z(l))\pi_j(\theta \mid z(l))d\theta} \tag{7}$$

which is unaffected by the arbitrary constants k_i. The $B_{ij}(l)$ is often referred to as partial Bayes factor. Expression (7) can be generally rewritten as

$$B_{ij}(l) = B_{ij}^N B_{ji}^N(z(l))$$

where B_{ij}^N is the full Bayes factor given by (6) and

$$B_{ji}^N(z(l)) = \frac{m_j^N(z(l))}{m_i^N(z(l))}.$$

Different versions of the intrinsic Bayes factor have been proposed (Berger and Pericchi, 1996; 1998). However, the most suitable for non-nested situation, as the one-sided hypothesis testing, seem to be the encompassing expected intrinsic Bayes factor (Berger and Mortera, 1999). The idea is to embed the competing models in a larger encompassing model, say M_0, so that M_1 and M_2 are both nested within it. The resulting Bayes factor is defined as

$$B_{ij}^{EEI} = B_{ij}^N \frac{\mathbf{E}_{\hat{\theta}}^{M_0}[B_{i0}(z(l))]}{\mathbf{E}_{\hat{\theta}}^{M_0}[B_{j0}(z(l))]}$$

where the expectations are computed under the encompassing model M_0, evaluated at the maximum likelihood estimate $\hat{\theta}$ under M_0.

An alternative formulation of the Bayes factor, a variant of the partial Bayes factor, is the so-called *fractional Bayes factor* (O'Hagan, 1995). When model M_i is compared with model M_j, the fractional Bayes factor is defined as

$$B_{ij}^{(F)} = \frac{m_i^*(z)}{m_j^*(z)}$$

where

$$m_i^*(z) = \frac{\int p(z;\theta)\pi_i^N(\theta)d\theta}{\int [p(z;\theta)]^b \pi_i^N(\theta)d\theta}.$$

The indeterminate constants k_i cancel out in $m_i^*(z)$.

In the fractional Bayes factor, b is the proportion of data used for training. There are no precise recipes for choosing b. In what follows we will always set b equal to the dimension of the minimal training sample divided by the number of the observations, as suggested by O'Hagan (1995).

Considerable insight into the behavior of a default Bayes factor can be provided by the corresponding 'intrinsic' prior (Berger and Pericchi, 1996), if it exists. It is defined as the prior that would give a Bayes factor (asymptotically) equivalent to the default Bayes factor. The analysis of an intrinsic prior helps in detecting the existence of biases in the Bayes factor towards one of the hypothesis.

4.2 Main Result

Turning back to the situation depicted in Section 1, and using the following reparameterization:

$$\xi = \frac{\alpha}{\beta}; \qquad \omega = \mathrm{sgn}(\beta)\sqrt{\alpha^2 + \beta^2}$$

the likelihood function may be written as

$$\ell(\xi,\omega) = \frac{n}{2\pi\sigma^2} \exp\left\{-\frac{n}{2\sigma^2}\left[\left(\omega - \frac{\overline{x}\xi + \overline{y}}{1+\xi^2}\right)^2 + \frac{(\overline{x} - \overline{y}\xi)^2}{1+\xi^2}\right]\right\}.$$

Notice that we have included the correct constant in the likelihood so that it can be interpreted as a probability distribution on the parameter space. Also, the resulting Jeffreys' prior is (Liseo, 2003)

$$\pi^J(\xi,\omega) = \frac{|\omega|}{(1+\xi^2)}.$$

We now derive default Bayes factors and intrinsic priors starting from the Jeffreys' prior and using the fractional approach by O'Hagan (1995) and the encompassing intrinsic approach, developed in Berger and Mortera (1999).

The choice of using the Jeffreys prior instead of the reference prior (Berger and Bernardo, 1992), which proved itself superior to Jeffreys prior in the point null hypothesis case, is mainly due to computational reasons: calculations with Jeffreys' prior are almost always possible in closed form and this is important in terms of comparisons with other approaches.

Each default Bayes factor B_{21}^D can be written as $B_{21}^D = B_{21}^N B^{cf}$, where B_{21}^N is the Bayes factor based on the usual noninformative prior and B^{cf} is the correction factor resulting from the specific used approach. In our problem B_{21}^N is defined as

$$B_{21}^N(\xi,\omega) = \frac{\int_{\xi>\xi_0} \ell(\xi,\omega)\pi^J(\xi,\omega)d\xi\,d\omega}{\int_{\xi\leq\xi_0} \ell(\xi,\omega)\pi^J(\xi,\omega)d\xi\,d\omega}. \tag{8}$$

Following Berger and Mortera (1999) we will also determine the intrinsic prior for (ξ,ω), say $\pi^I(\xi,\omega)$. The restrictions of $\pi^I(\xi,\omega)$ on M_1 and M_2 will be denoted by $\pi_1^I(\xi,\omega)$ and $\pi_2^I(\xi,\omega)$, respectively.

From Berger and Mortera (1999) it is known that $\pi_1^I(\xi,\omega)$ and $\pi_2^I(\xi,\omega)$ is the solution of the equation system

$$
\begin{cases}
B_2^*(\xi,\omega) = \dfrac{\pi_2^I(\xi,\omega)\pi_1^N(\psi_1(\xi,\omega))I_{M_2}(\xi,\omega)}{\pi_1^I(\psi_1(\xi,\omega))\pi_2^N(\xi,\omega)} \\[4mm]
B_1^*(\xi,\omega)^{-1} = \dfrac{\pi_1^I(\xi,\omega)\pi_2^N(\psi_2(\xi,\omega))I_{M_1}(\xi,\omega)}{\pi_2^I(\psi_2(\xi,\omega))\pi_1^N(\xi,\omega)}
\end{cases}
\tag{9}
$$

where, for $i = 1,2$,

$$
B_i^*(\xi,\omega) = \lim_{n\to\infty} B^{cf} \quad \text{under model } M_i,\ j\neq i
$$

and

$$
\psi_i(\xi,\omega) = \lim_{n\to\infty} (\hat{\xi},\hat{\omega}) \quad \text{under model } M_j\ (j\neq i; i,j=1,2)
$$

where $\hat{\xi}$ and $\hat{\omega}$ are the maximum likelihood estimator of ξ and ω under model M_i.

In the case of fractional Bayes factor, the correction factor, indicated with $\tilde{B}_{12}^F$, is given by

$$
\tilde{B}_{12}^F = \frac{\int_{\xi\leq\xi_0} \ell^b(\xi,\omega)\pi^J(\xi,\omega)d\xi\, d\omega}{\int_{\xi>\xi_0} \ell^b(\xi,\omega)\pi^J(\xi,\omega)d\xi\, d\omega}.
$$

Turning back to the original parameterization in terms of (α,β), and noticing that this way the posterior distribution is bivariate normal, the ratio may be written as

$$
\tilde{B}_{12}^F = \frac{1 - G(b,\xi_0,\overline{x},\overline{y})}{G(b,\xi_0,\overline{x},\overline{y})}
$$

where
$$
G(b,\xi_0,\overline{x},\overline{y}) = \int\limits_{\beta<0}\int\limits_{\alpha<\xi_0\beta} \frac{nb}{2\pi\sigma^2} \exp\left\{-\frac{nb}{2\sigma^2}[(\overline{x}-\alpha)^2 + (\overline{y}-\beta)^2]\right\} d\alpha d\beta
$$
$$
+ \int\limits_{\beta>0}\int\limits_{\alpha>\xi_0\beta} \frac{nb}{2\pi\sigma^2} \exp\left\{-\frac{nb}{2\sigma^2}[(\overline{x}-\alpha)^2 + (\overline{y}-\beta)^2]\right\} d\alpha d\beta.
$$

Using Theorem 5.1 one can see that

$$
G(b,\xi_0,\overline{x},\overline{y}) = \begin{cases}
1 + 2\left[T(h,r) - T(h_1,r_1)\right] & \text{if } \xi_0 \leq \dfrac{\overline{x}}{\overline{y}} \\[4mm]
2\left[T(h,r) - T(h_1,r_1)\right] & \text{if } \xi_0 > \dfrac{\overline{x}}{\overline{y}}
\end{cases}
\tag{10}
$$

where
$$
h = -\frac{\sqrt{n\,b}}{\sigma\sqrt{1+\xi_0^2}}(\overline{y}\xi_0 - \overline{x}), \qquad h_1 = -\frac{\sqrt{n\,b}}{\sigma}\overline{y}
$$
$$
r = -\frac{(\overline{y}+\xi_0\overline{x})}{(\overline{y}\,\xi_0 - \overline{x})}, \qquad\qquad r_1 = \frac{\overline{x}}{\overline{y}}
$$

and $T(\lambda,\gamma)$ is the well known T function proposed by Owen (1956) and it represents the volume of a bivariate standard normal density over the region

$$
\{(u,v) : \lambda < u < +\infty,\quad 0 < v < \lambda u\}.
$$

Also, B_{21}^N can be computed using the same approach, putting $b = 1$ and inverting numerator and denominator

$$B_{21}^N = \frac{G(1, \xi_0, \overline{x}, \overline{y})}{1 - G(1, \xi_0, \overline{x}, \overline{y})}$$

Then, with the usual choice of $b = 1/n$, the fractional Bayes factor for this problem is given by

$$B_{21}^F = \frac{G(1, \xi_0, \overline{x}, \overline{y})}{1 - G(1, \xi_0, \overline{x}, \overline{y})} \frac{1 - G(1/n, \xi_0, \overline{x}, \overline{y})}{G(1/n, \xi_0, \overline{x}, \overline{y})}. \tag{11}$$

To figure out what kind of intrinsic priors are associated to the last factor in (11), i.e. the correction factor associated to the fractional approach, $\tilde{B}_{12}^F$, we now study its asymptotic behavior.

Independently of the true model it is easy to see that the two T functions which appear in $G(b, \xi_0, \overline{x}, \overline{y})$ have the following limiting behavior

$$\lim_{n \to \infty} T(h, r) = \begin{cases} 0 & b = 1 \\ T\left(\frac{\omega(\xi - \xi_0)}{\sigma\sqrt{1+\xi^2}\sqrt{1+\xi_0^2}}, \frac{1+\xi\xi_0}{\xi_0 - \xi}\right) & b = \dfrac{1}{n} \end{cases}$$

and

$$\lim_{n \to \infty} T(h_1, r_1) = \begin{cases} 0 & b = 1 \\ T\left(-\frac{\omega}{\sigma\sqrt{1+\xi^2}}, \xi\right) & b = \dfrac{1}{n}. \end{cases}$$

Repeatedly using the above results and the fact that the sampling means $\overline{x}$ and $\overline{y}$ are consistent estimators of α and β, one can easily show that, in the (ξ, ω) parameterization,

$$\tilde{B}_{12}^F \xrightarrow{\text{under } M_1} B_1^*(\xi, \omega) = \left(d(\xi, \omega)^{-1} - 1\right).$$

where

$$d(\xi, \omega) = 2\left[T\left(\frac{\omega(\xi - \xi_0)}{\sigma\sqrt{1+\xi^2}\sqrt{1+\xi_0^2}}, \frac{1+\xi\xi_0}{\xi_0 - \xi}\right) - T\left(-\frac{\omega}{\sigma\sqrt{1+\xi^2}}, \xi\right)\right] \tag{12}$$

and

$$\tilde{B}_{12}^F \xrightarrow{\text{under } M_2} B_2^*(\xi, \omega) = \left(-d(\xi, \omega)^{-1} - 1\right)^{-1}.$$

Also

$$\psi_1(\xi, \omega) = \lim_{n \to \infty, M_2} (\hat{\xi}, \hat{\omega}) = (\xi_0, \omega)$$

and

$$\psi_2(\xi, \omega) = \lim_{n \to \infty, M_1} (\hat{\xi}, \hat{\omega}) = (\xi_0, \omega)$$

In the following theorem we give the closed form expression of the intrinsic priors for problem (2).

Theorem 4.1 *The fractional intrinsic priors for the calibration testing problem (2) are given by*

$$\pi_1^I(\omega) = \pi_2^I(\omega) = \text{ any prior}$$

$$\pi_1^I(\xi \mid \omega) = k_1(\xi_0, \omega) \frac{1 + \xi_0^2}{1 + \xi^2} \frac{1}{B_1^*(\xi, \omega)} \qquad \xi \le \xi_0 \tag{13}$$

$$\pi_2^I(\xi \mid \omega) = k_2(\xi_0, \omega) \frac{1 + \xi_0^2}{1 + \xi^2} B_2^*(\xi, \omega) \qquad \xi > \xi_0 \tag{14}$$

where the conditional priors of ξ are proper under both the hypotheses while k_1 and k_2 appearing in the above expressions are suitable normalizing functions.

Proof. One simply needs to solve the equation system (9) which, in our case is

$$
\begin{cases}
B_2^*(\xi,\omega) = \dfrac{\pi_2^I(\xi,\omega)(1+\xi^2)I_{M_2}(\xi,\omega)}{\pi_1^I(\xi_0,\omega)(1+\xi_0^2)} \\[2ex]
\dfrac{1}{B_1^*(\xi,\omega)} = \dfrac{\pi_1^I(\xi,\omega)(1+\xi^2)I_{M_1}(\xi,\omega)}{\pi_2^I(\xi_0,\omega)(1+\xi_0^2)}.
\end{cases}
$$

If we also assume that $\pi_1^I(\omega) = \pi_2^I(\omega)$ then the conditional intrinsic priors for $\xi \mid \omega$ must satisfy the following:

$$
\begin{cases}
\dfrac{\pi_2^I(\xi \mid \omega)I_{M_2}(\xi,\omega)}{\pi_1^I(\xi_0 \mid \omega)} = \dfrac{1+\xi_0^2}{1+\xi^2}B_2^*(\xi,\omega) \\[2ex]
\dfrac{\pi_1^I(\xi \mid \omega)I_{M_1}(\xi,\omega)}{\pi_2^I(\xi_0 \mid \omega)} = \dfrac{1+\xi_0^2}{1+\xi^2}\dfrac{1}{B_1^*(\xi,\omega)}
\end{cases}
$$

which directly leads to the priors stated in (13) and (14). To prove that the above prior is proper it suffices to show that, for any value of ω, the quantities $1/B_1^*(\xi,\omega)$ and $B_2^*(\xi,\omega)$ are bounded. Without loss of generality, assume that $\omega > 0$. Then, for all values of ξ, this is ensured by the boundedness of the T function, which always takes value in the interval $[-0.25, 0.25]$ and it is zero only when the first argument goes to infinity or the second argument goes to zero. From (12), this cannot happen for fixed ξ. ∎

Notice that the prior for the nuisance parameter is not crucial and any prior will do the job, provided that the same prior is used both under M_1 and M_2. The conditional prior for ξ is then a sort of Cauchy prior times a correction factor, which is expressed in terms of the Owen's T function. Since the T function takes values in the bounded support $(-0.25, 0.25)$, it easily follows that the intrinsic priors are proper.

On the other hand, the above expressions of the priors do not clarify whether they are well calibrated priors or not. To do that, one could check the sufficient condition that, for any given value of ω, the prior conditional probability of the two hypotheses is the same; we did it numerically for several values of σ and ξ_0. It turns out that the two priors are well calibrated only when $\xi_0 = 0$, while the prior mass over M_2 is larger than the prior mass over M_1 for $\xi_0 > 0$ (the reverse is true when $\xi_0 < 0$). Then, it seems that the fractional approach is not able to adequately weigh the two hypotheses. Although we check the condition on the conditional prior distributions rather than the joint priors, there are no marginal priors which can adjust the bias because the above relations hold for all ω.

Calculations for the expected encompassing intrinsic Bayes factor are quite similar in theory but they do not lead to a close form solution. One should first define the natural encompassing model which is given here by $M_0 : \xi \in R$. Then a minimal training sample is given by a single pair of observations (x_i, y_i) and, using the same sort of calculations, one can see that

$$
B_{10}^E(x_i, y_i) = 1 - G(1, \xi_0, x_i, y_i)
$$

and
$$
B_{20}^E(x_i, y_i) = G(1, \xi_0, x_i, y_i).
$$

Now one has to evaluate the expected values of $B_{10}^E(x_i, y_i)$ and $B_{20}^E(x_i, y_i)$ under the larger model and under the hypothesis that the parameter values are equal to their maximum likelihood estimates.

Unfortunately, it is not known how to compute the expected value of the Owen's T function and the correction factor needed to get B_{21}^{EEI} must be computed numerically. A first exploratory analysis shows that results are quite similar to those obtained via the fractional approach.

5.   Technical Details

Theorem 5.1   *Let $R(\gamma, \delta, \zeta)$ be the integral of the bivariate standard Normal distribution over the region between the lines $v = \gamma w + \delta$ and $w = \zeta$ (see Fig. 1), then*

$$R(\gamma, \delta, \zeta) = \begin{cases} 1 + 2[T(k, s) - T(k_1, s_1)] & \text{if } \frac{\delta}{\zeta} \leq 0 \\[2mm] 2[T(k, s) - T(k_1, s_1)] & \text{if } \frac{\delta}{\zeta} > 0 \end{cases} \tag{15}$$

where $k = -\dfrac{\delta}{\sqrt{1 + \gamma^2}}$, $s = \dfrac{\zeta(1 + \gamma^2)}{\delta} + \gamma$, $k_1 = \zeta$, $s_1 = \dfrac{\delta}{\zeta} + \gamma$ and T represents the Owen's function.

Proof. Assume, without loss in generality, that γ, δ and ζ are positive, so that $\frac{\delta}{\zeta} > 0$ and the integrated region has the form shown in Fig. 1.

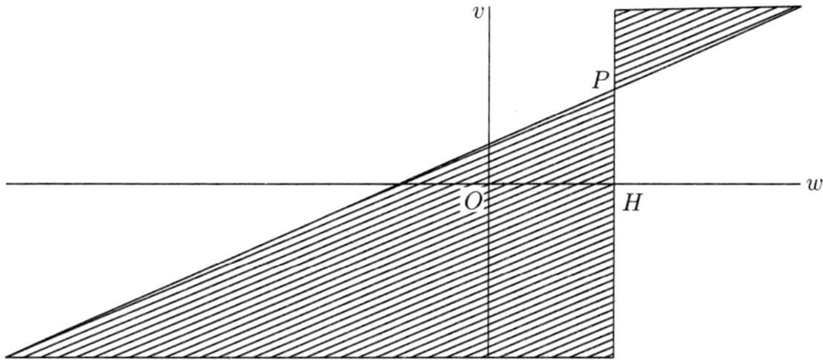

Fig. 1.   The shaded region is between the lines $v = \gamma w + \delta$ **and** $w = \zeta$ $(\delta/\zeta > 0)$**.**

Using the symmetry properties of the bivariate standard Normal distribution, the integral over the shaded region in Fig. 1 may be equivalently computed as the integral over the shaded region in Fig. 2.

In Fig. 2, point Q is the midpoint of the segment between P and P'' and it lies on the intersection between the line $v = \gamma w + \delta$ and the line orthogonal to it through the origin O. The distance between the points P and H is the same as the distance between the points P' and H. Thus $R(\gamma, \delta, \zeta)$ may be computed as the integral of the bivariate standard normal distribution over $\Re^2$ minus the integral over the region complementary to the shaded one which has been splitted into the four regions A, B, C and D, with similar shapes. Also, observe that, still exploiting the symmetry properties of the bivariate standard Normal distribution

$$\iint_A \frac{1}{2\pi} \exp\left\{\frac{1}{2}(w^2 + v^2)\right\} dv\, dw = \iint_B \frac{1}{2\pi} \exp\left\{\frac{1}{2}(w^2 + v^2)\right\} dv\, dw$$

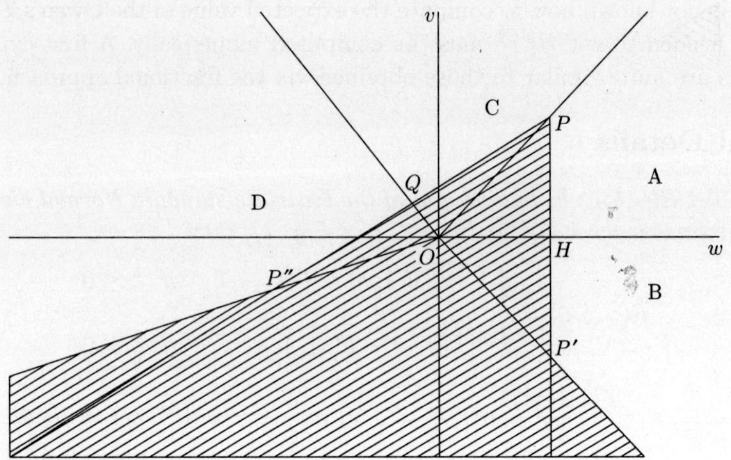

Fig. 2.   The integral over the shaded region gives $R(\gamma, \delta, \zeta)$ when $\delta/\zeta > 0$.

and

$$\iint_A \frac{1}{2\pi} \exp\left\{\frac{1}{2}(w^2 + v^2)\right\} dv\, dw = \int_\zeta^{+\infty} \int_0^{\frac{\delta}{\zeta}+\gamma} \frac{1}{2\pi} \exp\left\{\frac{1}{2}(w^2 + v^2)\right\} dv\, dw$$

$$= T\left(\zeta, \frac{\delta}{\zeta} + \gamma\right)$$

Also

$$\iint_C \frac{1}{2\pi} \exp\left\{\frac{1}{2}(w^2 + v^2)\right\} dv\, dw = \iint_D \frac{1}{2\pi} \exp\left\{\frac{1}{2}(w^2 + v^2)\right\} dv\, dw$$

and

$$\iint_C \frac{1}{2\pi} \exp\left\{\frac{1}{2}(w^2 + v^2)\right\} dv\, dw = \int_{-\frac{\delta}{\sqrt{1+\gamma^2}}}^{+\infty} \int_0^{\frac{\zeta(1+\gamma^2)}{\delta}+\gamma} \frac{1}{2\pi} \exp\left\{\frac{1}{2}(w^2 + v^2)\right\} dv\, dw$$

$$= -T\left(\frac{\delta}{\sqrt{1+\gamma^2}}, \frac{\zeta(1+\gamma^2)}{\delta} + \gamma\right).$$

The last equality holds because of the following properties of the Owen's T function, which is directly derived from its definition: $T(-h, r) = T(h, r)$ and $T(h, -r) = -T(h, r)$, with $h > 0$ and $r > 0$. This completes the proof of the first equality in (15).

Assume now, still without loss of generality, that $\gamma < 0$, $\delta > 0$ and $\zeta < 0$, so that $\frac{\delta}{\zeta} < 0$ and the integrated region may be represented as in Fig. 3.

Similar to the previous case, the integral over the shaded region in Fig. 3 may be equivalently computed as the integral over the shaded region in Fig. 4, where the letters have the same meaning as in the previous case.

Since now

$$\iint_A \frac{1}{2\pi} \exp\left\{\frac{1}{2}(w^2 + v^2)\right\} dv\, dw = \int_{-\infty}^{\zeta} \int_0^{\frac{\delta}{\zeta}+\gamma} \frac{1}{2\pi} \exp\left\{\frac{1}{2}(w^2 + v^2)\right\} dv\, dw = -T\left(\zeta, \frac{\delta}{\zeta} + \gamma\right)$$

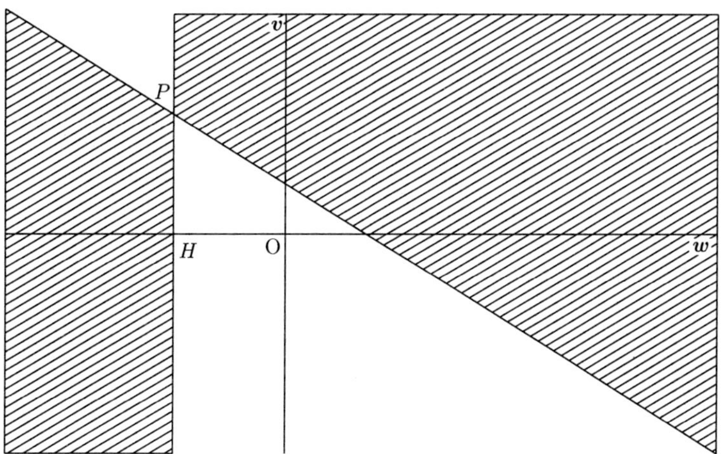

Fig. 3.   The shaded region is between the lines $v = \gamma w + \delta$ and $w = \zeta$ $(\delta/\zeta < 0)$.

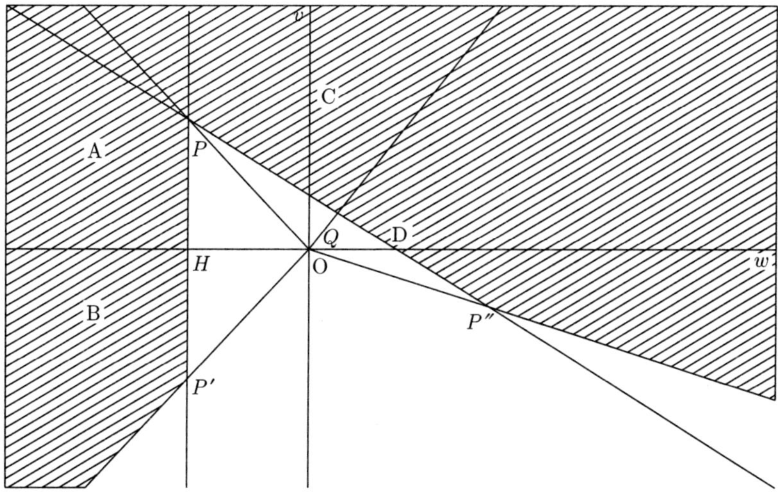

Fig. 4.   The integral over the shaded region gives $R(\gamma, \delta, \zeta)$ when $\delta/\zeta < 0$.

and

$$\iint_C \frac{1}{2\pi} \exp\left\{\frac{1}{2}(w^2 + v^2)\right\} dv\, dw = \int_{-\frac{\delta}{\sqrt{1+\gamma^2}}}^{+\infty} \int_0^{\frac{\zeta(1+\gamma^2)}{\delta} + \gamma} \frac{1}{2\pi} \exp\left\{\frac{1}{2}(w^2 + v^2)\right\} dv\, dw$$

$$= T\left(\frac{\delta}{\sqrt{1+\gamma^2}}, \frac{\zeta(1+\gamma^2)}{\delta} + \gamma\right)$$

the first equality in (15) is proved. ∎

6. Discussion

We have considered, from several different Bayesian perspectives, the problem of one side testing in linear calibration. The solutions we have found are not completely satisfactory since the intrinsic priors induced by the default Bayes factors are not calibrated between the two competing models. Therefore, the straight use of the intrinsic or fractional Bayes factors would imply a biased weight of the two hypotheses. As suggested by a referee, a simple recalibration can be however introduced, since, for each ξ_0 one can calculate the prior probability given to the alternative hypotheses.

We did all the computations using the Jeffreys' prior, since it allows closed form computation of the Bayes factor and the intrinsic priors. One can suspect that results could be different when starting from the Berger and Bernardo reference prior; we are currently considering this way. Finally, to be really useful and effective, our proposed approach should be generalized to the case where (X, Y) follow a bivariate normal distribution with unknown covariance matrix Σ. In this case, however, there are no closed form expressions for the default Bayes factor and comparisons with other methods are complicated. The case with $\Sigma = \sigma^2 I$, with σ^2 unknown, can be easily dealt with and, similarly to what happens in the point null case (Barbieri et al., 2000), the expressions of the default Bayes factors are in terms of Kummer's function (Abramowitz and Stegun, 1970; p. 509).

References

Abramowitz, M. and Stegun, I. (1970). *Handbook of Mathematical Functions.* Applied Mathematics Series, **55**, Washington D.C.: National Bureau of Standards.

Barbieri, M.M., Liseo, B. and Petrella, L. (2000). Bayes Factors for Fieller's problem. *Biometrika,* **87**, 3, 717-723.

Berger, J.O. and Bernardo, J.M. (1992). On the development of the reference prior. In *Bayesian Statistic IV,* eds. J.M. Bernardo, J.O. Berger, A.P. Dawid and A.F.M. Smith, 35-60, Oxford: Clarendon Press.

Berger, J.O., Liseo, B. and Wolpert, R.L. (1999). Integrated likelihood methods for eliminating nuisance parameters (with Discussion). *Statist. Sc.,* **14**, 1-28.

Berger, J.O. and Mortera, J. (1999). Default Bayes factor for one sided hypothesis testing. *J. Amer. Statist. Assoc.,* **94**, 542-554.

Berger, J.O. and Pericchi, L.R. (1996). The intrinsic Bayes factor for model selection and prediction. *J. Am. Statist. Assoc.,* **91**, 109-22.

Berger, J.O. and Pericchi, L.R. (1998). Accurate and stable Bayesian model selection: the median intrinsic Bayes factor. *Sankhyā B* **60**, 1-18.

Buonaccorsi, J.P. and Gatsonis, C.A. (1988). Bayesian inference for ratios of coefficients in a linear model. *Biometrics* **44**, 87-101.

Casella, G. and Berger, R.L. (1987). Reconciling Bayesian and frequentist evidence in the one-sided testing problem (with discussion). *J. Am. Statist. Assoc.,* **82**, 106-11.

Cheng, C.L. and Van Ness, J.W. (1994). On estimating linear relationships when both variables are subject to errors. *J. R. Statist. Soc. B,* **56**, 167-83.

Fieller, E.C. (1954). Some problems in interval estimation. *J. R. Statist. Soc. B,* **16**, 175-85.

Gleser, L.J. and Hwang, J.T. (1987). The non existence of $100(1 - \alpha)\%$ confidence sets of finite expected diameter in errors-in-variable and related models. *Ann. Statist.* **15**, 1351-62.

Kass, R.E. and Raftery, A.E. (1995). Bayes factors. *J. Am. Statist. Assoc.,* **90**, 773-95.

Krishnamoorthy, K, Kulkarni, P.M. and Mathew, T. (2001). Multiple use one-sided hypotheses testing in univariate linear calibration. *J. Statist. Plann. Inf.,* **93**, 211-223.

Kubokawa, T. and Robert, C.P. (1994). New prespective in linear calibration. *J. Mult. Anal.,* **51**, 178-200.

Lindley, D.V. and El-Sayyad, G.M. (1968). The Bayesian estimation of a linear functional relationship. *J. R. Statist. Soc. B,* **30**, 190-202.

Liseo B. (2003). Bayesian and Conditional Frequentist Analyses of the Fieller's Problem: A critical review. *Metron*, LXI, 133-152.

Moreno, E. (2005). Objective Bayesian methods for one-sided testing. *Test*, **14**, 1,181-198.

O'Hagan, A. (1995). Fractional Bayes factors for model comparison (with Discussion). *J. R. Statist. Soc. B*, **57**, 99-138.

Owen, D.B. (1956). Tables for Computing Bivariate Normal Probabilities. *Annals of Mathematical Statistics*, **27**, 1075-1090.

Bayesian Statistics and Its Applications
Edited by S.K. Upadhyay, U. Singh and D.K. Dey
Anamaya Publishers, New Delhi, India

Bayesian Hierarchical Spatio-Temporal Analysis of fMRI Data: A Case Study

Sandeep Basak[1] and Sumitra Purkayastha[2]

[1]Upper Ground Floor, Bay # 4, GE Towers, Sector Road
Sector 53, DLF Phase 5, Gurgaon-122 002, India

[2]Theoretical Statistics and Mathematics Unit, Indian Statistical Institute
203 B.T. Road, Kolkata-700 108, India

Abstract

Analysis of fMRI data has traditionally proceeded along the direction of deciding separately if each voxel is activated by a stimulus. Linear model formulation of this problem takes into account the temporal dependence that is likely to be present between fMR observations. However, associated spatial and spatio-temporal dependence have been receiving substantial attention for the last few years. Hierarchical Bayesian modeling appears to be quite useful in such studies. Some relevant references, by no means exhaustive, are Genovese (1999, 2000), Gössl et al. (2000, 2001a), Penny et al. (2005), Woolrich et al. (2004a, b, 2005). The approach proposed in Gössl et al. (2000, 2001a) appear to be relatively simpler than those proposed in the others, even though it must be said that there is considerable flexibility in them (in Genovese, 1999, for example). The work reported in this paper has been inspired by Gössl et al. (2000, 2001a). We have employed some of the models proposed in Gössl et al. (2001a) on a data-set studied by Worsley (Worsley et al., 2002; Worsley; 2003, http://www.math.mcgill.ca/keith/fmristat). This has been pursued with a view to seeing how well do the models proposed in Gössl et al. (2001a) perform when employed on a data-set where non-Bayesian method has already been tried. The data-set studied by Prof. Worsley came in handy for this purpose. We wanted to see to what extent do the results obtained by employing some of the models in Gössl et al. (2001a) agree to those obtained by Prof. Worsley and also wanted to have some understanding about the computational aspects, the running time, in particular. We have used the results obtained by Prof. Worsley as the benchmark. Our results agree to a large extent with those obtained by Prof. Worsley. Our computations reveal certain important aspects. In particular, for reducing computing time, introducing some simplifications seemed unavoidable.

---

1. Introduction

Functional magnetic resonance imaging (fMRI) is a technique to determine which parts of the brain are activated by different types of physical sensation or activity, such as sight, sound or the movement

---

AMS 2000 Subject Classifications: 62F15; 62P10; 62M99.

of a subject's fingers. In functional brain imaging studies, the signal changes of interest are caused by the neuronal activity, but such electrical activity is not directly detectable by fMRI. The signal being measured in fMRI experiments is called the *blood oxygenation level dependent* (BOLD) response, which is a consequence of the hemodynamic changes (including local changes in blood flow, volume and oxygenation level) occurring within a few seconds of changes in neuronal activity induced by external stimuli. The BOLD signal is usually used as a proxy for the underlying neuronal activity. Most fMRI studies concern the detection of sites of activation ('hot-spots') in the brain and their relationship to the experimental stimulation and we refer to these studies as the *activation* studies. Interested readers can refer to Clare (1997), Frackowiak et al. (2004), and Jezzard et al. (2001) for the details on the theory and application of fMRI in neurology, neuroscience, psychology and psychiatry.

Analysis of fMRI data usually proceeds by modelling the BOLD signal as the dependent variable in a linear model where suitable transform(s) of the external stimulus (stimuli), obtained by convolving the stimulus (stimuli) with the hemodynamic response function (HRF) (see, e.g., Glover, 1999), has (have) played the role of the independent variable(s). The associated error distribution is assumed to be an autoregressive process of an appropriate order. The question whether a voxel is activated by one of the stimuli being employed is posed as one of testing whether the corresponding regression coefficient is greater than zero against being equal to zero (see, e.g., Worsley et al. (2002), Worsley (2003)). Some important references on statistical analysis of fMRI data and related issues are Bullmore et al. (1996), Friston et al. (1995), Lange (1996), Lazar et al. (2001), Worsley and Friston (1995) etc. and this list is by no means exhaustive.

The approach described above takes into account the temporal dependence that is likely to be present between the BOLD signals. However, this does not take into account the spatial dependence that is likely to be present between the BOLD signals. Possible spatio-temporal dependence is also ignored. On the other hand, it is only natural to demand that our analysis should take these aspects into account. Motivated by these, several authors have tried to incorporate spatial and spatio-temporal aspects both into modeling of fMRI data and subsequent stage of inference. Some relevant references are Genovese (1999, 2000), Gössl et al. (2000, 2001a), Penny et al. (2005), Woolrich et al. (2004a, 2005). All these works are based on a Bayesian formulation. Moreover, hierarchical modeling is seen to be quite useful in this problem. For a detailed discussion on hierarchical modeling and associated Bayesian methods, we refer the reader to Banerjee et al. (2004). It needs to be mentioned here that apart from the purpose with which a Bayesian formulation has been employed, several researchers have also employed Bayesian formulation for the purpose of estimating the HRF. Some relevant references are Gössl et al. (2001b), Marrelec et al. (2003), Woolrich et al. (2004b). We would like to add here that Bayesian approach has received substantial attention in analysis of fMRI data and related issues and the references cited above are by no means exhaustive. Some relevant non-Bayesian references are Polzehl and Spokoiny (2001), Purdon et al. (2001) and Solo et al. (2001).

The work reported in this paper has been inspired by Gössl et al. (2001a) (henceforth, abbreviated to GAF) and also by Gössl et al. (2000). The approach proposed in GAF appears to be relatively simpler than those proposed in the others, even though it must be said that there is considerable flexibility in them (in Genovese, 1999, for example). We have employed some of the models proposed in GAF on a data-set studied by Worsley (Worsley, et al. 2002; Worsley, 2003; http://www.math.mcgill.ca/keith/fmristat) with a view to seeing to what extent do the results obtained by employing some of the models in GAF agree to those obtained by Prof. Worsley and also to have some understanding about the computational aspects, about the running time, in particular. We describe the experiment we have studied and the data-set we have analysed in the next two paragraphs.

The aim of the experiment has been to identify areas related to pain perception. We shall see that the experiment allows one to locate separately areas activated by painful heat stimulus and warm stimulus also. During the experiment, the subject was exposed to the following two stimuli: (a) a painful heat stimulus (49°C) to the left forearm and (b) a warm stimulus (35°C). The experiment began with a period of rest for 9 secs which was followed by each stimulus being alternately administered for 9 secs, intercepted by 9 secs of rest. The whole cycle was repeated 10 times for 6 mins. in total with TR = 3 secs. In other words, each type of scan lasted for 3 secs., and was replicated thrice.

The data-set consists of 13 slices of 128 × 128 images, for each scan. For every scan, the slices are interleaved, i.e. first images of the odd-numbered slices are scanned and then those of the even-numbered slices are obtained. Hence, there are 120 images corresponding to each slice. The work reported in this paper has been pursued with only the *third* slice.

Section 2 discusses the models of GAF we have considered and employed in our context. Section 3 describes our goals and also the methods we have employed. Section 4 pertains to certain issues associated with implementation of the methods. Section 5 presents the results and offers some relevant discussion. Section 6 concludes with some discussion on issues worth exploring in future.

2. Our Models

Suppose, for ease in notation, the *hot* stimulus is the first one and the *warm* stimulus is the second one. Denote by $s_j(t)$, the function representing the jth ($j = 1, 2$) stimulus. This function is usually taken to be a zero-one function (a *box-car* function). Fig. 1 shows the function $s_1(t)$. Its definition

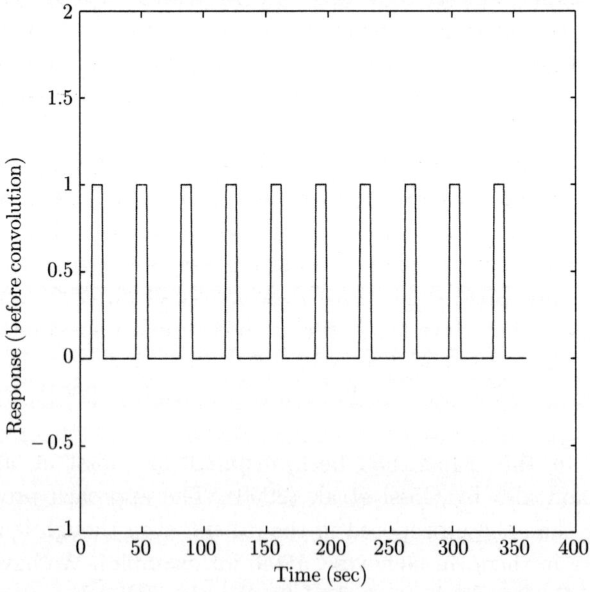

Fig. 1.   Box-car function $s_1(t)$ representing the first stimulus.

requires onset time, duration, height ($= 1$, usually). The experiment we are studying involves two stimuli. In this situation, the box-car function $s_2(t)$ for the second stimulus is a *suitable* translate of $s_1(t)$.

The corresponding fMRI response is not instantaneous; there is a blurring and a delay of the peak response by about 6 secs. A standard way of capturing this is to assume that the noise-free fMRI response depends on the external stimulus by convolution with a hemodynamic response function (HRF) $h(t)$. Thus, for each of the two stimuli, the corresponding noise-free fMRI response is given by

$$z_t^{(j)} = \int_0^\infty h(u)s_j(t-u)\,du, \quad j = 1, 2.$$

The function $h(t)$ we shall work with is the one due to Glover (1999):

$$h(t) = \left(\frac{t}{d_1}\right)^{a_1} \exp\left(-\frac{t-d_1}{b_1}\right) - c\left(\frac{t}{d_2}\right)^{a_2} \exp\left(-\frac{t-d_2}{b_2}\right),$$

where t is time in secs, $d_j = a_j b_j$ is the time to the peak, and $a_1 = 6, a_2 = 12, b_1 = b_2 = 0.9$ secs, and $c = 0.35$ (Fig. 2).

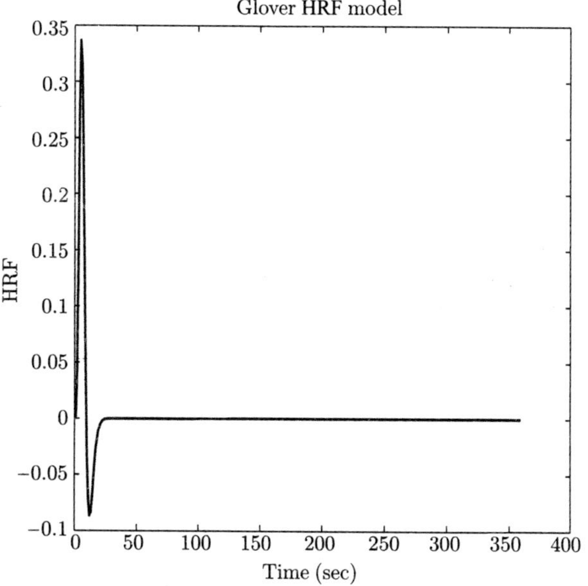

Fig. 2.   Glover's HRF $h(t)$.

The function $z_t^{(1)}$, obtained after convolution, is shown in Fig. 3. The function $z_t^{(2)}$ will obviously be a translate of $z_t^{(1)}$. Both the functions $z_t^{(1)}$ and $z_t^{(2)}$ are subsampled at $T = 120$ scan acquisition times. These quantities have been computed by using the MATLAB function **fmridesign.m**, which is one module of the software **fmristat**, available at http://www.math.mcgill.ca/keith/fmristat.

To describe the model let I be the number of voxels in the slice under consideration. In our problem, $I = 128 \times 128$. For the ith voxel, $i = 1, \ldots, I$, the time series $\{y_{it} : t = 1, \ldots, T\}$ of fMR signals is related to the quantities $\{z_t^{(1)} : t = 1, \ldots, T\}$ and $\{z_t^{(2)} : t = 1, \ldots, T\}$ as follows (GAF):

$$y_{it} = w_t' a_i + z_t^{(1)} b_{it} + z_t^{(2)} c_{it} + \epsilon_{it}, \tag{1.1}$$

$$\epsilon_{it} \overset{\text{i.i.d.}}{\sim} N(0, \sigma_i^2), \tag{1.2}$$

where $w_t' a_i$ is the drift, w_t a known design vector that models parametrically the drift, a_i the multiplicative constant (a vector) associated with the drift, b_{it} the multiplicative constant associated with the *hot* stimulus, c_{it} the multiplicative constant associated with the *warm* stimulus and ϵ_{it} the noise.

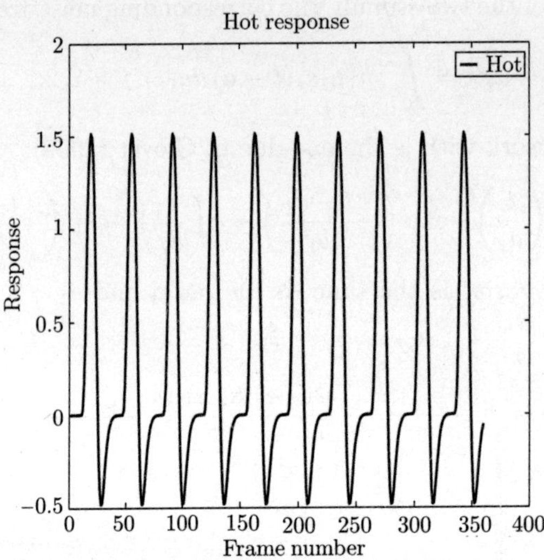

Fig. 3.   Convolution $z_i^{(1)}$ of stimulus $s_1(t)$ and Glover's HRF $h(t)$.

Our observation model is given by (1.1)-(1.2). Note that this model ignores temporal dependence present between fMRI observations. Keeping in mind this issue and also the issues of spatial and spatio-temporal dependence, GAF has replaced this model by those described in the following.

2.1 Model for Spatial Dependence

The model for the observations is assumed to be

$$y_{it} = a_i + z_t^{(1)} b_i + z_t^{(2)} c_i + \epsilon_{it},$$
$$\epsilon_{it} \overset{\text{i.i.d.}}{\sim} N(0, \sigma_i^2). \tag{2}$$

The prior for

$$\boldsymbol{b} = (b_1, \dots, b_I)'$$

is assumed to be an intrinsic autoregressive prior, also called pairwise difference prior (Besag et al., 1991; Banerjee et al. 2004, Section 3.3). For every voxel i, a neighbourhood denoted by ∂i is assumed to consist of n_i nearest neighbours. This number n_i equals 8 for all but the voxels remaining on the boundary of the slice, where this number equals 3 or 5 according as it is one of the four corners or not. The neighborhood relation between i and j is denoted by $i \sim j$. In other words, for every voxel i, ∂i is the 3×3 window around i without the voxel i. The prior is given by

$$p(\boldsymbol{b}|\lambda) \propto \exp\left\{ -\frac{\lambda}{2} \sum_{\substack{i \sim j \\ i > j}} (b_i - b_j)^2 \right\}. \tag{3.1}$$

It can also be written as

$$p(\boldsymbol{b}|\lambda) \propto \exp\left(-\frac{\lambda}{2}\boldsymbol{b}'Q^{b,\text{space}}\boldsymbol{b}\right),\tag{3.2}$$

where the precision matrix $Q_{b,\text{space}}$ (i.e., the inverse of the variance-covariance matrix) is of the form

$$Q_{ij}^{b,\text{space}} = \begin{cases} n_i & \text{if } i = j, \\ -1 & \text{if } i \sim j, \\ 0 & \text{otherwise.} \end{cases}$$

This model has the following conditional distributions:

$$b_i|b_{j\neq i}, \lambda \sim N\left(\frac{1}{n_i}\sum_{j\in\partial i} b_j, \frac{1}{\lambda n_i}\right).$$

Priors $p(\boldsymbol{a}|\lambda)$, $p(\boldsymbol{c}|\lambda)$ for

$$\boldsymbol{a} = (a_1,\ldots,a_I)' \quad \text{and} \quad \boldsymbol{c} = (c_1,\ldots,c_I)'$$

are similar. All these priors are assumed to be independent.

In the last stage of the hierarchy, we assume, following GAF, Gamma$(1,1)$ priors for the unknown hyperparameters, i.e., precision parameter λ and inverse variances $1/\sigma_i^2$ of the observation errors.

Notice that the model (2) is a special case of the one given by (1.1) and (1.2), with the activation coefficients b_i and c_i depending only on the voxel i. The prior given by (3.1) (or (3.2)) allows us to introduce spatial correlation. Compared to pixelwise parametric modeling, estimates are spatially smoothed.

2.2 Model for Temporal Dependence

The model for the observations is assumed to be the following (Gössl et al., 2000, 2001a):

$$y_{it} = a_{it} + z_t^{(1)}b_{it} + z_t^{(2)}c_{it} + \epsilon_{it},$$
$$\epsilon_{it} \overset{\text{i.i.d.}}{\sim} N(0,\sigma_i^2).\tag{4}$$

The priors for $\boldsymbol{a}_i = (a_{i,1},\ldots,a_{i,T})'$, $\boldsymbol{b}_i = (b_{i,1},\ldots,b_{i,T})'$, $\boldsymbol{c}_i = (c_{i,1},\ldots,c_{i,T})'$ are as follows:
We assume that for $t = 3,4,\ldots,T$,

$$a_{it} = 2a_{i,t-1} - a_{i,t-2} + \zeta_{it}, \quad \zeta_{it} \overset{\text{i.i.d.}}{\sim} N(0,\sigma_{\zeta,i}^2),\tag{5.1}$$

$$b_{it} = 2b_{i,t-1} - b_{i,t-2} + \eta_{it}, \quad \eta_{it} \overset{\text{i.i.d.}}{\sim} N(0,\sigma_{\eta,i}^2),\tag{5.2}$$

$$c_{it} = 2c_{i,t-1} - c_{i,t-2} + \xi_{it}, \quad \xi_{it} \overset{\text{i.i.d.}}{\sim} N(0,\sigma_{\xi,i}^2),\tag{5.3}$$

and assume appropriate initial priors for $b_{i,1}, b_{i,2}$ (and also for $a_{i,1}, a_{i,2}$ and $c_{i,1}, c_{i,2}$) so that the prior for

$$\boldsymbol{b}_i = (b_{i,1},\ldots,b_{i,T})'$$

can be written as

$$p(\boldsymbol{b}_i|\lambda_{\eta,i}) \propto \exp\left(-\frac{1}{2}\lambda_{\eta,i}\boldsymbol{b}_i'\boldsymbol{Q}^{b,\text{temp}}\boldsymbol{b}_i\right),\tag{6}$$

where
$$Q^{b,\text{temp}} = \begin{bmatrix} 2 & -2 & 1 & 0 & 0 & \cdots & 0 & 0 \\ -2 & 6 & -4 & 1 & 0 & \cdots & 0 & 0 \\ 1 & -4 & 6 & -4 & 1 & \cdots & 0 & 0 \\ \vdots & \vdots & \vdots & \vdots & \vdots & \vdots & \vdots & \vdots \\ 0 & \cdots & 0 & 1 & -4 & 6 & -4 & 1 \\ 0 & \cdots & 0 & 0 & 1 & -4 & 5 & -2 \\ 0 & \cdots & 0 & 0 & 0 & 1 & -2 & 1 \end{bmatrix}_{T \times T},$$

and $\lambda_{\eta,i}$ is the precision parameter of the prior, given by

$$\lambda_{\eta,i} \stackrel{\text{def}}{=} \frac{1}{\sigma_{\eta,i}^2}.$$

The expression for the prior given by (6) will be derived later. We need to mention only that the priors for $b_{i,1}$ and $b_{i,2}$ are assummed to be independent $N(0, \sigma_{\eta,i}^2)$ variables. Priors for $\boldsymbol{a}_i = (a_{i,1}, \ldots, a_{i,T})'$ and $\boldsymbol{c}_i = (c_{i,1}, \ldots, c_{i,T})'$ are similar. All these priors are assumed to be independent.

In the last stage of the hierarchy, we assume that

$$\sigma_{\eta,1} = \sigma_{\eta,2} = \cdots = \sigma_{\eta,I} = \sigma_{\xi,1} = \sigma_{\xi,2} = \cdots = \sigma_{\xi,T} = \sigma_{\zeta,1} = \sigma_{\zeta,2} = \cdots = \sigma_{\zeta,T} = \sigma, \qquad (7.1)$$

say. This implies

$$\lambda_{\eta,1} = \lambda_{\eta,2} = \cdots = \lambda_{\eta,I} = \lambda_{\xi,1} = \lambda_{\xi,2} = \cdots = \lambda_{\xi,T} = \lambda_{\zeta,1} = \lambda_{\zeta,2} = \cdots = \lambda_{\zeta,T} = \lambda, \qquad (7.2)$$

say. Also, we assume, following GAF, Gamma$(1, 1)$ priors for the unknown hyperparameters, i.e. precision parameter λ and inverse variances $1/\sigma_i^2$ of the observation errors.

Equations in (5.1)-(5.3) enforce smoothness, by imposing Gaussian prior restrictions, on the second-order differences

$$\zeta_{it} = a_{it} - 2a_{i,t-1} + a_{i,t-2}, \quad \eta_{it} = b_{it} - 2b_{i,t-1} + b_{i,t-2}, \quad \xi_{it} = c_{it} - 2c_{i,t-1} + c_{i,t-2}.$$

These differences may be interpreted as penalties. They penalize deviations from a straight line for which the penalty is zero. Thus, b_{it}'s (and similarly a_{it}'s and c_{it}'s) follow a second-order random walk.

One specific reason for extending (1.1)-(1.2) to (4), (5.1)-(5.3) is to allow flexibility in the form of baseline trend as opposed to a particular form of baseline trend (like a polynomial of degree 3) and also to allow time-dependent effects of covariates.

2.3 An Additive Spatio-Temporal Model

The model for the observations is assumed to be

$$y_{it} = a_{it} + z_t^{(1)} b_{it} + z_t^{(2)} c_{it} + \epsilon_{it},$$
$$\epsilon_{it} \stackrel{\text{i.i.d.}}{\sim} N(0, \sigma_i^2).$$

The previous two models took care, respectively, of the spatial aspect or the temporal aspect of the data separately by introducing appropriate smoothness priors. To obtain models that simultaneously consider these two aspects, a combination of the above properties is necessary. This can be done by different types of space-time interactions. Here we consider an additive model as a reasonable compromise between computational complexity and model complexity. This is obtained by splitting the activation effects for the *hot* and *warm* stimulus b_{it} and c_{it}, and the baseline trend a_{it} as follows:

$$a_{it} = \alpha_i + \beta_{it}, \quad b_{it} = \gamma_i + \delta_{it}, \quad c_{it} = \varphi_i + \theta_{it}.$$

We define

$$\boldsymbol{\beta}_i = (\beta_{i1}, \dots, \beta_{iT})', \quad \boldsymbol{\delta}_i = (\delta_{i1}, \dots, \delta_{iT})', \quad \boldsymbol{\theta}_i = (\theta_{i1}, \dots, \theta_{iT})',$$

and assume that each of the vectors $\boldsymbol{\beta}_i$, $\boldsymbol{\delta}_i$, $\boldsymbol{\theta}_i$ are centered at zero. This is to ensure identifiability.

We assume a spatial smoothness prior for the time-constant part γ_i (cf. (3.1) or (3.2)) and a voxelwise temporal second-order random walk prior (cf. (6)) for the time-varying effects δ_{it}. Similar priors are assumed for α_i, $\boldsymbol{\beta}_i$, φ_i and $\boldsymbol{\theta}_i$.

The priors for the precisions are assumed to be Gamma$(1, 1)$. Also, as before, all the priors are assumed to be independent.

3. Our Goal and Methods

We have mentioned earlier that the problem of deciding if a voxel is activated by a specific stimulus is formulated as one of testing if the relevant activation coefficient is zero. This is usually answered within the frequentist framework. Such an approach does not take into account the spatial or spatio-temporal dependence among the fMRI observations which are natural issues in our context. We are addressing such issues within the Bayesian framework.

3.1 Spatial Model

We are interested in the following probabilities:

$$\text{(i) } P(b_i > 0|\boldsymbol{Y}), \quad \text{(ii) } P(c_i > 0|\boldsymbol{Y}), \quad \text{(iii) } P(b_i > c_i|\boldsymbol{Y}), \; i = 1, \dots, I,$$

where $\boldsymbol{Y}$ is the vector of all $(= I \cdot T)$ observations. We shall plot the above values of the probabilities for all the pixels to get activation maps as before. These probabilities are obtained by employing Gibbs sampling (Gelfand and Smith, 1990). The final results are given by images of the posterior probabilities as above. Some initial calculations leading to the full conditionals for the steps of Gibbs sampling are presented below.

To begin with, we work out a more precise expression for $p(\boldsymbol{b}|\lambda)$. The total number of terms in the sum

$$\sum_{\substack{i \sim j \\ i > j}} (b_i - b_j)^2 \text{ is } A/2,$$

where $A = 4 \cdot 3 + (\sqrt{I} - 2) \cdot 4 \cdot 5 + (\sqrt{I} - 2)^2 \cdot 8 = 8I - 12\sqrt{I} + 4.$

Hence,

$$p(\boldsymbol{b}|\lambda) \propto (\sqrt{\lambda})^{4I - 6\sqrt{I} + 2} \cdot \exp\left\{ -\frac{\lambda}{2} \sum_{\substack{i \sim j \\ i > j}} (b_i - b_j)^2 \right\}.$$

The prior for λ is Gamma (a, b), with $a = b = 1$. Hence, it is given by

$$p(\lambda) = \exp(-\lambda), \quad \lambda > 0. \tag{8}$$

The prior for $1/\sigma_i^2$ is given by $\sigma_i^2 \sim IG(a, b)$ with $a = b = 1$. Write $t = 1/\sigma_i^2$. Then

$$p(t) = \exp(-t), \quad t > 0.$$

From this, we obtain

$$p(\sigma_i^2) = \exp\left(-\frac{1}{\sigma_i^2}\right) \cdot \left(\frac{1}{\sigma_i^2}\right)^2, \quad \sigma_i^2 > 0. \tag{9.1}$$

Define

$$\boldsymbol{\sigma} = (\sigma_1^2 \ \sigma_2^2 \ \ldots \ \sigma_I^2)^t.$$

Then

$$p(\boldsymbol{\sigma}) = \prod_{i=1}^{I} p(\sigma_i^2). \tag{9.2}$$

In view of (2), the likelihood is given by

$$l(\boldsymbol{b}, \boldsymbol{c}, \boldsymbol{a}, \boldsymbol{\sigma}|\boldsymbol{Y}) \propto \prod_{i=1}^{I} \frac{1}{\sigma_i^T} \exp\left(-\frac{1}{2\sigma_i^2} \sum_{t=1}^{T} (y_{it} - a_i - z_t^{(1)} b_i - z_t^{(2)} c_i)^2\right),$$

where $\boldsymbol{Y}$ is the vector of all $(= I \cdot T)$ observations. Posterior of the parameters is given by

$$p(\boldsymbol{b}, \boldsymbol{c}, \boldsymbol{a}, \boldsymbol{\sigma}, \lambda|\boldsymbol{Y}) \propto l(\boldsymbol{b}, \boldsymbol{c}, \boldsymbol{a}, \boldsymbol{\sigma}|\boldsymbol{Y}) \cdot p(\boldsymbol{b}|\lambda) \cdot p(\boldsymbol{c}|\lambda) \cdot p(\boldsymbol{a}|\lambda) \cdot p(\lambda) \cdot p(\boldsymbol{\sigma}).$$

The steps of Gibbs sampling is obtained from the expression of the posterior. In Appendix A.1, we present the expressions for the full conditionals.

3.2   Temporal Model

We are interested in the following probabilities

(i) $P(b_{it} > 0|\boldsymbol{Y})$,   (ii) $P(c_{it} > 0|\boldsymbol{Y})$,   (iii) $P(b_{it} > c_{it}|\boldsymbol{Y})$, $i = 1, \ldots, I, t = 1, \ldots, T,$

where $\boldsymbol{Y}$ is the vector of all $(= I \cdot T)$ observations. For a given t, we can plot the above values of the probabilities for all the pixels to get activation maps as before. These probabilities are obtained by employing Gibbs sampling. In what follows, we present some initial calculation leading to the full conditionals for the steps of Gibbs sampling.

To begin with, we derive the expression for the prior for $\boldsymbol{b}_i$ given by (6). Refer to (5.2), (7.1) and (7.2). Let us define

$$\eta_{it} = b_{it} - 2b_{i,t-1} + b_{i,t-2}, \quad t = 3, \ldots, T,$$

$$= b_{it}, \qquad\qquad\qquad t = 1, 2.$$

We assume that η_{it}, $t = 1, \ldots, T$, are i.i.d. $N(0, \sigma_{\eta,i}^2)$. Define now

$$\boldsymbol{\eta}_i \stackrel{\text{def}}{=} (\eta_{i,1}, \ldots, \eta_{i,T})',$$

and recall that

$$\boldsymbol{b}_i = (b_{i,1}, \ldots, b_{i,T})'.$$

Observe now that

$$\boldsymbol{\eta}_i = \boldsymbol{A}\boldsymbol{b}_i,$$

$$\text{where} \quad \boldsymbol{A} = \begin{bmatrix} 1 & 0 & 0 & 0 & 0 & \cdots & 0 & 0 \\ 0 & 1 & 0 & 0 & 0 & \cdots & 0 & 0 \\ 1 & -2 & 1 & 0 & 0 & \cdots & 0 & 0 \\ 0 & 1 & -2 & 1 & 0 & \cdots & 0 & 0 \\ \vdots & \vdots & \vdots & \vdots & \vdots & \vdots & \vdots & \vdots \\ 0 & \cdots & 0 & 1 & -2 & 1 & 0 & 0 \\ 0 & \cdots & 0 & 0 & 1 & -2 & 1 & 0 \\ 0 & \cdots & 0 & 0 & 0 & 1 & -2 & 1 \end{bmatrix}_{T \times T}$$

The pdf of $\boldsymbol{\eta}_i$ is given by

$$p(\boldsymbol{\eta}_i) \propto \lambda^{T/2} \exp\left(-\frac{1}{2}\lambda \boldsymbol{\eta}_i' \boldsymbol{\eta}_i\right).$$

Hence, the pdf of $\boldsymbol{b}_i$ is given by

$$p(\boldsymbol{b}_i|\lambda) \propto \lambda^{T/2} \exp\left(-\frac{1}{2}\lambda \boldsymbol{b}_i' \boldsymbol{A}' \boldsymbol{A} \boldsymbol{b}_i\right).$$

Let us write $\boldsymbol{Q}^{b,\text{temp}} = \boldsymbol{A}'\boldsymbol{A}$. It is easy to verify that

$$\boldsymbol{Q}^{b,\text{temp}} = \begin{bmatrix} 2 & -2 & 1 & 0 & 0 & \cdots & 0 & 0 \\ -2 & 6 & -4 & 1 & 0 & \cdots & 0 & 0 \\ 1 & -4 & 6 & -4 & 1 & \cdots & 0 & 0 \\ \vdots & \vdots & \vdots & \vdots & \vdots & \vdots & \vdots & \vdots \\ 0 & \cdots & 0 & 1 & -4 & 6 & -4 & 1 \\ 0 & \cdots & 0 & 0 & 1 & -4 & 5 & -2 \\ 0 & \cdots & 0 & 0 & 0 & 1 & -2 & 1 \end{bmatrix}_{T \times T}.$$

Notice that the algebraic calculation above implies the following:

$$\boldsymbol{Q}^{b,\text{temp}} = \boldsymbol{Q}^{c,\text{temp}} = \boldsymbol{Q}^{a,\text{temp}} = \boldsymbol{Q}, \text{ say.}$$

The prior for the common precision parameter λ (cf. (7.1) and (7.2)) is same as in (8).

The prior for

$$\boldsymbol{\sigma} = (\sigma_1^2 \ \sigma_2^2 \ \ldots \ \sigma_I^2)'$$

is same as in (9.1) and (9.2).

The likelihood is given by

$$l(\boldsymbol{b}_i, \boldsymbol{c}_i, \boldsymbol{a}_i, i = 1, \ldots, I, \boldsymbol{\sigma}|\boldsymbol{Y}) \propto \prod_{i=1}^{I} \frac{1}{\sigma_i^T} \exp\left(-\frac{1}{2\sigma_i^2} \sum_{t=1}^{T} (y_{it} - a_{it} - z_t^{(1)} b_{it} - z_t^{(2)} c_{it})^2\right),$$

where $\boldsymbol{Y}$ is the vector of all $(= I \cdot T)$ observations. Posterior of the parameters is given by

$$p(\boldsymbol{b}_i, \boldsymbol{c}_i, \boldsymbol{a}_i, i = 1, \ldots, I, \boldsymbol{\sigma}, \lambda|\boldsymbol{Y}) \propto l(\boldsymbol{b}_i, \boldsymbol{c}_i, \boldsymbol{a}_i, i = 1, \ldots, I, \boldsymbol{\sigma}|\boldsymbol{Y})$$

$$\times \prod_{i=1}^{I} p(\boldsymbol{b}_i|\lambda) \cdot \prod_{i=1}^{I} p(\boldsymbol{c}_i|\lambda) \cdot \prod_{i=1}^{I} p(\boldsymbol{a}_i|\lambda) \cdot p(\lambda) \cdot p(\boldsymbol{\sigma}).$$

The full conditionals associated with the Gibbs sampling are obtained from the expression of the posterior. They appear in Appendix A.2.

3.3 Spatio-Temporal Model

We are interested in the following probabilities:

$$\text{(i) } P(b_{it} > 0|\boldsymbol{Y}), \ \ \text{(ii) } P(c_{it} > 0|\boldsymbol{Y}), \ \ \text{(iii) } P(b_{it} > c_{it}|\boldsymbol{Y}), \ i = 1,\dots,I, \ t = 1,\dots,T,$$

where $\boldsymbol{Y}$ is the vector of all ($= I \cdot T$) observations. For a given t, we can plot the above values of the probabilities for all the pixels to get activation maps as before. These probabilities are obtained by employing Gibbs sampling. We skip the details. Interested readers can contact the authors to obtain these information.

4. Issues of Implementation

All our computations have been done on a typical 1.8 GHz Pentium IV desktop PC with 128MB of RAM. There are several simplifications we have introduced with the purpose of reducing running time. The most important is the following: We simulated λ and σ_i^2s directly from their prior distributions, instead of their *full conditional distributions*. Moreover, to reduce the running time of the Gibbs sampler to a reasonable time frame, and to reduce the working memory load in the computer, we have worked with only part of the data-set, i.e., we have taken only first 60 time points for the temporal model and first 40 time points for the spatio-temporal model (instead of the full 120 available to us in our data-set). This data reduction was done only for the models for temporal dependence and spatio-temporal dependence in which the large number of parameters forced us to apply the above step to make our program run within reasonable time limits. For the spatial model, however, we worked with the full data-set.

The Gibbs sampler and the codes for the calculation of the posterior probabilities, were all written in MATHEMATICA while the final plots of the probabilities (i.e., the activation maps) were prepared using MATLAB. For preserving independence among successive samples, we have selected every 5th value generated by the Gibbs sampler, thus discarding all but m of the $5m$ (see Raftery and Lewis, 1992) random variates it generates.

Table 1 gives the number of iterations and approximate running time of the Gibbs sampler.

Table 1. Approximate running times

Model	Iterations		Time taken (approx.)
	Burn-in	Convergence	
Spatial	1000	5000	2 hr 20 min
Temporal	1000	1000	5 hr 50 min
Spatio-temporal	500	500	15 hr 10 min

In order to make a model assessment, we carried out a preliminary analysis of the residuals (for a few random pixels) to check for heteroscedascity and non-normality. All the results obtained validated our assumptions.

5. Results and Discussion

For describing how we present our results, let us restrict our attention to the model for spatial dependence. We want following probabilities:

$$\text{(i) } P(b_i > 0|\boldsymbol{Y}), \quad \text{(ii) } P(c_i > 0|\boldsymbol{Y}), \quad \text{(iii) } P(b_i > c_i|\boldsymbol{Y}), \; i = 1, \dots, I,$$

where $\boldsymbol{Y}$ is the vector of all $(= I \cdot T)$ observations. For every voxel, denoted by i, of the slice under consideration, we present these quantities in plots. Fig. 4 (a, b) gives the first and third of these plots, for the model for spatial dependence. Such plots are known as *posterior probability maps*. Voxels with high probabilities constitute the region(s) of activation due to the stimulus.

Fig. 4 (a) shows some very highly activated regions (colored white) and a substantial amount of high activation regions (colored red) due to the *hot* stimulus. Due to the presence of such a high amount of activation in different regions of the brain, it is quite difficult to distinguish where exactly the primary activation takes place due to the *hot* stimulus. However, Fig. 4(b) very clearly distinguishes between the activated and non-activated regions. It is seen that the primary activation is located in the central part of the brain (colored white). Additionally small activated regions can be seen in the lower part of the brain. This figure essentially tells us which parts of the brain are responsible for perception of pain finding which is the aim of the experiment. Also, as a consequence, this has motivated us to concentrate on 'such' type of maps in the other models also.

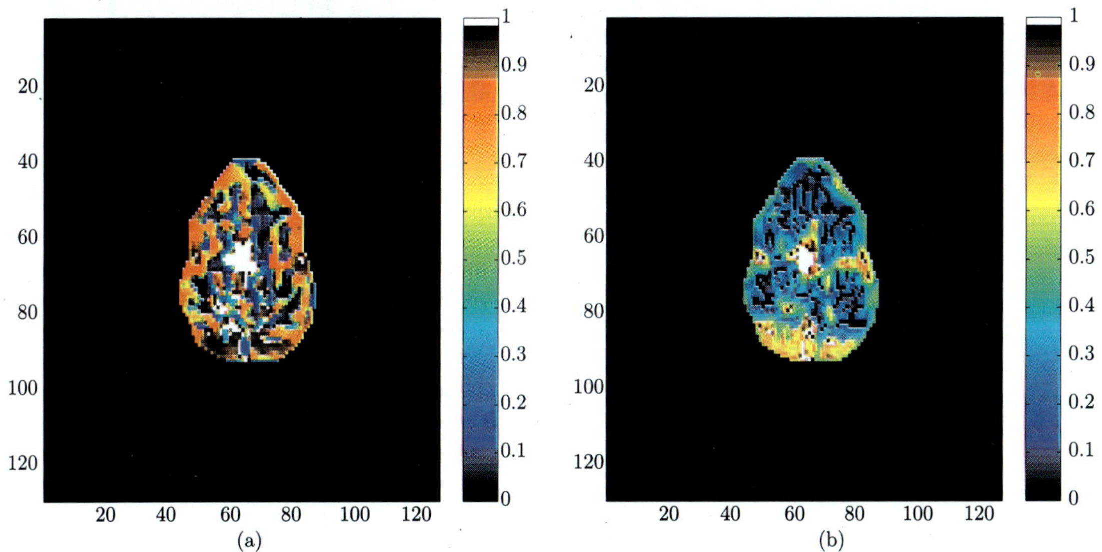

(a) (b)

Fig. 4. Plots of (a) $P(b_i > 0|Y)$ and (b) $P(b_i > c_i|Y)$ for the model for spatial dependence.

We now consider the models for temporal and spatio-temporal dependence. Let us recall that we are interested in the following probabilities:

$$\text{(i) } P(b_{it} > 0|\boldsymbol{Y}), \quad \text{(ii) } P(c_{it} > 0|\boldsymbol{Y}), \quad \text{(iii) } P(b_{it} > c_{it}|\boldsymbol{Y}), \; i = 1, \dots, I, \, t = 1, \dots, T,$$

where $\boldsymbol{Y}$ is the vector of all $(= I \cdot T)$ observations. For a given t, we can plot the above values of the probabilities for all the pixels to get posterior probability maps as before.

The figures given next and pertaining to (iii) above were obtained from the application of the temporal model. Being obtained from the temporal model, each such map corresponds to a specific time point of the experiment. Fig. 5 (a) corresponds to the 22nd time point of the experiment, while Fig. 5 (b) corresponds to the 27th time point of the experiment. Both these maps were thresholded at the probability level 1-10^{-3}, meaning that all those pixels for which the corresponding probabilities were greater than our threshold were assigned values above 0.8 in the colorbar. This sort of thresholding is necessary here to bring out the activated regions. (What we are doing here is what is known in image processing as *Contrast Stretching*. See Appendix A.3 for details.)

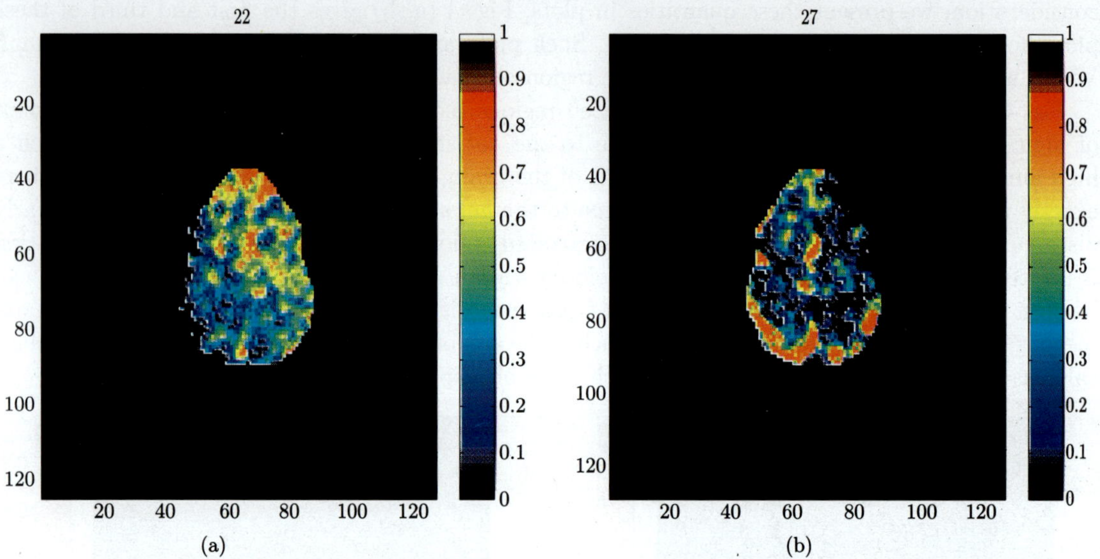

Fig. 5. Plots of $P(b_{it} > c_{it}|Y)$ at (a) $t = 22$ and (b) $t = 27$ for the model for temporal dependence.

What is readily seen to be *common* among the two plots is the activation around the central region of the brain. However, in Fig. 5 (a) we see significant activation in some parts of the top of the brain. Also, in Fig. 5 (b) parts of the lower left portion and lower right portion (near the boundary) show high activation. These observations suggest that there are temporal fluctuations in the activated regions of the brain which our temporal model can catch. This is because we had designed our priors in such a way so that we had a temporally varying activation profile (β_{it}). On the other hand, the spatial model, which assumes a time constant activation α_i, can detect only the primary activation areas (areas which get activated more or less throughout the whole duration of the experiment), but is unable to detect the temporally varying activation regions. However, it can be seen that the plots from the temporal model does not assign very high probabilities to the primary activation regions, as was the case for the spatial model. Also the activation is spread over a large part of the brain, instead of some specific regions. This may be due to the large number of parameters associated with the temporal model which may lead to a splitting of the information (thus widening the Bayesian credibility regions). Also to be noted is the fact that this analysis was carried out using half the data-set (as was mentioned earlier). So the inferences from this model cannot be as precise as those from the spatial model.

Finally, the plots given next and pertaining to (iii) above were obtained from the additive spatio-temporal model. As before, each such plot corresponds to a particular time point of the experiment. Fig. 6 (a) corresponds to the 10th time point of the experiment, while Fig. 6 (b) corresponds to the 18th time point of the experiment. Both these maps were thresholded at 1-10^{-6} level. *It should be noted that as the number of parameters in our model increase, we have to increase our thresholding limits in order to bring out the activated regions. Thresholding at lower levels do not give satisfactory results, i.e., we do not get pictures/images where we can clearly distinguish the activated regions from the non-activated ones.* If we threshold at lower levels, then we get large parts of the brain as activated regions. So, unless we threshold at higher probability levels, we are unable to decide which are the areas of the brain that actually stand out as activated compared to the other surrounding areas.

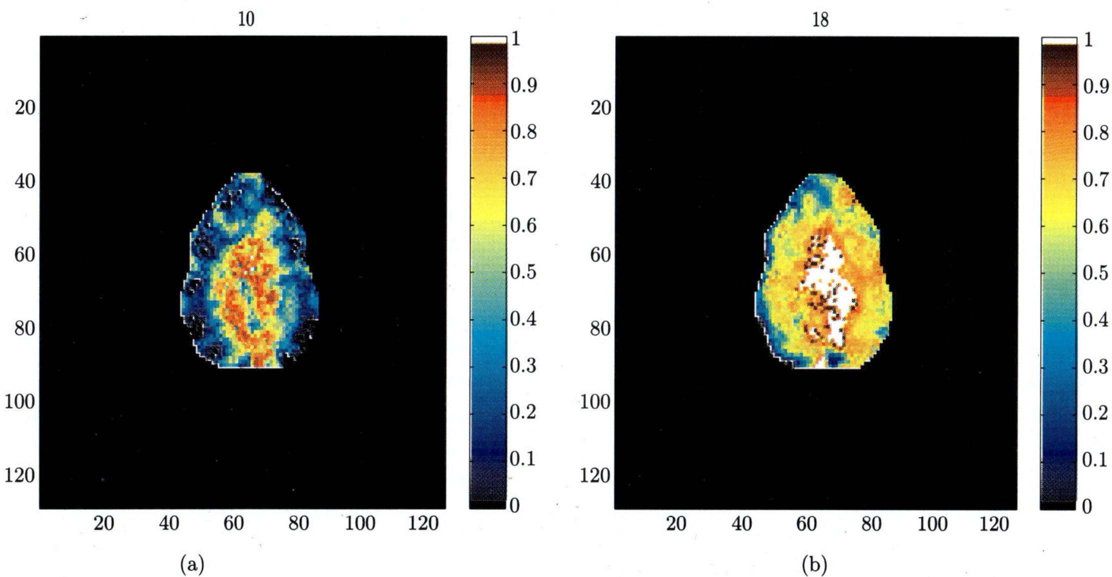

Fig. 6. Plots of $P(b_{it} > c_{it}|Y)$ at (a) $t = 10$ and (b) $t = 18$ for the model for spatio-temporal dependence.

Fig. 6 (a, b) shows quite a substantial part of the brain activated including the central part which we had already identified in the results of the spatial and temporal models. However, a comparison between these two plots shows an interesting feature. In Fig. 6 (a) the regions which are quite highly activated (deep red regions), a large part of those regions show very high activation (white regions) in Fig. 6 (b). Also those regions which show low levels of activation (blue and green regions) in Fig. 6 (a) show increased activation in Fig. 6 (b) (yellow and orange regions). This suggests that regions of the brain do not get activated all at once, the levels of activation vary smoothly over time. This is what one would expect, as BOLD signal depends on the blood flow inside the brain which is expected to vary smoothly over time. However, as has been said about results for the temporal model, the results obtained from the spatio-temporal model show activation over a large part of the brain (including the suspected central region). The large number of parameters and the small portion of the data-set we used for this model ($1/3^{\rm rd}$ part of the original data-set) may have caused this. Hence, the inferences drawn from the results of this model will obviously not be as precise as those of the spatial model. However, the results show that this model is able to capture the primary activated regions as well as capture the smooth temporal variation of different parts of the brain in response to the stimulus.

We conclude this section by observing that the region(s) activated by pain perception and picked up by the methods we have employed agree to a large extent with the one obtained in Worsley

(Worsley et al., 2002; Worsley, 2003; http://www.math.mcgill.ca/keith/fmristat) by application of non-Bayesian method. However, we get activation over other regions and larger regions.

6. Future Directions

In order to simplify our calculations, we have assumed that λ_i's are all equal (cf. (7.2)). We have also simplified calculations by simulating directly from the prior distributions of the hyperparameters λ's and σ_i^2's. We should do the whole exercise without these simplifications. However, in view of our experience, we may add that we shall require *some* simplifications for reducing computing time. We may also try with different hyperprior specifications, say by changing the values of γ_a and γ_b, to see how our results vary.

To reduce the computation time of our programs, we have worked with only half of our original data-set in case of the temporal model and one third of our data-set was used for fitting the spatio-temporal model. Besides this, our results and inference were obtained by running the analysis on a single slice. In future one might try running the analysis on all the slices and the full data-set, to draw more accurate inferences. Here also we have to address separately the issue of reducing computing time.

A non-additive spatio-temporal model has been suggested in GAF that takes into account the temporal variations in adjacent pixels, unlike the *additive* spatio-temporal model. They propose priors in which both the spatial and temporal dependencies are incorporated simultaneously. This is one direction in which some future work may be pursued. Also, models and methods proposed in Genovese (1999), Gössl et al. (2001b), Marrelec et al. (2003), Woolrich et al. (2004a) may be tried. Carlin and Banerjee (2003) also may be helpful in this context.

The referee has asked if there is any motivation for choosing ∂i, for every voxel i, the 3×3 window around i without the voxel i. In the present context, this choice has been driven by requirement of simplicity and consequential reduction in computing time. However, we also wish to address this issue by varying the window size and making a decision by employing some tool of model selection. In particular, we may try an empirical Bayes approach using ML-II method (Berger, 1985, pp. 99–101) in this context. Purdon et al. (2001) and Solo et al. (2001) also are relevant references here.

The analysis presented in GAF is preceded by application of suitable algorithm for correcting for the subject's motion. Usually an image registration algorithm is applied for this purpose. This is a common preprocessing step and is quite effective in practice (Genovese, 2000). We may also think of some sort of image processing on the scanned images, possibly to reduce noise and increase sharpness. Such techniques can be applied to our images. We have ignored these issues as our focus was more on successful implementation of some of the models proposed in GAF.

It has to be noted that we are employing several models on the same data-set and have also kept several others in mind. Issues of model selection need to be addressed here. Also, in addition to looking at certain posterior probabilities we need to develop and work out the appropriate Bayes tests for activation studies in analysis of fMRI data. See Genovese (1999) in this context.

A final extension might be the incorporation of further substantial prior information, e.g., anatomical priors obtained from structural images to constrain estimates of activity to be in or near the grey matter. This would both lend biological validity and restrict the number of voxels to be analysed and so speed up computation—both important issues.

Acknowledgements

The work reported in this paper has been supported partially by fund available through the project "Analysis of fMRI Data and Human Brain Mapping" of Division of Theoretical Statistics and

Mathematics, Indian Statistical Institute. This project was undertaken with a view to initiating research in statistical brain imaging at Indian Statistical Institute. Prof. J.K. Ghosh introduced Sumitra Purkayastha to this area and also encouraged him to pursue research in this area. Sumitra Purkayastha is grateful to him for these. The authors are also grateful to him for several helpful discussions and to Professors N.A. Lazar and W.D. Penny for their encouraging comments on an earlier version of this paper. Some of their suggestions have been incorporated in the section on future directions of work. They are also grateful to Prof. K.J. Worsley for bringing some important references to their notice. Finally, the authors wish to record their sincere gratitude to an anonymous referee for going through an earlier version of this paper and offering several helpful suggestions.

References

Banerjee, S., Carlin, B.P. and Gelfand, A.E. (2004). *Hierarchical modeling and analysis for spatial data*. Boca Raton: Chapman & Hall/CRC.

Berger, J.O. (1985). *Statistical Decision Theory and Bayesian Analysis* (2nd ed.). New York: Springer.

Besag, J., York, J. and Mollié, A. (1991). Bayesian image restoration with two applications in spatial statistics. *Annals of the Institute of Statistical Mathematics*, **43**, 1-59.

Bullmore, E., Brammer, M., Williams, S.C.R., Rabe-Hesketh, S., Janot, N., David, A., Mellers, J., Howard, R. and Sham, P. (1996). Statistical methods of estimation and inference for functional MR image analysis. *Magnetic Resonance in Medicine*, **35**, 261-277.

Carlin, B.P. and Banerjee, S. (2003). Hierarchical multivariate CAR models for spatio-temporally correlated survival data. In *Bayesian Statistics* **7**, eds. J.M. Bernardo, M.J. Bayarri, J.O. Berger, A.P. Dawid, D. Heckerman, A.F.M. Smith, and M. West. Oxford: Oxford University Press, 45-64.

Clare, S. (1997). *Functional Magnetic Resonance Imaging: Methods and Applications*, Unpublished Ph.D. thesis. [http://www.fmrib.ox.ac.uk/~stuart]

Frackowiak, R.S.J., Friston, K.J., Frith, C.D., Dolan, R.J., Price, C.J., Zeki, S., Ashburner, J. and Penny, W. (2004). *Human brain function* (2nd ed.). Elsevier, Academic Press: Amsterdam.

Friston, K.J., Holmes, A.P., Poline, J.-B., Grasby, P.J., Williams, S.C.R., Frackowiak, R.S.J. and Turner, R. (1995). Analysis of fMRI time-series revisited. *Neuroimage*, **2**, 45-53.

Gelfand, A.E. and Smith, A.F.M. (1990). Sampling-based approaches to calculating marginal densities. *Journal of the American Statistical Association*, **85**, 398-409.

Genovese, C.R. (1999). Imaging and spatio-temporal inference (with discussion). In *Bayesian Statistics* **6**, eds. J. Bernardo, J.O. Berger, A.P. Dawid and A.F.M. Smith, Oxford University Press, 255-274.

Genovese, C.R. (2000). A Bayesian time-course model for functional magnetic resonance imaging data (with discussion). *Journal of American statistical association*, **95**, 691-719.

Glover, G. H. (1999). Deconvolution of impulse function response in event-related BOLD fMRI. *NeuroImage*, **9**, 416-429.

Gössl, C., Auer, D.P. and Fahrmeir, L. (2000). Dynamic models in fMRI. *Magnetic Resonance in Medicine* **43**, 72-81.

Gössl, C., Auer, D.P. and Fahrmeir, L. (2001a). Bayesian spatiotemporal inference in functional magnetic resonance imaging. *Biometrics*, **57**, 554-562.

Gössl, C., Fahrmeir, L. and Auer, D.P. (2001b). Bayesian modeling of the hemodynamic response function in BOLD fMRI. *NeuroImage*, **14**, 140-148.

Jezzard, P., Matthews, P.M. and Smith, S.M. (2001). (Eds.) *Functional MRI: An introduction to methods*. New York: Oxford University Press.

Lange, N. (1996). Statistical approaches to human brain mapping by functional magnetic resonance imaging. *Statistics in Medicine*, **15**, 389-428.

Lazar, N.A., Eddy, W.F., Genovese, C.R. and Welling, J. (2001). Statistical issues in fMRI brain imaging. *International Statistical Review*, **69**, 105-127.

Marrelec, G., Benali, H., Ciuciu, P., Pélégrini-Issac, M. and Poline, J.-B. (2003). Robust Bayesian estimation of the hemodynamic response function in event-related BOLD MRI using basic physiological information. *Human Brain Mapping*, **19**, 1-17.

Penny, W.D., Trujillo-Barreto, N.J. and Friston, K.J. (2005). Bayesian fMRI time series analysis with spatial priors. *NeuroImage*, **24**, 350-362.

Polzehl, J. and Spokoiny, V.G. (2001). Functional and dynamic magnetic resonance imaging using vector adaptive weights smoothing. *Journal Royal Statist. Soc. C*, **50**, 485-501.

Purdon, P.L., Solo, V., Weisskoff, R.M. and Brown, E.N. (2001). Locally regularized spatiotemporal modeling and model comparison for functional MRI. *NeuroImage*, **14**, 912-923.

Raftery, A.E. and Lewis, S.M. (1992). How many iterations in the Gibbs sampler? In *Bayesian Statistics* **4**, eds. J. M. Bernardo, J. O. Berger, A. P. Dawid and A. F. M. Smith, 255-274. Oxford: Oxford University Press.

Solo, V., Purdon, P.L., Weisskoff, R.M. and Brown, E.N. (2001). A signal estimation approach to functional MRI. *IEEE Transactions on Medical Imaging*, **20**, 26-35.

Woolrich, M.W., Jenkinson, M., Brady, J.M. and Smith, S.M. (2004a). Fully Bayesian spatio-temporal modeling of fMRI data. *IEEE Transactions on Medical Imaging*, **23**, 213-231.

Woolrich, M., Behrens, T.E.J. and Smith, S.M. (2004b). Constrained linear basis set for HRF modelling using variational Bayes. *NeuroImage*, **21**, 1748-1761.

Woolrich, M.W., Behrens, T.E.J., Beckmann, C.F. and Smith, S. M. (2005). Mixture models with adaptive spatial regularization for segmentation with an application to fMRI data. *IEEE Transactions on Medical Imaging*, **24**, 1-11.

Worsley, K.J., Liao, C.H., Aston, J., Petre, V., Duncan, G.H., Morales, F. and Evans, A.C. (2002). A general statistical analysis for fMRI data. *NeuroImage*, **15**, 1-15.

Worsley, K. J. (2003). Detecting activation in fMRI data. *Statistical Methods in Medical Research*, **12**, 401-418.

Worsley, K.J. and Friston, K.J. (1995). Analysis of fMRI time series revisited—again. *NeuroImage*, **2**, 173-181.

Appendix A.1   Full Conditionals for the Steps of Gibbs Sampling for the Model for Spatial Dependence

- *Full conditional of each b_i*: Note that

$$p(b_i|\boldsymbol{b}_{-i}, \boldsymbol{c}, \boldsymbol{a}, \boldsymbol{\sigma}, \lambda, \boldsymbol{Y}) \propto \exp\left(-\frac{1}{2\sigma_i^2}\sum_{t=1}^{T}(y_{it} - a_i - z_t^{(1)}b_i - z_t^{(2)}c_i)^2 - \frac{\lambda}{2}\left(n_i b_i^2 - 2b_i\sum_{j\in\partial i} b_j\right)\right).$$

From this, it is possible to show that

$$p(b_i|\boldsymbol{b}_{-i}, \boldsymbol{c}, \boldsymbol{a}, \boldsymbol{\sigma}, \lambda, \boldsymbol{Y}) \sim N(\mu_1, \tau_1^2),$$

where $\quad \mu_1 = \dfrac{\dfrac{\sigma_i^2}{n_i}\sum_{j\in\partial i} b_j + \dfrac{1}{n_i\lambda}\sum_{t=1}^{T} z_t^{(1)}(y_{it} - a_i - z_t^{(2)}c_i)}{\sigma_i^2 + \dfrac{\sum_{t=1}^{T}(z_t^{(1)})^2}{n_i\lambda}}, \quad \tau_1^2 = \dfrac{\dfrac{\sigma_i^2}{n_i\lambda}}{\sigma_i^2 + \dfrac{\sum_{t=1}^{T}(z_t^{(1)})^2}{n_i\lambda}}.$

limit, the hypothetical infinite population of Fisher and the like, rest on a notion of probability which is frequentist in spirit. Spelling out rule (iv), he argues that our state of knowledge about scientific laws, including the relativity and quantum theories, provides strong reason to believe that there is no conclusive evidence to consider that our present laws are final. Since we may not be able to give a deductive proof or disproof for those theories, Jeffrey writes, "there is a valid primitive idea of expressing the degree of confidence [degree of belief] that we may have in a proposition." He goes on to state that *"the essence of the present theory* is that no probability... is simply frequency. The fundamental idea is that of a reasonable degree of belief, which satisfies certain rules of consistency and can in consequence of these rules be formally expressed by numbers..."

For Jeffreys, then, the notion of probability is an agent's degree of belief in a proposition/hypothesis where the agent's degree of belief must be coherent. For them, this coherence condition implies that the agent's degree of belief must satisfy the rules of the probability calculus along with the rule vi, which is known as the simplicity postulate. According to this postulate, "the simpler theories have the greater prior probabilities." Zeller has recently extended this simplicity postulate to various models of economics (Zellner, 2001). Given the JZA, the only plausible theory turns out to be the Bayesian theory of inference that makes central the agent's degree of belief in a parameter or a hypothesis satisfying some coherence conditions. So Royall is right in pointing out that the way Bayesianism has been practiced, given the JZA, one could only obtain the agent's degree of belief central to Bayesianism.

However, one key feature of Bayesianism needs to be mentioned.[2] A crucial aspect of statistics, possibly significantly more crucial than parameter estimation, is prediction. Bayesians are in a much better shape here than frequentists, because the former can incorporate the uncertainty due to estimation of parameters. Frequentists use only plug-in predictors which can't address the said problem.

3.   The Evidence Question

Now consider Lele's non-Bayesian approach to desiderata for a statistical theory of evidence. He is a non-Bayesian because he works within an evidentialist framework (that is, a generalized version of the likelihood framework) that eschews use of subjective prior probability. Lele begins with the law of likelihood and then defines a class of functions called "the evidence functions" to quantify the strength of evidence for one hypothesis over the other. He imposes some desiderata on this class of evidence functions, which could be regarded as epistemological conditions. Some of the conditions satisfied by the evidence function are

(i)   *The translation invariance condition*: If one translates an evidence function by adding a constant to it to change the strength of the evidence, then the evidence function should remain unaffected by that addition of a constant.

(ii)   *The scale invariance condition*: If one multiplies an evidence function by a constant to change the strength of evidence, then the evidence function should remain unaffected by that constant multiplier.

(iii)   *The reparameterization invariance*: The evidence function must be invariant under reparameterization. It means that if there is an evidence function, Ev1 and the latter is reparameterized to Ev2, then both Ev1 and Ev2 must provide the identical quantification of the strength of evidence.

---

[2]I owe this point to an anonymous referee.

- *Full conditional of each c_i:* Proceeding as above, we can show that

$$p(c_i | \boldsymbol{c}_{-i}, \boldsymbol{b}, \boldsymbol{a}, \boldsymbol{\sigma}, \lambda, \boldsymbol{Y}) \sim N(\mu_2, \tau_2^2),$$

where
$$\mu_2 = \frac{\dfrac{\sigma_i^2}{n_i} \sum_{j \in \partial i} c_j + \dfrac{1}{n_i \lambda} \sum_{t=1}^{T} z_t^{(2)}(y_{it} - a_i - z_t^{(1)} b_i)}{\sigma_i^2 + \dfrac{\sum_{t=1}^{T} (z_t^{(2)})^2}{n_i \lambda}}, \quad \tau_2^2 = \frac{\dfrac{\sigma_i^2}{n_i \lambda}}{\sigma_i^2 + \dfrac{\sum_{t=1}^{T} (z_t^{(2)})^2}{n_i \lambda}}.$$

- *Full conditional of each a_i:* It is possible to show that

$$p(a_i | \boldsymbol{a}_{-i}, \boldsymbol{b}, \boldsymbol{c}, \boldsymbol{\sigma}, \lambda, \boldsymbol{Y}) \sim N(\mu_3, \tau_3^2),$$

where
$$\mu_3 = \frac{\dfrac{\sigma_i^2}{n_i} \sum_{j \in \partial i} a_j + \dfrac{1}{n_i \lambda} \sum_{t=1}^{T} (y_{it} - z_t^{(1)} b_i - z_t^{(2)} c_i)}{\sigma_i^2 + \dfrac{T}{n_i \lambda}}, \quad \tau_3^2 = \frac{\dfrac{\sigma_i^2}{n_i \lambda}}{\sigma_i^2 + \dfrac{T}{n_i \lambda}}.$$

- *Full conditional of each σ_i^2:*

$$p(\sigma_i^2 | \boldsymbol{\sigma}_{-i}, \boldsymbol{a}, \boldsymbol{b}, \boldsymbol{c}, \lambda, \boldsymbol{Y}) \propto \left\{ \frac{1}{\sigma_i^T} \exp\left(-\frac{1}{2\sigma_i^2} \sum_{t=1}^{T} (y_{it} - a_i - z_t^{(1)} b_i - z_t^{(2)} c_i)^2 \right) \right\}$$
$$\times \ \exp\left(-\frac{1}{\sigma_i^2} \right) \times \left(\frac{1}{\sigma_i^2} \right)^2.$$

Therefore,

$$p(\sigma_i^2 | \boldsymbol{\sigma}_{-i}, \boldsymbol{a}, \boldsymbol{b}, \boldsymbol{c}, \lambda, \boldsymbol{Y}) \sim IG\left(1 + \frac{1}{2} \sum_{t=1}^{T} (y_{it} - a_i - z_t^{(1)} b_i - z_t^{(2)} c_i)^2, \frac{T+2}{2} \right).$$

- *Full conditional of λ:* It is given by

$$p(\lambda | \boldsymbol{a}, \boldsymbol{b}, \boldsymbol{c}, \boldsymbol{\sigma}, \boldsymbol{Y}) \sim \text{Gamma}\left(a^\star, b^\star \right),$$

where

$$a^\star = a + \frac{1}{2} \left\{ \sum_{i=1}^{I} n_i a_i^2 + \sum_{i=1}^{I} n_i b_i^2 + \sum_{i=1}^{I} n_i c_i^2 \right\} - \left(\sum_{i \sim j} a_i a_j + \sum_{i \sim j} b_i b_j + \sum_{i \sim j} c_i c_j \right),$$
$$b^\star = b + \frac{3}{2}(4I - 6\sqrt{I} + 2).$$

We need to mention here that for reducing computing time, at each step of iteration we have drawn sample from the priors of the σ_i^2's and also λ. However, for the other parameters, the samples are drawn from the full posteriors.

Appendix A.2   Full Conditionals for the Steps of Gibbs Sampling for the Model for Temporal Dependence

- *Full conditional of each b_i:*

$$
p(b_i|b_{j\neq i}, c_i, a_i, i=1,\ldots,I, \sigma, \lambda, Y) \propto \left\{ \frac{1}{\sigma_i^T} \exp\left(-\frac{1}{2\sigma_i^2} \sum_{t=1}^{T} (y_{it} - a_{it} - z_t^{(1)} b_{it} - z_t^{(2)} c_{it})^2 \right) \right\}
$$
$$
\times \exp\left(-\frac{1}{2} \lambda b_i' Q b_i \right).
$$

Let us write

$$
\begin{aligned}
y_i &= (y_{i1},\ldots,y_{iT})', \\
\mu_i &= (a_{i1} + z_1^{(1)} b_{i1} + z_1^{(2)} c_{i1}, \ldots, a_{iT} + z_T^{(1)} b_{iT} + z_T^{(2)} c_{iT})', \\
Z^{(1)} &= \mathrm{diag}(z_1^{(1)},\ldots,z_T^{(1)}), \\
Z^{(2)} &= \mathrm{diag}(z_1^{(2)},\ldots,z_T^{(2)}).
\end{aligned}
$$

It follows that

$$
\mu_i = a_i + Z^{(1)} b_i + Z^{(2)} c_i.
$$

Now note that

$$
p(b_i|b_{j\neq i}, c_i, a_i, i=1,\ldots,I, \sigma, \lambda, Y) \propto \exp\left(-\frac{1}{2\sigma_i^2}(y_i - \mu_i)'(y_i - \mu_i) \right) \exp\left(-\frac{1}{2}\lambda b_i' Q b_i \right).
$$

Observe that

$$
\begin{aligned}
&\frac{1}{2\sigma_i^2}(y_i - \mu_i)'(y_i - \mu_i) + \frac{1}{2}\lambda b_i' Q b_i \\
&= \frac{1}{2\sigma_i^2}[Z^{(1)} b_i - (y_i - a_i - Z^{(2)} c_i)]'[Z^{(1)} b_i - (y_i - a_i - Z^{(2)} c_i)] + \frac{1}{2}\lambda b_i' Q b_i \\
&= b_i'\left[\frac{1}{2\sigma_i^2} Z^{(1)\prime} Z^{(1)} + \frac{\lambda}{2} Q \right] b_i - \frac{(y_i - a_i - Z^{(2)} c_i)' Z^{(1)} b_i}{\sigma_i^2} + \frac{(y_i - a_i - Z^{(2)} c_i)'(y_i - a_i - Z^{(2)} c_i)}{\sigma_i^2}
\end{aligned}
$$

All these show that the conditional distribution to be found is a T-variate normal distribution with dispersion matrix and mean vector given by

$$
\left[\frac{1}{\sigma_i^2} Z^{(1)\prime} Z^{(1)} + \lambda Q \right]^{-1} \quad \text{and} \quad \left[\frac{1}{\sigma_i^2} Z^{(1)\prime} Z^{(1)} + \lambda Q \right]^{-1} \frac{Z^{(1)}(y_i - a_i - Z^{(2)} c_i)}{\sigma_i^2}.
$$

- *Full conditional of each c_i:* It is given by a T-variate normal distribution with dispersion matrix and mean vector given by

$$
\left[\frac{1}{\sigma_i^2} Z^{(2)\prime} Z^{(2)} + \lambda Q \right]^{-1} \quad \text{and} \quad \left[\frac{1}{\sigma_i^2} Z^{(2)\prime} Z^{(2)} + \lambda Q \right]^{-1} \frac{Z^{(2)}(y_i - a_i - Z^{(1)} b_i)}{\sigma_i^2}.
$$

- *Full conditional of each a_i:* It is given by a T-variate normal distribution with dispersion matrix and mean vector given by

$$
\left[\frac{1}{\sigma_i^2} I_T + \lambda Q \right]^{-1} \quad \text{and} \quad \left[\frac{1}{\sigma_i^2} I_T + \lambda Q \right]^{-1} \frac{(y_i - Z^{(1)} b_i - Z^{(2)} c_i)}{\sigma_i^2}.
$$

- *Full conditional of each σ_i^2:*

$$p(\sigma_i^2|\boldsymbol{\sigma}_{-i}, \boldsymbol{a}_i, \boldsymbol{b}_i, \boldsymbol{c}_i, \, i = 1, \ldots, I, \, \lambda, \boldsymbol{Y})$$

$$\propto \left\{ \frac{1}{\sigma_i^T} \exp\left(-\frac{1}{2\sigma_i^2} \sum_{t=1}^{T} (y_{it} - a_{it} - z_t^{(1)} b_{it} - z_t^{(2)} c_{it})^2 \right) \right\} \times \exp\left(-\frac{1}{\sigma_i^2} \right) \times \left(\frac{1}{\sigma_i^2} \right)^2.$$

Therefore,

$$p(\sigma_i^2|\boldsymbol{\sigma}_{-i}, \boldsymbol{a}_i, \boldsymbol{b}_i, \boldsymbol{c}_i, \, i = 1, \ldots, I, \, \lambda, \boldsymbol{Y}) \sim IG\left(1 + \frac{1}{2} \sum_{t=1}^{T} (y_{it} - a_{it} - z_t^{(1)} b_{it} - z_t^{(2)} c_{it})^2, \frac{T+2}{2} \right).$$

- *Full conditional of λ:*

$$p(\lambda|\boldsymbol{a}_i, \boldsymbol{b}_i, \boldsymbol{c}_i, \, i = 1, \ldots, I, \, \boldsymbol{\sigma}, \boldsymbol{Y})$$

$$\propto \prod_{i=1}^{I} \exp\left(-\frac{1}{2}\lambda \boldsymbol{a}_i' \boldsymbol{Q} \boldsymbol{a}_i - \frac{1}{2}\lambda \boldsymbol{b}_i' \boldsymbol{Q} \boldsymbol{b}_i - \frac{1}{2}\lambda \boldsymbol{c}_i' \boldsymbol{Q} \boldsymbol{c}_i \right) \cdot \exp(-\lambda) \cdot \lambda^{3T/2}.$$

Therefore, the required conditional distribution is given by

$$p(\lambda|\boldsymbol{a}_i, \boldsymbol{b}_i, \boldsymbol{c}_i, \, i = 1, \ldots, I, \, \boldsymbol{\sigma}, \boldsymbol{Y}) \sim \text{Gamma}\left(a^\star, b^\star \right),$$

where $\quad a^\star = 1 + \dfrac{\sum_{i=1}^{I}(\boldsymbol{a}_i'\boldsymbol{Q}\boldsymbol{a}_i + \boldsymbol{b}_i'\boldsymbol{Q}\boldsymbol{b}_i + \boldsymbol{c}_i'\boldsymbol{Q}\boldsymbol{c}_i)}{2}, \quad b^\star = \dfrac{3T+2}{2}.$

We need to mention here that for reducing computing time, at each step of iteration we have drawn sample from the priors of the σ_i^2's and also λ. However, for the other parameters, the samples are drawn from the full posteriors.

Appendix A.3   Details on Contrast Stretching

Denote $1\text{-}10^{-3}$ by T (threshold level). On the colorbar in MATLAB, we can assign values from 0 to 1, 0 representing black, 1 representing white. Note that each voxel value is a probability and hence is a number in $[0, 1]$. However, we want to enhance those voxels having values (i.e., probabilities) greater than T. This is achieved by employing the following piecewise linear transformation $f: [0, 1] \longrightarrow [0, 1]$, characterized by $f(0) = 0, f(T) = 0.8, f(1) = 1$, on each voxel value (i.e., probability) x. It is given by

$$f(x) = \frac{0.8 \times x}{T} \qquad \text{if } 0 \leq x \leq T,$$

$$= 0.8 + \frac{x - T}{1 - T} \times (1 - 0.8) \text{ if } T \leq x \leq 1.$$

Bayesian Statistics and Its Applications
Edited by S.K. Upadhyay, U. Singh and D.K. Dey
Anamaya Publishers, New Delhi, India

Intrinsic Point Estimation
of the Normal Variance

José M. Bernardo

Departmento de Estadística e I.O., Universitat de València,
Facultad de Matemáticas, 46100-Burjassot, València, Spain

Abstract

Point estimation of the normal variance is surely one of the oldest non-trivial problems in mathematical statistics and yet, there is certainly no consensus about its more appropriate solution. Formally, point estimation may be seen as a decision problem where the action space is the set of possible values of the quantity on interest; foundations then dictate that the solution must depend on both the utility function and the prior distribution. An estimator intended for general use should surely be invariant under one-to-one transformations, and this requires the use of an invariant loss function; moreover, an objective solution requires the use of a prior which does not introduce subjective elements. The combined use of an invariant information-theory based loss function, the *intrinsic discrepancy*, and an objective prior function, the *reference prior*, produces a general Bayesian objective solution to the problem of point estimation. In this paper, point estimation of the normal variance is considered in detail, and the behaviour of the solution found is compared with the behaviour of alternative conventional solutions from both a Bayesian and a frequentist perspective.

---

1. Introduction

Point estimation of the normal variance has a long fascinating history which is far from settled. As mentioned by Maata and Casella (1990) in their lucid discussion of the frequentist decision-theoretic approach to this problem, the list of contributors to the twin problems of point estimation of the normal mean and point estimation of the normal variance reads like a *Who's Who* in modern 20th century statistics.

In this paper, an objective Bayesian decision-theoretic solution to point estimation of the normal variance when the mean is unknown is presented. In marked contrast with most approaches, this solution is invariant under one-to-one reparametrization. The behaviour of this new solution is compared to the behaviour of known alternatives from both a Bayesian and a frequentist viewpoint.

1.1 Notation

A brief review of notation is needed to proceed. Probability distributions are described through their probability density functions, and no notational distinction is made between a random quantity and the particular values that it may take. Bold italic roman fonts are used for observable random vectors (typically data) and bold italic greek fonts for unobservable random vectors (typically parameters); lower case is used for variables and upper case calligraphic for their dominion sets. The standard mathematical convention of referring to functions, say $f_{\boldsymbol{x}}(\cdot)$ and $g_{\boldsymbol{x}}(\cdot)$ of $\boldsymbol{x} \in \mathcal{X}$, respectively, by $f(\boldsymbol{x})$ and $g(\boldsymbol{x})$ will often be used. Thus, the conditional probability density of observable data $\boldsymbol{x} \in \mathcal{X}$ given $\boldsymbol{\theta}$ will be represented by either $p_{\boldsymbol{x}}(\cdot \,|\, \boldsymbol{\theta})$ or $p(\boldsymbol{x} \,|\, \boldsymbol{\theta})$, with $p(\boldsymbol{x} \,|\, \boldsymbol{\theta}) \geq 0$, $\boldsymbol{x} \in \mathcal{X}$ and $\int_{\mathcal{X}} p(\boldsymbol{x} \,|\, \boldsymbol{\theta}) \, d\boldsymbol{x} = 1$, and the posterior density of a non-observable parameter vector $\boldsymbol{\theta} \in \Theta$ given $\boldsymbol{x}$ will be represented by either $\pi_{\boldsymbol{\theta}}(\cdot \,|\, \boldsymbol{x})$ or $\pi(\boldsymbol{\theta} \,|\, \boldsymbol{x})$, with $\pi(\boldsymbol{\theta} \,|\, \boldsymbol{x}) \geq 0$, $\boldsymbol{\theta} \in \Theta$ and $\int_{\Theta} \pi(\boldsymbol{\theta} \,|\, \boldsymbol{x}) \, d\boldsymbol{\theta} = 1$. Density functions of specific distributions are denoted by appropriate names. In particular, if x has a normal distribution with mean μ and variance σ^2, its probability density function will be denoted by $\mathrm{N}(x \,|\, \mu, \sigma)$. The maximum likelihood estimators of μ and σ^2, given a random sample $\boldsymbol{x} = \{x_1, \dots, x_n\}$ from $\mathrm{N}(x \,|\, \mu, \sigma)$ will, respectively, be denoted by

$$\hat{\mu} = \bar{x} = \frac{1}{n} \sum\nolimits_{j=1}^{n} x_j, \qquad \hat{\sigma}^2 = s^2 = \frac{1}{n} \sum\nolimits_{j=1}^{n} (x_j - \bar{x})^2.$$

The more common point estimators of the normal variance are members of the family of *affine invariant* estimators

$$\tilde{\sigma}_{\nu}^2 = \frac{n s^2}{\nu} = \frac{1}{\nu} \sum\nolimits_{j=1}^{n} (x_j - \bar{x})^2, \quad \nu > 0. \tag{1}$$

In particular, the maximum likelihood estimator (MLE) is $s^2 = \tilde{\sigma}_n^2$, and the commonly used unbiased estimator is $\tilde{\sigma}_{n-1}^2$.

1.2   Conventional Point Estimation of the Normal Variance

The basic facts on frequentist point estimation of the normal variance are well known (see, e.g., Pal, Ling and Lin, 1998). The MLE of σ^2 is s^2 and, since maximum likelihood estimation is invariant under one-to-one transformations, the MLE of σ is s. The uniformly minimum variance unbiased estimator (UMVUE) of σ^2 is $\tilde{\sigma}_{n-1}^2 = n s^2/(n-1)$, but the UMVUE of σ is not its squared root, but $s \sqrt{(n/2)}\, \Gamma((n-1)/2)/\Gamma(n/2)$ (see e.g., Lehmann and Casella, 1998, p. 91); for $n = 2$, these respectively yield $\sqrt{2}\, s$ and $\sqrt{\pi}\, s$, with $s = |x_1 - x_2|/2$, a 25% difference using precisely the same procedure; this is a good example of methodological inconsistency.

Despite many warnings on its inappropriate behaviour ("I find it hard to take the problem of estimating σ^2 with quadratic loss very seriously" Stein, 1964; see also Brown, 1968, 1990), decision theoretical approaches to the normal variance estimation are typically based on the standardized quadratic loss function

$$\ell_{\mathrm{squad}}\{\tilde{\sigma}^2, \sigma^2\} = [(\tilde{\sigma}^2/\sigma^2) - 1]^2 \tag{2}$$

where overestimation of σ^2 is much more severely penalized than underestimation, thus leading to presumably too small estimates. Indeed, the minimum risk equivariant estimator (MRE) of σ^2 under this loss, which is also minimax, is $\tilde{\sigma}_{n+1}^2$ (see e.g., Lehmann and Casella, 1998, p. 172), smaller than both the MLE and the unbiased estimator, and it is often considered the "straw man" to beat in this problem (George, 1990). By considering a larger class of estimators than (1) which are also scale invariant, namely those of the form $\phi_n(z)\, n s^2$, where ϕ_n is a real valued function and $z = \bar{x}/s$, Stein (1964) found that

$$\tilde{\sigma}_{\mathrm{stein}}^2 = \min\left\{ \tilde{\sigma}_{n+1}^2, \ \tilde{\sigma}_{(n+2)/(1+z^2)}^2 \right\} \tag{3}$$

dominates $\tilde{\sigma}_{n+1}^2$ under the standardized quadratic loss (2) and, thus, $\tilde{\sigma}_{n+1}^2$ is inadmissible under that loss. The intuition behind this is that small z values indicate that μ may be close to 0, and then $\tilde{\sigma}_{n+2}^2$, which would the best affine invariant estimator under the quadratic loss (2) if μ were known to be zero, might be better than $\tilde{\sigma}_{n+1}^2$. This prompted a whole class of so-called *preliminary test* estimators where the estimator takes one of typically two different forms, depending of the value of z (Brown, 1968; Brewster and Zidek, 1974). For a review of their performance, see Csörgö and Faraway (1996). Notice however that $\tilde{\sigma}^2_{\text{Stein}}$ must also be inadmissible, for admissible estimators are limits of Bayes estimators, and so must be analytic.

The results mentioned above are obtained under the mathematically convenient—but otherwise rather unsatisfactory—standardized quadratic loss. Stein (1964) suggested the use of far more appropriate *entropy loss*

$$\ell_{\text{entropy}}\{\tilde{\sigma}^2, \sigma^2\} = \int_{\Re} \mathrm{N}(x \mid \mu, \sigma) \log \frac{\mathrm{N}(x \mid \mu, \sigma)}{\mathrm{N}(x \mid \mu, \tilde{\sigma})} \, \mathrm{d}x = \frac{1}{2}\left[\frac{\tilde{\sigma}^2}{\sigma^2} - 1 - \log \frac{\tilde{\sigma}^2}{\sigma^2} \right] \tag{4}$$

and showed that the best invariant estimator for this loss is $\tilde{\sigma}^2_{n-1}$, the unbiased estimator. Brown (1968) proved that, from a frequentist decision theoretic viewpoint, this may be also improved by appropriate preliminary test estimators. In particular,

$$\tilde{\sigma}^2_{\text{brown}} = \min\left\{ \tilde{\sigma}^2_{n-1}, \ \tilde{\sigma}^2_{n/(1+z^2)} \right\} \tag{5}$$

dominates $\tilde{\sigma}^2_{n-1}$ under the entropy loss (4), and thus $\tilde{\sigma}^2_{n-1}$ is inadmissible under that loss. On the other hand, Lin and Pal (2005) recently found that, from a frequentist viewpoint, $\tilde{\sigma}^2_{n-1}$ may be proposed as a good compromise estimator, for it performs moderately well under several alternative criteria (risk, Pitman nearness and stochastic domination), while the performance of the other estimators considered dramatically depends on the criterion used.

It may certainly be argued that the frequentist emphasis of the concept described by the emotionally charged word "inadmissible" may well be inappropriate. Indeed, for some assumed loss function, a particular estimator is declared *inadmissible* (do not dare to use it!) only because its *average* loss under repeated sampling is larger than that of another estimator; whether or not the "inadmissible" estimator is actually better for most regions of the sampling space (which are often identifiable) is simply ignored. Yet, one would certainly expect that decent people would prefer a country where there is no poverty to one with a larger *average* income induced a small group of very rich people which over-compensates the small income of the poor people, thus producing a larger average. In more technical terms, one should certainly analyse the sampling properties of any statistical procedure, and avoid procedures which for some possible parameter values would give misleading conclusions most of the time, which is the weak repeated sampling principle of Cox and Hinkley (1974, p. 45); however, insisting on the average result over all possible samples as the overwhelming criterion is *not* to be recommended. The crucial properties of a statistical procedure are those *conditional* on the observed data, not those obtained by *averaging* over them: a sensible procedure should produce appropriate answers whatever the data obtained, with special attention to the parameter values supported by the observed data. But, of course, this type of analysis requires a Bayesian approach.

Conventional Bayesian point estimation typically consists of some location measure of the marginal posterior distribution of the quantity of interest. The solution naturally depends on the prior used. Objective Bayesian estimators, which do not involve any information about the parameters beyond that contained in the assumed model—and may, therefore, be meaningfully compared with their frequentist counterparts—require an objective prior, that is a positive prior function to be formally used in Bayes theorem, which only depends on the assumed model and on the quantity of interest. In the particular problem where data $x = \{x_1, \ldots, x_n\}$ consists of a random sample from a normal $\mathrm{N}(x \mid \mu, \sigma)$ distribution and the quantity of interest is either σ, or any one-to-one function of σ (say the variance σ^2, the precision $1/\sigma^2$, or the approximate location parameter $\log\sigma$), there is a clear consensus on the objective prior function to use, namely

$$\pi(\mu,\sigma) = \sigma^{-1} \tag{6}$$

a uniform prior on both μ and $\log\sigma$. Indeed, this was already suggested by Barnard (1952) on invariance arguments, and recommended by Jeffreys (1939, p. 138), Lindley (1965, p. 37) and Box and Tiao (1973, p. 49) in their pioneering books. As one would expect, this is also the relevant reference prior (Bernardo, 1979).

The corresponding marginal posterior of the standard deviation, which is the reference posterior distribution of σ, is

$$\pi(\sigma\,|\,\boldsymbol{x}) \propto \int_{-\infty}^{\infty} \prod_{j=1}^{n} \mathrm{N}(x_j\,|\,\mu,\sigma)\,\pi(\mu,\sigma)\,\mathrm{d}\mu \propto \sigma^{-n} e^{-\frac{n}{2}\frac{s^2}{\sigma^2}}. \tag{7}$$

Hence, the reference posterior of the variance σ^2 is

$$\pi(\sigma^2\,|\,\boldsymbol{x}) \propto (\sigma^2)^{-(n+1)/2} e^{-\frac{n}{2}\frac{s^2}{\sigma^2}} \tag{8}$$

an inverted gama $\mathrm{Ig}(\sigma^2\,|\,(n-1)/2, ns^2/2)$, and the reference posterior of $\tau = ns^2/\sigma^2$ is $\pi(\tau\,|\,n) = \chi_{n-1}^2$, a central chi-square with $n-1$ degrees of freedom. Naïve Bayesian estimators of σ and σ^2 are given by the corresponding posterior means

$$\mathrm{E}[\sigma\,|\,\boldsymbol{x}] = \sqrt{\frac{n}{2}}\frac{\Gamma[(n-2)/2]}{\Gamma[(n-1)/2]}\,s\,, \qquad \mathrm{E}[\sigma^2\,|\,\boldsymbol{x}] = \frac{n\,s^2}{n-3} = \tilde{\sigma}_{n-3}^2 \tag{9}$$

and posterior modes,

$$\mathrm{Mo}[\sigma\,|\,\boldsymbol{x}] = s\,, \qquad \mathrm{Mo}[\sigma^2\,|\,\boldsymbol{x}] = \frac{n\,s^2}{n+1} = \tilde{\sigma}_{n+1}^2. \tag{10}$$

Notice that these estimation procedures are *not* invariant under reparametrization. It may be appreciated that there are many direct relations between frequentist and naïve objective Bayesian estimators. For instance, the mode of the reference posterior of σ is s, its MLE, and the mode of the reference posterior of σ^2 is $\tilde{\sigma}_{n+1}^2$, its MRE.

A more formal Bayesian approach to point estimation is to consider point estimation as a decision problem where the action space is the set of possible values of the quantity of interest. This requires to specify a loss function. Naïve loss functions reproduce naïve estimators. Thus, the optimal Bayes estimator under quadratic loss is the posterior expectation, leading to the results in (9), and the optimal Bayes estimator under a zero-one loss is the posterior mode, leading to the results in (10). Bayesian decision-theoretic point estimation is considered in detail in Section 2.1.

In this paper, a particular objective Bayesian solution to point estimation of any one-to-one function of the normal variance is presented. Section 2 is a review of the methodology used, *intrinsic estimation* (Bernardo, 1999; Bernardo and Rueda, 2002; Bernardo and Juárez, 2003; Bernardo, 2005), which is an objective Bayesian decision-theoretic approach using an information-theory based loss function and a reference prior. Section 3 contains the derivation on the intrinsic point estimator of the normal variance with unknown mean and, by the invariance of the arguments used, that of any one-to-one transformation of the variance. In Section 4, the results obtained are compared with other solutions in the literature, from both a Bayesian and a frequentist viewpoint.

2. Intrinsic Estimation

2.1 Point Estimation as a Decision Problem

Let $\boldsymbol{x}$ be the available data, which are assumed to consist of one observation from model $\mathcal{M} \equiv \{p(\boldsymbol{x} \mid \omega), \boldsymbol{x} \in \boldsymbol{\mathcal{X}}, \omega \in \Omega\}$, and let $\boldsymbol{\theta} = \boldsymbol{\theta}(\omega) \in \Theta$ be the vector of interest. Often, but not necessarily, data $\boldsymbol{x}$ consist of a random sample $\boldsymbol{x} = \{x_1, \dots, x_n\}$ of some simpler model $\{q(\boldsymbol{x} \mid \omega), \boldsymbol{x} \in \boldsymbol{\mathcal{X}}, \omega \in \Omega\}$, in which case $p(\boldsymbol{x} \mid \omega) = \prod_{j=1}^{n} q(x_j \mid \omega)$. Without loss of generality, the original model $\mathcal{M}$ may be written as $\mathcal{M} \equiv \{p(\boldsymbol{x} \mid \boldsymbol{\theta}, \boldsymbol{\lambda}), \boldsymbol{x} \in \boldsymbol{\mathcal{X}}, \boldsymbol{\theta} \in \Theta, \boldsymbol{\lambda} \in \Lambda\}$, in terms of the vector of interest $\boldsymbol{\theta}$ and a vector $\boldsymbol{\lambda}$ of nuisance parameters. A *point estimator* of $\boldsymbol{\theta}$ is some function of the data $\tilde{\boldsymbol{\theta}}(\boldsymbol{x}) \in \Theta$ such that, for each possible set of observed data $\boldsymbol{x}$, $\tilde{\boldsymbol{\theta}}(\boldsymbol{x})$ could be regarded as an appropriate proxy for the actual, unknown value of $\boldsymbol{\theta}$.

For each given data set $\boldsymbol{x}$, to choose a point estimate $\tilde{\boldsymbol{\theta}}$ is a *decision problem*, where the action space is the class Θ of possible $\boldsymbol{\theta}$ values. Foundations dictate (see, e.g., Bernardo and Smith, 1994, Ch. 2 and references therein) that to solve this decision problem it is necessary to specify a *loss function* $\ell\{\tilde{\boldsymbol{\theta}}, (\boldsymbol{\theta}, \boldsymbol{\lambda})\}$ measuring the consequences of acting *as if* the true value of the quantity of interest were $\tilde{\boldsymbol{\theta}}$, when the actual parameter values are $(\boldsymbol{\theta}, \boldsymbol{\lambda})$. Given data $\boldsymbol{x}$, the loss to be expected if $\tilde{\boldsymbol{\theta}}$ were used as the true value of the quantity of interest is

$$l\{\tilde{\boldsymbol{\theta}} \mid \boldsymbol{x}\} = \int_{\Theta} \int_{\Lambda} \ell\{\tilde{\boldsymbol{\theta}}, (\boldsymbol{\theta}, \boldsymbol{\lambda})\} \, \pi(\boldsymbol{\theta}, \boldsymbol{\lambda} \mid \boldsymbol{x}) \, \mathrm{d}\boldsymbol{\theta} \, \mathrm{d}\boldsymbol{\lambda}$$

where $\pi(\boldsymbol{\theta}, \boldsymbol{\lambda} \mid \boldsymbol{x}) \propto p(\boldsymbol{x} \mid \boldsymbol{\theta}, \boldsymbol{\lambda}) \, \pi(\boldsymbol{\theta}, \boldsymbol{\lambda})$ is the joint posterior density of $\boldsymbol{\theta}$ and $\boldsymbol{\lambda}$, and $\pi(\boldsymbol{\theta}, \boldsymbol{\lambda})$ is the joint prior of the unknown parameters. Given data $\boldsymbol{x}$, the *Bayes estimate* is that $\tilde{\boldsymbol{\theta}}$ value which minimizes in Θ the (posterior) expected loss $l\{\tilde{\boldsymbol{\theta}} \mid \boldsymbol{x}\}$. The *Bayes estimator* is the function

$$\boldsymbol{\theta}^*(\boldsymbol{x}) = \arg \min_{\tilde{\boldsymbol{\theta}} \in \Theta} l\{\tilde{\boldsymbol{\theta}} \mid \boldsymbol{x}\}. \tag{11}$$

For any given model, the Bayes estimator depends on both the loss function $\ell\{\tilde{\boldsymbol{\theta}}, (\boldsymbol{\theta}, \boldsymbol{\lambda})\}$ and the prior distribution $\pi(\boldsymbol{\theta}, \boldsymbol{\lambda})$. In the case of the normal variance considered in this paper, data $\boldsymbol{x} = \{x_1, \dots, x_n\}$ are assumed to be a random sample from $\mathrm{N}(x \mid \mu, \sigma)$, the parameter of interest is either σ or some one-to-one function of σ, and μ is a nuisance parameter. It has already been mentioned that the undisputed objective prior for this problem is $\pi(\mu, \sigma) = \sigma^{-1}$, which leads to the reference posterior distributions (7) and (8).

2.2 The Loss Function

The loss function is context specific, and should be chosen in terms of the anticipated uses of the estimate; however, a number of conventional loss functions have been suggested for those situations where no particular uses are envisaged. The more common of these conventional loss functions (which often ignore the possible presence of nuisance parameters) is the ubiquitous *quadratic loss*, $\ell\{\tilde{\boldsymbol{\theta}}, (\boldsymbol{\theta}, \boldsymbol{\lambda})\} = (\tilde{\boldsymbol{\theta}} - \boldsymbol{\theta})^t (\tilde{\boldsymbol{\theta}} - \boldsymbol{\theta})$; the corresponding Bayes estimator is then the (marginal) *posterior mean* $\boldsymbol{\theta}^* = E[\boldsymbol{\theta} \mid \boldsymbol{x}]$, assuming that the mean exists (see, e.g., Bernardo and Smith, 1994, p. 257).

In the case of normal variance, the conventional quadratic loss $\ell(\tilde{\sigma}^2, \sigma^2) = c\,(\tilde{\sigma}^2 - \sigma^2)^2$ leads to the reference posterior expectation $E[\sigma^2 \mid \boldsymbol{x}] = \tilde{\sigma}^2_{n-3}$ quoted in (9). If the slightly more sophisticated standardized quadratic loss (2) is used, the Bayes estimator is

$$\arg \min_{\tilde{\sigma}^2 > 0} \int_0^\infty [(\tilde{\sigma}^2/\sigma^2) - 1]^2 \, \pi(\sigma^2 \mid \boldsymbol{x}) \, \mathrm{d}\sigma^2 = \frac{n\,s^2}{n+1} = \tilde{\sigma}^2_{n+1} \tag{12}$$

which is also the MRE of σ^2 under this loss. Using the entropy loss (4) yields

$$\arg\min_{\tilde{\sigma}^2>0} \int_0^\infty [(\tilde{\sigma}^2/\sigma^2) - 1 - \log(\tilde{\sigma}^2/\sigma^2)]\,\pi(\sigma^2\,|\,\boldsymbol{x})\,\mathrm{d}\sigma^2 = \frac{n\,s^2}{n-1} = \tilde{\sigma}^2_{n-1} \tag{13}$$

which is also both the MRE of σ^2 under this loss, and the unbiased estimator.

Conventional loss functions are typically *not* invariant under reparametrization. As a consequence, the Bayes estimator $\boldsymbol{\psi}^*$ of a one-to-one transformation $\boldsymbol{\psi} = \boldsymbol{\psi}(\boldsymbol{\theta})$ of the original parameter $\boldsymbol{\theta}$ is not necessarily $\boldsymbol{\psi}(\boldsymbol{\theta}^*)$. Yet, scientific applications require this type of invariance; indeed, it would be hard to argue that the best estimate of, say a galaxy speed, is θ^* but that the best estimate of the logarithm of that speed is *not* $\log(\theta^*)$. Invariant loss functions are required to guarantee invariant estimators.

With no nuisance parameters, *intrinsic loss functions* (Robert, 1996), of the general form $\ell(\tilde{\boldsymbol{\theta}}, \boldsymbol{\theta}) = \ell\{p_{\boldsymbol{x}}(.\,|\,\tilde{\boldsymbol{\theta}}), p_{\boldsymbol{x}}(.\,|\,\boldsymbol{\theta})\}$, shift attention from the discrepancy between the estimate $\tilde{\boldsymbol{\theta}}$ and the true value $\boldsymbol{\theta}$, to the more relevant discrepancy between the statistical *models* they label, and they are always invariant under one-to-one reparametrization. The *intrinsic discrepancy* (Bernardo and Rueda, 2002) is an intrinsic loss with very attractive properties. The intrinsic discrepancy between two models $p_{\boldsymbol{x}}(\cdot\,|\,\boldsymbol{\theta}_1)$ and $p_{\boldsymbol{x}}(\cdot\,|\,\boldsymbol{\theta}_2)$ for data $\boldsymbol{x} \in \boldsymbol{\mathcal{X}}$ is

$$\delta_{\boldsymbol{x}}(\boldsymbol{\theta}_1, \boldsymbol{\theta}_2) = \delta\{p_{\boldsymbol{x}}(.\,|\,\boldsymbol{\theta}_1), p_{\boldsymbol{x}}(.\,|\,\boldsymbol{\theta}_2)\} = \min\{\kappa(\boldsymbol{\theta}_1\,|\,\boldsymbol{\theta}_2), \kappa(\boldsymbol{\theta}_2\,|\,\boldsymbol{\theta}_1)\}$$

$$\kappa(\boldsymbol{\theta}_i\,|\,\boldsymbol{\theta}_j) = \int_{\boldsymbol{\mathcal{X}}} p_{\boldsymbol{x}}(\boldsymbol{x}\,|\,\boldsymbol{\theta}_j)\log\frac{p_{\boldsymbol{x}}(\boldsymbol{x}\,|\,\boldsymbol{\theta}_j)}{p_{\boldsymbol{x}}(\boldsymbol{x}\,|\,\boldsymbol{\theta}_i)}\,\mathrm{d}\boldsymbol{x} \tag{14}$$

that is, the minimum Kullback-Leibler logarithmic divergence between them. This is a proper discrepancy measure; indeed, (i) it is symmetric, (ii) it is non-negative and (iii) it is zero if, and only if, $p(\boldsymbol{x}\,|\,\boldsymbol{\theta}_1) = p(\boldsymbol{x}\,|\,\boldsymbol{\theta}_2)$ almost everywhere. The intrinsic discrepancy is invariant under one-to-one transformations of either the parameter vector $\boldsymbol{\theta}$ or the data set $\boldsymbol{x}$. Moreover, the intrinsic discrepancy is well defined in irregular models, where the sample space may depend on the parameter value and, therefore, the support of $p_{\boldsymbol{x}}(.\,|\,\boldsymbol{\theta}_i)$ may be strictly smaller than the support of $p_{\boldsymbol{x}}(.\,|\,\boldsymbol{\theta}_j)$.

The intrinsic discrepancy is *additive* with respect to conditionally independent observations. Thus, if the available data consist of a random sample $\boldsymbol{x} = \{x_1, \ldots, x_n\} \in \boldsymbol{\mathcal{X}} = \mathcal{X}^n$ from some distribution $q_x(\cdot\,|\,\boldsymbol{\theta}, \boldsymbol{\lambda})$, $x \in \mathcal{X}$, then the intrinsic discrepancy $\delta_{\boldsymbol{x}}(\boldsymbol{\theta}_1, \boldsymbol{\theta}_2)$ between the joint distributions $p_{\boldsymbol{x}}(\cdot\,|\,\boldsymbol{\theta}_1)$ and $p_{\boldsymbol{x}}(\cdot\,|\,\boldsymbol{\theta}_2)$ is simply n times the intrinsic discrepancy $\delta_{\boldsymbol{x}}(\boldsymbol{\theta}_1, \boldsymbol{\theta}_2)$ between $q_x(\cdot\,|\,\boldsymbol{\theta}_1)$ and $q_x(\cdot\,|\,\boldsymbol{\theta}_2)$. Formally,

$$\boldsymbol{x} = \{x_1, \ldots, x_n\}\ \text{i.i.d.} \quad \rightarrow \quad \delta_{\boldsymbol{x}}(\boldsymbol{\theta}_1, \boldsymbol{\theta}_2) = n\,\delta_{\boldsymbol{x}}(\boldsymbol{\theta}_1, \boldsymbol{\theta}_2). \tag{15}$$

The intrinsic discrepancy between two families of distributions is defined as the minimum intrinsic discrepancy between their members. In the context of point estimation, this leads to the (invariant) intrinsic loss function

$$\delta_{\boldsymbol{x}}\{\tilde{\boldsymbol{\theta}}, (\boldsymbol{\theta}, \boldsymbol{\lambda})\} = \min_{\tilde{\boldsymbol{\lambda}}\in\Lambda} \delta_{\boldsymbol{x}}\{(\tilde{\boldsymbol{\theta}}, \tilde{\boldsymbol{\lambda}}), (\boldsymbol{\theta}, \boldsymbol{\lambda})\} \tag{16}$$

which measures the discrepancy between the assumed model $p_{\boldsymbol{x}}(\cdot\,|\,\boldsymbol{\theta}, \boldsymbol{\lambda})$ and its closest approximation within the family $\{p_{\boldsymbol{x}}(\cdot\,|\,\tilde{\boldsymbol{\theta}}, \tilde{\boldsymbol{\lambda}}), \tilde{\boldsymbol{\lambda}} \in \Lambda\}$ of all models with $\boldsymbol{\theta} = \tilde{\boldsymbol{\theta}}$. Notice that the value of $\delta\{\tilde{\boldsymbol{\theta}}, (\boldsymbol{\theta}, \boldsymbol{\lambda})\}$ does *not* depend on the particular parametrization chosen to describe the problem.

Given data x generated by $p(x \mid \theta, \lambda)$ and no subjective prior information, the reference posterior expected intrinsic loss from using $\tilde{\theta}$ as a proxy for θ is the *intrinsic statistic function*

$$d\{\tilde{\theta} \mid x\} = \int_\Theta \int_\Lambda \delta_x\{\tilde{\theta}, (\theta, \lambda)\} \, \pi(\theta, \lambda \mid x) \, d\theta \, d\lambda \qquad (17)$$

where $\delta_x\{\tilde{\theta}, (\theta, \lambda)\}$ is the intrinsic discrepancy between the true model and the family of models with $\theta = \tilde{\theta}$, and $\pi(\theta, \lambda \mid x)$ is the joint posterior distribution which results from formal use of Bayes theorem with the reference prior $\pi(\theta) \, \pi(\lambda \mid \theta)$ associated to model $p(x \mid \theta, \lambda)$ when θ is the quantity of interest.

It immediately follows from (14), (16) and (17) that $d\{\tilde{\theta} \mid x\}$ is the posterior expectation of the minimum expected likelihood ratio between the true model and a model with $\theta = \tilde{\theta}$. This is an explicit measure of the expected posterior loss associated to any particular estimate $\tilde{\theta}$.

The *intrinsic estimate* is the value θ^* which minimizes $d\{\tilde{\theta} \mid x\}$, that is, the Bayes estimate with respect to the intrinsic discrepancy loss and the reference posterior. Formally, the intrinsic estimator is then

$$\theta^*(x) = \arg\min_{\tilde{\theta} \in \Theta} d\{\tilde{\theta} \mid x\} \qquad (18)$$

where $d\{\tilde{\theta} \mid x\}$ is given by (17). Since both the intrinsic loss function and the reference prior are invariant under one-to-one reparametrization, the intrinsic estimator $\phi^*(x)$ of any one-to-one function $\phi\{\theta\}$ of θ will simply be $\phi^*(x) = \phi\{\theta^*(x)\}$.

3. Intrinsic Point Estimation of Normal Variance

In this section, the intrinsic point estimator of a normal variance is derived and by the invariance of the arguments used, this also provides the intrinsic point estimator of any one-to-one function of the variance.

The intrinsic discrepancy $\delta_x\{p_1, p_2\}$ between the two normal densities $p_1(x)$ and $p_2(x)$, with $p_i(x) = \mathrm{N}(x \mid \mu_i, \sigma_i)$, is

$$\delta_x\{p_1, p_2\} = \min\{\kappa\{p_1 \mid p_2\}, \ \kappa\{p_2 \mid p_1\}\}$$

$$\kappa\{p_i \mid p_j\} = \int_{-\infty}^{\infty} p_j(x) \log \frac{p_j(x)}{p_i(x)} \, dx = \frac{1}{2} \left\{ \log \frac{\sigma_i^2}{\sigma_j^2} + \frac{\sigma_j^2}{\sigma_i^2} - 1 + \frac{(\mu_i - \mu_j)^2}{\sigma_i^2} \right\}. \qquad (19)$$

The behaviour of the intrinsic discrepancy is *very* different from the conventional quadratic distance $\ell_{\text{quad}}\{p_1, p_2\} = c\{(\mu_1 - \mu_2)^2 + (\sigma_1 - \sigma_2)^2\}$. In Fig. 1, the discrepancy between a normal distribution $\mathrm{N}(x \mid \mu, \sigma)$ and a standard normal $\mathrm{N}(x \mid 0, 1)$ is represented for both the intrinsic discrepancy and the quadratic distance as a function of μ and $\log \sigma$; a useful range of parameter values $\mu \in [-3, 3]$ and $\sigma \in [e^{-3}, e^3] \approx [0.05, 20.1]$ has been used and, to facilitate comparison, the constant c in the quadratic distance has been chosen such that both surfaces have the same value at the extreme point $(3, 3)$. It is apparent from Fig. 1 that, as one would expect from its analytical form, the quadratic distance essentially ignores discrepancies due to small σ values; the quadratic distance is simply not appropriate to describe the divergence between two normal distributions.

It may be verified that the minimum logarithmic discrepancy of $\{\mathrm{N}(x \mid \mu_i, \sigma_i), \mu_i \in \Re\}$ from $\mathrm{N}(x \mid \mu_j, \sigma_j)$ is achieved when $\mu_i = \mu_j$, so that

$$\min_{\mu_i \in \Re} \kappa\{p_i \mid p_j\} = \kappa\{p_i \mid p_j\} \Big|_{\mu_i = \mu_j} = \frac{1}{2} \left\{ \log \frac{\sigma_i^2}{\sigma_j^2} + \frac{\sigma_j^2}{\sigma_i^2} - 1 \right\}. \qquad (20)$$

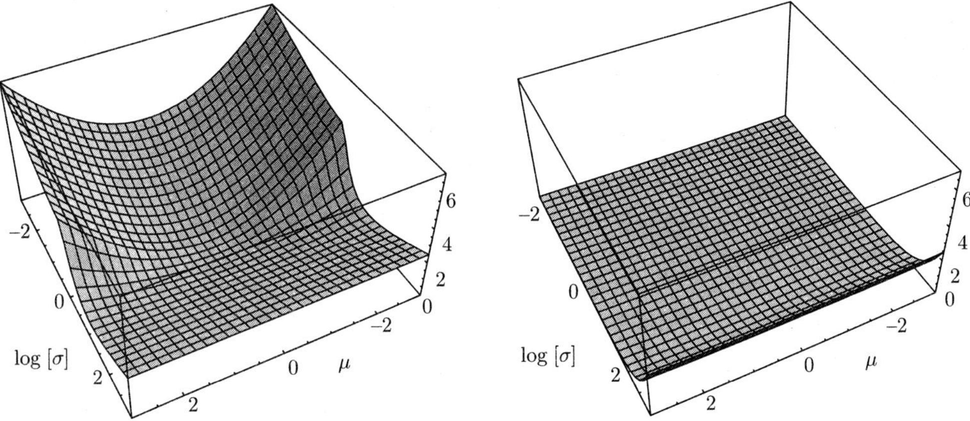

Fig. 1.   **Intrinsic discrepancy (left panel) and quadratic distance (right panel)
between $N(x \mid \mu, \sigma)$ and $N(x \mid 0, 1)$ as a function of μ and $\log \sigma$.**

Thus, the intrinsic discrepancy between the normal $N(x \mid \mu, \sigma)$ and the set of normals with standard deviation $\tilde{\sigma}$, $\mathcal{M}_{\tilde{\sigma}} \equiv \{N(x \mid \tilde{\mu}, \tilde{\sigma}), \ \tilde{\mu} \in \Re)\}$ is achieved when $\tilde{\mu} = \mu$, and, using (19) and (20)

$$\delta_x\{\mathcal{M}_{\tilde{\sigma}}, \ N(x \mid \mu, \sigma)\} = \delta_x(\theta) = \begin{cases} \frac{1}{2}[\log \theta^{-1} + \theta - 1], & \text{if } \theta < 1 \\ \frac{1}{2}[\log \theta + \theta^{-1} - 1], & \text{if } \theta \geq 1 \end{cases} \tag{21}$$

which only depends on $\theta = \tilde{\sigma}^2/\sigma^2$, the ratio of the two variances. Comparison with the entropy loss (4) immediately shows that the entropy loss is the same as the intrinsic loss for $\theta < 1$, i.e. for $\tilde{\sigma}^2 < \sigma^2$, but rather different for $\theta > 1$ (i.e. for $\tilde{\sigma}^2 > \sigma^2$).

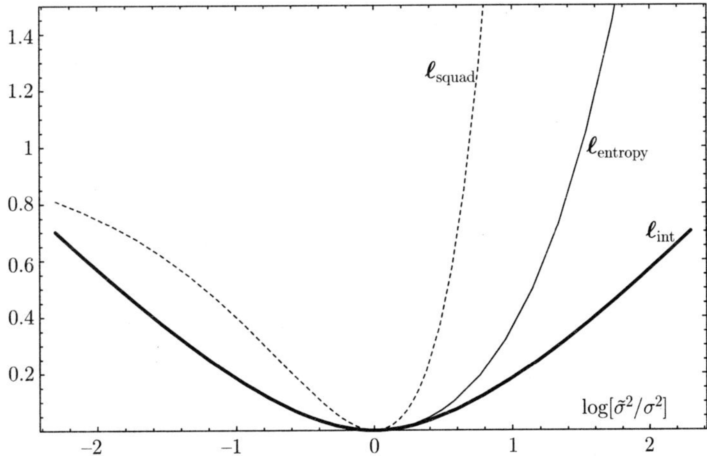

Fig. 2.   **Standardized quadratic loss $\ell_{\mathrm{squad}}(\phi)$, entropy loss $\ell_{\mathrm{entropy}}(\phi)$ and
intrinsic loss $\ell_{\mathrm{int}}(\phi)$ as a function of $\phi = \log(\theta) = \log\,[\tilde{\sigma}^2/\sigma^2]$.**

Thus the intrinsic loss for this problem, a loss function which grants a similar treatment to the cases $\tilde{\sigma}^2 < \sigma^2$ and $\tilde{\sigma}^2 > \sigma^2$, turns out to be a symmetrized version of the entropy loss which Stein apparently suggested for this problem. This is better seen in a logarithmic scale; Fig. 2 represents the standardized quadratic loss (2), the entropy loss (4) and the intrinsic loss (20) as a function of $\log[\tilde{\sigma}^2/\sigma^2]$. The dramatic overpenalization of large estimates by the standardized quadratic loss is immediately apparent; the entropy loss behaves better, but it still clearly overpenalizes large estimates.

Since the normal is a location-scale model, the reference prior when σ (or any one-to-one transformation of σ) is the parameter of interest is the universally recommended (improper) prior $\pi(\mu, \sigma) = \sigma^{-1}$. The corresponding reference posterior distribution of $\theta = \tilde{\sigma}^2/\sigma^2$, after a random sample $\boldsymbol{x} = \{x_1, \ldots, x_n\}$ of size $n \geq 2$ has been observed, is the (always proper) gamma density

$$\pi(\theta \mid \boldsymbol{x}) = \pi(\theta \mid n, s^2, \tilde{\sigma}^2) = \mathrm{Ga}\left(\theta \,\middle|\, \frac{n-1}{2}, \frac{ns^2}{2\tilde{\sigma}^2}\right), \qquad n \geq 2 \tag{22}$$

where, again, s^2 is the MLE of σ^2. Hence, the intrinsic estimator of the normal variance is that value σ_{int}^2 of $\tilde{\sigma}^2$ which minimizes the expected posterior loss; using the additivity property in (15), this is

$$\sigma_{\mathrm{int}}^2 = \arg \min_{\tilde{\sigma}^2} \ d(\tilde{\sigma}^2 \mid n, s^2)$$

$$d(\tilde{\sigma}^2 \mid n, s^2) = \int_0^\infty n\, \delta_x(\theta)\, \pi(\theta \mid n, s^2, \tilde{\sigma}^2)\, \mathrm{d}\theta \tag{23}$$

where $\delta_x(\theta)$ is given by (21) and $\pi(\theta \mid n, s^2, \tilde{\sigma}^2)$ is the gamma density (22). Since (22) implies that the reference posterior distribution of $\tau = ns^2/\sigma^2$ is a central χ_{n-1}^2, the expected posterior loss from using $\tilde{\sigma}^2$ may further be written as

$$d(\tilde{\sigma}^2 \mid n, s^2) = d(a \mid n) = \int_0^\infty n\, \delta(a\tau)\, \chi^2(\tau \mid n-1)\, \mathrm{d}\tau, \qquad a = \frac{\tilde{\sigma}^2}{ns^2}. \tag{24}$$

Thus, the intrinsic estimator is an affine equivariant estimator of the form

$$\sigma_{\mathrm{int}}^2(n, s^2) = c(n)\, s^2, \qquad c(n) = n\, a_n^* \tag{25}$$

where a_n^* is the value of a which minimizes $d(a \mid n)$ in (24). The exact value of a_n^*, and hence that of $c(n) = n\, a_n^*$, may be numerically found by one-dimensional numerical integration followed by numerical optimization. This is tabulated in Table 1, and represented in Fig. 3. It may be appreciated that $c(2) \approx 5$ and that, for moderately larger, $c(n)$ is reasonably well approximated by $n/(n-2)$.

Summarizing, $\sigma_{\mathrm{int}}^2(2, s^2) \approx 5\, s^2$, $\sigma_{\mathrm{int}}^2(3, s^2) \approx 2.4\, s^2$, $\sigma_{\mathrm{int}}^2(4, s^2) \approx 1.8\, s^2$ and, for moderate sample sizes,

$$\sigma_{\mathrm{int}}^2(n, s^2) \approx \frac{n\, s^2}{n-2} = \tilde{\sigma}_{n-2}^2 \tag{26}$$

which is larger than both the MLE (which divides by n the sum of squares) and that the conventional unbiased estimate (which divides the sum of squares by $n-1$). Since intrinsic estimation is consistent under one-to-one reparametrizations, the intrinsic estimator of the standard deviation is just σ_{int}, the squared root of σ_{int}^2, and the intrinsic estimator of, say, $\log \sigma$ is simply $\log \sigma_{\mathrm{int}}$.

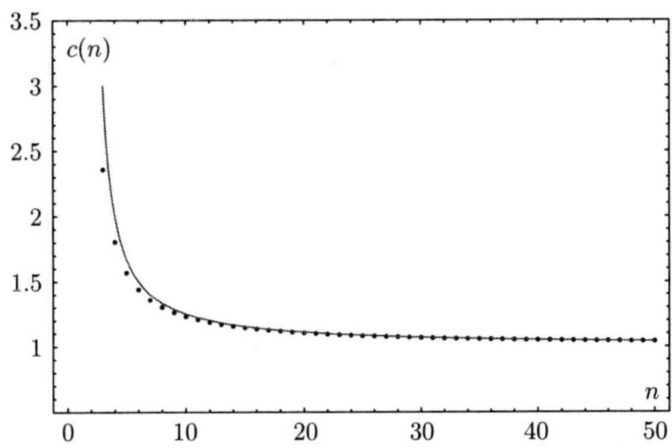

Fig. 3.   The intrinsic estimator of the normal variance is $\sigma^2_{\text{int}} = c(n)s^2$. The exact value of $c(n)$, $n > 2$, are represented by dots, and its approximation $n/(n-2)$ by a continuous line.

Table 1.   Exact value and approximation of the intrinsic estimator of the normal variance $\sigma^2_{\text{int}} = c(n)\,s^2$

n	$c(n)$	$n/(n-2)$
2	4.982	—
3	2.357	3.000
4	1.803	2.000
5	1.569	1.667
6	1.440	1.500
7	1.359	1.400
8	1.303	1.333
9	1.262	1.286
10	1.231	1.250
20	1.106	1.111
30	1.069	1.071
40	1.051	1.053
50	1.041	1.042
60	1.034	1.034
70	1.029	1.029
80	1.025	1.026
90	1.022	1.023
100	1.020	1.020

4.   Discussion

The (posterior) expected intrinsic loss from using an estimator $\tilde{\sigma}_i^2$ given a random sample $x = \{x_1, \ldots, x_n\}$ of Normal $N(x \mid \mu, \sigma)$ observations is

$$d(\tilde{\sigma}_i^2 \mid n, s^2) = \int_0^\infty n\, \delta_x\{\tilde{\sigma}_i^2, \sigma^2\}\, \pi(\sigma^2 \mid n, s^2)\, d\sigma^2 \tag{27}$$

which precisely measures, as a function of the relevant observed data (n, s^2), the expected loss to be suffered if $\tilde{\sigma}_i^2 = \tilde{\sigma}_i^2(n, s^2)$ where used as a proxy for σ^2, given a random sample of size n. Notice that since the intrinsic loss function δ is invariant under reparametrization, the value of the expected loss (27) is independent of the particular parametrization chosen; for instance, $d(\tilde{\sigma}_i \,|\, n, s) = d(\tilde{\sigma}_i^2 \,|\, n, s)$.

It immediately follows from (24) that the expected intrinsic loss $d(\tilde{\sigma}_i^2 \,|\, n, s^2)$ of any particular affine equivariant estimator $\tilde{\sigma}_i^2 = k_n\, s^2$ is actually independent of s^2 and only depends on the sample size n. Table 2 provides, for different n values, the expected posterior intrinsic losses which correspond to the estimators σ_{int}^2, $\tilde{\sigma}_{n-1}^2$ and $\tilde{\sigma}_{n+1}^2$ which are, respectively, the Bayes estimators under the intrinsic loss, the entropy loss, and standardized quadratic loss.

Table 2. Reference expected posterior intrinsic losses and intrinsic risks associated to the use of the intrinsic loss, the entropy loss and the quadratic loss Bayes estimators

n	$d(\tilde{\sigma}_{\text{int}}^2 \,\vert\, n, s^2)$ $r(\sigma_{\text{int}}^2 \,\vert\, n, \sigma^2)$	$d(\tilde{\sigma}_{n-1}^2 \,\vert\, n, s^2)$ $r(\tilde{\sigma}_{n-1}^2 \,\vert\, n, \sigma^2)$	$d(\tilde{\sigma}_{n+1}^2 \,\vert\, n, s^2)$ $r(\tilde{\sigma}_{n+1}^2 \,\vert\, n, \sigma^2)$
2	0.972	1.124	1.696
3	0.675	0.750	1.143
4	0.584	0.636	0.944
5	0.545	0.580	0.840
10	0.480	0.500	0.650
20	0.460	0.480	0.560
30	0.480	0.480	0.540
40	0.480	0.480	0.520
50	0.450	0.450	0.500

It may be appreciated that with $n = 2$ the posterior expected loss of the intrinsic estimator σ_{int}^2 is only 57% of that of the "straw man" $\tilde{\sigma}_{n+1}^2$ and 86% of the conventional estimator $\tilde{\sigma}_{n-1}^2$. As one would expect, those very large relative gains decrease with the sample size, since all those estimators converge to each other as $n \to \infty$.

The frequentist behaviour of each possible estimator is conventionally described by its expected loss under sampling, i.e. its *risk* function

$$r(\tilde{\sigma}_i^2 \,|\, n, \sigma^2) = \int_0^\infty n\, \delta_x\{\tilde{\sigma}_i^2, \sigma^2\}\, p(s^2 \,|\, n, \sigma^2)\, \mathrm{d}s^2. \tag{28}$$

In this problem σ^2 and s^2 play dual roles from, respectively, a Bayesian and a frequentist perspective. This follows from the interesting fact that the reference posterior distribution of $\tau = n\, s^2/\sigma^2$ is precisely the same as the sampling distribution of $t = n\, s^2/\sigma^2$, namely a χ_{n-1}^2 distribution. Since the intrinsic loss is symmetric, so that $\delta\{\tilde{\sigma}_i^2, \sigma^2\} = \delta\{\sigma^2, \tilde{\sigma}_i^2\}$, this implies that the intrinsic risk of any affine equivariant estimator (the expected intrinsic loss under repeated sampling) is precisely equal to its expected reference posterior loss, that is, for all affine equivariant estimators $\tilde{\sigma}_i^2 = k_n\, s^2$,

$$r(\tilde{\sigma}_i^2 \,|\, n, \sigma^2) = d(\tilde{\sigma}_i^2 \,|\, n, s^2), \qquad n \geq 2.$$

Hence, their risks for different sample sizes are also given by Table 2. This implies that, under intrinsic loss, the intrinsic estimator dominates all affine equivariant estimators.

It follows from the preceding discussion that if the arguments given for the general use of the intrinsic discrepancy loss are accepted, then the *optimal* estimator of the normal variance is the intrinsic estimator $\tilde{\sigma}_{\text{int}}^2$, which is given by (25) and quite well approximated by $\tilde{\sigma}_{n-2}^2$ for all but very small samples. Moreover, since intrinsic estimation is an invariant procedure, the optimal estimate of any one-to-one function $\psi[\sigma^2]$ of the normal variance is precisely $\tilde{\psi}_{\text{int}} = \psi[\tilde{\sigma}_{\text{int}}^2]$. Finally, the intrinsic estimator is also the best affine equivariant estimator under the frequentist decision-theoretic criterion of minimizing the (intrinsic) risk.

References

Barnard, G. A. (1952). The frequency justification of certain sequential tests. *Biometrika*, **39**, 155–150.

Bernardo, J. M. (1979). Reference posterior distributions for Bayesian inference. *J. Roy. Statist. Soc. B*, **41**, 113–147 (with discussion). Reprinted in *Bayesian Inference* (N.G. Polson and G.C. Tiao, eds.) Brookfield, VT: Edward Elgar, 1995, 229–263.

Bernardo, J. M. (1999). Nested hypothesis testing: The Bayesian reference criterion. *Bayesian Statistics 6* (J. M. Bernardo, J. O. Berger, A. P. Dawid and A. F. M. Smith, eds.) Oxford: University Press, 101–130 (with discussion).

Bernardo, J. M. (2005). Reference analysis. *Handbook of Statistics* **25** (D. K. Dey and C. R. Rao eds.). Amsterdam: Elsevier, 17–19.

Bernardo, J. M. and Juárez, M. (2003). Intrinsic estimation. *Bayesian Statistics 7* (J. M. Bernardo, M. J. Bayarri, J. O. Berger, A. P. Dawid, D. Heckerman, A. F. M. Smith and M. West, eds.) Oxford: University Press, 465–476.

Bernardo, J. M. and Rueda, R. (2002). Bayesian hypothesis testing: A reference approach. *Internat. Statist. Rev.*, **70**, 351–372.

Bernardo, J. M. and Smith, A. F. M. (1994). *Bayesian Theory*. Chichester: Wiley (Student edition in 2000).

Box, G. E. P. and Tiao, G. C. (1973). *Bayesian Inference in Statistical Analysis*. Reading, MA: Addison-Wesley.

Brewster, J. F. and Zidek, J. V. (1974). Improving on equivariant estimators. *Ann. Statist.*, **2**, 21–38.

Brown, L. (1968). Inadmissibility of the usual estimators of scale parameters in problems with unknown location and scale parameters. *Ann. Math. Statist.*, **39**, 24–48.

Brown, L. (1990). Comment on Maata and Casella (1990).

Csörgö, S. and Faraway, J. J. (1996). On the estimation of a normal variance. *Statistics and Decisions*, **14**, 23–34.

Cox, D. R. and Hinkley, D. V. (1974). *Theoretical Statistics*. London: Chapman and Hall.

George, E. I. (1990). Comment on Maata and Casella (1990).

Jeffreys, H. (1961). *Theory of Probability* (3rd ed.) Oxford: Oxford University Press.

Lehmann, E. L. and Casella, G. (1998). *Theory of Point Estimation* (2nd ed.) Berlin: Springer

Lin, J.-J. and Pal, N. (2005). Comparison of normal variance estimators under multiple criteria and towards a compromise estimator. *J. Statist. Computation and Simulation*, **75**, 645–666.

Lindley, D. V. (1965). *Introduction to Probability and Statistics from a Bayesian Viewpoint*. Volume 2: Inference. Cambridge: University Press.

Maata, J. M. and Casella, G. (1990). Developments in decision-theoretic variance estimation. *Statist. Sci.*, **5**, 90–120, (with discussion).

Pal, N., Ling, C, and Lin, J.-J. (1998). Estimation of a normal variance—a critical review. *Statistical Papers* **39**, 389-404.

Robert, C. P. (1996). Intrinsic loss functions. *Theory and Decision*, **40**, 192–214.

Stein, C. (1964). Inadmissibility of the usual estimator for the variance of a normal distribution with unknown mean. *Ann. Inst. Statist. Math.*, **16**, 155–160.

Bayesian Statistics and Its Applications
Edited by S.K. Upadhyay, U. Singh and D.K. Dey
Anamaya Publishers, New Delhi, India

Analyzing Financial Data Using Polya Trees

David G.T. Denison[1] and Bani K. Mallick[2]

[1]Department of Mathematics, Imperial College of Science, Technology and Medicine,
180 Queens Gate, London, SW7 2BZ, UK

[2]Department of Statistics, Texas A&M University, College Station, TX 77843-3143, USA

Abstract

We present a new approach to generalized autoregressive conditional heteroscedastic (GARCH) modeling for asset returns. Instead of attempting to choose a specific distribution for the errors, as in the usual GARCH model formulation, we use a nonparametric distribution to estimate these errors. This takes into account the common problems encountered in financial time series, for example, asymmetricity and excessive kurtosis, whilst maintaining the simple, and highly interpretable, GARCH model.

---

1. Introduction

Generalized autoregressive conditional heteroscedastic (GARCH) models are well established in the financial time series literature (Bollerslev, 1986; Bollerslev, Chou and Kroner, 1992; Bollerslev, Engle and Nelson, 1994). The focus of the analysis of a financial dataset is on the changing nature of the underlying volatility of the asset in question, and the standard GARCH (p,q) model (Bollerslev, 1986) is given by

$$y_t = \epsilon_t \sigma_2, \quad \sigma_t^2 = \alpha_0 + \sum_{i=1}^{p} \alpha_i y_{t-i}^2 + \sum_{i=1}^{q} \beta_i \sigma_{t-i}^2 \tag{1}$$

where y_t is the mean corrected return of the asset at time t ($> \max\{p,q\}$) given by

$$y_t = 11 \times \left\{ (\log x_t - \log x_{t-1}) - \frac{1}{T} \sum_{i=1}^{T} (\log x_i - \log x_{i-1}) \right\} \tag{2}$$

and x_t is the price of the asset at time $t = 1, \ldots, T$. The σ_t^2 are the volatilities of stock at each time point and ϵ_t are the errors which are assumed to be independent and drawn from a distribution with zero mean and unit variance. Restrictions on the coefficients $\beta = (\alpha_0, \ldots, \alpha_p, \beta_1, \ldots, \beta_q)$ are required to ensure that the volatilities are all positive ($\alpha_0 > 0, \alpha_1 \geq 0, \beta_j \geq 0$) and to ensure covariance stationarity ($\sum_i \alpha_i + \sum_j \beta_j < 1$). Although the error distribution in (1) is not necessarily chosen to be normal this is commonly the case with the most popular alternative being the Student-t distribution (Bollerslev, 1986).

The GARCH model, itself, has been generalized by many authors (see, for example, Nelson, 1991; Geweke, 1986; Drost and Nijman, 1993). The motivation for these generalizations is the inability of standard GARCH models to adequately model the true perceived nature of volatilities. It is believed that models should be able to respond more rapidly to falls than rises in the asset (Campbell and Hentschel, 1992) and, to further complicate matters, this asymmetry is believed to be more prevalent in equities than currencies (Nelson, 1991). As well as this skewness in the errors, financial datasets often exhibit excessive kurtosis and this needs to be taken into account to produce reliable models. In fact Kim, Shephard and Chib (1998), when analyzing some real financial datasets, find that the standard GARCH model can work very well in conjunction with an error distribution close to the truth.

In this paper we generalize the standard GARCH (l, l) model to analyze financial data by taking the error distribution to be nonparametric, in a similar spirit to Gallant, Hsieh and Tauchen (1991) and Linton (1993). This flexible distribution allows accurate modeling of the true skewness and kurtosis of the error distribution. We can also use the nonparametric distribution function found to compare the possible parametric distributions which could be used to repeat the analysis. For instance, we can check whether the error distribution is modeled better with a Student-t distribution, with some chosen degrees of freedom, a normal distribution or even a skewed distribution such as the Gumbel.

We use Polya trees (Lavine, 1992, 1994; Mauldin, Sudderth and Williams, 1992) as a basis to model the unknown error distribution. This involves assuming that ϵ_t come from an unknown distribution F which has a Polya tree representation. These have recently been used to estimate frailty distributions in survival analysis and random effects in hierarchical linear models (Walker and Mallick, 1997). We perform the analysis within a Bayesian framework, this provides a coherent means of inference and allows easy computation of Polya trees by standard Markov chain Monte Carlo (MCMC) techniques (Gelfand and Smith, 1990). Note that other Bayesian analysis of GARCH models with parametric error distributions has been undertaken by Kim et al. (1998), Müller and Pole (1997) and Bauwens and Lubrano (1998).

2. Polya Tree Distributions

Polya tree distributions have received much attention recently (Lavine, 1992, 1994; Maudlin et al., 1992) and have been shown to uncover new insights, which cannot be captured by usual parametric methods, in some data analysis (e.g. Walker et al., 1999). Polya trees can be used to define probability distributions on any space of interest Ω. To do this we must define a partition of Ω, say $\Pi = \{\mathcal{B}_\tau\}$, and a set of non-negative real numbers, say $\mathcal{A} = \{\gamma_\tau\}$, associated with each element of Π.

We can recursively define the partitions by letting $\{B_0, B_1\}$ be the set of level 1 partitions which exhaustively split Ω. The level 2 partitions B_{00} and B_{01} split the 'parent' B_0, with the other partitions B_{10}, B_{11} being descendents of B_1. In general, a parent B_τ has 'children' $B_{\tau 0}$ and $B_{\tau 1}$, where $B_{\tau 0} \cap B_{\tau 1} = \emptyset$ and $B_{\tau 0} \cup B_{\tau 1} = B_\tau$. This is an example of binary partitioning.

Once Π has been fixed we need to associate probabilities of a draw from the random distribution being in each partition. A good analogy to help understand this concept is that of a particle cascading through the tree. Consider the particle starting in Ω and then it can move to either B_0 or B_1 with probability C_0 and $C_1 = 1 - C_0$, respectively. In general, on entering B_τ, the particle can move to either $B_{\tau 0}$ or $B_{\tau 1}$ with probability $C_{\tau 0}$ or $C_{\tau 1} = 1 - C_{\tau 0}$, respectively. For Polya trees these probabilities are random quantities which follow Beta distributions, such that $C_{\tau 0} \sim \text{Beta}(\gamma_{\tau 0}, \gamma_{\tau 1})$ and $C_{\tau 1} = 1 - C_{\tau 0}$, where $\gamma_{\tau 0}$ and $\gamma_{\tau 1}$ are both non-negative. The last stage of defining a Polya tree distribution is defining the $\mathcal{A} = \{\gamma_\tau\}$.

It is possible to center the Polya tree prior on a particular probability (base) measure G on Ω by taking the partitions to coincide with the percentiles of G and then to take $\gamma_{\tau 0} = \gamma_{\tau 1}$ for each τ. This involves setting $B_0 = (-\infty, G^{-1}(1/2)), B_1 = [G^{-1}(1/2), \infty)$ and, at level m, setting, for

$j = 1, \ldots, 2^m$, $B_j = [G^{-1}((j-1)/2^m), G^{-1}(j/2^m))$, with $G^{-1}(0) = -\infty$ and $G^{-1}(1) = +\infty$, where $(B_j : j = 1, \ldots, 2^m)$ correspond, in order, to the 2^m partitions of level m. With this formulation we find that $E\{F(_\tau)\} = G(B_\tau)$ for all τ. We can use a hierarchical setup in assigning the Polya tree prior distribution G. In particular, G may depend on some random quantity ξ so we can use the relationship $P(F, \xi) = p(F|\xi)p(\xi)$, to assign the joint prior of F and ξ. We use this in our Bayesian model formulation described in Section 3.1 where we take median of the base measure G to be random.

We choose to use the prior on $\mathcal{A}$ suggested by Walker and Mallick (1997). Hence, we take all the γ_τ corresponding to level m to be $c_m = cm^2$, where c is a constant to be chosen and affects the prior influence on the posterior. This prior ensures that at low levels (m large) nearby bins have similar probability. In practice, we only update upto a fixed level M; this is known as a partially specified Polya tree distribution (Lavine, 1992). This distribution has similarities to histogram approaches (Lenk, 1993; Hartigan, 1996) but Polya trees allow prior information to be incorporated straightforwardly whereas histograms rely almost entirely on the data to provide inference. However, posterior inference using Polya trees is dependent on the partitions chosen to construct the tree. To learn about the distribution everywhere we need to ensure that very few, if any, observations lie in the end bins $B_{0\ldots0}$ and $B_{1\ldots1}$ as otherwise we get little idea of the distribution outside $(G^{-1}(2^{-M}), G^{-1}(2^M))$. The side-effect of this is that fewer bins are then concerned with modeling the center of the distribution but we have not found this to be a significant problem. Histograms can avoid this problem by allowing bins to be located randomly over some range. However, flexible partitioning destroys Polya trees' conjugate nature, removing much of their computational attractiveness.

In our GARCH model formulation we shall use the Polya tree to estimate the error distribution from which ϵ_τ have arisen. Thus, given the data, we assume that $\epsilon_t = y_t/\sigma_t$ ($t = p+1, \ldots, T$) are random draws from the Polya tree distribution F. Hence the probability of observing the tth observation is

$$f(\epsilon_\tau | F, \beta, \xi) = F(B_{\tau 1 \ldots \tau_M}(t)) \tag{3}$$

where $\epsilon_t \in B_{\tau 1 \ldots \tau_M}(t)$.

Sampling from $p(F|\beta, \xi, \mathcal{D})$ is straightforward using a Polya tree formulation because this posterior is just an updated version of the prior Polya tree, given the ϵ_t. The updated posterior Polya tree has the same partitions as before but the γ_τ are replaced by $\gamma_\tau + T_\tau$, where T_τ are the number of observations that are in B_τ.

3. The Bayesian Model

3.1 The Model

We must first assign prior distributions to all the parameters in the model. We assume a uniform prior over $[0, 1]$ for each of the coefficients in β. We do not restrict the model to be stationary ($\alpha_1 + \beta_1 < 1$) because, if the data is nonstationary we wish the method to inform us of this. To allow comparison of our method with a standard GARCH model we take σ_0^2 as the empirical variance of the first 20 datapoints, although assigning a prior to it would be straightforward.

The base probability distribution, or prior, for the Polya tree is chosen to be a normal distribution with unknown median, or mean, ξ and a standard deviation of three. This ensures that few observations lie in the end bins removing the problems highlighted in the last section. Similar results would be obtained by using a Student-t distribution with a low number of degrees of freedom but this prior choice is not explored further here.

The prior for the median of the Polya tree distribution ξ is assigned a normal distribution with mean zero and standard deviation 0.1. This allows the distribution to be skewed without contradicting the general arbitrage assumption in financial models that $E(\epsilon_t) = 0$. Theoretically, our model satisfies this constraint as $E(y_t) = 0$ (efficient market hypothesis) and $E(\sigma_t) > 0$ as σ_t is a strictly positive random variable, so by taking expectations of (1) we see that $E(\epsilon_t) = 0$. The set $\mathcal{A}$ is chosen so that the γ_τ at level m are given by $c_m = 0.1m^2$ for $m = 1, \ldots, 8$, i.e. we choose $c = 0.1$ and $M = 8$.

We model the errors $\epsilon_t = y_t/\sigma_2$ as independent draws from the nonparametric Polya tree distribution F. Thus, the likelihood of the data given the other parameters and the error distribution is given by

$$L(Y_T|F,\beta,\xi) = \prod_{t=1}^{T} p(y_t|Y_{t-1},F,\beta,\xi) = \prod_{t=1}^{T} \sigma_t^{-1} f(y_t/\sigma_t) \tag{4}$$

where $Y_t = (y_1, \ldots, y_t)$ and f is the density of F given in (3).

The standard GARCH model uses only unit variance error distributions to model the data to ensure identifiability in the coefficients β. Fixing the variance of a Polya tree distribution requires, at best, a very *ad hoc* procedure so we prefer to fix the scale of the coefficients to maintain identifiability. Following Linton (1993) we set $\alpha_0 = 1.0$ and allow the error distribution to have a non-fixed variance. Meaningful comparison of the standard GARCH model coefficient estimates and those produced by the Polya tree model can then only take place after suitable scaling of the Polya tree ones. This is achieved by the multiplication of the coefficient values by the posterior variance of the Polya tree distribution. This is acceptable as we can use the MCMC sample to make inference about any such linear transformations of the random quantities (e.g. Tierney, 1994).

3.2 The MCMC Algorithm

The Markov Chain Monte Carlo algorithm to simulate from the posterior of interest is, in essence, easy to construct. Each iteration consists of two steps: the first draws a sample from $p(F|\beta,\xi,\mathcal{D})$, as outlined at the end of Section 2, and the second from $p(\beta,\xi|F,\mathcal{D})$ where β is the vector of coefficients in the GARCH model, F is the error distribution, ξ is the median of F, and $\mathcal{D}$ is the data. To ease the following exposition we now augment β to include ξ as these parameters are all updated together and now describe how to update $p(\beta|F,\mathcal{D})$.

Due to the high correlation between the coefficients in the GARCH model great care must be taken to prevent the sampler from becoming trapped in local modes. From experience we find that a simple Metropolis-Hastings algorithm (Metropolis et al., 1953; Hastings, 1970) is inappropriate and much better results can be found using a generalization of this algorithm found in Holmes and Mallick (Technical Report, Imperial College). This method is based on a standard genetic algorithm (Holland, 1975) with suitable modifications to ensure that the proposal density preserves the detailed balance requirements. This method is particularly suited to drawing samples from spaces where the parameters are highly correlated, as in our case. Also, the strategy used prevents the sampler from becoming stuck in local modes, which is a particular problem when using such flexible models.

Standard genetic algorithms involve maintaining a population of individuals and then updating (or evolving) the population by *crossover* and *mutation* steps. In the MCMC approach we can think of the progress of each individual as being represented by a single Markov chain and updates to the individual are made using information from the other population members, whose status' are represented by other, parallelly run, Markov chains.

The crossover step involves randomly swapping 'gene' information (elements of the coefficient vectors) between individuals in the population at a given time step. The proposal for updating the individual k with coefficients $\beta^{(k)}$ is made using coefficients (i.e. gene) information at four other uniformly chosen chains (i.e. individuals), say $(\beta^{(p_1)}, \beta^{(p_2)})$ and $(\beta^{(q_1)}, \beta^{(q_2)})$, where p_i, q_i $(i = 1, 2)$ and k are distinct. Each of these pairs of individuals can be thought of as parents as they are used to produce crossover coefficient vectors $\beta^{(c_p)}$ and $\beta^{(c_q)}$. These are found by setting each element of $\beta^{(c_\bullet)}$ as the corresponding element from either of its parents $\beta^{(\bullet_1)}$ or $\beta^{(\bullet_2)}$, each chosen with equal probability. The proposed new coefficients for individual k are then chosen to be a distance, drawn from a $U[0, 1]$ distribution, along the unit vector between $\beta^{(c_p)}$ and $\beta^{(c_q)}$. Note that this step does not use the current coefficient values $\beta^{(k)}$ for the proposed move to simplify the acceptance probability.

The mutation step is more straightforward and mimics the usual single chain Metropolis-Hastings sampler. This just involves choosing, uniformly at random, one of the chains to update and proposing a new coefficient vector which is the same as the old one but with each coefficient, in turn, mutated with a zero-mean, Gaussian proposal distribution with standard deviation 0.1. Thus this step performs small 'local' moves around the posterior space whereas the crossover step makes more bold moves.

Thus, the step which updates $p(\beta|F, \mathcal{D})$ first chooses, uniformly at random, an individual from the population on which to perform the crossover step. Then the mutation step is performed on another randomly drawn individual. In all the work that follows we chose to include 100 individuals in the population, i.e. 100 parallel chains were run.

4.   Examples

4.1   Simulated Examples

To demonstrate the efficacy of our method in the analysis of financial data in this Section we apply the methodology to simulated datasets with known zero mean, unit variance error distributions. In each simulated example the underlying true model was a GARCH (1, 1) process with coefficients $\alpha_0 = 7; \alpha_1 = 0.4$ and $\beta_1 = 0.2$, and 801 points were generated.

We compare the results we found with those obtained using a standard GARCH (1, 1) model with an assumed standard normal error distribution. In each case we ran the sampler in an identical manner. This involved drawing the initial values of each coefficient from a $U[0, 1]$ distribution, then running the algorithm for 50,000 burn-in iterations followed by a further 50,000 iterations from which a sample is collected. Convergence of the chain after 50,000 iterations was checked using the R-statistic of Gelman and Rubin (1992).

The sample of coefficients was collected by, every ten iterations, picking at random one of the 100 parallel chains and then the coefficients from the chosen chain were taken to be in the (approximately) independent sample. This produces a sample with much lower auto correlations than one that might have been produced using a standard Metropolis-Hastings algorithm.

The three error distributions used are: (a) a standard normal distribution; (b) a Student-t distribution with unit variance and 4 degrees of freedom and (c) a Gumbel distribution (e.g. Kotz and Johnson, 1970) with unit variance and zero mean (i.e. parameters for scale and location given by 0.780 and -0.450, respectively).

The posterior mean Polya tree distribution was found by averaging the probability values in each of the partitions over the models in the posterior sample. From this mixture of Polya tree distributions we generated 5,000 realizations. To compare this density with the true ones, which all had unit variance, we scaled the simulated points from the error distribution so these too had unit variance. The densities shown in Fig. 1 were then found using the $S+$ function density with a standard bandwidth of one (i.e. width $= 1$).

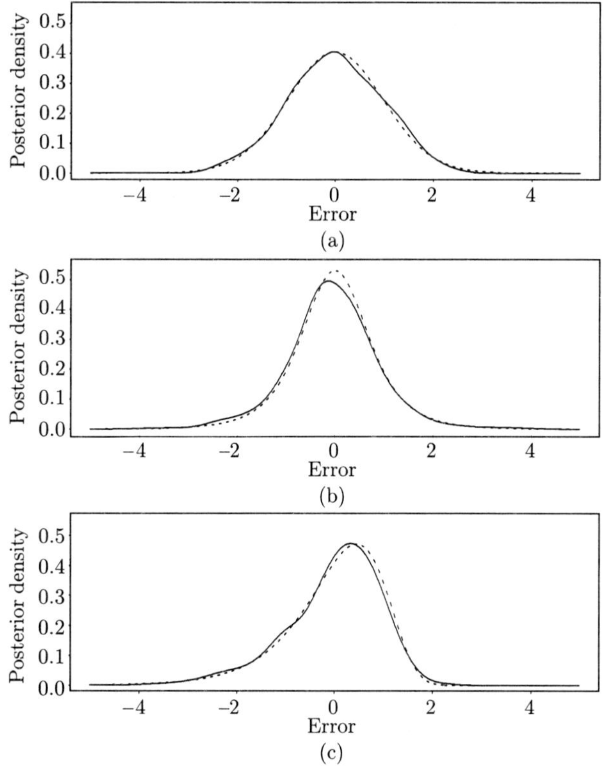

Fig. 1. **Posterior densities of error distributions (solid lines) with true error distributions (dotted lines) for the three simulated datasets: (a) $N(0,1)$ errors; (b) Student-$t(4)$ errors and (c) Gumbel errors.**

To test the 'closeness' of the true densities from the posterior Polya tree estimates we must use a suitable distance metric. We chose to use the Kullback-Leibler (KL) generalized distance measure (Kullback, 1959), that is

$$d(f_0, f_1) = \int f_0(x) \log \frac{f_0(x)}{f_1(x)} dx \qquad (5)$$

where f_0 is the true density and f_1 an estimate of f_0.

Table 1 gives the scaled coefficient values from the generated sample of models. As mentioned previously, the scaling took place to allow comparison between the posterior sample of coefficient values and the real coefficient values which assumed unit error variance. We see that, although the Polya tree method does not pinpoint the true coefficient values, in general the truth lies within the 95% Bayesian credible intervals. Note that these intervals are valid for finite sample sizes in contrast to those found using quasi-maximum likelihood estimation with the GARCH(1,1) model (e.g. Lee and Hansen, 1994) which are only asymptotically consistent.

Fig. 1 shows the posterior mean error distributions for the three datasets. In each case the Polya tree distribution captures the form of the error distributions even with these relatively small

Table 1.   **Scaled coefficient estimates of the GARCH (1, 1) model for the simulated datasets using the Polya tree model and maximum likelihood. The results using the Polya tree include the MCMC standard errors, given in brackets, and the Kullback-Liebler distance between the true and estimated density. Note that the true values are $\alpha_0 = 0.7$, $\alpha_1 = 0.4$ and $\beta_1 = 0.2$**

	Polya tree model			Maximum likelihood		
	(a)	(b)	(c)	(a)	(b)	(c)
$\hat{\alpha}_0$	0.592	0.580	0.827	03923	0.620	0.900
$\hat{\alpha}_1$	0.385(0.12)	0.315(0.14)	0.523(0.21)	0.321	0.370	0.351
$\hat{\beta}_1$	0.207(0.06)	0.157(0.07)	0.169(0.10)	0.138	0.277	0.092
log KL	−7.373	−7.670	−7.204	−	−	−

$(T = 801)$ datasets. Not only are the central parts of the distributions well estimated., but also the tails.

The KL-distance between the posterior distribution found using dataset (b) and Student-t distributions with various degrees of freedom is shown in Fig. 2. We calculated the KL-distance between our estimate and zero mean, unit variance Student-t distributions with degrees of freedom between 2.75 and 10 in steps of 0.25. The true number of degrees of freedom (i.e. four) is nearly picked out by the Polya tree, with the minimum of the KL-distance occurring at 4.5 degrees of freedom.

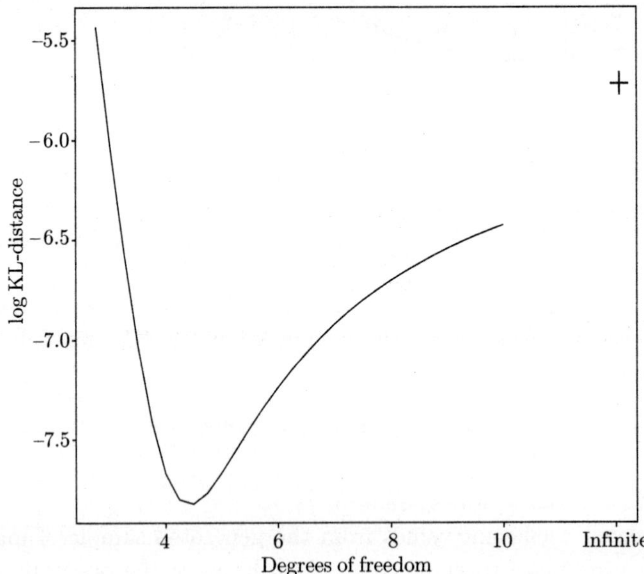

Fig. 2.   **Kullback-Liebler distance between the posterior mean Polya tree density, found using simulated dataset (b), and zero mean, unit variance Student-t distributions with varying degrees of freedom. The + indicates the KL-distance between the estimated distribution and the standard normal distribution.**

To check that the method was not overly affected by the choice of $c = 0.1$ in the prior for γ_τ we reran the algorithm using various values of c between 0.01 and 0.5. We found that the KL-distance from the $N(0,1)$ distribution for the first simulated dataset was between -6.90 (found taking $c = 0.5$) and -7.37 (with $c = 0.01$, 0.05, and 0.1). Thus $c = 0.1$ is among the better values for this constant for this particular dataset but the results are not unduly affected by other choices over a fairly wide range.

4.2   Real Data Examples

Now we use the proposed methodology to estimate the error distributions found when analyzing some real financial datasets. The datasets we shall use have previously been studied in Shephard (1996) and can be obtained from DATASTREAM.

 (i) The daily German deutschemark-sterling exchange rate from 1st January 1986 to 12th April 1994 ($T = 2160$).

(ii) The daily Japanese yen-sterling exchange rate from 1st January 1986 to 12th April 1994 ($T = 2160$).

(iii) The daily Nikkei 500 Price Index from 2nd April 1986 to 6th May 1994 ($T = 2113$).

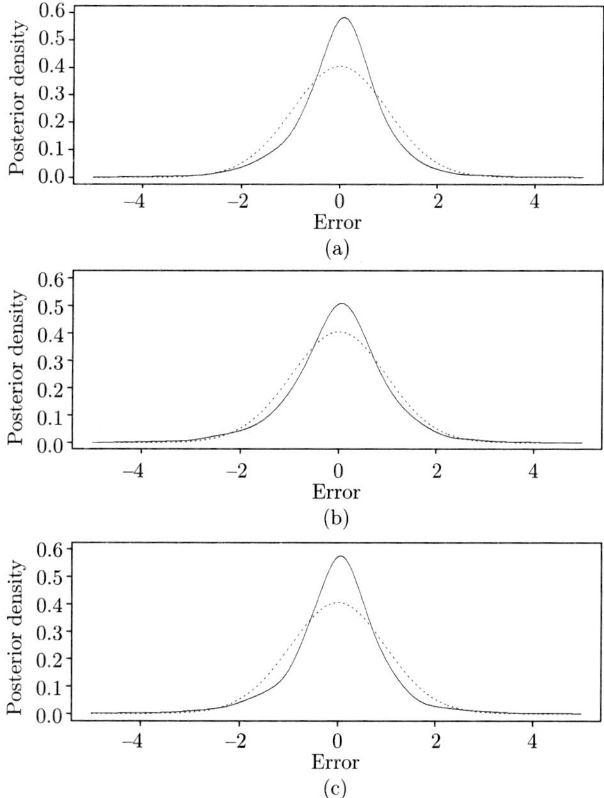

Fig. 3.   Posterior densities of error distributions (solid lines) compared to standard normal distributions (dotted lines) for the three real datasets: (a) deutschemark-sterling exchange rate; (b) yen-sterling exchange rate and (c) Nikkei 500.

Note that for each of these datasets the values were taken at the end of the weekdays and, when the market was closed on a weekday (e.g. a Bank holiday), the previous day's value was used. Also, the algorithm we use in this section was run identically to that described in Section 4.1.

Fig. 3 shows the posterior densities of F for the three examples together with the $N(0,1)$ density. The error distributions found for each of these examples are negatively skewed and exhibit much higher kurtosis than a normal distribution. The asymmetric error distributions suggest that large falls in the market are more likely than large rises.

5.   Discussion

We have presented a Bayesian nonparametric method which estimates the error distribution using a GARCH model formulation. In particular, this allows the simple interpretability of the GARCH (1, 1) model to be retained whilst gaining further insight into the nature of the errors from the highly flexible Polya tree error distribution.

The posterior Polya tree distribution has been shown to cope with all the most likely deviations from normality associated with financial time series; notably skewness and excessive kurtosis. It can also be used for model choice as the KL-distance between the mean error distribution and each member of a possible set of predetermined parametric error distributions can be found. Then the parametric distribution which is closest to the estimated error distribution can be used, if required. The main advantage of this would be to ensure the expectation of the error distribution is exactly, not just theoretically, zero. Also, using the output from the Polya tree methodology contains far more information than just the possible error distribution such as the posterior distributions of the coefficients.

Acknowledgements

The authors wish to thank Mike Pitt and Chris Holmes, as well as the referee and editors, for helpful comments. The work of the second author was supported by a grant from the National Cancer Institute (CA-104620).

References

Bauwens, L. and Lubrano, M. (1998). Bayesian inference on GARCH models using the Gibbs sampler. *Econometrics J.* **1**, C23-C46.

Bollerslev, T. (1986). Generalized autoregressive conditional heteroscedasticity. *J. Econometrics*, **51**, 307-327.

Bollerslev, T., Chou, R. and Kroner, K.F. (1992). ARCH modeling in finance. *J. Econometrics*, **52**, 5-59.

Bollerslev., T., Engle, R.F. and Nelson, D.B. (1994). ARCH models. In *The Hand Book of Econometrics, Volume 4*, Eds. R.F. Engle and D. McFadden, 2959-3038. Amsterdam: North-Holland.

Campbell, J. Y. and Hentschel, L. (1992). No news is good news: an asymmetric model of changing volatility in stock returns. *J. Finan. Economics*, **31**, 281-318.

Drost, F.C. and Nijman, T.E. (1993). Temporal aggregation of GARCH processes. *Econometrica*, **61**, 909-927.

Gallant, A.R., Hsieh, D.A. and Tauchen, G.E. (1991). On fitting a recalcitrant series: The pound/dollar exchange rate, 1974-1983. In *Nonparametric and Semiparametric Methods in Econometrics and Statistics*, eds. W.A. Barnett, J. Powell and G.E. Tauchen, 199-240. Cambridge: Cambridge University Press.

Gelfand, A.E. and Smith, A.F.M. (1990). Sampling-based approaches to calculating marginal densities. *J. Am. Statist. Assoc.*, **85**, 398-409.

Gelman, A. and Rubin, D.B. (1992). Inference from iterative simulation using multiple sequences. *Statist. Sci*, **7**, 457-472.

Geweke, J. (1986). Modeling the persistence of conditional variances: a comment. *Econometric Reviews*, **5**, 57-61.

Hartigan, J.A. (1996). Bayesian Histograms. In *Bayesian Statistics 5*, eds. J.M. Bernardo, J.D. Berger, A.P. Dawid & A.F.M. Smith, 211–222. Oxford: Clarendon Press.

Hastings, W.K. (1970). Monte Carlo sampling methods using Markov chains and their applications. *Biometrika*, **57**, 97–109.

Holland, J .H. (1975). *Adaption in Natural and Artificial Systems*. Ann Arbor, Michigan: University of Michigan Press.

Kim, S., Shephard, N. and Chib, S. (1998). Stochastic volatility: likelihood inference and comparison with ARCH models. *Review of Economic Studies*, **65**, 361–393.

Kotz, S. and Johnson, N .L. (1970). *Continuous Univariate Distributions-I*. New York: John Wiley & Sons.

Kullback, S. (1959). *Information Theory and Statistics*. New York: John Wiley.

Lavine, M. (1992). Some aspects of Polya tree distributions for statistical modeling. *Ann. Statist.*, **20**, 1222–1235.

Lavine, M. (1994). More aspects of Polya tree distributions for statistical modeling. *Ann. Statist.*, **22**, 1161–1176.

Lee, S. W. and Hansen, B.E. (1994). Asymptotic theory for the GARCH(1, 1) quasi-maximum likelihood estimator. *Econometric Theory*, **10**, 29–52.

Lenk, P. (1993). A nonparametric Bayesian density estimator. *J. Nonpar. Statist.*, **3**, 53–69.

Linton, O. (1993). Adaptive estimation in ARCH models. *Econometric Theory*, **9**, 539–569.

Maudlin, R.D., Sudderth, W.D. and Williams, S.C. (1992). Polya trees and random distributions. *Ann. Statist.*, **20**, 1203–1221.

Metropolis, N., Rosenbluth, A.W., Rosenbluth, M.N., Teller, A.H. and Teller, E. (1953). Equations of state calculations by fast computing machines. *J. Chem. Phys.*, **21**, 1087–1091.

Müller, P. and Pole, A. (1997). Monte Carlo posterior integration in GARCH models. *Sankhya B*, **60**, 127–144.

Nelson, D.B. (1991). Conditional heteroscedasticity in asset pricing: a new approach. *Econometrica*, **59**, 347–370.

Shephard, N. (1996). Statistical aspects of ARCH and stochastic volatility. In *Time Series Models in Econometrics, Finance and Other Fields*, eds. D.R. Cox, D.V. Hinkley and O.E. Barndorff-Nielsen, 1–67. London: Chapman & Hall.

Tierney, L. (1994). Markov chains for exploring posterior distributions (with Discussion). *Ann. Statist.*, **22**, 1701–1762.

Walker, S.G. and Mallick, B.K. (1997). Hierarchical generalized linear models and frailty models with Bayesian nonparametric mixing. *J. Roy. Statist. Soc. B*, **59**, 845–860.

Walker, S.G., Damien, P., Laud, P. W. and Smith, A.F.M. (1999). Bayesian nonparametric inference for random distributions and related functions. *J. Roy. Statist. Soc. B*, **61**, 485–527.

Appendix

The GA sampler is easily implemented by adding a few extra lines of code to a standard Metropolis-Hastings computer simulation program.

The state space must be augmented to accommodate the parallel states. This can be achieved by simply increasing by one the dimension of the array encoding the state of the Markov chain. The pseudo-code for the GA sampler is then as follows.

Let β_t^{ij} represent the jth component of the ith individual in the population at time t.

```
• Draw initial population, P, from over dispersed random values.
• FOR t=1 to total-number-of-iterations
```

- Select an individual state β_t^a from $\mathcal{P}$ for updating.
- IF $u < R$, where $u \sim U(0,1)$ and R is some mutation rate
  * Perform standard Metropolis-Hastings update:
  * Propose new state β_{t+1}^{a*} using proposal density $q \sim N_p(0, \sigma_m^2)$ so that $\beta_{t+1}^{a*} = \beta_{t+1}^a + q$.
- ELSE perform genetic crossover operator
  * Select two states, β_t^i and β_t^j, at random from the population such that $i \neq j \neq a$.
  * Calculate D, the distance between β_t^i and β_t^j so that $D = \|\beta_t^i - \beta_t^j\|$.
  * Create crossover state $\mathcal{C}_t^x$ using uniform scheme:
    - FOR k=1 to dimensionality of state
    - IF $u < 0.5$, where $u \sim U(0,1)$ THEN $\mathcal{C}_t^{xk} = \mathcal{C}_t^{ik}$
    - ELSE $\mathcal{C}_t^{xk} = \mathcal{C}_t^{jk}$
    - END. for loop.
  * /* Now create proposal state. */
  * IF $u < F$, where $u \sim U(0,1)$ and F is some flip rate, THEN perform FLIP step
    - FOR k=1 to dimensionality of state
    - $\beta_{t+1}^{a*k} = \beta_t^{ak} + 2 \times (\beta_t^{xk} - \beta_t^{ak})$
    - END. for loop.
  * ELSE perform crossover step
    - Calculate normalised vector $V = (\beta_t^x - \beta_t^a)/\|\beta_t^x - \mathcal{C}_t^a\|$.
    - Calculate jump size α where α is drawn from a Normal density with zero mean and unit variance scaled by D the distance between parent states, $\alpha = D \times N(0,1)$
    - FOR k=1 to dimensionality of state
    - $\beta_{t+1}^{a*k} = \beta_t^{ak} + (\alpha \times V^k)$
    - END. for loop.
  * END. flip or crossover stage
- END. metropolis or GA stage
- /* Update all states in the population */
- FOR m = 1 to number in population
  * $\beta_{t+1}^m = \beta_t^m$
- END. population update
- /* Now go back and check for acceptance of update to β^a */
- IF $u < Q$, where $u \sim U(0,1)$, and Q is from (3) THEN $\beta_{t+1}^a = \beta_{t+1}^{a*}$
- END. for loop iterations.

Bayesian Statistics and Its Applications
Edited by S.K. Upadhyay, U. Singh and D.K. Dey
Anamaya Publishers, New Delhi, India

Shape Classification Procedures with Application to Schizophrenia Diagnosis

Dipak K. Dey[1], Rosalee E. Zackula[2] and Athanasios C. Micheas[3]

[1]Department of Statistics, University of Connecticut, Storrs, CT 06269
[2]University of Kansas, School of Medicine-Wichita, USA
[3]Department of Statistics, University of Missouri-Columbia, USA

Abstract

This article discusses the classification procedures in a shape analysis context. We derive discriminants in shape space, while considering a complex Watson shape distribution for the data, as well as in a tangent space to shape space. Both frequentist and Bayesian approaches are considered. Using MAP (maximum a posteriori) estimates of parameters involved, we derive discriminant rules, and calculate misclassification probabilities using Monte Carlo methods. The methods are exemplified through an example, where we are interested in classifying patients into the normal or schizophrenic groups, based on shapes created by MRI's (magnetic resonance images) of their brain. Hence, the methods provide us with a new way of diagnosis of this medical condition while controlling the error of misallocation.

1. Introduction

The study of shapes that appear in experiments, has drawn much of the attention of researchers in many disciplines, including image analysis of magnetic resonance images (MRI), psychology, genetics, biology and a host of others. New methodologies are developed in an effort to provide better understanding of the shape of an object. Some of the major goals in shape analysis include the development of methods that help obtain a modal (average) shape that summarizes the shape information contained in a population of shapes; comparing populations of shapes, for example, through comparison of modal shapes; employ an appropriate probability distribution that would describe shape stochastically; capture shape variability through credible/confidence sets and finally classify a new shape into one of two or more populations of shapes via discriminant rules. There are several methods that can be employed to answer these questions. Some of the major contributions in the literature include Bookstein (1991, 1996), Dryden and Mardia (1998), Kendall (1977, 1989), Kent (1994), Kendall et al. (1999), Mardia and Dryden (1999), Rohlf (1999) and Micheas and Dey (2005).

In particular, shape classification procedures are of great interest, as it is very often the case where shape information is available for two or more populations and classification of a new shape is desired. Perhaps the most important application of shape classification can be found in medicine, where physicians are often called upon to diagnose and consequently assign a treatment for a disease or disability, solely based on shapes that appear in MRI scans. For example, a radiologist would observe an MRI brain scan of a schizophrenic patient, and try to detect which areas are affected by certain treatments or differ from a normal brain and so forth. Areas, for example, such as the corpus callosum, cerebellum and the ventricles have been shown to be affected by schizophrenia, using postmortem examinations, with respect to both size and shape (Shenton et al., 2001). For instance, there appears to be a ventricular enlargement in schizophrenic brains, as well as abnormal asymmetry of the planum temporale (there is a left asymmetry in normals that is much reduced in patients with schizophrenia). In this context, shapes are created by selecting anatomical landmarks, points of correspondence that match between shapes of objects, that enjoy an interpretation from a biological or medical point of view, directly from the MRI brain scan of a patient. Then based on these points on the plane, we would create the shape configuration (raw landmarks) by representing any point with a complex number with real part the x-coordinate and imaginary part the y-coordinate originally observed.

In order to study shape, we need to remove the effects of translation, dilation and rotation contained in the data. Thus, if $\mathbf{z}^c = (z_1^c, z_2^c, ..., z_k^c)$, denotes the configuration of a shape in k-landmarks on the complex plane, and let $\mathbf{z} = \frac{\mathbf{H}\mathbf{z}^c}{\|\mathbf{H}\mathbf{z}^c\|}$, denotes the corresponding pre-shape on the $(k-1)$-dimensional complex sphere $\mathcal{CS}^{k-1} = \{\mathbf{z} \in \mathcal{C}^{k-1} : \mathbf{z}^*\mathbf{z} = 1\}$, where $\mathcal{C}$ is the complex plane, and $\mathbf{H}$ the sub-Helmert matrix, that is, the last $(k-1)$ rows of the Helmert matrix. The space of all shapes $\mathbf{z}$ and its rotation $e^{i\theta}\mathbf{z}$, for any angle θ, is called the pre-shape space. Assuming that we have two or more populations of pre-shapes, our goal is to derive discriminant rules that would allow us to classify a new pre-shape into one of the populations under consideration. The classification rules are derived based on modelling shapes on the complex sphere. An alternative, and popular approach, is tangent space inference. In this case the original landmarks are transformed into a linear space, tangent to the complex sphere at some pole, usually chosen to be the Full Procrustes mean (see, e.g., Dryden and Mardia, 1999, p. 72, for Kent's partial Procrustes tangent coordinates). Since this is a linear space, it is a very attractive to work with, especially for data that are not highly dispersed, although the approach heavily depends on the selection of the pole on the complex sphere. Standard discriminant analysis techniques can then be used to create classification rules, but they still depend on the pole selected. This article derives rules using both approaches, working directly on the complex sphere, where even for highly dispersed data the general methodology will yield trustworthy results.

It should be noted however, that the approaches we consider are applied to two-dimensional data. Medical research of many diseases is still based on investigation of two-dimensional MRI brain scans and with this in mind, the classification of shapes based on two dimensional data is addressed herein.

In particular, Section 2 discusses models used to describe shape stochastically on the complex sphere, and the estimation of parameters involved from a frequentist and Bayesian perspective. Section 3 explores tangent space in inference and briefly discusses methods of classification in this linear space. Section 4 develops discriminant rules in shape space with implementation via Monte Carlo methods for computing misclassification probabilities. An important application of the methodology is developed in Section 5 while Section 6 contains some concluding remarks.

2. Modeling Shape on the Complex Sphere and Estimation

The complex Bingham shape distribution was proposed and studied by Kent (1994), which provided an appealing framework for the analysis of two-dimensional shape data. The density of the distribution is given by

$$f(\mathbf{z}|\mathbf{A}) = \left(2\pi^{k-1}\sum_{j=1}^{k-1}\frac{\exp(\lambda_j)}{\prod_{i\neq j}(\lambda_i - \lambda_j)}\right)^{-1} \exp(\mathbf{z}^*\mathbf{A}\mathbf{z}) \tag{2.1}$$

where $\mathbf{z}^*$ is the conjugate-transpose of $\mathbf{z}$, $\mathbf{A}$ is negative definite and Hermitian matrix, i.e., $\mathbf{A} = \mathbf{A}^*$, and $\lambda_1 < ... < \lambda_{k-2} < \lambda_{k-1} = 0$ are the eigenvalues of $\mathbf{A}$. The matrices $\mathbf{A}$ and $\mathbf{A} + a\mathbf{I}$ define the same complex Bingham distribution, since $\mathbf{z}^*\mathbf{z} = 1$. We remove this non-identifiability by setting $\lambda_{k-1} = 0$. We write $CB_{k-1}(\mathbf{A})$ to denote the complex Bingham distribution on the complex sphere of $k - 1$ dimension. The modal shape here is γ_{k-1}, the dominant eigenvector of $\mathbf{A}$, and is the most likely pre-shape.

Notice that $f(e^{i\psi}\mathbf{z}) = f(\mathbf{z})$, i.e., it is invariant under rotations of the pre-shape $\mathbf{z}$, and hence the distribution is suitable for shape analysis as rotational effects are removed automatically by the distribution. Moreover, if $\mathbf{z} \sim CN_{k-1}(\mathbf{0}, -\mathbf{A}^{-1})$ then $\mathbf{z}|\{||\mathbf{z}|| = 1\} \sim CB_{k-1}(\mathbf{A})$. This relationship is useful for generating samples from the complex Bingham distribution through generated values of the complex Normal distribution. See Kent (1994) and Dryden and Mardia (1998) for more on the properties of the complex Bingham distribution.

In the case where an isotropic structure is reasonable for the data, we consider the complex Watson shape distribution, defined and explored by Mardia and Dryden (1999), the density of which can be written as

$$f(\mathbf{z}|\xi,\mu) = \left(2\pi^{k-1}\xi^{2-k}[\exp(\xi) - \sum_{r=0}^{k-3}\frac{\xi^r}{r!}]\right)^{-1} \exp(\xi\,||\mathbf{z}^*\mu||^2) \tag{2.2}$$

where $\xi \in \mathcal{R}$ is the concentration parameter and μ the mean pre-shape. We denote the complex Watson distribution as $CW_{k-1}(\xi, \mu)$. Notice that if $\xi > 0$ the modes are obtained at $e^{i\theta}\mu, 0 \leq \theta < 2\pi$, which is useful in shape analysis for representing population shape. If $\xi < 0$ the distribution has modes at all vectors orthogonal to μ and for $\xi = 0$ the distribution is reduced to the uniform distribution on $\mathcal{CS}^{k-1}$. The complex Watson shape distribution is a special case of the complex Bingham distribution if we assume $\mathbf{A}$ has two distinct eigenvalues, i.e., if we let $\mathbf{A} = -\xi(\mathbf{I}_{k-1}-\mu\mu^*)$.

2.1 Frequentist Approach

Assume that $\mathbf{z}_1, ..., \mathbf{z}_n$ is a sample from $CB_{k-1}(\mathbf{A})$, then it is easy to show (Kent, 1994) that the MLE's of $\gamma_1, ..., \gamma_{k-1}$, the eigenvectors of $\mathbf{A}$ are $\widehat{\gamma}_j = \mathbf{v}_j$, $j = 1, 2, ..., k - 1$, where $\mathbf{v}_1, ..., \mathbf{v}_{k-1}$ the eigenvectors of $\mathbf{S} = \sum_{i=1}^{n}\mathbf{z}_i\mathbf{z}_i^*$, the complex sum of squares and products matrix. Under high concentrations, i.e., when the landmarks are not highly dispersed, $\widehat{\lambda}_j \approx -\frac{n}{l_j}, j = 1, 2, ..., k - 2$, where $0 < l_1 < ... < l_{k-1}$ are the eigenvalues of $\mathbf{S}$.

Based on a sample $\mathbf{z}_1, ..., \mathbf{z}_n$ from $CW_{k-1}(\xi, \mu)$, we can estimate the parameters by their MLE's as $\widehat{\mu} = \mathbf{v}_{k-1}$ and under high concentrations $\widehat{\xi} \approx \frac{2n(k - 2)}{n - l_{k-1}}$.

2.2 Bayesian Perspective

Let us consider modeling the data according to a complex Watson distribution, that is, $\mathbf{z}_i \sim CW_{k-1}(\xi, \boldsymbol{\mu}) \equiv CB_{k-1}(\mathbf{A} = -\xi(\mathbf{I}_{k-1} - \boldsymbol{\mu}\boldsymbol{\mu}^*))$, $i = 1, ..., n$, for some concentration parameter ξ and modal shape $\boldsymbol{\mu}$ and under high concentrations we will estimate ξ by its corresponding MLE for simplicity, and hence $\widehat{\xi} \approx \dfrac{2n(k-2)}{n - l_{k-1}}$. A test for isotropy is of course required in this case, in order to assess goodness of fit of the model to the data.

Since we are primarily interested in $\boldsymbol{\mu}$, the modal shape, we consider several prior distributions in order to employ the Bayesian paradigm. A natural choice, that is attractive from a mathematical as well as from a modeling point of view, is the conjugate prior distribution which is a complex Bingham, i.e., $\boldsymbol{\mu} \sim CB_{k-1}(\mathbf{A}_{\mathrm{prior}})$, for some known negative definite and Hermitian matrix $\mathbf{A}_{\mathrm{prior}}$. We can easily show that the posterior distribution is $\boldsymbol{\mu}|\mathbf{z} \sim CB_{k-1}(\mathbf{A}_{\mathrm{post}}^{B})$, where $\mathbf{A}_{\mathrm{post}}^{B} = \widehat{\xi}\mathbf{S} + \mathbf{A}_{\mathrm{prior}}$, $\mathbf{S} = \sum_{i=1}^{n} \mathbf{z}_i \mathbf{z}_i^*$, $\mathbf{z} = (\mathbf{z}_1, ..., \mathbf{z}_n)^T$. In this case, the MAP estimator of $\boldsymbol{\mu}$ is the dominant eigenvector of $\widehat{\xi}\mathbf{S} + \mathbf{A}_{\mathrm{prior}}$.

An alternative modeling scheme, would be to consider an isotropic prior distribution as well, $\boldsymbol{\mu} \sim CW_{k-1}(\xi_{\mathrm{prior}}, \boldsymbol{\mu}_{\mathrm{prior}})$, for some known hyper-parameters ξ_{prior} and $\boldsymbol{\mu}_{\mathrm{prior}}$. Under isotropy the posterior distribution becomes $\boldsymbol{\mu}|\mathbf{z} \sim CB_{k-1}(\mathbf{A}_{\mathrm{post}}^{W})$, where $\mathbf{A}_{\mathrm{post}}^{W} = \widehat{\xi}\mathbf{S} - \xi_{\mathrm{prior}}(\mathbf{I}_{k-1} - \boldsymbol{\mu}_{\mathrm{prior}}\boldsymbol{\mu}_{\mathrm{prior}}^*)$. This scheme is somewhat restrictive although, as isotropy is required for both the data and the prior distribution. The dominant eigenvector of $\mathbf{A}_{\mathrm{post}}^{W}$ is once again the MAP estimator of modal shape.

The a priori modal beliefs can be modeled in both cases by the same $\boldsymbol{\mu}_{\mathrm{prior}}$, that is, $\mathbf{A}_{\mathrm{prior}}$ in the conjugate prior depends on $\boldsymbol{\mu}_{\mathrm{prior}}$. The complex vector $\boldsymbol{\mu}_{\mathrm{prior}}$ is often referred to as the ideal shape. The ideal a priori information contained in $\boldsymbol{\mu}_{\mathrm{prior}}$ will be updated by the data to form the posterior distribution.

3. Tangent Space to Shape Space Formulation

Dryden and Mardia (1998) illustrate that standard multivariate statistical techniques in tangent space are equivalent to non-Euclidean shape methods provided the data are highly concentrated. This homomorphism allows the use of all standard multivariate techniques of classification.

Consider Kent's partial Procrustes tangent plane coordinates (Kent, 1994). Assume that we observe pre-shapes $\mathbf{z}_1, ..., \mathbf{z}_n$ and define $\mathbf{S} = \sum_{i=1}^{n} \mathbf{z}_i \mathbf{z}_i^*$. The pole $\boldsymbol{\gamma}$ of the complex sphere where the tangent plane is taken, is chosen to be the full Procrustes mean pre-shape defined as the dominant eigenvector of $\mathbf{S}$. The partial Procrustes tangent coordinates for a pre-shape $\mathbf{z}$, are defined by

$$\mathbf{u} = e^{i\theta}[\mathbf{I}_{k-1} - \boldsymbol{\gamma}\boldsymbol{\gamma}^*]\mathbf{z}_S, \ \mathbf{u} \in \mathcal{T}(\boldsymbol{\gamma}) \tag{3.1}$$

where $\theta = \arg(-\boldsymbol{\gamma}^*\mathbf{z})$, and $\mathcal{T}(\boldsymbol{\gamma})$ denotes the tangent plane at $\boldsymbol{\gamma}$. Note that $\mathbf{u}$ involves only rotation to match the pre-shape $\mathbf{z}$, but not scale effects, followed by a projection onto the tangent plane at $\boldsymbol{\gamma}$. The matrix $[\mathbf{I}_{k-1} - \boldsymbol{\gamma}\boldsymbol{\gamma}^*]$ is the matrix of complex projection into the space orthogonal to $\boldsymbol{\gamma}$, i.e., $\mathbf{u}^*\boldsymbol{\gamma} = 0$. In order to obtain the pre-shape coordinates from the tangent coordinates we use the inverse transformation

$$\mathbf{z}e^{i\theta} = [1 - \mathbf{u}^*\mathbf{u}]^{\frac{1}{2}}\boldsymbol{\gamma} + \mathbf{u}, \ \mathbf{z} \in \mathcal{CS}^{k-1}. \tag{3.2}$$

Clearly, since $\mathcal{T}(\gamma)$ is $(k-1)$ dimensional complex linear space, we can perform classification in $\mathcal{T}(\gamma)$ without any difficulty. Thus, since distances on this tangent plane can approximate (see, e.g. Dryden and Mardia, 1999, p. 73) distances on pre-shape space, provided there is not much variability in the data (which is very often a valid assumption), we can use standard multivariate techniques in tangent space and approximate very well, non-Euclidean shape methods, i.e., methods based on the complex sphere.

Now the usual multivariate classification techniques can be applied to perform classification. Some of the methods include Fisher discriminants, Likelihood ratio, Predictive Bayesian discriminant rules and more. These methods are standard under multinormal populations and, hence, are not presented here (see Mardia et al., 1979, for more details).

4. Classification Methods in Shape Space

Only but a few methods may carry over to the complex sphere. Methods include Likelihood ratio, Bayes discriminant rules and a decision theoretic approach. Classification procedures on the complex sphere require the construction of discriminant rules that are created by dividing the space where the pre-shape $\mathbf{z}$ is defined (the complex sphere $\mathcal{CS}^{k-1}$ of $k-1$ dimension) into disjoint regions. For simplicity and without loss of generality, we consider two populations, and let regions R_1 and R_2 be such that $R_1 \cup R_2 = \mathcal{CS}^{k-1}$. Then a discriminant rule d is defined when we allocate $\mathbf{z}_0$ to Π_i if $\mathbf{z}_0 \in R_i$, $i = 1, 2$.

Classification will be more accurate if Π_i has most of its probability concentrated in region R_i for each $i = 1, 2$.

For what follows, we note that the distributions of the populations Π_1 and Π_2 are assumed to be complex Watson. Then the likelihood that $\mathbf{z}_0$ belongs to either population Π_1 or Π_2 is

$$L_1(\mathbf{z}_0|\widehat{\xi}_1, \boldsymbol{\mu}_1) = f(\mathbf{z}_0|\widehat{\xi}_1, \boldsymbol{\mu}_1) = c(\widehat{\xi}_1)\exp(\widehat{\xi}_1 \|\mathbf{z}_0^*\boldsymbol{\mu}_1\|^2) \tag{4.1}$$

or

$$L_2(\mathbf{z}_0|\widehat{\xi}_2, \boldsymbol{\mu}_2) = f(\mathbf{z}_0|\widehat{\xi}_2, \boldsymbol{\mu}_2) = c(\widehat{\xi}_2)\exp(\widehat{\xi}_2 \|\mathbf{z}_0^*\boldsymbol{\mu}_2\|^2) \tag{4.2}$$

with normalizing constants given by

$$c(\xi) = \left(2\pi^{k-1}\xi^{2-k}[\exp(\xi) - \sum_{r=0}^{k-3}\frac{\xi^r}{r!}]\right)^{-1} \tag{4.3}$$

where $\widehat{\xi}_1$ and $\widehat{\xi}_2$ are the MLE's of ξ_1 and ξ_2, respectively, the concentration parameters of the two populations. The likelihoods above are known (up to estimation of parameters).

An intuitively appealing procedure is to assign a pre-shape to be classified $\mathbf{z}_0$ to Π_1 if the likelihood function corresponding to this population is large enough, compared to the likelihood function corresponding to population Π_2. Moreover, assume that we have a priori information about the two populations, and hence we might regard a pre-shape as more likely to come from population Π_1 instead of Π_2. For example, consider classification of a pre-shape created from the MRI of the brain of a patient that has been tested for a certain disease or disability by some medical means. There are certain conditions that would be more likely for the patient to have based on the initial medical tests, i.e., a patient diagnosed with schizophrenia by a medical doctor, should have higher probability to be classified into the schizophrenic group instead of the normal group. Thus, the need for procedures that incorporate this prior information naturally arises. Let $\pi_1 > 0$ and $\pi_2 > 0$ denote the prior probabilities of populations Π_1 and Π_2, respectively. Then a first classification procedure is suggested when

$$\mathbf{z}_0 \text{ is allocated to population } \Pi_1 \text{ if } \pi_1 L_1(\mathbf{z}_0|\widehat{\xi}_1, \boldsymbol{\mu}_1) > \pi_2 L_2(\mathbf{z}_0|\widehat{\xi}_2, \boldsymbol{\mu}_2) \tag{4.4}$$

otherwise $\mathbf{z}_0$ is allocated to population Π_2, where $\boldsymbol{\mu}_1$, $\boldsymbol{\mu}_2$ are the modal shapes for each population. When $\pi_1 L_1(\mathbf{z}_0|\widehat{\xi}_1, \boldsymbol{\mu}_1) = \pi_2 L_2(\mathbf{z}_0|\widehat{\xi}_2, \boldsymbol{\mu}_2)$, the rule would allocate $\mathbf{z}_0$ to either population with equal weight (cannot decide).

4.1   Decision Theoretic Discriminant Rule

Consider the loss function

$$L(i,j) = c_{ij}, \text{ when } \pi_1 L_1(\mathbf{z}_0|\widehat{\xi}_1, \boldsymbol{\mu}_1) > \pi_2 L_2(\mathbf{z}_0|\widehat{\xi}_2, \boldsymbol{\mu}_2)$$

the loss of allocating an observation as coming from Π_i, when it actually comes from Π_j, with $L(i,i) = 0$, i.e., no loss for correctly classifying the pre-shape. The Bayes rule then is easily shown to be

$$\mathbf{z}_0 \text{ is allocated to population } \Pi_1 \text{ if } c_{12}\pi_1 L_1(\mathbf{z}_0|\widehat{\xi}_1, \boldsymbol{\mu}_1) > c_{21}\pi_2 L_2(\mathbf{z}_0|\widehat{\xi}_2, \boldsymbol{\mu}_2) \tag{4.5}$$

otherwise $\mathbf{z}_0$ is allocated to population Π_2. Under the complex Watson shape distribution for the data and for high concentrations we have, $\widehat{\xi}_1 \approx \dfrac{2n(k-2)}{n - l^1_{k-1}}$, $\widehat{\xi}_2 \approx \dfrac{2n(k-2)}{n - l^2_{k-1}}$, where l^1_{k-1} and l^2_{k-1} are the dominant eigenvectors of $\mathbf{S}_1 = \displaystyle\sum_{i=1}^{n} \mathbf{z}_i \mathbf{z}_i^*$ and $\mathbf{S}_2 = \displaystyle\sum_{i=1}^{n} \mathbf{w}_i \mathbf{w}_i^*$, respectively, based on samples from the two populations $\mathbf{z}_i \in \Pi_1$ and $\mathbf{w}_i \in \Pi_1$, $i = 1, ..., n$.

After some algebra, the decision theoretic discrimination rule becomes:

$$\mathbf{z}_0 \text{ is allocated to population } \Pi_1 \text{ if } \widehat{\xi}_1 \mathbf{z}_0^* \boldsymbol{\mu}_1 \boldsymbol{\mu}_1^* \mathbf{z}_0 > \widehat{\xi}_2 \mathbf{z}_0^* \boldsymbol{\mu}_2 \boldsymbol{\mu}_2^* \mathbf{z}_0 + \ln \frac{c_{21}\pi_2 c(\widehat{\xi}_2)}{c_{12}\pi_1 c(\widehat{\xi}_1)} \tag{4.6}$$

otherwise allocate to Π_2. Note that for $c_{12} = c_{21} = 1$, we obtain the Bayes discriminant rule, and if furthermore $\pi_1 = \pi_2$ $(= \frac{1}{2})$ we obtain the Likelihood ratio discriminant rule. The estimated discriminant rule is easily obtained by replacing $\boldsymbol{\mu}_1$ and $\boldsymbol{\mu}_2$ above by their corresponding ML or MAP estimators.

4.2   Posterior Expected Loss (PEL) Discriminant Rule

A natural Bayesian approach to classification is found in the PEL discrimination rule, using the loss function introduced in the previous section. First, the PEL's can be shown to be

$$\text{PEL}_1 = E^{\pi(\boldsymbol{\mu}_1, \boldsymbol{\mu}_2|\mathbf{z}_1, \mathbf{z}_2)}(L(1,2)) = c_{12}p_{12}$$

and

$$\text{PEL}_2 = E^{\pi(\boldsymbol{\mu}_1, \boldsymbol{\mu}_2|\mathbf{z}_1, \mathbf{z}_2)}(L(2,1)) = c_{21}p_{21}$$

where

$$p_{ij} = P^{\pi(\boldsymbol{\mu}_1, \boldsymbol{\mu}_2|\mathbf{z}_1, \mathbf{z}_2)}\left(\widehat{\xi}_i \mathbf{z}_0^* \boldsymbol{\mu}_i \boldsymbol{\mu}_i^* \mathbf{z}_0 > \widehat{\xi}_j \mathbf{z}_0^* \boldsymbol{\mu}_j \boldsymbol{\mu}_j^* \mathbf{z}_0 + \ln \frac{\pi_j c(\widehat{\xi}_j)}{\pi_i c(\widehat{\xi}_i)} \right) \tag{4.7}$$

with $p_{12} + p_{21} = 1$, $\pi(\boldsymbol{\mu}_1, \boldsymbol{\mu}_2|\mathbf{z}_1, \mathbf{z}_2)$ the joint posterior distribution of $\boldsymbol{\mu}_1, \boldsymbol{\mu}_2$ given $\mathbf{z}_1, \mathbf{z}_2$, the observed matrices of pre-shapes from populations Π_1 and Π_2, respectively. Finally the rule becomes

$$\mathbf{z}_0 \text{ is allocated to population } \Pi_1 \text{ if } c_{12}p_{12} < c_{21}p_{21} \tag{4.8}$$

otherwise allocate into Π_2. Notice that the PEL rule is accounting for variability in the parameters of the model by averaging over the joint posterior distribution.

4.3 Misclassification Probabilities

The misclassification probabilities are the estimated measures of performance of any discriminant rule, and play a vital role in the selection of the best procedure we should use for future samples that we wish to allocate into one of the populations. The misclassification probabilities are defined in general as

$$\Pr\left(\text{classify } \mathbf{z}_0 \text{ into } \Pi_1 | \mathbf{z}_0 \in \Pi_2\right) = \alpha \tag{4.9}$$

and

$$\Pr\left(\text{classify } \mathbf{z}_0 \text{ into } \Pi_2 | \mathbf{z}_0 \in \Pi_1\right) = \beta \tag{4.10}$$

The probabilities α and β should be small, and are often different in medical applications. For instance, it could be argued that the cost of classifying a normal subject as schizophrenic is less than classifying a schizophrenic as normal, since the latter patient may pose an immediate threat to society once released from care.

Under the decision theoretic discrimination rule we have

$$\widehat{\alpha} = P(\widehat{\xi}_1 \mathbf{z}_0^* \widehat{\boldsymbol{\mu}}_1 \widehat{\boldsymbol{\mu}}_1^* \mathbf{z}_0 > \widehat{\xi}_2 \mathbf{z}_0^* \widehat{\boldsymbol{\mu}}_2 \widehat{\boldsymbol{\mu}}_2^* \mathbf{z}_0 + C | \mathbf{z}_0 \sim CW_{k-1}(\widehat{\xi}_2, \widehat{\boldsymbol{\mu}}_2)) \tag{4.11}$$

and
$$\widehat{\beta} = P(\widehat{\xi}_1 \mathbf{z}_0^* \widehat{\boldsymbol{\mu}}_1 \widehat{\boldsymbol{\mu}}_1^* \mathbf{z}_0 < \widehat{\xi}_2 \mathbf{z}_0^* \widehat{\boldsymbol{\mu}}_2 \widehat{\boldsymbol{\mu}}_2^* \mathbf{z}_0 + C | \mathbf{z}_0 \sim CW_{k-1}(\widehat{\xi}_1, \widehat{\boldsymbol{\mu}}_1))$$

where
$$C = \ln \frac{c_{21} \pi_2 c(\widehat{\xi}_2)}{c_{12} \pi_1 c(\widehat{\xi}_1)} \tag{4.12}$$

Note that $\widehat{\alpha}$ and $\widehat{\beta}$ depend on the selection of the constant C. Typically, for application of the rule, we search for C that would give us misclassification probabilities $\widehat{\alpha}$ and $\widehat{\beta}$ approximately equal.

For the PEL discrimination rule the estimated misclassification probabilities are computed as follows:

$$\widehat{\alpha} = P\left(c_{12}p_{12} < c_{21}p_{21} | \mathbf{z}_0 \sim CW_{k-1}(\widehat{\xi}_2, \widehat{\boldsymbol{\mu}}_2)\right) \tag{4.13}$$

and
$$\widehat{\beta} = P\left(c_{12}p_{12} > c_{21}p_{21} | \mathbf{z}_0 \sim CW_{k-1}(\widehat{\xi}_1, \widehat{\boldsymbol{\mu}}_1)\right)$$

Here, $\widehat{\boldsymbol{\mu}}_1$ and $\widehat{\boldsymbol{\mu}}_2$ are the MAP estimates (or the MLE's) of $\boldsymbol{\mu}_1$ and $\boldsymbol{\mu}_2$, respectively.

5. Application of Methods to Diagnosing Schizophrenia

Schizophrenia is a mental disorder that affects an estimated 1% of the world's population, with approximately 1.5 to 2 million people affected in the United States alone. Characteristics of this disorder include from affective flattening, alogia (poverty of speech), auditory disturbances, delusions, distorted thoughts, to grossly disorganized or catatonic behavior, hallucinations, and many more. To date, schizophrenia remains poorly understood and the cause is largely unknown.

Current theories suggest that schizophrenia may include as many as five subtypes including: paranoid, disorganized, catatonic, undifferentiated, and residual (Diagnostic and Statistical Manual of Mental Disorders, fourth edition, or DSM-IV, 1994). Moreover, no laboratory findings have been identified to be a diagnostic tool for schizophrenia (DSM-IV, p. 280). Therefore, three things that can potentially be used to help determine the true nature of this mental disorder are: an accurate diagnostic tool and a reliable method to identify different types of schizophrenia, as well as a rigorous statistical procedure that would help physicians classify a patient into the schizophrenic or normal group.

With regard to the epidemiology of schizophrenia, much of what we know comes from postmortem studies, and more recently, structural analysis of brain-imaging studies such as computed tomography (CT) and magnetic resonance imaging (MRI). In brief, findings from structural analysis of brain imaging include such things as ventricular enlargement, gray matter volume reduction of medial and neocortical temporal lobe regions, abnormal asymmetry of the planum temporale of the temporal lobe, volume reduction in frontal lobe and parietal lobe, and abnormalities of the corpus callosum, hippocampus, and cerebellum (Shenton et al., 2001).

However, DeQuardo et al. (1996) argue that these structural analyses have generally been limited to a single region of interest considered in relation to other intracranial structures, or to brain size as a whole, which has often led to opposing findings. Moreover, the authors argue that region of interest methods have limited usefulness due to the difficulty in demarcating the same region in brains of different shapes and sizes. Therefore, even with these modern techniques, researchers in the field of schizophrenia continue to re-define, re-classify, and explore causality of schizophrenia.

In order to help answer these important questions, we conducted an analysis of midsagittal slices of MRI brain scans from 28 subjects who were identified by an expert as either schizophrenic or normal,

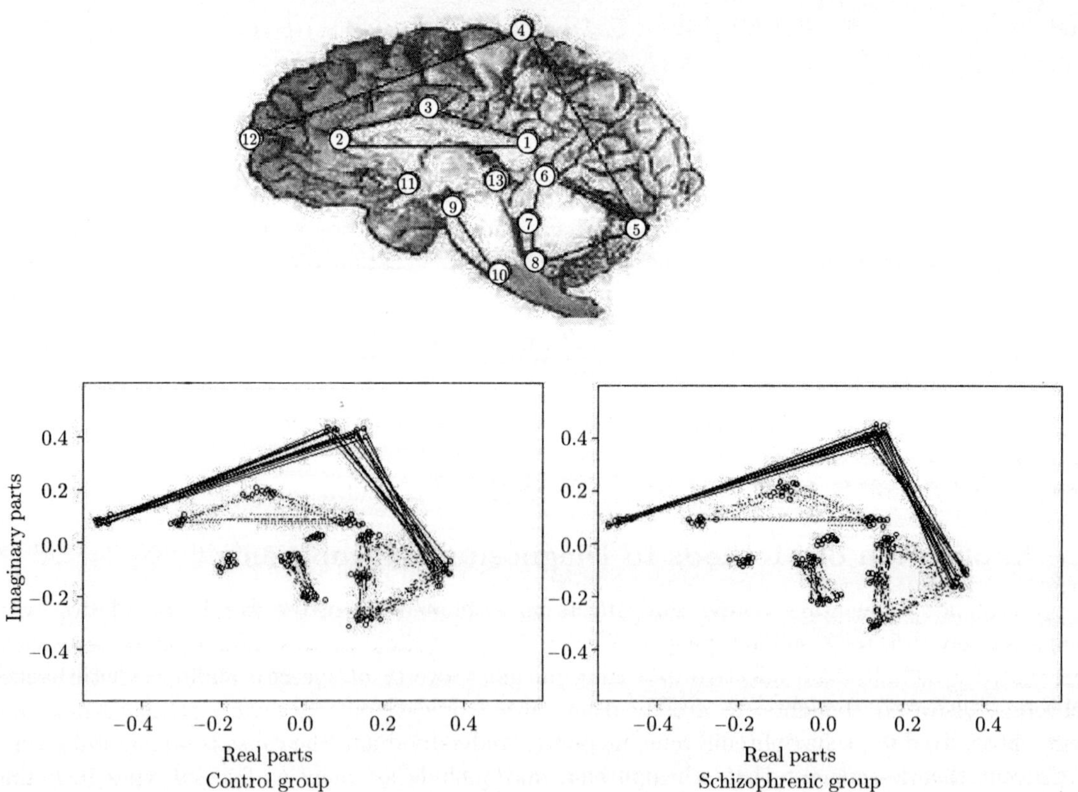

Fig. 1.   Midsagittal image.

with 14 subjects per group. The data comes from DeQuardo et al. (1996) where doctors identified thirteen structural points that were potentially involved in the disease process of schizophrenia. Fig. 1 outlines thirteen points as they would appear on a normal midsagittal brain image. Starting from the leftmost point the landmarks include: (12) frontal pole (extension of a line from (1) through (2) until it intersects the dura), (4) vertex (top of the head, a point relaxed from a standard landmark along the apparent margin of the dura), (5) tentorium of cerebellum at dura; (2) genu (anteriormost point on corpus callosum), (3) top of corpus callosum (uppermost point on arch of callosum) and (1) splenium (posteriormost point of corpus callosum); (11) optic chiasm, (13) superior colliculus; (9) top of pons (anterior margin), (10) bottom of pons (anterior margin); (8) bottom of cerebellum, (7) tip of fourth ventricle, and (6) top of cerebellum. Fig. 1 also illustrates icons for the two groups.

We used these structural points to describe the shape of the brains from each group, examined a variety of possible methods for the analysis, and introduced new methods of analyzing MRI data. More importantly, we outlined a classification process to first separate and then to allocate subjects into one of the two groups, thereby establishing a diagnostic tool for schizophrenia.

A few points are in order about the selection of these structural points as well as the sample size of the study. First, certainly, it is possible that other areas not described above are affected by this condition. Our goal is to assess whether or not these areas are indeed affected by schizophrenia, and use them to help us potentially create a classification rule. Secondly, larger studies are needed that include more areas that medical research is currently concentrating on. With this in mind, we should be cautious about the interpretation of the results for this dataset.

5.1 Tangent Space Classification

Using (3.1), we obtain Kent's partial procrustes tangent plane coordinates of the observed pre-shapes and proceed to analyze the data using standard multivariate analysis techniques. First, we assess bivariate normality at each landmark and for each group, using inspection of 50% confidence ellipses. For the control group, landmarks 1, 3, 5, 8 and 10 have more than 50% of the sample observations falling outside the ellipse, whereas only landmark 10 has observations greater than 50% in the schizophrenic group. Therefore, the normality assumption is suspect, especially in the control group. However, the sample size is small (only 14 per group), and it could be that for larger samples of MRI data the assumption of multivariate normality would not be suspect.

Next, we explore several covariance structures for each group, including heteroscedastic, equal correlation, proportional, group spherical, homoscedastic and spherical. Covariance structures were selected on the basis of significant results from likelihood ratio tests, AIC and BIC scores. Two covariance structures appeared to be appropriate for the schizophrenia data: homoscedastic and proportional. Tables 1 and 2 show results from the discriminant analyses with regard to the two types of structures for a variety of discriminant rules. All Bayes discriminant rules incorporated prior probabilities associated with lifetime prevalence of schizophrenia.

Interestingly, using the Bayes discriminant rule with a homoscedastic covariance structure resulted in classifying all subjects, both schizophrenic and control, into the normal group. In other words, we were unable to properly diagnose schizophrenia with this method. However, an investigation of Box's M-test revealed homoscedasticity between groups for only 6 out of 13 landmark pairs.

In obvious contrast, notice in Table 2 that with a proportional structure the likelihood based discriminant rule yields the lowest error rates over all analyses performed. None of the original control subjects were misclassified into the schizophrenic group, while only three of the schizophrenic patients were misclassified into the normal group.

Table 1.   Discriminant analysis with homoscedastic covariance structure

Discriminant function	Prior probabilities	Actual normal	Error for schizo	Overall error	Post overall	Cross validation overall
Likelihood	Equal	.214	.286	.25	.225	.857
	.99/.01	0	1.0	.01	.03	.787
Predictive	Equal	.214	.286	.25	.328	.857
Bayesian	.99/.01	0	1.0	.01	.014	.787
Fisher	Equal	.214	.286	.25	.225	.857
	.99/.01	0	1.0	.01	.03	.787

Table 2.   Discriminant analysis with proportional covariance structure

Discriminant function	Prior probabilities	Actual normal	Error for schizo	Overall error	Post overall
Likelihood	Equal	.643	0	.321	.174
	.99/.01	0	.214	.002	.092
Predictive	Equal	0	.929	.464	.252
Bayesian	.99/.01	0	1.0	.01	.004

Perhaps a classification rule that was based on a larger sample size would reduce this probability of misclassification by improving the covariance estimates, thereby allowing for greater insight into the structure. It may be that the normality assumption is violated making multivariate statistical results invalid (recall that bivariate normality was rejected in several of the landmarks). Finally, more accurate classification methods can be obtained when conducting the analysis in pre-shape space utilizing distributions that have been shown to be effective in modeling the pre-shapes. We explore this approach next.

5.2   Shape Space Classification

First we begin by assessing the complex Watson model for the data. We notice that at each landmark site there are tight clusters of points that are roughly circular in nature (a requirement for an isotropic model); notice also that the width of each circular cloud is similar (indications that the complex Watson distribution may be appropriate). Moreover, the eigenvalues from the complex sum of squares and products matrix for the control and schizophrenic groups were: (13.935, 0.0222, 0.0162, 0.009, 0.006, 0.004, 0.003, 0.002, 0.001, 0.0004, 0.0003, 0.0002) and (13.925, 0.0274, 0.0191, 0.009, 0.008, 0.004, 0.003, 0.002, 0.001, 0.0007, 0.0005, 0.0001). Notice that the first eigenvalue is approximately equal to the sample size, while the remaining values are all close to zero, and thus an indication of high concentrations as well as isotropic structure. Finally, we test for isotropy in the tangent space to shape space for each group, using multivariate normal models. Using the likelihood ratio test with Bartlett's approximation the test statistics and p-values are 83.39, $p = .675$ (controls) and 84.38, $p = .647$ (schizophrenics). There are 90 degrees of freedom for the approximate χ^2 statistic under isotropy, and therefore we have no evidence against isotropy in each group, and consequently no evidence against the complex Watson distribution.

For what follows we will model the two groups as: $\mathbf{z}_i^{\text{schizo}} \sim CW_{k-1}(\widehat{\xi}_{\text{schizo}}, \boldsymbol{\mu}_{\text{schizo}})$, where $\widehat{\xi}_{\text{schizo}} \approx \dfrac{2n(k-2)}{n - l_{k-1}^{\text{schizo}}} = 4113.432$, and $\mathbf{z}_i^{\text{cont}} \sim CW_{k-1}(\widehat{\xi}_{\text{cont}}, \boldsymbol{\mu}_{\text{cont}})$, where $\widehat{\xi}_{\text{cont}} \approx \dfrac{2n(k-2)}{n - l_{k-1}^{\text{cont}}} = 4754.752$. We consider the conjugate prior distributions: $\boldsymbol{\mu}_{\text{schizo}}, \boldsymbol{\mu}_{\text{control}} \sim CB_{k-1}(\mathbf{A}_{\text{prior}})$, for some negative definite and Hermitian matrix $\mathbf{A}_{\text{prior}}$ that is created by the eigenvectors of $\mathbf{I}_{k-1}$ and the ideal a priori modal pre-shape (given here in Helmertized form):

$$\boldsymbol{\mu}_{\text{prior}} = (-0.26 + 0.15i, 0.044 + 0.08i, 0.28 + 0.13i, 0.15 - 0.42i, 0.04 - 0.15i, -0.05 - 0.23i,$$

$$-0.11 - 0.33i, -0.13 - 0.024i, -0.15 - 0.16i, -0.24 + 0.06i, -0.4 + 0.33i, 0.03 + 0.0i)$$

The posterior distributions are easily shown to be: $\boldsymbol{\mu}_{\text{schizo}} | \mathbf{z}_{\text{schizo}} \sim CB_{k-1}(\mathbf{A}_{\text{post}}^{\text{schizo}})$, where $\mathbf{A}_{\text{post}}^{\text{schizo}} = \widehat{\xi}_{\text{schizo}}\mathbf{S}_{\text{schizo}} + \mathbf{A}_{\text{prior}}$, $\mathbf{S}_{\text{schizo}} = \displaystyle\sum_{i=1}^{n} \mathbf{z}_i^{\text{schizo}} \left(\mathbf{z}_i^{\text{schizo}}\right)^*$, and $\boldsymbol{\mu}_{\text{cont}} | \mathbf{z}_{\text{cont}} \sim CB_{k-1}(\mathbf{A}_{\text{post}}^{\text{cont}})$, where $\mathbf{A}_{\text{post}}^{\text{cont}} = \widehat{\xi}_{\text{cont}}\mathbf{S}_{\text{cont}} + \mathbf{A}_{\text{prior}}$, $\mathbf{S}_{\text{cont}} = \displaystyle\sum_{i=1}^{n} \mathbf{z}_i^{\text{cont}} \left(\mathbf{z}_i^{\text{cont}}\right)^*$. Note that $\boldsymbol{\mu}_{\text{cont}} | \mathbf{z}_{\text{cont}}$ and $\boldsymbol{\mu}_{\text{schizo}} | \mathbf{z}_{\text{schizo}}$ are independent. The MAP estimates for each population are given by the corresponding dominant eigenvectors of $\mathbf{A}_{\text{post}}^{\text{cont}}$ and $\mathbf{A}_{\text{post}}^{\text{schizo}}$.

First, we apply the decision theoretic rule given in (4.6), with misclassification probabilities computed in (4.11), and constant C as defined in (4.12). We estimated $\boldsymbol{\mu}_{\text{cont}}$ and $\boldsymbol{\mu}_{\text{schizo}}$ in the classification rule, by their corresponding MAP estimators. There were 10000 generated shapes from each population used, in order to conduct the analysis. Table 3 summarizes the Monte Carlo simulations depending on the value of C we choose. Note that we take Π_1 the normal and Π_2 the schizophrenic groups.

Table 3.   Misclassification probabilities for schizophrenia data

	Value of actual probabilities from data				
	$\widehat{\alpha}$	$\widehat{\beta}$	C	α	β
Fixing	.1	0	636.89	.14	.07
the value	.05	.000333	637.77	.14	.07
of α	.01	.00233	639.9	.14	.21
Common	.004333	.004333	640.65	.14	.21
Fixing	.002	.01	641.465	.14	.35
the value	0	.05	643.009	.07	.5
of β	0	.1	644.0638	.07	.5

Notice that the actual misclassification probabilities vary significantly from the estimated ones. There are several reasons that cause this discrepancy. First, we have very small sample sizes, 14 from each group, that heavily affect the actual probabilities. Moreover, we investigated differences in average modal shapes in order to assess how close the two populations were. Using two sample Hotelling T^2 tests between landmarks (on tangent space to shape space and using Kent's partial Procrustes tangent coordinates), we found that landmarks 6, 9 and 13 do not differ significantly, which may account for the actual misclassification probabilities we see in Table 1. It is clear that this method has difficulty in discriminating between populations of pre-shapes since the mean pre-shapes are very similar (and their tangent Procrustes tangent coordinates are very close), but this is a problem that most discriminant rules face and the situation might be different from dataset to dataset.

In order to eliminate any possible effect of the a priori information we employed to the analysis, we consider the PEL discriminant rule. Not surprisingly, the PEL discriminant rule applied to the actual data performs quite well with misclassification probabilities equal to zero. Here we had costs $c_{12} = 1 = c_{21}$ and $\pi_1 = .99$, $\pi_2 = .01$. Hence, it seems that the PEL method is able to capture variability in the priors that might affect the results.

6. Concluding Remarks

We investigated several classification rules for classifying shapes into one or more populations of shapes. We developed a decision theoretic discriminant rule as well as discriminant rules based on PEL. Although exact misclassification probabilities are difficult to obtain due to the nature of the probability distributions involved, we can apply Monte Carlo methodology to estimate them. Moreover, we discussed an important application to identifying schizophrenia based on shape created by MRI brain scans. We concluded that it is possible to accurately diagnose schizophrenia utilizing landmarks from MRI data, although one must proceed with caution due to the small sample sizes. Further studies are needed of larger datasets, that would incorporate more information about the subject, such as gender, age, medical treatment they have been under, and so forth.

We discussed methods on the complex sphere and working directly with pre-shapes as well as a tangent space approach, where the data is transformed into a suitable linear space, like a tangent plane at some complex pole, and then performed standard multivariate techniques on the new shape coordinates. As illustrated, the methodologies derived on the shape space outperformed those on the tangent plane. We should not discard tangent plane approaches however, since the sample sizes are small and this is perhaps the reason for the bad performance of rules on the tangent space.

References

Bookstein, F.L. (1991). *Morphometric tools for landmark data: Geometry and Biology.* Cambridge University Press, Cambridge.

Bookstein, F.L. (1996). Biometrics, Biomathematics and the Morphometric Synthesis. *Bulletin of Mathematical Biology,* **58**, 313-365.

DeQuardo, J.R., Bookstein, F.L., Green, W.D.K., Brunberg, J.A. and Tandon, R. (1996). Spatial relationships of neuroanatomic landmarks in schizophrenia. *Psychiatry Research: Neuroimaging,* **67**, 81-95.

Dryden, I.L. and Mardia, K.V.(1998). *Statistical Shape Analysis.* John Wiley & Sons.

Kendall, D.G. (1977). The diffusion of shape. *Advances in Applied Probability,* **9**, 428-430.

Kendall, D.G. (1989). A Survey of the Statistical Theory of Shape. *Statistical Science,* **4**, 87-120.

Kendall, D.G., Barden, D., Carne, T.K. and Le, H. (1999). *Shape & Shape Theory.* John Wiley & Sons.

Kent, J.T. (1994). The complex Bingham distribution and shape analysis. *Journal of the Royal Statistical Society, Series B,* **56**, 285-299.

Kent, J.T. (1997). Data analysis for shapes and images. *Journal of Statistical Planning and Inference,* **57**, 181-193.

Mardia, K.V. and Dryden, I.L. (1999). The complex Watson distribution and shape analysis. *Journal of the Royal Statistical Society, Series B,* **61**, 913-926.

Mardia, K.V., Kent, J.T. and Bibby, J.M. (1979). *Multivariate Analysis.* Academic Press.

Micheas, A.C. and Dey, D.K. (2005). Modeling shape distributions and inferences for assessing differences in shapes. *Journal of Multivariate Analysis,* **92**, 257-280.

Rohlf, F.J. (1999). Shape Statistics: Procrustes Superimpositions and Tangent Spaces. *Journal of Classification,* **16**, 197-223.

Shenton, M.E., Dickey, C.C., Frumin, M. and McCarley, R.W. (2001). A review of MRI findings in schizophrenia. *Schizophrenia Research,* **49**, 1-52.

Bayesian Statistics and Its Applications
Edited by S.K. Upadhyay, U. Singh and D.K. Dey
Anamaya Publishers, New Delhi, India

Checking for Prior-Data Conflict with Hierarchically Specified Priors

Michael Evans and Hadas Moshonov

Department of Statistics, University of Toronto, Canada

Abstract

Priors are often specified component-wise. This may entail placing independent priors on parameter components or specifying the prior in a sequential or hierarchical fashion. We consider methods for checking for the source of any prior-data conflict in the individually specified components of the prior.

1. Introduction

Virtually all statistical analyses are dependent on choices made by the analyst. In Bayesian analyses these choices take the form of the sampling model and the prior. If inferences drawn from the model, prior and data are to have validity, then it is important that we feel confident that the choices made make sense in light of the observed data.

Model checking in this context has been discussed in Guttman (1967), Box (1980), Rubin (1984), Gelman, Meng and Stern (1996), Bayarri and Berger (2000) and Johnson (2004). Most of these papers considered the effect of both the sampling model and the prior simultaneously while Bayarri and Berger (2000) focused on the sampling model.

As pointed out in Evans and Moshonov (2005), however, there are two possible ways in which the Bayesian model can fail. The sampling model may fail, in the sense that the observed data is surprising for each of the assumed distributions in the model or, when the sampling model is appropriate, the prior may place its mass primarily on distributions in the sampling model for which the observed data is surprising. We refer to the latter failure as *prior-data conflict*. It is also argued in Evans and Moshonov (2005) that it is important to check for these possible failures separately. For, when there is sufficient data, prior-data conflict can be ignored as the impact of the prior on inference disappears with increasing amounts of data.

The developments in Evans and Moshonov (2005) lead to basing the assessment of whether or not a prior-data conflict exists, on the prior predictive distribution for the minimal sufficient statistic T for the model. By this we mean that, if the observed value of T is a surprising (out in the tails) value for the prior predictive distribution of T, then we have evidence that a prior-data conflict exists. Part of the justification for this is that the prior predictive distribution of the full data given T does not depend on the prior and so can tell us nothing about the existence of prior-data conflict.

Further, it was shown in Evans and Moshonov (2005) that when there is an ancillary statistic $U(T)$, i.e., an ancillary that is a function of T, then it is necessary to replace the prior predictive of T by the conditional prior predictive of T given $U(T)$. This removes variation in the prior predictive that does not depend on the particular prior used and so results in a more accurate assessment of a prior-data conflict.

We denote the sample space for the data s by S, the parameter space by Ω, the sampling model by the collection of densities $\{f_\theta : \theta \in \Omega\}$ with respect to some support measure μ, and the prior probability measure on Ω by Π, with density π with respect to a support measure υ. The developments in Evans and Moshonov (2005) then lead to the factorization of the full joint distribution of (θ, s) as

$$P\left(\cdot \mid T\right) \times P_{U(T)} \times M_T\left(\cdot \mid U(T)\right) \times \Pi\left(\cdot \mid T\right) \tag{1}$$

where $P\left(\cdot \mid T\right)$ is the conditional distribution of the data given T, $P_{U(T)}$ is the marginal distribution of $U(T)$, $M_T\left(\cdot \mid U(T)\right)$ is the conditional prior predictive of T given $U(T)$ and $\Pi\left(\cdot \mid T\right)$ is the posterior distribution of the parameter θ. As $P\left(\cdot \mid T\right)$ and $P_{U(T)}$ depend only on the model $\{f_\theta : \theta \in \Omega\}$ they are available for checking the sampling model, $M_T\left(\cdot \mid U(T)\right)$ is available for checking for prior-data conflict and finally $\Pi\left(\cdot \mid T\right)$ is available for inference about θ. In effect each component of this factorization plays a role in a statistical analysis quite separate from the others. First we check for the correctness of the sampling model, if no evidence is found that the sampling model is incorrect, then we proceed to check for prior-data conflict, and finally proceed to inference about θ if no evidence of prior-data conflict is obtained.

The end result of deciding that there is evidence that the model is not appropriate for the data, or that there is evidence of a prior-data conflict can, however, vary depending on the circumstances. For example, we may decide that, even though there is evidence for a prior-data conflict, the amount of data is sufficient so that the conflict can be ignored and we can proceed to inference about θ using $\Pi\left(\cdot \mid T\right)$. Diagnostics are developed in Evans and Moshonov (2005) to assess this. A similar comment applies to checking for correctness of the sampling model. Throughout the remainder of the paper we assume that the appropriate checks have been applied to the sampling model and the outcome of this is such that we are prepared to proceed to check for prior-data conflict. We note that if our conclusion from the first stage is that the sampling model is definitely not appropriate, then it would not make any sense to proceed to checking for prior-data conflict.

As discussed in Evans and Moshonov (2005), it is also apparent that, if a so-called noninformative prior can result in a prior-data conflict, then the prior is indeed adding some information into the analysis. Thus, the lack of any possibility for a prior-data conflict to exist, is at least a partial characterization of what it means for a prior to be noninformative. Since the characterization of prior-data conflict depends on the prior predictive of T, and this is only a proper probability distribution when π is proper, our discussion of this is initially limited to proper priors. This is extended, however, to improper priors by looking at the limits of sequences of proper priors and determining when these sequences are noninformative. As shown in Evans and Moshonov (2005), it matters how we take limits in a sequence if we want to say that the sequence is noninformative.

In this paper we are concerned with not just determining whether or not a prior-data conflict exists, but in determining the source of that conflict. For example, suppose that the prior π is specified sequentially, or hierarchically, as

$$\pi\left(\theta\right) = \pi_1\left(\theta_1\right)\pi\left(\theta_2 \mid \theta_1\right)\cdots\pi\left(\theta_p \mid \theta_1, \theta_2, \ldots, \theta_{p-1}\right) \tag{2}$$

where $\theta = (\theta_1, \theta_2, \ldots, \theta_p)$. Then if we decide that a prior-data conflict exists, it is reasonable to ask if it is some particular component, or components, of π that are causing this and not others. The methods discussed in Evans and Moshonov (2005) deal with the full prior. Further, we extend the criterion for noninformativity to the component parameter case.

The methods we develop here for checking for individual components will be seen not to apply to a general decomposition of a prior, as written in (2). In particular, we will restrict our attention to exponential statistical models and group statistical models and even in those contexts only certain decompositions will be seen to be amenable to our methodology. Also, we will restrict our attention to the case where $p = 2$. Note, however, that this does not require that the θ_i be 1-dimensional parameters. For example θ_1 could be k-dimensional and θ_2 l-dimensional. We will consider more general decompositions in a further paper.

Although there is a certain natural logic to the methods developed here, we can't say that there isn't an approach capable of handling a general decomposition. It is also possible, however, that the methods developed here do place some restrictions on how we decompose a prior for hierarchical specification, at least when we want to assess individual components for conflict with the data. In any case, the methods developed in Evans and Moshonov (2005) are always available for checking the full prior.

2.   Checking Prior Components

Suppose that the prior has been specified as in Eq. (2). Rather than assessing whether or not the full prior is in conflict with the data we consider instead the problem of assessing whether or not each component, as specified in Eq. (2), is in conflict. This approach results in a more refined understanding of how a prior conflicts with the data when such a conflict exists.

When assessing prior-data conflict for the full prior, we first reduced to the prior predictive distribution of the minimal sufficient statistic $T : S \to \mathcal{T}$. This reduction, as argued in Evans and Moshonov (2005), was based in part on the fact that the prior predictive distribution of the data given T does not depend on the prior and so the data, beyond the value of $T(s)$, can tell us nothing about whether a prior-data conflict exists. To determine whether or not a component of the prior is in conflict it seems sensible then to first reduce to the prior predictive distribution of T, denoted as

$$M_T(B) = \int_B \int_\Omega f_{\theta T}(t)\pi(\theta)\, \upsilon(d\theta)\, \lambda(dt) = \int_B m_T(t)\, \lambda(dt)$$

where B is any measurable subset of $\mathcal{T}$ and λ is a support measure on $\mathcal{T}$.

To assess whether or not the full prior is in conflict we then compare the observed value $T(s)$ with M_T to see if it is a surprising value and, if so, conclude that a prior-data conflict exists. Intuitively, $T(s_0)$ is a surprising value if M_T places relatively little mass at $T(s_0)$ compared to other possible values for $T(s)$. We have chosen to follow Box (1980) and measure this by comparing the value of the density m_T at $T(s_0)$ with other possible values via $M_T(m_T(t) \leq m_T(T(s_0)))$ and conclude that $T(s_0)$ is surprising when this probability is small. While this approach has imperfections, it has worked well in the examples we have considered where λ is Euclidean volume in the continuous case or counting measure in the discrete case. Further, Evans and Moshonov (2005) discusses the use of diagnostics to assess whether or not an assessment of a prior-data conflict via the P-value can be ignored.

In addition, if $U(T)$ is ancillary for θ, then we replace M_T in this comparison by the conditional distribution of T given $U(T)$, namely $M_T(\cdot\,|\,U(T))$. This has the effect of removing variation in M_T that is only dependent on the sampling model and so makes for a more accurate assessment of whether or not $T(s)$ is surprising. Further we take $U(T)$ to be a maximal ancillary so that the maximum amount of unrelated variation is removed from the comparison. More details on this, and many examples, can be found in Evans and Moshonov (2005).

We generalize checking for prior-data conflict for the full prior to components by generalizing the concept of ancillarity according to the following definition.

Definition 1. A function V of the minimal sufficient statistic T is said to be *ancillary* for θ_2 if the distribution of $V(T)$ depends on θ_1 but not θ_2.

Note that if $U(T)$ is ancillary for θ and $V(T)$ is ancillary for θ_2, then the conditional distribution of $V(T)$ given $U(T)$ also does not depend upon θ_2. We have also required that the distribution of $V(T)$ depends on θ_1 so that $V(T)$ is not ancillary for the full parameter.

So suppose $\theta = (\theta_1, \theta_2)$ and $\pi(\theta) = \pi_1(\theta_1)\pi_2(\theta_2 \mid \theta_1)$. This kind of decomposition typically arises in hierarchical modeling when θ_2 is the parameter of interest (not necessarily 1-dimensional) and θ_1 is comprised of either nuisance parameters or hyperparameters. In many situations θ_2 is of central interest so this decomposition is not arbitrary.

Initially we consider checking for any prior-data conflict with the prior π_2 on θ_2. If $V(T)$ is ancillary for θ_2 then, just as with an ancillary for the full parameter, we could decide that $T(s)$ is a surprising value from the prior predictive distribution $M_T(\cdot \mid U(T))$, simply because $V(T(s))$ is a surprising value from its conditional prior predictive distribution $M_{V(T)}(\cdot \mid U(T))$. While this is appropriate in assessing whether any prior-data conflict exists, such a conflict cannot be caused by the component π_2, because the prior predictive distribution (conditional on $U(T)$ or otherwise) of $V(T)$ does not depend on π_2. Accordingly, to assess whether or not there is any prior-data conflict caused by the choice of π_2 we must compare $T(s)$ with $M_T(\cdot \mid U(T), V(T))$, as this removes the variation in $M_T(\cdot \mid U(T))$ due to $V(T)$.

Again we want to remove the maximum amount of variation in making this comparison, so we choose $V(T)$ to be a maximal ancillary for θ_2, namely, require that $V(T)$ not be a function of any statistic that is also ancillary for θ_2. As discussed in Lehmann and Scholz (1992) for full ancillaries, there is currently no general method for constructing maximal ancillaries for θ_2 or even proving that a given ancillary is maximal. Further, it is conceivable that there is more than one maximal ancillary for θ_2. We note, however, as in Evans and Moshonov (2005), the lack of a unique maximal ancillary for θ_2 does not cause a problem in this context, because two different maximal ancillaries $V_1(T)$ and $V_2(T)$ simply represent different methods of removing variation from the full prior predictive and so provide different assessments. In other words, if the assessment via $M_T(\cdot \mid U(T), V_1(T))$ provides evidence of a conflict and that via $M_T(\cdot \mid U(T), V_2(T))$ does not (or conversely), this is not a problem as we would conclude that a prior-data conflict due to π_2 exists. In such a situation these tests do not contradict one another, they simply represent different methods of slicing up the conditional prior predictive $M_T(\cdot \mid U(T))$ to see if $T(s)$ looks anomalous.

We also note that the conditional prior predictive distribution for $V(T)$, namely, $M_{V(T)}(\cdot \mid U(T))$, is available for checking whether or not there is any conflict due to π_1. We have the following result.

Theorem 1. If $V(T)$ is ancillary for θ_2 then $M_{V(T)}(\cdot \mid U(T))$ does not depend on π_2.

Proof: Let $\Omega_2(\theta_1) = \{\theta_2 : (\theta_1, \theta_2) \in \Omega\}$, then we have that

$$M_{V(T)}(B \mid U(T)) = \int_{\Omega} P_{V,\theta}(B \mid U(T))\pi(\theta)\, \upsilon(d\theta)$$

$$= \int_{\Omega_1} \int_{\Omega_2(\theta_1)} P_{V,\theta_1}(B\,|\,U(T))\pi_1(\theta_1)\,\pi_2(\theta_2\,|\,\theta_1)\,\upsilon_2(d\theta_2)\,\upsilon_1(d\theta_1)$$

$$= \int_{\Omega_1} P_{V,\theta_1}(B\,|\,U(T))\pi_1(\theta_1)\left(\int_{\Omega_2(\theta_1)}\pi_2(\theta_2\,|\,\theta_1)\,\upsilon_2(d\theta_2)\right)\upsilon_1(d\theta_1)$$

$$= \int_{\Omega_1} P_{V,\theta_1}(B\,|\,U(T))\pi_1(\theta_1)\,\upsilon_1(d\theta_1)$$

which establishes the result. $\square$

We see then that checking for the individual prior components has lead to a further factorization of $M_T(\cdot\,|\,U(T))$ as

$$M_T(\cdot\,|\,U(T)) = M_{V(T)}(\cdot\,|\,U(T)) \times M_T(\cdot\,|\,U(T),V(T)) \tag{3}$$

The availability of this decomposition is dependent on the existence of a statistic $V(T)$ that is ancillary for the parameter of interest θ_2 and in general there is nothing to guarantee this. In Sections 3 and 4, however, we examine some quite general contexts where this is the case.

It is notable that the existence of an appropriate factorization is not dependent on the form of the prior, only on how we specify the prior sequentially. For example, if we specify that θ_1 and θ_2 are a priori independent there is no simplification and we still need to specify that θ_2 is the parameter of primary interest. Further, if instead we specify the prior as $\pi(\theta) = \pi_2(\theta_2)\,\pi_1(\theta_1\,|\,\theta_2)$, then the analysis will be different. Overall the methods we are developing here are meant to be applicable when we have specified the prior according to a specific hierarchical structure and we have reasons for choosing this. If we do not have such a structure, then we can still use the methods of Evans and Moshonov (2005) to assess the full prior.

Suppose now that we have a statistic $V(T)$ that is ancillary for θ_2 and sufficient for θ_1, i.e., $V(T)$ is ancillary for θ_2 and the conditional distribution of T given $V(T)$ does not depend on θ_1. Such a statistic is called sufficient-ancillary for (θ_1,θ_2) in Fraser (1979). In this case, when θ_1 and θ_2 are a priori independent, we have the following result.

Theorem 2. Suppose that $\Omega = \Omega_1 \times \Omega_2$, the prior density is given by $\pi(\theta_1,\theta_2) = \pi_1(\theta_1)\,\pi_2(\theta_2)$ and that $V(T)$ is sufficient-ancillary for (θ_1,θ_2). Then $M_T(B\,|\,U(T),V(T))$ is independent of π_1.

Proof: Denoting the range space for V by $\mathcal{V}$, and the support measure on Ω_i by υ_i, we have that $M_T(\cdot\,|\,U(T),V(T))$ satisfies,

$$M_T(B\,|\,U(T)) = \int_{\mathcal{V}} M_T(B\,|\,U(T),V(T)=v)\,M_{V(T)}(dv\,|\,U(T))$$

$$= \int_{\Omega} P_{T,\theta}(B\,|\,U(T))\pi(\theta)\,\upsilon(d\theta)$$

$$= \int_{\Omega_2}\int_{\Omega_1} P_{T,\theta_1,\theta_2}(B\,|\,U(T))\pi_1(\theta_1)\,\pi_2(\theta_2)\,\upsilon_1(d\theta_1)\,\upsilon_2(d\theta_2)$$

$$= \int_{\Omega_2}\int_{\Omega_1}\left(\int_{\mathcal{V}} P_{T,\theta_2}(B\,|\,U(T),V(T)=v)\,P_{V(T),\theta_1}(dv\,|\,U(T))\right)$$

$$\times \pi_1(\theta_1)\,\pi_2(\theta_2)\,\upsilon_1(d\theta_1)\,\upsilon_2(d\theta_2)$$

$$= \int_{\mathcal{V}} \int_{\Omega_2} P_{T,\theta_2}(B \,|\, U(T), V(T) = v)\pi_2\,(\theta_2)\,\upsilon_2\,(d\theta_2)$$

$$\times \int_{\Omega_1} P_{V(T),\theta_1}(dv \,|\, U(T))\pi_1\,(\theta_1)\,\upsilon_1\,(d\theta_1)$$

$$= \int_{\mathcal{V}} \int_{\Omega_2} P_{T,\theta_2}(B \,|\, U(T), V(T) = v)\pi_2\,(\theta_2)\,\upsilon_2\,(d\theta_2)\, M_{V(T)}(dv \,|\, U(T))$$

Therefore

$$M_T(B \,|\, U(T), V(T) = v) = \int_{\Omega_2} P_{T,\theta_2}(B \,|\, U(T), V(T) = v)\pi_2\,(\theta_2)\,\upsilon_2\,(d\theta_2)$$

almost surely and this is independent of π_1 as claimed. $\square$

From Theorem 2 we see that the check for π_2 is independent of how we specify π_1 whenever the conditions of the theorem obtain. As noted in Fraser (1979), however, situations where we have a sufficient-ancillary statistic for (θ_1, θ_2) are relatively hard to find although we provide one context in Example 1. Of course, the lack of dependence of the check for π_2 on π_1 is also dependent on specifying independent priors for θ_1 and θ_2.

So, in general, when we specify $\pi\,(\theta) = \pi_1\,(\theta_1)\,\pi_2\,(\theta_2 \,|\, \theta_1)$ the check for π_2 is really checking π_1 in part, even though conditioning on an ancillary for θ_2 is helping to remove this dependence somewhat. For this reason it seems that there is a natural order to the checking unless the conditions of Theorem 2 hold. First we check π_1 and we note, by Theorem 1, that this does not depend on how we specify π_2. If we obtain no evidence of any prior-data conflict for π_1, then we proceed to check π_2. This is similar to the restriction that we first check for the correctness of the sampling model, and only proceed to check for prior-data conflict, when we have no evidence against the sampling model.

3.   Checking Prior Components with Exponential Models

Consider the following examples where we can apply the methodology discussed in Section 2.

Example 1. Multinomial Model

Suppose we observe $(f_1, f_2) \sim$ Multinomial$(n, \theta_1, \theta_2, 1 - \theta_1 - \theta_2)$ with

$$\Omega = \{(\theta_1, \theta_2) : \theta_1, \theta_2 \geq 0, 0 \leq \theta_1 + \theta_2 \leq 1\}$$

Then $T = (f_1, f_2)$ is a minimal sufficient statistic. If we prescribe the prior as $\pi\,(\theta_1, \theta_2) = \pi_1\,(\theta_1)\,\pi_2\,(\theta_2 \,|\, \theta_1)$, then $V(T) = f_1 \sim$ Binomial(n, θ_1) is ancillary for θ_2. So we can check for any prior-data conflict with π_1 by using the prior predictive of f_1 and then check for any prior-data conflict with π_2 by using the conditional prior predictive of (f_1, f_2) given f_1.

For example, suppose that θ_1 has a Beta(α_1, β_1) distribution and

$$\frac{\theta_2}{1 - \theta_1} \,|\, \theta_1 \sim \text{Beta}\,(\alpha_2, \beta_2)$$

Note that the joint prior distribution of (θ_1, θ_2) is Dirichlet$(\alpha_1, \alpha_2, \beta_2)$ if and only if $\beta_1 = \alpha_2 + \beta_2$.

The joint prior predictive of (f_1, f_2) is given by

$$m\,(f_1, f_2) = \left\{ \binom{n}{f_1} \frac{\Gamma\,(\alpha_1 + \beta_1)}{\Gamma\,(\alpha_1)\,\Gamma\,(\beta_1)} \frac{\Gamma\,(f_1 + \alpha_1)\,\Gamma\,(n - f_1 + \beta_1)}{\Gamma\,(n + \alpha_1 + \beta_1)} \right\}$$

$$\times \left\{ \binom{n - f_1}{f_2} \frac{\Gamma\,(\alpha_2 + \beta_2)}{\Gamma\,(\alpha_2)\,\Gamma\,(\beta_2)} \frac{\Gamma\,(f_2 + \alpha_2)\,\Gamma\,(n - f_1 - f_2 + \beta_2)}{\Gamma\,(n - f_1 + \alpha_2 + \beta_2)} \right\}. \tag{4}$$

By considering a simple binomial model with a beta prior it is easy to see that each of the factors in Eq. (4) is a probability function and so the first factor is the prior predictive for f_1 and the second factor is the conditional prior predictive for f_2 given f_1.

Notice that if we reparameterize this model as $(\psi_1, \psi_2) = (\theta_1, \theta_2/(1-\theta_1))$ then f_1 is sufficient-ancillary for (ψ_1, ψ_2). Further the above prior on (θ_1, θ_2) induces independent priors on ψ_1 and ψ_2 so that Theorem 2 applies. This tells us that the conditioning on f_1 to check for conflict with π_2 has completely removed the dependence on π_1.

Suppose now that $n = 20, (\alpha_1, \beta_1) = (3, 15), (\alpha_2, \beta_2) = (3, 4)$ and we observe $(f_1, f_2) = (7, 12)$. Then the prior predictive for f_1 is plotted in Fig. 1. We see that $f_1 = 7$ is not very extreme for this distribution. In fact the probability of getting a value at least as far out in the tails as this value is 0.103. Therefore we have no evidence of any prior-data conflict with π_1.

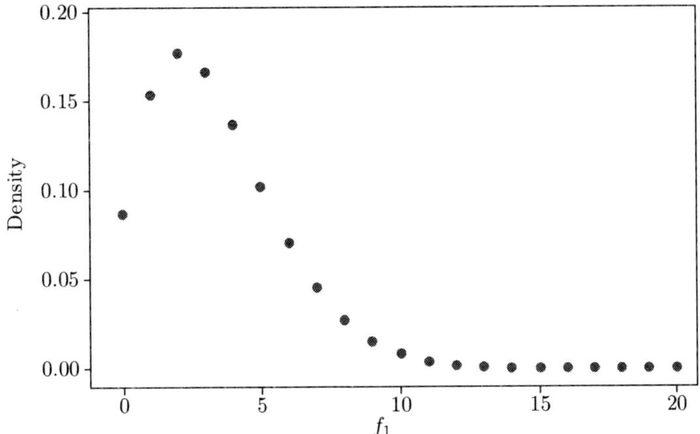

Fig. 1. Prior predictive density for f_1 in Example 1.

We now proceed to assess whether or not $f_2 = 12$ is a reasonable value from its conditional prior predictive given $f_1 = 7$. The conditional prior predictive is plotted in Fig. 2. The probability of getting a value at least as far out in the tails as $f_2 = 12$ is 0.017. Therefore, we have evidence of prior-data conflict with π_2.

Now suppose we choose a uniform prior for θ_2 given the value of θ_1. This entails choosing $\alpha_2 = \beta_2 = 1$. From Eq. (4) we have that the conditional prior predictive for f_2 given f_1 is uniform on $\{0, \ldots, n - f_1\}$. The implication of this is that we would never obtain evidence of a prior-data conflict. As discussed in Evans and Moshonov (2005) this is what we would expect from a prior that is noninformative. Similarly, when we choose $\alpha_1 = \beta_1 = 1$, we get that the prior predictive for f_1 is uniform on $\{0, \ldots, n\}$ and so the component prior is noninformative for θ_1. $\square$

Example 2. Location-scale Normal Model

Suppose that $x = (x_1, ..., x_n)$ is a sample from a $N(\mu, \sigma^2)$ distribution where $\theta_1 = \sigma > 0$ and $\theta_2 = \mu \in R^1$ are unknown. With $s^2 = (n-1)^{-1} \sum (x_i - \overline{x})^2$, then $(\overline{x}, s^2)$ is a minimal sufficient statistic for $\theta = (\sigma^2, \mu)$ with $\overline{x} \sim N(\mu, \sigma^2/n)$ independent of $s^2 \sim (\sigma^2/(n-1))\chi^2_{(n-1)}$. We see immediately that s^2 is ancillary for μ. Note that s^2 is not sufficient-ancillary for this model so Theorem 2 is not applicable.

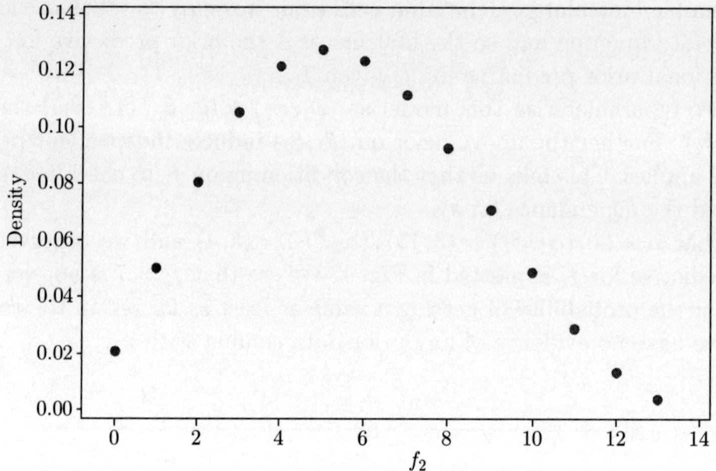

Fig. 2.   Conditional prior predictive of f_2 given that $f_1 = 7$.

Suppose we put the conjugate prior on (σ^2, μ) given by

$$\frac{1}{\sigma^2} \sim \text{Gamma}(\alpha_0, \beta_0)$$

$$\mu \,|\, \sigma^2 \sim N(\mu_0, \tau_0^2 \sigma^2)$$

Hence, we have that $\theta = (\sigma^2, \mu)$ and π_1 is an inverse gamma density while $\pi_2\left(\mu \,|\, \sigma^2\right)$ is the $N(\mu_0, \tau_0^2 \sigma^2)$ density.

The joint prior predictive distribution of $(\overline{x}, s^2)$ is given by

$$m(\overline{x}, s^2) = \int_{-\infty}^{\infty} \int_0^{\infty} f(\overline{x}, s^2 \,|\, \sigma^2, \mu) \pi(\mu \,|\, \sigma^2) \pi(\sigma^2) \, d\sigma^2 \, d\mu$$

$$= \frac{\Gamma\left(\frac{n}{2} + \alpha_0\right)}{\Gamma\left(\alpha_0\right)} \left(n + \frac{1}{\tau_0^2}\right)^{-1/2} \frac{(2\pi)^{-n/2} \beta_0^{\alpha_0}}{\tau_0} \left(\beta_x\right)^{-(n/2) - \alpha_0} \tag{5}$$

where

$$\beta_x = \beta_0 + \frac{1}{2} \frac{\mu_0^2}{\tau_0^2 \left(n\tau_0^2 + 1\right)} + \frac{n-1}{2} s^2 + \frac{n}{2} \frac{1}{n\tau_0^2 + 1} \left(\overline{x} - \mu_0\right)^2. \tag{6}$$

We can easily determine that the marginal prior predictive of s^2 is given by $s^2 \sim (\beta_0/\alpha_0) F_{(n-1, 2\alpha_0)}$. From Eqs. (5) and (6) we see that the conditional prior predictive of $\overline{x}$ given s^2 is distributed as $\mu_0 + \tilde{\sigma} t$ where

$$\tilde{\sigma}^2 = \frac{1}{n\tau_0^2 \left(n + 2\alpha_0 - 1\right)} \left\{ \left(n\tau_0^2 + 1\right) \left(2\beta_0 + (n-1)s^2\right) \right\}$$

and $t \sim \text{Student}(n + 2\alpha_0 - 1)$.

For example, suppose that $\mu = 0$, $\sigma^2 = 1$ and that for a sample of size $n = 20$ from this distribution we obtain $\overline{x}_0 = 0.0358324$ and $s_0^2 = 0.836563$. For the prior we specify $\tau_0^2 = 1$, $\mu_0 = 50$, $\alpha_0 = 1$ and $\beta_0 = 5$.

To assess if there is any conflict with π_1 we compare $s^2/5 = 0.1673126$ with the $F(19, 2)$ distribution. Computing the probability of obtaining a value from the $F(19, 2)$ distribution with density smaller than that obtained at the observed value, leads to the P-value

$$P\left(F(19,2) \leq 0.1673126\right) + P\left(F(19,2) > 1.5295\right) = .47832$$

which does not indicate any prior-data conflict.

In Evans and Moshonov (2005) it was determined that, for a sequence of priors in this example to be noninformative, when checking the full prior for any prior-data conflict, it was necessary that $\alpha_0/\beta_0 \to 0$. This implies that the relevant P-value converges to 1 no matter what data is observed and so no evidence of any conflict will ever be obtained in the limit. When checking for any prior-data conflict with π_1 we see that $P((\beta_0/\alpha_0)F_{(n-1,2\alpha_0)} \leq s_0^2) \to 0$ and $P((\beta_0/\alpha_0)F_{(n-1,2\alpha_0)} \geq s_0^2) \to 1$ as $\alpha_0/\beta_0 \to 0$. Therefore such a sequence of priors will satisfy the natural requirement of non-informativity for θ_1, namely the sequence does not conflict with any data in the limit.

To assess if there is any conflict with π_2 we compare

$$\frac{\overline{x} - \mu_0}{\tilde{\sigma}} = \frac{0.0358324 - 50}{1.1378642} = -43.910485$$

to the Student(21) distribution. This is clearly a very extreme value and in fact the two-sided P-value is 0 to 7 decimals. This check has appropriately detected the discrepancy between the prior and the location of the data.

Evans and Mosohonov (2005) also considered assessing whether any prior-data conflict existed, with respect to the full prior, by comparing the observed value of $\overline{x}$ with its marginal prior predictive. With this data, this comparison resulted in a P-value of .0021 using the marginal prior predictive (a Student (2) distribution). Intuitively this check seemed to be assessing whether the prior for μ was in conflict, but there was no clear rationale for saying this. This article is concerned with developing methods appropriate to this kind of assessment and we note that the conditional check has given a much more extreme P-value reflective of just how extreme the choice of the prior is when we put $\mu_0 = 50$.

As $\tau_0^2 \to \infty$ we have that $\tilde{\sigma}^2 \to \infty$ for every value of s^2. This implies that

$$P\left(|t| \geq \left|\frac{\overline{x}_0 - \mu_0}{\tilde{\sigma}}\right|\right) \to 1$$

as $\tau_0^2 \to \infty$. Therefore, provided that $\tau_0^2 \to \infty$ we have that such a sequence of priors will be noninformative for θ_2.

As previously mentioned, it is necessary that $\alpha_0/\beta_0 \to 0$ for a sequence of priors to be noninformative for the full parameter (σ^2, μ). The analysis here shows that, if we require that the sequence of priors be noninformative for μ by itself, then we need to require that $\tau_0^2 \to \infty$. If we only require noninformativity for μ, there is no need to require that $\alpha_0/\beta_0 \to 0$. If we require noninformativity for σ^2, however, then we do need that $\alpha_0/\beta_0 \to 0$. So the situation is somewhat different depending on what we want the sequence of priors to be noninformative for. As noted in Evans and Moshonov (2005) the requirement of no prior-data conflict is only a partial characterization for noninformativity. Our analysis here indicates that in addition to the requirement that $\alpha_0/\beta_0 \to 0$ we really do want $\tau_0^2 \to \infty$ as well, at least when the prior is specified hierarchically, and we want the sequence to be noninformative with respect to both parameters. $\square$

Both examples can be generalized in the sense that it is possible to find a function V of the minimal sufficient statistic T so that V is ancillary for θ_2. For example, Example 1 can be generalized to the Multinomial$(n, p_1, \dots, p_k)$ case where we put $\theta_1 = (p_1, \dots, p_l)$ for $l < k-1$ and $\theta_2 = (p_{l+1}, \dots, p_{k-1})$. Example 2 can be generalized in several ways. For example, if we have a linear model $y = \beta_1 x_1 + \cdots + \beta_k x_k + z$ with $z \sim N(0, \sigma^2)$, then we can take $\theta_1 = \sigma^2$ and $\theta_2 = (\beta_1, \ldots, \beta_k)$. If we are sampling from a p-dimensional normal $N_p(\mu, \Sigma)$, we can take $\theta_1 = \Sigma$ and $\theta_2 = \mu$.

All of the above examples are models that have exponential form. Consider then a model whose density takes the form

$$f_{(\psi_1, \ldots, \psi_p)}(s) = c(\psi_1, \ldots, \psi_p) h(s) \exp\{\psi_1 T_1(s) + \cdots + \psi_k T_k(s)\}$$

where $\psi = (\psi_1, \ldots, \psi_p) \in \Psi$ with Ψ an open subset of R^p and suppose that the functions $T_1, \ldots, T_p$ are linearly independent. Then $T = (T_1, \ldots, T_p)$ is a complete minimal sufficient for $(\psi_1, \ldots, \psi_p)$. Note that by Basu's theorem any ancillary U for ψ is independent of T and so we can ignore conditioning on U when checking for prior-data conflict.

To apply the approach of Section 2 to some function θ_2 of $(\psi_1, \ldots, \psi_p)$ we need to find a function V of T ancillary for θ_2. In general it isn't obvious how to do this. Indeed it is probably not the case that such a V will exist for an arbitrary θ_2 and currently we lack a characterization of such parameters. The following section gives another general context where it is easier to find such a decomposition.

4.   Checking Prior Components with Group Models

Consider the following example where we apply the methodology discussed in Section 2.

Example 3. Location-scale Cauchy

Suppose we have a sample $s = (s_1, \ldots, s_n)$ from a distribution with density at x equal to

$$\frac{1}{\pi\sigma\left(1 + (x - \mu)^2 / \sigma^2\right)}$$

where $\theta = (\sigma, \mu)$ and $\mu \in R^1$, $\sigma > 0$ are unknown. It is known that the order statistic $T(s) = \left(s_{(1)}, \ldots, s_{(n)}\right)$ is a minimal sufficient statistic for this model.

Putting $r = s_{(n)} - s_{(1)}$, we have that

$$U(T(s)) = \left(\frac{s_{(2)} - s_{(1)}}{r}, \cdots, \frac{s_{(n-1)} - s_{(1)}}{r}\right) \tag{7}$$

is ancillary for θ, since U is invariant under location-scale transformations, and we note that $U \in R^{n-2}$. The conditional distribution of T given U can be expressed as the conditional distribution of $(s_{(1)}, r)$ given $U(T)$. To obtain this we make the following transformation $(s_{(1)}, ..., s_{(n)}) \rightarrow (v_1, ..., v_n)$ defined as follows:

$$v_1 = s_{(1)}, v_2 = r = s_{(n)} - s_{(1)}, v_{i+1} = \left(s_{(i)} - s_{(1)}\right)/r,$$

for $i = 2, \ldots, n - 1$. Expressing the order statistic in terms of these new variables we have $s_{(1)} = v_1, s_{(2)} = v_1 + v_2 v_3, ..., s_{(i)} = v_1 + v_2 v_{i+1}$, for $i = 2, ..., n - 1$ and $s_{(n)} = v_1 + v_2$. The Jacobian of this transformation is then given by $v_2^{n-2} = r^{n-2}$. The joint density of $(s_{(1)}, r, U) = (v_1, \ldots, v_n)$ is then the joint density of the order statistic at the above values times r^{n-2}. This implies that the conditional distribution of $(s_{(1)}, r)$ given U satisfies

$$f_{\sigma,\mu}(s_{(1)}, r \,|\, U) \propto \sigma^{-2} \left(\frac{r}{\sigma}\right)^{n-2} \left(1 + \left(\frac{s_{(1)} - \mu}{\sigma}\right)^2\right)^{-1} \left(1 + \left(\frac{r}{\sigma} + \frac{s_{(1)} - \mu}{\sigma}\right)^2\right)^{-1}$$

$$\times \prod_{i=3}^{n} \left(1 + \left(\frac{r}{\sigma} v_i + \frac{s_{(1)} - \mu}{\sigma}\right)^2\right)^{-1}. \tag{8}$$

If we integrate $s_{(1)}$ from Eq. (8) we see that what remains does not involve μ and so we have immediately that r is ancillary for μ.

Denoting the prior on (σ, μ) by $\pi(\sigma, \mu) = \pi_1(\sigma)\, \pi_2(\mu \,|\, \sigma)$ we then have that the conditional prior predictive density of $(s_{(1)}, r)$ given U satisfies

$$m(s_{(1)}, r \,|\, U) = \int_0^\infty \int_{-\infty}^\infty f_{\mu,\sigma}(s_{(1)}, r \,|\, U)\pi_1(\sigma)\, \pi_2(\mu \,|\, \sigma)\, d\mu\, d\sigma. \tag{9}$$

In Eq. (9) we transform from (σ, μ) to $x = \left(s_{(1)} - \mu\right)/\sigma$ and $y = r/\sigma$ and use (8) to obtain,

$$m(s_{(1)}, r \,|\, U) = \int_0^\infty \int_{-\infty}^\infty \left(r^2/y^3\right) f_{s_{(1)} - xr/y, r/y}(s_{(1)}, r \,|\, U)\pi\left(s_{(1)} - xr/y, r/y\right) dx\, dy$$

$$\propto \int_0^\infty \int_{-\infty}^\infty y^{n-2} \left(1 + x^2\right)^{-1} \left(1 + (y + x)^2\right)^{-1} \prod_{i=3}^{n} \left(1 + (yv_i + x)^2\right)^{-1}$$

$$\times\, y^{-1}\pi_1(r/y)\, \pi_2\left(s_{(1)} - rx/y \,|\, r/y\right) dx\, dy. \tag{10}$$

Integrating $s_{(1)}$ and r out of Eq. (10), using the fact that $\pi_2\left(s_{(1)} - rx/y \,|\, r/y\right)$ is a density in $s_{(1)}$ and $y^{-1}\pi_1(r/y)$ is a density in r, we obtain the inverse normalizing constant as

$$c = \int_0^\infty \int_{-\infty}^\infty y^{n-2} \left(1 + x^2\right)^{-1} \left(1 + (y + x)^2\right)^{-1} \prod_{i=3}^{n} \left(1 + (yv_i + x)^2\right)^{-1} dx\, dy$$

$$= \int_0^\infty \int_{-\infty}^\infty k(x, y)\, dx\, dy.$$

Observe that the marginal prior predictive density for r is obtained by integrating $s_{(1)}$ out of Eq. (10), to obtain

$$m_1(r \,|\, U) = \int_0^\infty g(y \,|\, U)y^{-1}\pi_1(r/y)\, dy \tag{11}$$

where

$$g(y \,|\, U) = c^{-1} \int_{-\infty}^\infty k(x, y)\, dx \tag{12}$$

is the conditional density of y given U. From Eq. (11) we see that $m_1(r \,|\, U)$ is a scale probability mixture of the prior on σ. Now note that

$$m_2(s_{(1)} \,|\, r, U) = (m_1(r \,|\, U))^{-1} \int_0^\infty \left(c^{-1} \int_{-\infty}^\infty k(x, y)\pi_2\left(s_{(1)} - rx/y \,|\, r/y\right) dx\right) y^{-1}\pi_1(r/y)\, dy$$

$$= (m_1(r \,|\, U))^{-1} \int_0^\infty h(s_{(1)}, y \,|\, U)y^{-1}\pi_1(r/y)\, dy.$$

In terms of organizing the computations we first tabulate $g(\cdot\,|\,U)$, which also requires the computation of c, next we tabulate $m_1(\cdot\,|\,U)$ and use this, together with the observed value of r, to assess whether or not there is any prior-data conflict with π_1. Finally, we need to tabulate $\int_0^\infty h(\cdot, y\,|\,U)y^{-1}\pi_1(r/y)\,dy$ to obtain the conditional prior predictive of $s_{(1)}$ and use this, together with the observed value of $s_{(1)}$, to assess whether or not there is any prior-data conflict with π_2.

To evaluate a P-value associated with this conditional prior predictive we must integrate numerically. For example, consider the following ordered sample of $n = 10$ from the Cauchy distribution with $\mu = 0$ and $\sigma = 1$.

-4.4829	-2.9692	-0.8915	-0.7164	-0.5501
-0.2805	0.0474	2.1665	4.1467	18.7272

Then the observed values of the relevant statistics are $s_{(1)} = -4.48290, r = 23.2101$ and the values of $v_3, \ldots, v_{10}$ are given as follows:

0.0652163	0.1547339	0.1622763	0.1694449
0.1810571	0.1951884	0.2864865	0.3718015

First we need to assess whether or not $r = 23.2101$ is a surprising value from $m_1(\cdot\,|\,U)$. For this we first compute the conditional density of y given $v_3, \ldots, v_{10}$. We use Eq. (12) for this and note that $v_3, \ldots, v_{10}$ are all positive. Thus, based on the expression for $k(x, y)$, for each $y > 0$ we can integrate over x by summing over values that lie within the effective range of the Cauchy. In Fig. 3 we have plotted the conditional density of y given U. From this we can see that tabulating y in the range $(0, 100)$ will be adequate.

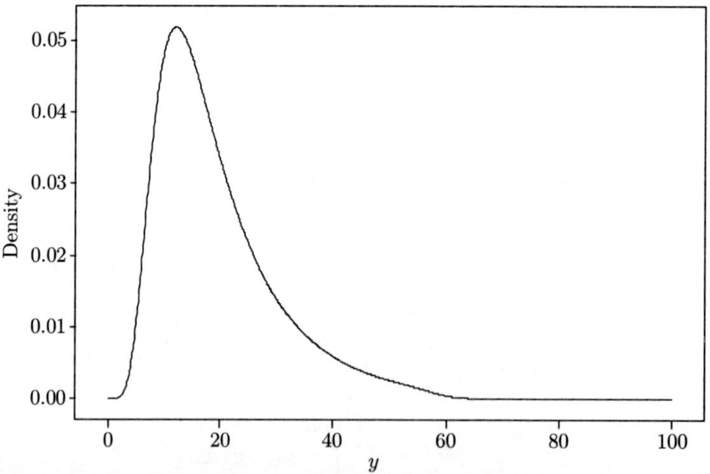

Fig. 3.   **Conditional density** $g(\cdot|U)$.

Now suppose that we use the prior as given in Example 2 but with $\tau_0^2 = 1$, $\mu_0 = 0$, $\alpha_0 = 2$ and $\beta_0 = 3$. The conditional prior predictive of r given U, namely $m_1(\cdot\,|\,U)$, is displayed in Fig. 4. We see from this that the observed value of $r = 23.2101$ is in the central hump of the distribution and so we have no evidence of any prior-data conflict with π_1.

Fig. 5 shows the plot $m_2(\cdot\,|\,r, U)$. From this we can see that the observed value of $s_{(1)} = -4.48290$ does not lead to any prior-data conflict with π_2. $\square$

Example 3 is a particular example of a group model. For a group model we have a group G and a family of transformations $\{W_g : g \in G\}$ acting on the sample space $\mathcal{T}$ for the minimal sufficient

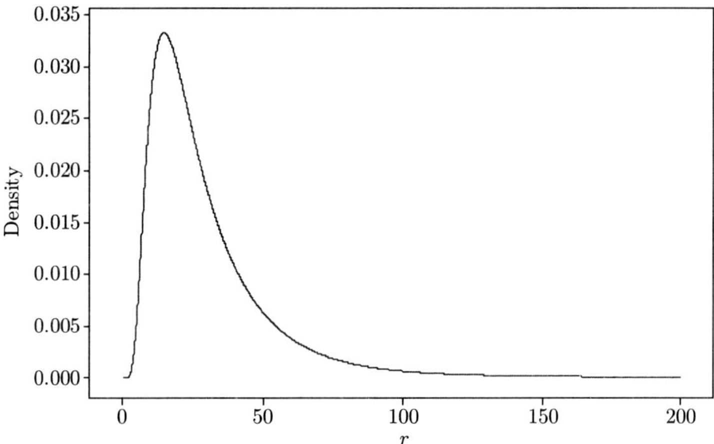

Fig. 4. The conditional prior predictive density $m_1(\cdot\,|\,U)$.

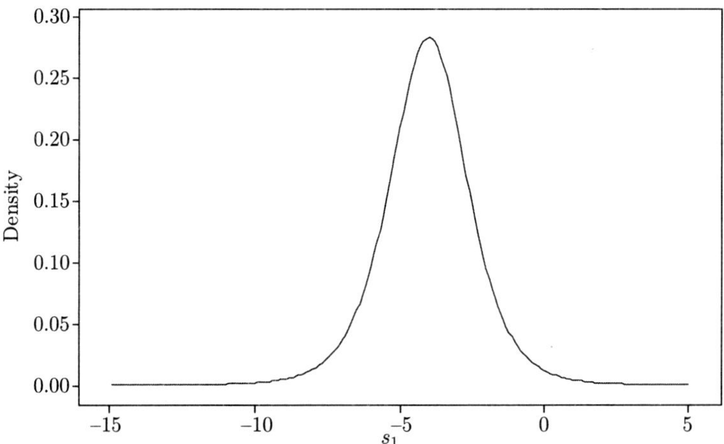

Fig. 5. The conditional prior predictive $m_2(\cdot\,|\,r,U)$.

statistic T. We assume that this action is free, namely it satisfies $W_{g_1}(t) = W_{g_2}(t)$ if and only if $g_1 = g_2$ for any t (perhaps after removing a set of measure 0 from $\mathcal{T}$). Further we have a distribution with density f with respect to some support measure μ on $\mathcal{T}$. The sampling model is then given by the family of distributions, indexed by $g \in G$, generated via $t = W_g(z)$ where $z \sim f$. So we have that $\theta = g$ and $\Omega = G$.

In Example 3 we have the location-scale group $G = \{(a,c) : a \in R^1, c > 0\}$, with product $(a_1, c_1)(a_2, c_2) = (a_1 + c_1 a_2, c_1 c_2)$, acting on R^n via $t = W_{(a,c)}(z) = a1_n + cz$, where 1_n is the n-dimensional vector of 1's. The distribution of z is that of the order statistic based on a sample of n from the standard Cauchy.

Now note that we can write the order statistic $t = \left(s_{(1)}, \ldots, s_{(n)}\right)$ as $t = s1_n + rU = W_{(s_{(1)}, r)}(U(t))$. Since $U(t) = (t - s_{(1)})/r$ is a maximal invariant under the action of this group, we have that U is ancillary and this result holds no matter what distribution we take for z. Note that once we have specified a maximal invariant U this immediately specifies the data-dependent element in G that restores the full data from U. In Example 3, choosing U as in (7) specifies $(s_{(1),r}) \in G$. So once we have specified a maximal invariant U, and there are many possible choices, there is a unique element $[t] \in G$ such that $t = W_{[t]}(U(t))$.

In the location-scale group G we can write $(a, c) = (a, 1)(0, c)$ and the elements $(a, 1)$ belong to the location group $G_2 = \{(a, 1) : a \in R^1\}$ while the elements $(0, c)$ belong to the scale group $G_1 = \{(0, c) : c > 0\}$. In effect the location-scale group is a semidirect product of the location and scale groups since we can write $G = G_2 G_1$ and G_2 is a normal subgroup of G. If a group G can be written as the semidirect product $G = G_2 G_1$, then any element $g \in G$ can be written uniquely as $g = g_2 g_1$ where $g_1 \in G_1$ and $g_2 \in G_2$. For details on this see, for example, Robinson (1980). Further we have that $W_g(t) = W_{g_2 g_1}(t) = W_{g_2} \circ W_{g_1}(t)$ and $[t] = [t]_2 [t]_1$ with $[t]_1 \in G_1$ and $[t]_2 \in G_2$.

Using the fact that, for any maximal invariant U we have $U(t) = U(W_g(t))$, observe that

$$W_g = W_{g_2 g_1} \circ W_{[t]_2 [t]_1}(U(t)) = W_{g_2 g_1 [t]_2 [t]_1}(U(t))$$

$$= W_{g_2 g_1 [t]_2 g_1^{-1} g_1 [t]_1}(U(t)) = W_{g_2 g_1 [t]_2 g_1^{-1}} \circ W_{g_1 [t]_1}(U(t)) \tag{13}$$

and the normality of G_2 implies that $g_1 [t]_2 g_1^{-1} \in G_2$ and so $g_2 g_1 [t]_2 g_1^{-1} \in G_2$. For the location-scale group, using the maximal ancillary equal to (7), we have that $[t]_1 = (0, r)$ and $[t]_2 = (s_{(1)}, 1)$ and so when $g = (a, c) = (a, 1)(0, c)$

$$g_2 g_1 [t]_2 g_1^{-1} = (a, 1)(0, c)(s_{(1)}, 1)(0, c^{-1}) = (a + c s_{(1)}, 1)$$

$$g_1 [t]_1 = (0, c)(0, r) = (0, cr)$$

Now $t = W_{[t]}(U(t))$ and so

$$W_g(t) = W_{[W_g(t)]_2 [W_g(t)]_1}(U(W_g(t))) = W_{[W_g(t)]_2 [W_g(t)]_1}(U(t))$$

together with (13) implies that $[W_g(t)]_2 = g_2 g_1 [t]_2 g_1^{-1}$, $[W_g(t)]_1 = g_1 [t]_1$. In particular this implies that $V(t) = [W_g(t)]_1$ is invariant under G_2 and equivariant under G_1. From this we have the following result.

Theorem 3. Suppose that the statistical model for the minimum sufficient statistic T is given by $\{f_{\theta T} : \theta \in G\}$ where G is a group acting freely on $\mathcal{T}$ via the class of transformations $\{W_g : g \in G\}$ and $f_{\theta T}$ corresponds to the distribution of t when we write $t = W_\theta(z)$ with $z \sim f$. Further suppose that G is a semidirect product $G = G_2 G_1$ and we write $\theta = \theta_2 \theta_1$ with $\theta_i \in G_i$. For a particular choice of maximal invariant U, and thus corresponding transformation $[\cdot] : \mathcal{T} \to G$ such that $t = W_{[t]}(U(t))$, then $V(T)$ given by $V(t) = [t]_1$ is ancillary for θ_2.

Theorem 3 gives a very large class of models for which the methods of this paper are applicable for checking the prior-data conflict for the components π_1 and π_2. For example, any regression model with nonnormal error falls in this class. For a discussion of a wide variety of group models in the frequency context see Fraser (1979). For many of these models the group involved can be decomposed as a semidirect product and so the analysis of this section will apply. Of course, this requires that we specify the components of the full prior distribution in a way that conforms to this decomposition.

5. Conclusions

We have been concerned here with checking for prior-data conflict by checking individual components of the prior when the prior is specified hierarchically in two stages. Thus, the parameter θ is decomposed into two parts θ_1 and θ_2 where θ_2 is the parameter of interest (not necessarily

1-dimensional) and θ_1 is comprised of nuisance parameters and perhaps hyperparameters. Then a prior π_1 is specified for θ_1 and a prior $\pi_2\left(\cdot\,|\,\theta_1\right)$ is specified for θ_2, conditional on θ_1. Of course the prior $\pi_2\left(\cdot\,|\,\theta_1\right)$ can also be independent of θ_1. The methodology developed is a generalization of the methods discussed in Evans and Moshonov (2005). The key requirement is the existence of statistic $V(T)$ that is ancillary for θ_2. We have shown that, in a wide variety of models, such a statistic exists provided the decomposition of the prior conforms to a certain structure.

The necessity of the existence of the statistic $V(T)$ is of course a restriction on the approach taken here. While we can't say that methods cannot be developed for general decompositions, there does appear to be a natural logic to our approach. This might be interpreted as a restriction on how we should specify priors hierarchically, at least when we want to check the individual components separately for prior-data conflict. When such a decomposition does not exist, or is not deemed suitable for the hierarchical specification, we can always resort to the methods of Evans and Moshonov (2005) for checking the full prior.

There are several aspects of our discussion here where more work is indicated. In particular we need to develop methods appropriate for the general decomposition given by Eq. (2) where $p > 2$. These will naturally be a generalization of the methods we have discussed in this paper. Further the computations involved for general models with general priors will obviously be more complicated than those we have presented. Strictly speaking straightforward Monte Carlo methods are not available. The computational problems are more similar to those involved in computing inverse normalizing constants than posterior calculations. These aspects of the problem are currently under investigation.

Acknowledgements

The authors thank a referee for constructive comments and Professor Satyanshu Upadhyay for his work on this volume.

References

Bayarri, M.J. and Berger, J.O. (2000). P values for composite null models. *Journal of the American Statistical Association*, **95, (452)**, 1127-1142.

Box, G.E.P. (1980). Sampling and Bayes' inference in scientific modelling and robustness. *Journal of the Royal Statistical Society*, **A,143**, 383-430.

Evans, M. and Moshonov, H. (2005). Checking for prior-data conflict. To appear in Bayesian Analysis. Technical Report 0413.

Fraser, D.A.S. (1979). Inference and Linear Models. McGraw-Hill.

Gelman, A., Meng, X.L. and Stern, H.S. (with discussion) (1996). Posterior predictive assessment of model fitness via realized discrepancies. *Statistica Sinica*, **6**, 733-808.

Guttman, I. (1967). The use of the concept of a future observation in goodness-of-fit problems. *Journal of the Royal Statistical Society*, **B, 143,** 383-430.

Johnson, V. (2004). A Bayesian χ^2 test for goodness-of-fit. *Annals of Statistics*, **32, (6)**, 2361-2384.

Lehmann, E.L. and Scholz, F.W. (1992). Ancillarity. Current Issues in Statistical Inference: Essays in Honor of D.Basu. Malay Ghosh and Pramod K. Pathak, Editors. IMS Lecture notes-monograph series, Hayward, CA, 32-51.

Robinson, D.J.S. (1980). A Course in the Theory of Groups. Springer-Verlag, New York.

Rubin, D.B. (1984). Bayesianly justifiable and relevant frequency calculations for the applied statistician. *Annals of Statistics*, **12**, 1151-1172.

Bayesian Statistics and Its Applications
Edited by S.K. Upadhyay, U. Singh and D.K. Dey
Anamaya Publishers, New Delhi, India

Bayesian Cross-Sectional Analysis of the Conditional Distribution of Earnings of Men in USA (1967-1996)

John Geweke[1] and Michael Keane[2]

[1]Departments of Economics and Statistics, University of Iowa
[2]Department of Economics, Yale University

Abstract

This study develops practical methods for Bayesian nonparametric inference in regression models. The emphasis is on extending a nonparametric treatment of the regression function to the full conditional distribution. It applies these methods to the relationship of earnings of men in the USA to their age and education over the period 1967 through 1996. Principal findings include increasing returns to both education and experience over this period, rising variance of earnings conditional on age and education, a negatively skewed and leptokurtic conditional distribution of log earnings, and steadily increasing inequality with asymmetric and changing impacts on high- and low-wage earners. These results are insensitive to several alternative nonparametric specifications of the distribution of earnings conditional on age and education.

1. Introduction

Much of applied statistics and econometrics is concerned with the measurement and interpretation of conditional distributions. In statistics the core curriculum devotes substantial time to regression, a topic that practically defines elementary econometrics and is the main point of departure in advanced treatments. Typically the conditional distribution is that of a univariate random variable y conditional on a random vector $\mathbf{x}$. Regression, narrowly defined, is concerned only with $E\left(y \mid \mathbf{x}\right)$, yet this topic alone is the basis of a huge literature in mathematical statistics and theoretical econometrics. Elementary treatments typically assume that the regression is linear in $\mathbf{x}$, perhaps after a logarithmic, Box-Cox, or other nonlinear transformation of y. This assumption is rarely justified on theoretical grounds and as an empirical matter often can be overturned. Therefore much of this literature has taken nonparametric or semiparametric approaches to regression. The textbooks by Hardle (1989) and Green and Silverman (1994) provide introductory yet comprehensive treatments of non-Bayesian approaches; for Bayesian treatments, see, for example Erkanli and Gopalan (1994), Wong and Kohn (1996), Smith and Kohn (1996), Koop and Tobias (2006) and Koop and Poirier (2004).

In general, substantive interest in the conditional distribution goes well beyond conditional expectations to embrace the entire distribution $p(y \mid \mathbf{x})$. Well-known examples include the pricing of derivatives of financial assets and the study of inequality and mobility of earnings across individuals. Nonparametric approaches are much less common at this level of generality, due mainly to well-understood practical problems, but even semiparametric treatments of both the regression function and the conditional distribution simultaneously are rare. In this study we take nonparametric and semiparametric Bayesian approaches to this question. The specific statistical question we address is Bayesian inference for a functional of $p(y \mid \mathbf{x})$. This includes not only the expectation of y corresponding to given value of $\mathbf{x}$, but also functionals of the conditional distribution such as the coefficient of kurtosis and measures of inequality like the Gini coefficient. The emphasis is on methods that can be carried out quickly and reliably using software that is widely available. We achieve our objectives by combining, alternatively, polynomial basis functions or Wiener process smoothness priors for regression, with normal mixture modeling for regression residuals $y - E(y \mid \mathbf{x})$. Section 3 describes our methods in detail. The closest precedent for our approach appears to be that of Smith and Kohn (1996). That study, however, is concerned with outliers and confines consideration to scale mixture of normals distributions. The findings in Section 4 show that assumption clearly would be inappropriate in our application.

2. Earnings and the PSID Data

Our substantive concern is with the distribution of the earnings of individual men conditional on age and education, whose primary importance for earnings emerges from a rich theoretical and empirical literature in economics beginning with Mincer (1958). To introduce notation used throughout this study, we have available for each of 30 years $t = 1967, \ldots, 1996$ a sample of the age a_{ti}, education e_{ti} and logarithm of earnings y_{ti} for each of $n(t)$ men $(i = 1, \ldots, n(t))$. We are interested in functionals of the density $p(y_{ti} \mid \mathbf{x}_{ti})$, where $\mathbf{x}_{ti} = (a_{ti}, e_{ti})'$. Due to the structure of the data, described in Section 2.2, we treat each year as a separate sample and do not model dependence between these samples.

2.1 Modelling Earnings

There is a long and well established literature studying the relationship between earnings and the determinants of earnings suggested by life-cycle human capital models. Going back at least to the seminal work of Mincer (1958) the essence of these models is that an individual's productivity, or human capital, is an increasing function of formal education and work experience. Heckman et al. (2003) review this work. By far the most common measure of formal education is years of schooling, and the most common measure of experience is age for individuals who have limited or no spells of labor force nonparticipation following the completion of their formal education. Since our study is limited to men, the latter assumption is reasonable.

A fully specified relationship between earnings, on the one hand, and age and education, on the other, has several properties that are of substantive interest. One is the impact of education on earnings, sometimes termed returns to education. We use that terminology here, recognizing that returns so measured do not distinguish between the impact of education on the earnings of a given individual and the fact that individuals for whom further education is more likely to provide economic benefits are more likely to choose more years of schooling. Another is the distribution of earnings over the life cycle. As individuals age the relative benefits of working and leisure change because of increasing financial wealth (on average), a changing planning horizon, and changing family circumstances.

This relationship need not be constant over time. On the demand side, changes in technology can change the relative productivity of more and less highly educated individuals, and may make it more advantageous for a given number of hours of work over the life cycle to be more concentrated in fewer years or spread out over more years. On the supply side, demographic changes driven by changes in fertility and immigration shift the age distribution of the labor force, and to the extent that older and younger workers are imperfect substitutes in the workforce their relative wages will change for this reason as well. Changes in economic policy including the taxation of earnings, public subsidies for higher education, unemployment benefits, and public pension benefits, will also affect wages and decisions about hours of work, thereby changing the relationship. There is no reason that the effect of these changes should be limited to expectations of earnings (or log earnings) conditional on age and education: the entire distribution may be affected. The extent to which this distribution has changed, and especially the implications for inequality in the distribution of earnings, has been a topic of intense interest in both the academic literature and public policy forums. In this study we characterize the distribution and its change over the period 1967 through 1996.

2.2 Data

Our findings are all based on the panel study of income dynamics (PSID). The PSID is a household-based panel that has collected information on earnings and other aspects of household economic activity. From 1968 through 1997 the survey was fielded annually, collecting data pertaining to the previous year. We use these data, identifying each wave of the panel by the year previous to the interview since that is the year to which the information pertains. The survey was not fielded in 1998, and data has been conducted biannually since 1999. Death or divorce can lead to one or two new households, and in these cases the PSID tracks the new households. It also follows households formed eventually by children in households, and from time to time the sample is refreshed with new households as required to preserve the stratification of the sample. For each year we assemble data on age, education and earnings of these individuals. We model the distribution of earnings conditional on age and education separately for each year. Thus we do not exploit the panel structure of the data in this study.

We restrict the full PSID sample in several ways in this study. First, we include only male household heads. Primarily this eliminates women, for whom the distribution of earnings conditional on age and education is distinct from that of men, and for whom the identification of individual earnings in a male-headed household is difficult in the PSID in any event. Second, we include male household heads only between the ages of 25 and 65 inclusive, because the conditional distribution of earnings is likely to change sharply for younger men, many of whom are marginal labor force participants still in school, and for older men, many of whom are retired or partially retired. Third, we exclude black males and the 1989-1993 Latino source sample because earnings distributions differ by race. Finally, we include these men only if they are labor force participants, as indicated by earnings of \$1,000 or more in a year. Our results are essentially unchanged if the criterion is changed to \$500 or \$2,000, because there are very few men with positive earnings below \$2,000.

Fig. 1 shows some properties of the sample that are important in understanding the findings reported in Section 4. Because the PSID follows households the sample has grown steadily since its inception. Our data from 1994 through 1996 is pre-release and does not include new, young households. Through the period 1967-1996 white male labor force participants became steadily better educated, as a group. There has been a strong tendency to leave the labor force earlier and, since 1976, a tendency to enter later. Conclusions about the distribution of earnings conditional on age and education will, therefore, most strongly reflect the data for men between the ages of 30 and

50 with 12 or more years of education. At the other extreme, if education is less than 8 years, then conclusions will strongly reflect prior information about the similarity of the conditional distribution for poorly educated men to the conditional distribution for well educated men, together with the information in the data about well educated men. These differences are conveyed in greater posterior uncertainty about the distribution of earnings conditional on combinations of age and education that are poorly represented in the sample, as compared with combinations that are well represented.

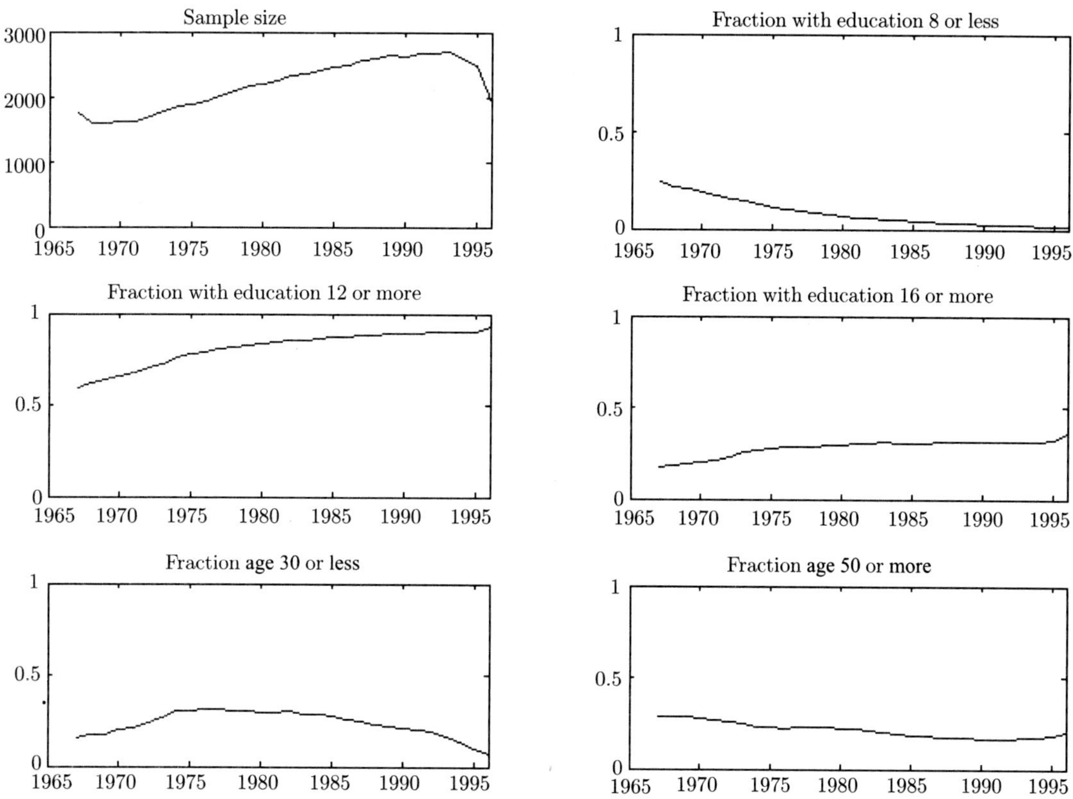

Fig. 1.   **Some aspects of the PSID data sets.**

3.   Methodology

We are interested in learning about $p_t\,(y_{ti} \mid \mathbf{x}_{ti})$. This notation reflects the fact that we model each year separately. We do not exploit the panel features of the data. It is important to bear in mind that earnings are not conditionally independent across years. Beginning with the textbook Bayesian linear model we first weaken the assumption that $E_t\,(y_{ti} \mid \mathbf{x}_{ti})$ is a linear function of $\mathbf{x}_{ti}$ (Section 3.1) and then weaken the assumption that the conditional distribution of y_{ti} given $\mathbf{x}_{ti}$ is conditionally Gaussian (Section 3.2).

3.1   Nonlinear Regression

We begin by weakening the assumption that the regression function is linear in $\mathbf{x}_t$ in favor of the specification that it is a smooth function of $\mathbf{x}_t$, while maintaining the assumption that $y_{ti} - E_t\,(y_{ti} \mid \mathbf{x}_{ti})$ is i.i.d. Gaussian. As with all subjective conditions, "smoothness" can be characterized in different ways. This section takes up two different approaches. Each leads to a posterior distribution identical to that of a normal model linear in unknown coefficients, but with a different matrix of covariates and with a different interpretation of the coefficient vector in each case. That nonlinear regression is thus isomorphic to linear regression has two desirable consequences. On the practical level, many Bayesian computational methods for the normal linear model can be applied in normal nonlinear regression. On the conceptual level, many of the rich elaborations of the normal linear model that have been taken up in Bayesian analysis can be applied directly in nonlinear regression. These include the extension to non-normality in Section 3.2.

3.1.1 *Polynomial Basis Functions*

A sequence of normal linear models A_j $(j = 1, 2, \ldots)$ for each of the period t models captures the essentials of nonlinear regression with basis functions. In model A_j

$$y_{ti} = f_{tj}(\mathbf{x}_{ti}) + \varepsilon_{ti} = \sum_{\ell=1}^{k_j} \beta_{tj\ell}\phi_{j\ell}(\mathbf{x}_{ti}) + \varepsilon_{ti} = \beta'_{tj}\phi_j(\mathbf{x}_{ti}) + \varepsilon_{ti}; \quad \varepsilon_{ti} \overset{\text{i.i.d.}}{\sim} N\left(0, h^{-1}\right) \tag{1}$$

Model A_j specifies basis functions ϕ_{ji} $(i = 1, \ldots, k_j; j = 1, 2, \ldots)$ so that as j increases, the function f_{tj} is, loosely speaking, more flexible. We take the basis functions to be monomials; other possibilities include Fourier functions (Gallant, 1981) and the Muntz-Szatz series (Barnett and Jonas, 1983). In the *interactive polynomial model* $\phi_{j\ell}(\mathbf{x}_{ti})$ consists of all terms of the form $a_{ti}^m e_{ti}^n$ $(m = 0, \ldots, k_a(j); n = 0, \ldots, k_e(j))$. In the *separable polynomial model* $\phi_{j\ell}(\mathbf{x}_{ti})$ consists of all terms of the form a_{ti}^m or e_{ti}^n $(m = 0, \ldots, k_a(j); n = 1, \ldots, k_e(j))$. In both cases the relation between j and $(k_a(j), k_e(j))$, for $j = 1, \ldots, 9$ is as follows:

	$k_e = 0$	$k_e = 1$	$k_e = 2$
$k_a = 0$	$j = 1$	$j = 3$	$j = 6$
$k_a = 1$	$j = 2$	$j = 4$	$j = 8$
$k_a = 2$	$j = 5$	$j = 7$	$j = 9$

The pattern continues for larger values of j. The Weirstrass Theorem guarantees that the interactive polynomial model provides an arbitrarily close approximation of $E(y_{ti} \mid \mathbf{x}_{ti})$ for j sufficiently large, but at the cost of a vector of basis functions of high dimension. There is no such justification for the separable polynomial model, but separability is of substantive interest.

In formulating prior distributions of the $k_j \times 1$ coefficient vector β_{tj}, it is useful to think in terms of the regression function f_{tj} defined in (1). This is especially important in comparing variants with different numbers of basis functions, because it ensures comparable priors. For example, the prior distribution consisting of the components $f_{tj}(\mathbf{x}_i^*) \overset{\text{i.i.d.}}{\sim} N(\mu, \sigma^2)$ for selected covariate values $\mathbf{x}_i^*$ $(i = 1, \ldots, n)$ implies the prior distribution consisting of the components $\beta'_{tj}\phi_j(\mathbf{x}_i^*) \overset{\text{i.i.d.}}{\sim} N(\mu, \sigma^2)$ $(i = 1, \ldots, n)$ when the order of expansion is j. If J is the highest order of expansion considered and $n \geq k_J$ points are chosen appropriately, this approach will provide comparable and proper prior distributions for the coefficients in all orders of expansion considered. It also facilitates comparison of polynomial models with models based on smoothness priors described in the next section.

For year t we take the points $\mathbf{x}_i^*$ to be the observed covariates $\mathbf{x}_{ti}$ and choose $\mu = 10.5$, which is close to the sample mean of y_{ti} for all years. Then denoting the matrix of covariates in year t and order of expansion j by $\mathbf{X}_{tj}$, the prior distribution of the $k_j \times 1$ vector β_{tj} is

$$\beta_{tj} \sim N\left[\theta, \lambda n(t)\left(\mathbf{X}'_{tj}\mathbf{X}_{tj}\right)^{-1}\right]$$

where $\theta' = (\mu, 0, \dots, 0)$ and λ is a hyperparameter to be chosen. The prior distribution of h in (1) is gamma,

$$\underline{s}^2 h \sim \chi^2 (\underline{\nu})$$

Experimentation with the three hyperparameters λ, $\underline{s}^2$ and $\underline{\nu}$ showed that findings are robust over changes of nearly an order of magnitude. Based on marginal likelihoods for several years and several variants of the models, we settled on $\lambda = 200$, $\underline{s}^2 = 12$ and $\underline{\nu} = 1$. Marginal likelihoods were then computed corresponding to all orders of polynomial expansion k_a and k_e for both the interactive polynomial and separable polynomial models. For the thirty separate regressions corresponding to the 30 years in the sample, average marginal likelihood was maximized with the choice $k_a = 4$, $k_e = 1$ in the interactive polynomial model, and the choice $k_a = 4$, $k_e = 2$ in the separable polynomial model. These values are used in all of the results reported subsequently.

3.1.2   Wiener Process Smoothness Priors

There is a complementary approach when the regression function is separable, so that

$$y_{ti} = f_{ta}(a_{ti}) + f_{te}(e_{ti}) + \varepsilon_{ti}; \quad \varepsilon_{ti} \sim N\left(0, h^{-1}\right) \tag{2}$$

Models of this form have been widely studied in the non-Bayesian semiparametric estimation literature. In the Bayesian literature the same problems have been addressed by smoothness priors. We take a Bayesian smoothness prior approach that emphasizes two properties. The first is that the prior distribution should be capable of allowing the investigator to study the behavior of f_{ta} or f_{te} at points not in the data set: for example, in some years t not all levels of education are represented in our data set as indicated in Section 2.2. The second is that the prior distribution should be formulated in a fashion that assures comparability between models using smoothness priors and models using basis functions, so that Bayes factors are not driven by arbitrary assumptions that are implicit in priors but opaque to the investigator.

The essentials of nonlinear regression with smoothness priors are captured in the simpler model with a single covariate,

$$y_t = f(x_t) + \varepsilon_t, \varepsilon_t \overset{\text{i.i.d.}}{\sim} N\left(0, h^{-1}\right) \quad (t = 1, \dots, T) \tag{3}$$

The function $f(\tau)$ is defined on a closed interval $\tau \in [\tau_1, \tau_2]$. The prior must incorporate the idea that f is a smooth function, in the sense that it is differentiable and $df(\tau)/d\tau$ changes slowly with τ. By formulating a prior that pertains to all points $\tau \in [\tau_1, \tau_2]$ we guarantee coherence if the investigator decides to incorporate a point that is of interest but for which there is no data.

A convenient and powerful analytic tool for expressing these beliefs is the Wiener process $W(\tau)$, defined on $\tau \in [0, \infty)$ with $W(0) = 0$. A standard representation is $W(\tau) = \int_0^\tau dW(u)$, it being understood that the orthogonal increments $dW(u)$ are normally distributed. One important property of a Wiener process is $W(\tau + s) - W(\tau) \sim N(0, s)$, for all $\tau \geq 0$ and all $s > 0$: this limits the rapidity with which W can move as a function of τ, and this feature can in turn be controlled by appropriate scaling of W. Another important property is that any pair of increments $W(\tau + s) - W(\tau)$ and $W(\tau' + s') - W(\tau')$ has a bivariate normal distribution. Each increment has mean zero. If $[\tau, \tau + s]$ and $[\tau', \tau' + s']$ do not overlap then the increments are uncorrelated, whereas if the intervals do overlap then their covariance is the length of the overlap: in general,

$$\mathrm{cov}\left[W(\tau + s) - W(\tau), \ W(\tau' + s') - W(\tau')\right] = \int_0^\infty I_{[\tau, \tau+s]}(u)\, I_{[\tau', \tau'+s']}(u)\, du$$

for all positive τ, τ', s and s'.

A Wiener process has the properties ascribed to the function $f'(\tau) = df(\tau)/d\tau$, and thus we pursue the idea that in the prior distribution $f(\tau)$ is the integral of such a process. The approach is that of Shiller (1984). Thus, $f(\tau) = \underline{h}^{-1/2} \int_0^\tau W(u)\, du$, where the precision hyperparameter $\underline{h}$ controls smoothness. For any two points τ and s,

$$f(\tau) = \underline{h}^{-1/2} \int_0^\tau W(u)\, du = \int_0^\tau \int_0^u dW(r)\, du \tag{4}$$

$$f(s) = \underline{h}^{-1/2} \int_0^s W(v)\, dv = \int_0^s \int_0^v dW(p)\, dv \tag{5}$$

have a joint normal distribution, with $E[f(\tau)] = E[f(s)] = 0$. If $s \geq \tau$ then from (4) and (5),

$$E[f(\tau) f(s)] = \underline{h}^{-1} \int_0^\tau \int_0^s \min(u,v)\, dv du = \underline{h}^{-1} \int_0^\tau \left[\int_0^u v\, dv + \int_u^s u\, dv \right] du$$

$$= \underline{h}^{-1} \int_0^\tau \left[\frac{u^2}{2} + u(s-u) \right] du = \frac{\tau^2}{6}(3s - \tau) \tag{6}$$

Since $\operatorname{var}[f(\tau)] = \underline{h}^{-1/2} \tau^3/3$, the prior variance ascribed to $f(\tau)$ at a point $\tau = s_1$ will depend strongly on the idea that $\tau = 0$ is a special point at which it is known *a priori* that $f'(0) = 0$. This is an artificial assumption. It arises not from prior ideas about smoothness (the reason for introducing the Wiener process as a model for the prior) but rather from the analytical necessity of an initial condition for $f'(\tau)$. There is a similar problem with the slope of the function $f(\tau)$ between two points s_1 and s_2 $(s_2 > s_1)$, $[f(s_2) - f(s_1)]/(s_2 - s_1)$. From (6)

$$\operatorname{var}\left\{ [f(s_2) - f(s_1)]/(s_2 - s_1) \right\}$$

$$= \frac{1}{6\underline{h}} \begin{bmatrix} -(s_2 - s_1)^{-1} \\ (s_2 - s_1)^{-1} \end{bmatrix}' \begin{bmatrix} 2s_1^3 & s_1^2(3s_2 - s_1) \\ s_1^2(3s_2 - s_1) & 2s_2^3 \end{bmatrix} \begin{bmatrix} -(s_2 - s_1)^{-1} \\ (s_2 - s_1)^{-1} \end{bmatrix}$$

$$= (4s_1 + 2s_2)/6\underline{h} = [6s_1 + 2(s_2 - s_1)]/6\underline{h} \tag{7}$$

which depends not only on the length of the interval $s_2 - s_1$, but also on the size of s_1. However for any three points $s_1 < s_2 < s_3$, we find that for the change in the slope of $f(\tau)$,

$$\operatorname{var}\left[\frac{f(s_3) - f(s_2)}{s_3 - s_2} - \frac{f(s_2) - f(s_1)}{s_2 - s_1} \right] = \frac{(s_3 - s_1)}{3\underline{h}} \tag{8}$$

which can be derived from (6) in the same way that (7) was derived. Thus the distribution of any change in slopes does not depend on the artifice of an initial condition for $f'(\tau)$. This fact is not surprising, given that $f'(\tau)$ is a Wiener process, and changes in the level of a Wiener process over an interval depend only on the length of the interval and not on the distance of the interval from the origin $\tau = 0$. Given $s_4 > s_3$,

$$\operatorname{cov}\left[\frac{f(s_3) - f(s_2)}{s_3 - s_2} - \frac{f(s_2) - f(s_1)}{s_2 - s_1}, \ \frac{f(s_4) - f(s_3)}{s_4 - s_3} - \frac{f(s_3) - f(s_2)}{s_3 - s_2} \right] = \frac{(s_3 - s_2)}{6\underline{h}} \tag{9}$$

If $s_6 > s_5 > s_4 \geq s_3 > s_2 > s_1$, then

$$\operatorname{cov}\left[\frac{f(s_3) - f(s_2)}{s_3 - s_2} - \frac{f(s_2) - f(s_1)}{s_2 - s_1}, \ \frac{f(s_6) - f(s_5)}{s_6 - s_5} - \frac{f(s_5) - f(s_4)}{s_5 - s_4} \right] = 0 \tag{10}$$

Without loss of generality suppose that in a sample of size T there are m distinct values of $x_1, \ldots, x_T$. Denote the ordered distinct values by s_i $(i = 1, \ldots, m)$, and define $\mathbf{s} = (s_1, \ldots, s_m)'$

and

$$\beta = [f(s_1), \ldots, f(s_m)]' \tag{11}$$

Then (3) may be written $\mathbf{y} = \mathbf{X}\beta + \varepsilon$ with $x_{ti} = 1$ if $x_t = s_i$ and $x_{ti} = 0$ otherwise, and $\beta_i = f(s_i)$. The information in the smoothness prior of the form (8)-(10) may be expressed

$$\mathbf{R}\beta \sim N(\mathbf{0}, \mathbf{G}) \tag{12}$$

The matrix $\mathbf{R}$ is $(m-2) \times m$, with

$$r_{ii} = (s_{i+1} - s_i)^{-1}$$

$$r_{i,i+1} = -\left[(s_{i+1} - s_i)^{-1} + (s_{i+2} - s_{i+1})^{-1}\right]$$

$$r_{i,i+2} = (s_{i+2} - s_{i+1})^{-1} \quad (i = 1, \ldots, m-2)$$

and all other elements 0. The matrix $\mathbf{G}$ is $(m-2) \times (m-2)$ with

$$g_{ii} = (s_{i+2} - s_i)/3\underline{h}, \quad g_{i,i+1} = g_{i+1,i} = (s_{i+2} - s_{i+1})/6\underline{h} \quad (i = 1, \ldots, m-2)$$

and all other elements 0.

Because (12) provides a distribution of $m-2$ linear combinations of m coefficients, more information is needed for a proper prior distribution for β. To construct a proper prior that is comparable with those used in the previous section, we amend (3) slightly, by writing

$$y_t = \alpha_1 + \alpha_2 x_t + f(x_t) + \varepsilon_t, \quad \varepsilon_t \overset{\text{i.i.d.}}{\sim} N(0, h^{-1}) \quad (t = 1, \ldots, T) \tag{13}$$

Then (13) has the form $y = \mathbf{X}_1\alpha + \mathbf{X}_2\beta + \varepsilon$, for the suitably arranged $T \times 2$ matrix $\mathbf{X}_1$ and $T \times m$ matrix $\mathbf{X}_2 = \mathbf{X}$. The two restrictions

$$\sum_{i=1}^{m} f(s_i) = 0, \quad f(s_1) = f(s_m) \tag{14}$$

identify α_1, α_2 and f in (13) without imposing any additional restrictions. They can be imposed by writing

$$\beta = \mathbf{Q}\beta^*, \text{ with } \mathbf{Q}' = \left[\iota_{m-2}(-1/2) \vdots \mathbf{I}_{m-2} \vdots \iota_{m-2}(-1/2)\right]. \tag{15}$$

The restrictions (15) plus the prior information (12) in $y = \mathbf{X}_1\alpha + \mathbf{X}_2\beta + \varepsilon$ are equivalent to $\beta^* \sim N(\mathbf{0}, \underline{\mathbf{H}}_2^{-1})$ in $y = \mathbf{X}_1\alpha + \mathbf{X}_2^*\beta^* + \varepsilon$, where $\mathbf{X}_2^* = \mathbf{X}_2\mathbf{Q}$ and $\underline{\mathbf{H}}_2 = \mathbf{Q}'\mathbf{R}'\mathbf{G}^{-1}\mathbf{R}\mathbf{Q}$. If $\alpha \sim N(\underline{\alpha}, \underline{\mathbf{H}}_1^{-1})$ is independent of β in the prior distribution, then as $\underline{h} \to \infty$ in $\mathbf{G}$, the marginal likelihood of the model must approach that of the linear model $y_t = \alpha_1 + \alpha_2 x_t + \varepsilon_t$. By using the same prior distribution for α and h that we used in the special case of the linear model in the polynomial expansions of the previous section, we guarantee comparability of those models with models based on smoothness priors.

Extension from the case of a single covariate (3) to the model at hand (2) is straightforward so long as f_{ta} and f_{te} in (2) are independent *a priori*, which we take them to be. The extension may be expressed as

$$y_{ti} = \beta_{t1} + \beta_{t2}a_{ti} + \beta_{t3}e_{ti} + f_{ta}(a_{ti}) + f_{te}(e_{ti}) + \varepsilon_{ti}; \quad \varepsilon_{ti} \sim N(0, h^{-1})$$

In the prior distribution the components h, $(\beta_{t1}, \beta_{t2}, \beta_{t3})'$, f_{ta} and f_{te} are mutually independent. The prior distributions of h and $(\beta_{t1}, \beta_{t2}, \beta_{t3})'$ are the same as those for the particular case $j = 3$ of the linear model. The prior distributions of f_{ta} and f_{te} are given by (12) and (14). The hyperparameter $\underline{h}^{-1/2}$, which is the prior standard deviation of $f(\tau) - f(\tau - 1)$ controlling smoothness, is set to 0.01 for f_{ta} and 0.10 for f_{ea}, values that produce large marginal likelihood values relative to most other choices in most of the samples. Experimentation showed that varying these values by a factor of 10 had little effect on the results that we report in Section 4.

3.1.3 Computation

Both of these approaches to nonlinear regression lead to regression models of the form

$$\mathbf{y} = \mathbf{X}\beta + \varepsilon \tag{16}$$

$$\varepsilon \sim N\left(\mathbf{0},\, h^{-1}\mathbf{I}_T\right) \tag{17}$$

where the matrix of covariates $\mathbf{X}$ is $T \times k$, and independent prior distributions of the form

$$\beta \sim \mathbf{N}\left(\underline{\beta}, \underline{\mathbf{H}}^{-1}\right), \quad \underline{s}^2 h \sim \chi^2\left(\underline{\nu}\right) \tag{18}$$

In both cases the number of observations T is the number in the sample for the year to which the regression model is being applied: 1751 in the smallest sample and 2698 in the largest. In the interactive polynomial model $k_a = 4$ and $k_e = 1$, so $k = 10$, and in the separable polynomial model $k_a = 4$ and $k_e = 2$, so $k = 7$. In the separable models with smoothness priors there are 41 distinct values of a_{ti} (ages 25 through 65); the corresponding vector β (see (11)) has 41 elements and β^* (see (15)) has 39 elements. There are 17 distinct values of e_{ti}, so the corresponding vector β^* has 15 elements. Thus $k = 57$ in the separable models with smoothness priors.

In the posterior distribution corresponding to (16)-(18),

$$\beta|\,(h,\mathbf{y},\mathbf{X}) \sim N\left(\overline{\beta}, \overline{\mathbf{H}}^{-1}\right), \text{ with } \overline{\mathbf{H}} = \underline{\mathbf{H}} + h\mathbf{X}'\mathbf{X} \text{ and } \overline{\beta} = \overline{\mathbf{H}}^{-1}\left(\underline{\mathbf{H}}\underline{\beta} + h\mathbf{X}'\mathbf{y}\right)$$

$$\overline{s}^2 h|\,(\beta,\mathbf{y},\mathbf{X}) \sim \chi^2\left(\overline{\nu}\right), \text{ with } \overline{s}^2 = \underline{s}^2 + (\mathbf{y} - \mathbf{X}\beta)'(\mathbf{y} - \mathbf{X}\beta) \text{ and } \overline{\nu} = \underline{\nu} + T$$

The corresponding Gibbs sampling algorithm produces a sequence of draws $\left\{\beta^{(m)}, h^{(m)}\right\}$ from the posterior distribution with little or no detectable serial correlation. Computations were carried out using the Bayesian Analysis, Computation and Communications (BACC) software[1] running under Matlab 6.5. The results reported here are based on 1,000 iterations following 10 warm-up iterations. Computation time is less than one second for the polynomial models and less than five seconds for the separable models with smoothness priors.

3.2 Non-Gaussian Conditional Distributions

Based on our earlier work (Geweke and Keane, 2000) the assumption of Gaussian disturbances is bound to be poor in this context. In measuring the evolution of inequality in earnings over the period 1967 through 1996, it is therefore essential to employ a more flexible distribution of earnings conditional on age and education. In measuring changes in returns to education, the age profile of earnings, and the conditional variance of earnings, a more flexible distribution should provide a more reliable indication of uncertainty about these changes.

---

[1] The Bayesian Analysis, Computation and Communication Extension for Matlab, Gauss, and Splus; see http://www.cirano.qc.ac/~bacc.

3.2.1 *Mixture of Normals Distributions*

The normal mixture linear model begins with (16) and then introduces the latent state vector $\widetilde{\mathbf{s}} = (\widetilde{s}_1, \ldots, \widetilde{s}_T)'$. Conditional on $\mathbf{X}$, the $\widetilde{s}_t$ are i.i.d. with $P(\widetilde{s}_t = j) = \pi_j$, and thus

$$p(\widetilde{\mathbf{s}}\,|\,\mathbf{X}) = \prod_{t=1}^{T}\pi_{\widetilde{s}_t} = \prod_{j=1}^{m}\pi_j^{T_j} \tag{19}$$

where $T_j = \sum_{t=1}^{T}\delta(\widetilde{s}_t, j)$ is the number of observations t for which $\widetilde{s}_t = j$.

Corresponding to each of the m states j there is a mean parameter α_j and a positive precision parameter h_j; let $\alpha = (\alpha_1, \ldots, \alpha_m)'$, $\mathbf{h} = (h_1, \ldots, h_m)'$ and $\pi = (\pi_1, \ldots, \pi_m)'$. Conditional on $\widetilde{s}_t = j$, $\varepsilon_t \sim N\left[\alpha_j, (h\cdot h_j)^{-1}\right]$. Thus

$$p\left[y_t\mid \beta, h, \pi, \propto, \mathbf{h}, \widetilde{s}_t = j, \mathbf{X}\right] \propto (h\cdot h_j)^{1/2}\cdot\exp\left[-h\cdot h_j\,(y_t - \alpha_j - \beta'\mathbf{x}_t)^2/2\right] \quad (t = 1, \ldots, T) \tag{20}$$

The disturbances ε_t are i.i.d. and follow a full discrete normal mixture distribution:

$$p(\varepsilon_t\mid h, \pi, \alpha, \mathbf{h}, \mathbf{X}) \propto h^{1/2}\sum_{j=1}^{m}\pi_j h_j^{1/2}\exp\left[-h\cdot h_j\,(\varepsilon_t - \alpha_j)^2/2\right]$$

The mixture of normals distribution is very flexible. Fig. 2 provides several examples. For the special case in which the means α_j are all the same the normal mixture distribution is known as the scale mixture of normals distribution. That distribution is symmetric, unimodal, and must be leptokurtic, that is, the coefficient of kurtosis $K = E\left[\varepsilon_t - E(\varepsilon_t)\right]^4/\mathrm{var}(\varepsilon_t)^2 > 3$, its value if ε_t is normally distributed. Panels (a) and (f) of Fig. 2 provide examples. If the means α_j are not all the same then the normal mixture distribution can be skewed, as illustrated in panels (c) and (d). It can also be platykurtic (i.e. $K < 3$), as is the case in panels (b) and (e). Of course, these distributions can be multimodal (panel (e)). With a sufficient number of components, the normal mixture distribution can mimic distributions that are quite different from the normal, like the uniform (panel (b)).

The conditionally conjugate prior densities in the normal mixture linear model are (18) for β and h. The prior distribution of π is Dirichlet

$$p(\pi) = \Gamma(mr)\,\Gamma(r)^{-m}\prod_{j=1}^{m}\pi_j^{r-1} \tag{21}$$

The components of $\mathbf{h}, \underline{\nu}^2 h_j \overset{\text{i.i.d.}}{\sim} \chi^2(\underline{\nu})$ $(j = 1, \ldots, m)$ have independent gamma distributions,

$$p(\mathbf{h}) = 2^{m\underline{\nu}/2}\Gamma(\underline{\nu}/2)^{-m/2}\left(\underline{\nu}^2\right)^{m\underline{\nu}/2}\prod_{j=1}^{m}h_j^{(-\underline{\nu}-2)/2}\exp\left(-\underline{\nu}^2 h_j/2\right) \tag{22}$$

Finally, $\alpha\mid h \sim N\left[\mathbf{0}, (\underline{h}_\alpha\cdot h)^{-1}\mathbf{I}_m\right]$, so that

$$p(\alpha\mid h) = (2\pi)^{-m/2}(\underline{h}_\alpha h)^{m/2}\exp\left(-\underline{h}_\alpha h\alpha'\alpha/2\right) \tag{23}$$

These prior distributions are conditionally conjugate, symmetric across states, and require the specification of just three hyperparameters: r, $\underline{\nu}$ and $\underline{h}_\alpha$. The specification $E(\alpha) = \mathbf{0}$ resolves the identification issues with respect to α and β. The prior variance in β conveys uncertainty about the location of the distribution of $\mathbf{y}$ given $\mathbf{X}$. The prior distribution of α is scale dependent on $h^{-1/2}$, i.e. it states prior beliefs about the shape of the distribution. Keeping in mind that $E(\mathbf{h}) = \mathbf{e}_m$, a prior

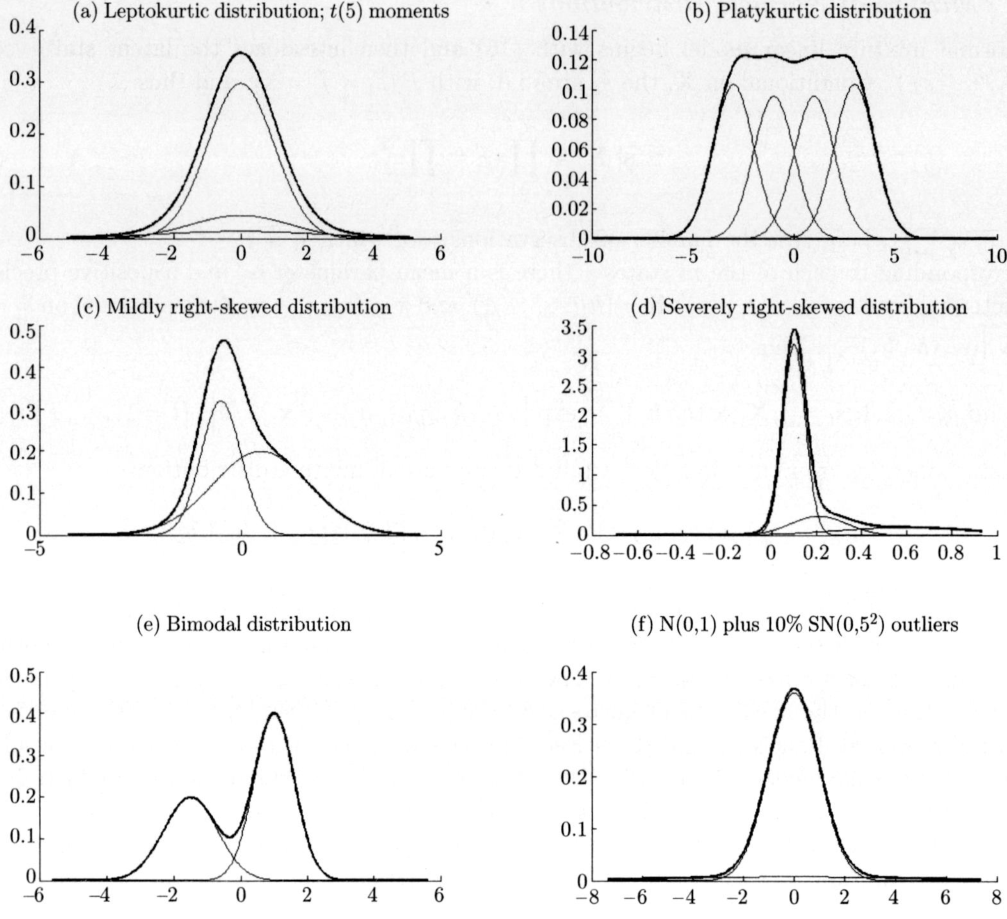

Fig. 2.   Several mixture of normals probability density functions.

distribution with $\underline{h}_\alpha^{-1/2} = 5$ implies a prior probability of multimodality that is near 1, whereas $\underline{h}_\alpha^{-1/2} = 1/5$ makes this probability negligibly small. Keeping in mind that $E(\alpha) = \mathbf{0}$, choice of $\underline{\nu}.$ governs the prior probability of tail thickness in the mixture normal density relative to the normal. In the prior distribution the ratio $h_j/h_k \sim F(\underline{\nu}.,\underline{\nu}.)$ for all $j \neq k$. If $\underline{\nu}. = 1$ the prior probability of component variance ratios at least as great as those shown in Fig. 2(f) is significant, whereas if $\underline{\nu}. = 5$ it is negligible. With these considerations in mind, and after examining the implications of different choices for marginal likelihood in the three regression models and several samples, we settled on the choices $\underline{h}_\alpha^{-1/2} = 1.58$, $\underline{\nu}. = 0.2$, and $r = 1$.

The states are unidentified in this model: that is, a relabelling that interchanges two or more states leaves the posterior distribution unaffected. In our application this lack of identification is harmless, because the states are simply a modeling device to provide a flexible conditional distribution and have no independent substantive interpretation. None of the questions of interest depend in any way on identification of the states. Moreover, leaving the states unidentified provides some advantages in computation.

3.2.2   *Computation*

Posterior inference in the normal mixture model utilizes five blocks: (α, β), h, π, $\mathbf{h}$, and $\widetilde{\mathbf{s}}$. It is useful to define

$$\underset{T\times m}{\widetilde{\mathbf{Z}}\,(\widetilde{\mathbf{s}})} = \widetilde{\mathbf{Z}} = [\widetilde{\mathbf{z}}_1, \dots, \widetilde{\mathbf{z}}_T]' = [\delta\,(\widetilde{s}_t, j)], \quad \underset{T\times(m+k)}{\widetilde{\mathbf{W}}} = \begin{bmatrix} \widetilde{\mathbf{Z}} & \mathbf{X} \end{bmatrix}$$

$$\underset{(m+k)\times 1}{\underline{\gamma}} = \begin{pmatrix} \mathbf{0} \\ \underline{\beta} \end{pmatrix}, \quad \underset{(m+k)\times(m+k)}{\underline{\mathbf{H}}_\gamma\,(h)} = \underline{\mathbf{H}}_\gamma = \begin{bmatrix} \underline{h}_\alpha h \mathbf{I}_m & \mathbf{0} \\ \mathbf{0} & \underline{\mathbf{H}}_\beta \end{bmatrix}$$

and $\underset{T\times T}{\widetilde{\mathbf{Q}}\,(\widetilde{\mathbf{s}})} = \widetilde{\mathbf{Q}} = \mathrm{diag}\,(h_{\widetilde{s}_1}, \dots, h_{\widetilde{s}_T})$. With this notation (20) is equivalent to

$$p\,(\mathbf{y} \mid \gamma, h, \pi, \mathbf{h}, \widetilde{\mathbf{s}}, \mathbf{X}) \propto h^{T/2} \left| \widetilde{\mathbf{Q}} \right|^{1/2} \cdot \exp\left[-h\,\left(\mathbf{y} - \widetilde{\mathbf{W}}\gamma\right)' \widetilde{\mathbf{Q}} \left(\mathbf{y} - \widetilde{\mathbf{W}}\gamma\right)/2\right] \tag{24}$$

The kernel of the conditional posterior density of γ is the product of (18), (23), and (24), from which the conditional posterior distribution is

$$\gamma \sim \mathbf{N}\,(\overline{\gamma}, \overline{\mathbf{H}}_\gamma); \quad \overline{\mathbf{H}}_\gamma = \underline{\mathbf{H}}_\gamma + h\widetilde{\mathbf{W}}'\widetilde{\mathbf{Q}}\widetilde{\mathbf{W}}, \quad \overline{\gamma} = \overline{\mathbf{H}}_\gamma^{-1}\left[\underline{\mathbf{H}}_\gamma\underline{\gamma} + h\widetilde{\mathbf{W}}'\widetilde{\mathbf{Q}}\widetilde{\mathbf{y}}\right] \tag{25}$$

The conditional posterior density of h is the product of (18), (23) and (20). This kernel corresponds to the conditional posterior distribution

$$\left[\underline{s}^2 + \underline{h}_\alpha \alpha'\alpha + \sum_{t=1}^{T} h_{(t)}\,(y_t - \alpha'\widetilde{\mathbf{z}}_t - \beta'\mathbf{x}_t)^2\right] h \sim \chi^2\,(\underline{\nu} + m + T) \tag{26}$$

The conditional posterior density kernel of π is the product of (19) and (21), $\prod_{j=1}^{m} p_j^{r+T_j-1}$, and thus the conditional posterior distribution is Dirichlet with parameters $r + T_j$ $(j = 1, \dots, m)$. The conditional posterior density kernel of $\mathbf{h}$ is the product of (22) and (20), which implies

$$\left[\underline{s}^2 + \sum_{t=1}^{T} \delta\,(s_t, j)\,(y_t - a_j - \beta'\mathbf{x}_t)^2\right] h_j \sim \chi^2\,(\underline{\nu} + T_j) \quad (j = 1, \dots, m).$$

The conditional posterior density kernel for the state assignments $\widetilde{\mathbf{s}}$ is the product of (19) and (20) taken over $t = 1, \dots, T$. Thus the states $\widetilde{s}_t$ are conditionally independent, with

$$P\,(\widetilde{s}_t = j) \propto p_j h_j \exp\left[-h \cdot h_j\,(y_t - a_j - \beta'\mathbf{x}_t)^2/2\right] \quad (j = 1, \dots, m) \tag{27}$$

Draws from these multinomial distributions are straightforward.

Computations follow a Gibbs sampling algorithm based on these five blocks. Since the states are left unidentified, the Markov chain is characterized by label switching. We observed this regularly with mixtures of three components and occasionally in models with two components, using 20,000 iterations in each case. The results presented in the next section are all based on every tenth value of 20,000 iterations following 1,000 warm-up iterations. Fig. 3 illustrates MCMC output using the 1986 data, the two component mixture model, and two of the regression functions. The illustration labels as state 1 the state with the smaller probability: the top panels display π_1, the middle panels show $\alpha_2 - \alpha_1$, and the lower panels $(h_1/h_2)^{1/2}$. The MCMC output is nearly serially uncorrelated in all cases: relative numerical efficiencies are between 0.65 and 0.75 for the polynomial models (left-hand panels), and between 0.25 and 0.50 for the separable smooth function models. All computations were carried out using BACC running under Matlab 6.5, and require about 4 min. for each polynomial model and 10 minutes for each separable smooth function model.

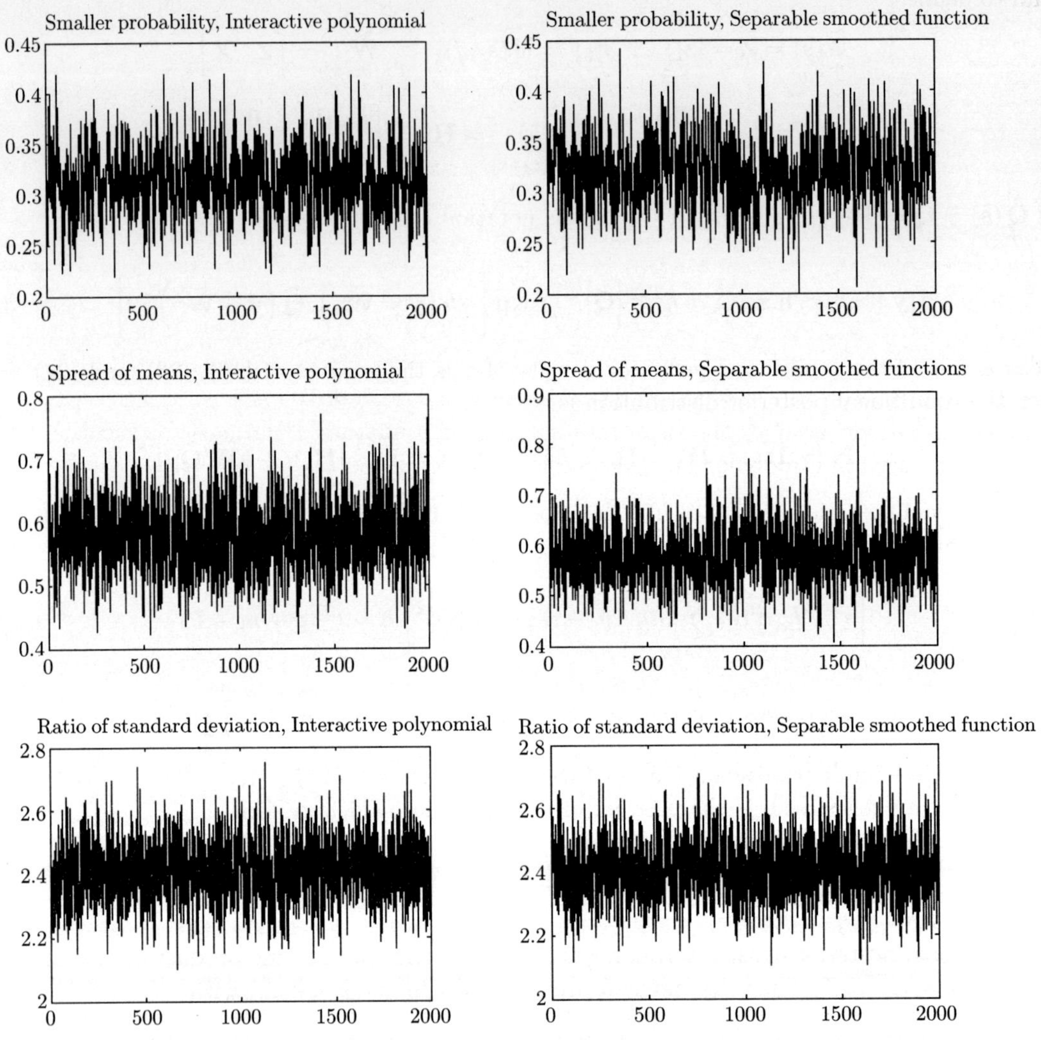

Fig. 3. **Markov chain Monte Carlo for model with mixture of two normals, 1985 sample.**

4. Findings

This section presents the evidence on the substantive questions of interest motivating our application. We begin by summarizing the relative performance of the alternative approaches to nonlinear regression and the distribution of regression residuals (Section 4.1). Next we take up returns to education and the age profile of earnings, which are functionals of the regression (Section 4.3). Lastly, we turn to measures of inequality in the distribution of earnings, which are functionals of the entire conditional distribution (Section 4.4). The principal vehicles for conveying the results are conditional medians and interquartile ranges for these functionals for each of the thirty years in the sample. Tables in an appendix[2] provide corresponding posterior means and standard deviations.

4.1   Model Comparison

The conclusion of Section 3.1.1 describes the selection of orders of polynomial expansion and prior hyperparameters in the basis function approach to regression, the conclusion of Section 3.1.2 indicates how the smoothness hyperparameters were chosen in the Wiener process prior approach, and the end of Section 3.2.1 does the same for the three prior distribution hyperparameters for the normal mixture models of regression residuals. Given these choices, we now compare these approaches to modeling $p(y \mid \mathbf{x})$ using Bayes factors.

Fig. 4 provides histograms for the posterior probabilities across the three alternative specifications of the regression function assuming a Gaussian distribution for disturbances, in the three left panels. It does the same assuming a mixture of two normal distributions in the three right panels. In both cases interactive polynomials are favored in the early samples (through about 1980) but not thereafter. In the later samples (after about 1980) the models with Wiener process priors are more highly favored if one assumes Gaussian residuals. Given a mixture of normals specification for the residuals, the separable polynomial specification is favored more often than the Wiener process priors after about 1980. Perhaps more important than any of these results, however, is the fact that no single model has posterior probability near one for all years—note that the figures convey small positive probabilities for quite a few models and years.

Appendix Table 1 provides log Bayes factors for the models, all taken relative to an interactive polynomial model of order $k_a = 3$ in age and $k_e = 1$ in education, with Gaussian residuals. (We take this as a benchmark because it is the closest to a consensus of earnings model specifications in the literature.) The most important information conveyed by this table is that models with mixtures of two normal distributions completely dominate the Gaussian models. For any combination of specifications of the regression function the log Bayes factor in favor of a mixture of two normals model versus a Gaussian model is never less than 170 and exceeds 300 in some cases. Comparison of this table with Fig. 1 shows that the log Bayes factors comparing normal mixture and Gaussian models are roughly proportional to sample size, as might be expected. Bayes factors comparing mixtures of two normals with mixtures of three normals (not presented) weakly favor a mixture of two normals. The reason for this will become apparent when we inspect conditional distributions in detail in Section 4.4.

Because no one mixture of normals model dominates in many of the years, it is necessary to average across models for those functionals whose posterior distribution is sensitive to the specification of the regression function. In Sections 4.3 and 4.4 we report results for all three specifications, and it turns out that these distributions are insensitive to specification. In the case of conditional means and variances, we report findings for all six specifications (the product of three for regressions and two for residual specification) in order to assess the sensitivity of the posterior distribution to the incorrect assumption of normality of residuals.

4.2   Model Evaluation

Bayes factors provide relative comparisons of alternative models, but they do not reveal the adequacy or inadequacy of these models in describing interesting aspects of the data. We address this question through the structured but less formal method of posterior predictive analysis. To describe this

---

[2] http://www.biz.uiowa.edu/faculty/jgeweke/papers/paper89/appendix.pdf

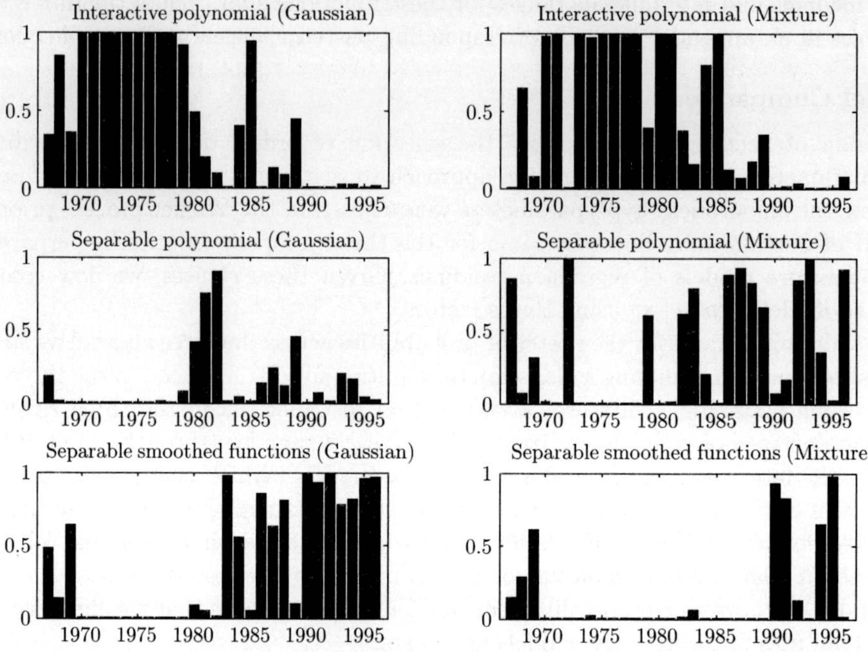

Fig. 4. The left [right] panels show posterior probabilities of regression functions given Gaussian [mixture of normals] regression residuals.

method succinctly, denote the parameters of the model by θ, the observables in the sample by $\mathbf{y}$, and the entire observed data set by $\mathbf{y}^0$. Let $g\left(\mathbf{y}\right)$ be an interesting scalar feature that can be observed in the data, such as a sample moment or quantile. Then the posterior distribution of θ induces a distribution on $g\left(\mathbf{y}\right)$ by means of the model density $p\left(\mathbf{y}\mid\theta\right)$: it is the predictive distribution for $g\left(\mathbf{y}\right)$ corresponding to a hypothetical repetition of the experiment of drawing a sample of the same size from the same population. A finding that with extremely high probability $g\left(\mathbf{y}\right)$ would exceed or fall short of the observed $g\left(\mathbf{y}^0\right)$ in this hypothetical experiment casts doubt on the specification of the model. This idea goes back to the notion of surprise discussed by Good (1956) and its essentials were further developed by Rubin (1984) in what he termed "model monitoring by posterior predictive checks." The mechanics of model evaluation entail generating an artificial sample $\mathbf{y}^{(m)}$ corresponding to each draw $\theta^{(m)}$ from the posterior simulator and computing $g\left(\mathbf{y}^{(m)}\right)$ $(m=1,\dots,M)$, and then finding the position of $g\left(\mathbf{y}^0\right)$ relative to the empirical c.d.f., which can be expressed

$$p_g^* = M^{-1} \sum_{m=1}^{M} I_{(-\infty, g(\mathbf{y}^0))} \left[g\left(\mathbf{y}^{(m)}\right) \right].$$

If p_g^* is close to zero, then the posterior distribution overpredicts $g\left(\mathbf{y}\right)$; if it is close to one, then the posterior distribution underpredicts. For further details see Lancaster (2004, Section 2.5) or Geweke (2005, Section 8.3).

We evaluate the Gaussian and mixture models using six different features of the data $g\left(\mathbf{y}\right)$. We include the Gaussian models in spite of their inferior performance in the model comparison exercise because we wish to see if the mixture models perform at least as well as the Gaussian models with respect to all these features. Figs. 5 through 10 provide the results of this analysis. Each figure corresponds to a different function g, to be described shortly. The six panels in each figure correspond to the three Gaussian and three mixture models, just as in Fig. 4. The value p_g^* was computed for each of the 30 samples corresponding to the years 1967-1996, and each panel provides a histogram of the p_g^* values over the 30 samples. Values of p_g^* are sorted into five bins of equal size, except that if $g\left(\mathbf{y}^0\right) < g\left(\mathbf{y}^{(m)}\right)$ for all iterations, then p_g^* is assigned to a bin just below zero (see, e.g. Fig. 9) and if $g\left(\mathbf{y}^0\right) > g\left(\mathbf{y}^{(m)}\right)$ for all iterations, then p_g^* is assigned to a bin just above one (see, e.g., Fig. 10); all results are based on $M = 2,000$ iterations. Thus if a histogram is concentrated to the left, the posterior distribution tends to overpredict the observed feature of the data, and if concentrated to the right tends to underpredict. Appendix Tables 2 through 7 provide the values p_g^* for the individual years.

Figs. 5, 6 and 7 provide evidence on the ability of the models to describe systematic differences in earnings by age and education. In Fig. 5 the function g is the difference in average log earnings between men with 16 years of education and men with 12 years of education, in each of the 30 sample years. In Fig. 6 it is this difference computed for men age 45 and age 25, and in Fig. 7 it is for men age 60 and men age 45. There is a mild tendency for all models to underpredict the observed returns to college education (Fig. 5). The tendency is slightly more pronounced for mixture models than for Gaussian models. The difficulty is most pronounced in the mixture models with interactive polynomials, in which the sample returns to education fell below the posterior median in 29 out of 30 years, and in one year (1977) $p_g^* = 1.0$. All models do better in capturing average log earnings by age: Figs. 6 and 7 show little, if any, tendency for models to overpredict or underpredict. While the difficulties in capturing average log earnings between men with 16 years of education and men

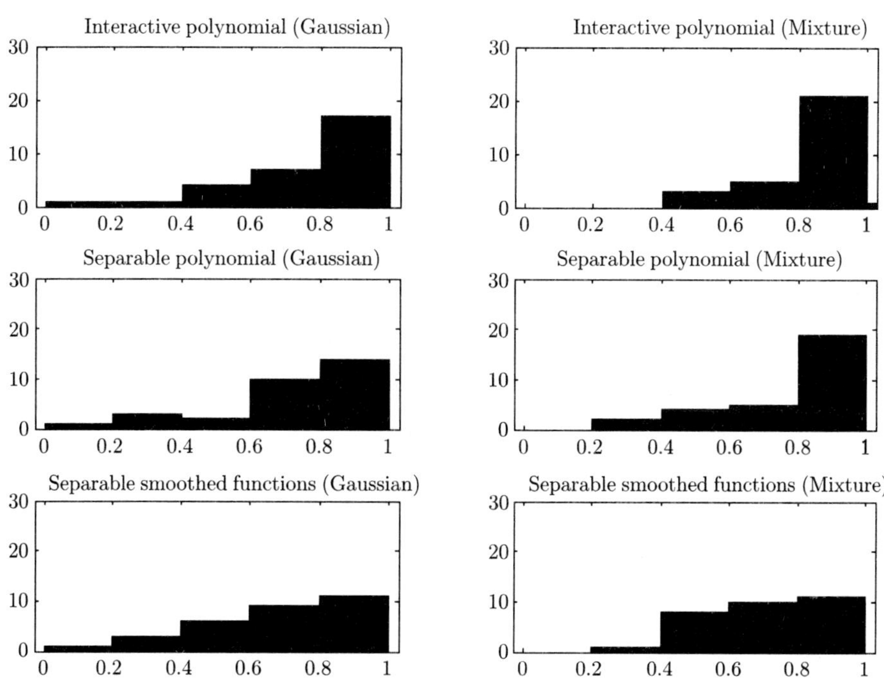

Fig. 5.   **Posterior predictive p_g^* for difference between average sample log earnings of men with 16 and 12 years of education.**

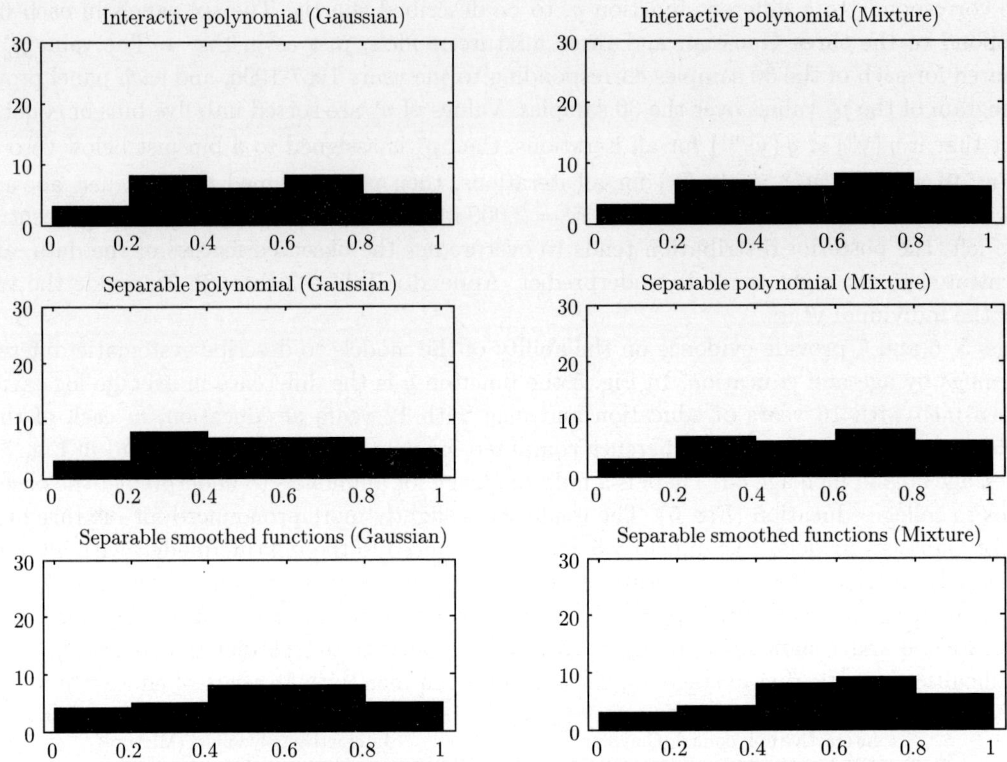

Fig. 6.   **Posterior predictive** p_g^* **for difference between average sample log earnings of men age 45 and men age 25.**

with 12 years of education in the sample are not extreme, we find them somewhat puzzling in view of the fact that the evidence supports a very nearly linear relationship between education and log earnings in the regression function. The fact that there are many more men with exactly 16 years and exactly 12 years of education than there are men of any one age (in particular, ages 25, 45 and 60) means that there is less sampling variation in observed returns to education, and this may contribute to the evident mild difficulties in fitting these returns.

Figs. 8, 9 and 10 show how well each of the models describes aspects of the distribution of log earnings conditional on age and education. In the case of Fig. 8, the function $g(\mathbf{y})$ was computed by finding least squares estimates of the regression function (e.g., a polynomial of order 4 in age and 2 in education in the case of the separable polynomial models), and then computing the usual least squares estimate of the standard deviation of the disturbance. In Fig. 9 the procedure was the same except that the final object $g(\mathbf{y})$ is the coefficient of skewness of the least squares residuals, and in Fig. 10 it is the coefficient of kurtosis of these residuals. The first of these figures stands in marked contrast to the latter two. For the Gaussian models, the sample conditional standard deviation $g(\mathbf{y}^0)$ in Fig. 8 is always close to the median of the distribution. This reflects the well-known fact that a normal linear model recovers the population linear projection and the variance about that projection whether the model is correctly specified or not (see Geweke, 2005, Example 3.4.3). For similar reasons $g(\mathbf{y}^0)$ is also near the median in the mixture models. Fig. 9 dramatically demonstrates the failure of the Gaussian models to describe the negative skewness in the distribution of log earnings conditional

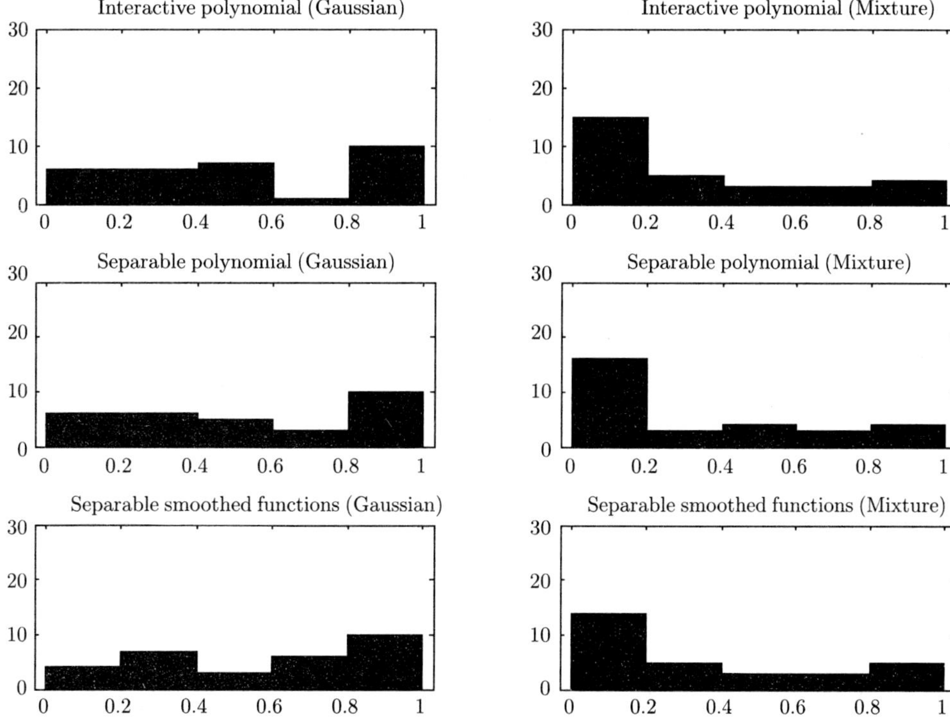

Fig. 7. Posterior predictive p_g^* for difference between average sample log earnings of men age 60 and men age 45.

on age, and Fig. 10 does the same for the leptokurtosis in this distribution. The mixture models account for these features rather well, although there is a mild tendency for the posterior distribution to underpredict the degree of conditional kurtosis observed in the sample.

Posterior predictive analysis can be conducted with other interesting functions of the observables $g(\mathbf{y})$. Appendix Tables 8, 9 and 10 report results related to aspects of conditional distributions studied in Section 4.4.2. In many of these cases the failure of the normal models was as dramatic as that in Figs. 9 and 10, and in all cases the mixture models had stronger posterior predictive performance.

4.3 Conditional Means and Variances

Each of the six models provides the functionals $E(y \mid \mathbf{x})$ and $var(y \mid \mathbf{x})$; the former depends on $\mathbf{x}$ but the latter does not. Fig. 11 shows posterior medians and interquartile ranges for $f_1(a) = E_{1986}(y \mid a, e = 12)$ and Fig. 12 does the same for $f_2(e) = E_{1986}(y \mid a = 40, e)$. A change in the conditioning $e = 12$ or $a = 40$ produces only level shifts in the medians and interquartile ranges for the separable polynomial and Wiener process prior specifications, whereas for the interactive polynomial regressions both the shapes of the curves and the sizes of the interquartile ranges will be affected. The values of $f_1(a)$ for different values of a are highly dependent, via the polynomial restriction for the first two specifications and the Wiener process prior for the third. The posterior medians and interquartile ranges in Fig. 11, and the posterior means and standard deviations in

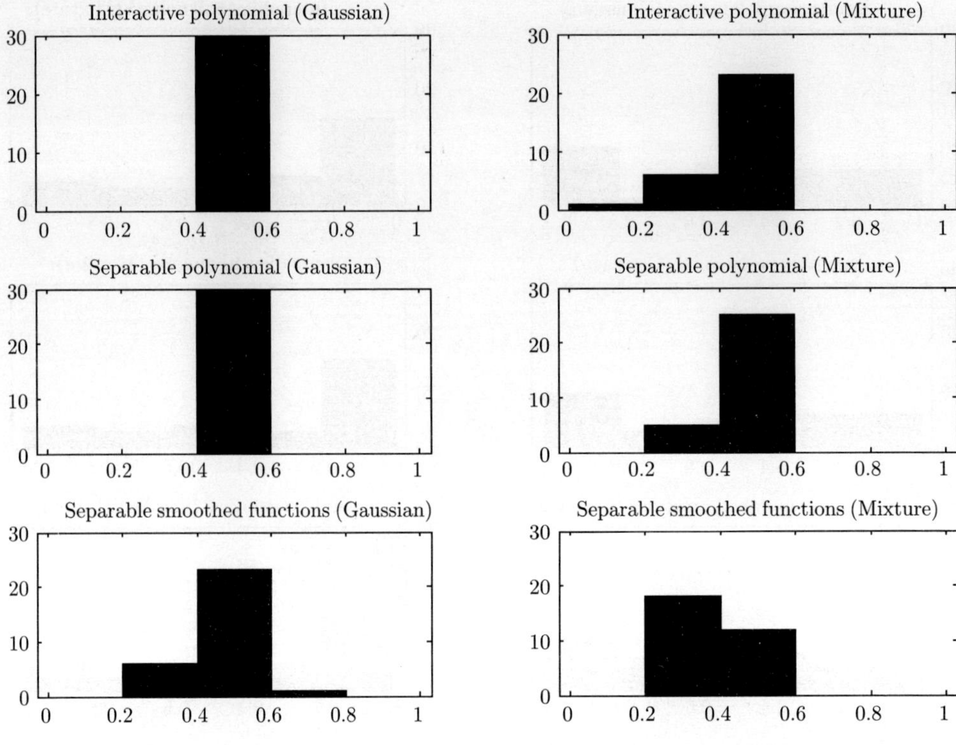

Fig. 8.   **Posterior predictive p_g^* for standard deviation of least squares residuals.**

Appendix Table 11, pertain to the marginal posterior distribution of $f_1(a)$ at each value of a, and therefore do not provide an indication of uncertainty about the regression function as a whole. A similar precaution is in order for the interpretation of Fig. 12 and Appendix Table 12.

The posterior medians and means of $f_1(a)$ are remarkably insensitive to the specification of the model. Differences in posterior means are always substantially less than corresponding posterior standard deviations, and differences in posterior medians are less than the corresponding interquartile ranges. The same cannot be said of $f_2(e)$. We observe a similar congruence across specifications for those values of e on which the sample is concentrated, $e = 12$ and above (Fig. 1). For lower values of e discrepancies markedly increase, especially for the interactive polynomial model in which $f_2(e)$ is linear. The models have substantially different implications for $f_2(e)$, $e \leq 4$, but because of the dearth of data for such low levels of education different implications have very little impact on the posterior distribution: these are cases of almost pure extrapolation.

The sizes of the posterior interquartile ranges and standard deviations tell a more complex, but quite informative, story. Consider first the case of $f_2(e)$, which is somewhat simpler. For all specifications the posterior interquartile range and standard deviation increase as e decreases. This is due to the lack of observations on poorly educated men, and is familiar from elementary econometrics. For any given specification of the regression function and value of a, the interquartile ranges and posterior standard deviations of $f_2(e)$ are smaller for most mixture of normal specifications of the disturbance than for the Gaussian specification. (This feature is somewhat more transparent in Appendix Table 12 than it is in Fig. 12.) This is due to the fact that under the

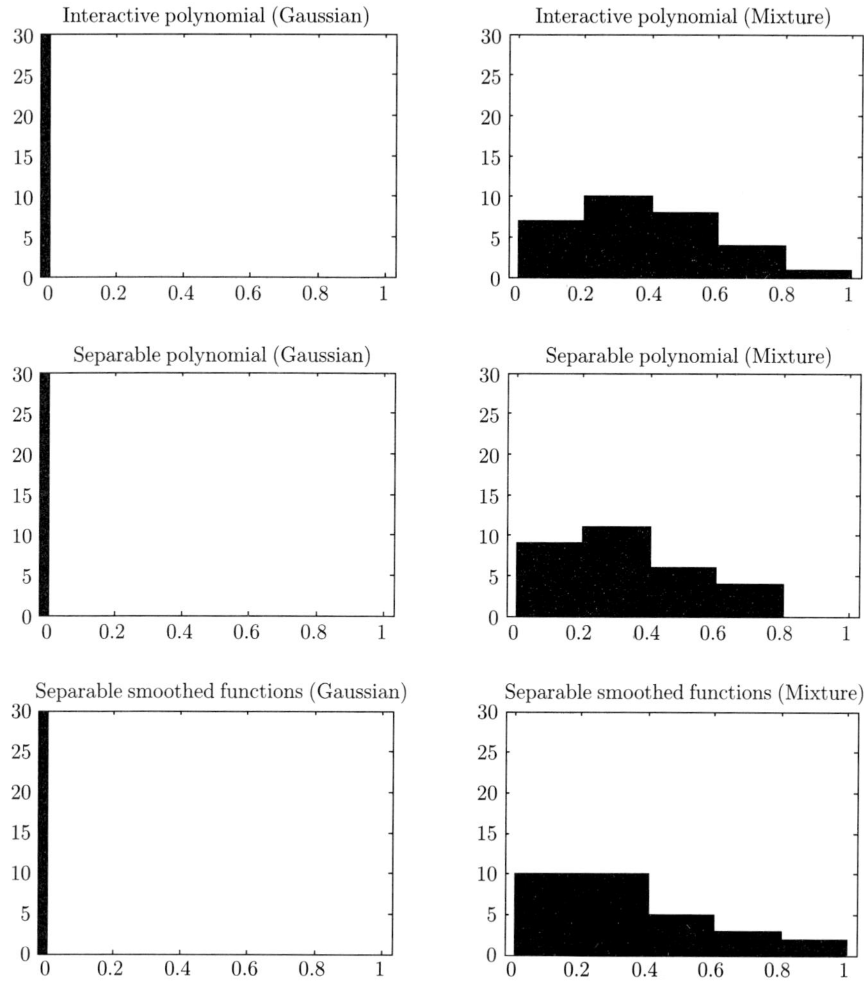

Fig. 9. Posterior predictive p_g^* for coefficient of skewness of least squares residuals.

mixture of normals specification the conditional distribution is substantially leptokurtic, as described below in Section 4.4.1. The Gaussian and mixture of normals specification lead to similar inferences about conditional variance, a fact familiar from the analysis of misspecification of regression residual distributions in linear regression and demonstrated below in Section 4.3.3 for this application. In this situation the mixture of normals distribution is more informative about the regression function than is the Gaussian distribution due to its higher concentration of residuals near the mean of the distribution, illustrated below in Section 4.4.1 for this application. This is perhaps most easily seen in the limiting case in which kurtosis increases without bound while variance remains constant: Chebychev's inequality implies that the probability density function of the residuals collapses about the point zero, and as this limit is approached the posterior distribution conveys the value of $E\left(y \mid \mathbf{x}\right)$ with certainty given a fixed specification of the model and prior.

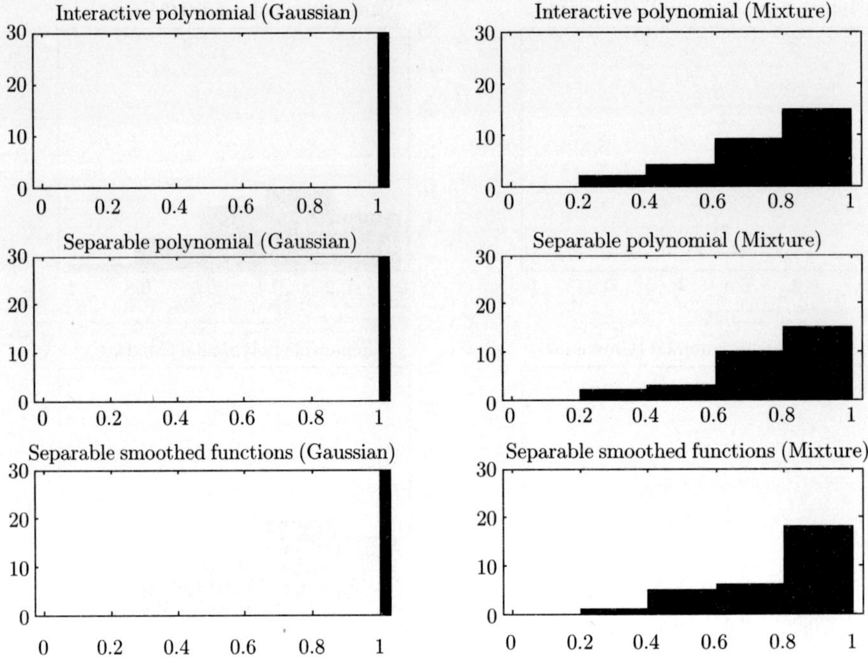

Fig. 10. Posterior predictive p_g^* **for coefficient of kurtosis of least squares residuals.**

Interquartile ranges and posterior standard deviations for $f_1(a)$ increase as a increases, due to the fact that sample size is a decreasing function of age in 1986. For the same specification of the regression function, the ratio of the posterior standard deviation in the mixture of normals specification to that in the Gaussian specification is a monotone increasing function of age, as may be confirmed from Appendix Table 11. The same is true of the interquartile range, although this admittedly requires a very keen eye inspecting Fig. 11. This can be traced to the fact that in the mixture of normals model, draws of s_t from (27) occur more often from the high-variance state than from the low-variance state for older men relative to the frequency for younger men. This reflects a misspecification of the model, which implies that the distribution of these states should be the same at all ages. We return to this point when discussing directions for future research in the concluding section.

4.3.1   Returns to Education

Changes in education imply changes in expected log earnings. We focus here on the difference in expected log earnings between college graduates ($e = 16$) and high school graduates ($e = 12$). In the separable polynomial and Wiener process prior models these differences do not depend on age. For the interactive polynomial regression specification we condition on age 40, at which point expected log earnings approach their peak over the life cycle. Fig. 13 provides the posterior medians, upper and lower quartiles of the posterior distribution of this functional for all six models and each of the 30 data sets for the years 1967,..., 1996, and Appendix Table 13 indicates the corresponding posterior means and standard deviations. Recall that posterior distributions are constructed separately for each year, and there is no functional or prior dependence between years in Fig. 13 (or in any of the succeeding figures) as there was between years of education in Fig. 11 or ages in Fig. 12. The

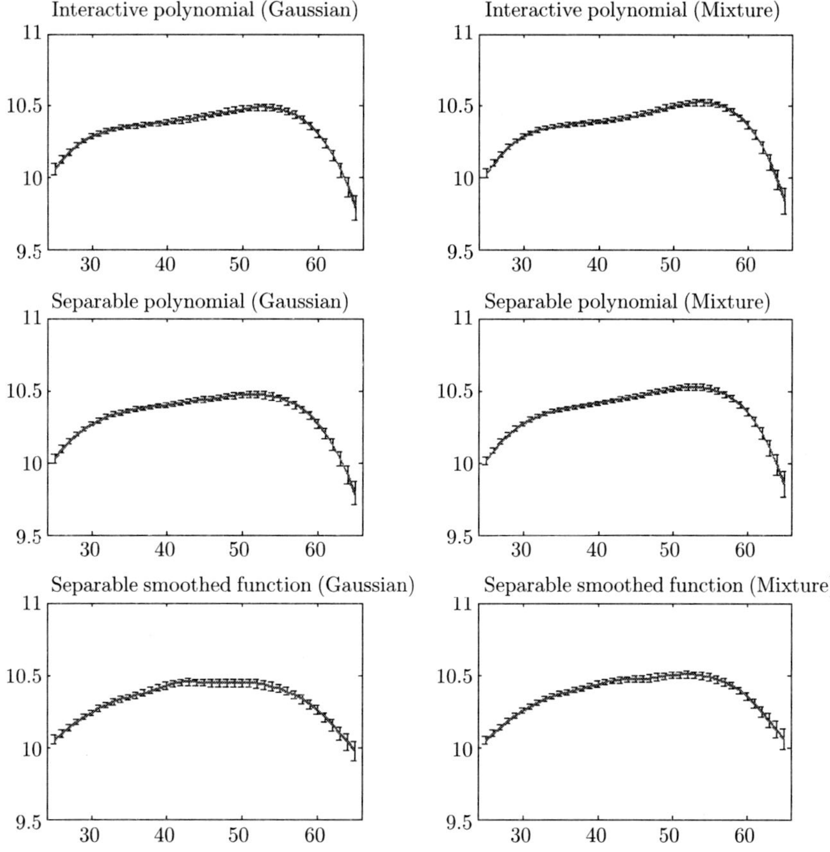

Fig. 11.   Posterior medians, upper and lower quartiles for expected log earnings conditional on age and 12 years of education, 1986 sample.

results are not independent across years, since that data are taken from a panel. While our models do not address this dependence, regarding results as independent across years is probably a good approximation in interpreting these figures and similar figures that follow.

Posterior means and medians for returns to education are sensitive to the specification of the regression function, and almost completely insensitive to the specification of the distribution of regression residuals. However, there is substantially less posterior uncertainty about returns in the mixture of normals models, for the reasons just discussed in considering $f_2(e)$: posterior standard deviations and interquartile ranges are about 25% smaller. Across regression specifications and for the same specification of the residual distribution, models that have the highest posterior probability (Fig. 4) tend to provide the smallest posterior standard deviations and interquartile ranges.

The separable regression functions indicate stagnating returns to college education early in the sample, declining from about 0.35 in 1967 to about 0.25 in 1979. By 1990 returns have nearly doubled, to about 0.48, where they remain through 1996. In the context of the mixture of normals specification the early downward movement amounts to about four posterior standard deviations, and the later increase is the equivalent of about eight posterior standard deviations. Conditional on these models, there is substantial confidence in the "U" shape of returns to college education

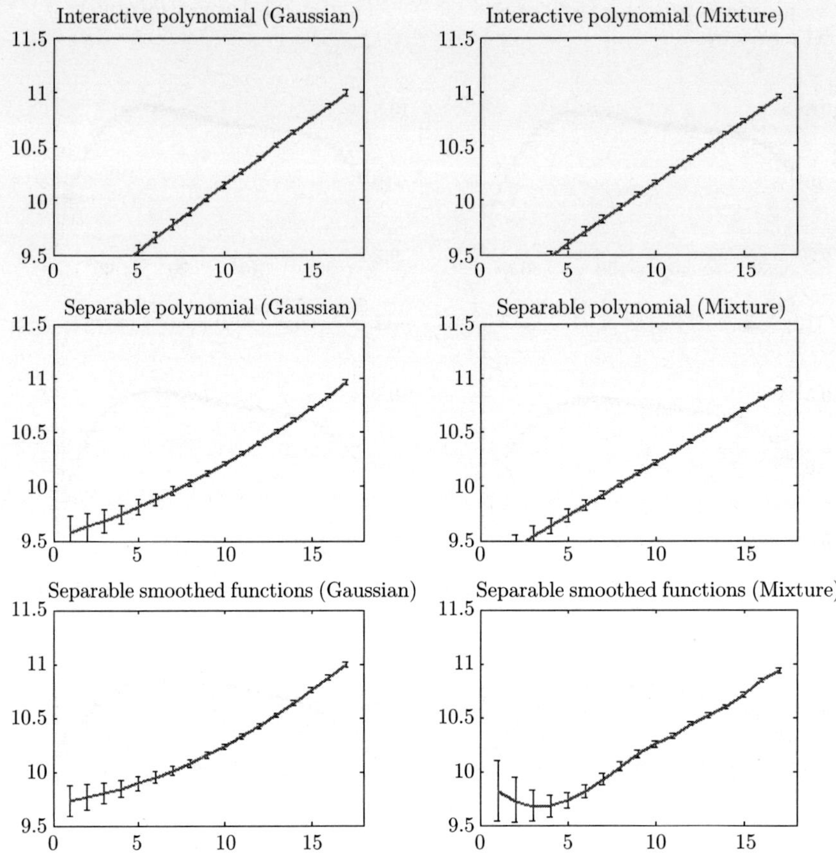

Fig. 12. **Posterior medians, upper and lower quartiles for expected log earnings conditional on education at age 40, 1986 sample.**

over the period 1967-1996. Given the interactive polynomial specification, returns to college change substantially less over the period. They remain in the range 0.30 to 0.35, which has roughly the same length as the interquartile range, through 1979. They then increase to about 0.48, but not until the early 1990's. All models identify 1979 through 1996 as a period in which returns to college education increased substantially, with an upward shift on the order of 0.20.

Recall that the interactive polynomial specification was favored in many of the years prior to 1980 (Fig. 4), although the log-odds ratio never exceeds about 15 (Appendix Table 1). In the interactive polynomial models, returns to education can depend on age. Fig. 14 and Appendix Table 14 show that in these models, returns to college education are lower for younger men than for older men, in roughly the first half of the sample. Through about 1985, returns to college are between 0.1 and 0.2 for men age 25, as opposed to the substantially higher levels for men age 40 or 60. After 1985 returns to college are similar for men of all ages, consistent with the Bayes factors in favor of the separable models for those years. In all specifications the sample provides considerably more information about returns to college at age 40 than at age 25 or 60. This is consistent with the distribution of age and education in the sample (Fig. 1). In particular, in the last years of the sample there are very few very young men, and this is reflected in large interquartile ranges in Fig. 14.

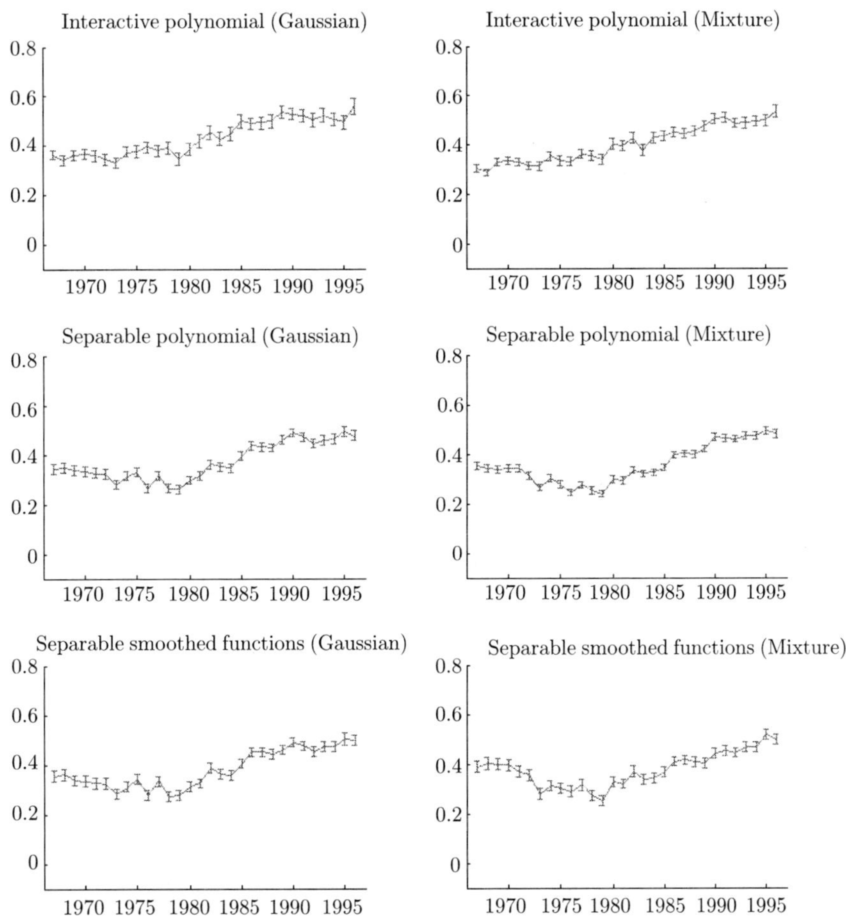

Fig. 13.   Posterior medians, upper and lower quartiles for the difference in expected log earnings given 16 years of education versus 12 years of education at age 40.

4.3.2   Age Profile of Earnings

The hump displayed in the age profile of earnings for 1986 in Fig. 11 is characteristic of all years in the sample, but the shape changes markedly over the period 1967-1996. We measure the change by considering two functionals of $E\left(y \mid \mathbf{x}\right)$

$$E\left(y \mid a = 45, e = 12\right) - E\left(y \mid a = 25, e = 12\right)$$

and

$$E\left(y \mid a = 60, e = 12\right) - E\left(y \mid a = 45, e = 12\right)$$

The first reflects mainly returns to experience, since virtually all men are labor market participants between the ages of 25 and 45, and most of them work full-time during almost all of these years. The second reflects mainly changes in labor market participation that accelerate between the ages of 45 and 60. Conditioning on 12 years of education matters only in the interactive polynomial models of regression.

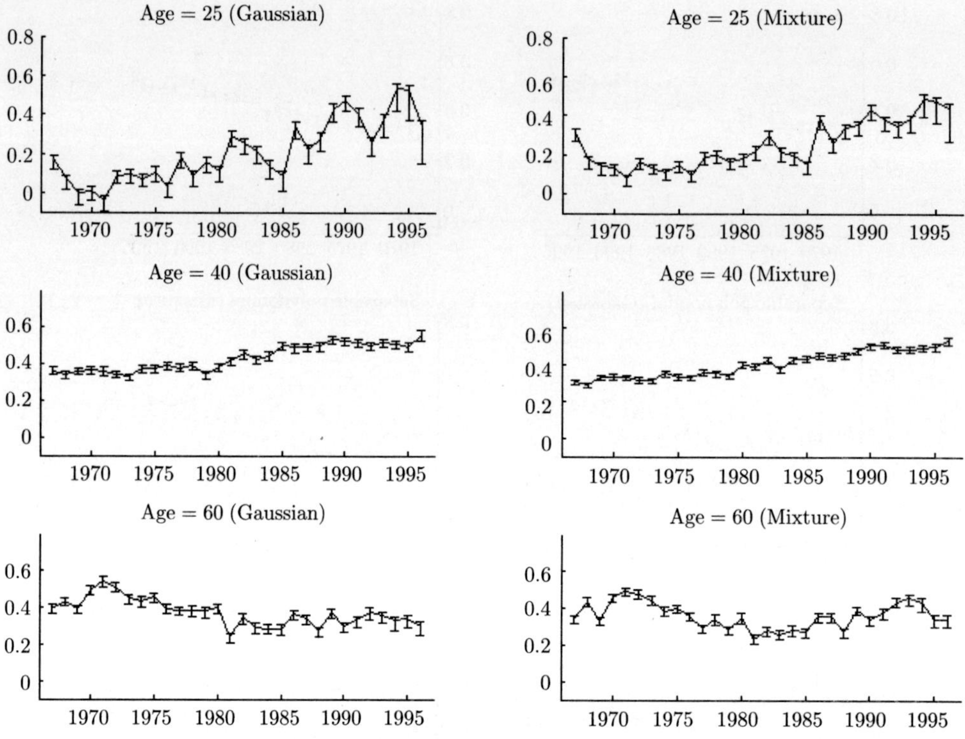

Fig. 14. Posterior medians, upper and lower quartiles for the difference in expected log earnings given 16 years of education vs 12 years of education, conditional on alternative ages, interactive polynomials models.

As was the case with returns to education, posterior means and medians for the age profiles shown in Figs. 15 and 17 and Appendix Tables 15 and 17 depend on the model for the regression function, but are insensitive to whether the specification of the regression residual is Gaussian or a mixture of normals. On the other hand, posterior interquartile ranges and standard deviations are smaller in the mixture models, for reasons discussed above. Across the years changes in the size of posterior interquartile ranges and standard deviations are due to changes in the size of the sample. In the case of returns to experience between the ages of 25 and 45 these sizes increase substantially in 1994, 1995 and 1996, due to the dearth of 25 to 30 year olds in the samples for those years, above and beyond the decline in overall sample size indicated in Fig. 1.

Referring to Fig. 15 and Appendix Table 15, all models indicate that returns to experience rose between 1967 and 1975, but the amount depends on the model: the increase is about 0.15 for the separable regression functions but only about 0.06 in the interactive polynomial regression function. From 1975 through 1991 return to the 20 years of experience between the ages of 25 and 45 remains about 0.40, in all of the models. Over the period 1991 to 1996, the polynomial regression functions show a sharp increase in these returns, reaching about 0.55 by 1996. The smoothed separable models, on the other hand, indicate only a modest increase, from about 0.40 to about 0.43. We believe this difference can be ascribed to the thinning of the sample of very young men in the later years,

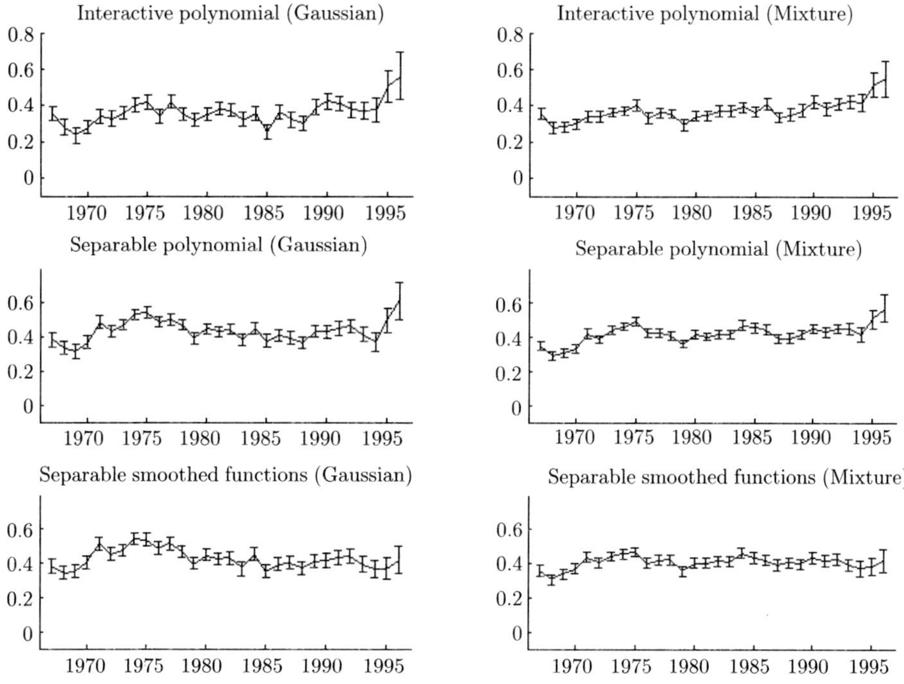

Fig. 15. Posterior medians, upper and lower quartiles for the difference in expected log earnings at age 45 versus age 25, given 12 years of education.

which implies that expected earnings for 25-year-olds in these samples is largely extrapolation. Since the polynomial regression functions and the Wiener process prior approach extrapolation in fundamentally different ways, the different conclusions are not surprising. This difference in conclusions should be resolved when we update the pre-release sample with the full data.

Fig. 16 and Appendix Table 16 indicate how returns to experience varies by level of education in the interactive polynomial models. Overall, the expected difference in log earnings between ages 45 and 25 is greater, the higher the level of education, but this difference steadily narrowed throughout the thirty-year period and practically vanished by 1996. Through 1985, the difference for poorly educated men is only about half what it is for high school graduates. After 1985 there is evidence that returns to experience for these men rises to returns for high school graduates, but by the 1990's the point is moot since there are almost no poorly educated young men in the sample. This accounts for the huge interquartile ranges in the top two panels of Fig. 16 in the 1990's. College graduates average about 15% greater growth in earnings than do high school graduates, through about 1985. After that, earnings growth drops for the college graduates and increases for the high school graduates. The scarcity of young men in the last years of the sample makes it hard to assess returns to experience in the 1990's.

Fig. 17 and Appendix Table 17 exhibit findings about the difference in expected log earnings at age 60 and age 45. As expected, the difference is always negative. All of the models indicate that the algebraic difference increased by about .10 to .15 between the late 1960's and 1988, after which it declined sharply to the values of the late 1960's. These changes are all modest absolutely and in

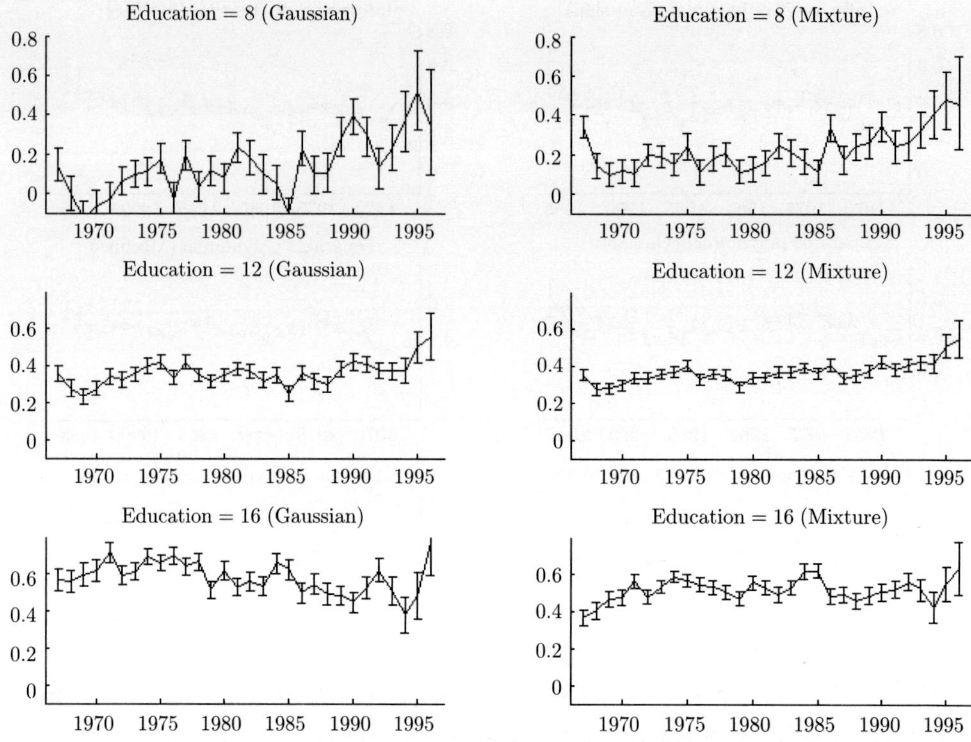

Fig. 16. **Posterior medians, upper and lower quartiles for the difference in expected log earnings at age 45 vs age 25, conditional on alternative levels of education, interactive polynomials models.**

comparison with changes over the 30-year period found for returns to education and to experience between ages 25 and 45. A striking feature of these results is that the difference in expected earnings between men of ages 60 and 45 is systematically more negative in the Gaussian models than it is in the mixture models. We believe that this can be traced to the evident misspecification of the assumption that state probabilities are independent of age. Older men are classified more frequently into the less probable state. As detailed in Section 4.4.1, the less probable normal distribution has a lower mean and a higher standard deviation than the more probable state. With more men classified into the less probable state at age 60 than at age 45, some of the decline in expected log earnings between ages 45 and 60 is accounted for by the high probability of a normal mixture component with a lower mean at age 60 than at age 45, leaving less to be explained by the regression function. We suggest ways that future research can address this issue in the concluding section.

In the interactive polynomial models, the expected difference in log earnings between ages 60 and 45 can depend on the level of education. Fig. 18 and Appendix Table 18 show that education does not have quite as strong an impact on expected earnings growth between the ages of 45 and 60 as it does between the ages of 25 and 45, but this impact changed in a systematic way between 1967 and 1996. Early in the sample, expected log earnings fell most rapidly between the ages of 45 and 60 for the most poorly educated men. For example, in 1970 the difference is about −0.30 for men with 8 years of education, -0.15 for high school graduates, and about zero for college graduates. In 1977

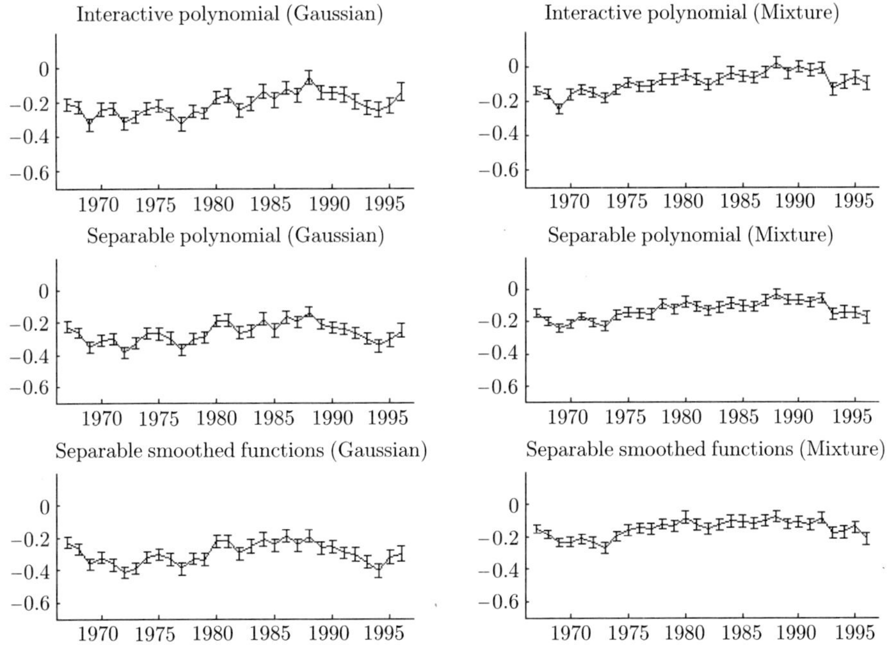

Fig. 17.   Posterior medians, upper and lower quartiles for the difference in expected log earnings at age 60 vs age 45, given 12 years of education.

the pattern reversed, and the reversal became stronger over time. By the 1990's, men with 8 years of education have almost no change in earnings between the ages of 45 and 60, whereas expected log earnings decline by about 0.10 for high school graduates and 0.20 for college graduates. It is important to keep in mind that in the latter years, there is considerable uncertainty about the impact of education on the age profile of earnings, as indicated by the Bayes factors summarized in Fig. 4. We conjecture these changes over the years, like many of the others discussed in this section, can be assessed more accurately in a longitudinal model in which the entire sample can be studied at once.

4.3.3   Conditional Variances

The standard deviation of log earnings conditional on age and education increased sharply over the period, from about 0.6 in the late 1960's to about 0.75 in the early 1990's: (see Fig. 19 and Appendix Table 19). In the misspecified Gaussian models the standard deviation is still the standard deviation of the actual distribution of regression residuals, so it is not surprising that the posterior means and medians of the standard deviation of this distribution are about the same: those for the mixture models are on the order of 1% to 2% higher than those in the Gaussian models. Differences across the specification of the regression function are even smaller.

Not surprisingly, posterior interquartile ranges and standard deviations of the regression residual standard deviations are substantially larger for the normal mixture model than for the Gaussian model, in most years by about a factor of two. Nevertheless, the change in the standard deviation of the conditional distribution over the 1967-1996 period amounts to about six normal mixture model posterior standard deviations.

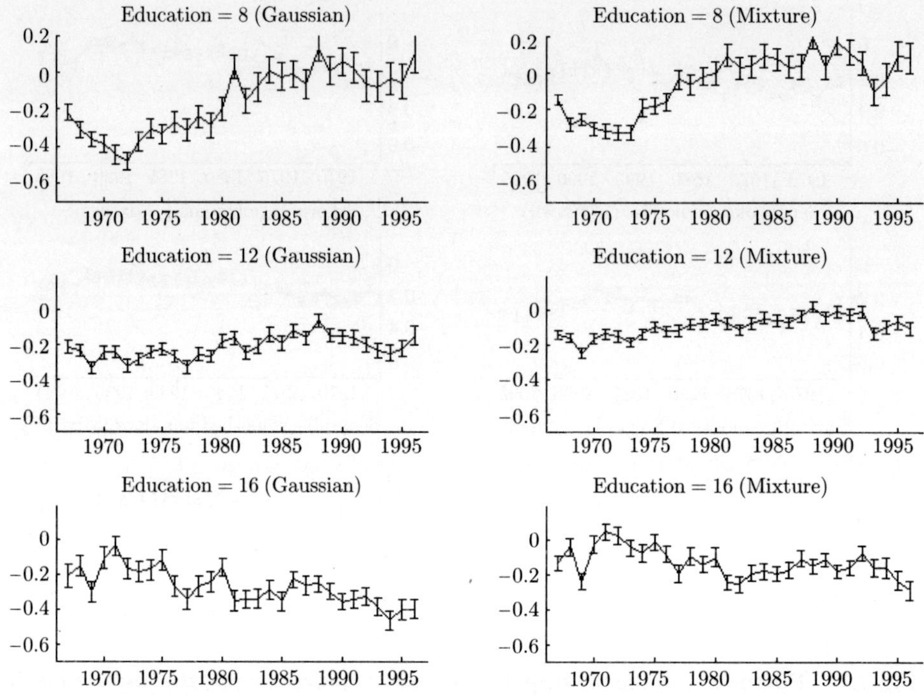

Fig. 18. **Posterior medians, upper and lower quartiles for the difference in expected log earnings at age 60 vs age 45, conditional on alternative levels of education, interactive polynomials models.**

4.4 Conditional Distributions

As discussed in Section 3.2.1, the normal mixture models accommodate a rich variety of distributions. With separate models for each of the 30 years in the sample, we can track changes in the conditional distribution in the same way that changes in the regression function were tracked in Section 4.3. This can be done through a wide variety of functionals. We concentrate here on functionals familiar from distribution theory (Section 4.4.1) and those related to the measurement of inequality (Section 4.4.2).

First, however, we present some evidence on the distribution of log earnings regression residuals. Each panel of Fig. 20 provides the posterior mean of a mixture of normals density (heavier line) together with the posterior mean of a normal density with the same mean (zero) and standard deviation (about 0.68) as the mixture of normals density. The posterior means are taken with respect to the 1986 sample. Comparing top, middle and bottom panels, mixture of normals densities are insensitive to the specification of the regression function. Comparing the left and right panels, the mixture of normals density with three components is very similar to that with two components. A mixture of three normals requires three more parameters than a mixture of two normals. Since both distributions provide about the same fidelity to the data, marginal likelihoods favor a mixture of two normals over a mixture of three normals. This happens for all years, and hence we conduct all further analysis using the two-component mixture models. Given the similarity of the conditional mixture of normals densities across regression functions, all further results in the paper pertain to

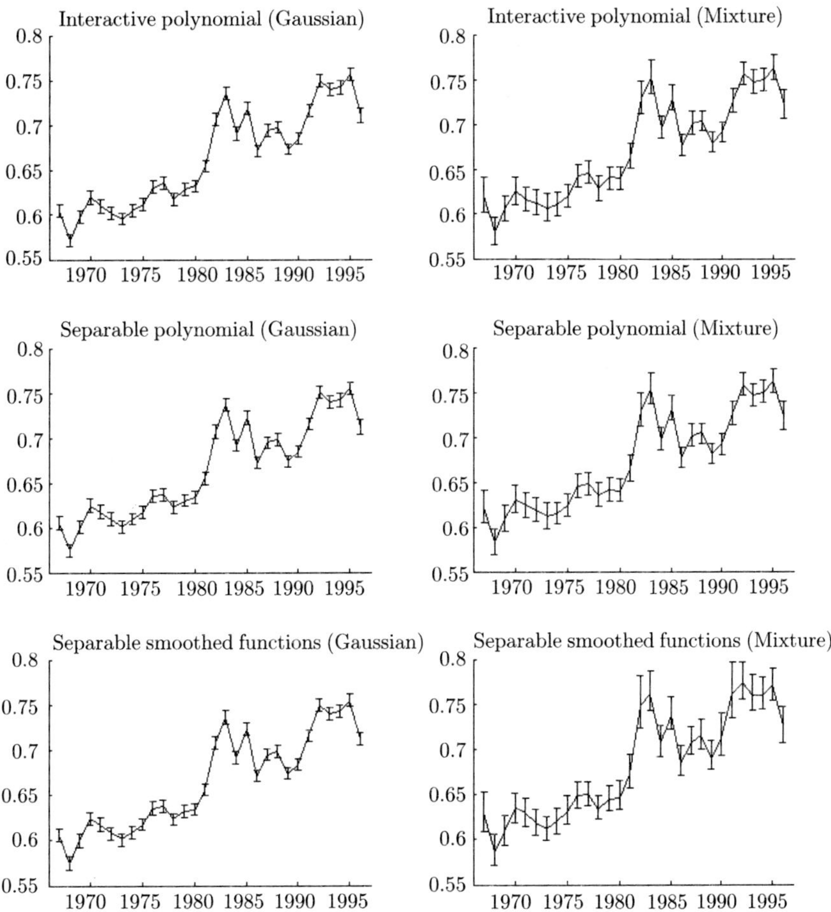

Fig. 19. Posterior medians, upper and lower quartiles for the standard deviation of the distribution of log earnings conditional on age and education.

the Wiener process prior separable regression function. Detailed results for other regression functions may be found in Appendix Tables 20 through 25.

These mixture of normal distributions are strongly negatively skewed: the mode is almost one-half standard deviation larger than the mean and the left tail is much thicker than the right tail. The excess kurtosis of the mixture of normals distribution is evident in the fact that the mode of its density is substantially higher than the mode of the corresponding normal density. Because the skewness coefficient is negative, the thicker left tail of the normal mixture density is evident in Fig. 20 whereas the thicker right tail is not.

Fig. 21 exhibits the two-component normal mixture probability density function posterior means for six years evenly spaced through the sample. The increasing standard deviation, documented in Section 4.3.3, is evident. The shape as well as the scale of the distribution change, a feature that we document more closely in the next subsection.

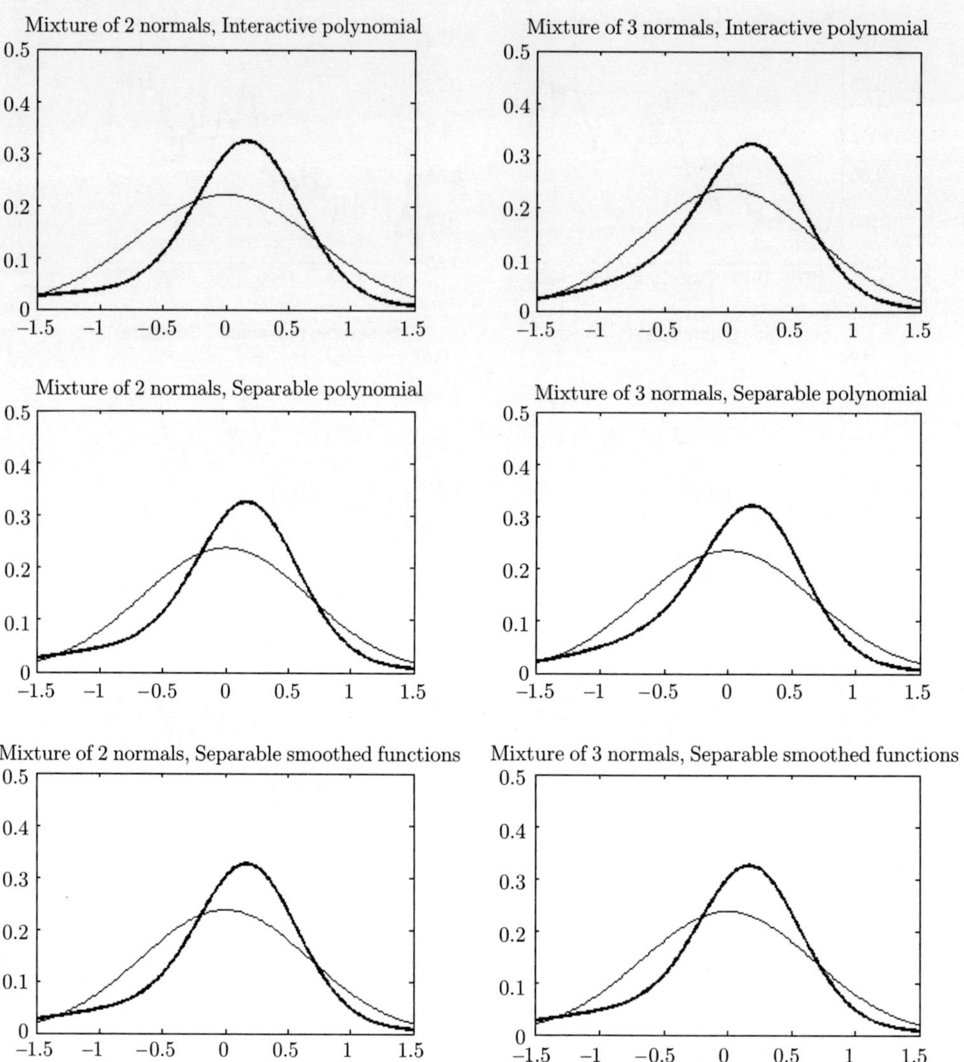

Fig. 20. Posterior mean of the mixture of normals conditional p.d.f. for the 1985 sample (heavy line) together with the posterior p.d.f. of the corresponding normal p.d.f. (light line).

4.4.1 *The Evolution of the Conditional Distribution of Earnings*

There are many ways of summarizing the shape of a probability density. Two of the most familiar are skewness and kurtosis. The top panels of Fig. 22, and Appendix Table 20, show how these measures have changed for the distribution of log earnings conditional on age and education over the sample. Skewness in the distribution has diminished, moving from about -1.25 in the late 1960's to around -1.15 in the 1970's to about -1 by 1990. Thus the difference between the left and right tail of the distribution has diminished; but, of course, the tails have also expanded with the growth in standard deviation, and we return to the implications of both movements for low earnings in Section 4.4.2.

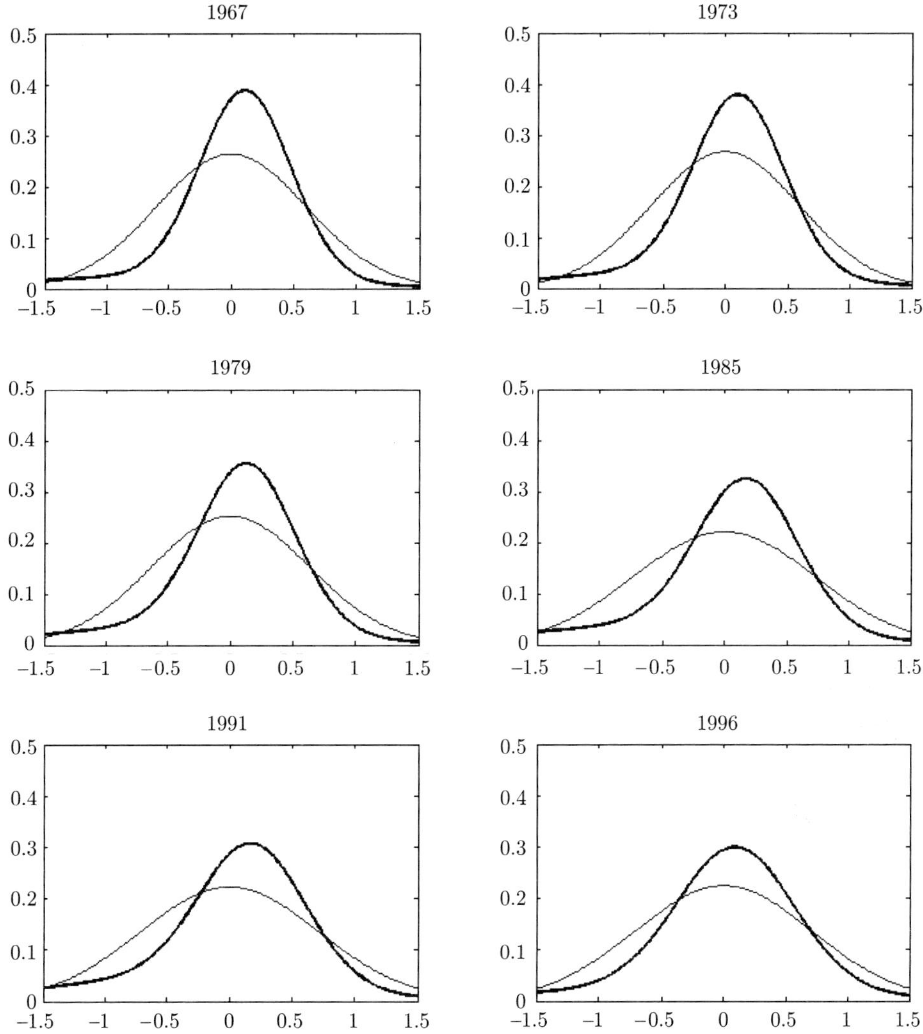

Fig. 21.   Posterior mean of the two-component mixture of normals conditional p.d.f. for each of several samples (heavy line) together with the posterior mean of the corresponding normal p.d.f. (light line).

Kurtosis declined sharply through the mid-1980's, after which it rose again, returning to 1970 levels by 1996. We conclude that the shape of the distribution has evolved slowly and systematically, even as standard deviation was increasing. However these changes are small relative to the absolute magnitude of skewness and excess kurtosis.

The shape may also be summarized in terms of the components of the normal mixture model. In all years, the component of the mixture model with the higher probability has a positive mean and a smaller standard deviation; the component with the lower probability has a negative mean and a larger standard deviation. The ratio of the larger to the smaller standard deviation declined until about 1980 and then rose again beginning about 1990. The pattern is similar to the movement in kurtosis, and accounts for most of its change. The spread in the means of the components was about the same in the first and last years in the sample, but declined somewhat in the years before 1983 and rose again thereafter. The fact that the difference in the component means was about the same in the early 1990's as in the late 1960's, together with the increase in the standard deviation, accounts for movement in the skewness coefficient towards zero over the thirty-year period. The probabilities of the mixture components change somewhat over the period: the component with the negative mean and larger standard deviation has probability roughly 0.2 in the earlier and later years, but is between 0.25 and 0.3 during the 1980's.

The lower right panel of Fig. 22 tracks the proportion of the probability density that is within one standard deviation of the mean, in each year (see also Appendix Table 22). For a Gaussian density this proportion is 0.68. Note that the change in this probability closely follows the change in kurtosis shown in the top right panel: for the same standard deviation, the more leptokurtic the

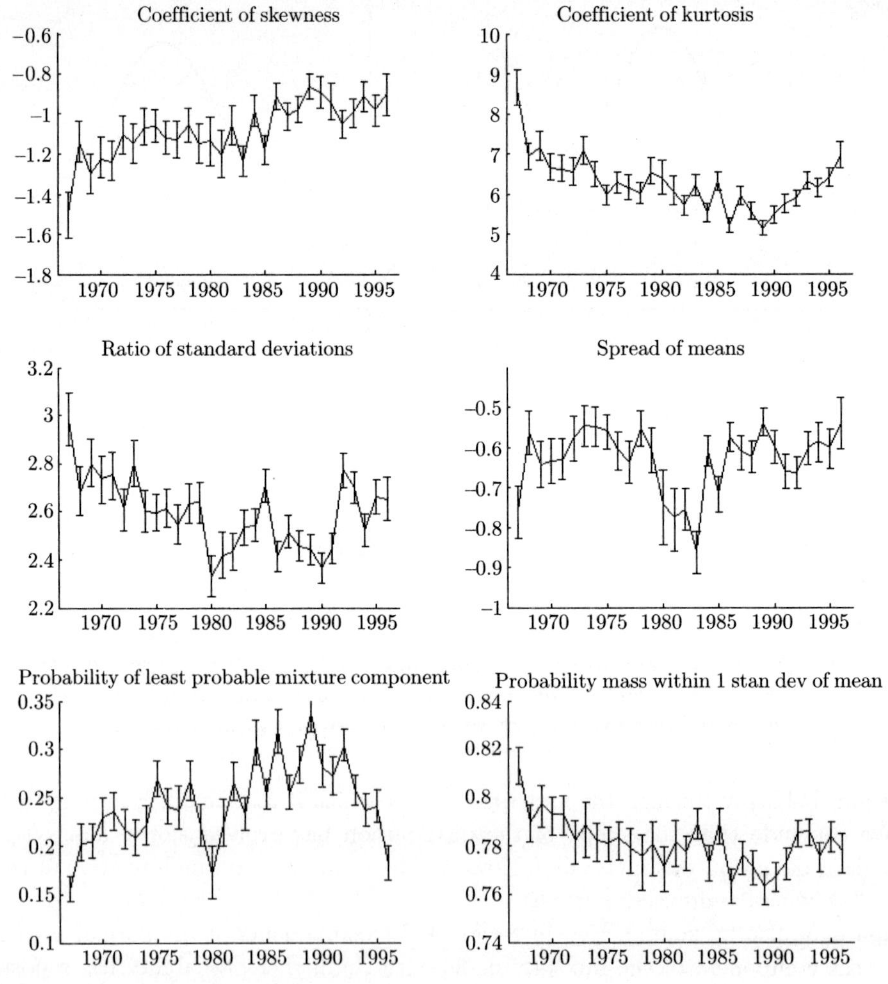

Fig. 22.   **Posterior medians, upper and lower quartiles for several aspects of the two-component mixture of normals distribution of log earnings conditional on age and education.**

distribution, the more probability must be concentrated near the mean in order to preserve the standard deviation. Since 1976, changes in this probability have been small.

4.4.2 The Evolution of Inequality in Earnings

Inequality in the distribution of earnings, or of any other non-negative variable associated with a population of individuals, can be measured in a variety of ways. Perhaps the most familiar is the Gini coefficient, which derives from the Lorenz curve. The Lorenz curve $L(p)$ is defined on the interval $[0, 1]$ and is the fraction of total earnings accruing to individuals in earnings quantile p or lower. If all individuals have the same earnings then $L(p) = p$ and in general $L(p) < p$. The Gini coefficient is $G = 2 \int_0^1 [p - L(p)] dp$; $G \in [0, 1]$, with $G = 1$ if and only if all individuals have the same earnings and $G = 0$ if and only if all earnings accrue to one individual. We consider two other measures of inequality: P, the fraction of men with earnings less than one-half of median earnings, and R, the fraction of earnings accruing to men in the top decile of the earnings distribution.

These measures depend on the population, and we consider two such populations. The first is a hypothetical population of men of the same age and education, which leads to measures of inequality conditional on age and education. Given the specifications of our models, this measure of inequality will be the same for all ages and education, but it changes from year to year. For example, it will be affected by the change in the conditional standard deviation conveyed in Fig. 19, but it will not be affected by the increasing return to education identified in Fig. 13. The second population we consider is the distribution of age and education in our sample for 1986, selected only as a convenient benchmark, and leads to measures of inequality unconditional on age and education. Measures of inequality for this population will be affected by changes in the regression of log earnings on age and education, such as the increasing return to education. We expect unconditional measures of inequality of earnings to be greater than conditional measure of inequality. (If inequality were measured by variance, this would be a consequence of the Rao-Blackwell Theorem.) It is important to keep in mind that all of the changes considered here condition on a fixed distribution of age and education, and abstract from the important demographic changes that are evident in Fig. 1.

Fig. 23 shows the evolution of each measure of inequality over the 1967-1996 period; unconditional measures are in the left panels, and conditional measures are in the right panels. Each panel provides a measure of inequality based on the mixture model by means of the darker lines and, for comparison, the measure based on the Gaussian model by means of the lighter lines. Appendix Tables 23, 24 and 25 provide detail for posterior means and standard deviations of these measures based on the mixture models.

The discrepancy between inequality measures based on the Gaussian and mixture distributions is striking. In all cases, the Gaussian specification leads to a higher measure of inequality than does the mixture specification. This occurs because the Gaussian specification reliably captures the mean and variance in the leptokurtic mixture distributions, but in so doing moves most points in the distribution farther from the mean. This is documented in the lower right panel of Fig. 22, which shows that the mixture specification places about 78% of the probability within one standard deviation of the mean, whereas for a Gaussian distribution this fraction is always 68%. The difference is largest in the fraction of earnings accruing to men in the top decile, and lowest in the fraction of men with earnings below half the median. This occurs because the mixture distributions are negatively skewed, meaning that the overdispersion in the Gaussian specification is less for earnings below the conditional mean than above.

Most measures of inequality have risen steadily through the period, and in every case the increase is large relative to posterior interquartile ranges or standard deviations. Inequality arises primarily

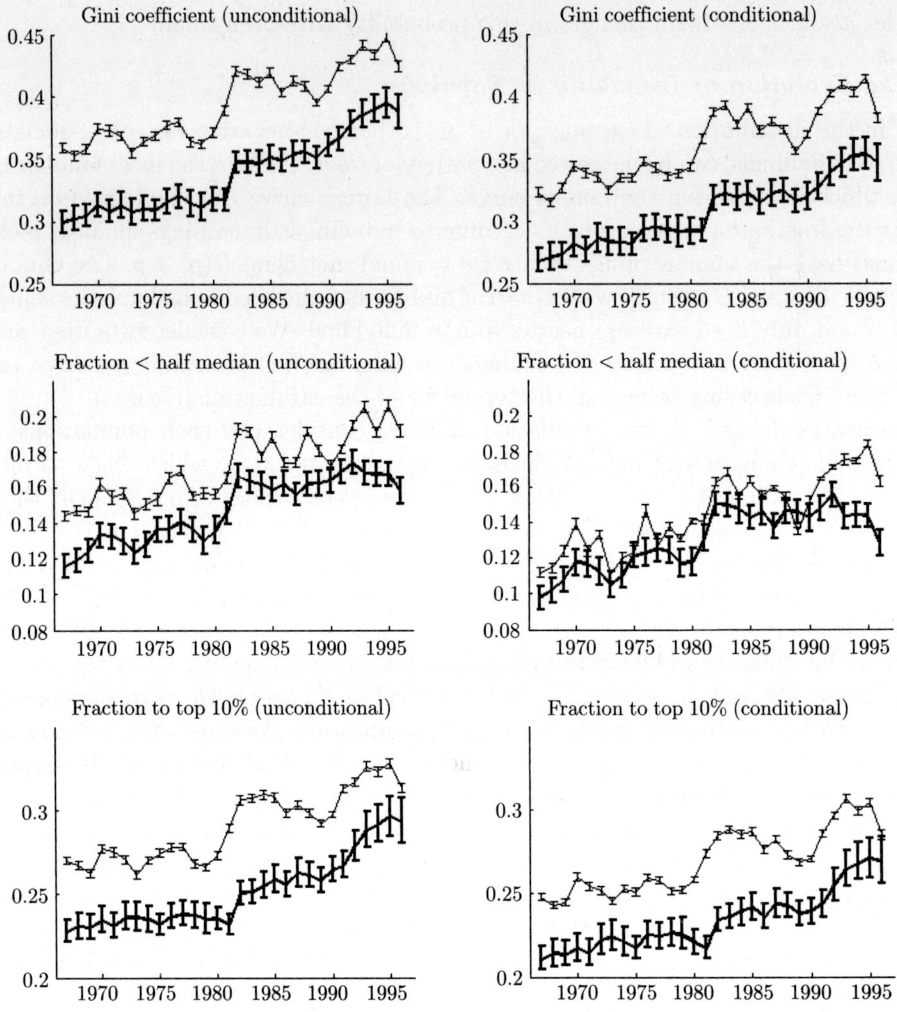

Fig. 23.   Posterior medians, upper and lower quartiles for several measures of inequality, using a mixture of normals distribution (heavy line) and a Gaussian distribution (light line) of the residuals.

from the distribution of earnings conditional on age and education, rather than from differences in expected earnings for different levels of age and education, as indicated by the fact that the conditional measure is nearly as large as the unconditional measure in each case.

The rise in the Gini coefficient has been steady, with plateaus in the 1970's and 1980's and a sharp jump between 1981 and 1982. The pattern of changes in the unconditional and conditional Gini coefficients are quite similar. The change in the Gini coefficient is similar to that in the standard deviation of the conditional distribution of log earnings (Fig. 19). The change in the Gini coefficient can be ascribed primarily to the increase in these standard deviations.

The fraction of men with earnings below half the median rose from 11.5% to 13.5% between 1967 and 1980, and jumped to 16.5% by 1982 where it remained to the end of the sample. Conditional on age and education the pattern is similar. The movement in the standard deviation documented in Fig. 19 accounts in part for these changes. But the impact of the increase in standard deviations is tempered by the decline in the probability of the low-mean state, documented in the lower left panel of Fig. 22, and this accounts for the plateau in this measure of inequality in the late 1980's and early 1990's.

The fraction of earnings accruing to men in the top decile shows the same slow increase through the 1970's followed by a sharp increase over the 1980-1982 period, but then it continues to rise steadily through the rest of the sample period. In fact, this measure grows nearly twice as much after 1982 as it does before 1980. Comparison of Fig. 19 and Appendix Table 19 with the lower panels of Fig. 23 and Appendix Table 25 indicates that like the Gini coefficient, the movement in this measure is driven by the change in scale of the distribution of log earnings conditional on age and education.

5. Conclusions and Directions for Future Research

We undertook this study using a Bayesian approach because we find it natural. It allows us to address questions the way they are asked, conditioning on data and assumptions, and to answer them by providing probability distributions for interesting unobservables like returns to education and measures of inequality. The results are exact and can be traced to explicit assumptions, which can be varied deliberately in order to assess the sensitivity of conclusions to different assumptions. We have done that for several important aspects of the specification of the model.

The study has advanced this methodology in two ways, viz.

1. It shows that state-of-the-art Bayesian methods permit the simultaneous non-parametric modeling of the regression function, employing either basis functions or smoothness priors, and the conditional distribution, using mixture of normals distributions.
2. The computations required are modest, both absolutely and in comparison with what is often required in non-Bayesian nonparametric methods. They amount to special cases of simple models already incorporated in extensions of popular mathematical applications software.

We began by examining the evidence in the data on the suitability of alternative modelling assumptions, and reached two main conclusions as follows:

1. Three alternative expansions of the regression function—interactive polynomials, separable polynomials, and separable functions with Wiener process smoothness priors—are highly competitive in the sense that different formulations are favored in different years, and the Bayes factors in any one year are rarely extreme. Moreover, substantive conclusions are generally insensitive to choice among these three specifications and consequently averaging over specifications is not essential.
2. The traditional specification of a Gaussian conditional distribution is decidedly inferior to a mixture of two normals. Mixing three normals does not improve the fit.

In the context of this specification we reached following conclusions about the evolution of the distribution of the earnings of men in the U.S. over the thirty-year period 1967 through 1996.

1. The ratio of expected earnings of college graduates to high school graduates declined from about 1.4 in 1967 to around 1.3 in 1979, and then rose to approximately 1.6 by 1990 where it remained through 1996.

2. The ratio of expected earnings at age 45 to those at age 25 grew from about 1.4 in 1967 to almost 1.6 by 1975, where it remained until 1991, after which there is some evidence of further growth in this ratio.

3. The conditional variance of earnings increased steadily over the period. The standard deviation rose from about 0.6 in 1967 to around 0.75 by the mid-1990's.

4. The conditional distribution of earnings was negatively skewed, but the coefficient of skewness rose from -1.25 to -1 over the sample period. The coefficient of kurtosis is between 5 and 7 for most of the sample

5. Inequality in earnings rose steadily through the period. Conditional on age and education, the Gini coefficient steadily rose from about 0.25 to 0.35. Returns to education and experience contribute further to inequality, and the unconditional Gini coefficient steadily rose from about 0.30 to 0.40 over the sample period. This increase in inequality is reflected in growth in the proportion of men with low earnings in the first half of the sample, and in the fraction of income accruing to the top decile in the latter half of the sample.

This study is part of our ongoing research on the evolution of earnings in the U.S. We note several extensions of this work, e.g.

1. The analysis here can be repeated, organizing by age rather than by year. That is, we can construct 41 samples of a-year-olds, and examine $p_a\left(y_{ai} \mid t_{ni}, e_{ni}\right)$. This approach explicitly models the impact of the evolution of earnings, and drops any assumption about smoothness in age. One could organize by education or cohort, as well, but these do not lead to cross-sectional analyses.

2. In the model used in this work, the density $p_t\left(y_{ti} \mid a_{ti}, e_{ti}\right)$ need not be Gaussian, and the regression $E_t\left(y_{ti} \mid a_{ti}, e_{ti}\right)$ is very flexible. Yet the distribution of the disturbance term $y_t - E_t\left(y_{ti} \mid a_{ti}, e_{ti}\right)$ does not depend on a_{ti} or e_{ti}: for example, there is no conditional heteroskedasticity. A natural extension of the approach taken in this work is to permit the state probabilities to depend on a_{ti} and e_{ti}. This extension removes this restriction on the distribution of the disturbance term. (see Geweke and Keane (2006) for this extension and its application to a small subset of the data used in this study).

3. It should be possible to incorporate the extension in Geweke and Keane (2005) in the conventional life-cycle model earnings model for longitudianl earnings data (Lillard and Willis, 1978), building on our earlier work (Geweke and Keane, 2000).

Acknowledgement

Grant R01-HD37060-01 from the National Institutes of Health provided financial support for this work.

References

Barnett, W.A. and Jonas, A. (1983). The Muntz-Szatz demand system: An application of a globally well balanced series expansion. Economics Letters, **11**, 337-342.

Erkanli, A. and Gopalan, R. (1994). Bayesian nonparametric regression: Smoothing using Gibbs sampling, in: D. Berry, K. Chaloner and J. Geweke (eds.), Bayesian Statistics and Econometrics: Essays in honor of Arnold Zellner. Wiley, New York.

Gallant, A.R. (1981). On the bias in flexible functional forms and an essentially unbiased form: The Fourier flexible form. *Journal of Econometrics*, **15**, 211-245.

Geweke, J. (2005). Contemporary Bayesian Econometrics and Statistics. New York: Wiley.

Geweke, J. and Keane, M. (2000). An empirical analysis of earnings dynamics among men in the PSID: 1968-1989. *Journal of Econometrics*, **96**, 293-356.

Geweke, J. and Keane, M. (2005). Smoothly mixing regressions. Working paper.

Good, I.J. (1956), The surprise index for the multivariate normal distribution. *Annals of Mathematical Statistics*, **27**, 1130-1135.

Green, P. and Silverman, B. (1994). Nonparametric regression and generalized linear models. Chapman and Hall, London.

Hardle, W. (1989), Applied Nonparametric Regression. Cambrdge University Press, Cambridge.

Heckman, J.J., Lochner, L.J. and Todd, P. (2003). Fifty years of Mincer earnings regressions. IZA Discussion Paper 775.

Koop, G. and Poirier, D.J. (2004). Bayesian variants of some classical semiparametric regression techniques. *Journal of Econometrics*, (forthcoming).

Koop, G. and Tobias, J.L. (2004). Semiparametric Bayesian regression in smooth coefficient models. *Journal of Econometrics*, (forthcoming).

Lancaster, T. (2004). An Introduction to Modern Bayesian Econometrics. Malden MA: Blackwell Publishing.

Lillard, L. and Willis, R. (1978). Dynamic Aspects of Earnings Mobility. *Econometrica*, **46**, 985–1012.

Mincer, J. (1958). Investment in human capital and personal income distribution. *Journal of Political Economy*, **66**, 281-302.

Rubin, D.B. (1984). Bayesianly justifiable and relevant frequency calculations for the applied statistician. *Annals of Statistics*, **12**, 1151-1172.

Shiller, R.J. (1984). Smoothness priors and nonlinear regression. *Journal of the American Statistical Association* **79:** 609-615.

Smith, M. and Kohn, R. (1996). Nonparametric regression using Bayesian variable selection. *Journal of Econometrics*, **75**, 317-344.

Wong, C. and Kohn, R. (1996). A Bayesian approach to additive semiparametric regression. *Journal of Econometrics*, **74**, 209-236.

Bayesian Statistics and Its Applications
Edited by S.K. Upadhyay, U. Singh and D.K. Dey
Anamaya Publishers, New Delhi, India

Role of Randomization in Bayesian Analysis: An Expository Overview

Jayanta K. Ghosh

Department of Statistics, Purdue University, West Lafayette, USA

Abstract

The evolving role of randomization in Bayesian analysis as well as arguments for and against randomization is discussed. We note that Bayesian analysis is moving towards a substantially reduced role for randomization, but more work of this kind is needed in sample surveys.

1. Introduction

We provide a relatively non-technical overview of how the Bayesian approach to randomization has changed over time. The overview depends primarily on the cited papers of Basu (1988), Kadane and Seidenfeld (1999), Rubin (1978) and some extracts from Savage, quoted in Kadane and Seidenfeld (1999). These references are supplemented by more recent papers of Berry (2004) and Spiegelhalter (2004).

Most of the paper deals with the role of randomization in Bayesian analysis of clinical trials (Section 5), but there is some discussion of random samples (Section 4) and randomization tests (Section 3). Section 6 contains concluding remarks.

2. Bayesian Views on Randomization: An Imaginary Conversation

We present an imaginary conversation by suitably arranging written views of the distinguished Bayesians mentioned in the Introduction. For example, the first two paragraphs below are from Savage (1961,1962), quoted at the beginning of the cited article of Kadane and Seidenfeld (1999).

Savage: "Applying the theory (of personal probability) naively one quickly comes to the conclusion that randomization is without value for statistics. This conclusion does not sound right...

The need for randomization presumably lies in the imperfection of actual people and, perhaps, in the fact that more than one person is ordinarily concerned with an investigation."

Basu: "The mere artifact of randomization cannot generate any information that is not there already. However, in survey practice situations will arise where it will be necessary to insist upon a random sample. But this will be only to safeguard against some unknown biases.

The inner consistency of the Bayesian point of view is granted...(But) who can be a true Bayesian and live with thousands of parameters. Survey statistics is more an art than a science."

Kadane and Seidenfeld: "Many have criticized randomization analysis for failing the likelihood principle: see Basu (1988). This reply to Fisher, Kempthorne and others, who defend randomization analysis, is we think, what Savage means by the "naive" Bayesian rejection of randomization.... When only one decision maker is relevant we would not randomize."

After considering Bayesian alternatives to randomization, Kadane and Seidenfeld go on to observe: "Thus even a sophisticated (rather than naive) Bayesian defense of randomization....fails to establish it as a sine qua non of sound experimental methods."

Rubin: "Classical randomized designs stand out as especially appealing mechanisms designed to make inference for causal effects straightforward by limiting the sensitivity of a valid Bayesian Analysis."

Kadane and Seidenfeld: "Randomization is one way to accomplish this but it is not unique in having the virtue. We join with Savage and many others, however, in his respect for randomization as a statistical tool for enhancing interpersonal communication."

At first reading there seems to be major differences. A second or third reading of the original papers shows a basic unity along with important differences. As Kadane and Seidenfeld (1999) point out, there are two kinds of experiments—experiments to learn and experiments to prove. The second kind of experiment involves at least two decision makers, of which one is the Bayesian who designed the experiment and analyzed the data and the other represents other Bayesians who have to be convinced. There, i.e. in the second set of experiments, randomization helps ensure trust in the analysis. Most Bayesians would agree with the sophisticated view expressed by Kadane and Seidenfeld (1999) in the last two extracts from their paper.

This paper is about learning—my learning. It is not a paper to prove. Hopefully, it will help other Bayesians who have not made up their mind about randomization but would like to do so.

We explore these basic issues in some more detail as follows.

3. Randomization and Permutation Tests

Randomization appears in classical inference and design of experiment—either to reduce the effect of unknown biases by some form of averaging over randomizations or to justify permutation tests by introducing exchangeability. Randomization also appears in classical inference via devices like randomized tests.

While the idea of using randomization to cope with unknown biases seems reasonable, each of the other two applications is contrary to principles of Bayesian analysis with a single decision maker. The single decision maker is supposed to maximize his posterior expectation of utility. He may, but need not, randomize if two or more actions are optimal.

If randomization is used at the design stage, it violates the likelihood principle since the likelihood is the same no matter what randomization is used at the design stage to draw a sample from a population or allocate different treatments to different sampling units. In particular, randomization will not have any effect on the posterior. Therefore any method of analysis, in which randomization plays an important role, cannot be based solely on likelihood or posterior. We examine below the logical difficulties associated with such tests. Logical difficulties arise from the fact that certain observed or observable quantities are ignored even though they may contain relevant information.

We consider Example B of Kadane and Seidenfeld (1999) on the permutation t-test for the difference of means of two normal populations with possibly unequal variances. Let (x_i, y_i), $i = 1, 2 \ldots, n$, be n observations. Use a random permutation of y's to generate data with a new pairing,

(x_i, y_{j_i}), $i = 1, 2 \ldots, n$. The differences $\{z_i = (x_i - y_{j_i}), \; i = 1, 2 \ldots, n\}$ are equally likely over the permutations. Let

$$t = \sqrt{(n-1)}\,\overline{z}/S_z,$$

$$S_z^2 = \sum_1^n (z_i - \overline{z})^2/(n-1).$$

For large n, the permutation distribution of t is approximately $N(0, 1)$ under the null hypothesis of same mean of X and Y. When we use this distribution obtained by averaging over $n!$ permutations, we ignore the random pairing and hence, S_z, which are not ancillary and contain information.

Kadane and Seidenfeld show how odd the argument is by examining the case of $n = 2$. There are only two values of the t-statistic since $n! = 2$. The values are

$$[(x_1 + x_2) - (y_1 + y_2)]/[|x_1 - x_2| - |y_1 - y_2|]$$

and

$$[(x_1 + x_2) - (y_1 + y_2)]/[|x_1 - x_2| + |y_1 - y_2|].$$

It is clear that the first value of $|t|$ is larger, i.e. more significant because of the smaller value of the ignored S_z in the denominator of t. On the other hand, the conditional distribution of t given S_z is degenerate. Thus the permutation t distribution has an unclear logical status. Similar objections apply to permutation t or F-tests for randomized block experiments. None of these are acceptable to a Bayesian.

A second way in which randomization may enter analysis is via a randomized decision function which, given data, puts a probability distribution over the space of actions and chooses an action at random. This would be acceptable to a Bayesian only if this distribution sits on actions that minimize the posterior risk.

4. Randomization and Sample Surveys

Random sampling from a finite population with or without replacement, is supposed to ensure exchangeability of x_i's that are sampled (and hence seen) and remaining x_i's that are not sampled (and hence not seen). Exchangeability leads to a natural estimate of population total, based on the assumption that the mean of the unseen x_i's can be estimated by the mean of seen x_i's

$$\sum_{i \,\in\, \text{sample}} x_i + \left(\frac{1}{n} \sum_{i \,\in\, \text{sample}} x_i \right) (N - n) = N\overline{x}$$

where $N - n$ is the number of unseen units.

The catch here is that labels of the units are ignored. If labels are not ignored exchangeability does not hold.

Even more difficult to justify are pps (probability proportional to size) sampling designs and the corresponding Horvitz-Thompson (HT) estimate, which are routinely used but remain repugnant to a Bayesian. The HT estimate for the population total is

$$\sum_{i=1}^n x_i/p_i$$

where p_i is the probability of choosing the ith unit. In case $p_i = \frac{1}{n}$ this reduces to the estimate for random samples with equal probability. Basu (1988) presents a hilarious example of how absurd the estimate can be if the p_i's are poorly chosen.

Basu's Circus Example: A circus owner wants a rough estimate of the total weight of fifty adult elephants. He chooses Sambo as typical (based on past measurements), measures Sambo's current weight and multiplies by fifty. That is his estimate.

The circus statistician is "horrified" by this ad hoc estimate. He chooses a sampling plan "that allots a selection probability 99/100 to Sambo and 1/4900 to the rest." The object is to choose Sambo with a high probability and yet have a valid estimate of variance. Naturally, Sambo is selected and the statistician produces the HT estimate, namely, (Sambo's weight) which is approximately about (1/50)th of the owner's rough estimate (read Basu's own description of the problem). This is a clever example of an unbiased estimate with a very large variance. Hence, even a classical statistician would be shocked. Basu seems to be saying that this is what would often happen if one uses a sampling design with unequal probabilities. If a random sample with equal probabilities were used, the result would still be bad compared with the owner's common sense estimate but not as absurd as above.

The common sense estimate makes sense to a subjectivist Bayesian. He too may want to choose Sambo as typical, based on the prior information consisting of the past measurements. But what would be a simple, generally applicable Bayesian procedure for drawing a sample based on available information? What would be a Bayes estimate for the population total and an estimate of its variance? The parametric super population based approaches have not survived competition with the design based approaches. It is possible that one needs to be a nonparametric Bayesian, even in these relatively simple examples, i.e., sampling problems with small sample size.

We consider another simple example to indicate that we seem to lack simple Bayesian alternatives to randomization.

A Presidential candidate in the USA wants an estimate of the proportion of voters who will vote for him. This is an experiment to learn, the usual sample size is small (1000 to 1200). Can a Bayesian do better than a simple random sample?

If the sample size is quite large, one ought to be able to come up with good Bayesian non-parametric answers. But I have not seen any.

5. Bayesian Approach to Clinical Trials with or without Randomization

Most of this section is based on Rubin (1978) and Kadane and Seidenfeld (1999). Rubin assumes there is a target population and the data arise by random sampling from the target population and random experiment of treatments (as in a randomized block assignment).

Rubin notes that "intuitively the causal effect of one treatment relative to another for a particular experimental unit is the difference between the result, if instead, the unit had been exposed to a second treatment." Clearly, both observed and unobserved random variables are relevant.

Clinical trials are experiments to prove, with various groups, namely, the statisticians, doctors and FDA, who are to be convinced by the study undertaken by the pharmaceutical company. In addition, there are ethical concerns relating to patients.

Data Table

	Covariate (X)	Which treatment (W)	Post-treatment value (Y)
1	(X_{11}, X_{01})	W_1	$Y_{11}, \ldots, Y_{T1}$
2	(X_{12}, X_{02})	W_2	$Y_{12}, \ldots, Y_{T2}$
$\vdots$			
N	$\ldots$		

X_{11} = observed part of x for first unit; X_{01} = unobserved part of x for first unit; Y_{11} = value of Y if unit 1 gets the first treatment; Y_{T1} = value of Y if unit 1 gets the Tth treatment; N = # units in target population; $W_i = 0$ indicates ith unit was not selected and so not exposed to any treatment; $W_i = t$ indicates ith unit was selected and exposed to treatment t.

For the ith unit with $W_i = t$, $(Y_{1i}, \ldots, Y_{t-1,i}, Y_{t+1,i}, \ldots) =$ unobserved $\left(Y_{(0)}\right)$, $Y_{ti} =$ observed $\left(Y_{(1)}\right)$.

Model for data

$\left(X_{(1)}, Y_{(1)}, W\right)$ are observed random vectors. $\left(X_{(0)}, Y_{(0)}\right)$ are unobserved random vectors. Joint density is

$$f(X, Y | \Pi) k(W | X, Y, \Pi)$$

where Π is what Rubin calls parameters in the model, the factor k models assignment and the first factor is

$$f(X, Y | \Pi) = \prod_{i=1}^{n} f(X_i, Y_i | \Pi).$$

The object is to make inference on Π.

Definition: The assignment mechanism $k(W | X, Y, \Pi)$ is ignorable if k depends on (X, Y, Π) only through the observed part $\left(X_{(1)}, Y_{(1)}\right)$.

Rubin remarks as follows, "The more involved the assignment mechanism, the more complex must be the recording mechanism if the assignment mechanism is to be ignorable".

Ignorability has an important implication. In general, i.e, without ignorability,

$$P\{Y_{(0)} | X_{(1)}, Y_{(1)}, W\} = \frac{\int \int k(W | X, Y, \Pi) f(X, Y | \Pi) p(\Pi) d\Pi dX_{(0)}}{\int \int \int k(W | X, Y, \Pi) f(X, Y | \Pi) p(\Pi) d\Pi dX_{(0)} dY_{(0)}}.$$

Under ignorability the factor k comes out of the integrals in the numerator and denominator and so gets canceled. This implies $k(W | X, Y, \Pi)$ need not be modeled.

We can now list the advantages of randomization as suggested by Rubin.

1. It leads to ignorable k. Thus k need not be modeled, nor would one have to model unobserved values in terms of observed values. This makes the Bayesian analysis robust.

2. It "yields data having more than one treatment condition for any distinct value of covariate." This also implies robustness. Rubin points out if two units with identical $X_{(1)}$ represent two treatments, then randomization must have been used. This allows balancing of covariates for different treatments and hence achieves robustness in modeling effect of covariates.

Various ignorable alternatives to randomization have been suggested in recent years. One of these, due to Kadane and Sedransk, has been discussed by doctors, lawyers, and philosophers and implemented at Johns Hopkins, vide Kadane (1996) and Kadane and Seidenfeld (1999). The main features are summarized as follows:

1. Appoint a small number of experts on the disease and treatments.

2. The group chooses a single indicator of outcome "of most reasonably of concern to a patient."

3. The group agrees on a few diagnostic variables.

4. Each expert's opinion is elected as a dynamic probabilistic function $A(3)$ (see Kadane and Seidenfeld (1999) for the definition of the function $A(3)$) about the outcome indicator. This may involve modeling.

5. Based on (4), each expert has a preferred treatment as a function of the diagnostic variables. Given the values of these variables, determine the preferred treatments and choose one at random.

This is ignorable and more ethical than randomization. Also it would not be easy to deviate from this protocol. As observed by Kadane and Seidenfeld (1999), "The desire that the analysis of an experiment not involve a judgement of the motivation of the experimenter seems natural for science. ...This consideration independent of the prior and the utility function of the experimenter can be conducted under any design that puts the experiment on "automatic pilot" once it is started ... Randomization is one way to accomplish this but it is not unique in having this virtue."

Can such protocols yield data which are as robust as those arising from experiments with randomization? Berry (2004) argues with considerable theoretical, and numerical evidence that we learn just as much. Of course, it would be good to subject real data from new studies to critical evaluation. Berry also points out that ethical considerations for patients may increase the number of volunteers and thus lead to more reliable studies. Apparently, FDA has accepted some Bayesian recommendations. Other studies of the same kind include Spiegelhalter (2004), who explores many other aspects of health care and Christen et al. (2004), who provide the technical details for an implementable Bayesian protocol.

6. Concluding Remarks

Our review focuses on three applications of randomization—permutation tests, clinical trials and sample surveys—all of which disturb Bayesians. The first of these, namely, permutation tests are simple, but the simplicity comes with a price, one has to ignore part of the information. The second application, namely, clinical trials seem to be moving away from randomization to well planned, ethical, but more complex alternatives. In the case of the third application of randomization, namely, design based sample surveys, Basu (1988) had pointed out various logical difficulties. However, except for the use of post-sampling stratification, small area estimation and ingenious similar ideas, the basic philosophical structure of survey sampling has not apparently changed much, see, for example, Rao (1999) and Ghosh (1999). This remains a promising area neglected by Bayesians.

References

Basu, D. (1988). Statistical Information and Likelihood—A Collection of Critical Essays by D. Basu, editor J.K. Ghosh, Springer, New York.

Berry, D. A. (2004). Bayesian Statistics and the Efficiency and Ethics of Clinical Trials. *Statistical Science*, **19**, 175-187.

Christen, J.A., Müller, P., Wathen, K. and Wolf, J. (2004). Bayesian Randomized Clinical Trials: A Decision Theoretic Sequential Design, *The Canadian Journal of Statistics*, **32**, 4, 1-16.

Ghosh, J.K. (1999). Discussion of Some Current Trends in Sample Survey Theory and Methods, *Sankhyā B*, **61**, 36-40.

Kadane, J.B. (1996 Ed). Bayesian Methods and Ethics in a Clinical Trial Design, J. Wiley & Sons, New York.

Kadane, J.B. and Seidenfeld, T.(1999). Randomization in a Bayesian Perspective, in Rethinking the Foundations of Statistics. J.B. Kadane, M.J. Schervish, T. Seidenfeld, Eds, Cambridge University Press, Cambridge, U.K.

Rao, J.N.K. (1999). Some Current Trends in sample Survey Theory and Methods. *Sankhyā B*, **61**, 1-25.

Rubin, D.B. (1978). Bayesian Inference for Causal Effects: The Role of Randomization. *Ann Statist.*, **6**, 34-58.

Savage, L.J. (1961). The Foundations of Statistics Reconsidered. Proc. 4th Berkeley Sympos. Math. Statist. and Prob., **I**, 575–586, J. Neyman, Ed., Univ. California Press, Berkeley, Calif.

Savage, L.J. (1962). Subjective Probability and Statistical Practice, in The Foundations of Statistical Inference. M.S. Bartlett, Ed., Methuen, London.

Spiegelhalter, D.J. (2004). Incorporating Bayesian Ideas into Health-Care Evaluation. *Statistical Science*, **19**, 156-174.

Bayesian Statistics and Its Applications
Edited by S.K. Upadhyay, U. Singh and D.K. Dey
Anamaya Publishers, New Delhi, India

Bayesian Neural Nets for Survival Analysis

Malay Ghosh

University of Florida, USA

Abstract

The paper introduces parametric and semiparametric Bayesian neural network models which are potentially useful in survival analysis. These models are extremely useful for prediction purposes, and are particularly suitable when the linear relationship between a monotone function of the survival time and the available covariates does not hold.

1. Introduction

The celebrated proportional hazards model of Cox (1972) has been the cornerstone of research in survival analysis for more than three decades. However, over the years, many alternate models have been proposed to accomodate situations where the proportional hazards model has been found inappropriate.

One such model which has drawn attention from various authors is the accelerated failure time model where typically the logarithm of the failure time is modeled as a linear function of covariates plus an error term. A classical analysis of such models began with Kalbfleisch (1978) who assumed normality of the errors. Christensen and Johnson (1988) introduced a semi-Bayesian analysis by assuming the errors to be iid from a certain distribution F, and then assigning a Dirichlet process prior to F. The approach is semi-Bayesian in the sense that they obtained a marginal estimate of the regression parameter. Later, Kuo and Mallick (1997) and Walker and Mallick (1999) introduced a semiparametric but fully Bayesian approach to handle this problem. A detailed account of these methods is available in Ibrahim et al. (2001).

However, quite often the assumed linear relationship between the logarithm of the time to failure or for that matter any other monotone function of the same does not hold. Accordingly, alternate models to accomodate situations of this type are called for.

The objective of this article is to use Bayesian neural network models which are quite versatile to handle these situations. Originally intended as abstract models of the brain, the scope of neural nets has now enormously extended, and they are now used in a variety of applications involving prediction and classification. Our main goal is to use these techniques for prediction. For introduction to Bayesian neural networks, we refer to the monographs of Neal (1996) and Lee (2004).

The motivation of our study stems from the need to develop Bayesian procedures for predicting ten year survival probabilities for men diagnosed with clinically localized prostate cancer, i.e., those for whom the cancer was detected not to spread outside the organ. The important covariates that one can use are: (1) treatment type (Watchful Waiting, Radiotherapy and Radical Prostatectomy), (2) age, (3) race, (4) PSA type, (5) grade of tumor (1, 2 or 3), and (5) Charlson score (0-1 and > 1).

The current method of prediction is based on Cox's (1972) proportional hazards model. However, when *prediction* is the main objective, plug-in predictors based on Cox's model are not usually adequate. In particular, a simple application of the proportional hazards model does not provide any associated measures of precision. Besides, when it comes to prediction, one needs to have an estimate of the baseline hazard as well, and Cox's (1975) partial likelihood which is meant for inference about the regression parameters by eliminating the baseline hazard, alone will not suffice. Also, the Cox model assumes that the logarithm of the ratio of the current hazard rate with respect to the baseline hazard rate is a linear function of the covariates. Such an assumption may not hold under all situations. As pointed out already, the usual accelerated failure time models are also not always appropriate, since the log-linearity of failure times with the given covariates is too much to expect in any complex situation.

In Section 2 we introduce parametric Bayesian accelerated failure time neural network models and in Section 3 we consider semiparametric versions of the same. Some concluding remarks are made in Section 4.

Due to some constraint, it was not possible to carry out a full fledged numerical analysis. Only I have been able to do is lay out the models and discuss the inference procedures for these models.

2.   Bayesian Accelerated Failure Time Neural Network Models

Consider n subjects $1, \cdots, n$. The failure times (times to death) are denoted by $t_1, \cdots, t_n$. The corresponding censoring times are denoted by $\nu_1, \cdots, \nu_n$, where $\nu_i = 1$ if t_i is a failure time, while $\nu_i = 0$, if t_i is right-censored, i.e., the person has not died in the time-period considered. Let $\boldsymbol{y} = (y_1, \cdots, y_n)^T$ and $\boldsymbol{\nu} = (\nu_1, \cdots, \nu_n)^T$.

In an accelerated failure time model, $y_i = \log t_i$ are modeled as

$$y_i = \sum_{j=1}^{M} \beta_j \psi(\boldsymbol{x}_i^T \boldsymbol{\gamma}_j) + e_i = \boldsymbol{\beta}^T \boldsymbol{\eta}_i(\boldsymbol{\gamma}) + e_i, \tag{1}$$

where $\boldsymbol{\beta}^T = (\beta_1, \cdots, \beta_M)$, $\boldsymbol{x}_i^T = (x_{i1}, \cdots, x_{ip})$, $\boldsymbol{\eta}_i^T(\boldsymbol{\gamma}) = (\psi(\boldsymbol{x}_i^T \boldsymbol{\gamma}_1), \cdots, \psi(\boldsymbol{x}_i^T \boldsymbol{\gamma}_M))^T$, and the e_i are the errors. The errors e_i will be modeled both parametrically and nonparametrically.

The regression part of the model can be identified with a single layer feedforward neural network with fixed number of hidden nodes M. The $\boldsymbol{x}_i$ are the inputs, β_j are the weights attached to the hidden nodes j $(j = 1, \cdots, M)$. The $\boldsymbol{\gamma}_j$ are the weights attached to the inputs for node j $(j = 1, \cdots, M)$, and ψ is the activation function. Typically ψ is taken as the logistic function, or the hyperbolic tangent function, or the sigmoid function. In related neural network studies, we have found that the posterior predictive probabilities are quite insensitive to the choice of the activation function.

We first assume the errors e_i to be i.i.d. $N(0, \sigma^2)$. Throughout, we assume noninformative censoring, which seems appropriate here since other than regular PSA monitoring, the patients do not undergo any other drug treatment or hormonal therapy. Accordingly, while writing the joint likelihood for $\boldsymbol{\beta}$, $\boldsymbol{\gamma}_j$ $(j = 1, \cdots, M)$ and σ^2, we omit the factor which involves the joint distribution of the censoring variables.

Suppose t_0 is the specified number of years after surgery where we want to find the probability of death. Let $y_0 = \log(t_0)$. The likelihood function for $\boldsymbol{\beta}, \boldsymbol{\gamma}_1, \cdots, \boldsymbol{\gamma}_M, \sigma^2$ based on $\boldsymbol{y}$ and $\boldsymbol{\nu}$ is given by

$$L(\boldsymbol{\beta}, \boldsymbol{\gamma}_1, \cdots, \boldsymbol{\gamma}_M, \sigma^2) \propto \Pi_{i=1}^n \left[\exp\{-\frac{\nu_i}{2\sigma^2}(y_i - \sum_{j=1}^M \beta_j \psi(\boldsymbol{x}_i^T \boldsymbol{\gamma}_j))^2\} \Phi^{1-\nu_i} \left(\frac{\sum_{j=1}^M \beta_j \psi(\boldsymbol{x}_i^T \boldsymbol{\gamma}_j) - y_0}{\sigma} \right) \right].$$

$$(2)$$

At the first stage of the hierarchical prior, $\boldsymbol{\beta}, \boldsymbol{\gamma}_1, \cdots, \boldsymbol{\gamma}_M$ are mutually independent with $\boldsymbol{\beta} \sim \mathrm{N}(\mu_\beta \mathbf{1}_\beta, \sigma_\beta^2 \boldsymbol{I}_\beta)$, and $\boldsymbol{\gamma}_1, \cdots, \boldsymbol{\gamma}_M$ are iid $\mathrm{N}(\boldsymbol{\mu}_\gamma, \boldsymbol{S}_\gamma)$. At the second stage, the prior parameters $\mu_\beta, \boldsymbol{\mu}_\gamma, \sigma^2, \sigma_\beta^2$ and $\boldsymbol{S}_\gamma$ are mutually independent with $\mu_\beta \sim \mathrm{N}(a_\beta, A_\beta)$, $\boldsymbol{\mu}_\gamma \sim \mathrm{N}_p(\boldsymbol{a}_\gamma, \boldsymbol{A}_\gamma)$, $\sigma^2 \sim \mathrm{IG}(c_\sigma/2, c_\sigma C_\sigma/2)$, $\sigma_\beta^2 \sim \mathrm{IG}(c_\beta/2, c_\beta C_\beta/2)$ and $\boldsymbol{S}_\gamma \sim \mathrm{IW}(c_\gamma, c_\gamma^{-1} \boldsymbol{C}_\gamma^{-1})$. Here IG and IW denote respectively the inverse gamma and inverse Wishart distributions. Specifically, σ_β^2 has pdf

$$f(\sigma_\beta^2) \propto \exp(-\frac{c_\beta C_\beta}{2\sigma_\beta^2})(\sigma_\beta^2)^{-c_\beta/2-1}$$

and $\boldsymbol{S}_\gamma$ has pdf

$$f(\boldsymbol{S}_\gamma) \propto |\boldsymbol{S}_\gamma|^{-\frac{1}{2}(c_\gamma + p + 1)} \exp[-\frac{1}{2} tr(\boldsymbol{S}_\gamma^{-1} c_\gamma \boldsymbol{C}_\gamma)].$$

A major objective is to predict the probability of death (time to failure) of a patient with covariate $\boldsymbol{x}$. Let $\boldsymbol{\theta}^T = (\boldsymbol{\beta}^T, \boldsymbol{\gamma}_1^T, \cdots, \boldsymbol{\gamma}_M^T, \sigma^2, \mu_\beta, \sigma_\beta^2, \boldsymbol{\mu}_\gamma, \boldsymbol{S}_\gamma)$. Writing $\Pi(\boldsymbol{\theta}|\boldsymbol{y}, \boldsymbol{\nu})$ as the joint posterior pdf of $\boldsymbol{\theta}$ given $\boldsymbol{y}$ and $\boldsymbol{\nu}$, the posterior predictive survival probability of a new, say $(n+1)$th patient at time t_0 is given by

$$P(t_{n+1} > t_0|\boldsymbol{y}, \boldsymbol{\nu}) = P(y_{n+1} > y_0|\boldsymbol{y}, \boldsymbol{\nu})$$

$$= \int P(y_{n+1} > y_0|\boldsymbol{\theta}, \boldsymbol{y}, \boldsymbol{\nu}) \Pi(\boldsymbol{\theta}|\boldsymbol{y}, \boldsymbol{\nu}) d\boldsymbol{\theta}$$

$$= \int \Phi \left(\frac{\sum_{j=1}^M \beta_j \psi(\boldsymbol{x}_{n+1}^T \boldsymbol{\gamma}_j) - y_0}{\sigma} \right) \Pi(\boldsymbol{\theta}|\boldsymbol{y}, \boldsymbol{\nu}) d\boldsymbol{\theta}, \qquad (3)$$

where $y_{n+1} = \log(t_{n+1})$ and $y_0 = \log(t_0)$.

The posterior $\Pi(\boldsymbol{\theta}|\boldsymbol{y}, \boldsymbol{\nu})$ is given by

$$\Pi(\boldsymbol{\theta}|\boldsymbol{y}, \boldsymbol{\nu}) \propto L(\boldsymbol{\beta}, \boldsymbol{\gamma}_1, \cdots, \boldsymbol{\gamma}_M, \sigma^2)(\sigma_\beta^2)^{-\frac{M}{2}} \exp \left[-\frac{||\boldsymbol{\beta} - \mu_\beta \mathbf{1}_M||^2}{2\sigma_\beta^2} \right]$$

$$\times |\boldsymbol{S}_\gamma|^{-\frac{M}{2}} \exp \left[-\frac{\sum_{j=1}^M (\boldsymbol{\gamma}_j - \boldsymbol{\mu}_\gamma)^T \boldsymbol{S}_\gamma^{-1} (\boldsymbol{\gamma}_j - \boldsymbol{\mu}_\gamma)}{2} \right]$$

$$\times \exp \left[-\frac{(\mu_\beta - a_\beta)^2}{2A_\beta^2} - \frac{(\boldsymbol{\mu}_\gamma - \boldsymbol{a}_\gamma)^T \boldsymbol{A}_\gamma^{-1} (\boldsymbol{\mu}_\gamma - \boldsymbol{a}_\gamma)}{2} \right]$$

$$\times \exp \left(-\frac{c_\sigma C_\sigma}{2\sigma^2} \right) (\sigma^2)^{-\frac{c_\sigma}{2}-1} \exp \left(-\frac{c_\beta C_\beta}{2\sigma_\beta^2} \right) (\sigma_\beta^2)^{-\frac{c_\beta}{2}-1}$$

$$\times \exp[-tr(\boldsymbol{S}_\gamma^{-1} c_\gamma \boldsymbol{C}_\gamma)] |\boldsymbol{S}_\gamma|^{-\frac{c_\gamma + p + 1}{2}}. \qquad (4)$$

The posterior given in (4) is analytically intractable as its evaluation requires high dimensional numerical integration. Thus, an alternative Markov chain Monte Carlo (MCMC) numerical integration approach is undertaken which requires generating samples from the posterior and making inferences based on the generated samples.

Specifically, this alternative approach, known as the Gibbs sampler (Gelfand and Smith, 1990) requires generating samples from the full conditionals of the different parameters given the remaining parameters and the data. At the end of this section, we will find that some of the conditional distributions for the different parameters given the remaining parameters and the data, cannot be obtained in a closed form, and one needs to adopt the Metropolis-Hastings algorithm (Metropolis et al., 1953) to generate samples from these distributions. Chib and Greenberg (1995) gave a very succinct account of this algorithm for ready use by statisticians.

It may be noted that the parameters β_j's and the γ_j's are themselves non-identifiable. However, this does not cause much problem for the proposed inferential procedure since our objective is not estimation of the β_j's or γ_j's, but finding the predictive probabilities of future outcomes. Moreover, as a special case of Lemma 1 of Ghosh et al. (2000), it follows that with a proper prior, the posterior is also proper even if the likelihood is non-identifiable in some of the parameters.

The MCMC approximation of the posterior predictive survival probability is given by $B^{-1} \sum_{k=1}^{B} \Phi\left(\frac{\sum_{j=1}^{M} \beta_j^{(k)} \psi(\boldsymbol{x}_{n+1}^T \boldsymbol{\gamma}_j^{(k)}) - y_0}{\sigma^{(k)}}\right)$, where $\beta_j^{(k)}, \gamma_j^{(k)}$, $j = 1, \cdots, M$ and σ^2 are the kth MCMC samples generated from the posterior $\pi(\boldsymbol{\theta}|\boldsymbol{y}, \boldsymbol{\nu})$. Here, B denotes the number of such samples used to estimate the parameters.

There is a potential advantage of employing hierarchical Bayesian methods in this context over any frequentist or naive empirical Bayesian estimation procedure. According to the latter, denoting by $\widehat{\beta}_j$'s and $\widehat{\gamma}_j$'s and $\widehat{\sigma}^2$ the estimates of the β_j's, γ_j's, and σ^2, the prediction probabilities are estimated by $\Phi\left(\frac{\sum_{j=1}^{M} \widehat{\beta}_j \psi(\boldsymbol{x}_{n+1}^T \widehat{\boldsymbol{\gamma}}_j) - y_0}{\widehat{\sigma}}\right)$. These are the so-called "plug-in" estimators. Such an estimation procedure ignores uncertainty due to estimation of the β_j's and γ_j's and σ^2. Accordingly, although the point estimates are reasonable, any associated measures of uncertainty are typically underestimates. The hierarchical Bayesian procedure accounts for this uncertainty by assigning distributions to the hyperparameters. Thus the hierarchical Bayesian method will not only yield the point estimates, but will also provide the associated measures of precision via posterior standard deviations.

An alternate Bayesian procedure is to select the β_j's and γ_j's adaptively. Rather than iid priors, we assign independent $N(0, \tau_j^2)$ priors to the β_j's and independent $N(0, \boldsymbol{V}_j)$ priors to the γ_j's at the first stage of the hierarchical model. At the second stage, we assign independent inverse gamma priors to the τ_j^2's and independent inverse Wishart priors to the $\boldsymbol{V}_j$'s. If now the posterior of any τ_j^2 is near degenerate at zero, there is some evidence that the corresponding β_j is *a priori* near degenerate at zero, and, in such cases, it may be possible to delete it from the model. Similarly, by examining the posterior of $\boldsymbol{V}_j$, it may be appropriate to drop some of the components of γ_j. This process introduces an adaptive selection of the weights β_j's and γ_j's.

In the remainder of this section, we provide the full conditionals based on (4) needed for the implementation of MCMC algorithm. In what follows, for each conditional distribution of a parameter or a parameter vector, we will use $|\cdot$ as the generic symbol for the other parameters and the data $(\boldsymbol{y}, \boldsymbol{\nu})$. Let $\overline{\beta} = M^{-1} \sum_{j=1}^{M} \beta_j$ and $\overline{\gamma} = M^{-1} \sum_{j=1}^{M} \gamma_j$. The full conditionals are

(i) $\mu_\beta|\cdot \sim \mathrm{N}(M\sigma_\beta^{-2} + A_\beta^{-1})^{-1}(M\sigma_\beta^{-2}\overline{\beta} + A_\beta^{-1}a_\beta)(M\sigma_\beta^{-2} + A_\beta^{-1})^{-1}$

(ii) $\sigma_\beta^2|\cdot \sim \text{IG}\left(\frac{c_\sigma+M}{2}, \frac{c_\sigma C_\sigma+\|\boldsymbol{\beta}-\mu_\beta \mathbf{1}_M\|^2}{2}\right)$

(iii) $\boldsymbol{\mu}_\gamma|\cdot \sim \text{N}(\boldsymbol{S}_\gamma^{-1} + \boldsymbol{A}_\gamma^{-1})^{-1}(M\boldsymbol{S}_\gamma^{-1}\overline{\boldsymbol{\gamma}} + \boldsymbol{A}_\gamma^{-1}\boldsymbol{a}_\gamma), (\boldsymbol{S}_\gamma^{-1} + \boldsymbol{A}_\gamma^{-1})^{-1}$

(iv) $\boldsymbol{S}_\gamma^{-1}|\cdot \sim \text{W}(c_\gamma + M, c_\gamma \boldsymbol{C}_\gamma + \{\sum_{j=1}^{M}(\boldsymbol{\gamma}_j - \boldsymbol{\mu}_\gamma)(\boldsymbol{\gamma}_j - \boldsymbol{\mu}_\gamma)^T\})$

(v) $\sigma^2|\cdot \sim \text{IG}\left(\frac{c_\beta+M}{2}, \frac{c_\beta C_\beta+\|\boldsymbol{\beta}-\mu_\beta \mathbf{1}_M\|^2}{2}\right)$

(vi) $\pi(\boldsymbol{\beta}|\cdot) \propto L(\boldsymbol{\beta}, \boldsymbol{\gamma}_1, \cdots, \boldsymbol{\gamma}_M, \sigma^2)\exp\left[-\frac{\|\boldsymbol{\beta}-\mu_\beta \mathbf{1}_M\|^2}{2\sigma_\beta^2}\right]$

(vii) $\boldsymbol{\gamma}_j|\cdot \sim L(\boldsymbol{\beta}, \boldsymbol{\gamma}_1, \cdots, \boldsymbol{\gamma}_M, \sigma^2)\exp\left[-\frac{(\boldsymbol{\gamma}_j-\boldsymbol{\mu}_\gamma)^T \boldsymbol{S}_\gamma^{-1}(\boldsymbol{\gamma}_j-\boldsymbol{\mu}_\gamma)}{2}\right].$

Generation of samples from (i)-(v) is standard. This is not so for (vi) and (vii) which require the use of Metropolis-Hastings algorithm.

As pointed out by a reviewer, the full conditional for β is nonstandard because of censoring which results in the appearance of Φ in the likelihood. For actual data analysis, as in a probit model, one can treat the ν_i as binary variables, where $\nu_i = 1$ or 0 according as $t_i > t_0$ or $t_i \leq t_0$. This is equivalent to $y_i > y_0$ or $y_i \leq y_0$. Now the full conditional for β, which includes the ν_i's as well as conditioning variables is multivariate normal involving only those y_i for which $\nu_i = 1$. We also need a full conditional for the ν_i's which involves β and other conditioning variables. This is how WINBUGS hands censoring.

3. Semiparametric HB Neural Network Models

The models considered so far are fully parametric. It is possible though to enrich the proposed class of models by adopting a semiparametric hierarchical Bayesian approach. As a result, parts of the modeling are captured parametrically, in particular, the neural network structure for the canonical parameters β_j's, and γ_j's. Also, the likelihood remains the same as before. However, instead of normality, one can specify nonparametrically the distribution of the random effects e_i. In particular, one may use Dirichlet-normal mixture priors for the random effects. Such priors were first introduced by Ferguson (1973, 1974) and Antoniak (1974). Later work includes Ferguson (1983), Lo (1984), Brunner and Lo (1989, 1994), Escober (1994, 1995), West, Müller and Escober (1994), MacEachern and Müller (1998), Escober and West (1995), Newton, Czado and Chappell (1996), Kleinman and Ibrahim (1998), Müller and Rosner (1997) and Mukhopadhyay and Gelfand (1997) among others. Clearly, the nonparametric approach is intended to provide greater flexibility and robustness in Bayesian inference.

We begin with the definition of Dirichlet process priors introduced by Ferguson (1973, 1974). A probability measure $G \in \mathcal{G}$ on $\mathcal{X}$ is said to follow a Dirichlet process with parameter αG_0, symbolically written as $G \sim DP(\alpha G_0)$, if for any measurable partition $B_1, \cdots, B_m$ of $\mathcal{X}$, $G(B_1), \cdots, G(B_m) \sim \text{Dirichlet}(\alpha G_0(B_1), \cdots, \alpha G_0(B_m))$. Here G_0 is a specified probability measure, usually referred to as the base measure, and α is the precision parameter. This name for α is justified since $V[G(B_k)] = G_0(B_k)[1 - G_0(B_k)]/(\alpha + 1)$ which decreases in α for every k.

Computationally, it is more convenient to work with a family of Dirichlet mixture distributions. Let $\{f(\cdot|\boldsymbol{\xi}), \boldsymbol{\xi} \subset R^p\}$ be a parametric family of densities with respect to some dominating measure μ. Consider the family of probability distributions $\mathcal{F}_G$, $\{G : G \in \mathcal{G}\}$ with densities

$$f(e|G) = \int f(\cdot|\boldsymbol{\xi})dG(\boldsymbol{\xi}).$$

Here $G(\boldsymbol{\xi})$ is viewed as the conditional distribution of $\boldsymbol{\xi}$ given G. It is assumed that $G \sim DP(\alpha G_0)$, whence $f(e|G)$ arises by mixing with respect to a distribution having a Dirichlet process. It is often convenient to assume that G has a Dirichlet process prior with a normal base measure. Symbolically, $G \sim DP(\alpha G_0)$ and G_0 is $N(\zeta_0, \sigma_0^2)$, where α is the concentration parameter. Using the result of Antoniak (1974), it follows that the conditional distribution of e_j given $e_1, \cdots, e_{j-1}, e_{j+1}, \cdots, e_n$ is

$$e_j | e_1, \cdots, e_n \sim \frac{\alpha}{\alpha + n - 1} G_0(\cdot) + \frac{1}{\alpha + n - 1} \sum_{k=1(k \neq j)}^{n} I_{e_k}(\cdot) \tag{5}$$

where I is the usual indicator function. We may also note that for very large values of α relative to n, the Bayesian semiparametric method produces essentially the $N(\zeta_0, \sigma_0^2)$ conditional distribution for e_j, while very small values of α relative to n amounts to the fact that the conditional distribution of e_j essentially equals the "empirical" distribution of $e_1, \cdots, e_n$. We also note that as $\alpha \to \infty$ the Dirichlet process model reduces to specifying a parametric model on the e_i, namely $e_i \overset{i.i.d.}{\sim} G_0$ whereas $\alpha = 0$ simply implies a parametric model with a constant stratum effect, namely, $e_i = e_0$ and $e_0 \sim G_0$.

In our computations we assume a Gamma prior on α and resample from the full conditional distribution of α using a latent Beta variable, as described in Escobar and West (1995).

With the above model and prior specifications, one can obtain the full conditional distributions for the parameters.

In order to implement this semiparametric model, the first step is to integrate out G, and begin with the joint distribution $f(e_1, \cdots, e_n | G_0, \alpha)$. The e_i are no longer independent, but the joint distribution can be written explicitly (see e.g. Escobar and West, 1995 or Mukhopadhyay and Gelfand, 1997).

To complete the prior specification, one needs to specify G_0 and α. One option is to use a $N(0, V)$ prior and, subsequently an inverse gamma prior for V to describe G_0. This is the usual normal-gamma prior, although other priors seem possible also. Escobar and West (1995) suggested a gamma prior for α. However, others have found that even assignment of some specific value to α does not make much difference.

The implementation of the proposed Bayesian procedure is once again accomplished via MCMC numerical integration technique. This is similar to what is given in Escober and West (1995) or Mukhopadhyay and Gelfand (1997). In addition to the full conditionals for the β_j's, γ_j's, μ_β, μ_γ, σ_β^2, S_γ, this requires also generating samples from the conditionals of the e_i given the remaining parameters and the data, and a similar conditional for α. From the work of Antoniak (1974), it follows that the conditional pdf for the e_i given the remaining parameters (including e_j, $j \neq i$) and the data is a weighted average of a normal pdf, and some degenerate pdf $\delta_{e_j}(e_i)$, where $\delta_{e_j}(e_i) = 1$ or 0 according as $e_i = e_j$ or not. The full conditional for α can be expressed as a weighted average of two gamma distributions as in Escobar and West (1995) and Mukhopadhyay and Gelfand (1997).

4. Summary and Conclusion

The paper proposes both parametric and semiparametric Bayesian neural network models for the prediction of survival probabilities. There are many potential applications. I want to apply the proposed methodology to the prediction of survival probabilities of prostate cancer patients who were diagnosed with clinically localized prostate cancer.

Acknowledgement

Thanks are due to a reviewer for careful reading of the manuscript, and making many helpful suggestions.

References

Antoniak, C.E. (1974). Mixtures of Dirichlet processes with applications to nonparametric problems. *The Annals of Statistics*, **2**, 1152-1174.

Brunner, L.J. and Lo, A.Y. (1989). Bayes methods for a symmetric unimodal density and its mode. *The Annals of Statistics*, **17**, 1550-1566.

Brunner, L.J. and Lo, A.Y. (1994). Nonparametric Bayes methods for directional data. *The Canadian Journal of Statistics*, **22**, 401-412.

Chib, S., and Greenberg, E. (1995). Understanding the Metropolis-Hastings algorithm. *The American Statistician*, **49**, 327-335.

Christensen, R. and Johnson, W.O. (1988). Modelling accelerated failure time with a Dirichlet process. *Biometrika*, **75**, 693-704.

Cox, D.R. (1972). Regression models and life tables (with discussion). *Journal of the Royal Statistical Society, B*, **34**, 187-220.

Cox, D.R. (1975). Partial likelihood. *Biometrika*, **62**, 269-276.

Escobar, M.D. (1994). Estimating normal means with Dirichlet process priors. *Journal of the American Statistical Association*, **89**, 268-277.

Escobar, M.D. (1995). Nonparametric Bayesian methods in hierarchical models. *Journal of Statistical Planning and Inference*, **43**, 97-106.

Escobar, M.D. and West, M. (1995). Bayesian density estimation and inference using mixtures. *Journal of the American Statistical Association*, **90**, 577-588.

Ferguson, T.S. (1973). A Bayesian analysis of some nonparametric problems. *The Annals of Statistics*, **1**, 209-230.

Ferguson, T.S. (1974). Prior distributions on spaces of probability measures. *The Annals of Statistics*, **2**, 615-629.

Ferguson, T.S. (1983). Bayesian density estimation in mixtures of normal distributions. *Recent Advances in Statistics*. Eds. H. Rizvi and J. Rustagi. Academic Press, New York, 287-302.

Gelfand, A.E. and Smith, A.F.M. (1990). Sampling-based approaches to calculating marginal densities. *Journal of the American Statistical Association*, **90**, 398-409.

Ghosh, M., Ghosh, A., Chen, M.H. and Agresti, A. (2000). Noninformative priors for one-parameter item response models. *Journal of Statistical Planning and Inference*, **88**, 99-115.

Ibrahim, J.G., Chen, M-H. and Sinha, D. (2001). *Bayesian Survival Analysis*. Springer, New York.

Kalbfleisch, J.D. (1978). Nonparametric Bayesian analysis of survival time data. *Journal of the Royal Statistical Society, B*, **40**, 214-221.

Kleinman, K.P. and Ibrahim, J.G. (1998). A semi-parametric approach to generalized linear mixed models. *Statistics in Medicine*, **17**, 2579-2596.

Kuo, L. and Mallick, B.K. (1997). Bayesian semiparametric inference for the accelerated failure-time model. *The Canadian Journal of Statistics*, **25**, 457-472.

Lee, H.K.H. (2004). Bayesian Nonparametrics via Neural Networks. Siam, Philadelphia, Pennsylvania.

Lo, A.Y. (1984). On a class of Bayesian nonparametric density estimation: I, density estimates. *The Annals of Statistics*, **12**, 351-357.

MacEachern, S. and Müller, P. (1998). Estimating mixture of Dirichlet process models. *Journal of Computational and Graphical Statistics*, **7**, 223-238.

Metropolis, N., Rosenbluth, A.W., Rosenbluth, M.N., Teller, A.H. and Teller, E. (1953). Equations of state calculations by fast computing machines. *Journal of Chemical Physics*, **21**, 1087-1092.

Mukhopadhyay, S. and Gelfand, A.E. (1997). Dirichlet process mixed generalized linear models. *Journal of the American Statistical Association*, **92**, 633-639.

Müller, P. and Rosner, G. (1997). A Bayesian population model with hierarchical mixture priors applied to blood count data. *Journal of the American Statistical Association*, **92**, 1279-1292.

Neal, R.M. (1996). *Bayesian Learning for Neural Networks*, Springer-Verlag, New York.

Newton, M., Czado, C. and Chappell, R. (1996). Bayesian inference for semiparametric binary regression. *Journal of the American Statistical Association*, **91**, 142-153.

West, M., Müller, P. and Escober, M.D. (1994). Hierarchical priors and mixture models. *In Aspects of Uncertainty: A Tribute to D.V. Lindley*. Eds. A.F.M. Smith and P. Freeman. Wiley, U.K., 363-385.

Bayesian Statistics and Its Applications
Edited by S.K. Upadhyay, U. Singh and D.K. Dey
Anamaya Publishers, New Delhi, India

Semiparametric Accelerated Failure Time Models for Censored Data

Sujit K. Ghosh and Subhashis Ghosal

Department of Statistics, North Carolina State University, Raleigh, NC 27695-8203

Abstract

An accelerated failure time (AFT) semiparametric regression model for censored data is analyzed as an alternative to the widely used proportional hazards survival model. The statistical inference is based on a nonparametric Bayesian approach that uses a Dirichlet process prior for the mixing distribution. Consistency of the posterior distribution of the regression parameters in the Euclidean metric is established under certain conditions. Finite sample parameter estimates along with associated measure of uncertainties can be computed by a MCMC method. Simulation studies are presented to provide empirical validation of the new method. Analysis of data from two studies are provided to show the easy applicability of the proposed method.

1. Introduction

The framework of generalized linear models (GLM) popularized by McCullagh and Nelder (1989) is arguably one of the most flexible statistical tools and is used extensively in almost all fields of applications. However, such a flexible framework is not widely used in regression models for survival data with censoring. A key motivation of this article is to develop a flexible class of regression models that are similar in spirit to the GLM but retain the semiparametric structure of the widely used proportional hazard (PH) models. It is known that such an objective may be achieved by a class of mixture models where the mixing distribution is left unspecified (see Heckman and Singer, 1984, Ishwaran, 1996).

The Cox proportional hazard regression model (Cox, 1972) and the associated partial likelihood theory of estimation was a breakthrough in developing a flexible method of regression for censored data. The huge success of PH models testify to the many needs for this type of semiparametric regression models. The structure of PH is quite different from the usual GLM for regression, in that the link function is not specified via the mean but rather through the hazard function. The proportionality structure is interesting but it may be hard to interpret the regression coefficients. Sir D. Cox himself once remarked (Reid, 1994) that:

"Of course, another issue is the physical or substantive basis for the proportional hazards model. I think that's one of its weakness, that accelerated life models are in many ways more appealing because of their quite direct physical interpretation, particularly in an engineering context."

An accelerated failure time (AFT) model is characterized by specifying the conditional survival function, $S(t|\mathbf{Z} = z)$ of survival time, T given the covariates, $\mathbf{Z} = z$, as $S(t|\mathbf{z}) = S_0(tg(\mathbf{z}^T\boldsymbol{\beta}))$, where $S_0(\cdot)$ is an unspecified baseline survival function. If $S_0(\cdot)$ is specified parametrically (with possibly some unknown finite-dimensional parameter), then we get a parametric AFT model. Under such parametric models, estimates of the regression coefficient $\boldsymbol{\beta}$ can be obtained by the maximum likelihood or Bayesian method, for a given completely known link function $g(\cdot)$. The main contributions of this article are (a) a semiparametric formulation of the accelerated failure time model and (b) asymptotic justifications of the resulting estimators.

Among the various extensions of the traditional linear model, AFT models and the method of least squares to accommodate censored data seems very appealing, simply because the model is well known, widely used, well understood and well tested (Wei, 1992). Using a U-statistic representation, Koul et al. (1981) showed that their estimators are consistent and asymptotically normal under some regularity conditions. Following the simple idea of using synthetic data, several extensions of the method have appeared in the literature that use more efficient ways to obtain estimated responses (Zheng, 1984, Lai et al., 1995, Zhou, 1992 and references therein). These developments have been very exciting but generally lack stability of the estimators and hence are not as widely used as the PH model.

From a Bayesian perspective a wide class of semiparametric AFT regression models have been developed by, among others, Christensen and Johnson (1988), Johnson and Christensen (1989), Kuo and Mallick (1998), Walker and Mallick (1999), Campolieti (2001) and Hanson and Johnson (2004). However, to the best of our knowledge, none of the previous authors have attempted to formally prove the posterior consistency based on censored data from semiparametric AFT models. For the case of no censoring, Ishwaran (1998) presented some interesting results for Weibull semiparametric mixture under Dirichlet process and Ishwaran (1996) points out an important relationship between the constraint on mixing distribution and rates of estimation for the parameters in the model.

The proposed method of parameter estimation is quite similar in spirit to the previous approaches in terms of using the popular Markov Chain Monte Carlo (MCMC) methods. Moreover, we demonstrate that such Monte Carlo iterative methods can be easily implemented (see Appendix for a code) using freely available softwares like **WinBUGS** (Spiegelhalter et al., 1999). In addition to the computational stability and simplicity of the proposed method, we show that the posterior distribution is consistent under certain compactness assumptions on the parameter space.

In Section 2, the semi-parametric AFT model is presented using a mixture of Dirichlet processes. Section 3 discusses how an MCMC method can be easily implemented to obtain estimates based on the posterior distribution of the parameters. The posterior consistency of the proposed model is discussed in Section 4. Simulation studies are presented in Section 5. Finally, Section 6 presents couple of real data examples to illustrate our method. Proofs and codes are presented in the Appendices.

2. Semiparametric AFT Regression Model

Mixture models have been used for a variety of inference problems including density estimation, clustering analysis and robust estimation; see Lindsay (1995), McLachlan and Basford (1988), Banfield and Raftery (1993), Robert (1996) and Roeder and Wasserman (1997). In the Bayesian context, mixture models for density estimation were introduced by Ferguson (1983) and Lo (1984) who used a Dirichlet process prior on the mixing distribution and obtained expressions for Bayes estimates. Escobar and West (1995) developed this idea further and provided MCMC algorithms for the computation of the posterior distribution of parameters of a normal mixture model.

Consistency and rates of convergence issues for Bayesian density estimation in mixture models have been studied by Ghosal et al. (1999), Ghosal and van der Vaart (2001), Ghosal (2001) and Petrone and Wasserman (2002). For survival models without covariates, consistency has been studied by Ghosh and Ramamoorthi (1994), Ghosh et al. (1999) and Kim and Lee (2001). In this article we propose using an infinite mixture of standard parametric distributions (such as the Weibull distributions) to model the conditional distribution of a survival time given a set of covariates.

For each subject $i = 1, \ldots, n$, let T_i denote the failure time and C_i denote the censoring time. The observed survival data are $X_i = \min(T_i, C_i)$ and $\Delta_i = I(T_i \leq C_i)$, where $I(\cdot)$ is the indicator function; these and all other variables are independent across i. Let $\boldsymbol{Z}_i$ denote a p-dimensional vector of covariates associated with subject i. Assume that for each subject i the conditional survival function of T given $\boldsymbol{Z} = \boldsymbol{z}$ (suppressing the subscript i) is given by

$$S(t|\boldsymbol{z}, \boldsymbol{\beta}) = \int_0^\infty S_b(t/\mu g(\boldsymbol{z}^T \boldsymbol{\beta})) dH(\mu) \tag{1}$$

where $S_b(\cdot)$ is a survival function with a specified functional form, but may involve additional unknown parameters. The link function $g : \mathbb{R} \to [0, \infty)$ is completely specified. The mixing distribution function $H(\cdot)$ is left unspecified, which will be estimated nonparametrically with the constraint $H(0) = 0$. It follows that the density of T given $\boldsymbol{Z} = \boldsymbol{z}$ is then given by

$$p(t|\boldsymbol{z}, \boldsymbol{\beta}) = \int_0^\infty p_b(t/\mu g(\boldsymbol{z}^T \boldsymbol{\beta}))(\mu g(\boldsymbol{z}^T \beta))^{-1} dH(\mu) \tag{2}$$

where p_b is the density function corresponding to the survival function S_b. Note that (1) leads to an AFT model with $S_0(t) = \int_0^\infty S_b(\frac{t}{\mu}) dH(\mu)$. In other words, the baseline survival function is modeled as a mixture of parametric survival function with an unknown mixing distribution.

Let $m = \int_0^\infty t p_b(t) dt = \int_0^\infty S_b(t) dt$ stands for the mean of the distribution with survival function S_b. Then

$$E(T|\boldsymbol{z}, \alpha, \boldsymbol{\beta}) = m g(\boldsymbol{z}^T \boldsymbol{\beta}) \int_0^\infty \mu dH(\mu). \tag{3}$$

In most applications a logarithmic link is chosen for g^{-1}, that is, $g(w) = e^w$. Putting in (3), we obtain

$$\log \frac{E(T|\boldsymbol{z}_1, \boldsymbol{\beta})}{E(T|\boldsymbol{z}_2, \boldsymbol{\beta})} = (\boldsymbol{z}_1 - \boldsymbol{z}_2)^T \boldsymbol{\beta}$$

which provides an easy interpretation of the parameter $\boldsymbol{\beta}$ as the unit change in the logarithm of the mean response with respect to a unit change in the covariate. Henceforth, we shall work with the logarithmic link function. Note that the mixing distribution $H(\cdot)$ has been left completely unspecified.

We shall use a Weibull survival function with an unknown shape parameter as a choice for S_b, i.e., we assume that $S_b(t, \alpha) = e^{-t^\alpha}$. The distribution with survival function $e^{-t^\alpha/\lambda^\alpha}$ will be referred to as Weibull (α, λ) distribution, the Weibull distribution with scale parameter λ and shape parameter α. One may use other parametric families such as log normal or gamma. More generally, any parametric family supported on the positive half line may be considered. However, unlike the gamma or the log normal family, the Weibull distribution enjoys the advantage of the survival function being free of transcendental functions, and hence it will be easier to explicitly write down the likelihood for censored data. It may be noted that a mixture of Weibull distribution (as in (2)) can be seen as a location mixture model on a log-scale. More specifically, for Weibull mixtures, when we choose a log-link for g^{-1}, the conditional density of $Y = \log T$ given $\boldsymbol{Z} = \boldsymbol{z}$ is given by

$$p(y|\boldsymbol{z}, \boldsymbol{\beta}) = \int_0^\infty e^y p_b(e^y / \mu g(\boldsymbol{z}^T \boldsymbol{\beta}))(\mu g(\boldsymbol{z}^T \boldsymbol{\beta}))^{-1} dH(\mu)$$

$$= \int_0^\infty \alpha f_b(\alpha(y - \log \mu - \boldsymbol{z}^T \boldsymbol{\beta}))(\mu g(\boldsymbol{z}^T \boldsymbol{\beta}))^{-1} dH(\mu)$$

where $f_b(t) = \exp(t - e^t)$ is Gumbel's extreme value density. Thus $\log \mu$ takes the role of a location parameter and α (the reciprocal of) a scale parameter in the above location mixture. Therefore, letting $\alpha \to \infty$, mixtures over μ can approximate any arbitrary density in the total variation distance. As the total variation distance is invariant under one-to-one measurable transformations, the same conclusion holds in the original scale. Thus, Weibull mixture can adequately represent an arbitrary density in (2).

A likelihood based method to estimate the parameter $\boldsymbol{\beta}$ would require maximization of over the space of all distributions $H(\cdot)$ with the property $H(0) = 0$ and the shape parameter α. Such an approach may yield efficient estimates for $\boldsymbol{\beta}$, and under regularity conditions, an asymptotic estimate of the standard error of the estimate can be obtained. Optimizations over such large space of parameters may be computationally challenging. Alternatively, one may use a set of estimating equations (see Jin et al. 2003) to obtain parameter estimates. However, it turns out that for semiparametric AFT models, such estimating equations can be non-monotone and can lead to multiple solutions. In our nonparametric Bayesian approach, we use MCMC method to obtain the marginal posterior distribution of $\boldsymbol{\beta}$ by integrating out the nuisance parameters $H(\cdot)$ and α.

Prior distributions play a crucial role in Bayesian inference. Informative priors can be used if deemed plausible. However, in the present context, it is desirable to choose the prior distribution by some default mechanism. A priori, it is assumed that $(\alpha, \boldsymbol{\beta})$ independent of $H(\cdot)$, and that $H(\cdot)$ has a Dirichlet Process (DP) Prior, denoted by $\mathrm{DP}(M, H_b(\cdot))$ with base measure $H_b(\cdot)$ and precision parameter M. The proposed model is equivalently written as a Bayesian hierarchical model as follows:

$$T_i | \boldsymbol{Z}_i, \alpha, \mu_i, \boldsymbol{\beta} \;\sim\; \text{Weibull} \left(\alpha, \mu_i \exp \left\{ \boldsymbol{Z}_i^T \boldsymbol{\beta} \right\} \right)$$

$$\mu_i | H(\cdot) \stackrel{\text{i.i.d.}}{\sim} H(\cdot)$$

$$H(\cdot) \;\sim\; \mathrm{DP}(M, H_b(\cdot)) \tag{4}$$

$$(\alpha, \boldsymbol{\beta}) \;\sim\; \pi(\alpha, \boldsymbol{\beta})$$

where the joint density $\pi(\alpha, \boldsymbol{\beta})$ is usually chosen to be of the product type and often taken to be the diffuse uniform prior on $[a_0, A_0] \times [-B_0, B_0]^p$, where the fixed positive constants a_0, A_0 and B_0 are chosen to approximate a non-informative prior. A small value of $a_0 > 0$ and large values for A_0 and $B_0 > 0$ are usually chosen. A moderate value of $M > 0$ is usually used. In practice, a sensitivity study is generally needed to elicit these constants (see Section 4). Let Π denote the joint prior distribution of $(\alpha, \boldsymbol{\beta}, H)$. Note that from the properties of a DP it follows that

$$E(H(\cdot)|\mu_1, \ldots, \mu_n) = \frac{M}{M + n} H_b(\cdot) + \frac{n}{M + n} H_n(\cdot)$$

where $H_n(\cdot)$ is the empirical distribution function of $\mu_1, \ldots, \mu_n$. However, these latent variables $\mu_1, \ldots, \mu_n$ are not directly observable, and hence the posterior distribution may only be computed by MCMC methods. In the next section we discuss how to obtain the posterior distribution of the parameters given the observed data $\{(X_i, \Delta_i, \mathbf{Z}_i), i = 1, \ldots, n\}$.

3. Model Fitting

For the proposed AFT model, the posterior distribution of the parameters cannot be obtained in a closed form. Gibbs sampling with a data augmentation step will be used to obtain MCMC samples.

First assume that T_i's are indeed observed. In this case, a simple strategy to obtain samples from the posterior distribution of $(\alpha, \boldsymbol{\beta}, H(\cdot))$ can be described as follows (see MacEachern, 1998):

Gibbs Sampling: Initialize the parameters $\boldsymbol{\beta}^{(0)}$ and $\alpha^{(0)}$. At the k-th iteration of the Gibbs sampler, the full conditional density of μ_i given $(\boldsymbol{\beta}^{(k-1)}, \alpha^{(k-1)}, \mu_l^{(k-1)}, l \neq i, T_i, \mathbf{Z}_i)$ is given by (upto a proportionality constant),

$$p_b(T_i | Z_i, \mu, \alpha^{(k-1)}, \boldsymbol{\beta}^{(k-1)}) h_b(\mu), \qquad \text{with probability } q_{0i}^{(k)}$$

$$\mu_l^{(k-1)} I(l > i) + \mu_l^{(k)} I(l < i), \qquad \text{with probability } q_{li}^{(k)}, \quad l \neq i$$

where

$$q_{li}^{(k)} \propto \begin{cases} \mu_l^{(k)} p_b(T_i | \mathbf{Z}_i, \mu_l^{(k)}, \alpha^{(k-1)}, \boldsymbol{\beta}^{(k-1)}), & \text{if } l < i \\ \mu_l^{(k-1)} p_b(T_i | \mathbf{Z}_i, \mu_l^{(k-1)}, \alpha^{(k-1)}, \boldsymbol{\beta}^{(k-1)}), & \text{if } l > i \end{cases}$$

and

$$q_{0i}^{(k)} \propto M \int_0^\infty p_b(T_i | \mathbf{Z}_i, \mu, \alpha^{(k-1)}, \boldsymbol{\beta}^{(k-1)}) h_b(\mu) d\mu$$

such that $q_{0i}^{(k)} + \sum_{l \neq i} q_{li}^{(k)} = 1$ and $h_b(\mu)$ denotes the density corresponding to $H_b(\mu)$. Note that if we choose a conjugate base measure $H_b(\cdot)$, the above integral can be evaluated analytically and hence samples can be obtained by an inversion method. Otherwise a numerical one-dimensional integral can be used to compute $q_{0i}^{(k)}$'s. For our model we may achieve conjugacy by choosing an appropriate gamma distribution for μ_i^α. The full conditional density of (α, β) given $\{(\mu_i^{(k)}, T_i, Z_i), i = 1, \ldots, n\}$ is given by (upto a proportionality constant)

$$\prod_{i=1}^n p_b(T_i | \mathbf{Z}_i, \mu_i^{(k)}, \alpha, \boldsymbol{\beta}) h_b(\mu_i^{(k)})$$

which is a log-concave density and hence an adaptive rejection sampling (see Gilks, 1992) can be used to sample from the above density. Thus, a Gibbs sampler, as described above, can be used to obtain approximate samples from the joint posterior distribution of the parameters given the observed data. If desired, the posterior mean of $H(\cdot)$ can be obtained from the MCMC samples using (5).

When censoring is present, an imputation method can be used to "replace" the censored data by the imputed values. This is done as follows. If $\Delta_i = 1$, set $T_i = X_i$; otherwise set $T_i = (U_i + X_i^{\frac{1}{\alpha}})^\alpha$, where U_i's are independently exponentially distributed with means $\mu_i \exp\{\mathbf{Z}_i^T \boldsymbol{\beta}\}, i = 1, \ldots, n$.

Alternatively, a finite approximation for DP (Ishawaran and Zarepour, 2002) can be used within the software **BUGS** to implement the Gibbs sampling. In fact, for our simulation study we use a finite approximation technique to implement the required Gibbs sampling using **BUGS**. This is achieved by introducing latent variables $\mathbf{L} = (L_1, \ldots, L_n)$ which indicate the group membership for the hidden variables μ_i's along with a probability vector $\boldsymbol{w} = (w_1, \ldots, w_N)^T$. More precisely, (4) can be written as,

$$T_i | L_i, \boldsymbol{\alpha}, \boldsymbol{\beta} \sim \text{Weibull } (\alpha, \mu_{L_i} \exp\{\boldsymbol{Z}_i^T \boldsymbol{\beta}\})$$

$$L_i | \boldsymbol{w} \overset{\text{i.i.d.}}{\sim} \text{Multinomial } (\{1, \ldots, N\}, \boldsymbol{w})$$

$$\mu_l \overset{\text{i.i.d.}}{\sim} H_b(\cdot), l = 1, \ldots, N$$

$$\boldsymbol{w} \sim \text{Dirichlet } \left(\frac{M}{N}, \ldots, \frac{M}{N} \right)$$

$$(\alpha, \beta) \sim \pi(\alpha, \beta)$$

where N is a large integer. Notice that $\Pr[L_i = l] = w_l$ for $l = 1, \ldots, N$, where $\sum_{l=1}^{N} w_l = 1$. The above hierarchical framework which uses a finite dimensional Dirichlet distribution (with N possibly depending on n) is more convenient for programming in **BUGS**. For instance, Ishwaran and Zarepour (2002) suggested to use $N = \sqrt{n}$ for large n and $N = n$ for small n (see Section 5 of their article). However, more analytical work is necessary to choose the right order of N.

4.   Consistency of Posterior Distribution

Consistency is an important desirable large sample property of the posterior distribution which provides a useful validation of a particular Bayesian method in use. For discussion and examples of inconsistency, the readers are referred to Diaconis and Freedman (1986) and Ghosh and Ramamoorthi (2003). A celebrated theorem of Schwartz (1965) gives sufficient conditions for posterior consistency in terms of conditions involving the existence of appropriate tests and the prior positivity of a neighborhood defined by the Kullback-Leibler divergence. Useful extensions of Schwartz's theorem are given by Barron et al. (1999) and Ghosal et al. (1999). The goal of the present section is to justify our Bayesian analysis of the AFT model for censored data by posterior consistency. Ishwaran (1998) has studied the consistency of similar posterior distributions for uncensored data.

We shall assume that the domains of $\boldsymbol{Z}$, $\boldsymbol{\beta}$ and α, and the support of H are compact. The compactness assumption is agreeably somewhat restrictive. Also, we assume, without loss of generality, that $\boldsymbol{0}$ is a possible value of the covariate $\boldsymbol{Z}$. If not, we may shift the covariates to satisfy this condition. We further assume that the true density p_0 of T given $\boldsymbol{Z} = \boldsymbol{z}$ is actually a scale mixture of Weibull:

$$p_0(t|\boldsymbol{z}) = \int \alpha_0 (\mu \exp\{\boldsymbol{\beta}_0^T \boldsymbol{z}\})^{-\alpha_0} t^{\alpha_0 - 1} \exp[-(t/\mu \exp\{\boldsymbol{\beta}_0^T \boldsymbol{z}\})^{\alpha_0}] dH_0(\mu)$$

so that α_0, $\boldsymbol{\beta}_0$ and H_0 are, respectively, the true values of the parameters α, $\boldsymbol{\beta}$ and H. The covariates are assumed to be i.i.d. with an absolutely continuous distribution supporting the vector $\boldsymbol{0}$ on $\mathbb{R}^p$. The density of $\boldsymbol{Z}$ at $\boldsymbol{z}$ will be denoted by $q(\boldsymbol{z})$. Let $f_{\alpha, \boldsymbol{\beta}, H}(x, \delta, \boldsymbol{z})$ stand for the joint density of $(X, \Delta, \boldsymbol{Z})$, that is

$$f_{\alpha, \boldsymbol{\beta}, H}(x, \delta, \boldsymbol{z}) = \begin{cases} p(x|\alpha, \boldsymbol{\beta}, \boldsymbol{z}) q(\boldsymbol{z}), & \text{if } \Delta = 1 \\ S(x|\alpha, \boldsymbol{\beta}, \boldsymbol{z}) q(\boldsymbol{z}), & \text{if } \Delta = 0 \end{cases}$$

where $p(x|\alpha, \boldsymbol{\beta}, \boldsymbol{z})$ and $S(x|\alpha, \boldsymbol{\beta}, \boldsymbol{z})$ are as defined in equations (2) and (1), respectively. Note that the class of distributions that are supported in a given compact domain is also compact with respect to the weak topology on the space of probability measures. Hence the parameter space of $(\alpha, \beta, \boldsymbol{H})$ with respect to the product of Euclidean, Euclidean and weak topology, is also compact.

The following is the main theorem.

Theorem 1. *Suppose that the prior density* $\pi(\alpha, \beta)$ *for* (α, β) *has compact support containing* (α_0, β_0) *and the base measure* H_b *of the Dirichlet process has compact support that contains the support of* H_0. *Then the posterior distribution* $\Pi((\alpha, \beta, H) \in .|(X_1, \Delta_1), \dots (X_n, \Delta_n))$ *of* (α, β, H) *given* $(X_1, \Delta_1), \dots (X_n, \Delta_n)$ *is consistent with respect to the Euclidean distances on* β *and* α *and the weak topology on* H, *that is, given any* $\varepsilon > 0$ *and a weak neighborhood* $\mathcal{N}$ *of* H_0,

$$\Pi\{(\alpha, \beta, H) : |\alpha - \alpha_0| < \varepsilon, |\beta - \beta_0| < \varepsilon, H \in \mathcal{N}|(X_1, \Delta_1), \dots, (X_n, \Delta_n)\} \to 1$$

almost surely in $P^\infty_{(\alpha_0, \beta_0, H_0)}$*-probability.*

This gives a large sample justification of our procedure. The proof of Theorem 1 is given in Appendix A.

If the covariates are not random, but arise deterministically from a design, then by largely a similar analysis and using posterior consistency results for independent, non-identically distributed observations as in Amewou-Atisso et al. (2003), consistency will follow if the values of the covariate gradually fill up its range space as the sample size increases. The details are omitted here.

5. A Simulation Study

We performed extensive simulations to explore the sampling properties of the Bayes estimates obtained by the proposed method. We present only some of the significant findings from our simulation experiments. In order to perform the simulation experiments, we generate the data from model (2). In particular we use the following data generation process (DGP):

DGP for simulation: Fix true values $\beta_0 = (-0.5, 0.5)$, $\alpha_0 = 1.0$ and the sample size $n = 50, 100$ and 200.

1. Generate $U_i \stackrel{\text{i.i.d.}}{\sim}$ Beta $(3, 3)$ and set $\boldsymbol{Z}_i = 6U_i - 3$. This ensures that support of $\boldsymbol{Z}_i$'s is compact.

2. Generate $\mu_i \stackrel{\text{i.i.d.}}{\sim}$ Weibull $(\alpha^* = 1, \mu^* = 2)$. This means that the mixing distribution is also a Weibull distribution.

3. Generate $T_i \stackrel{\text{ind}}{\sim}$ Weibull $(\alpha_0 = 1, \mu_i \exp\{\boldsymbol{Z}_i^T \beta_0\})$.

4. Generate $C_i \stackrel{\text{i.i.d.}}{\sim}$ Weibull $(\alpha^*, 2\mu^*)$. We also perform simulation where C_i's are allowed to depend on $\boldsymbol{Z}_i$'s. In particular we generate $C_i \stackrel{\text{ind.}}{\sim}$ Weibull $(\alpha^*, 4.5 \exp\{\boldsymbol{Z}_i^T \beta\})$. These choices of censoring variable ensure that the censoring rate is about 30% on average.

5. Set $X_i = \min(T_i, C_i)$ and $\Delta_i = I(T_i \leq C_i)$, and output the observed data as: $\{(X_i, \Delta_i, \boldsymbol{Z}_i); i = 1, \dots, n\}$.

We fit the semiparametric Bayesian model to the data generated (at step 5) from the model and repeat (steps 1-5 above) 500 times with a fixed burn-in time of 500 iterations followed by a sample of size 1000 from the posterior distribution of β and α. We have tried several combinations of true values, mixing distributions and censoring variables for our simulations. In order to have a compactly supported prior distribution we have used fixed values of the constants $a_0 = 0.01$, $A_0 = 100$, $B_0 = 10$ and $M = 1$ (see description right after (5)). Also, we have used $N = \sqrt{n}$ for all our applications to approximate a DPP. A sensitivity study with $N = n$ revealed no significant difference in the posterior estimates. In this article we present results based on only one set of true values for three different sample sizes and two different censoring mechanisms.

Table 1. Sampling properties of Bayes estimates when the censoring mechanism does not depend on covariates. True values: $\beta_1 = -0.5$, $\beta_2 = 0.5$, $\alpha = 1.0$. The standard deviation of the corresponding posterior summaries is denoted by s.e. (e.g., mean, sd etc.) over 500 replications

	mean	sd	2.5%L	97.5%U	median	cp
			$n = 50$			
β_1	−0.505	0.244	−1.000	−0.036	−0.501	0.927
s.e.	0.273	0.053	0.313	0.288	0.271	
β_2	0.516	0.248	0.055	1.032	0.534	0.928
s.e.	0.253	0.055	0.269	0.289	0.253	
α	1.116	0.241	0.721	1.641	1.134	0.948
s.e.	0.224	0.051	0.148	0.289	0.233	
			$n = 100$			
β_1	−0.491	0.166	−0.821	−0.165	−0.483	0.935
s.e.	0.172	0.027	0.192	0.170	0.173	
β_2	0.494	0.167	0.172	0.828	0.491	0.938
s.e.	0.177	0.027	0.178	0.197	0.176	
α	1.032	0.188	0.716	1.430	1.021	0.954
s.e.	0.176	0.042	0.115	0.235	0.183	
			$n = 200$			
β_1	−0.502	0.121	−0.753	−0.279	−0.504	0.949
s.e.	0.123	0.009	0.113	0.098	0.102	
β_2	0.497	0.118	0.261	0.728	0.496	0.948
s.e.	0.120	0.013	0.131	0.139	0.129	
α	0.996	0.152	0.727	1.305	0.967	0.951
s.e.	0.145	0.034	0.072	0.168	0.128	

Average observed censoring 29.25% (range 10–40%).

The results are presented in Tables 1 and 2 by increasing sample size. In each table we present the average five-number posterior summary values, viz. posterior mean, posterior sd, posterior 2.5% and 97.5% percentiles and the posterior median with the associated standard error (s.e.) from 500 Monte Carlo repetitions. In addition, we also present the coverage probability (cp) based on a 95% equal-tail posterior interval obtained by posterior 2.5% and 97.5% percentiles. For example, in Table 1, for $n = 100$, the entry −0.491 for β_1 is the average of 500 posterior means with s.e. 0.172 (which is the standard deviation of 500 posterior means). Similarly the entry 0.166 is the average of 500 posterior sd's with s.e. 0.027. Finally, a cp of 0.935 means that 93.5% of 500 equal-tail posterior 95% intervals contained the true value of $\beta_1 = -0.5$.

We observe that at an average censoring rate of about 30%, the Bayes estimates (posterior mean and median) are nearly unbiased as sample size increases and also the coverage probability of a 95% posterior interval approaches the nominal value of 95%. In addition, we see that the average length of a 95% posterior interval decreases with increasing sample size, thus making the intervals tighter. For instance, for β_2 the average length of 95% interval drops from 0.977 (for $n = 50$) to an average length of 0.467 (for $n = 200$). Similar observations can be made based on Table 2. These results numerically assert the consistency of posterior distribution of β and α. We repeated similar studies for other mixing distributions and for lower and higher censoring rates, the results were very similar to what we see in Tables 1 and 2.

Table 2. Sampling properties of Bayes estimates when the censoring mechanism depends on covariates. True values: $\beta_1 = -0.5$, $\beta_2 = 0.5$, $\alpha = 1.0$. The standard deviation of the corresponding posterior summaries is denoted by s.e. (e.g., mean, sd etc.) over 500 replications

	mean	sd	2.5%L	97.5%U	median	cp
			$n = 50$			
β_1	−0.487	0.234	−0.953	−0.028	−0.487	0.925
s.e.	0.258	0.053	0.283	0.282	0.258	
β_2	0.502	0.233	0.037	0.956	0.503	0.928
s.e.	0.248	0.052	0.266	0.278	0.248	
α	1.121	0.244	0.723	1.654	1.137	0.945
s.e.	0.239	0.053	0.153	0.309	0.247	
			$n = 100$			
β_1	−0.489	0.159	−0.819	−0.175	−0.489	0.939
s.e.	0.165	0.024	0.188	0.168	0.169	
β_2	0.495	0.163	0.179	0.817	0.493	0.940
s.e.	0.167	0.021	0.172	0.190	0.171	
α	1.029	0.179	0.710	1.422	1.018	0.953
s.e.	0.168	0.038	0.108	0.230	0.177	
			$n = 200$			
β_1	−0.502	0.115	−0.746	−0.280	−0.503	0.950
s.e.	0.123	0.009	0.113	0.098	0.102	
β_2	0.501	0.105	0.287	0.709	0.500	0.949
s.e.	0.110	0.010	0.126	0.133	0.124	
α	1.004	0.144	0.719	1.294	0.990	0.949
s.e.	0.140	0.028	0.067	0.160	0.121	

6. Applications to Data from Two Studies

We now apply our proposed models to some popular real data sets that have been analyzed by other authors. In particular we fit our semi-parametric models to ovarian cancer data set (with $n = 26$, small sample size) and multiple myeloma data set (with $n = 65$, moderate sample size). For both data sets we have used a compactly supported prior with $a_0 = 0.1$, $A_0 = B_0 = 10$ and $M = 1$. Also for Dirichlet prior approximation we have used $N = \sqrt{n}$.

6.1 Analysis of Ovarian Cancer Data

Consider a study on ovarian cancer as reported in Edmunson et al. (1979). In this study $n = 26$ patients were monitored and age for each patient were also recorded. Let X_i denote the number of days patient i was on the study. For each patient, let A_i denotes the age (recorded as number of days/365.25). It was of interest to find a relation between the X_i's and the A_i's using statistical regression methods. However in this study 53.84% of the observations were censored. We obtain the posterior distribution of β, under the model $\log[E(T|A)] = A\beta$.

Using the proposed semiparametric method based on Weibull mixture models, we obtain the summary of the posterior distribution of (β, α) using the Gibbs sampling algorithm described in Section 3. In order to maintain numerical stability, we transformed the X_i to $X_i/500$ and also standardized the covariate age to $Z_i = (A_i - \overline{A})/\text{sd}(A)$, where $\overline{A}$ and $\text{sd}(A) = 10.1$ denote the sample mean and standard deviation of A_i's. The results are summarized in Table 3. In addition, we can obtain the entire posterior density of α and β based on MCMC samples. In Fig. 1, we plot the trace and kernel density estimates of (α, β) based on 2000 burn-ins and 5000 samples from three independent chains. From the plots it appears that the chains mixed well and there were no apparent problems with convergence to the stationary distributions. The MCMC diagnostic software **CODA** available in Splus and R was also used to check convergence. The Gelman-Rubin 50% and 97.5% shrink factors (based on three dispersed starting values) were found to be 1.00 and 1.01, respectively for β, which indicate good mixing (as also evident from the trace plots in Fig. 1).

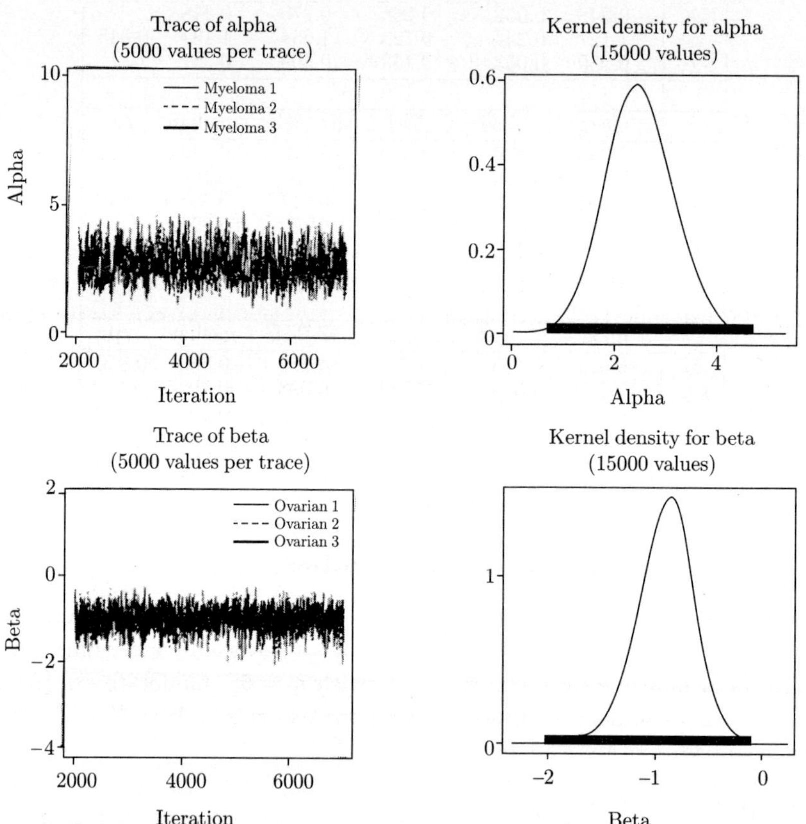

Fig. 1. Trace and density plots based on MCMC samples from the posterior distribution of (α, β) for the ovarian data.

From Table 3, it is apparent that age has a significant negative effect on the mean survival time. More precisely, for unit increase in standardized age (i.e., days/365.25), the expected survival time decreases by a factor of 1.09 ($= \exp(0.9/10.1)$) for patients with ovarian cancer. This type of straightforward interpretation is possible only for these kind of AFT models.

Table 3. Posterior summaries of the regression parameter (β) and the shape parameter (α)

	mean	sd	2.5%L	97.5%U	median
β	−0.922	0.243	−1.460	−0.487	−0.911
α	2.470	0.640	1.330	3.760	2.440

Table 4. Posterior summaries of the regression parameter (β) and the shape parameter (α). Within $(\cdot)$ corresponding frequentist estimates obtained by Jin et al. (2003) are presented

	mean	sd	2.5%L	97.5%U	median
β_1	-0.534	0.175	-0.870	-0.190	-0.531
freq est.	(-0.532)	(0.146)	(-0.818)	(-0.246)	
β_2	0.202	0.168	-0.128	0.528	0.199
freq est.	(0.292)	(0.169)	(-0.039)	(0.623)	
α	1.040	0.074	0.885	1.170	1.050

7.   Analysis of Multiple Myeloma Data

We analyze a data set on multiple myeloma reported by Krall et al. (1975). This data has recently been analyzed by Jin et al. (2003) using a semiparametric AFT model. We compare our findings to that of Jin et al. (2003). In this study $n = 65$ patients were treated with alkylating agents and survival time (in months) were monitored along with several prognostic variables.

In order to compare our results to that of Jin et al. (2003) we use the logarithm of the blood urea nitrogen measurement and the hemoglobin measurement, both measured at diagnosis as the covariates and standardize them to bring numerical stability. In other words, we fit a model such that $\log[E(T|\boldsymbol{Z})] = \boldsymbol{Z}^T \boldsymbol{\beta}$, where $\boldsymbol{Z} = (Z_1, Z_2)^T$ is the vector of standardized covariates and $\boldsymbol{\beta} = (\beta_1, \beta_2)^T$ the vector of regression coefficients to be estimated. We ran three independent parallel chains to draw MCMC samples from the posterior distribution of $(\alpha, \boldsymbol{\beta})$. Fig. 2, shows the trace plots and density estimates of $(\alpha, \boldsymbol{\beta})$ based on 5000 samples from three parallel chains after a burn-in of 2000 samples. The Gelman-Rubin 50% and 97.5% shrink factors (based on three dispersed starting values) were found to be 1.00 and 1.01, respectively for $\boldsymbol{\beta}$, which indicate good mixing (as also evident from the trace plots in Fig. 2). It appears that there is no problem of convergence of the samples to the stationary (posterior) distribution.

Based on the final 15000 samples we compute the posterior summaries of $(\alpha, \boldsymbol{\beta})$ and present these values in Table 4. First note that the posterior mean and median of $\boldsymbol{\beta}$ are quite similar to those obtained by Jin et al. (2003), which are presented in parenthesis. Also from the posterior MCMC samples of $\boldsymbol{\beta}$ we can easily obtain the posterior correlation coefficient of $\boldsymbol{\beta}$ to be $\text{Corr}(\beta_1, \beta_2|\text{data}) = -0.187$. This is clearly an advantage of a Bayesian method over a frequentist method, as almost any posterior summary can be computed from the posterior samples generated by the MCMC method. Note that Jin et al. (2003) had to use a resampling method to compute the asymptotic variance covariance matrix of $\hat{\boldsymbol{\beta}}$. In contrast, our method produces the entire posterior density of $\boldsymbol{\beta}$ using a relatively simple code in **BUGS** (see Appendix B). Further, our interval estimates of $\boldsymbol{\beta}$ as presented in Table 4, are based on the exact (finite sample) posterior percentiles as compared to large sample normal percentiles of Jin et al. (2003), and hence our interval estimates seem to be wider (accounting for the true uncertainty) than those obtained by Jin et al. (2003).

From this study, it appears that the logarithm of the blood urea nitrogen measurement has a significant negative effect on the logarithm of the mean survival time, whereas hemoglobin does not seem to have a significant effect. We have not done a formal Bayesian testing (say using Bayes factor). These conclusions are based on the concentration of posterior distribution around zero. Based on our study we may conclude that, for unit increase in the logarithm of blood urea nitrogen (in the standardized scale), the expected survival time decreases by a factor of 5.53 $(= \exp{(0.534/0.312)})$

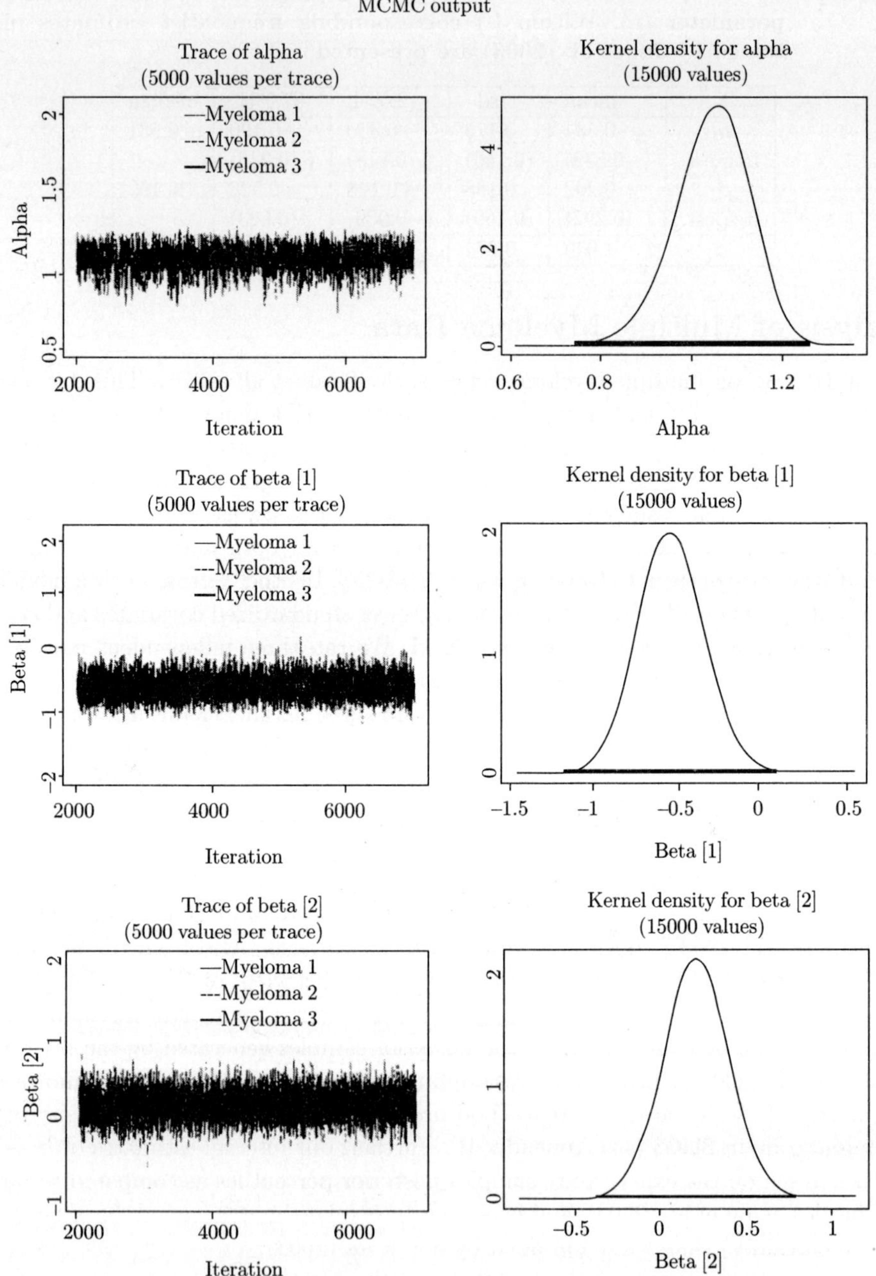

Fig. 2.   **Trace and density plots based on MCMC samples from the posterior distribution of (α, β) for the myeloma data.**

for patients with multiple myeloma. This again illustrates the straightforward interpretation of the regression coefficients based on semiparametric AFT models.

Acknowledgements

The work of the second author was partially supported by the NSF Grant, DMS-0349111.

References

Amewou-Atisso, M., Ghosal, S., Ghosh, J. K. and Ramamoorthi, R.V. (2003). Posterior consistency for semiparametric regression problems. *Bernoulli*, **9**, 291–312.

Banfield, J.D. and Raftery, A.E. (1993). Model based Gaussian and non-Gaussian clustering. *Biometrics*, **49**, 803-821.

Barron, A., Schervish, M.J. and Wasserman, L. (1999). The consistency of posterior distributions in nonparametric problems. *Ann. Statist.*, **27**, 536–561.

Campolieti, M. (2001). Bayesian semiparametric estimation of discrete duration models: an application of the Dirichlet process prior. *Journal of Applied Econometrics*, **16**, 1–22.

Christensen, R. and Johnson, W.O. (1988). Modelling accelerated failure time with a Dirichlet process. *Biometrika*, **75**, 693–704.

Cox, D. R. (1972). Regression models and life tables (with discussion). *J. Roy. Statist. Soc., Ser. B*, **34**, 187–202.

Diaconis, P. and Freedman, D. (1986). On the consistency of Bayes estimates (with discussion). *Ann. Statist.*, **14**, 1–67.

Edmunson, J. H., Fleming, T. R., Decker, G. D., Malkasam, G. D., Kvols, L. K. (1979). Different chemotherapeutic sensitivities and host factors affecting prognosis in advanced ovarian carcinoma versus minimal disease residual. *Cancer Treatment Reports*, **63**, 241–247.

Escobar, M.D. and West, M. (1995). Bayesian density estimation and inference using mixtures. *J. Amer. Statist. Assoc.*, **90**, 577–588.

Ferguson, T. S. (1983). Bayesian density estimation by mixtures of normal distributions. In *Recent Advances in Statistics* (M. Rizvi, J. Rustagi, and D. Siegmund, eds.) 287–302.

Ghosal, S. (2001). Convergence rates for density estimation with Bernstein polynomials. *Ann. Statist.*, **29** 1264–1280.

Ghosal, S., Ghosh, J. K. and Ramamoorthi, R. V. (1999). Posterior consistency of Dirichlet mixtures in density estimation. *Ann. Statist.*, **27**, 143–158.

Ghosal, S. and van der Vaart, A. W. (2001). Entropies and rates of convergence of maximum likelihood and Bayes estimation for mixtures of normal densities. *Ann. Statist.* **29**, 1233–1263.

Ghosh, J. K. and Ramamoorthi, R. V. (1994). Consistency of Bayesian inference for survival analysis with or without censoring. In *Analysis of Censored Data* (Pune, 1994/1995), 95–103, IMS Lecture Notes Monogr. Ser., **27**, Inst. Math. Statist., Hayward, CA.

Ghosh, J. K. and Ramamoorthi, R. V. (2003). *Bayesian Nonparametrics*, Springer-Verlag, New York.

Ghosh, J. K., Ramamoorthi, R. V. and Srikanth, K. R. (1999). Bayesian analysis of censored data. Special issue in memory of V. Susarla. *Statist. Probab. Lett.*, **41**, 255–265.

Gilks, W.R. (1992). Derivative free adaptive rejection sampling for Gibbs sampling. In Bayesian Statistics, **4** (J.M. Bernardo, J.O. Berger, A.P. Dawid and A.F.M. Smith, eds.). Oxford Univesity Press, Oxford, 641–649.

Hanson, T.E. and Johnson, W. O. (2004). A Bayesian semiparametric AFT model for interval-censored data. *Journal of Computational Graphics and Statistics*, **13**, 341–361.

Heckman, J. and Singer, B. (1984). A method for minimizing the impact of distributional assumptions in econometric models for duration data. *Econometrika*, **52**, 271–320.

Ishwaran, H. (1996). Uniform rates of estimation in the semiparametric Weibull mixture model. *Ann. Statist.*, **24**, 1572–1585.

Ishwaran, H. (1998). Exponential posterior consistency via generalized Plya urn schemes in finite semiparametric mixtures. *Ann. Statist.*, **26**, 2157–2178.

Ishwaran, H. and Zarepour, M. (2002). Dirichlet prior sieves in finite normal mixtures. *Statistica Sinica*, **12**, 941–963.

Jin, Z. Lin, D.Y., Wei, L.J. and Ying, Z. (2003). Rank-based inference for the accelerated failure time model. *Biometrika*, **90**, 341–353.

Johnson, W.O. and Christensen, R. (1989). Nonparametric Bayesian analysis of the accelerated failure time model. *Statistics and Probability Letters*, **8**, 179–184.

Kim, Y. and Lee, J. (2001). On posterior consistency of survival models. *Ann. Statist.*, **29**, 666–686.

Koul, H., Sursula, V. and Van Ryzin, J. (1981). Regression analysis with randomly right censored data. *Ann. Statist.*, **9**, 1276–1288.

Krall, J.N., Utho, V.A. and Harley, J. B. (1975). A set-up procedure for selecting variables associated with survival. *Biometrics*, **31**, 49–57.

Kuo, L. and Mallick, B.K. (1998). Variable selection for regression models. *Sankhyā B*, **60** 65–81.

Lai, T.L., Ying, Z. and Zheng, Z. (1995). Asymptotic normality of a class of adaptive statistics with applications to synthetic data methods for censored regression. *J. Multivariate Anal.*, **52**, 259–279.

Lindsay, B. (1995). *Mixture Models: Theory, Geometry and Applications*, NSF- CBMS Regional Conference Series in Probability and Statistics **5**, Institute of Mathematical Statistics, Hayward, CA.

Lo, A. Y. (1994). On a class of Bayesian nonparametric estimates I: Density estimates. *Ann. Statist.*, **12**, 351–357.

McCullagh P. and Nelder, J. A. (1989). *Generalized Linear Models* (2nd Edition), Chapman & Hall, London.

McLachlan, G. and Basford, K. (1988). *Mixture Models: Inference and Applications to Clustering.* Marcel-Dekker, New York.

Petrone, S. and Wasserman, L. (2002). Consistency of Bernstein polynomial posteriors. *J. Roy. Statist. Soc., Ser. B*, **64**, 79–100.

Reid, N. (1994). A conversation with Sir David Cox. *Statist. Sci.*, **9**, 439–455.

Robert, C.P. (1996). Mixtures of distributions: inference and estimation. In *Markov Chain Monte Carlo in Practice* (W. Gilks, S. Richardson and D. Spiegelhalter, eds.), Chapman and Hall, London, 441–464.

Roeder, K. and Wasserman, L. (1997). Practical Bayesian density estimation using mixtures of normals. *J. Amer. Statist. Assoc.*, **92**, 894–902.

Schwartz, L. (1965). On Bayes procedures. *Z. Wahr. Verw. Gebiete*, **4**, 10–26.

Spiegelhalter, D.J., Thomas, A. and Best, N.G. (1999). *WinBUGS Version 1.2 User Manual*, MRC Biostatistics Unit. http://www.mrc-bsu.cam.ac.uk/bugs/welcome.shtml

Walker, S.G. and Mallick, B. K. (1999). A Bayesian semiparametric accelerated failure time model. *Biometrics*, **55**, 477–483.

Wei, L. J. (1992). The accelerated failure time model: A useful alternative to the Cox regression model in survival analysis. *Statistics in Medicine*, **11**, 1871–1879.

Zheng, Z. (1984). *Regression with Randomly Censored Data.* Ph.D Thesis, Columbia University, NY.

Zhou, M. (1992). Asymptotic normality of the 'synthetic data' regression estimator for censored survival data. *Ann. Statist.*, **20**, 1002–1021.

Appendix A: Proof of the Main Result

Now we prove the main theorem by verifying the conditions of the following general result.

Theorem 2 *Let $Y_1, \ldots, Y_n$ be i.i.d. with density p lying in a class $\mathcal{P}$ of probability densities which is compact under the total variation metric. Let Π be a prior on $\mathcal{P}$ and p_0 be the true value of the*

density and let P_0 be the probability measure corresponding to p_0. If for all $\varepsilon > 0$,

$$\Pi\left\{ p \in \mathcal{P} : \int p_0 \log \frac{p_0}{p} < \varepsilon \right\} > 0$$

then the posterior distribution is consistent at p_0 under the total variation (or the L_1 distance a.s. $[P_0^\infty]$), that is, for all $\varepsilon > 0$,

$$\Pi\left\{ p \in P : \int |p - p_0| < \varepsilon | Y_1, \dots, Y_n \right\} \to 1 \qquad a.s.\ P_0^\infty$$

Proof. Observe that a compact set has finite metric entropy. Taking $\mathcal{P}$ itself as the sieve at each stage, (ii) of Theorem 2 of Ghosal et al. (1999) is satisfied. Alternatively, as $\mathcal{P}$ is compact under the variation distance, which gives a topology stronger than the weak one, the two topologies actually coincide on $\mathcal{P}$. Thus, total variation neighborhoods are also weak neighborhoods. As Schwartz's (1965) testing condition is satisfied for the weak topology (see Theorem 4.4.2 of Ghosh and Ramamoorthi, 2003), the result follows. $\square$

To verify the conditions of Theorem 2, we shall need the following three lemmas.

Lemma 1. *The model is identifiable.*

Proof. This result follows from the discussion presented in Section 1 of Heckman and Singer (1984). $\square$

Lemma 2. *Consider the product topology on (α, β, H), where α and β are given the usual Euclidean topology and H the weak topology. On the densities $f_{\alpha,\beta,H}(x, \delta, z)$, put the total variation (or the L_1) distance defined as*

$$\|f_{\alpha_1,\beta_1,H_1} - \hat{f}_{\alpha_2,\beta_2,H_2}\| = \int\int |f_{\alpha_1,\beta_1,H_1}(x,\delta,z) - f_{\alpha_2,\beta_2,H_2}(x,\delta,z)| dx dz \tag{6}$$

Then

$$\|f_{\alpha_\nu,\beta_\nu,H_\nu} - f_{\alpha,\beta,H}\| \to 0$$

if and only if $(\alpha_\nu, \beta_\nu, H_\nu) \to (\alpha, \beta, H)$. In other words, the variation topology on the densities is equivalent to the product topology on the indexing parameters.

Proof. For the "if" part, it suffices to show that the densities converge pointwise and then apply Scheffe's theorem. Fix (x, δ, z). We shall show the proof only for $\delta = 1$.

As μ has a fixed compact range, the integrand

$$\alpha x^{\alpha-1}(\mu \exp\{z^T \beta\})^{-\alpha} \exp[-(x/\mu \exp\{z^T \beta\})^\alpha]$$

in the definition of $f_{\alpha,\beta,H}$, as a family of functions of (α, β) indexed by μ, is equicontinuous. For a given $\varepsilon > 0$, find ν large enough so that the integrands are ε-close for all μ. For such a ν, in the decomposition,

$$|f_{\alpha_\nu,\beta_\nu,H_\nu} - f_{\alpha,\beta,H}| \leqslant |f_{\alpha_\nu,\beta_\nu,H_\nu} - f_{\alpha,\beta,H_\nu}| + |f_{\alpha,\beta,H_\nu} - f_{\alpha,\beta,H}| \tag{7}$$

the first term on the RHS is clearly less than ε. The second term on the RHS converges to 0 as $\nu \to \infty$ since H_ν converges weakly to H and the integrand is a bounded continuous function of μ.

To prove the "only if" part, consider any subsequence of the original sequence. By the compactness of the domains of α, β and μ, we can extract a further subsequence along which α_ν, β_ν and H_ν converge in the respective topologies to α^*, β^* and H^*, say. Thus by the "if" part, along that subsequence, $\|f_{\alpha_\nu,\beta_\nu,H_\nu} - f_{\alpha^*,\beta^*,H^*}\| \to 0$. Hence $f_{\alpha,\beta,H} = f_{\alpha^*,\beta^*,H^*}$, and so by Proposition 1, $(\alpha,\beta,H) = (\alpha^*,\beta^*,H^*)$. Therefore the original sequence must converge to (α,β,H). $\square$

Lemma 3. *For all $\varepsilon > 0$,*

$$\Pi\left\{(\alpha,\beta,H): \int\int f_{\alpha_0,\beta_0,H_0} \log \frac{f_{\alpha_0,\beta_0,H_0}}{f_{\alpha,\beta,H}} dx dz < \varepsilon \right\} > 0$$

Proof. The log likelihood ratio is given by

$$\Lambda(\alpha,\beta,H) = \begin{cases} \log \dfrac{p_{\alpha,\beta,H}(x,z,\delta)}{p_{\alpha_0,\beta_0,H_0}(x,z,\delta)}, & \text{if } \delta = 1 \\[2em] \log \dfrac{s_{\alpha,\beta,H}(x,z,\delta)}{s_{\alpha_0,\beta_0,H_0}(x,z,\delta)}, & \text{if } \delta = 0. \end{cases}$$

Thus, the Kullback-Leibler divergence breaks up into two terms corresponding to $\delta = 1$ and $\delta = 0$. We shall bound the contribution corresponding to $\delta = 1$, the other case being simpler. Note that β, α, z are all bounded. Thus the integrand defining $f_{\alpha,\beta,H}$ is bounded above and below by functions of the form $Cx^{k_1}e^{cx_2^k}$. Taking ratio and then logarithm, Λ in the tails (in x) is bounded by a multiple of a power of x. This is integrable with respect to the true density, and hence we can find a central region, outside which the contribution to the integral $\int\int f_{\alpha_0,\beta_0,H_0} \log \frac{f_{\alpha_0,\beta_0,H_0}}{f_{\alpha,\beta,H}} dx dz$ is small. In the central region, the family indexed by (x,z) is equicontinuous in the parameters. Rest of the proof follows as in that of Theorem 3 of Ghosal et al. (1999). $\square$

Proof of Theorem 1. Lemma 2 shows that it suffices to consider neighborhoods with respect to the L_1-distance. Also, the space is compact. The condition of prior positivity has been verified in Lemma 3. The proof therefore follows. $\square$

Appendix B: BUGS Code for AFT Model

```
model myeloma;
const
n=65, N=8, p=2, b0=-10, B0=10, M=1, a0=0.1, A0=10;
var
X[n], alpha, lambda[n], Cen[n], eta[N], Z[n, p], delta[p], beta[p],
latent[n], prob[N], mu[N], a[N];
data X, Cen, Z in "myeloma.dat";
inits in "myeloma.in";
{
for(i in 1:n){
X[i] ~ dweib(alpha, lambda[i])I(Cen[i], );
log(lambda[i]) <- eta[latent[i]] + inprod(Z[i, ], delta[]);
latent[i] ~ dcat(prob[]);}
for(j in 1:p){
delta[j] ~ dunif(b0, B0);
beta[j] <- -delta[j]/alpha; }
prob[1:N] ~ ddirch(a[]);
```

```
for(k in 1:N){
eta[k] ~ dunif(b0, B0);
mu[k] <- exp(-eta[k]/alpha);
a[k] <- M/N;}
alpha ~ dunif(a0, 0);
}
```

We have used the unix version of **classic BUGS** to fit our models. This code has not been tested on latest version of **WinBUGS**

Bayesian Statistics and Its Applications
Edited by S.K. Upadhyay, U. Singh and D.K. Dey
Anamaya Publishers, New Delhi, India

Sequential Monte Carlo in Bayesian Inference for Dynamic Models: An Overview

Prem K. Goel[1], Lixin Lang[1] and Bhavik Bakshi[2]

[1]Department of Statistics, The Ohio State University, 404C Cockins Hall,
1958 Neil Avenue, Columbus, OH 43210
[2]Department of Chemical and Biomolecular Engineering, The Ohio State University,
121 Koffolt Labs, 140 West 19th Avenue, Columbus, OH 43210

Abstract

Sequential Monte Carlo (SMC), also known as Sequential Importance Resampling (SIR), or Particle Filtering (PF) has been applied successfully to perform Bayesian inference in engineering problems. However, a naive choice of importance function can lead to various complications, including large Monte Carlo errors. The key problem that occurs quite often with a long sequence of prior-posterior updates as new data arrive sequentially, is that importance weights can have a large variance, which can lead to estimates with high Monte Carlo variance. This article presents an overview of SMC methodology for both static and dynamic models. A variety of important functions, resampling and particle-moving strategies are also discussed that improve the basic SMC algorithm. A couple of examples illustrate the applications of these strategies.

1. Introduction

Numerical and Monte Carlo simulation based methods have been used for a long time for developing a critical understanding of complex scientific and industrial modeling problems that are not amenable to analytic solutions. Bayesian approach to solve engineering problems, while incorporating realistic prior information based on expert knowledge and judgment, requires the use of non-conjugate prior distributions in the analysis. For most of these problems, properly normalized posterior distributions could not be expressed in analytical form. Therefore, sophisticated numerical integration techniques were developed to integrate complicated functions involving posterior probability distribution of interest (Smith et al., 1987). However, since these methods were appropriate only for moderate dimensions, Markov chain Monte Carlo (MCMC) methods to solve complex Bayesian integration problems (Gelfand and Smith, 1990) became popular in early 1990s. For the theory underlying the MCMC methods and basic exposition on applications of these results (see, e.g. Tierney, 1994, Athreya et al., 1996, Robert and Casella, 1998). MCMC via Gibbs samplers involved drawing a chain of samples from full conditional distributions of one variable, given all others. Under certain conditions, the stationary distribution of this chain would be the joint posterior distribution. It was quickly realized that Gibbs sampling represents a special type of

MCMC algorithms based on sampling using Metropolis method developed in 1950s (Metropolis et al., 1953, Hammersley and Morton, 1954, Geman and Geman, 1984). Now general MCMC algorithms are based on the Metropolis-Hastings (MH) method, which builds a Markov chain with initial state sampled according to a convenient proposal distribution, and whose stationary distribution is the desired posterior distribution, or the so called *target distribution* (Hastings, 1970). Once the chain is thought to have converged to its stationary distribution, samples are drawn from the chain to make statistical inferences. Usually, the target distribution under this context is fixed and the simulation is done off-line in a batch mode. A typical example is a static problem with a fixed parameter space (state space), with all its data available in one batch.

However, usual MCMC methods are not appropriate for problems in which decision-making requires prior to posterior updating at each time new data (measurement) becomes available. This holds for static problems in which the data arrive sequentially over time, as well as for dynamic problems, with time varying parameters, whose states evolve over time. In some problems requiring real-time decisions/actions, the duration between two adjacent time steps may be too short for the convergence of the Markov chain so that valid samples from the posterior can be drawn. The main difficulty arises because the usual MCMC requires the prior specification in an analytic form, but in the sequential Bayesian setup, the prior distribution before the second set of data arrives, i.e., the posterior distribution after the first prior to posterior update, is represented by samples. If one were to start from scratch with the initial prior every time, as if in a batch mode, previous knowledge of the target distribution is not reused at its next updating, thus resulting in an increasing waste of computing resources.

Unlike MCMC, Sequential Monte Carlo (SMC) is capable of recursively adjusting previous estimates upon arrival of additional data with minimum computational efforts. Thus SMC is quite appropriate for real-time applications. Essentially a Bayesian filter with Monte Carlo simulation, SMC simultaneously uses samples and importance weights to implement such adjustment. In recent engineering literature, Monte Carlo samples from the posterior distribution are called "particles" in order to distinguish them from the measurements representing samples from the underlying population of observables. The inference based on these particles is called 'particle filtering.' Details are presented in the next few sections.

In fact, the use of SMC can be traced back to the late 1960s in the control engineering literature, e.g., Handschin and Mayne (1969). The rebirth of these techniques can be credited to the dramatic increase in computing power. Ever since Gordon et al. (1993), who developed a practical Bayesian filtering method for non-Gaussian, non-linear models, there has been a surge of interest in improving SMC methods and applying them to a large variety of processes. Doucet et al. (2000) present a general framework unifying many methods proposed earlier. Doucet et al. (2001) is a useful reference for readers interested in a variety of applications.

In Section 2, we review the generic SMC algorithm for static models as well as the dynamic models in the state-space equation form. As is well known, variance of importance weights plays a crucial role in determining size of the simulation error over different runs; the smaller the variance of the weights, the lower the simulation error. In Section 3, the discussion on particle degeneracy and resampling covers important issues about the dynamics of changes in importance weight attached to the particles and how to decrease the variance of these weights. Resampling lowers the variance of the importance weights but slowly leads to a very small number of distinct particles to represent the distribution. This phenomenon is called 'particle depletion,' and it can lead to large Monte Carlo errors in the estimates. In Section 4, a particle's moving strategy is described to avoid particle depletion. The SMC usage is illustrated for some examples in Section 5, and conclusion/future directions are presented in Section 6.

2.   Generic Sequential Monte Carlo Method

2.1   State Space Models for Dynamic Systems

State space models are usually used to represent a dynamic system, whose unknown states at each time point can be estimated to some extent through noisy measurements of some function(s) of the states. Mathematically, a state equation (1) and a measurement equation (2) are used to describe a discrete time dynamic model for $t = 1, 2, \cdots$

$$x_t = f(x_{t-1}, \omega_t) \tag{1}$$

$$y_t = h(x_t, \nu_t) \tag{2}$$

At each time point t, x_t is the state of system (possibly vector valued), ω_t the process noise, y_t the measurement(s), and ν_t denotes the measurement noise. System states are assumed to consist of a hidden Markov chain with transition density, $p(x_t|x_{t-1})$, with a prior distribution, $p(x_1)$, on the initial state. The measurements $y_{1:t} = (y_1, y_2, \cdots, y_t)$ are assumed to be conditionally independent given $x_{1:t} = (x_1, x_2, \cdots, x_t)$. Therefore the likelihood function of the states $x_{1:t}$, given the measurements $y_{1:t}$, is proportional to the product of the conditional probability densities $p(y_i|x_i)$, of the measurements $y_i, i = 1, 2, \cdots, t$. Of course, the problem of interest is to obtain the joint posterior distribution, $p(x_{1:t}|y_{1:t})$, of $x_{1:t}$, as well as the marginal posterior distribution, $p(x_t|y_{1:t})$, of x_t at each time point t.

It is generally impossible to obtain an analytic form of the above posterior, except in few limited cases like linear state space model, with additive Gaussian noises in both the system and measurement equations. In this situation, the posterior distributions are also Gaussian, with mean given by the famous Kalman Filter. For nonlinear/non-Gaussian system, a popular approach is the Extended Kalman Filter (EKF), which involves taking a first order Taylor series expansion of the system and/or measurement equation, and applying Kalman filter to this approximate linear system (see, e.g., Anderson and Moore, 1979). However, for many practical systems, EKF does not perform, in that the estimates do not come close to the true values of the states. Therefore, intense research has focused on approximating the target distribution(s) via simulations of a recursive system. One recursive scheme for the joint posterior distributions is given below (see, e.g., Doucet et al., 2000):

$$p(x_{1:t+1}|y_{1:t+1}) = \frac{p(x_{1:t}|y_{1:t})p(x_{t+1}|x_t)p(y_{t+1}|x_{t+1})}{p(y_{t+1}|y_{1:t})} \tag{3}$$

This interpretation shows that the prior distribution at time $(t + 1)$ is proportional to the product of the posterior distribution at time t and the transition density $p(x_{t+1}|x_t)$.

2.2   Sequential Importance Sampling

Importance sampling is a useful technique to simulate complex distributions. Suppose that one needs to evaluate the expectation of a function $g(x)$, where x is distributed as $p(x)$. If it is difficult to draw samples from $p(x)$, but we can easily simulate samples $x^{(1)}, x^{(2)}, \cdots, x^{(N)}$, from another density function $\pi(x)$ (see, Geweke (1989) for appropriate conditions), then a Monte Carlo estimate of $E[g(x)]$ is given by

$$E[g(x)] \approx \frac{1}{N} \sum_{i=1}^{N} w^{(i)} g\left(x^{(i)}\right) \tag{4a}$$

where, the importance weights $w^{(i)}$ are given by

$$w^{(i)} = p(x^{(i)})/\pi(x^{(i)}), \ i = 1, 2, \cdots N \tag{4b}$$

The convenient sampling density $\pi(x)$ is called the importance function. In order to implement recursive estimation, not all importance functions can be used. Noting the form of (4b), it is easy to see that an importance function would be suitable, if the following factorization holds:

$$\pi(x_{1:t}|y_{1:t}) = \pi(x_{1:t-1}|y_{1:t-1})\pi(x_t|x_{1:t-1}, y_{1:t})$$

Correspondingly, the unnormalized importance weights can be computed as

$$\tilde{w}_t^{(i)} = \frac{p(x_{1:t}^{(i)}|y_{1:t})}{\pi(x_{1:t}^{(i)}|y_{1:t})} = w_{t-1}^{(i)} \frac{p(y_t|x_t^{(i)})p(x_t^{(i)}|x_{t-1}^{(i)})}{\pi(x_t^{(i)}|x_{1:t-1}^{(i)}, y_{1:t})} \tag{5}$$

In equation (5), a constant of proportionality has been ignored. Therefore, the final normalized importance weights are given by

$$w_t^{(i)} = \frac{\tilde{w}_t^{(i)}}{\sum_{j=1}^{N} \tilde{w}_t^{(j)}}, \ i = 1, ..., N$$

The general Sequential Importance Sampling (SIS) algorithm is as follows:

> Draw samples from initial prior, $x_1^{(i)} \sim p(x_1)$, $i = 1, ..., N$
> Set $w_1^{(i)} = 1/N$, for each $i = 1, ..., N$
> For time $t = 2, 3, ...$
>   For each $i = 1, ..., N$
>   Draw sample $x_t^{(i)} \sim \pi(x_t|x_{1:t-1}^{(i)}, y_{1:t})$
>   Calculate weight $\tilde{w}_t^{(i)}$
>   End
>   Normalize weights, $w_t^{(i)}$, $i = 1, ..., N$
> End

2.3 Importance Function

Among all choices of importance functions, the density $\pi(x_t|x_{1:t-1}^{(i)}, y_{1:t}) = p(x_t|x_{t-1})$ is the most commonly used due to its simplicity and generality. Also known as a weighted bootstrap filter (WBF), its practical usage can be traced back to Gordon et al. (1993). A graphical illustration of this method is shown in Fig. 1.

Even though this importance function is easy to implement, it is usually not capable of generating efficient particles, since the importance function does not depend on observations. When encountering outliers in measurements, it would give imprecise empirical prediction density (Pitt and Shephard, 1999). Consequently, several importance functions were proposed to improve the performance. The optimal importance function $\pi(x_{1:t}|y_{1:t}) = p(x_t|x_{t-1}, y_t)$ is claimed to have the least variance among importance weights. However, since its usage is limited to Gaussian state space model with linear measurement equation, local-linearized importance function (Doucet et al., 2000) was developed to deal with problems involving non-linear measurement. For much broader non-linear, non-Gaussian models, Unscented Particle Filter based importance function is shown to outperform other types of importance densities (Merwe et al., 2000) for some simple examples.

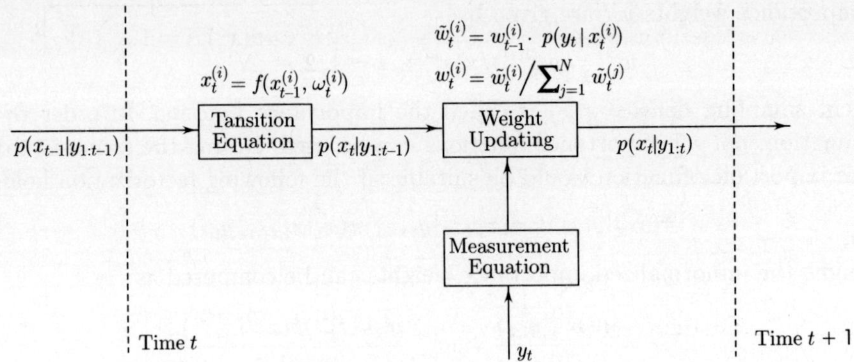

Fig. 1. SIS Algorithm with a WBF importance function.

2.4 Static Model

Static model can be viewed as a special case of dynamic model, i.e., the states (parameters) are assumed to be constant over time, and given the parameters, the measurements are believed to be conditionally i.i.d. Usually, the static estimation problem is written as:

$$y_k|\theta \ \sim \ p(.|\theta) \tag{6}$$

$$\theta \sim p(\theta) \tag{7}$$

The key problem of interest is the approximation of $p(\theta|y_{1:k})$. In practice, the observations could be available all at once. SMC can still be applied here, assuming that data arrive sequentially, especially if the data is massive, see Ridgeway and Madigan (2003), and Balakrishnan and Madigan (2004). The special form of model (6) allows a natural choice of the prior, $p(\theta)$, as the importance function. Therefore, only the importance weights are updated. However, such algorithm could lead to particle depletion, in that the weights are associated with a small number of distinct particles, as described in Section 4. One solution to get around this problem is to use a MCMC step in a particle moving strategy (see Section 4).

3. Particle Degeneracy and Resampling

As shown in Doucet et al. (2000), repeated updating of importance weights will inevitably lead to a situation of particle degeneracy, where all but few weights are nearly zero. Thus a great deal of computational effort is wasted on updating particles that make very little contribution toward the estimation. To overcome this problem, a resampling procedure with probability proportional to the weights is performed after the normalization of importance weights. However unnecessary resampling, which tends to bring multiple identical particles, would hurt the performance especially for the static model. To minimize the effect of unnecessary resampling, Liu (1996) proposed the Effective Sample Size (ESS), essentially, inversely proportional to the coefficient of variation of the weights, as a measure of degeneracy among the particles. Whenever ESS gets below a specified threshold value, resampling step is implemented. A generic SIS/Resampling algorithm is given as follows.

For variations of the following algorithm and other resampling schemes, interested readers are referred to Liu and Chen (2000), and Chapter 4 in Doucet et al. (2001).

Draw random sample from initial prior, $x_1^{(i)} \sim p(x_1), i = 1, \ldots, N$
Set $w_1^{(i)} = 1/N$, for each $i = 1, \ldots, N$
For time $t = 2, 3, \ldots$
    For each $i = 1, \ldots, N$
        Draw random sample $\tilde{x}_t^{(i)} \sim \pi(\tilde{x}_t | x_{1:t-1}^{(i)}, y_{1:t})$
        Calculate weight $\tilde{w}_t^{(i)}$
    End
    Normalize weights, $w_t^{(i)}, i = 1, \ldots, N$
    Evaluate effective sample size $\widehat{N}_{\text{ess}}$
    if $\widehat{N}_{\text{ess}}$ less than a threshold value,
        Draw random sample $x_t^{(i)}, i = 1, 2, \ldots, N$ from
            $P[x_t^{(i)} = \tilde{x}_t^{(j)}] = w_t^{(j)}, j = 1, 2, \ldots, N$
        Reset $w_t^{(i)} = 1/N$ for each i
    else
        Set $x_t^{(i)} = \tilde{x}_t^{(i)}$ for each i
    end
End

Fig. 2 shows a multinomial resampling scheme, with 10 particles, indexed from 1 to 10, and the respective importance weights represented by the step size at each index. The points along y-axis are 10 uniform $(0, 1)$ random draws, which are used to pick the particles as indicated by the dotted lines. The resampling outcomes are shown with vertical bars along x-axis, the minimal positive

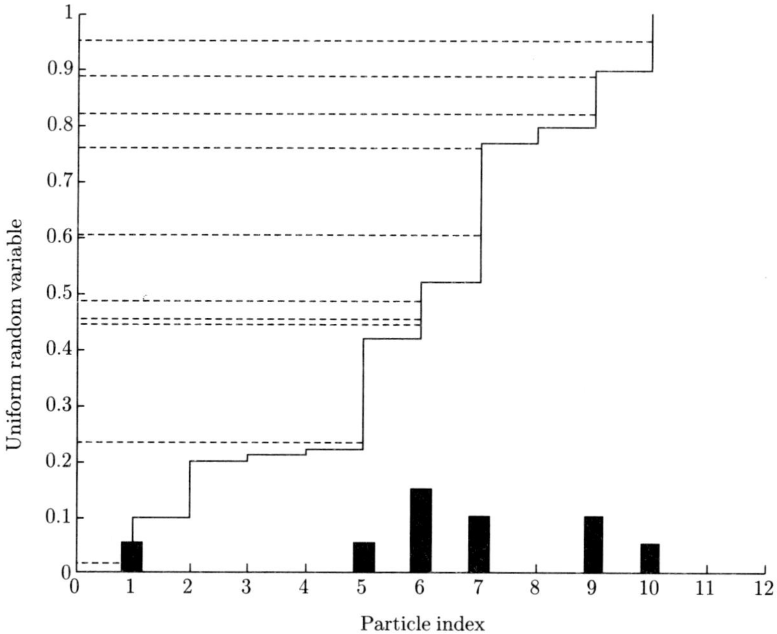

Fig. 2. Multinomial resampling illustrated.

height being 1. Thus, particles 2, 3, 4, 8 are not selected, particles 1, 5 and 10 are drawn once, particles 7 and 9 are drawn twice, and particle 6 is selected three times.

4.   Particle Depletion and Moving Strategies

Repeated resampling would decrease the number of distinct particles, since particles with larger weights tend to be selected more often. This problem can become much more severe for static problems, as particles in dynamic models are able to evolve into distinct ones through system equation. As shown by our simulation for a static model, the particles could collapse into very few distinct ones after a few iterations. An example is included where Fig. 3(a) indicates the situation when degeneracy happens due to an outlier in the measurement. All but two importance weights are nearly zero. Fig. 3(b) presents the situation when depletion happens after repeated resampling.

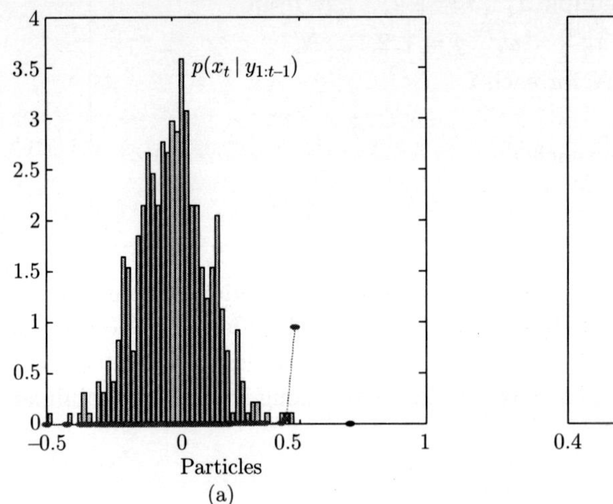

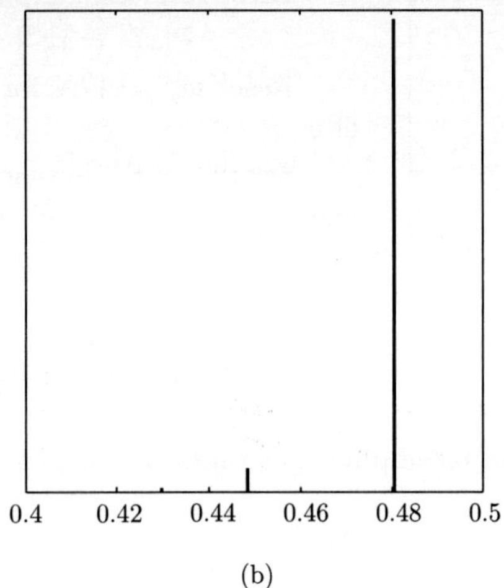

Fig. 3.   **Particle degeneracy and depletion.**

To reduce particle depletion (or impoverishment) after the resampling step, particle moving via an MCMC step based on an appropriately chosen transition kernel is intuitively appealing, in that it brings variations to identical particles without changing the underlying distributions. This step is also known as particle rejuvenation. Birzuini and Gilks (2001) discussed this novel technique inside their claimed unified framework of resample-move. In brief, each particle after resampling is subject to be moved to a new position by a transition kernel in the MCMC step. Thus, duplicated particles have a chance to become distinct.

Inspired by the above method, Chopin (2002) prefers a normal distribution as the transition kernel. He argued that an independent normal kernel with the empirical sample mean and variance is a reasonable choice since posterior tends to be Gaussian asymptotically. Compared with random walk transitions around each particle, independent kernel behaves more like a black box without worrying about the selection of walking distance. However, for dynamic models, with usually one observation at each time point, the posteriors may not be close to a Gaussian, and one needs to carefully study the impact of the Gaussian transition kernels.

Ridgeway and Madigan (2003) also integrated MCMC moving into the importance resampling for static model with massive data sets in a similar fashion, in addition to using ESS to further control the resampling times.

Compared to the independent normal kernel of Chopin (2002), a newly proposed transition kernel is also normal, but it moves a particle around itself instead of around the sample mean. To move a particle, say $p_t^{(j)}$, one candidate particle is sampled from the Gaussian transition kernel, i.e., $N(p_t^{(j)}, \sigma^2)$, where σ^2 is the sample variance of all particles. Based on several simulations, the proposed method runs very similar to that of Chopin, but it needs to be investigated further for possible advantages.

5.   Examples

5.1   Multivariate Probit Model

A GLM example in Chopin (2002) described below was implemented to illustrate the use of SMC in static models. In order to demonstrate the performance of SMC, results obtained from batch MCMC were also included as a reference. Furthermore, the SMC with moving was also compared with SMC without moving. For each method, 10 simulation runs were performed to get an averaged estimation which are listed in Table 1 (2000 particles were used). In addition to the point estimates, Fig. 4 also shows the marginal posterior empirical *cdf* of β_1 for batch MCMC and SMC-moving. Table 1 and Fig. 4 clearly indicate that SMC with moving has a similar performance with MCMC method, while the SMC without moving does not come close to the target distribution as indicated by the MCMC. This example indicates that for static problems, SMC should be implemented carefully in order to avoid incorrect estimates.

Static model to illustrate the use of SMC

$Y_i \sim Bernoulli\,(\rho_i), \quad i = 1, 2, \dots, 200.$

Probit link function: $\rho_i = \Phi(X_i\beta)$, where Φ is the standard normal *cdf*

$\beta = [\beta_1 \ \beta_2 \ \beta_3 \ \beta_4 \ \beta_5]^T,$

Prior distribution: $N([0 \ \ 0 \ \ 0 \ \ 0 \ \ 0]^T, 25I_{5\times5})$

True value of $\beta = [-1 \ \ 0.7 \ \ -0.5 \ \ -0.1 \ \ 0.3]$

Table 1.   Estimation results for the multivariate probit model, Example 5.1

True value	MCMC		SMC with moving		SMC without moving	
	mean	std	mean	std	mean	std
$\beta_1 \ = \ -1$	-1.0973	0.0089	-1.0906	0.0023	-3.0951	1.7763
$\beta_2 \ = \ 0.7$	0.8947	0.0086	0.8893	0.0040	2.9414	2.2571
$\beta_3 \ = \ -0.5$	-0.8704	0.0082	-0.8642	0.0026	0.1765	1.2036
$\beta_4 \ = \ -0.1$	-0.1039	0.0049	-0.1057	0.0026	0.7143	1.1674
$\beta_5 \ = \ 0.3$	-0.2798	0.0066	-0.2790	0.0025	-0.0572	1.2066

5.2   Nonlinear Dynamic Model

A non-linear dynamic model in Gordon et al. (1993) is used to illustrate WBF. This model is also discussed in Kotecha and Djurić (2003).

Non-linear dynamic model to illustrate WBF

System equation: $x_t = 0.5 * x_{t-1} + 25 * x_{t-1}/[1 + (x_{t-1})^2] + 8\cos(1.2 * (t-1)) + \omega_t,$

Measurement equation: $y_t = x_t^2/20 + \nu_t,$ where

$\omega_t \sim N(0,10), \nu_t \sim N(0,1), \ t = 1, 2, \cdots, 50,$

Prior $N(0,2),$ and

True initial value: 0.1

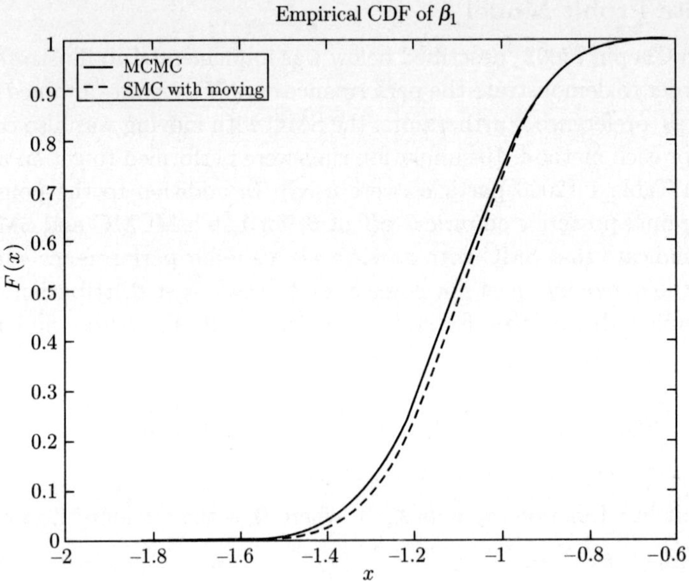

Fig. 4.   **Empirical CDF of β_1 obtained from MCMC and SMC with moving.**

For this model, we apply the EKF and the SMC using 500 particles. Fig. 5 provides EKF and the SMC estimates. Note that the measurement is a noisy version of a quadratic function of the state. Therefore, for some observations the posterior distribution is bimodal. In this Example, EKF estimates have an incorrect sign at several time points, even though the true state is quite far away from zero. In general, it is well known that for many non-linear systems, SMC will beat the EKF in performance.

6.   Conclusions and Future Directions

In this overview, we illustrate the application of SMC for Bayesian inference in both static and dynamic models. As demonstrated, fairly accurate estimation through the simulated posterior at any time point and reasonably fast computational implementation via SMC, make this a highly promising approach for marginal posterior inference for complex stochastic systems. Of course, due to the curse of dimensionality, the density estimation in higher dimensions would not be reliable, even if one uses a large number of particles. As a result, inference made on the estimated joint distribution at a fixed time point is not justifiable.

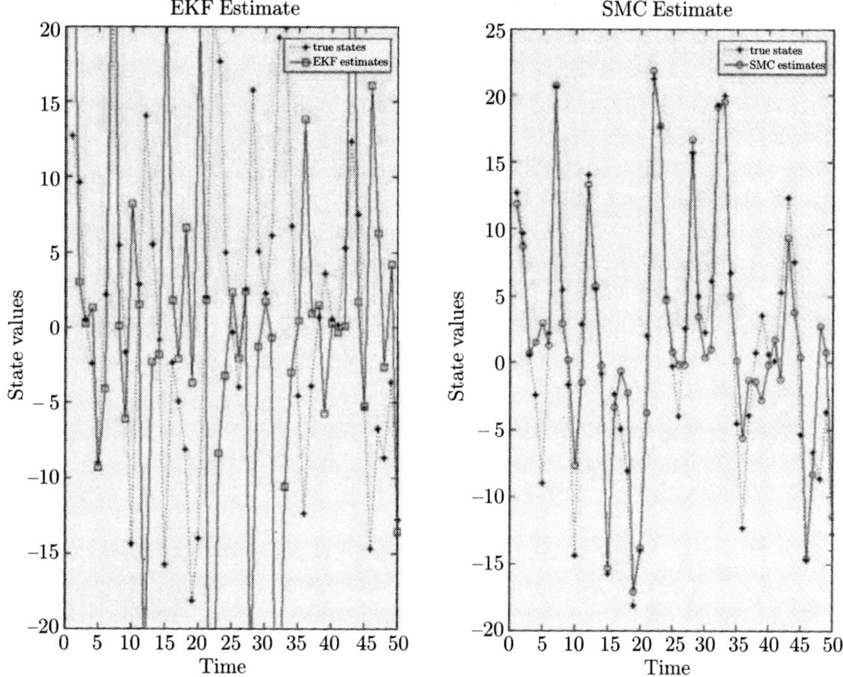

Fig. 5. EKF and SMC estimates and the true state values for the dynamic model in Example 5.2.

In addition, due to the strong correlation of particles sampled between adjacent times, SMC approach is not suitable for joint inference over time, even for just one-dimensional state variables. A very important factor that has been largely ignored in the research is the impact of prior information incompatible with the likelihood. Being a simulation based method, SMC for dynamic systems may suffer in the presence of incompatible prior information, in that particles drawn from the prior are not sufficient to represent posterior particles adequately and system equation fails to propagate the sampled particles to appropriate locations. It would be of great practical importance to make SMC robust under incompatible priors or a bad initial guess for the state.

Acknowledgements

We would like to thank the referees for useful comments that have led to improvements in the article. This work was supported by the National Science Foundation, Grant # CTS-0321911.

References

Anderson, B.D. and Moore, J.B. (1979). *Optimal Filtering.* Prentice Hall, New Jersey.

Athreya, K.B., Doss, H. and Sethuraman, J. (1996). On the Convergence of the Markov Chain Simulation Method. *The Annals of Statistics*, **24**, 69-100.

Balakrishnan, S. and Madigan, D. (2004). One-Pass Sequential Monte Carlo Method for Bayesian Analysis of Massive Datasets. www.stat.rutgers.edu/~madigan/PAPERS/pfpaper-03-08-04.pdf

Berzuini, C. and Gilks, W. R. (2001). RESAMPLE-MOVE filtering with cross-model jumps. In Doucet et al, *Sequential Monte Carlo Methods in Practice*. New York: Springer, 117-138.

Chopin, N. (2002). A sequential particle filter method for static models. *Biometrika*, **89**(3), 539-552.

Doucet, A., de Freitas, J.F.G ., and Gordon, N.J. (eds) (2001). *Sequential Monte Carlo Methods in Practice*, New York: Springer.

Doucet, A., Godsill, S.J., and Andrieu, C. (2000). On sequential Monte Carlo sampling methods for Bayesian filtering. *Statistics and Computing*, **10**, #3, 197-209.

Gelfand, A.E. and Smith, A.F.M. (1990). Sampling based approaches to calculating marginal densities. *Journal of the American Statistical Association*, **85**, 398-408.

Geman, S. and Geman, D. (1984). Stochastic relaxation, Gibbs distributions and the Bayesian restoration of images, *IEEE Transaction on Pattern Analysis and Machine Intelligence*, **6**, 721-741.

Geweka, J. (1989). Bayesian inference in econometric models using Monte Carlo integration. *Econometrica*, **57**, 1317-1339.

Gordon N.J., Salmond D. and Smith A.F.M. (1993). Novel Approach to Nonlinear and Non-Gaussian Bayesian State Estimation, *Radar and Signal Processing, IEE Proceedings-F*, **140**, 107-113.

Hammersley, J.M. and Morton, K.W. (1954). Poor man's Monte Carlo. *Journal of the Royal Statistical Society, Series B*, **16**, 23-28.

Handschin, J.E. and Mayne D.Q. (1969). Monte Carlo techniques to estimate the conditional expectation in multi-stage non-linear filtering. *International Journal of Control*, **9**, 547-559.

Hastings, W. K. (1970). Monte Carlo sampling methods using Markov chains and their applications. *Biometrika*, **57**, 97-109.

Kotecha, J.H. and Djurić, P.M. (2003). Gaussian Particle Filtering, *IEEE Transactions on Signal Processing*, **51,** #10, 2592-2601.

Liu, J.S. (1996). Metropolized independent sampling with comparisons to rejection sampling and importance sampling, *Statistics and Computing*, **6**(2), 113-119.

Liu, J.S. and Chen, R. (1998) Sequential Monte Carlo methods for dynamic systems. *Journal of the American Statistical Association*, **93**, 1032-1044.

Merwe, R. van der, Doucet, A., Freitas, N. de, and Wan E. (2000). The Unscented Particle Filter. *Technical report CUED/F-INFENG/TR-380*, Dept. of Eng., Cambridge University

Metropolis, N., Rosenbluth, A. W., Rosenbluth, M. N., Teller, A. H., and Teller, E. (1953). Equation of state calculations by fast computing machines, *Journal of Chemical Physics*, **21**, 1087-1092.

Pitt M. and Shephard N. (1999). Filtering via Simulation: Auxiliary Particle Filters. *Journal of the American Statistical Association*, **94**, 590-599.

Ridgeway, G. and Madigan, D. (2003). A Sequential Monte Carlo Method for Bayesian Analysis of Massive Datasets, *Data Mining and Knowledge Discovery*, **7**, 301-319.

Robert, C.P. and Casella, G. (1998). *Monte Carlo Statistical Methods*, New York: Springer.

Smith, A.F.M., Skene, A.M., Shaw, J.E.H., and Naylor, J.C. (1987). Progress with numerical and graphical methods for practical Bayesian statistics, *Statistician*. **36**, 75-82.

Tierney, L. (1994). Markov chains for exploring posterior distributions with discussion. *The Annals of Statistics*, **22**, 1701-1762.

Bayesian Statistics and Its Applications
Edited by S.K. Upadhyay, U. Singh and D.K. Dey
Anamaya Publishers, New Delhi, India

Estimation of Threshold Time Series Models Using Efficient Jump MCMC

Elena Goldman[1] and Terence D. Agbeyegbe[2]

[1]Department of Finance and Economics, Lubin School of Business, Pace University, New York, NY 10038
[2]Department of Economics, Hunter College, CUNY, 695 Park Avenue, New York, NY 10021

Abstract

This article discusses how a Metropolis-Hastings algorithm with efficient jump can be constructed for the estimation of multiple threshold time series of the U.S. short term interest rates. The results show that interest rates are persistent in a lower regime and exhibit weak mean reversion in the upper regime. For model selection and specification several techniques are used such as marginal likelihood and information criteria, as well as estimation with and without truncation restrictions imposed on thresholds.

1. Introduction

Many time series exhibit non-linear dynamics that can be characterized by regime changes in persistence and volatility. Recently many papers used Threshold Autoregressive (TAR) models in economic and financial time series to model the dynamics of short-term interest rates, real exchange rates, unemployment rate, stock prices, production, and inventories.

In spite of popularity of TAR models the problem of efficient estimation of thresholds is still unsolved. The likelihood function is discontinuous in the thresholds and has multiple peaks in case of more than two regimes. There is enormous literature using classical methods where a threshold is estimated via grid search or residual scatter plots (see, e.g., Tong and Lim (1980), Tong (1983, 1990), Tsay (1989, 1998), Hansen (1997), Lanne and Saikkonen (2002), etc.). We face a problem in estimating simultaneously multiple thresholds by these methods since accurate estimation with many dimensions of grid search takes a long time. To solve this problem, Gonzalo and Pitarakis (2002) used sequential estimation on a grid of points and found that their method works well only when regimes samples are equally spaced. The Bayesian analogue of a grid search is a Griddy Gibbs sampler within MCMC and it was done in Phann, Schotman and Tschering (1996). Other Bayesian methods were suggested for a two-regime simple autoregressive specification with one threshold by Geweke and Terui (1993) using Monte Carlo integration for the case of noninformative priors. This method requires numerical multiple integration. Chen and Lee (1995) used a Metropolis algorithm within a Gibbs sampler and noted that their procedure avoids sophisticated integration or use of plots, which may result in imprecise values of estimated threshold parameter. However, they admit that the MCMC algorithm that they suggested is restrictive to a simple two-regime TAR model and much remains to be done for a general TAR model with multiple regimes. Koop and Potter (1999) extended Monte Carlo integration method suggested in Geweke and Terui (1993) for a three regime autoregressive model with informative priors. Forbes, Kalb and Kofman (1999) extended numerical integration for multiple thresholds using Rao-Blackwellized estimates of marginal posterior densities. In this paper we propose a Metropolis-Hastings algorithm with efficient jumps and estimate multiple thresholds jointly in a general class of ARMA-GARCH models. Our model can be extended to other models with many parameters. This algorithm is more efficient than existing methods; it avoids numerical integration or grid search and achieves convergence in reasonable computational time.

In this paper we analyze behavior of the U.S. short-term interest rates using a threshold model. The short-term interest rate is a key variable in many finance models, including term structure models and models that use risk-free rate as an input. Switches in regimes for short-term interest rates were studied among others by Hamilton (1988), Cai (1994), Gray (1996), etc. using Markov switching model; Lanne and Saikkonen (2002) and Phann, Schotman and Tschering (1996) used threshold models. Hamilton (1988) analyzed the term structure of interest rates in response to FED's change in monetary policy in October 1979, which resulted in a period of a high level and high volatility of interest rates. He suggested to use a two-regime Markov-switching model instead of a constant coefficient autoregressive model for short rates since the former provided a better description of the univariate process for the short rate and was more consistent with historical correlation between long and short rates. Phann, Schotman and Tschering (1996) (PST) modeled the dynamics of short-term interest rates as an autoregressive threshold model with autoregressive parameters and volatility depending on the level of interest rates. They noted that the traditional linear models of interest rate dynamics, emphasized in Chan, Karolyi, Longstaff and Sanders (CKLS) (1992), are not able to explain the empirical regularity that long term rates are not as volatile as short-term rates. The general stochastic differential equation used by CKLS to model the adjustment process of short-term interest rates can be represented by

$$dr(t) = (\mu + \beta r(t))dt + \sigma r(t)^{\lambda}dW \tag{1}$$

where $r(t)$ is the interest rate level at time t, W is a standard Brownian motion and μ, β, σ and λ are parameters[1]. CKLS shows that with appropriate restrictions on the parameters, many popular interest rate models can be obtained. For example, setting $\lambda = 0.5$ yields the Cox, Ingersoll and Ross (1980) model, while setting $\lambda = 1.0$ yields the Brennan and Schwartz (1980) model. The model of Cox, Ingersoll and Ross (1980) is obtained when λ is set equal to 1.5. The model also requires that $\mu = \beta = 0$.

In this model $\mu + \beta r(t)$ is the drift and $\sigma^2 r(t)^{2\lambda}$ is the variance of unexpected interest rate changes. The β is interpreted as the mean reversion parameter—it measures the speed of mean reversion in rate levels; σ^2 as the volatility parameter and λ as the elasticity of volatility with respect to the level of interest rates. At higher λs, the volatility is more sensitive to interest rates. Since the CKLS model is inadequate in explaining the empirical regularity that long term rates are not as volatile as short-term rates PST considered a non-linear framework and obtained results that are consistent with this empirical fact.

This article discusses how a Metropolis-Hastings algorithm with efficient jump can be applied to the estimation of multiple threshold time series models. The algorithm that we suggest is more efficient than existing methods for threshold models. The model is then estimated for the U.S. 3 month Treasury-bill rate. Our first result is that the interest rates are persistent in a lower regime

---

[1] The model proposed by CKLS has been the subject of several investigations and extensions. See for example the papers by (among others) Broze, Scaillet and Zakoian (1995), Bliss and Smith (1998), Brenner, Harjes and Kroner (1996), Gray (1996), Nowman (1997), Ball and Torous (1996, 1999), Dahlquist and Gray (2000) and Yu and Phillips (2001).

and exhibit weak mean reversion in the upper regime. Second, volatility is more persistent in the lower regime than in the upper regime, while the level of volatility is higher when interest rates are high. Third, allowing for both GARCH effect and level effect (dependence of volatility on the level of interest rate) in TAR model results in smaller elasticity λ estimates, significant GARCH effects, and smaller standard errors of threshold parameter estimates. These results have important implications for measuring risk-free rate and for the term structure of interest rates. The empirical analysis to be presented corroborates earlier results obtained in Gray (1996) using Markov-switching model and Phann, Schotman and Tschering (1996) using TAR model. However, our specification is more general, since we allow multiple regimes and thresholds, ARMA, GARCH, and all parameters are flexible to change with regime. The model we suggest generalizes other models available in the literature so that CKLS, GARCH, ARMA, and multiple threshold models for interest rates are nested in our model as special cases.

For model selection we used several techniques such as marginal likelihood and Bayesian information criteria, as well as estimation with and without truncation restrictions imposed on thresholds. For testing non-linearity we suggest to use posterior distribution of difference in parameters in different regimes.

Section 2 presents our non-linear framework for multiple regime modelling and Metropolis-Hastings algorithm. Section 3 gives empirical estimates of the model applied to the U.S. Treasury-bill rate and Section 4 concludes.

2. The Threshold Model

Our generalized threshold model nests the threshold autoregressive model (TAR or SETAR), ARMA, GARCH, and CKLS models. Volatility of interest rates is assumed to have both level and GARCH effects, following earlier interest rate models suggested by Brenner et al. (1996) and Koedijk et al. (1997), who found that if GARCH is omitted it results in a biased estimate of level effect. In what follows, we will let y_t represent the level of interest rate at time t and define new parameters, for different regimes, extending equation (1) to the following threshold model:

$$\Delta y_t = \gamma_1^j + \gamma_2^j y_{t-1} + |y_{t-1}|^{\lambda^j} u_t \tag{2}$$

$$u_t = \frac{\Theta^j(B)}{\Phi^j(B)}\,\epsilon_t \tag{3}$$

$$\epsilon_t \sim N(0, \sigma_t^2) \tag{4}$$

$$\sigma_t^2 = \alpha_0^j + \sum_{i=1}^{a} \alpha_i^j \epsilon_{t-i}^2 + \sum_{i=1}^{s} \beta_i^j \sigma_{t-i}^2; \quad \alpha_0^j > 0,\ \alpha_i^j \geq 0,\ (j = 1, \cdots, a),\ \beta_i^j \geq 0,\ (i = 1, \cdots, s) \tag{5}$$

where y_t belongs to regime j given by

$$\begin{cases} j = 1 & y_{t-d} < r_1 \\ j = 2 & r_1 \leq y_{t-d} < r_2 \\ \dots & \dots \\ j = k+1 & y_{t-d} \geq r_k \end{cases} \tag{6}$$

We assume in the model above that there are k thresholds $r_1, r_2, ..., r_k$ and $k+1$ regimes for y_t.

$\Theta(B)$ and $\Phi(B)$ are q-th and p-th order polynomials in the backward shift operator B, respectively:

$$\Theta(B) = 1 + \theta_1 B + \cdots + \theta_q B^q \quad \text{and} \quad \Phi(B) = 1 - \phi_1 B - \cdots - \phi_p B^p$$

Each parameter in this model $\{\gamma^j, \phi^j, \theta^j, \alpha^j, \beta^j, \lambda^j\}$ takes $k+1$ values depending on the regime j where y_{t-d} belongs. The delay parameter d will be selected together with the orders (p, q, a, s) of $\text{ARMA}(p,q)\text{-GARCH}(a,s)$ process in each regime to fit the best model according to a Bayesian information criterion based on estimated marginal likelihood.

Let the prior pdf of the parameters γ', ϕ', θ', α' and β' be given by

$$\pi(\gamma, \phi, \theta, \alpha, \beta) = \prod_{i=1}^{k+1} N(\gamma_{0i}, \Sigma_{\gamma_i}) \times N(\phi_{0i}, \Sigma_{\phi_i}) \times N(\theta_{0i}, \Sigma_{\theta_i}) \times N(\alpha_{0i}, \Sigma_{\alpha_i})$$

$$\times N(\beta_{0i}, \Sigma_{\beta_i}) \times I\left(0 \leq \lambda^{(i)} \leq \lambda^{up}\right) \times I\left(r_i \in [r_{\text{low}}^{(i)}, r_{\text{up}}^{(i)}]\right)$$

We assume a uniform prior for parameters λ and r with necessary constraints imposed on these parameters: (i) $\lambda \geq 0$ and less than certain upper bound[2]; (ii) each r_i is constrained so that minimum $\delta\%$ of observations are within each regime. We need sufficient sample size in each regime in order to estimate all parameters of the model. We use $\delta = 5\%$ of total number of observations as a minimum sample size in each regime.[3]

Given threshold values $r_1, ..., r_k$ the samples $\{y_t\}$ and $\{x_t\}$ are separated into regimes given by (6). We will denote $y_t^j = y_t$ and $x_t^j = x_t$ if $\{y_t, x_t\}$ belong to regime j.

Let us rescale the variables y_t^j and x_t^j by the level heteroscedasticity component $|y_{t-1}|^{\lambda^j}$, $(j = 1, ..., k+1)$:

$$\widetilde{y}_t^j = y_t^j / |y_{t-1}|^{\lambda^j} \tag{7}$$

$$\widetilde{x}_t^j = x_t^j / |y_{t-1}|^{\lambda^j}$$

The posterior distribution is given by

$$p(\gamma, \phi, \theta, \alpha, \beta | \text{data}) = \pi(\gamma, \phi, \theta, \alpha, \beta) \prod_{j=1}^{k+1} \prod_{t \in T_j} \frac{1}{\sigma_t |y_{t-1}|^{\lambda^j}} \phi\left(\frac{\widetilde{y}_t^j - g(Z_t)}{\sigma_t}\right) \tag{8}$$

where for every $t \in T_j = \{t : r_{j-1} \leq y_{t-d} \leq r_j\}$

$$e_t = \widetilde{y}_t^j - \widetilde{x}_t^j \gamma^j$$

$$\epsilon_t = \widetilde{y}_t^j - g(Z_t)$$

$$g(Z_t) = \widetilde{x}_t \gamma^j - \sum_{i=1}^{p} \phi_i^j e_{t-i} - \sum_{i=1}^{q} \theta_i^j \epsilon_{t-i}$$

$$\sigma_t^2 = \alpha_0^j + \sum_{i=1}^{r} \alpha_i^j \epsilon_{t-i}^2 + \sum_{i=1}^{s} \beta_i^j \sigma_{t-i}^2$$

---

[2] This in practice is never greater than 1.5. Some models make stationarity of variance restriction $\lambda < 1$. However, most studies that used CKLS model (1) without GARCH effect found that λ is between 1 and 1.5 for US Treasury-bills.

[3] δ depends on number of observations, if the sample size is small, higher δ is recommended. For example, Koop and Potter (1999) used 15% as a minimum sample size in each regime.

Phann, Schotman and Tschering (1996, PST) considered a two regime threshold autoregressive model with CKLS heteroscedasticity for the U.S. 3 month Treasury-bill rates. Their model is a special case of our model if parameters of GARCH and ARMA are set to zero and level effect λ is not allowed to change with regime. We also allow possibility of more than two regimes in our model. In Section 3 we show that a GARCH component is important since as we found the interest rate data shows strong conditional heteroscedasticity. Moreover, the persistence of volatility changes with regime exhibiting higher persistence in a lower regime. We allow more flexibility for the model error (ARMA(p,q)-GARCH(a,s) rather than white noise) where the orders p, q, a, s in each regime[4] are selected using a Bayesian information criterion. The delay parameter d and the threshold variable are also part of the model selection. PST use the threshold variable y_{t-1} with $d = 1$. Tsay (1998) suggests to use a stationary z_{t-d} variable as a threshold. It could be modelled, e.g., as Δy_{t-d} as is suggested in Hansen (1997). Another choice of a threshold variable could be a moving average $\Sigma_{i=0}^{d}|y_{t-i}|/(d+1)$ used in Tsay (1998) and in Franses and van Dijk (2000). We have chosen the threshold variable as y_{t-d}, which makes intuitive sense for interest rate models, as regimes are separated by high and low interest rates. In the next section we explain an efficient Metropolis-Hastings algorithm for the threshold model that works much faster compared with estimation on a grid of points, as is suggested in PST and other literature.

2.1 Estimation of the Threshold Model

The way we estimate threshold parameters $r = (r_1, ..., r_k)$ is different from the existing methods in the literature. We employ an efficient Metropolis jumping rule which solves the problem of efficient simultaneous estimation of multiple thresholds.

We describe the algorithm for the case of two and three regimes. For higher number of regimes the estimation procedure is similar. We will estimate parameters in blocks: (i) regression parameters γ; (ii) AR coefficients ϕ; (iii) MA coefficients θ; (iv) ARCH coefficients α; (v) GARCH coefficients β; (vi) threshold parameters r and (vii) volatility parameters λ.

We describe the procedure as follows:

(i) Choose initial values for γ, ϕ, θ, α, β, r, and λ. Start from crude estimates of mean or mode of the posterior distribution. Generally it is hard to find a good approximation for mean or mode of threshold distribution because of unconventional shape of the likelihood function. For starting values we simply divide the sample into regimes with equal samples. In case we have two regimes we use the mean value of y_t as a starting point $r^{(0)}$, if there are three regimes we divide sample into three equal subsamples to find starting values $r_1^{(0)}$ and $r_2^{(0)}$.

   Let the started points be denoted by $\gamma^{(0)}$, $\phi^{(0)}$, $\theta^{(0)}$, $\alpha^{(0)}$, $\beta^{(0)}$, $r^{(0)}$, and $\lambda^{(0)}$. Given initial values $r_1^{(0)}$, $r_2^{(0)}$, and d the samples $\{y_t\}$ and $\{x_t\}$ are separated into regimes based on y_{t-d}:

$$\begin{cases} j = 1 & y_{t-d} < r_1 \\ j = 2 & r_1 \leq y_{t-d} < r_2 \\ j = 3 & y_{t-d} \geq r_2 \end{cases}$$

(ii) Once the data are separated into regimes given thresholds r and level λ the model is transformed into arranged ARMA-GARCH model,[5] estimation of which using MCMC is standard (see, e.g.,

---

[4] Orders can be different in different regimes, so we have choice of p, q, a, s in each regime separately.

[5] We construct arranged ARMA-GARCH model in a similar way as Tsay (1989) and others constructed arranged autoregressive model sorting data y_t by regimes.

Nakatsuma (2000) who used independent chains or Goldman and Tsurumi (2005) who used random walk algorithm). We draw parameters of ARMA-GARCH in each regime block by block using random walk Markov Chain, e.g., parameters of a regression block are drawn for all regimes separately (using arranged by regimes data) and then are accepted or rejected jointly in one block. The acceptance rates are controlled by multiplying the variance of the proposal density with a scaling constant.

(iii) **Threshold parameters:** We found that standard random walk Metropolis algorithm with constant variance wanders between thresholds and it is very hard to control the acceptance rate; as a result it takes long time to converge. Alternatively, using estimation on a grid of points between upper and lower bounds (Griddy Gibbs sampling) is also not efficient (in our model with single threshold and 200 grid points MCMC with griddy Gibbs takes 20 times longer than using efficient jump MCMC, explained below). Estimation with number of regimes higher than 2 using griddy Gibbs is impractical.

We suggest following procedure with efficient jump:

Let the superscript (i) denotes the i-th draw. Each threshold parameter $r_j^{(i)}$ $(1 \leq j \leq k)$ can be drawn either in a separate block given other threshold parameters, or all thresholds $\{r^{(i)} = (r_1^{(i)}, ..., r_k^{(i)})\}$ could be drawn in one block, in the latter case acceptance rate will be lower.

We generate

$$r_j^{(i)} \sim N(r_j^{(i-1)}, stdr_j^{(i-1)})$$

where $stdr_j^{(i-1)}$ is initially selected as a constant C_0, such that the proposal normal distribution covers all threshold parameter space (e.g, C_0=half-distance between upper and lower bound for each regime). After sufficient number of draws mmm we set $stdr_j^{(i-1)}$ equal to the standard deviation of the sample of accepted draws $\{r_j^{(l)}, l = 1, ..i - 1\}$ multiplied by a scaling constant C. The variance of the proposal density is therefore proportional to the variance matrix estimated from the simulation

$$stdr^{(i)} = C * stdr(\{r^{(l)}\}), \qquad l = n_0, ..i - 1$$

Variance is adjusted using a scaling constant C, so that acceptance rate is reasonable. Gelman et al. (2004) suggest optimal acceptance rate of 44% for 1 parameter and 23% for many parameters.

If $r_j^{(i)}$ does not satisfy the condition

$$r_j^{\text{low}} < r_j^{(i)} < r_j^{\text{up}}$$

where r_j^{low} and r_j^{up} are defined so that the regimes below and above $r_j^{(i)}$ have at least $\delta\%$ of observations, then generate $r_j^{(i)}$ again until it falls within upper and lower bounds. We set $\delta = 5\%$.

We accept $\{r^{(i)} = (r_1^{(i)}, ..., r_k^{(i)})\}$ with probability

$$a_6 = \min\left\{ \frac{p(\gamma^{(i)}, \phi^{(i)}, \theta^{(i)}, \alpha^{(i)}, \beta^{(i)}, r^{(i)}, \lambda^{(i-1)}|\text{data})}{p(\gamma^{(i)}, \phi^{(i)}, \theta^{(i)}, \alpha^{(i)}, \beta^{(i-1)}, r^{(i-1)}, \lambda^{(i-1)}|\text{data})}, 1 \right\}.$$

Otherwise set $r^{(i)} = r^{(i-1)}$.

Alternatively, we can construct a separate block and acceptance rate for each threshold.

(iv) **Elasticity λ blocks**: Draws of $\lambda_j^{(i)}$ are done similarly to draws of $r_j^{(i)}$ using efficient jump with the exception that we set the constraint: $\lambda_j^{(i)} > 0$. In this block we use the same procedure as Qian, Ashizawa and Tsurumi (2005) for estimation of CKLS parameter λ in a linear model (1).[6]

Generate

$$\lambda_j^{(i)} \sim N(\lambda_j^{(i-1)}, stdl_j^{(i-1)})$$

where $stdl_j^{(i-1)}$ is initially selected as a constant and after sufficient number of draws mmm we set $stdl_j^{(i-1)}$ proportional to the standard deviation of the sample of accepted draws $\{\lambda_j^{(l)}, l = 1, ..i-1\}$.

If $\lambda_j^{(i)} > 0$, we accept $\lambda_j^{(i)}$ with probability

$$a_7 = \min\left\{\frac{p(\gamma^{(i)}, \phi^{(i)}, \theta^{(i)}, \alpha^{(i)}, \beta^{(i)}, r^{(i)}, \lambda_j^{(i)}, \lambda_{-j}^{(i)}|\text{data})}{p(\gamma^{(i)}, \phi^{(i)}, \theta^{(i)}, \alpha^{(i)}, \beta^{(i-1)}, r^{(i)}, \lambda^{(i-1)}, \lambda_{-j}^{(i)}|\text{data})}, 1\right\}$$

Otherwise set $\lambda_j^{(i)} = \lambda_j^{(i-1)}$ given previous draws of other parameters and λ's in other regimes $(\lambda_{-j}^{(i)})$. After all λ's are drawn we set $\lambda^{(i)} = (\lambda_1^{(i)}, ..., \lambda_3^{(i)})$.

Alternatively, all λ_j can be drawn in one block.

As with any standard MCMC procedure, we make N draws of the parameters in each of the blocks, and we burn the first m draws. Out of the remaining $N - m$ draws, we keep every h-th draw. We check convergence by testing that the draws attain mean and covariance stationarity.

2.2   Choice of the Model

Estimation of a threshold model involves choice of: (i) number of regimes, (ii) the threshold variable (e.g., y_{t-d}, $\Delta(y_{t-d})$ or $\Sigma_{i=0}^{d}|y_{t-i}|/(d+1)$) and the delay parameter d and (iii) orders of ARMA(p, q)-GARCH(r, s) process in each regime.

We use several criteria, like significance of coefficients, marginal likelihood, and a modified Bayesian Information Criterion (MBIC) discussed in Goldman and Tsurumi (2005). This criterion is a Bayesian analogue of Akaike Information Criterion (AIC) given by

$$\text{AIC}(p, q, r, s, d, n_{\text{regimes}}) = -2\Sigma_{j=1}^{n_{\text{regimes}}} \ln(L_j(p_j, q_j, r_j, s_j, d)) + 2(\nu + 1)$$

where $\ln(L_j())$ is a log-likelihood function for regime j and ν are degrees of freedom. For the case of $n_{\text{regimes}} = 2$: $\nu = 2 * k + p_1 + p_2 + q_1 + q_2 + a_1 + a_2 + s_1 + s_2 + 1$, where k is the dimension of x_t and is assumed to be the same in two regimes, $p_1, p_2, q_1, q_2, a_1, a_2, s_1, s_2$ are orders of ARMA-GARCH parameters in regimes 1 and 2 correspondingly and 1 degree of freedom is given to the choice of delay parameter d. In a similar way we can find ν for higher number of regimes.

The modified Bayesian information criterion is given by

$$\text{MBIC} = -2\ln m(x) + 2(\nu + 1)$$

---

[6] The focus of this article is not on CKLS model, but we use it as one of the features of interest rate models. The main contribution of this article is efficient jump for estimation of threshold parameters that eliminates problems associated with other algorithms.

where the marginal likelihood $m(x)$ is computed by the Laplace-Metropolis estimation and evaluated at either posterior mean or mode.[7]

For the choice of number of regimes we find the smallest MBIC as well as perform sensitivity analysis where estimation of thresholds r_j is done without restriction $r^{\text{low }(j)} < r_j < r^{\text{up }(j)}$. We look at the sensitivity of posterior densities of r_j to imposing $\delta\%$ restriction. If r_j is closer than $\delta\%$ of observations to one of the neighbouring thresholds, or either upper or lower bound for the sample $\{y_t\}$ we suspect that number of regimes is less than it was assumed.

To test whether the dynamics changes with regime we look at posterior distributions of differences in parameters of interest in upper and lower regimes. If there is considerable difference in some parameter's distributions it supports the hypothesis of non-linearity of series y_t.

Finally, the formal test is based on minimizing the modified Bayesian information criterion when models with 1 regime, 2 regimes, and 3 regimes are compared.

3. Empirical Estimation of Interest Rate Dynamics

This section presents the results of estimation of the generalized threshold model (2)-(5) for the U.S. short-term interest rates. The 3 month Treasury-bill monthly data for the period 1962-2005 was downloaded from the Federal Reserve Board of Governor's website.[8] As was mentioned we have chosen the threshold variable as y_{t-d}, which makes sense for interest rate models, as regimes are separated by high and low interest rates. Using model selection criteria identified in the previous section we found that the best model has 2 regimes compared to 3 regimes or 1 regime. We estimated model with and without restrictions of minimum $\delta = 5\%$ of observations in each regime and results were identical for the two-regime model. The best threshold variable turned out to be y_{t-1} with delay parameter $d = 1$.

All results of estimation are given in Table 1 and Figs. 1 to 4. The data are presented in Fig. 1(a). We can identify the period of 1979-1982 which is characterized by high levels of interest rates and high volatility.[9] The posterior distribution of the estimated threshold is given in Fig 1(b) and the mean value of estimated threshold distribution (10%) is shown on the graph with data separating y_t into two regimes; the upper regime coincides with historical period of change in monetary policy.

Some of the results given in Table 1 are similar to results obtained in PST for the period 1962-1990, while certain results are different.[10] One of the similar results is that the interest rate follows a unit root process in the lower regime ($y_{t-1} < r$) and the process has slow mean reversion once the interest rate is above the threshold. The same result was obtained earlier by Gray (1996) who used a Markov-switching GARCH model. The pdfs of persistence parameter γ_2 in two regimes are presented in Fig. 2(a) and the posterior distribution of the difference in persistence for two regimes ($\gamma_2^{(1)} - \gamma_2^{(2)}$) is given in Fig. 2(b). We see that in the lower regime (regime 1) the pdf is tightly distributed around zero. In the upper regime (regime 2), the distribution is centered around -0.21 and is more disperse. The posterior distribution of the difference shows that 85% highest posterior density interval (HPDI) corresponds to positive values of differences in γ_2, i.e., persistence is higher in lower regime. Using any reasonable significance level, e.g., the 95% highest posterior density interval, we clearly do not reject the null hypothesis of a unit root for γ_2 in the lower regime. Although for the upper regime

---

[7] Alternatively one can use Chib and Jeliazkov (2001) estimator of marginal likelihood.

[8] http://federalreserve.gov/releases/h15/data.htm

[9] Many authors included the period of FED's experiment 1979-1982 as part of the sample, e.g., PST considered the period 1962-1990.

[10] We also estimated our model with the same sample as PST (1962-1990) and results are very similar to the full sample 1962-2005 results presented in the text.

using the 95% HPDI we would also fail to reject unit root (95% HPDI is $(-0.468, 0.060)$), the variance of this distribution is high and the upper limit of the 95% HPDI is very close to zero. Using, for example 87% HPDI $(-0.402, -0.002)$ we would reject the unit root hypothesis for the upper regime.

Let us test stationarity of interest rates. In each regime the null hypothesis of a unit root is given by either

$$H_0 : \gamma_2 = 0 \text{ versus } H_1 : \gamma_2 < 0 \tag{9}$$

or
$$H_0 : \rho = 1 \text{ versus } H_1 : \rho < 1 \tag{10}$$

where ρ is the maximum absolute value of the inverse roots of the AR parameters in the error term u_t. Rejection of both null hypotheses above implies stationarity. However, if we find a unit root in the lower regime, but mean reversion in the upper regime, it would still imply stationarity since interest rates exhibit mean reversion once they are higher than the threshold.[11]

Table 1 shows that the max absolute value of inverted roots of AR parameters (ρ) in the error term is less than 1, so the error term is stationary in both regimes. Overall, we conclude that interest rates are weakly stationary, since the unit root hypothesis in the upper regime was rejected using 87% HPDI, but was not rejected with 95% HPDI.

Table 1.   Threshold ARMA model with CKLS-GARCH volatility

	Regime 1: $(y_{t-1} < r)$			Regime 2: $(y_{t-1} \geq r)$		
	mean	(std)	Corr	mean	(std)	Corr
γ_1	0.047	(0.030)	0.447	2.405	(1.545)	0.491
γ_2	-0.006	(0.007)	0.442	-0.208	(0.131)	0.507
ϕ_1	0.507	(0.138)	0.918	-0.123	(0.432)	0.886
θ_1	-0.129	(0.166)	0.927	0.740	(0.403)	0.876
α_0	0.001	(0.000)	0.771	0.029	(0.024)	0.777
α_1	0.142	(0.030)	0.796	0.328	(0.215)	0.657
β_1	0.835	(0.030)	0.883	0.329	(0.229)	0.826
λ	0.391	(0.130)	0.799	0.638	(0.137)	0.791
ρ=max AR root	0.507	(0.138)	0.918	0.396	(0.212)	0.665
r	10.034	(0.246)	0.315			

Notes: (1) ARMA(1, 1)–GARCH(1, 1) model for each regime was selected based on MBIC, significance of parameters and convergence patterns.
(2) Estimated model has $n_1 = 483$ (93%) observations in the lower regime and $n_2 = 34$ (7%) observations in the upper regime.
(3) The figures in parentheses are posterior standard deviations.
(4) Corr is the first order autocorrelation of the MCMC draws.
(5) ρ is the maximum absolute value of the inverse roots of the AR parameters.

The posterior pdf of the threshold parameter presented in Fig. 1 (b) with mean value $r = 10.03$ (see Table 1) is close to PST but the standard error is significantly smaller. Since we included a GARCH component (which turns out to be significant in the model) we increased the precision of the threshold estimator.

---

[11] For a formal definition of stationarity of a two-regime TAR model see, e.g., Franses and van Dijk (2000). They explain that a unit root in lower regime and stationarity in the upper regime imply stationarity of non-linear time series.

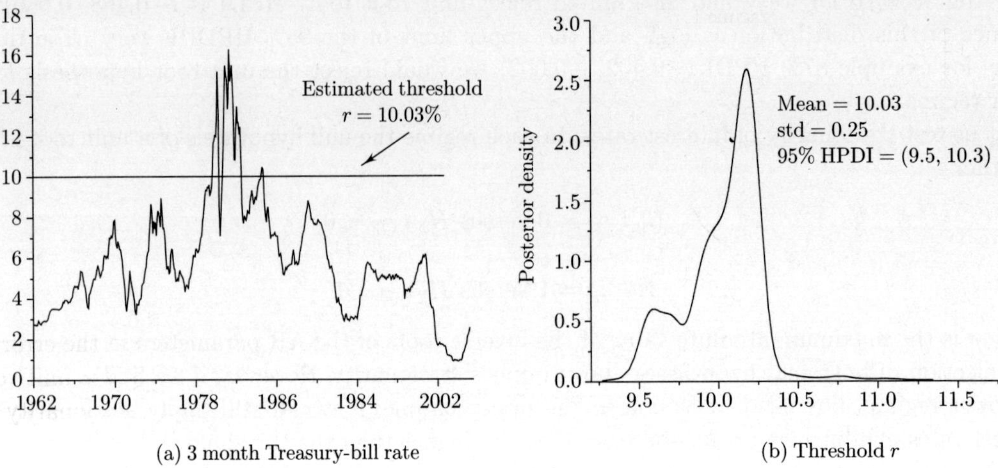

(a) 3 month Treasury-bill rate   (b) Threshold r

Fig. 1. (a) U.S. three-month Treasury-bill rate, monthly data, 1962-2005; (b) Posterior pdf of the threshold parameter r.

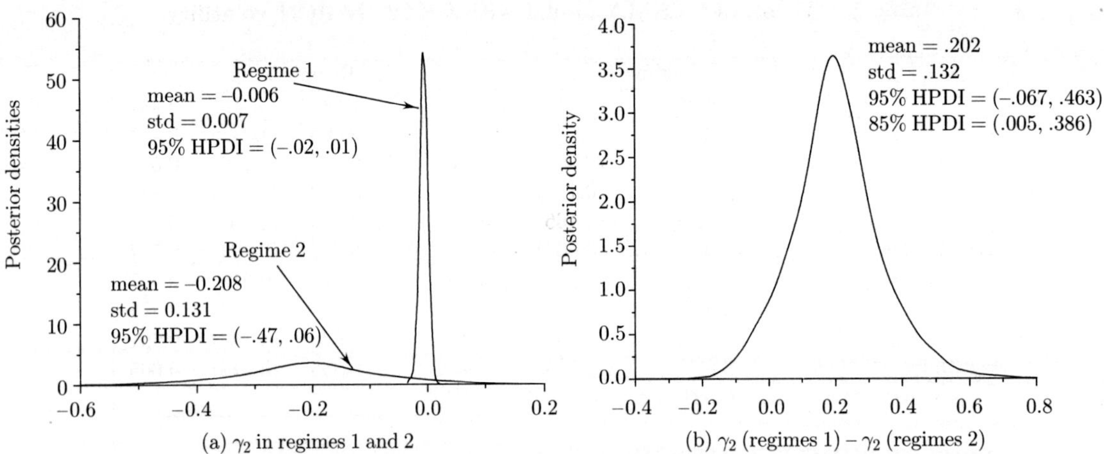

(a) γ_2 in regimes 1 and 2   (b) γ_2 (regimes 1) − γ_2 (regimes 2)

Fig. 2. (a) Posterior pdfs of persistence parameter γ_2 in regimes 1 and 2; (b) Posterior pdf of the difference in persistence parameters.

Fig. 4(a) shows the posterior pdf of a GARCH parameter ab, which measures the persistence of volatility in each regime:

$$ab = \Sigma_{i=1}^{l}(\alpha_i + \beta_i) \tag{11}$$

where $l = \max(a, s)$, $\alpha_i = 0$ for $i \geq a$ and $\beta_i = 0$ for $i \geq s$. The posterior pdf for ab can be used to test a null hypothesis whether a GARCH component is present in the data. The null and alternative hypotheses are given by $H_0 : ab = 0$ (i.e., the error process does not have GARCH) versus $H_1 : ab > 0$ (i.e., the error process has GARCH). In Fig. 4(a) we show the pdfs for ab in each regime and find ab significant in both regimes. The persistence in the lower regime is close to one, while in the upper regime it is less than one (using any HPDI). We also show the posterior distribution of the

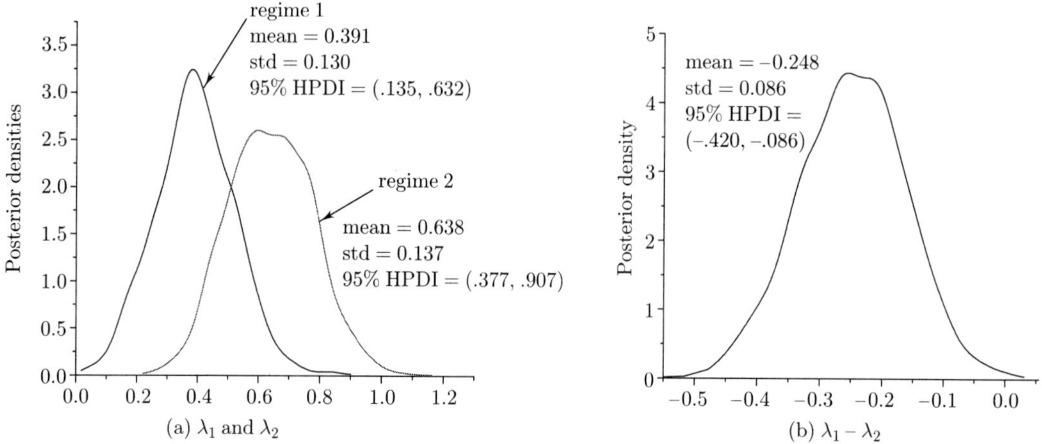

(a) λ_1 and λ_2

(b) $\lambda_1 - \lambda_2$

Fig. 3. (a) **Posterior pdfs of elasticity of volatility λ in regimes 1 and 2; (b) Posterior pdf of the difference in elasticity of volatility parameters.**

difference in the persistence in volatility in lower and upper regimes and find lower persistence in the upper regime using 85% HPDI. If we look at a constant term in the GARCH model we find that when interest rates fall they become less volatile (if we compare magnitudes of α_0 in regimes 1 and 2 from Table 1), however, at lower interest rates volatility is more persistent. The property that level of volatility is proportional to interest rates is well-known, however, change in persistence is an interesting result. We may conclude that overall volatility process is stationary, since it is mean-reverting when interest rates are high and volatility is high.

Finally, we find two similar distributions for level heteroscedasticity parameter λ in two regimes (Fig. 3). HPDI for both distributions include $\lambda = 0.5$. Our estimates of λ are smaller than estimates given in CKLS and PST and are consistent with results of other studies that omitting the GARCH component results in overestimating the level effect in volatility of interest rates.

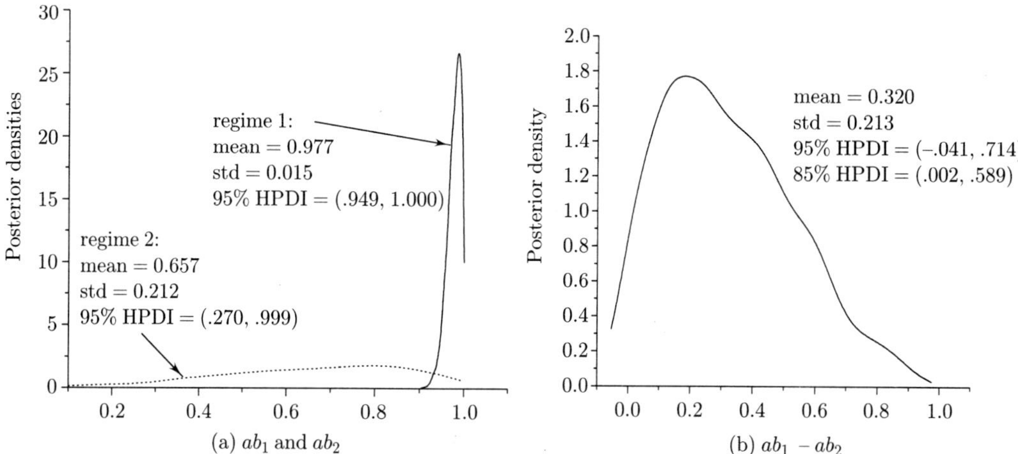

(a) ab_1 and ab_2

(b) $ab_1 - ab_2$

Fig. 4. (a) **Posterior pdfs of persistence in volatility ab in regimes 1 and 2; (b) Posterior pdf of the difference in persistence of volatility parameters.**

4. Conclusion

We applied MCMC algorithm with efficient jump for a general class of threshold time series models with multiple regimes that work much faster than estimation on a grid of points and is simpler than numerical integration techniques suggested in literature.

We used as example U.S. Treasury-bills data and found that the data can be characterized by two regimes; the interest rate data process follows a unit root in a lower regime and a weak mean reversion process above a certain threshold; the volatility is more persistent in lower regime; the modification to allow for GARCH error results in more efficient threshold parameter estimates and lower level effects in volatility.

Acknowledgement

We are indebted to Prof. Hiroki Tsurumi, Rutgers University, for his invaluable comments.

References

Ball, C.A and Torous, W.N. (1996). Unit roots and estimation of interest rate dynamics. *Journal of Empirical Finance*, **3**, 215-238.

Ball, C.A and Torous, W.N. (1999). The stochastic volatility of short-term interest rates: some international evidence. *Journal of Finance*, **54**, 2339-2359.

Bliss, R.J., and Smith, D.C. (1998). The Elasticity of Interest Rate Volatility: Chan, Karolyi, Longstaff and Sanders Revisited, Federal Reserve Bank of Atlanta, Working Paper 97:13a.

Brennan, M. J. and Schwartz, E.S. (1980). Analyzing Convertible Bonds. *Journal of Financial and Quantitative Analysis*, **15**, 907-929.

Brenner, R.J., Harjes, R.H. and Kroner, K.F. (1996). Another Look at Models of the Short-Term Interest Rates, *Journal of Financial and Quantitative Analysis*, **31**, 1, 85-107.

Broze, L., Scaillet, O. and Zakoian, J. (1995). Testing for Continuous-time Models of Short-term Interest Rate. *Journal of Empirical Finance*, **2**, 199-223.

Cai, J. (1994). A Markov Model of Switching-Regime ARCH. *Journal of Business & Economic Statistics*, **12**, 3, 309-316.

Chan, K.C, Karolyi, G.A., Longstaff, F.A. and Sanders, A.B. (1992). An Empirical Comparison of Alternative Models of the Short-Term Interest Rate. *Journal of Finance*, **48**, 1209-1227.

Chen, C.W.S. and Lee, J.C. (1995). Bayesian inference of threshold autoregressive models. *Journal of Time Series Analysis*, **16**, 483-492.

Chib, S. and Jeliazkov, I. (2001). Marginal Likelihood from the Metropolis-Hastings Output. *Journal of the American Statistical Association*, **96**, 270–281.

Cox, J., Ingersoll, J. and Ross, S. (1980). An Analysis of Variable Rate Loan Contracts. *Journal of Finance*, **35**, 389-403.

Cox, J., Ingersoll, J. and Ross, S. (1985). A Theory of the Term Structure of Interest Rates. *Econometrica*, **53**, 385-407.

Dahlquist, M. and Gray, S.F. (2000). Regime-switching and interest rates in the European Monetary System, *Journal of International Economics*, **50**, 399-419.

Forbes, C.S., Kalb, G.R.J. and Kofman, P. (1999). Bayesian Arbitrage Threshold Analysis. *Journal of Business & Economics Statistics*, **17**, 3, 364-372.

Franses, P. H. and van Dijk, D. (2000). *Non-linear Times Series Models in Empirircal Finance*. Cambridge: Cambridge University Press.

Gelman, A., Carlin, J.B., Stern, H.S. and Rubin, D.B. (2004). *Bayesian Data Analysis*. 2nd edition, Chapman & Hall/CRC.

Geweke, J. and Terui, N. 1993. Bayesian threshold autoregressive models for nonlinear time series. *Journal of Time Series Analysis*, **14**, 441-454.

Goldman, E. and Tsurumi, H. 2005. Bayesian Analysis of a Doubly Truncated ARMA-GARCH Model. *Studies in Nonlinear Dynamics & Econometrics*, **9** (2).

Gonzalo, J. and Pitarakis, J. (2002). Estimation and Model Selection Based Inference in Single and Multiple Threshold Model. *Journal of Econometrics*, **110**, 319-352.

Gray, S.F. (1996). Modelling the conditional distribution of interest rates as a regime-switching process. *Journal of Financial Economics*, **42**, 27-62.

Hamilton, J. (1988). Rational Expectations Econometric Analysis of Changes in Regime: An Investigation of the Term Structure of Interest Rates. *Journal of Economic Dynamics and Control*, **12**, 385-423.

Hansen B.E. (1997). Inference in TAR models. *Studies in Nonlinear Dynamics and Econometrics*, **2**.

Koedijk, K.G., Nissen, F.G.J.A., Schotman, P.C. and Wolff, C.C.P. (1997). The dynamics of short-term interest rate volatility reconsidered. *European Finance Review*, **1**, 105130.

Koop, G. and Potter S.M. (1999). Dynamic Asymmetries in U.S. Unemployment. *Journal of Business and Economic Statistics*, **17**, 298-312.

Lanne, M. and Saikkonen, P. (2002). Threshold Autoregressions for Strongly Autocorrelated Time Series. *Journal of Business and Economic Statistics*, **20**, 282-289.

Nakatsuma, T. (2000). Bayesian analysis of ARMA-GARCH models: A Markov chain sampling approach. *Journal of Econometrics*, **95**, 57-69.

Nowman, K.B. (1997). Gaussian Estimation of Single-Factor Continuous-Time Models of the Term Structure of Interest Rates. *Journal of Finance*, **52**, 1695-1706.

Phann, G.A., Schotman, P.C. and Tschering, R. (1996). Non-linear interest rate dynamics and implications for the term structure. *Journal of Econometrics*, **74**, 149-176.

Qian, X., Ashizawa, R. and Tsurumi, H. (2005). Estimation of short term interest rate models: comparison of Bayesian and non-Bayesian estimators. Forthcoming in the *Communications in Statistics-Theory and Methods*.

Tong, H and Lim, K.S. (1980). *Threshold Autoregression, Limit Cycles and Cyclical data. J. Roy. Stat. Soc.* B, **42**, 24592.

Tong, H. (1983). *Threshold models in non-linear time series analysis*. Springer-Verlag Inc (Berlin; New York).

Tong, H. (1990). *Nonlinear time series: a synamic system approach*. Oxford University Press (Oxford; U.K.).

Tsay, R.S. (1989). Testing and modelling threshold autoregressive processes. *Journal of the American Statistical Association*, **84**, 231-240.

Tsay, R.S. (1998). Testing and modelling multivariate threshold models. *Journal of the American Statistical Association*, **93**, 1188-1202.

Yu, J. and Phillips, P.C.B. (2001). A Gaussian approach for continuous-time models of the short-term interest rate. *Econometrics Journal*, **4**, 211-225.

Bayesian Statistics and Its Applications
Edited by S.K. Upadhyay, U. Singh and D.K. Dey
Anamaya Publishers, New Delhi, India

A Semiparametric Accelerated Failure Time Model for Survival Data with Time Dependent Covariates

Timothy E. Hanson[1], Wesley O. Johnson[2] and Purushottam W. Laud[3]

[1]Division of Biostatistics, University of Minnesota School of Public Health, Minneapolis, MN 55455

[2]Department of Statistics, University of California, Irvine, CA 92697

[3]Division of Biostatistics, Medical College of Wisconsin Milwaukee, WI 53226

Abstract

This article presents Bayesian methods for fitting a newly developed generalization of the accelerated failure time model (AFTD) and discusses the formulation and some properties of the AFTD. The prior for the baseline distribution in the model is taken to be a mixture of Polya trees and posterior inference is obtained through standard Markov chain Monte Carlo methods. We demonstrate the implementation of this model first on a simulated dataset and then apply the techniques presented here to the study of the timing of cerebral edema diagnosis during emergency room treatment of diabetic ketoacidosis in children.

1. Introduction

Semiparametric Bayesian methods for survival data in the regression context have been developed by many authors over the past three decades. The review article on Bayesian survival analysis by Sinha and Dey (1997) summarizes this work as well as that in the context of exchangeable observations. A recent book by Ibrahim, Chen and Sinha (2001) provides an update and more details.

Generally, from the Bayesian viewpoint of employing a full likelihood, two main components are relevant: a nonparametric prior on the space of all baseline distributions considered in the model and a parametric form specifying how the covariates modify the baseline distribution. The latter is a modeling issue common to both Bayesian and frequentist inference schemes. The former is a requirement of the Bayesian paradigm.

For the regression specification, the proportional hazards (PH) model of Cox (1972) is the most widely used, followed by the accelerated failure time (AFT) model considered, for example, by Kalbfleisch and Prentice (1980). A less frequently used alternative is the additive hazards model of Aalen (1980). We consider the second in this article.

Priors on baseline distributions can be approached from various angles. The Dirichlet process of Ferguson (1973) provides one building block for considering priors on distribution or survival functions. Another is the Polya tree (PT) prior as described by Ferguson (1974), and Lavine (1992, 1994). A different approach considers priors on cumulative hazard functions. Here, some prominent choices have been the gamma process (Kalbfleisch, 1978), neutral to the right processes (Doksum, 1974) and the beta process of Hjort (1990). Authors who have considered priors on hazard functions include Dykstra and Laud (1981), Lo and Weng (1989), Arjas and Gasbarra (1994) and Nieto-Barajas and Walker (2002).

Bayesian inference for the PH model has been implemented in the past few years via Markov chain Monte Carlo (MCMC) techniques for various choices of priors. Among these are Laud, Damien and Smith (1996) for the hazard function prior of Dykstra and Laud (1981), Laud, Damien and Smith (1998) for Hjort's (1990) beta process, Arjas and Gasbarra (1994) for their correlated gamma process and Ishwaran and James (2004) for the weighted gamma priors of Lo and Weng (1989). Many have also successfully used piecewise constant (see, for example, Walker and Mallick, 1997; Wang and Taylor, 2001; and Brown and Ibrahim, 2003) random functions as data-driven approximations to the above priors.

The Bayesian AFT model with covariates was first developed by modeling the underlying baseline distribution for that model as a Dirichlet process (Christensen and Johnson, 1988; Johnson and Christensen, 1989). These efforts were only partially successful since it was not possible at that time to give a fully Bayesian approach to the semiparametric AFT model due to mathematical intractability. However, Kuo and Mallick (1997) and Kottas and Gelfand (2001) were able to overcome these difficulties by using Dirichlet process mixture (DPM) models in conjunction with MCMC techniques that were introduced by Escobar (1994). In addition, Walker and Mallick (1999) developed methods based on MCMC sampling using a finite Polya tree prior for the baseline survivor distribution. Hanson and Johnson (2002) extended their method to mixture of Polya tree priors (MPT). Hanson and Johnson (2004) developed MCMC methods with interval censored data using mixture of Dirichlet process priors (MDP). These latter two approaches provide direct semiparametric extensions of standard parametric AFT models by starting with a parametric family of distributions (like the log-normal regression model for example). The approach then builds a prior distribution on the space of all possible distributions that is in some sense centered at this parametric family. So the parametric family is one of the possibilities, but serious departures from that family are allowed in the construction.

Bayesian analysis of survival data with time-dependent covariates (TDC's) has received some attention, in Ibrahim, Chen and Sinha (2001) and Brown and Ibrahim (2003) for example, under the heading of joint modeling of longitudinal and survival data. Frequentist analysis in this context dates back to Prentice (1982) who pointed out that the usual partial likelihood needs modification if the covariates are measured with error. More recent work of Tsiatis et al. (1992), Tsiatis, DeGruttola and Wolfsohn (1995) and others focuses on modeling the covariate process and justifying proposed estimators via asymptotic considerations. From the Bayesian viewpoint, Faucett and Thomas (1996), Wang and Taylor (2001) and Brown and Ibrahim (2003) show how conditional models can be augmented by marginals to accommodate both covariate function modeling and survival modeling. An excellent summary of much of this work can be found in Chapter 7 of Ibrahim, Chen and Sinha (2001).

This article considers the AFT regression model for survival data, and generalizes it to handle TDC's. The model is denoted as AFTD. We take the covariate function to be fixed and observed without (or with negligible) error as this is the case in many applications (e.g., time of transplant, fluids administered to a patient, onset of a well defined condition). Moreover, as addressed in Section 6, this conditional treatment can be readily extended to include covariate process models via a multiplicative factor in the likelihood. As for the baseline distribution underlying the model, the nonparametric prior we employ is the MPT of Hanson and Johnson (2002). Besides computational simplicity, this prior provides flexibility, ease of elicitation, continuous survival functions and multimodality accommodation.

The AFTD model is developed below by direct analogy with the comparable development for incorporating TDC's into the PH model. We condition on all time dependent process information and, for initial simplicity, only consider TDC's that correspond to jump processes. Thus, as is standard practice, each TDC will take on a finite number of values that change at discrete times over the course of the experiment. We refer to the times at which a TDC changes as TDC changepoints. An extension to continuously varying TDC's is discussed later.

Similar to the case of modeling TDC's in the context of the Cox model, where the resulting model is no longer a PH model, the AFTD model is no longer an AFT model. In the AFTD model, the hazard function is in the usual AFT form at time zero with an acceleration factor (AF) that depends on the values of the covariate information at time zero. The AF remains constant until the first TDC changepoint at which time there is a new AF. The hazard function now is in the same form as an AFT model with the new AF until the next TDC changepoint, and so on. Section 3 makes this precise and provides the machinery to make MCMC inferences. This model is analogous to the PH model where hazards are proportional for two individuals over periods of time between TDC changepoints.

Section 2 gives the necessary background material for the AFT model and the Bayesian semiparametric approach to it. Section 3 introduces the AFTD model, develops its properties, and then develops MCMC methods for statistical inference. These methods involve only straightforward extension of those required for parametric models. Section 4 addresses diagnostics. Section 5.1 presents an example with simulated data and Section 5.2 gives an analysis of data collected to study the effect of several covariates on the time of onset of cerebral edema, a very serious condition, in children admitted to the emergency department with diabetic ketoacidosis. Section 6 discusses our approach with some concluding remarks.

2. Background Material

This section discusses the AFT model for survival analysis with fixed covariates as background to modeling with time dependent covariates, and the mixture of Polya trees prior that we employ.

Hanson and Johnson (2002) present a Bayesian semiparametric AFT model in which the baseline survival distribution has a mixture of finite Polya trees prior. Let T be a survival time for an individual with covariates $x = (x_1, \ldots, x_p)'$. Let $F_0(t) = P(T_0 \leq t)$, $S_0(t) = P(T_0 > t)$, $f_0(t) = \frac{d}{dt}F_0(t)$, and $h_0(t) = f_0(t)/S_0(t)$ be the baseline cdf, survival function, pdf, and hazard function respectively. Hanson and Johnson (2002) consider the model

$$T = e^{-x\beta}T_0, \quad T_0 \sim S_0, \quad S_0 \sim \int PT_M(w, G_\theta)p(\theta)d\theta, \tag{1}$$

where β is a p-dimensional vector of regression coefficients. Given S_0, the survivor function for such a T is

$$S(t|x, \beta) = e^{-e^{x\beta}\int_0^t h_0(se^{x\beta})ds} = e^{-H_0(e^{x\beta}t)},$$

where $H_0(t) = \int_0^t h_0(s)ds$. The notation $S_0 \sim \int PT_M(w, G_\theta)p(\theta)d\theta$ is shorthand for a particular prior that "centers" S_0 around a parametric family $\{G_\theta : \theta \in \Theta\}$ with weight $w > 0$. We will refer to $c \equiv e^{x\beta}$ as the *acceleration factor* and say that $T \sim AFT(c, S_0)$. Note that the cumulative hazard is $H(t|x, \beta) = H_0(ct)$.

To give a brief description of the Polya tree prior as used here and by many others (Lavine 1992, 1994; Walker and Mallick, 1997, 1999; Berger and Guglielmi, 2001; Hanson and Johnson, 2002; Hanson, Bedrick, Johnson, and Thurmond, 2003; Paddock, Ruggeri, Lavine, and West 2003), let M be a positive integer. Let G_θ denote a parametric family of cumulative distribution functions, such as the log-normal or log-logistic, indexed by θ. A Polya tree (PT) prior is constructed from a set of partitions Π_M^θ and a family $\mathcal{A}_M$ of positive reals. Here G_θ is the centering distribution of the Polya tree prior. Following Walker and Mallick (1999) and Hanson and Johnson (2002), we consider the baseline distribution fixed such that $G_\theta(1) = 0.5$, yielding a generalization of a standard median regression model on the log scale. Consider the canonical partition $\Pi_M^\theta = \{B_\epsilon^\theta : \epsilon \in \bigcup_{l=1}^M \{0,1\}^l\}$. If j is the base-10 representation of the binary $\epsilon = \epsilon_1 \cdots \epsilon_k$ at level k, then $B_{\epsilon_1 \cdots \epsilon_k}^\theta$ is defined to be the interval $(G_\theta^{-1}(j/2^k), G_\theta^{-1}((j+1)/2^k))$. For example, with $k = 3$, and $\epsilon = 000$, then $j = 0$ and $B_{000} = (0, G_\theta^{-1}(1/8))$, and with $\epsilon = 001$, then $j = 1$ and $B_{001} = (G_\theta^{-1}(1/8), G_\theta^{-1}(2/8))$, etc.

Note then that at each level k, the class $\{B_\epsilon : \epsilon \in \{0,1\}^k\}$ forms a partition of the positive reals and furthermore $B_{\epsilon_1 \cdots \epsilon_k}^\theta = B_{\epsilon_1 \cdots \epsilon_k 0}^\theta \bigcup B_{\epsilon_1 \cdots \epsilon_k 1}^\theta$ for $k = 1, 2, \ldots, M - 1$. We take the family $\mathcal{A}_M = \{\alpha_\epsilon : \epsilon \in \bigcup_{j=1}^M \{0,1\}^j\}$ to be defined by $\alpha_{\epsilon_1 \cdots \epsilon_k} = wk^2$ for some $w > 0$ (Walker and Mallick, 1999; Hanson and Johnson, 2002). The parameter w acts much like the precision in a Dirichlet process (Ferguson, 1973). As w tends to zero the posterior baseline is almost entirely data-driven. As w tends to infinity we obtain a fully parametric analysis.

Given Π_M^θ and $\mathcal{A}_M$, the Polya tree prior is defined up to level M by the class of random vectors $\mathcal{Y}_M = \{(Y_{\epsilon 0}, Y_{\epsilon 1}) : \epsilon \in \bigcup_{j=1}^{M-1} \{0,1\}^j\}$ through the product

$$S_0(B_{\epsilon_1 \cdots \epsilon_k}^\theta | \mathcal{Y}_M, \theta) = \prod_{j=1}^k Y_{\epsilon_1 \cdots \epsilon_j}, \tag{2}$$

for $k = 1, 2, \ldots, M$, where we define $S_0(A)$ to be the baseline measure of any set A. Vector (Y_0, Y_1) is set to $(0.5, 0.5)$ to ensure $S_0(1|\mathcal{Y}_M, \theta) = 0.5$; the remaining vectors $(Y_{\epsilon 0}, Y_{\epsilon 1})$ are independent Dirichlet:

$$(Y_{\epsilon 0}, Y_{\epsilon 1}) \sim \text{Dirichlet } (\alpha_{\epsilon 0}, \alpha_{\epsilon 1}), \epsilon \in \bigcup_{j=2}^{M-1} \{0,1\}^j. \tag{3}$$

Beyond sets at the level M in Π_M we assume $S_0|\mathcal{Y}_M, \theta$ follows the baseline G_θ. Hanson and Johnson (2002) show that this assumption yields predictive distributions that are the same as from a fully specified (infinite) Polya tree for large enough M; this assumption also avoids a complication involving infinite probability in the tail of $S_0|\mathcal{Y}_M, \theta$ that arises from taking $S_0|\mathcal{Y}_M, \theta$ to be flat on these sets. Note that $S_0(B_{\epsilon_1 \cdots \epsilon_M}^\theta | B_{\epsilon_1 \cdots \epsilon_{M-1}}^\theta, \mathcal{Y}_M, \theta) = Y_{\epsilon_1 \cdots \epsilon_M}$ which converges to 0.5 in probability. This implies that as M grows $S_0(A|B, \mathcal{Y}_M, \theta) \approx G_\theta(A|B)$ for sets B with small Lebesgue measure and $A \subset B$.

Define the vector of probabilities $p = p(\mathcal{Y}_M) = (p_1, p_2, \ldots, p_{2^M})'$ as $p_{j+1} = S_0(B_{\epsilon_1 \cdots \epsilon_M}^\theta | \mathcal{Y}, \theta) = \prod_{i=1}^M Y_{\epsilon_1 \cdots \epsilon_i}$, where j is the base-10 representation of $\epsilon_1 \cdots \epsilon_M$. After simplification, the baseline survival function is

$$S_0(t|\mathcal{Y}_M, \theta) = p_N [N - 2^M G_\theta(t)] + \sum_{j=N+1}^{2^M} p_j, \tag{4}$$

where N denotes the integer part of $2^M G_\theta(t) + 1$ and $g_\theta(\cdot)$ is the density corresponding to G_θ. The density associated with $S_0(t|\mathcal{Y}_M, \theta)$ is given by

$$f_0(t|\mathcal{Y}_M, \theta) = \sum_{j=1}^{2^M} 2^M p_j g_\theta(t) I_{B^\theta_{\epsilon_M(j-1)}}(t) = 2^M p_N g_\theta(t), \tag{5}$$

where $\epsilon_M(i)$ is the binary representation $\epsilon_1 \cdots \epsilon_M$ of the integer i. Note that the number of elements of $\mathcal{Y}_M$ may not be prohibitively large. This number is $\sum_{j=1}^M 2^j = 2^{M+1} - 2$. For $M = 5$, a typical level, this is 62. To obtain p, $\sum_{j=2}^M 2^j = 2^{M+1} - 4$ multiplications are required; for $M = 5$ this is 60.

The mixture of Polya trees (MPT) prior provides an intermediate choice between a strictly parametric analysis and allowing S_0 to be completely arbitrary. In some ways it provides the best of both worlds. In areas where data are sparse, such as the tails, the MPT prior places relatively more posterior mass on the underlying parametric family $\{G_\theta : \theta \in \Theta\}$. In areas where data are plentiful the posterior is more data driven; and features not allowed in the strictly parametric model, such as left-skew and multimodality, become apparent. The user-specified weight w controls how closely the posterior follows $\{G_\theta : \theta \in \Theta\}$ with larger values of w yielding inference closer to that obtained from the underlying parametric model.

Since the tails are essentially parametric, one can use survival models with an MPT prior in reliability settings, where extrapolation is common and questions concerning the upper quantiles are of interest. However, inferences requiring the whole of the distribution, such as estimated hazard functions and associated credible intervals, are not required to have a simple parametric form.

Also, since the MPT model is essentially a richly parameterized but fully specified probability model, one can check the adequacy of a parametric regression model versus a nonparametric MPT alternative using Bayes factors, pseudo Bayes factors, or other model selection criteria that rely on fully specified probability models. For the problem of testing whether data arise from a specified parametric family of densities, Berger and Guglielmi (2001) employ a MPT prior and examine Bayes factors over a range of weights w. We consider a different mixture that somewhat smooths over the partitioning effects inherent in both a simple Polya tree and the mixture in Berger and Guglielmi (2001).

The Dirichlet process mixture (DPM) of Antoniak (1974) and Escobar and West (1995) has provided Bayesians with a remarkably flexible alternative to fully parametric models for over a decade. This model has gained much favor in part due to its tractability; unlike most simple Dirichlet process models or mixtures of Dirichlet processes (MDP) models, predictive inference is easily obtained through a simple Gibbs sampler. Historically, inferences based on the DPM have been obtained with the Dirichlet process marginalized. For inferences involving the mixing distribution, the Dirichlet process must be approximately sampled (Gelfand and Kottas, 2002) after the marginalized model is fit.

In these models we require the full likelihood of a vector of regression parameters and the baseline survival distribution and so, given our approach, marginalization is not an option. Thus our approach lends itself to straightforward estimation of posterior quantiles of a survival distribution for example, or credible intervals for hazard, survival, or density functions. As we sample the MPT distribution within the Gibbs sampler, inferences regarding any aspect of the model are immediately obtainable. Our approach also lends itself to the incorporation of real prior information, if available. Moreover, as MPT models are often generalizations of existing parametric models and readily fit via maximum likelihood, parameter estimates provide good starting values and asymptotic covariance matrices can be used for Metropolis-Hastings proposal distributions for the more general MPT models.

Hanson (2004) shows that the random density $g(\cdot|\mathcal{Y}_M) = \int g(\cdot|\mathcal{Y}_M,\theta)p(\theta)d\theta$ is almost surely differentiable. This implies that, in contrast to simple Polya tree and piecewise exponential models, all prior mass is placed on a subset of differentiable hazard functions $\lambda_0(\cdot)$.

3. A Model for Time Dependent Covariates: Properties and Methods

Section 2 defined the basic AFT regression model in expression (1) for a single observation. Assume that there are p covariates and that any or all of these can be time dependent. The observed processes are assumed to be jump processes in discrete time. Define the observed vector of p processes as $x \equiv \{x(s) : s \leq u\}$, where u is the largest time at which the covariates are "observed," and is less than the survival time T or a censoring time.

As in the fixed covariate case, we assume a baseline survival function S_0. This distribution now corresponds to the survival function for an individual with constant, zero covariates for all t, that is, if T_0 corresponds to covariates $x(s) = 0$ for all s, then $T_0 \sim S_0$. Covariates may be transformed so that this baseline is interpretable. In this case, the full model proposed in Sections 3.1 to 3.3 allows for the inclusion of real prior information for S_0 in terms of, say, the mean or median baseline survival time and variance or interquartile range.

3.1 Model Specification Through the Hazard Function

Define the hazard function for an individual with covariate $x(s)$ to be

$$h(s|x,\beta) = e^{x(s)\beta}h_0(se^{x(s)\beta}), \quad s \leq u, \tag{6}$$

where $h_0(\cdot)$ is an arbitrary "baseline" hazard function. Thus for periods between TDC changepoints, namely periods when $x(s)\beta$ remains constant, the hazard is same as the hazard for an AFT model with covariates fixed to be those values at the beginning of the interval. Now let $\{r_j : j = 1,\dots,m\}$ denote the ordered TDC changepoints in $(0,u)$ and let $r_0 = 0$. We assume that $s \in [r_{j-1},r_j)$ implies that $x(s) = x(r_{j-1})$ for $j = 1,\dots,m$ and $x(s) = x(r_m)$ for $s \geq r_m$. Define $c_j = e^{x(r_{j-1})\beta}$, $j = 1,\dots,m+1$.

Let $j^*(t) = \max_j\{r_j \leq t\}$. Then $j^*(u) = m$. The cumulative hazard for this individual is thus

$$H(t|x,\beta) = \int_0^t e^{x(s)\beta}h_0(se^{x(s)\beta})ds$$

$$= \sum_{j=1}^{j^*(t)} c_j \int_{r_{j-1}}^{r_j} h_0(sc_j)ds + c_{(j^*(t)+1)} \int_{j^*(t)}^t h_0(sc_{(j^*(t)+1)})ds,$$

and the survivor function is $S(t|x,\beta) = \exp\{-H(t|x,\beta)\}$.

Define $S_j(t) = P(T > t|T > r_{j-1})$ and let $p_j = S_j(r_j)$. Then for $t \in [r_{j-1},r_j)$,

$$S_j(t) = e^{-c_j \int_{r_{j-1}}^t h_0(c_j s)ds} = e^{-\{H_0(c_j t)-H_0(c_j r_{j-1})\}} = \frac{S_0(c_j t)}{S_0(c_j r_{j-1})}.$$

We have

$$S(t|x,\beta) = \left\{\prod_{j=1}^{j^*(t)} p_j\right\} S_{(j^*(t)+1)}(t) = \left\{\prod_{j=1}^{j^*(t)} \frac{S_0(c_j r_j)}{S_0(c_j r_{j-1})}\right\} \frac{S_0(c_{(j^*(t)+1)}t)}{S_0(c_{(j^*(t)+1)}r_{j^*(t)})} \tag{7}$$

for all $t \leq u$. This characterizes our model for all $t > 0$, but with finite number of TDC changepoints. We thus proceed to develop an alternative specification of the model which allows for countably infinite number of such changepoints.

3.2   Alternative Representation and Characterization of the Model

A slightly more elaborate specification than what was given above is required. Thus, instead of thinking about the actual observed process, we expand our definition to include the possibility of a conceptual covariate process, $\{x(s) : s \in (0, \infty)\}$, that is defined for all times. We now have the possibility of a countably infinite sequence of changepoints $\{r_j : j = 1, 2, \ldots\}$ and allow for the possibility that for some $k \geq 0$, $r_k = r_{k+1} = \ldots$, there may be no TDC's or, if there are, they need only change a finite number of times. Define $c_k = e^{x(r_{k-1})\beta}$, $k = 1, 2, \ldots$. Our model specification will now be conditional on this process.

For a given baseline distribution S_0, let $R_k \sim$ AFT $(c_k, S_0)I[r_{k-1}, r_k)$ where the notation suggests that R_k is distributed as an AFT model with acceleration factor c_k that is restricted or truncated to the set $[r_{k-1}, r_k)$. So if we started with some $\tilde{R}_k \sim$ AFT(c_k, S_0), then the conditional survivor function is

$$P(\tilde{R}_k > t | \tilde{R}_k \in [r_{k-1}, r_k)) = \frac{S_0(c_k t) - S_0(c_k r_k)}{S_0(c_k r_{k-1}) - S_0(c_k r_k)}.$$

Thus, $[\tilde{R}_k | \tilde{R}_k \in [r_{k-1}, r_k)] = [R_k]$, where $[\cdot]$ is now standard notation for the probability density for the indicated variable. It can be verified that the above conditional survivor function is identical to the one obtained using our modeled survivor function (7), with $S_0(\cdot) \equiv e^{-H_0(\cdot)}$, for all times that are defined for (7).

We require the analogue of (7) for the extended TDC process. Define this to be $S(t|x, \beta)$, $t > 0$, just as it is defined there, only now with the full sequence of changepoints $\{r_j : j = 1, 2, \ldots\}$. For this to be a proper survivor function, we require $\lim_{m \to \infty} \sum_{j=1}^{m} \{H_0(c_j r_j) - H_0(c_j r_{j-1})\} = \infty$, or equivalently that $\lim_{m \to \infty} \prod_{j=1}^{m} p_j = 0$ in the case of infinitely many changepoints, and we require $\lim_{t \to \infty} H_0(c_m t) = \infty$, in the case of, say, m changepoints. The latter condition is guaranteed provided $c_m > 0$. The former is guaranteed provided there exists some $c > 0$ such that for all sufficiently large j, $c_j > c$ and $\lim_{t \to \infty} H_0(t) = \infty$. This is satisfied, for example, if $|x(s)| \leq K < \infty$ for all s and S_0 is a proper survivor function. If these requirements are not met it is theoretically possible for an individual to "live forever."

Now let $\tilde{T}_k = c_k^{-1} V$, $V \sim S_0$. Then $\tilde{T}_k \sim$ AFT(c_k, S_0), and if we restrict $\tilde{T}_k$ to $[r_{k-1}, r_k)$, the resulting distribution is AFT $(c_k, S_0)I[r_{k-1}, r_k)$. Now let $q_k = S(r_{k-1}|x, \beta) - S(r_k|x, \beta) = \{(1 - p_k) \prod_{j=1}^{k-1} p_j\}, k = 1, 2, \ldots$, and note that $p_j = S_0(c_j r_j)/S_0(c_j r_{j-1})$. So if we know (S_0, β), the p_j's and consequently q_k's are all known numbers and $\sum_k q_k = 1$.

Finally, for given (S_0, β), define a random variable T to be $\tilde{T}_k$ with probability q_k, $k \in \{1, 2, 3, \ldots\}$. Then T has the survivor function (7) expanded to the full covariate process on all t described above. Specifically, let p_K be the probability mass function for a random variable K such that $P(K = k) = p_K(k) = q_k$. Then

$$T|K = k \sim \text{AFT}(c_k, S_0)I[r_{k-1}, r_k), \quad K \sim p_K(\cdot).$$

Thus, a complete probability model is constructed for given (S_0, β) that is more general than the partial specification given in Section 3.1 and which is consistent with that specification. Our full Bayesian probability model can now be specified.

3.3 The Full Probability Model

Combining the AFTD specification of how covariates modify the baseline survival function through time with an MPT prior for S_0, we can construct the full likelihood. Our introduction of the MPT is reflected by the use of notation $(\mathcal{Y}_M, \theta)$. In particular, the contribution to it from an individual with covariate process $x(\cdot)$ and survival time $T = t$, where N denotes the integer part of $2^M G_\theta(c_{m+1}t) + 1$, is

$$L_x(\beta, \mathcal{Y}_M, \theta | T = t) = \left\{ \prod_{j=1}^{m} p_j \right\} \frac{2^M p_N f_0(c_{m+1}t | \mathcal{Y}_M, \theta) c_{m+1}}{S_0(c_{m+1}r_m | \mathcal{Y}_M, \theta)}$$

where we note that $j^*(t) = m$ and $p_j = S_0(c_j r_j | \mathcal{Y}_M, \theta) / S_0(c_j r_{j-1} | \mathcal{Y}_M, \theta)$. The likelihood contribution for an observation right-censored at time t is

$$L_x(\beta, \mathcal{Y}_M, \theta | T > t) = \left\{ \prod_{j=1}^{m} p_j \right\} \frac{S_0(c_{m+1}t | \mathcal{Y}_M, \theta)}{S_0(c_{m+1}r_m | \mathcal{Y}_M, \theta)}.$$

The complete data involve n independent event times, $\{t_i\}_{i=1}^n$, that are the observed survival times ($T_i = t_i$) or are right-censoring times ($T_i > t_i$), and n covariate processes $\{x_i(\cdot)\}_{i=1}^n$. There are thus n likelihood contributions like the one above. Define the contribution for individual i with covariate process $x_i(\cdot)$ as $L_i(\beta, \mathcal{Y}_M, \theta)$. Then the complete likelihood is $\mathcal{L}(\beta, \mathcal{Y}_M, \theta) = \prod_{i=1}^n L_i(\beta, \mathcal{Y}_M, \theta)$. There are additional notational changes. For example, m_i is the number of TDC changepoints for individual i occurring at or before time u_i and r_{ij} is the time of the jth changepoint for individual i, etc.

The random quantities in the prior are thus $\mathcal{Y}_M$, θ, and β, and all are assumed *a priori* independent. In the Gibbs sampler we alternate between sampling $\beta, \theta | \mathcal{Y}_M$ and $\mathcal{Y}_M | \beta, \theta$ (where dependence on the data is suppressed). The former can be sampled via a Metropolis-Hastings step (Tierney, 1994) or slice sampling (Neal, 2003).

The full conditional $\mathcal{Y}_M | \beta, \theta$ will not have a recognizable closed form. A simple Metropolis-Hastings step for updating the components $(Y_{\epsilon 0}, Y_{\epsilon 1})$ first samples a candidate $(Y_{\epsilon 0}^*, Y_{\epsilon 1}^*)$ from a Dirichlet$(mY_{\epsilon 0}, mY_{\epsilon 1})$ distribution, where $m > 0$, typically $m = 20$ or 30. This candidate is accepted as the "new" $(Y_{\epsilon 0}, Y_{\epsilon 1})$ with probability

$$\rho = \min \left\{ 1, \frac{\Gamma(mY_{\epsilon 0})\Gamma(mY_{\epsilon 1})(Y_{\epsilon 0})^{mY_{\epsilon 0}^* - wj^2}(Y_{\epsilon 1})^{mY_{\epsilon 1}^* - wj^2} \mathcal{L}(\beta, \mathcal{Y}_M^*, \theta)}{\Gamma(mY_{\epsilon 0}^*)\Gamma(mY_{\epsilon 1}^*)(Y_{\epsilon 0}^*)^{mY_{\epsilon 0} - wj^2}(Y_{\epsilon 1}^*)^{mY_{\epsilon 1} - wj^2} \mathcal{L}(\beta, \mathcal{Y}_M, \theta)} \right\}.$$

where j is the number of digits in the binary number $\epsilon 0$ and $\mathcal{Y}_M^*$ is the set $\mathcal{Y}_M$ with $(Y_{\epsilon 0}^*, Y_{\epsilon 1}^*)$ replacing $(Y_{\epsilon 0}, Y_{\epsilon 1})$. This may be done in blocks or all at once with the acceptance probability being changed accordingly, although we have had good luck simply sampling the components of $\mathcal{Y}_M$ one at a time.

For interval censored data, where $T \in [a, b)$, for $b < \infty$ we assume that the observed TDC's are constant over the interval $[a, b)$. This would be the case for events like death since if vital signs were taken at some time during the censoring interval, then it would be known that the individual was still alive at the time of taking the information. We assume that $u = a$ and that there is no new measurement on any TDC in $[a, b)$. If this is not the case, our formulas are easily modified, though at the expense of notational simplicity, to account for the extra information. The likelihood contribution is

$$L_x(\beta, \mathcal{Y}_M, \theta | T \in [a, b)) = \left\{ \prod_{j=1}^{m} p_j \right\} \frac{S_0(c_{m+1}a | \mathcal{Y}_M, \theta) - S_0(c_{m+1}b | \mathcal{Y}_M, \theta)}{S_0(c_{m+1}r_m | \mathcal{Y}_M, \theta)}.$$

The likelihood contribution for right truncated data is similarly obtained.

Note that while our model is related to the AFT model, it is not an AFT model unless all the covariates are fixed at the beginning of the study.

4. Diagnostics

We define simple residuals (or more accurately posterior error estimates) to roughly assess model fit. Several authors have advocated the use of "Bayesian residuals" to check model fit and for outlier detection (e.g. Zellner, 1975; Zellner and Moulton, 1985; Chaloner and Brant, 1988). Such plots can be especially useful for censored data (Chaloner, 1991). Let $\mathcal{D}$ be the data $\{(x_i, t_i, \delta_i)\}_{i=1}^{n}$ and $\mathcal{D}_{-i}$ denote that data $\mathcal{D}$ with the ith case (x_i, t_i, δ_i) omitted. We consider the posterior mean survival function for each individual evaluated at their observed event time $\hat{S}_i(t_i) = E\{S(t_i | x_i, \mathcal{D})\}$ and define Cox-Snell residuals as $r_i = -\log \hat{S}(t_i)$. For a given (S_0, β), $-\log S(T | x_i) \sim \exp(1)$ and a plot of the empirical integrated hazard plot (Nelson, 1972) for the censored residuals $\{(r_i, \delta_i)\}$, where δ_i is a noncensoring indicator, should be relatively straight with slope one if the model is "true." Baltazar-Aban and Pena (1995) argue that Cox-Snell residual plots can be uninformative in richly parameterized and nonparametric models; i.e., they tend to be straight even when the model fits poorly. Similarly, we have found that marked curvature in these plots indicates gross departures from modeling assumptions but that there still may be considerable differences among models that all have relatively straight lines.

As the baseline survival function S_0 and β are modeled simultaneously, one can define a standard residual of "observed-expected under the model." For example, let T be independent of the data and let $T \sim \text{AFT}D(x_i, S_0)$. A residual based on the posterior expected survival time is defined as $e_i = E\{T_i - T | x_i, \mathcal{D}\}$. For uncensored data this is $e_i = T_i - E\{T | x_i, \mathcal{D}\}$. These residuals can be plotted against the ranked event times or expected event times to determine whether there is systematic over or under fitting of the model for early versus late survival. They can also be plotted versus fixed effects to determine inadequate model fit. Since survival distributions are often markedly skewed, a more useful residual is the observed minus the posterior expected median survival time $e_i^* = E\{T_i - S^{-1}(0.5 | x_i) | \mathcal{D}\}$, which equals $e_i^* = T_i - E\{S^{-1}(0.5 | x_i) | \mathcal{D}\}$ when $\delta_i = 1$. These residuals can be standardized in some fashion, however, it is informative to examine a raw residuals plot to get a feel for how well the model does globally in predicting each observed datum. Section 5 examines plots of e_i^* versus the "fitted" posterior expected median survival times $E\{S^{-1}(0.5 | x_i) | \mathcal{D}\}$ for the cerebral edema data set.

For $i = 1, \ldots, n$, we also obtain the conditional predictive ordinate (CPO) (Geisser, 1993) statistic:

$$\text{CPO}_i = [S(t_i | x_i, \mathcal{D}_{-i})]^{\delta_i - 1} [f(t_i | x_i, \mathcal{D}_{-i})]^{\delta_i}.$$

For an uncensored observation CPO_i is the posterior predictive density given the subset $\mathcal{D}_{-i}$ evaluated at the observed survival time t_i. For a censored observation CPO_i is $P(T_i > t_i | \mathcal{D}_{-i})$. Chen, Shao and Ibrahim (2000) discuss MCMC computation of CPO_i. See Sahu and Dey (2004) for a recent application of CPO statistics for comparing survival models.

5. Examples

Now we first simulate AFTD model data, fit the model to these data, and examine the fit of the model to these data, and then analyze a new data set involving cerebral edema in children with diabetic ketoacidosis.

5.1 Simulated Data

We simulated data with two distinct covariates and with a log-normal $(\log(2), 0.2^2)$ baseline. Group one followed the baseline and thus had $x_i(t) \equiv 0$ for $i = 1, \ldots, 10$. Group 2 had the TDC $x_i(t) = I_{[1,\infty)}(t)$ for $i = 11, \ldots, 100$. The regression coefficient β_1 was fixed at $\log(2) \approx 0.69$. In all three models a MPT prior was assumed for the baseline with $c = 1$ and $M = 4$. The log-logistic family $G_\theta(t) = t^{1/\theta-1}e^{-\beta_0/\theta}/(1 + t^{1/\theta}e^{-\beta_0/\theta})$ centered S_0 and the flat prior/hyperprior $p(\theta, \beta_0, \beta_1) = 4B^{-3}$ was assumed with support $(0, B] \times [-B, B] \times [-B, B]$ for large B.

Integrated hazard plot of the Cox-Snell residuals (Fig. 1) shows a good fit of the AFTD model. The plot of $e_i^* = T_i - E\{S^{-1}(0.5|x_i)|\text{data}\}$ versus $E\{S^{-1}(0.5|x_i)|\text{data}\}$ shows nothing unusual for the model (not shown).

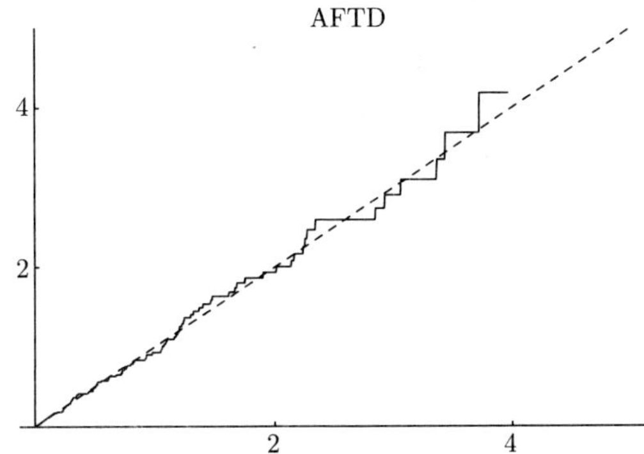

Fig. 1. Simulated data: integrated hazard plot of Cox-Snell residuals of the AFTD model.

Based on this limited sample size, the AFTD model estimates β_1 quite well (Table 1) with a posterior median value of $\hat{\beta} = 0.65$ and an equal-tailed credible interval of $(0.48, 0.96)$. Fig. 2 shows that the posterior baseline density from the AFTD model comes close to the true density.

Table 1. Posterior inferences for simulated data

Parameter	Posterior Mean	95% Credible Interval
β_0	0.63	(0.50, 0.81)
β_1	0.65	(0.48, 0.96)
θ	0.12	(0.08, 0.16)

5.2 Cerebral Edema Data

Here we develop a survival model for data collected by Glaser et al. (2001), who assessed risk factors associated with the onset of cerebral edema (CE) in children with diabetic ketoacidosis. Cerebral edema is a dangerous complication associated with emergency department and in-patient hospital care of children with diabetic ketoacidosis. Children with symptoms of diabetic ketoacidosis are

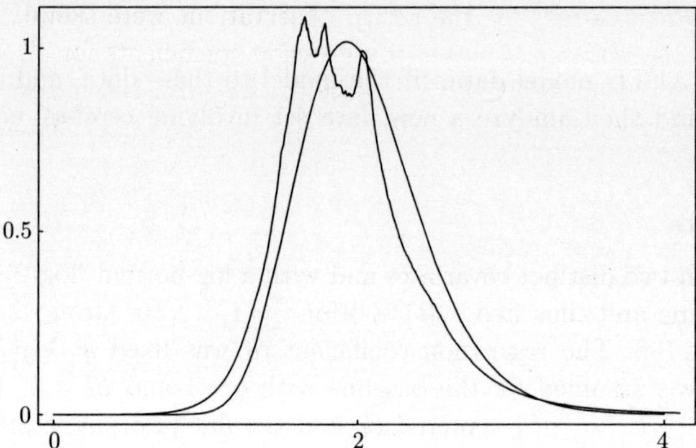

Fig. 2. Simulated data: baseline density estimate; the smoother function is the true density.

initially treated in the emergency department, then moved to the hospital, typically the pediatric intensive care unit, over the course of 24 hr. The main purpose of treatment is to normalize blood serum chemistry and acid-base abnormalities. A major, but infrequent complication of children associated with diabetic ketoacidosis and its treatment is CE, or swelling in the brain, which may result in death or permanent neurological damage. While Glaser et al. (2001) developed models for the probability of onset as a function of potential risk factors and confounders, here we consider only the children in that study who developed CE; $n = 58$. Our goal is to ascertain the effect of treatment procedures, which are changing in time, and the effect of fixed covariates on the timing of CE. For example, what are the factors associated with early versus late onset, if any?

At the time of initial admission to the emergency department, several measurements were taken, and at the same time, various treatments applied, continuously for up to 24 hr. The original data include 19 fixed covariates and 12 time dependent covariates. Rather than attempt a complete analysis with so many variables, we considered only a subset of the covariates for our illustration, thus the analysis should only be considered as a demonstration of how our modeling works rather than a definitive analysis of these data.

The only fixed variable considered here is age (in years) at the time of admission to the emergency department. Time dependent covariates were recorded hourly rather than continuously. Two types of TDC's are considered. The first involves simply monitoring biochemical variables over time. Two of these included here are serum bicarbonate (BIC) (concentration in the blood measured in mmol per liter) and blood urea nitrogen (BUN) (concentration measured in mg/deciliter). The second type involves actions by the physicians over time. The two used here are fluids administered (FL) (volume of fluids in ml/kg/hr) and sodium administered (NA) (measured in mEq/kg/hr). The ith patient in the data has a covariate process given by $\{(r_{ij}, x(r_{ij}))\}_{i=1}^{m_i}$, $i = 1, \ldots, 58$, and is diagnosed with CE at time t_i. Note that then $r_{m_i} < t_i$. None of the event times are censored.

We used the log-logistic family with unknown scale to center the MPT survival model with TDC's with $M = 4$ and $c = 1$. After selecting candidate generating distributions, we were able to establish convergence of our chain by standard techniques after 10,000 iterations.

Integrated Cox-Snell residual plots did not show radical departures from the assumption of a correct model. A plot of log CPO_i values versus the index i showed that the data are supported by the model; and a plot of the residuals e_i^* versus predicted values also showed no obvious lack of fit for the model. Prediction intervals for the actual observations were skewed, but all contained the observed values. Table 2 provides estimates of regression coefficients for all variables in the model. There is a 99% posterior probability that the coefficient for Adm-NA is positive and at least a 96% posterior probability that the coefficient for the interaction is negative. The Serum-BIC variable has a 94% probability of being positive. Other variables have posterior probability between 80% and 91% of being positive, or negative.

Table 2. Posterior inferences for cerebral edema data

Parameter	Posterior Mean	95% Credible Interval
Age (Fixed)	0.028	$(-0.01,\ 0.08)$
Serum-BIC (TD)	0.04	$(-0.01,\ 0.13)$
Serum-BUN (TD)	-0.005	$(-0.02,\ 0.01)$
Adm-FL (TD)	-0.03	$(-0.09,\ 0.03)$
Adm-NA (TD)	0.60	$(0.16,\ 0.93)^*$
FL×NA	-0.011	$(-0.03,\ -0.00)^*$
BIC^2	-0.005	$(-0.01,\ 0.006)$

Thus, larger values of Serum-BIC are associated with earlier diagnosis of CE. For example, comparing two children who are otherwise being treated the same over the same period of time and are of the same age, the hazard of CE for a child with a larger value for BIC will be accelerated relative to the one with a lower value for BIC.

Fig. 3 presents an estimated relative hazard comparing the two subjects having the same characteristics (age 10, BUN = 35, fluids constant at 3.6, NA constant at 0.7, and BIC increasing from 5 to 22) except that subject 1 has constant NA=0.7 and subject 2 has constant NA = 0.35. Observe that subject 2 is at lower risk of CE, but the estimated risk varies considerably over the first 18 hours.

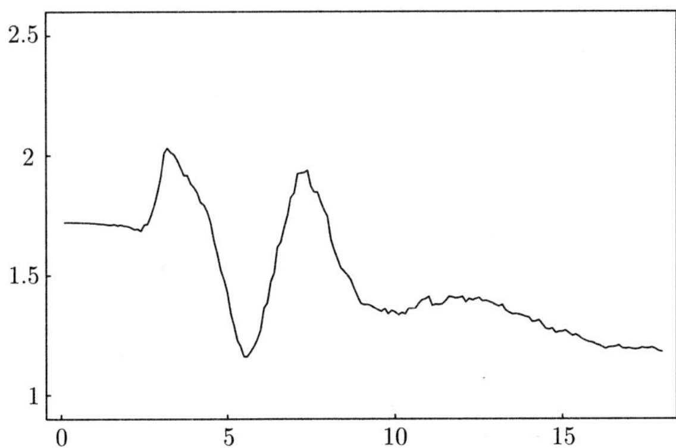

Fig. 3. CE data hazard ratio for subject with NA = 0.7 versus NA = 0.35.

Our analysis is far from complete; including other variables in the model could change interpretations. For example, we do not adjust the sodium infusion rate for the baseline sodium concentration in our analysis because baseline sodium was nonsignificant in the model with an exponential baseline. Moreover, the sample size is very small, and the precise timing of CE is actually uncertain. The diagnosis could well have come after the actual onset of CE, in which case our data should be considered to be left censored. And finally, one must be very careful to not over interpret the associations found between increasing fluids and sodium and the onset of CE since it may well be the case that the physicians were simply reacting to a child who was not doing well, in which case it could be the onset of CE that was bringing on the higher levels of these factors rather than the other way around. A more careful analysis of these data is under consideration.

6.   Discussion

As mentioned in Section 1, Bayesian joint models for longitudinal and survival data are described in some detail in Ibrahim, Chen and Sinha (2001). Similar to the development there, it is possible to accommodate covariate modeling by extending the model in this article. In general, in place of observing the true covariate $x(\cdot)$ in (6), suppose we observe $x^*(t) = x(t) + \epsilon(t)$, where $x(t)$ is either stochastic or deterministic and $\epsilon(t)$ is stochastic error. For example, consider $x(t) = \gamma_0 + \gamma_1 t + \cdots + \gamma_k t^k$. Then (6) can be viewed as a conditional model with true covariate $x(t)$ that is related to the observable $x^*(t)$ as modeled. Let γ denote the parameters associated with $x(t)$ and let τ denote any parameters associated with the error process $\epsilon(t)$. The full model can be written by multiplying the conditional model by $[x^*(\cdot)|\gamma, \tau]$. Now, in the Gibbs sampling scheme, the full conditionals for β and $S_0(\cdot)$ remain the same as given in previous sections while an additional draw from $[\gamma, \tau|x^*(\cdot), \beta, S_0(\cdot)]$ is needed. We defer the details to another article.

Results obtained here are easily extended to right truncated data by straightforward modification of the likelihood functions, while left truncated data create difficulties due to the necessity of observing covariate processes from time zero.

In theory, the model can be fit to arbitrary continuous $x(t)$ subject to $\|x(t)\|$ being bounded. The main obstacle in fitting the model is performing the integrations involved in the cumulative hazards. However, these can be computed using (4) and (5) in conjunction with any number of numerical integration techniques. The Bayesian approach to these models has typically assumed the baseline hazard to be a step function with jumps at quantiles suggested by the data (Ibrahim, Chen and Sinha, 2001). Although this is a flexible approach we believe that the MPT is perhaps a more realistic model for the baseline survival function that allows for the straightforward inclusion of prior information on baseline survival.

In regard to the dependence of inference on the choice of baseline centering family, Hanson and Johnson (2002) note that posterior density estimates were practically indistinguishable when fitting MPT models centered at the normal and logistic families to a particular data set. However, posterior inference will inherit some characteristics of the centering family, especially in the tails of S_0. When the MPT is employed as a means of generalizing a given parametric family, we view such inheritance as an advantage. Nonetheless, the magnitude of the impact of the chosen parametric family can be controlled via M, which defines how fine a mesh is used on $\mathbf{R}^+$, and c, which probabilistically affects how "close" S_0 is to members of the centering family. A choice of small c and large M diminishes the role of the parametric family but results in a model that is much harder to fit, mainly because then the posterior density has many peaked modes and the MCMC algorithm does not mix well. Reasonable choices strike a good balance between computational ease and relaxation of the parametric form. Also, although we fixed the parameter $c = 1$ in all models, there seems to be real information in the data for c (Hanson, 2004). It is certainly possible to place a vague prior on c, if desired.

It has been pointed out to us by a referee that the development of AFTD appears similar to what is called Step-Stress Accelerted Life Testing in the reliability literature. Parametric models in this area go back to DeGroot and Goel (1979, 1988) and Shaked and Singpurwalla (1983). More recent work includes van Dorp et al. (1996) and Tyoskin and Krivolapov (1996).

Interpretation of the AFT model, with fixed covariates, is designed to be meaningful when seen as specifying time scale changes via covariate values. Attempts at interpretation via hazards lead to difficulties if baseline hazard is arbitrary. Such properties are inherited by the AFTD from the AFT.

Acknowledgements

The authors thank Drs. Nathan Kupperman and Nicole Glaser of the University of California at Davis Medical School for permission to use their CE data and for providing medical insight for the analysis presented here.

References

Aalen, O.O. (1980). A model for nonparametric regression analysis of counting processes. In *Springer Lecture Notes in Statistics*, W. Klonecki, A. Kozek and J. Rosinski, editors, **2**, 1-25.

Antoniak, C.E. (1974). Mixtures of Dirichlet processes with applications to Bayesian nonparametric problems. *Annals of Statistics* **2**, 1152-1174.

Arjas, E. and Gasbarra, D. (1994). Nonparametric Bayesian inference from right censored survival data using the Gibbs sampler. *Statistica Sinica*, **4**, 505-524.

Baltazar-Aban, I. and Pena, E.A. (1995). Properties of hazard based residuals and implications in model diagnostics. *Journal of the American Statistical Association*, **90**, 185-197.

Berger, J.O. and Guglielmi, A. (2001). Bayesian testing of a parametric model versus nonparametric alternatives. *Journal of the American Statistical Association*, **96**, 174-184.

Brown E.R. and Ibrahim J.G. (2003). A Bayesian Semiparametric Joint Hierarchical Model for Longitudinal and Survival Data. *Biometrics*, **59**, 221-228.

Chaloner, K. (1991). Bayesian residual analysis in the presence of censoring. *Biometrika*, **78**, 637-644.

Chaloner, K. and Brant, R. (1988). A Bayesian approach to outlier detection and residual analysis. *Biometrika*, **75**, 651-659.

Chen, M.H., Shao, Q.M. and Ibrahim, J.G. (2000). *Monte Carlo Methods in Bayesian Computation*. Springer-Verlag: New York.

Christensen, R. and Johnson, W.O. (1988). Modeling accelerated failure time with a Dirichlet process. *Biometrika*, **75**, 693-704.

Cox, D.R. (1972). Regression models and life tables (with discussion). *Journal of the Royal Statistical Society, Series B*, **34**, 187-208.

DeGroot, M.H. and Goel, P.K. (1979). Bayesian estimation and optimal designs in partially accelerated life testing, *Naval Research Logistics Quarterly*, **26**, 223-235.

DeGroot, M.H. and Goel, P.K. (1988). Bayesian design and analysis of accelerated life testing with step-stress. In: Proc. S.I.P. Course CII, Accelerated Life Testing and Experts' Opinion in Reliability, C.A. Clarotti and D.V. Lindley editors, 193-2021, North Holland.

Doksum, K.A. (1974). Tailfree and neutral random probabilities and their posterior distributions. *Annals of Probability*, **2**, 183-201.

Dykstra, R.L. and Laud, P.W. (1981). A Bayesian nonparametric approach to reliability. *Annals of Statistics*, **9**, 356-367.

Escobar, M.D. (1994). Estimating normal means with a Dirichlet process prior. *Journal of the American Statistical Association*, **89**, 268-277.

Escobar, M.D. and West, M. (1995). Bayesian density estimation and inference using mixtures. *Journal of the American Statistical Association*, **90**, 577-588.

Faucett, C.J. and Thomas, D.C. (1996). Simultaneously modeling censored survival data and repeatedly measured covariates: A Gibbs sampling approach. *Statistics in Medicine*, **15**, 1663-1685.

Ferguson, T.S. (1973). A Bayesian analysis of some nonparametric problems. *Annals of Statistics*, **1**, 209-230.

Ferguson, T.S. (1974). Prior distributions on spaces of probability measures. *Annals of Statistics*, **2**, 615-629.

Geisser, S. (1993). *Predictive Inference: An Introduction.* London: Chapman & Hall.

Gelfand, A.E. and Kottas, A. (2002). A computational approach for full nonparametric Bayesian inference under Dirichlet process mixture models. *Journal of Computational and Graphical Statistics*, **11**, 289-305.

Glaser, N., Barnett, P., McCaslin, I., Nelson, W., Trainor, J., Louie, J., Kaufman, F., Quayle, K., Roback, M., Malley, R. and Kuppermann, N. (2001). Risk factors for cerebral edema in children with diabetic ketoacidosis. *New England Journal of Medicine*, **344**, 264-269.

Hanson, T.E. (2004). Inference for mixtures of finite Polya tree models. Technical Report, University of New Mexico Department of Mathematics and Statistics.

Hanson, T., Bedrick, E.J., Johnson, W.O. and Thurmond, M.C. (2003). Modeling fetal survival in dairy cattle. *Statistics in Medicine*, **22**, 1725-1739.

Hanson, T. and Johnson, W.O. (2004). A Bayesian semiparametric AFT model for interval censored data. *Journal of Computational and Graphical Statistics*, **13**, 341-361.

Hanson, T.E. and Johnson, W.O. (2002). Modeling regression error with a mixture of Polya trees. *Journal of the American Statistical Association*, **97**, 1020-1033.

Hjort, N.L. (1990). Nonparametric Bayes estimators based on beta processes in models of life history data. *Annals of Statistics*, **18**, 1259-1294.

Ibrahim, J.G., Chen, M.H., Sinha, D. (2001). *Bayesian Survival Analysis.* Springer-Verlag, New York.

Ishwaran, H. and James, L.F. (2004). Computational methods for multiplicative intensity models using weighted gamma processes : Proportional hazards, marked point processes and panel count data. *Journal of the American Statistical Association*, **99**, 175-190.

Johnson, W.O. and Christensen, R. (1989). Bayesian nonparametric survival analysis for the accelerated failure time model. *Statistics and Probability Letters*, **8**, 179-184.

Kalbfleisch, J.D. (1978). Nonparametric Bayesian analysis of survival time data. *Journal of the Royal Statistical Society B*, **40**, 214-221.

Kalbfleisch, J.D. and Prentice, R.L. (1980). *The Statistical Analysis of Failure Time Data.* Wiley, NY.

Kottas, A. and Gelfand, A.E. (2001). Bayesian semiparametric median regression modeling. *Journal of the American Statistical Association*, **95**, 1458-1468.

Kuo, L. and Mallick, B.K. (1997). Bayesian semiparametric inference for the accelerated failure-time model. *Canadian Journal of Statistics*, **25**, 457-472.

Laud, P.W., Damien, P. and Smith, A.F.M. (1998). Bayesian nonparametric and covariate analysis of failure time data. In *Practical Nonparametric and Semiparametric Bayesian Statistics*, D.Dey, P.Müller and D.Sinha editors. *Lecture Notes in Statistics*, **133**, 213-225.

Laud, P.W., Damien, P. and Smith, A.F.M. (1996). Monte Carlo methods for approximating a posterior hazard rate process. *Statistics and Computing*, **6**, 77-83.

Lavine, M. (1992). Some aspects of Polya tree distributions for statistical modeling. *Annals of Statistics*, **20**, 1222-1235.

Lavine, M. (1994). More aspects of Polya tree distributions for statistical modeling. *Annals of Statistics*, **22**, 1161-1176.

Lo, A.Y. and Weng, C.S. (1989). On a class of Bayesian nonparametric estimates : II. Hazard rates estimates. *Annals of the Institute for Statistical Mathematics*, **41**, 227-245.

Neal, R.M. (2003). Slice sampling (with discussion). *Annals of Statistics*, **31**, 705-767.

Nelson, W. (1972). Theory and applications of hazard plotting for censored failure data. *Technometrics*, **14**, 945-965.

Nieto-Barajas, L.E. and Walker, S.G. (2002). Bayesian nonparametric survival analysis for Lévy driven Markov processes. Technical Report, University of Bath.

Paddock, S.M., Ruggeri, F., Lavine M. and West, M. (2003). Randomized Polya tree models for nonparametric Bayesian inference. *Statistica Sinica*, **13**, 443-460.

Prentice, R.L. (1982). Covariate measurement errors and parameter estimation in a failure time regression model. *Biometrika*, **69**, 331-342.

Sahu, S.K. and Dey, D.K. (2004). On multivariate survival models with a skewed frailty and a correlated baseline hazard process. *Skew-Elliptical Distributions and Their Applications: A Journey Beyond Normality*, M.G. Genton (ed). CRC/Chapman & Hall, Boca Raton, FL, 321-338.

Sinha, D. and Dey, D.K. (1997). Semiparametric Bayesian analysis of survival data. *Journal of the American Statistical Association*, **92**, 1195-1212.

Shaked, M. and Singpurwalla, N.D. (1983). Inference for step-stress accelerated life tests. *J. Statistical Planning & Inference*, **7**, 295-306.

Tsiatis, A.A., Dafni, U., DeGruttola, V., Propert, K., Strawderman, R.L. and Wulfsohn, M. (1992). The relationship of CD4 counts over time to survival in patients with AIDS: Is CD4 a good surrogate marker? In *AIDS Epidemiology: Methodological Issues*, N. Jewell, K. Keitz and V. Farewell editors, 256-274, Birkhauser, Boston.

Tierney, L (1994). Markov chains for exploring posterior distributions. *Annals of Statistics*, **22**, 1701-1762.

Tsiatis, A.A., DeGruttola, V. and Wolfsohn, M. (1995). Modeling the relationship of survival to longitudinal data measured with error: Applications to survival and CD4 counts in patients with AIDS. *Journal of the American Statistical Association*, **90**, 27-37.

Tyoskin, O.I. and Krivolapov, S.Y. (1996). Nonparametric model for step-stress accelerated life testing. *IEEE Transactions on Reliability*, **45**, 346 - 350.

van Dorp, J.R., Mazzuchi, T.A., Fornell, G.E. and Pollock, L.R. (1996). A Bayes approach to step-stress accelerated life testing. *IEEE Transactions on Reliability*, **45**, 491-498.

Walker, S.G. and Mallick, B.K. (1997). Hierarchical generalized linear models and frailty models with Bayesian nonparametric mixing. *Journal of the Royal Statistical Society B*, **59**, 845-860.

Walker, S.G. and Mallick, B.K. (1999). Semiparametric accelerated life time model. *Biometrics*, **55**, 477-483.

Wang, Y. and Taylor, J.M.G. (2001). Jointly modeling longitudinal and event time data with application to acquired immunodeficiency syndrome. *Journal of the American Statistical Association*, **96**, 895-905.

Zellner, A. (1975). Bayesian Analysis of Regression Error Terms. *Journal of the American Statistical Association*, **70**, 138-144.

Zellner, A. and Moulton, B.R. (1985). Bayesian regression diagnostics with applications to international consumption and income data. *Journal of Econometrics*, **29**, 187-211.

Bayesian Statistics and Its Applications
Edited by S.K. Upadhyay, U. Singh and D.K. Dey
Anamaya Publishers, New Delhi, India

Biological Monitoring: A Bayesian Model for Multivariate Compositional Data

Devin S. Johnson[1], Jennifer A. Hoeting[2] and N. LeRoy Poff[3]

[1]National Marine Mammal Laboratory, Alaska Fisheries Science Center, NOAA, 7600 Sand Point Way NE, Seattle WA, 98115, USA

[2]Department of Statistics, Colorado State University, Fort Collins, CO 80523-1877, USA

[3]Department of Biology, Colorado State University, Fort Collins, CO 80523, USA

Abstract

This article develops a model to relate a multivariate compositional response to a number of covariates and proposes a new graphical model, called the random effects discrete regression (REDR) model, which allows for examination of the complex conditional relationships between a set of covariates and multiple discrete response variables. The approach offers a number of advantages over previous approaches and allows for a wide range of inferences. Relationships between compositional observations can be evaluated through a set of interaction parameters and inference about the influence of covariates is possible through a set of regression coefficients. The model also allows for examination of relationships between the covariates via another set of interactions. Parameter inference via Bayesian methods and MCMC is discussed. The proposed model and MCMC methods are used to examine the relationship between compositional observations of two characteristics of fish species and a number of covariates. These relationships are of interest to the U.S. Environmental Protection Agency for stream monitoring.

1. Introduction

This article proposes a class of models for compositional data based on traditional graphical models. A compositional observation $\mathbf{P} = (P_1, \ldots, P_D)$ possesses the two constraints: $\sum_j P_j = 1$ and $P_j \geq 0$ for $j = 1, \ldots, D$. Compositional analysis is preferred to the unconstrained positive multivariate observations if relative size is a more appropriate measure than the observed counts for each vector element. A discrete multivariate compositional observation arises when numerous sampled individuals at a site are cross classified according to a number of categorical variables forming a (possibly) high dimensional contingency table. The composition of interest is the probability that a randomly selected individual falls into a particular cross classification. The goal herein is to relate a multivariate compositional response to a number of covariates. Our approach offers a number of advantages over previous approaches and allows for a wide range of inferences including inferences on the relationship between various compositional response variables as well as inferences on the relationships between compositions and the explanatory variables.

Aitchison (1986) provides an overview of models for compositional data. There are a number of challenges in modeling compositional data. First, there is a necessary correlation structure due to the sum-to-one constraint. In fact, Pearson (1897) used compositional data as an example of spurious correlation. Secondly, there is an absence of an interpretable correlation structure. Not all positive definite matrices are valid covariance matrices for compositional random vectors. Finally, many existing models impose a rigid correlation structure that is due solely to the sum-to-one constraint. The Dirichlet distribution possesses this inflexibility.

The logistic-normal distribution has been proposed as a suitable distribution for compositional data in that it has a sensible covariance structure and offers a suitably rich family of distributions with which to model compositional data. If the random vector $\mathbf{X} = (X_1, \ldots, X_{D-1})$, where $X_j = \log(P_j/P_D)$, follows a multivariate normal distribution, then $\mathbf{P} = (P_1, \ldots, P_D)$ follows a logistic normal distribution (Aitchison and Shen, 1980; Aitchison, 1982). This distribution is not fully suitable for the problem considered here, however, because our data are discrete compositional data with extra variability due to sampling error. Also, the logistic normal density is not defined when the proportion in a particular class P_j equals 0, a common occurrence when the raw data are counts. A solution to this second issue was proposed by Ghosh et al. (2002) who adopt a mixture distribution consisting of a logistic normal distribution and a distribution that is degenerate at 0.

Billheimer (1995), Billheimer and Guttorp (1997), and Billheimer et al. (2001) proposed a model for a univariate compositional response which eliminates the problems in the logistic normal model caused by zeroes and low abundance counts when modeling species compositions. We build upon and extend their model to allow for multivariate compositions.

Graphical models are distributions for analyzing the conditional relationships of a Markov random field (Lauritzen, 1996). We propose a two component graphical chain model in which a set of categorical (or discrete) random variables is modeled as a response to a set of categorical and continuous covariates. This proposed model, called the random effects discrete regression (REDR) model, offers a number of advantages to analyzing compositional data. For the univariate composition considered in the Billheimer-Guttorp model, it is challenging to extend the results to more than three levels of response in a given category. We offer a graphical model framework which overcomes this difficulty and use the REDR model to construct a Bayesian hierarchical model for inference. Using the REDR model and some new results on graphical models, the complex conditional relationships between the covariates and multiple response variables as a whole system through inference from a graphical chain model can be examined.

To illustrate analysis with the REDR model, we examine the relationship between two characteristics of fish species and a number of environmental covariates. These relationships are of interest to the U.S. Environmental Protection Agency for stream monitoring. Our focus is on an application related to ecological modeling, but compositional data are prevalent in many other areas including geology, biology, economics and chemistry.

Section 2 describes the problem of interest while Section 3 introduces the model and a Bayesian approach to parameter estimation. The results of the fish species abundance analysis are discussed in Section 4.

2. Statistical Analysis of Species Composition

Relating the presence or absence of organisms and their biological characteristics to local environmental conditions is a challenging problem for ecologists (Legendre et al., 1997). Distributions of specific species are usually of limited interest as they are often biogeographically constrained, thereby restricting ecological inference to one geographic range. Recoding taxonomic species in terms of their membership in a number of functional trait categories allows for modeling of organism-environment relationships that can transcend biogeographic boundaries and may even apply across ecosystem types (Poff and Allan, 1995).

2.1   Previous Work

The problem of relating species traits to environmental conditions is of interest to ecologists trying to understand basic natural processes. Beyond this, scientists and policy makers concerned with monitoring natural resources also find such analyses to be useful. The U.S. Environmental Protection Agency (EPA) monitors the chemical, physical, and biological quality of streams in the United States, in part by counting the number of fish species at a number sample locations and relating this to environmental conditions at the sites. Since fish species vary over the landscape due to a large number of factors, it is useful to monitor the traits of fish instead of the specific species themselves. For example, species differ in their tolerance for high organic pollutant loads that reduce dissolved oxygen in the water; characterizing the relative proportion of species possessing such tolerance across the landscape can provide insight into water pollution. Because all species can be characterized in terms of possessing this trait, the functional approach allows for broad, interregional comparisons.

Various approaches have been used to describe the relationship between species traits and environmental conditions. One approach, canonical correspondence analysis, attempts to ordinate each species along a set of environmental axes (ter Braak, 1985). Dolèdec et al. (1996) continued the ordination approach by developing methods for marginally and jointly analyzing so called R, L, and Q tables, where R is a table with data on environmental variables at each sampling site, L is a table of species occurrences at each site, and Q is a table of trait classifications for each species.

A more direct approach was introduced by Legendre et al. (1997), called "a solution to the fourth corner problem." For a single trait with multiple levels, the four corners represent four matrices: (i) a matrix of environmental variables by site, (ii) an indicator matrix of species presence by site, (iii) an indicator matrix of functional trait levels by species and (iv) a matrix of parameters relating environmental variables to the trait. The parameters in matrix (iv) are product moment correlations between the trait counts and environmental variables and are estimated by a method of moments approach.

There are three main problems with the previous methodologies. First, these approaches only consider a single response variable at a time, i.e., multiple traits cannot be analyzed simultaneously. Secondly, both of the previous approaches measure only marginal associations between the environment and traits in question. Conditional relationships can give a more detailed measure of association between variables. For example, variables that are marginally correlated may in fact be independent upon conditioning on a third variable. This may provide evidence of possible mitigation by the third variable. Finally, the previous methods provide no predictive ability. If a researcher wants to predict the functional composition of a biological community at a site using remotely sensed environmental measurements, the previous methods provide no means to accomplish this task. The methodology proposed in Section 3 addresses all three of these issues. In the following section the data set of interest is described.

2.2   Fish Traits in the MAHA Region

The EPA and other agencies are interested in assessing the ecological condition of streams and identifying any factors that might be associated with any degradation in stream conditions (U.S. Environmental Protection Agency, 2000). During 1993–1996 the EPA, along with the U.S. Fish and Wildlife Service and other contractors, surveyed 309 wadeable streams in the Mid-Atlantic Highlands as part of the Environmental Monitoring and Assessment Program (EMAP). Here, we will consider the data from 1994. Streams in the Mid-Atlantic Highlands Assessment (MAHA) were sampled during a 12-week period from April to July. Chemical samples and several physical habitat variables were measured at the sampled sites. Fish were sampled by electro-fishing and each fish species was classified according to several taxonomic and ecological categories. McCormick et al. (2001) provides a list of all fish species in the MAHA region along with their trait classifications. A set of watershed scale environmental variables was also calculated for these sites from a GIS (Geographic Information System) model. These variables include metrics such as watershed area and average amount of precipitation in the watershed. Sites that were considered to be incapable of maintaining a fish population and sites for which environmental covariates were not recorded were eliminated from the analysis (McCormick et al., 2001). After removing these sites, 91 sites remained with between 1 and 23 fish species observed at each site.

In order to perform the functional trait analysis of species occurrence, each observed species is categorized according to two categorical variables, habit and tolerance. The habit variable describes where species live in two levels: benthic species live on or near the stream bottom and column species inhabit water depths between the surface of a stream and the bottom. Species tolerance refers to a species' ability to withstand degraded environmental conditions caused by heavy silt loads or low dissolved oxygen. Species classified as intolerant are sensitive to human induced stream degradation, whereas tolerant species are relatively insensitive. Species between the two extreme tolerance classifications are classified as having intermediate tolerance. The proportion of benthic and intolerant species are important metrics used by the EPA to measure stream degradation (McCormick et al., 2001). Table 1 shows the cell counts for two sites. These species distributions at each site are the response in our models. Note that there is one contingency table for every sampled site. The distribution of cell counts for all sites is shown in Fig. 1.

Table 1. Cross tabulation of the response for two sites: the number of species observed in each combination of habit and tolerance category

			Tolerance	
	Habit	Intolerant	Intermediate	Tolerant
Site 1	Column	0	4	3
	Benthic	0	1	5
Site 53	Column	1	3	3
	Benthic	3	3	3

In our analysis, we are interested in the associations between the species distribution at each site and covariates measuring local environmental conditions and landscape setting (Table 2). The local covariates include stream site sulfate concentration (log μeq/L), which measures atmospheric acid deposition or acid mine drainage; chloride concentration (log μeq/L), which is associated with human activity; and water turbidity (log NTU), a measure of water cloudiness caused by very fine

Table 2. Summary of environmental covariates for 1994 MAHA streams

Covariate	Mean	St. Dev.	Min.	Max.
Area (km^2)	3.04	1.19	0.78	6.39
Chloride (μeq/L)	4.60	1.14	2.64	6.98
Elevation (kilometers)	0.07	0.03	0.0015	0.14
Precipitation (meters)	1.08	0.10	0.85	1.33
Sulfate (μeq/L)	5.42	0.88	3.78	8.42
Turbidity (NTU)	1.01	0.77	−0.92	3.40

suspended sediments. The landscape covariates include watershed area (log km^2), elevation (km), a measure of stream area, and mean annual watershed precipitation (meters), a measure of stream volume.

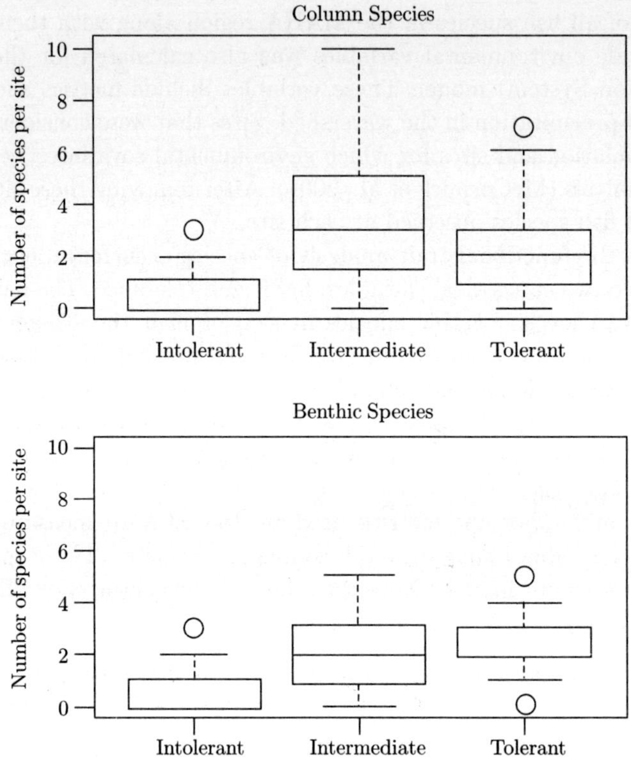

Fig. 1. **Distribution of fish species abundance for all 91 sites for the three different tolerance levels. Abundances are separated by habit type, column species and benthic species.**

3. Model Formulation

We consider models for multivariate compositional data where sampled individuals are classified according to two or more different categorical variables. The focus is on the proportion of counts of a particular category of possible joint outcomes, or equivalently, the probability that a randomly selected individual will belong to a certain cross-classification (cell).

The independence structure of the discrete variables used for cross-classification is also examined. For example, for a two-way cross-classification of individuals according to the discrete variables I and J, we examine whether the classification of a random individual to category i of variable I is independent of the event that the individual is classified to category j of variable J. In other words, we may be interested in whether separate probabilities should be modeled for each cell, or whether a model of the form $p_{ij} = p_i p_j$ would be more appropriate, where p_{ij}=Pr{individual in cell (i,j)} and p_i=Pr{individual in category i of I}. Even if independence of the classification variables is not a primary research concern, it may still be appropriate to include some independence structure to reduce the number of parameters and provide a more parsimonious model.

The term "site" is used to index each multivariate discrete compositional observation. If only the fully saturated model (complete dependence among all classification variables) is of interest, then a simple model that combines all of the cells into a single composition can be used. This is the approach used by Dominici (2000) to combine several contingency tables, some with missing dimensions. However, if one is interested in introducing some independence structure to the cell probabilities, then an extension of the simpler model must be considered and is proposed here.

The models proposed below are motivated by the class of models known as graphical models. In standard graphical modeling every sampled individual generates a multivariate observation of categorical and continuous variables. However, there is only one observation of the covariate vector for all of the individuals observed at a particular site. To achieve our goal of developing a class of graphical models to address this problem, the model developed here requires a more sophisticated model structure than either the standard graphical model or linear model structures.

3.1 Developing a Single Site Model

Let Φ denote the set of categorical response variables, of which there are D possible cross-classifications on the product space of the response variable levels. For each site, we are interested in the dependence relationships *among* the covariates measured at the site and the relationship *between* the covariates and the event that a randomly selected individual is cross-classified into one of the D cells $\mathbf{y}_\Phi$. Let $\mathbf{Y}_\Phi = \{Y_\phi : \phi \in \Phi\}$ denote the response vector for a single individual, which takes one of the D possible vectors, say $\mathbf{y}_\Phi$, as a realization. In addition, let $\mathbf{X} = (X_1, \ldots, X_p)$ denote a vector of observable covariates at the randomly selected site. The vector $\mathbf{X}$ can also be written $(\mathbf{X}_\Gamma, \mathbf{X}_\Delta)$, where Γ indexes the continuous covariates and Δ indexes the discrete covariates. In the analysis of fish species occurrence described in Section 2.2, we are interested in the event that a randomly selected fish species at a particular site belongs to a certain cross-classification of life-history traits. Table 3 gives such an example for a single "individual" (a single species of fish) observed at site 53. Section 4.1 further develops the model described below for the analysis of fish species occurrence.

Table 3. **Notation of Section 3.1 for the fish species occurrence example described in Section 2.2. Response $\mathbf{y}_\Phi$ is the response for a single species at site 53. This species has a benthic habit and is categorized as a tolerant species. Site specific indices are added in Section 3.2**

Notation	Example
Φ	Habit (column or benthic) and tolerance (intolerant, intermediate and tolerant)
D	$6 = (2 \text{ habit levels} \times 3 \text{ tolerance levels})$
$\mathbf{y}_\Phi$	(2,3) for 2nd level of habit variable and 3rd level of tolerance variable
$\mathbf{x}$	The values of the 6 continuous environment covariates at site 53

Typically many individuals are sampled at any given site. We now define the likelihood model for sampling N individuals at a site and subsequently develop the specific terms of the likelihood, including the probability that a single individual will be in a single cell at a given site. We first condition on the realization of a site, which amounts to conditioning on the covariates. Sampling N individuals at a site provides N realizations of the variable $\mathbf{Y}_\Phi$. These N realizations can be summarized into a D vector of counts $\mathbf{c} = \{c(\mathbf{y}_\Phi)\}$, where $c(\mathbf{y}_\Phi)$ represents the number of individuals that were cross-classified into cell $\mathbf{y}_\Phi$. The count vector $\mathbf{c}$ represents a complete and sufficient summarization of the N individual responses, so we can model the counts in order to make inference to $\mathbf{Y}_\Phi$. Using a multinomial model for the count vector $\mathbf{c}$, the joint distribution for the counts and covariates, for a fixed sample size N, is

$$f(\mathbf{c}, \mathbf{x}) = f_M(\mathbf{c}|\mathbf{x}) f_{CG}(\mathbf{x}) = \frac{N!}{\prod_{\mathbf{y}_\Phi} c(\mathbf{y}_\Phi)!} \left\{ \prod_{\mathbf{y}_\Phi} f(\mathbf{y}_\Phi|\mathbf{x})^{c(\mathbf{y}_\Phi)} \right\} \times f_{CG}(\mathbf{x}), \qquad (1)$$

where $f(\mathbf{y}_\Phi|\mathbf{x}) f_{CG}(\mathbf{x})$ represent the joint density of an individual classified to cell $\mathbf{y}_\Phi$ and covariates observed at the randomly selected site where the individual is observed. These terms are defined below.

The model in (1) includes terms for the probability that a single individual will be in a particular cell at a single site, $f(\mathbf{y}_\Phi|\mathbf{x}) f_{CG}(\mathbf{x})$. In the analysis of fish species occurrence described in Section 2.2, this corresponds to the probability that a randomly selected fish species at a particular site (which has a particular set of site-level covariates) belongs to a certain cross-classification of life-history traits. Using a log-linear model, we now specify a joint density for $(\mathbf{Y}_\Phi, \mathbf{X})$, which we call the discrete regression (DR) model,

$$f(\mathbf{y}_\Phi|\mathbf{x}) = \exp\left\{ \alpha_\Phi(\mathbf{x}) + \sum_{f \subseteq \Phi} \sum_{c \subseteq \Gamma} \sum_{d \subseteq \Delta} \beta_{fcd}(\mathbf{y}_\Phi, \mathbf{x}_\Delta) \prod_{\gamma \in c} x_\gamma + \sum_{\gamma \in \Gamma} \sum_{d \subseteq \Delta} \sum_{m=2}^{M} \omega_{\gamma dm}(\mathbf{y}_\Phi, \mathbf{x}_\Delta) x_\gamma^m \right\} \qquad (2)$$

and

$$f_{CG}(\mathbf{x}) = \exp\left[\sum_{d \subseteq \Delta} \left\{ \lambda_d(\mathbf{x}_\Delta) + \eta_d(\mathbf{x}_\Delta)' \mathbf{x}_\Gamma - \frac{1}{2} \mathbf{x}_\Gamma' \mathbf{\Psi}_d(\mathbf{x}_\Delta) \mathbf{x}_\Gamma \right\} \right]. \qquad (3)$$

In the response model (2), $\alpha_\Phi(\mathbf{x})$ is a normalizing constant with respect to $\mathbf{Y}_\Phi|\mathbf{x}$. The regression coefficients $\beta_{fcd}(\mathbf{y}_\Phi, \mathbf{x}_\Delta)$, and $\omega_{\gamma dm}(\mathbf{y}_\Phi, \mathbf{x}_\Delta)$ correspond to interaction terms which depend on $\mathbf{y}_\Phi$ and $\mathbf{x}_\Delta$ only through the variables associated with the sets $f \subseteq \Phi$ and $d \subseteq \Delta$, respectively. The summation from $m=2$ to M denotes the number of higher-order terms desired in the model for covariates x_γ. For example, if $M = 3$ the summation is over x_γ^2 and x_γ^3. The first order terms enter into the model in the β terms with c equal to a singleton set. In the response portion of the model (2), interaction terms for which $f = \emptyset$ can, without loss of generality, be set to zero (e.g. $\beta_{\emptyset cd} \equiv 0$ for any $c \subseteq \Gamma$ and $d \subseteq \Delta$) as they do not depend on $\mathbf{y}_\Phi$ and will cancel with the normalizing term $\alpha_\Phi(\mathbf{x})$. The conditional Gaussian (CG) density (3) is a joint distribution for continuous and discrete variables. As with the interaction terms in the response model $\lambda_d(\mathbf{x}_\Delta)$, $\eta_d(\mathbf{x}_\Delta)$, and $\mathbf{\Psi}_d(\mathbf{x}_\Delta)$ depend on $\mathbf{x}_\Delta$ only through the subset of variables associated with the set $d \subseteq \Delta$. The CG distribution is constructed by assuming that $\mathbf{X}_\Delta$ follows a log-linear model and $\mathbf{X}_\Gamma|\mathbf{x}_\Delta$ follows a multivariate normal distribution.

To complete the DR model we must impose some constraints to ensure identifiability of the model parameters. To accomplish this, first select a reference cell of $\mathbf{Y}$, say $\mathbf{y}_\Phi^*$, and a reference cell for the categorical covariates, say $\mathbf{x}_\Delta^*$. Without loss of generality, henceforth, we assume that $\mathbf{y}_\Phi^*$ and $\mathbf{x}_\Delta^*$ are appropriately sized vectors of ones, indicating the reference cells are those indexed by the first level of all the variables associated with Φ and Δ. Now that the reference cells are defined, set all interaction terms in (2) and (3) equal to zero if $y_\phi = 1$ for any $\phi \in f$ or $x_\delta = 1$ for any $\delta \in d$. These zero constraints are analogous to the zero constraints of interaction terms in classic ANOVA models. By using these constraints we can interpret the interaction terms as measuring interactions relative to the selected values $\mathbf{y}_\Phi^*$ and $\mathbf{x}_\Delta^*$. For example, given any response variable $\phi \in \Phi$ and any two covariates $\gamma \in \Gamma$, and $\delta \in \Delta$, a positive value for the interaction term $\beta_{\phi\gamma\delta}(\mathbf{y}_\Phi, \mathbf{x}_\Delta)$ implies that an increase in x_γ increases the probability that a randomly selected individual will be cross-classified according to a cell where $Y_\phi = y_\phi$ over a cell for which $Y_\phi = 1$ and the amount of increase depends on the categorical covariate X_δ.

Here we consider only the homogeneous CG distribution, where $\mathbf{\Psi}_d(\mathbf{x}_\Delta) = \mathbf{0}$ for $d \neq \emptyset$. This restriction is identical to the assumption that the covariance matrix of the continuous variables is constant over all of the cells defined by the product space of $\mathbf{X}_\Delta$. The model can be extended to be non-homogeneous, if desired.

3.2   Random Effects Discrete Regression

Section 3.1 described a model for a single randomly sampled site. Now, we extend this model to account for possibly hundreds of randomly selected sites. For each site, a separate model could be constructed, but this would increase the number of parameters to be estimated to an unmanageable level. In addition, the differences in non-zero parameter values are not usually of primary interest. Therefore, we propose a global random effects model for all sites that allows site-to-site flexibility in some of the non-zero parameter values. In order to add this flexibility as well as to model the randomness in site selection, we introduce a random error term to the response model (2).

The addition of a random effect to the response model (2) produces a full model for $\mathbf{Y}_\Phi$, $\mathbf{X}$, and the random effects $\boldsymbol{\epsilon}$ of the form

$$f(\mathbf{y}_\Phi, \mathbf{x}, \boldsymbol{\epsilon}) = f_{RE}(\mathbf{y}_\Phi | \mathbf{x}, \boldsymbol{\epsilon}) f_{CG}(\mathbf{x}) f(\boldsymbol{\epsilon}). \tag{4}$$

Since there is only one observation of the explanatory variables at each site we adopt the model for the covariates, $f_{CG}(\mathbf{x})$, that is given in (3). The response portion $f_{RE}(\mathbf{y}_\Phi | \mathbf{x}, \boldsymbol{\epsilon})$ of the random effects discrete regression (REDR) model is modified by the addition of a random intercept term to give

$$f_{RE}(\mathbf{y}_\Phi | \mathbf{x}, \boldsymbol{\epsilon}) = \exp \left\{ \alpha_\Phi(\mathbf{x}) + \sum_{f \subseteq \Phi} \sum_{c \subseteq \Gamma} \sum_{d \subseteq \Delta} \beta_{fcd}(\mathbf{y}_\Phi, \mathbf{x}_\Delta) \prod_{\gamma \in c} x_\gamma \right.$$
$$\left. + \sum_{f \subseteq \Phi} \sum_{\gamma \in \Gamma} \sum_{d \subseteq \Delta} \sum_{m=2}^{M} \omega_{f\gamma dm}(\mathbf{y}_\Phi, \mathbf{x}_\Delta) x_\gamma^m + \sum_{f \subseteq \Phi} \epsilon_f(\mathbf{y}_\Phi) \right\}, \tag{5}$$

where $\epsilon_f(y_\Phi) = 0$, if $y_\phi = 1$ for any $\phi \subseteq f$, to ensure identifiability. In order to allow modeling of a given independence structure for the multivariate response, we also introduce one other constraint on the random effects. For any set $f \subseteq \Phi$, if $\beta_{fcd}(\mathbf{y}_\Phi, \mathbf{x}_\Delta)$ and $\omega_{f\gamma dm}(\mathbf{y}_\Phi, \mathbf{x}_\Delta)$ are set to zero for all $\mathbf{y}_\Phi$ and $\mathbf{x}_\Delta$ then $\epsilon_f(\mathbf{y}_\Phi)$ is defined to be a zero vector. The remaining random interactions $\boldsymbol{\epsilon}_f = \{\epsilon_f(\mathbf{y}_\Phi) : y_\phi \neq 1 \text{ for any } \phi \subseteq f\}$ are given a multivariate distribution with mean $\mathbf{0}$ and covariance matrix (or scale parameter) $\mathbf{\Sigma}_f$.

The inclusion of random error terms in (5) has three benefits. First, the model can adjust for site-to-site variability. Secondly, the model will account for some level of over-dispersion to cell counts. Finally, every realization of the random effects provides cell probabilities that maintain the desired independence relationships among the response variable. Johnson, Hoeting and Fadely (2006) provide proof of this fact using a Möbius inversion calculation similar to Lauritzen (1996, p. 174). The site-based REDR model provides an improvement over the approach of Aitchison (1986) which only examines the average composition independence structure across sites.

In the REDR model description we did not specify the error distribution, as different situations may necessitate different error structures. If it is reasonable to assume that the error structure is symmetric with few outliers, then a multivariate-normal (MVN) distribution may be reasonable. In this case, the cell compositions will have a logistic-normal distribution. However, other distributions could be used. For example a multivariate t distribution with k degrees of freedom could be used if it is desirable to have an error with heavier tails or if there is a high level of over-dispersion in the cell counts. For the remaining discussion of the REDR model we will assume a MVN distribution for the random effects (i.e., $f(\epsilon_f) = f_N(\epsilon_f; \mathbf{0}, \mathbf{\Sigma}_f)$).

Now that we have added random effects to the response portion of the model, the likelihood for the response variable cell counts given the covariates $\mathbf{x}$ changes slightly from that given in (1). The conditional likelihood model for the response variable cell counts $\mathbf{c}$ given the covariates $\mathbf{x}$, the total number of individuals observed at a site, and the random effects is

$$f_M(\mathbf{c}|\mathbf{x}, \epsilon) = \frac{N!}{\prod_{\mathbf{y}_\Phi} c(\mathbf{y}_\Phi)!} \prod_{\mathbf{y}_\Phi} f_{RE}(\mathbf{y}_\Phi|\mathbf{x}, \epsilon)^{c(\mathbf{y}_\Phi)}, \tag{6}$$

where $f_{RE}(\mathbf{y}_\Phi|\mathbf{x}, \epsilon)$ is given by (5).

Now, we focus on the multiple site likelihood for the explanatory variables. Assuming that the covariate observations are independently distributed and follow a homogeneous CG distribution, we obtain the multiple site explanatory density

$$f(\mathbf{x}_1, \ldots, \mathbf{x}_S) = \prod_{i=1}^{S} f_{CG}(\mathbf{x}_i|\boldsymbol{\lambda}, \boldsymbol{\eta}, \boldsymbol{\Psi}_\emptyset), \tag{7}$$

where $\mathbf{x}_i$ denotes the set of observed covariates for sites $i = 1, \ldots, S$ and $\boldsymbol{\lambda}$ and $\boldsymbol{\eta}$ represent the collected parameter sets $\{\lambda_d(\mathbf{x}_\Delta) : d \subseteq \Delta\}$ and $\{\boldsymbol{\eta}_d(\mathbf{x}_\Delta) : d \subseteq \Delta\}$.

We now re-parameterize the homogeneous CG density in (7) into a more useful form. First, we break the CG density into a marginal model for the categorical components of the explanatory variable set and a conditional model for the continuous components. We then re-parameterize the conditional Gaussian distribution into an ANOVA-like form. This re-parameterization gives the following form for the homogeneous CG density,

$$f_{CG}(\mathbf{x}) = f(\mathbf{x}_\Delta) f(\mathbf{x}_\Gamma|\mathbf{x}_\Delta)$$

$$= \exp\left\{\sum_{d \subseteq \Delta} \lambda_d(\mathbf{x}_\Delta)\right\} \times \frac{1}{\sqrt{2\pi}} |\boldsymbol{\Psi}_\emptyset|^{1/2} \exp\left\{\frac{1}{2}\left(\mathbf{x}_\Gamma - \sum_{d \subseteq \Delta} \boldsymbol{\tau}_d(\mathbf{x}_\Delta)\right)' \boldsymbol{\Psi}_\emptyset \left(\mathbf{x}_\Gamma - \sum_{d \subseteq \Delta} \boldsymbol{\tau}_d(\mathbf{x}_\Delta)\right)\right\}, \tag{8}$$

where $\boldsymbol{\Psi}_\emptyset$ represents the inverse covariance matrix for the continuous variables, which have a MVN distribution, $\boldsymbol{\tau}_d(\mathbf{x}_\Delta) = \boldsymbol{\Psi}_\emptyset^{-1} \boldsymbol{\eta}_d(\mathbf{x}_\Delta)$, and $\lambda_\emptyset(\mathbf{x}_\Delta)$ represents a normalizing constant in the log-linear model for $\mathbf{x}_\Delta$.

Define the vector of cell counts $\mathbf{c}_\Delta = [c(\mathbf{x}_\Delta)]$, where $c(\mathbf{x}_\Delta)$ is the number of sites for which the categorical covariates $\mathbf{X}_\Delta = \mathbf{x}_\Delta$. Using the re-parameterization of the CG density in (8) we can write the joint density of the covariates over all sites (7) as

$$f(\mathbf{x}_1, \ldots, \mathbf{x}_S) = \left\{\prod_{i=1}^{S} f(\mathbf{x}_{\Delta i})\right\} \times \left\{\prod_{i=1}^{S} f_N(\mathbf{x}_{\Gamma i}|\mathbf{x}_{\Delta i})\right\} \propto f_M(\mathbf{c}_\Delta|\boldsymbol{\lambda}) \prod_{i=1}^{S} f_N\left(\mathbf{x}_{\Gamma i}; \sum_{d \subseteq \Delta} \boldsymbol{\tau}_d(\mathbf{x}_{\Delta i}), \boldsymbol{\Psi}_\emptyset\right), \tag{9}$$

where the explanatory observation at the ith site is given by $\mathbf{x}_i = (\mathbf{x}_{\Delta i}, \mathbf{x}_{\Gamma i})$ and $f_M(\mathbf{c}_\Delta|\boldsymbol{\lambda})$ is the multinomial density

$$f_M(\mathbf{c}_\Delta|\boldsymbol{\lambda}) = \frac{S!}{\prod_{\mathbf{x}_\Delta} c(\mathbf{x}_\Delta)!} \prod_{\mathbf{x}_\Delta} \exp\left\{\sum_{d \subseteq \Delta} \lambda_d(\mathbf{x}_\Delta)\right\}^{c(\mathbf{x}_\Delta)}. \tag{10}$$

The full likelihood for parameter estimation in the REDR model (4) is obtained by combining the likelihood for response variable cell counts at each site (6), the random effects density, and the explanatory variable likelihood (7).

$$\begin{aligned}
f\left(\{\mathbf{c}_i\}, \{\mathbf{x}_i\}, \{\boldsymbol{\epsilon}_i\}\right) &= \prod_{i=1}^{S} f_M(\mathbf{c}_i|\mathbf{x}_i, \boldsymbol{\epsilon}_i) f_{CG}(\mathbf{x}_i) f_N(\boldsymbol{\epsilon}_i) \\
&\propto \prod_{i=1}^{S} f_M(\mathbf{c}_i|\mathbf{x}_i) \times f_M(\mathbf{c}_\Delta|\boldsymbol{\lambda}) \times \prod_{i=1}^{S} f_N\left(\mathbf{x}_{\Gamma i}; \sum_{d \subseteq \Delta} \tau_d(\mathbf{x}_{\Delta i}), \boldsymbol{\Psi}_\emptyset\right) \tag{11} \\
&\quad \times \prod_{i=1}^{S} \prod_{f \subseteq \Phi} f_N\left(\boldsymbol{\epsilon}_{f,i}; \mathbf{0}, \boldsymbol{\Sigma}_f\right),
\end{aligned}$$

where $\mathbf{c}_i$ is a D vector of response variable cell counts for site i, $\mathbf{x}_i$ is a vector of observed covariates, c_Δ is a vector of cell counts for the categorical covariates, and $\boldsymbol{\epsilon}_i$ represents the collection of random effects vectors $\{\boldsymbol{\epsilon}_{f,i} : f \subseteq \Phi\}$ for the ith site.

3.3   Multivariate Compositions and Graphical Models

The models proposed in the previous sections were motivated by a class of models known as *graphical models* or Markov random fields. A graphical model, or more loosely a conditional independence model, is a probability density function for a multivariate vector that is parameterized in such a way that a complex independence structure can be characterized by a mathematical graph. A mathematical graph involves a set of vertices, one for each element of the vector, and a set of edges connecting some of the vertices. Edges can either be *undirected* or *directed*. An undirected edge between two vertices indicates bi-directional dependence between the elements while a directed edge signifies a causal or influential effect from one element to another. If two vertices are not connected, one can infer that the two variables which they represent are conditionally independent given their neighbors (those vertices which are connected to the pair in question). Lauritzen (1996) provides a thorough overview of graphical modeling.

The REDR model (5) and the corresponding CG model (8) for the covariates together define a chain graph model. The interaction parameters correspond to edges between response and covariate vertices. Edges between the covariates and response vertices are directed, whereas, within covariate and response vertices, the edges are undirected. A given graph represents the conditional independencies for a specific model. Edges between any two vertices are absent if and only if all interaction parameters with subscripts containing the two variable set are zero for all values of the covariates and response. Johnson and Hoeting (2003) provide a rigorous description of the graphical model properties of the REDR model.

There is one difference between the REDR model and standard graphical models. The joint models in (1) for counts and covariates are different than the standard sampling scheme for a graphical model. Usually, every individual sampled generates a multivariate observation of categorical and continuous variables. Here, however, there is only one observation of the covariate vector for all of the individuals observed at a particular site. The present sampling scheme is analogous to replication of an experiment at the same factor levels at each site.

3.4 Parameter Inference

In order to make inference about the parameters in the REDR model (4), we adopt a Bayesian approach for parameter estimation. The hierarchical structure of the REDR model makes Bayesian procedures particularly attractive. There are a large number of unobserved random effects that may or may not be considered nuisance parameters. If one is interested in dependence relationships between the observable covariates and the response variables, or dependence relationships between the response variables given the covariates, the random effects are usually considered nuisance parameters. As discussed in Johnson et al. (2006), these random effects can be marginalized over in some cases without affecting the conditional independencies of the joint distribution of the response and covariates. If, however, the unobserved compositions at all, or some, sites are of interest then estimates of the random effects are necessary for each site to calculate an estimate of the true site composition. When the goal is to predict compositions at sites for which only the explanatory variables are observed, estimates of the random effects for that site are also necessary. Modern Bayesian computational techniques can handle all of these goals with little modification. Full development of the Bayesian approach to parameter estimation is provided in the Appendix.

4. Graphical Analysis of Fish Species Occurrence

We adopt the model discussed above to analyze the fish trait data described in Section 2.2. Using the full REDR model, we can examine the complex conditional relationships between the environmental covariates, the habit response variable, and the tolerance response variable as a whole system through inference from a graphical chain model. The proportions of benthic and intolerant species are important metrics used by the EPA to measure stream degradation (McCormick et al., 2001). We are interested in inference concerning species occurrence, or the number of species observed in each cell.

4.1 Model Specification

We consider a main-effects-only model for the analysis of the multivariate composition of habit (H) and tolerance (T) species occurrence including only the first-order linear interaction terms. Consider the following multinomial model for the cell counts

$$f_M(\mathbf{c}_i|\mathbf{x}_i, \boldsymbol{\epsilon}_i) = \frac{N!}{\prod_{\mathbf{y}_\Phi} c(\mathbf{y}_\Phi)_i!} \prod_{\mathbf{y}_\Phi} f_{RE}(\mathbf{y}_\Phi|\mathbf{x}_i, \boldsymbol{\epsilon}_i)^{c(\mathbf{y}_\Phi)_i}, \tag{12}$$

where

$$f_{RE}(\mathbf{y}_\Phi) = \exp\left\{\alpha_\Phi(\mathbf{x}_i) + \sum_{f \subseteq \Phi}\sum_{\gamma=0}^{6} \beta_{f\gamma}(\mathbf{y}_\Phi)(x_{\gamma,i} - \overline{x}_\gamma) + \epsilon_{f,i}(\mathbf{y}_\Phi)\right\}. \tag{13}$$

Each of the environmental covariates was centered by subtracting its mean $\overline{x}_\gamma$. This was done to improve Markov chain convergence to the posterior distribution. We chose the reference cell $\mathbf{y}_\Phi^*$ to be column species with intermediate tolerance level. Thus, $\beta_{f\gamma}(\mathbf{y}_\Phi) \equiv \epsilon_{f,i}(\mathbf{y}_\Phi) \equiv 0$ for $\mathbf{y}_\Phi = (1,\ 2)$ and $f \subseteq \{H,\ T\}$ in equation (13).

We consider three different models. In the *independent* model, the habit variable is independent of the tolerance variable, so $\beta_{f\gamma}(\mathbf{y}_\Phi)$ and $\epsilon_{f,i}(\mathbf{y}_\Phi)$ are set to zero for $f = \{B,\ T\}$ for all covariates

γ, sites i, and cells $\mathbf{y}_\Phi$. In the *dependent* model with uncorrelated errors all interactions between cells are estimated. We assume the random effects vectors ϵ_f are independently distributed from one another for all sites. The *dependent correlated errors* model is identical to the previous model except that the random effects vectors are correlated within each site. This is equivalent to applying a single composition model to all six cells.

Since all of the environmental covariates are continuous, the homogeneous CG model reduces to a MVN distribution and we assumed $f_{CG}(\mathbf{x}_{\gamma,i} - \overline{\mathbf{x}}_{\gamma,i}) = MVN(\mathbf{x}_{ip} - \overline{\mathbf{x}}_p;\ \mathbf{0}, \boldsymbol{\Psi}_\emptyset)$. The centering here allows the elimination of the nuisance parameter $\boldsymbol{\tau}_\emptyset$ in (8), which is irrelevant for determining conditional independencies. Note that in this case the posterior distribution of $\boldsymbol{\Psi}_\emptyset$ is a Wishart distribution. However, since we are interested in the off-diagonal elements of $\boldsymbol{\Psi}_\emptyset$, using an MCMC sampling technique allows straightforward inference of these elements.

4.2 Model Estimation and Performance

MCMC procedures were performed with the hierarchically centered version described in the Appendix as well as with the model as originally parameterized in (13). The hierarchically centered version reached satisfactory convergence with substantially fewer MCMC iterations than (13).

The program WinBUGS was used to run the Gibbs sampler (Spiegelhalter et al., 2000). The Gibbs sampling algorithm was run for an initial 4000 iterations in which a MVN proposal density was tuned so that the Metropolis-within-Gibbs step for the parameters would have an acceptance rate of around 30%. The first 4000 iterations were discarded, after which the sampler was run for 20,000 iterations as a burn-in period. Finally, an additional 800,000 iterations was used for model inference. Standard diagnostics suggested that the Markov chains had converged to their corresponding posterior distributions (Givens and Hoeting, 2005). Every twentieth iteration was saved for parameter inferences in order to reduce storage constraints.

The Bayesian posterior predictive p-value method of Gelman et al. (1996) was used to assess model fit for each of the three models. With this method a goodness-of-fit statistic $T(y)$, which can be a function of the observed data y and model parameters, is used to compute the Bayesian predictive p-value $P_b = Pr\{T(y^{\text{rep}}) > T(y) \mid y\}$. The data y^{rep} represent a hypothesized replicate data set that could have resulted from the model and the interpretation is essentially the same as the classic p-value with the addition that the null distribution is the distribution of the statistic given only the observed data y. In an MCMC setting P_b is particularly easy to approximate by generating a replicate data set at each iteration in the Markov chain. The goodness-of-fit statistic used for this analysis was the Freeman-Tukey statistic (Freeman and Tukey, 1950)

$$T(\mathbf{c}_1, \ldots, \mathbf{c}_S) = \sum_{i=1}^{S} \sum_{\mathbf{y}_\Phi} \left(\sqrt{c(\mathbf{y}_\Phi)_i} - \sqrt{N_i f_{RE}(\mathbf{y}_\Phi|\mathbf{x}_i, \epsilon_i)} \right)^2, \qquad (14)$$

where $\mathbf{c}(\mathbf{y}_\Phi)_i$ is the number of species belonging to cell $\mathbf{y}_\Phi$ at site i, N_i is the total number of species observed at site i, and $f_{RE}(\mathbf{y}_\Phi|\mathbf{x}_i, \epsilon_i)$ is given by (13) and represents the cell composition.

This naive method of approximating the posterior predictive p-value can be conservative when rejecting the null hypothesis of model fit in that the MCMC approximated p-value can be too large when used without calibration (Robbins et al., 2000). Draper and Krnjajič (2005) propose a calibration method which could be included in a future analysis. The method is somewhat computationally intensive so we will omit it here and use the p-value obtained as an index of apparent model fit.

We used the deviance information criterion (DIC) for model selection (Spiegelhalter et al., 2003). DIC is composed of two competing elements, a measure of goodness-of-fit and a measure of the number of effective parameters. For example, random effect parameters may contribute less than one parameter to the number of effective parameters. The model that minimizes the DIC criterion is optimal under this criterion.

The measure of effective parameter dimension used in DIC can be sensitive to the shape of the marginal posterior distributions of the likelihood components (David Draper, personal communication). If the marginal distributions are far from Gaussian, the calculation can be very unstable. We examined the posterior distributions of the parameters and they all appeared to be close to Gaussian in shape. While a full examination would also require inspection of the random effect distributions as well, we omitted this for convenience. After fitting the models, we examined the estimated effective number of parameters for each model. The results for all models considered were in the expected order and appeared reasonable in size, so, the calculated DIC values were judged to be sufficient for our purposes.

4.3   Results

Table 4 lists the considered models and their DIC values. The best model as measured by DIC is the independent response model followed by the uncorrelated errors model. There is a difference of 10.1 in DIC score between the independent response model and the nearest dependent response model suggesting a sizeable improvement in model parsimony by selecting the independent response model. The Bayesian p-value, P_b, was greater than 0.9 for all three of the considered models. This suggests there is little evidence of lack-of-fit for any of the models. Although the dependent response models fit the data well, they seem to contain too many parameters to be parsimonious.

Table 4.   DIC and model complexity for multivariate fish species occurrence models. Models are listed in increasing DIC order. Column ΔDIC represents the difference in DIC from the model with the lowest DIC value and the column p_D represents the model complexity or effective number of parameters

Model	DIC	ΔDIC	p_D
Independent	1111.1	–	66.1
Dependent (Uncorrelated errors)	1117.8	6.7	106.1
Dependent (Correlated errors)	1166.8	55.7	162.5

The 95% highest posterior density (HPD) intervals for the β interaction coefficients in the independent response model are given in Table 5. The HPD intervals in Table 5 indicate that chloride concentration and sulfate concentration are related to the pollution tolerance response and that elevation is related to the benthic response variable. The HPD intervals for the interaction terms of the remaining environmental variables, watershed area, precipitation, and turbidity, contain zero for both the habit and tolerance response variables, therefore, the analysis does not provide strong evidence that they are related to either response. The HPD intervals for the off-diagonal elements of $\mathbf{\Psi}_\emptyset$ are given in Table 6. Again, by examining Table 6 the intervals that do not contain zero provide strong evidence that there exists an undirected edge between the two associated variables in the marginal graph for the covariates.

Fig. 2 illustrates the chain graph for the model that is suggested by the DIC criterion. To construct this figure, edges between vertices only were included when all of the 95% HPD intervals for the

Table 5. **95% HPD intervals for covariate interaction parameters in the independent response model for the analysis of fish species occurrence**

| | | Habit | | Tolerance | |
Covariate	Column*	Benthic	Intolerant	Intermediate*	Tolerant
Precipitation	–	(−2.17, 0.71)	(−1.64, 3.53)	–	(−2.68, 6.73)
Elevation	–	(0.17, 1.19)	(−0.73, 1.24)	–	(−0.97, 0.22)
Turbidity	–	(−0.25, 0.15)	(−0.37, 0.45)	–	(−0.19, 0.25)
Sulfate	–	(−0.21, 0.16)	(0.06, 0.71)	–	(−0.07, 0.35)
Chloride	–	(−0.17, 0.17)	(−0.77, −0.08)	–	(−0.07, 0.35)
Area	–	(−0.25, 0.02)	(−0.11, 0.42)	–	(−0.29, 0.03)

*In this analysis, the column habit and intermediate tolerance types were used as the reference cell. Therefore, the interaction coefficients are set to zero for those interaction terms referencing that cell.

parameters that correspond to the two vertices did not contain zero. For example, in Table 6, all of the intervals for precipitation contain 0, so this vertex has no edges to other vertices. The DIC criterion suggested that the independent response model is more parsimonious than a dependent response model. As noted in Section 3.2, the independent response model is a preservative model in the sense that relationships are preserved after marginalizing over the random effects (Johnson et al., 2006). Therefore, Fig. 2 shows the subgraph for the response and covariates only.

Table 6. **HPD intervals for the elements of the inverse covariance matrix Ψ_0 for the MAHA environmental variables. The intervals are presented on a correlation matrix scale**

	Elevation	Turbidity	Chloride	Sulfate	Area
Precipitation	(−0.293, 0.099)	(−0.152, 0.248)	(−0.122, 0.273)	(−0.131, 0.263)	(−0.149, 0.247)
Elevation		(−0.052, 0.336)	(0.391, 0.672)	(−0.381,−0.001)	(−0.528,−0.181)
Turbidity			(−0.392,−0.016)	(0.089, 0.456)	(−0.109, 0.284)
Chloride				(−0.623,−0.318)	(−0.485,−0.128)
Sulfate					(−0.028, 0.359)

The findings can be interpreted in light of basic understanding of stream ecology. Elevation in this dataset is a surrogate for headwater streams that are shallow and are occupied largely by benthic species. Intolerant species occurrence declines with elevated chloride levels, a surrogate for human disturbance. The positive association between high sulfate levels and intolerant species is unexpected and at odds with a previous study that used a univariate approach (McCormick et al., 2001). We note, however, that the 95% HPD interval for this association almost includes 0, and that there is a strong negative correlation of sulfate with chloride which suggests the possibility that these covariates are drawing from the same latent disturbance process to which sulfate concentration is negatively related. And, it is possible that a disturbance process is influencing species occurrence. Similarly the addition of latent process to model tolerance could enhance the model. The ordering in the tolerance variable, intolerant to moderate to tolerant species was ignored in this analysis; inclusion of a latent process measuring tolerance on the continuous scale but observed on a discrete 3-point scale could be used to better address this issue.

The model developed here allows for predictions at locations where only the covariates are observed. These results were not included in this analysis due to the relatively small sample size. An example of predictions for a similar model is given in Johnson (2003).

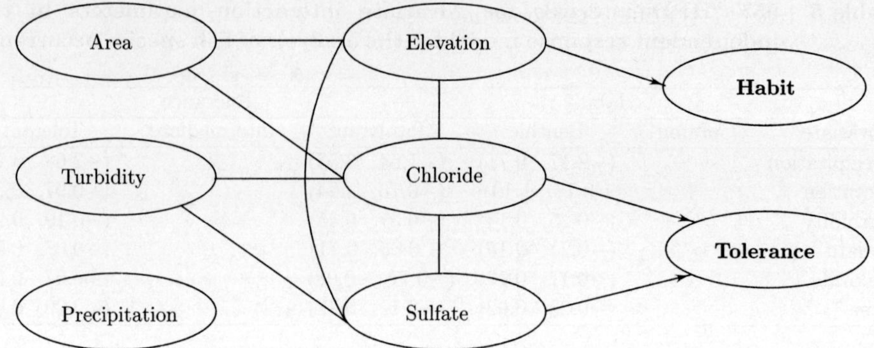

Fig. 2. **Data suggested chain graph for the multivariate composition of habit and tolerance. Arrows show that the covariates influence the responses, tolerance and habit. Undirected edges between covariates suggest that at least one of the HPD intervals for the covariance estimates between these covariates does not include 0. Lack of edge between any of the covariates or between the response and the covariates indicates that the corresponding intervals in Tables 5 and 6 include 0.**

Acknowledgements

The authors would like to thank David Draper for his suggested improvements to this paper and Scott Urquhart for helpful discussions. Research supported by the U.S. Environmental Protection Agency as part of the STAR Research Assistance Agreement CR-829095 (Johnson and Hoeting) and STAR grant R8286301 (Poff). The views expressed here are solely those of the authors. EPA does not endorse any products or services mentioned here.

References

Aitchison, J. (1982). The statistical analysis of compositional data (with discussion). *Journal of the Royal Statistical Society-B*, **44**, 139–177.

Aitchison, J. (1986). *The Statistical Analysis of Compositional Data*. Chapman and Hall, New York.

Aitchison, J. and Shen, S. M. (1980). Logistic-normal distribution: Some properties and uses. *Biometrika*, **67**, 261–272.

Billheimer, D. (1995). *Statistical Analysis of Biological Monitering Data: State-Space Models for Species Compositions*. PhD thesis, Department of Statistics, University of Washington.

Billheimer, D. and Guttorp, P. (1997). Natural variablity in benthis species compotiion in the Delaware Bay. *Environmental and Ecological Statistics*, **4**, 95–115. 24

Billheimer, D., Guttorp, P. and Fagen, W. F. (2001). Statistical interpretation of species composition. *Journal of the American Statistical Association*, **96**, 1205–1214.

Chen, M. H., Shao, Q. M. and Ibrahim, J. G. (2000). *Monte Carlo Methods in Bayesian Statistics*. Springer-Verlag, New York.

Dolèdec, S., Chessel, D., ter Braak, C. and Champely, S. (1996). Matching species traits to environmental variables: A new three-table ordination method. *Environmental and Ecological Statistics*, **3**, 143166.

Dominici, F. (2000). Combining contingency table data with missing observations. *Biometrics*, **56**, 546–553.

Draper, D. and Krnjajić, M. (2005). Bayesian model specication. Technical report, Department of Applied Mathematics and Statistics, University of California, Santa Cruz.

Freeman, M. and Tukey, J. (1950). Transformations related to the angular and square root. *Annals of Mathematical Statistics*, **21**, 607–611.

Gelman, A., Meng, X.-L. and Stern, H. (1996). Posterior predictive assessment of model tness via realized discrepancies. *Statistica Sinica*, **6**, 733–807.

Ghosh, J. K., Bhanja, J., Purkayastha, S., Samanta, T. and Sengupta, S. (2002). A statistical approach to geological mapping. *Mathematical Geology*, **34**(5), 505–528.

Givens, G. H. and Hoeting, J. A. (2005). *Computational Statistics*. Wiley, New York.

Johnson, D.S. (2003). *Random Effects Graphical Models for Discrete Compositional Data*. PhD thesis, Department of Statistics, Colorado State University.

Johnson, D.S. and Hoeting, J.A. (2003). Random effects graphical models for multiple site sampling. Technical Report 15, Department of Statistics, Colorado State University, USA.

Johnson, D.S., Hoeting, J.A. and Fadely, B.S. (2006). Random effects graphical models for multidimensional categorical data. Technical Report 2006/13, Department of Statistics, Colorado State University, USA.

Johnson, R. and Wichern, D. (1992). *Applied Multivariate Statistical Analysis*. Prentice-Hall, New Jersey. 642 pp.

Lauritzen, S. L. (1996). *Graphical Models*. Claredon Press.

Legendre, P., Galzin, R. and Harmelin-Vivien, M. (1997). Relating behavior to habitat: Solutions to the fourth-corner problem. *Ecology*, **78**, 547–562.

McCormick, F., Hughes, R., Kaufmann, P., Peck, D., Stoddard, J. and Herlihy, A. (2001). Development of an index of biotic integrity for the Mid-Atlantic Highlands Region. *Transactions of the American Fisheries Society*, **130**, 857–877.

Pearson, K. (1897). Mathematical contributions to the theory of evolution on a form of spurious correlation which may arise when indices are used in the measurment of organs. *Proceedings of the Royal Society*, 60:489–498.

Poff, N. and Allan, J. (1995). Functional-organization of stream fish assemblages in relation to hydrological variability. Ecology, **76**, 606–627.

Robbins, J.M., van der Vaart, A. and Ventura, V. (2000). Asymptotic distribution of P values in composite null models. *Journal of the American Statistical Association*, **95**, 1143–1156.

Spiegelhalter, D.J., Thomas, A. and Best, N.G. (2000). WinBUGS version 1.3, User Manual. MRC Biostatistics Unit, Institute of Public Health, Cambridge UK.

Spiegelhalter, D.J., Best, N.G., Carlin, B. P. and van der Linde, A. (2003). Bayesian measures of complexity and fit. *Journal of the Royal Statistical Society-B*, **64**, 583–639.

ter Braak, C. (1985). Correspondence analysis of incidence and abundence data: Properties in terms of a unimodal response model. *Biometrics*, **41**, 859–873.

U.S. Environmental Protection Agency (2000). Environmental Monitoring and Assessment Program, Mid-Atlantic Highlands Streams Assessment, EPA-903-R-00-015, US Environmental Protection Agency, National Health and Environmental Effects Research Laboratory, Western Ecology Division, Corvallis, OR.

Whittaker, J. (1990). *Graphical Models in Applied Multivariate Statistics*. Wiley, Chichster.

Appendix: Parameter Inference

Bayesian inference for graphical composition models proceeds by first defining a prior distribution for the parameters of the model $\pi(\boldsymbol{\beta}, \boldsymbol{\omega}, \boldsymbol{\lambda}, \boldsymbol{\tau}, \boldsymbol{\Psi}_\emptyset, \boldsymbol{\Sigma})$. Here, subscripts are removed in order to ease notational burden. For example, $\boldsymbol{\beta}$ refers to the entire set of parameters $\{\beta_{fcd}(\mathbf{y}_\Phi, \mathbf{x}_\Delta) : f \subseteq \Phi, c \subseteq \Gamma,$ and $d \subseteq \Delta\}$. Assuming that the observations at each site are independent, the posterior distribution of the parameters and the random effects is given by

$$f_{\text{post}}(\beta, \omega, \lambda, \tau, \Psi_\emptyset, \Sigma, \{\epsilon\} \mid \{\mathbf{c}\}, \{\mathbf{x}\}) \propto \prod_{i=1}^{S} f_M(\mathbf{c}_i | \beta, \omega, \mathbf{x}_i, \epsilon_i) \times f_M(\mathbf{c}_\Delta | S, \lambda)$$

$$\times \prod_{i=1}^{S} f_N(\mathbf{x}_{\Gamma i} | \mathbf{x}_{\Delta i}, \tau, \Psi_\emptyset) \times \prod_{i=1}^{S} \prod_{f \subseteq \Phi} f_N(\epsilon_{f,i} | \Sigma_f)$$

$$\times \, \pi(\beta, \omega, \lambda, \tau, \Psi_\emptyset, \Sigma), \tag{15}$$

where $\{\epsilon\} = \{\epsilon_i : i = 1, \ldots, S\}$, $\{\mathbf{c}\} = \{\mathbf{c}_i : i = 1, \ldots, S\}$, and $\{\mathbf{x}\} = \{\mathbf{x}_i : i = 1, \ldots, S\}$.

The posterior distribution (15) is a non-standard distribution; therefore analytical inference for posterior objects of interest such as expected values and credible intervals is not possible. We will draw a sample from this distribution using a Gibbs sampling approach (e.g., Givens and Hoeting (2005)).

Assuming the CG parameters are independent of the remaining parameters simplifies the analysis with $\pi(\beta, \omega, \lambda, \tau, \Psi_\emptyset, \Sigma) = \pi(\beta, \omega, \Sigma) \times \pi(\lambda, \tau, \Psi_\emptyset)$ so the posterior distribution is given by

$$f_{\text{post}}(\beta, \omega, \lambda, \tau, \Psi_\emptyset, \Sigma, \{\epsilon\} \mid \{\mathbf{c}\}, \{\mathbf{x}\}) \propto f_{\text{post}}(\beta, \omega, \Sigma, \{\epsilon\} \mid \{\mathbf{c}\}, \{\mathbf{x}\}) \times f_{\text{post}}(\lambda, \tau, \Psi_\emptyset \mid \{\mathbf{x}\}).$$

The parameters of the explanatory portion of the graphical model and the parameters of the response portion of the graphical model are *a posteriori* independent. This simplifies the analysis of the posterior distribution because two separate MCMC analyses can be performed, one for each chain component. This type of sequential estimation is often used to estimate chain model parameters (Whittaker, 1990, p. 310).

A.1   Hierarchical Centering Parameterization

Now we present a modification to the response model parameterizations presented in (5). When using a Gibbs MCMC procedure, the Markov chains for regression coefficient parameters in random effects generalized linear models, such as β and ω, are often slow to converge to the marginal posterior distribution (Chen et al., 2000, p. 40). It has been our experience that this is also the case for the proposed composition models. Therefore, we will use a hierarchical centering parameterization, as suggested by Chen et al. (2000), to help reduce the problem of poorly mixing chains for the regression coefficients β and ω.

In order to describe the hierarchical centering parameterization, first recall the response portion of the REDR distribution (5). We introduce a shortened notation for the fixed effects portion of the response model for each $f \subseteq \Phi$, with

$$\mu_f(\mathbf{y}_\Phi) = \sum_{c \subseteq \Gamma} \sum_{d \subseteq \Delta} \beta_{fcd}(\mathbf{y}_\Phi, \mathbf{x}_\Delta) \prod_{\gamma \in c} x_\gamma + \sum_{\gamma \in \Gamma} \sum_{d \subseteq \Delta} \sum_{m=2}^{M} \omega_{f\gamma dm}(\mathbf{y}_\Phi, \mathbf{x}_\Delta) x_\gamma^m. \tag{16}$$

Then we propose the following hierarchically centered re-parameterization of (5),

$$f_{RE}^{(h)}(\mathbf{y}_\Phi | \varphi) = \exp\left\{ \sum_{f \subseteq \Phi} \varphi_f(\mathbf{y}_\Phi) \right\}, \tag{17}$$

where $\varphi_f(\mathbf{y}_\Phi) = \mu_f(\mathbf{y}_\Phi) + \epsilon_f(\mathbf{y}_\Phi)$, $f \neq \emptyset$ and $\varphi_\emptyset$ represents the log-normalizing constant with respect to $\mathbf{Y}_\Phi$ given φ,

$$\varphi_\emptyset = -\log\left[\sum_{\mathbf{y}_\Phi}\exp\left\{\sum_{f\subseteq\Phi}\varphi_f(\mathbf{y}_\Phi)\right\}\right], \quad f\neq\emptyset. \tag{18}$$

If the assumption is made that $\boldsymbol{\epsilon}_f \sim f_N(\mathbf{0},\ \boldsymbol{\Sigma}_f)$ for $f\subseteq\Phi$ that are not set to zero, as portrayed in (11), then $\boldsymbol{\varphi}_f = \{\varphi_f(\mathbf{y}_\Phi): y_\phi\neq 1 \text{ for any } \phi\subseteq f\} \sim f_N(\boldsymbol{\mu}_f,\boldsymbol{\Sigma}_f)$, where $\boldsymbol{\mu}_f$ is the vector $[\mu_f(\mathbf{y}_\Phi)]$. Here we have simply changed the random effects $\boldsymbol{\epsilon}_f$ from a zero mean process to a process $\boldsymbol{\varphi}_f$, centered at the fixed effects $\boldsymbol{\mu}_f$. In the case of generalized linear mixed models there is no theoretical result to show that this will improve mixing of the MCMC procedure. It has been our experience, however, that the re-parameterization often greatly improves mixing for these models.

The general parameterization of the full posterior distribution is given by

$$f_{\text{post}}^{(h)}(\boldsymbol{\beta},\boldsymbol{\omega},\boldsymbol{\lambda},\boldsymbol{\Psi}_\emptyset,\{\boldsymbol{\varphi}_i\}|\{\mathbf{c}_i\},\{\mathbf{x}_i\})\propto\left[\prod_{i=1}^{S}f_M^{(h)}(\mathbf{c}_i|\boldsymbol{\varphi}_i)\left\{\prod_{f\subseteq\Phi}f_N(\boldsymbol{\varphi}_{fi}|\boldsymbol{\beta}_f,\boldsymbol{\omega}_f,\boldsymbol{\Sigma}_f,\mathbf{x}_i)\right\}\right]$$

$$\times\pi(\boldsymbol{\beta},\boldsymbol{\omega},\boldsymbol{\Sigma})\left[\prod_{i=1}^{S}f_N(\mathbf{x}_{\Gamma i}|\mathbf{x}_{\Delta i},\boldsymbol{\tau},\boldsymbol{\Psi}_\emptyset)\right]f_M(\mathbf{c}_\Delta|\boldsymbol{\lambda})\pi(\boldsymbol{\lambda},\boldsymbol{\tau},\boldsymbol{\Psi}). \tag{19}$$

A.2   Implementing the Gibbs Sampler

In order to implement the Gibbs sampler to draw a sample from (19) we need to obtain the full conditional distribution for each parameter. The full conditional distribution is the conditional distribution of the parameter in question given all remaining parameters as well as the observed data. A (non-independent) sample from the posterior is then drawn by iteratively drawing from each full conditional distribution. We derive the full conditional densities in two separate groups due to the fact that the full conditional densities for the response model parameters $\boldsymbol{\beta}$, $\boldsymbol{\omega}$, $\{\boldsymbol{\varphi}_i\}$ and $\boldsymbol{\Sigma}$, will not be functions of the explanatory model parameters $\boldsymbol{\lambda}$, $\boldsymbol{\tau}$ and $\boldsymbol{\Psi}_\emptyset$, as can be observed from (19). Therefore, we can make inferences about the response model parameters using only the first factor on the left hand side of the proportionality, while inferences about the explanatory portion of the chain graph uses only the second factor.

A.2.1   *Response Model Conditional Densities*

Before deriving the full conditional densities we introduce some notation. Let $\mathbf{E}$ refer to a matrix that has S rows and r columns including a column of ones, a column corresponding to each of the explanatory variables, and a column for each interaction and powers of the continuous covariates as given in (5). If a covariate X_δ, $\delta\subseteq\Delta$, is a categorical variable with b levels, then it will be represented by $b-1$ columns of indicator variables in $\mathbf{E}$, where each column indicates, with a one or zero, if X_δ takes the associated level at site i. The column associated with the reference level $X_\delta = 1$ is not included. The vector $\mathbf{E}_i$, $i = 1,\dots,S$ will denote an r-vector formed from the ith row of $\mathbf{E}$. In addition, let D_f denote the length of $\boldsymbol{\varphi}_f(\mathbf{y}_\Phi)$ and let $\mathbf{B}_f$ represent an $r\times D_f$ matrix of all the interaction coefficients $\{\beta_{fcd}(\mathbf{y}_\Phi): c\subseteq\Gamma,\ d\subseteq\Delta\}$ and $\{\omega_{f\gamma dm}(\mathbf{y}_\Phi,\mathbf{x}_\Delta):\gamma\in\Gamma,\ d\subseteq\Delta,\ m = 1,\dots,M\}$ such that the expected value (16) of the site i random effect $\boldsymbol{\varphi}_{f,i}$ is given by $\boldsymbol{\mu}_f = \mathbf{B}_f'\mathbf{E}_i$. The stacked version of $\mathbf{B}_f$ will be represented by $\mathbf{B}_{fs}$. The stacked version is a $rD_f\times 1$ vector where the columns of $\mathbf{B}_f$ have been concatenated in order. Although previously described as the collection of all random effects, $\boldsymbol{\varphi}_f$ will now specifically represent a $S\times D_f$ matrix of these random effects and $\varphi_{f,i}$ is a D_f vector formed from the ith row of φ_f. Finally, we will make use of the inverse of the $D_f \times D_f$ random effects covariance matrix $\mathbf{T}_f = \boldsymbol{\Sigma}_f^{-1}$.

Now we can derive full conditional distributions for the parameters of the response model, the coefficients in the matrices $\{\mathbf{B}_f : f \subseteq \Phi\}$, the site random effect matrices $\{\varphi_f : f \subseteq \Phi\}$, and the random effects inverse covariance matrices $\{\mathbf{T}_f : f \subseteq \Phi\}$. In addition to the increased rate of convergence, the hierarchical centering provides a Gibbs sampler that is easier to implement due to the fact that the interaction coefficients as well as the random effects covariance matrices will have standard full conditional densities. In the non-centered parameterization only the covariance matrices have standard full conditional densities. Here, we will also make the assumption that for each $f \subseteq \Phi$, the interaction coefficients and the covariance matrices are *a priori* mutually independent across all f, so $\pi(\boldsymbol{\beta}, \boldsymbol{\omega}, \boldsymbol{\Sigma}) = \prod_{f \subseteq \Phi} \pi(\boldsymbol{\beta}_f, \boldsymbol{\omega}_f)\pi(\boldsymbol{\Sigma}_f)$.

We begin with the interaction coefficients in $\mathbf{B}_f$ for any $f \subseteq \Phi$. First, note that due to the centering parameterization, given the random effects φ_{fi} at each site and the random effect inverse covariance matrix T_f, the interaction coefficients are independent of the cell counts. If we let $\pi(\boldsymbol{\beta}_f, \boldsymbol{\omega}_f) = \pi(\mathbf{B}_{fs}) = f_N(\mathbf{B}_{fs}; \boldsymbol{\mu}_{B_{fs}}, \mathbf{V}_{B_{fs}}^{-1})$ and $\hat{\mathbf{B}}_f = (\mathbf{E}'\mathbf{E})^{-1}\mathbf{E}'\varphi_f$ (correspondingly $\hat{\mathbf{B}}_{fs}$ represents the stacked version) then the full conditional distribution of the interaction coefficients is given as

$$
f(\mathbf{B}_{fs}|...) \propto \exp\left\{-\frac{1}{2}\sum_{i=1}^{S}\left(\varphi_{f,i} - \mathbf{B}_f'\mathbf{E}_i\right)'\mathbf{T}_f\left(\varphi_{f,i} - \mathbf{B}_f'\mathbf{E}_i\right)\right\}
$$

$$
\times \exp\left\{-\frac{1}{2}(\mathbf{B}_{fs} - \boldsymbol{\mu}_{B_{fs}})'\mathbf{V}_{B_{fs}}(\mathbf{B}_{fs} - \boldsymbol{\mu}_{B_{fs}})\right\}
$$

$$
\propto \exp\left\{-\frac{1}{2}tr\left[\mathbf{T}_f(\mathbf{B}_f - \hat{\mathbf{B}}_f)'\mathbf{E}'\mathbf{E}(\mathbf{B}_f - \hat{\mathbf{B}}_f)\right]\right\}\exp\left\{-\frac{1}{2}(\mathbf{B}_{fs} - \boldsymbol{\mu}_{B_{fs}})'\mathbf{V}_{B_{fs}}(\mathbf{B}_{fs} - \boldsymbol{\mu}_{B_{fs}})\right\}
$$

$$
= \exp\left\{-\frac{1}{2}(\mathbf{B}_{fs} - \hat{\mathbf{B}}_{fs})'(\mathbf{T}_f \otimes \mathbf{E}'\mathbf{E})(\mathbf{B}_{fs} - \hat{\mathbf{B}}_{fs})\right\}\exp\left\{-\frac{1}{2}(\mathbf{B}_{fs} - \boldsymbol{\mu}_{B_{fs}})'\mathbf{V}_{B_{fs}}(\mathbf{B}_{fs} - \boldsymbol{\mu}_{B_{fs}})\right\},
$$

$$(20)$$

where $\otimes$ represents the Kronecker product. The second proportionality statement for the random effects likelihood is due to Johnson and Wichern (1992, p. 322). Completing the square leads to

$$
f(\mathbf{B}_{fs}| \dots) = f_N(\mathbf{B}_{fs}; \boldsymbol{\mu}_{f,1}, \mathbf{V}_{f,1}^{-1}), \tag{21}
$$

where the mean and covariance are given by

$$
\boldsymbol{\mu}_{f,1} = \left[(\mathbf{T}_f \otimes \mathbf{E}'\mathbf{E}) + \mathbf{V}_{B_{fs}}\right]^{-1}\left[(\mathbf{T}_f \otimes \mathbf{E}'\mathbf{E})\hat{\mathbf{B}}_{fs} + \mathbf{V}_{B_{fs}}\boldsymbol{\mu}_{B_{fs}}\right]
$$

$$
\text{and} \qquad \mathbf{V}_{f,1} = (\mathbf{T}_f \otimes \mathbf{E}'\mathbf{E}) + \mathbf{V}_{B_{fs}}. \tag{22}
$$

Therefore, in the Gibbs sampler, drawing samples of the interaction coefficients is a relatively simple draw from a multivariate normal distribution.

We now derive the conditional distribution for the inverse covariance matrix $\mathbf{T}_f$ of the random effects φ_f. We assume, *a priori*, that $\mathbf{T}_f$ has a Wishart distribution, $f_W(\mathbf{T}_f; a_f, \mathbf{K}_f)$, with prior parameters $a > D_f - 1$, $D_f \times D_f$ positive definite matrix $\mathbf{K}_f$, and density

$$
\pi(\mathbf{T}_f) = f_W(\mathbf{T}_f; a_f, \mathbf{K}_f) \propto |\mathbf{T}_f|^{(a-D_f-1)/2} \exp\left\{-\frac{1}{2}tr\left[\mathbf{K}_f\mathbf{T}_f\right]\right\}. \tag{23}
$$

This is equivalent to specifying an inverse Wishart prior distribution for $\boldsymbol{\Sigma}_f$. Now, $\mathbf{T}_f$ only depends on φ_f and $\mathbf{B}_f$ through the random effects distribution, which is a MVN distribution. Therefore, we obtain the following full conditional distribution,

$$f(\mathbf{T}_f|\dots) \propto |\mathbf{T}_f|^{S/2} \exp\left\{-\frac{1}{2}\sum_{i=1}^{S}(\varphi_{f,i}-\mathbf{B}'_f\mathbf{E}_i)'\mathbf{T}_f(\varphi_{f,i}-\mathbf{B}'_f\mathbf{E}_i)\right\}$$

$$\times |\mathbf{T}_f|^{(a-D_f-1)/2}\exp\left\{-\frac{1}{2}tr\left[\mathbf{K}_f\mathbf{T}_f\right]\right\} = |\mathbf{T}_f|^{(a+S-D_f-1)/2}$$

$$\times \exp\left\{-\frac{1}{2}tr\left[\mathbf{T}_f\left\{\mathbf{K}_f+\sum_{i=1}^{S}(\varphi_{f,i}-\mathbf{B}'_f\mathbf{E}_i)'(\varphi_{f,i}-\mathbf{B}'_f\mathbf{E}_i)\right\}\right]\right\}. \qquad (24)$$

It follows, then, upon examination of (23), the full conditional distribution of $\mathbf{T}_f$ is given by

$$f(\mathbf{T}_f|\dots) = f_W(\mathbf{T}_f;\ a_{f,1},\mathbf{K}_{f,1}) \qquad (25)$$

where the full conditional parameters are $a_{f,1} = a_f + S$ and

$$\mathbf{K}_{f,1} = \mathbf{K}_f + \sum_{i=1}^{S}(\varphi_{f,i}-\mathbf{B}'_f\mathbf{E}_i)'(\varphi_{f,i}-\mathbf{B}'_f\mathbf{E}_i). \qquad (26)$$

Therefore, just like the interaction coefficients, the inverse covariance matrix $\mathbf{T}_f$ is relatively straightforward to sample from in the Gibbs algorithm.

The vector of site random effects $\varphi_{f,i}$ does not have a standard full conditional distribution. The full conditional density is given by

$$f(\varphi_{f,i}|\dots) \propto f_M^{(h)}(\mathbf{c}_i|\varphi_i)f_N(\varphi_{f,i};\ \mathbf{B}'_f\mathbf{E}_i,\mathbf{T}_f^{-1}). \qquad (27)$$

Since the full conditional density is non-standard, we employ a Metropolis-within-Gibbs step to sample from this full conditional distribution.

A.2.2 *Conditional Distributions for the Explanatory Variables Model*

To derive the conditional distributions for the parameters in the explanatory variable CG model (8), we first note that the categorical and continuous explanatory variable parameters are functionally independent. Therefore, we can perform separate posterior analyses for the discrete and continuous partitions of the explanatory model. The derivation of the required conditional distributions is similar to the derivations given above, so these are not given here. Complete derivations are provided in Johnson (2003).

Bayesian Statistics and Its Applications
Edited by S.K. Upadhyay, U. Singh and D.K. Dey
Anamaya Publishers, New Delhi, India

Semiparametric Multivariate Survival Models with Random Effects

Sungduk Kim[1], Dipak K. Dey[1] and Debajyoti Sinha[2]

[1]Department of Statistics, University of Connecticut, Storrs, CT 06269
[2]Department of Biometry, Medical University of South Carolina, Charleston, SC

Abstract

Often the dependence in multivariate survival data is modeled through an individual level effect called the frailty. Due to its mathematical simplicity the gamma distribution is often used as the frailty distribution for hazard modeling. However, it is well known that the gamma distribution for frailty has many drawbacks. For example, it weakens the effect of covariates. In addition, in presence of multilevel model overall frailty comes from several levels. To overcome such drawbacks more heavy tailed distributions are needed to model the frailty distribution in order to incorporate extra variability. In this article we develop a class of log-skew-t distributions for the frailty. This class includes the log-normal distribution along with many other heavy tailed distributions, e.g., log-Cauchy, log normal and log-t as special cases.

Conditional on the frailty, the survival times are assumed to be independent with proportional hazard structure. The modeling process is then completed by assuming multilevel frailty-effects. Instead of tuning a strict parameterization of the baseline hazard function, we consider a correlated prior process, which offers a great deal of flexibility. A case study involving a CGD data set in Fleming and Harrington (1991) is discussed in detail to demonstrate the proposed methodology.

1. Introduction

In studies of survival, the hazard function for each individual may depend on a set of risk factors or explanatory variables but usually not all such variables are known or measurable. This unknown and unobservable risk factor of the hazard function is often termed as the individual's heterogeneity or frailty. Frailty models are becoming increasingly popular in multivariate survival analysis since they allow us to model the association between the individual survival times within subgroups or clusters of subjects. Such data hierarchies may be present naturally in observation studies or may be due to the design of the experiment in experimental studies. In a chronic granulomatous disease (CGD) study, the effectiveness of a new treatment (γ-IFN) in reducing the rate of serious infections is investigated. Since each patient may experience multiple failure events, a commonly considered random effect survival model is the two-level model that assumes an independent identically distributed frailty term for each patient. Such a frailty term represents the variation due to the heterogeneity of patients. However, as each patient belongs to one of the 13 hospitals, the variation may possibly be due to a random hospital effect as well. In this case, a three-level survival model should be considered. According to the hierarchical structure of the data, infections are defined as level 1 units, patients as level 2 units, and hospitals as level 3 units. The purpose of the study is, therefore, to look for the effect of γ-IFN in reducing the rate of infection as well as estimate the variances of the clustered random effects. Ignoring such random effects may result in overlooking the importance of certain cluster effects and call into question the validity of traditional statistical techniques used for studying data relationships (Goldstein, 1995). Early work on frailty models for survival data by Hougaard (1984, 1986a, b) considered the parametric modelling of the heterogeneity between individuals in the population using the gamma, inverse Gaussian, and positive stable distribution. See for example the recent book by Ibrahim, Chen and Sinha (2001) for general introduction and early references. A convenient choice of the frailty distribution is the gamma distribution since it provides conjugate sampling distributions for Gibbs sampling (Gelfand and Smith, 1990). Gustafson (1997) discusses Bayesian hierarchical frailty models for multilevel multivaraiate survival data. The hierarchical modelling given by Gustafson (1997) has elements in common with the work of Clayton (1991), Gray (1994), Sinha (1993), Stangl (1995), and Stangl and Greenhouse (1995). Yau (2001) proposed the multilevel models for survival analysis with random effects for GCD data. Sahu and Dey (2005) developed a class of log-skew-t distribution for the frailty in multivariate survival model with a correlated baseline hazard process. Kim and Dey (2004) applied the skew-t distribution for multilevel survival analysis with random effects by using partial likelihood.

This article proposes the multilevel survival models to analyze survival data with nested random effects in the Bayesian framework. We consider log-normal, log-skew-normal and log-skew-t frailty models to setup the multilevel survival model. In the multilevel survival model, the class before the transformation reduces to the family of normal distributions for particular values of the parameters. On the other extreme it behaves as a half-normal (half-t or half-Cauchy) distribution by placing all its mass on the positive side of the real line. Thus, the family of distribution is quite flexible and general, and the existence of the mean and variance of the frailty distribution is not assumed, since the degrees of freedom of the t-distribution can be less than two. Piecewise exponential models provide a very flexible framework for modelling survival data. Although, in a strict sense, it is a parametric model, a piecewise exponential hazard can approximate any shape of a nonparametric baseline hazard. Therefore, in practice, defining a prior for a piecewise exponential hazard is the same as defining a prior process for the nonparametric hazard function. Therefore, a suitable stochastic process needs to be considered for the baseline hazard function. Several parametric and non-parametric models are available, see, e.g. Sinha and Dey (1997) for a review. Also we adopt a piecewise constant baseline hazard function. This choice is popular for modelling univariate survival data, see for example, Gamerman (1991), Arjas and Gasberra (1994) and McKeague and Tighiouart (2000). The correlated prior process imposes smoothness on the baseline hazard function in adjacent intervals. In particular, we generalize a first order autoregressive process considered in Sahu et al. (1997). In addition, we assume that the endpoints of the interval themselves form a time-homogeneous Poisson process. This introduces further flexibility since the number of endpoints where jumps are allowed to occur is left unknown. The full Bayesian model is rather complex and does not allow fitting and comparison using analytical methods. The straightforward Gibbs sampler is also not able to handle the computations since the parameter space is of varying dimension. Thus, we develop Bayesian computation methods using the reversible jump MCMC method, see for example, Green (1995). Consequently we extend the algorithm of Sahu and Dey (2005). A natural next step after Bayesian model fitting is to investigate the issues relating to model adequacy and model choice. In this perspective for model comparison, we adopt deviance information criterion (DIC). The DIC introduced by Spiegelhalter et al. (2002) is directly inspired by linear and generalized linear models, but it is not so naturally defined for missing (latent) data models. Celeux et al. (2003) reassessed the criterion for such models, testing the behavior of various extension in the cases of mixture and random effect models. As our proposed model has random effects, we use the complete DIC criterion of Celeux et al. (2003).

The article is organized by an outline of the developments of the estimation for a multilevel-level hierarchical survival model. Section 2 introduces the conditional proportional hazard models with frailty (Clayton and Cuzick, 1985; Oakes, 1989; Clayton, 1991) and smooth prior process for the baseline hazard function. The likelihood and prior specifications are discussed in Section 3. In Section 4, as an illustration, such a model is applied to the CGD data on recurrent infections of Fleming and Harrington (1991), with infection observations, patients and hospitals being considered as the three levels. In Section 5 we provide numerical examples. We conclude with a few summary remarks in Section 6.

2.   The Model

2.1   Frailty Models

In this section we consider three-level hierarchial survival model. For the CGD study, multiple failure time data are collected for the enrolled patients. Each patient is enrolled with one among several hospitals. The failure time observations, patients, and hospitals are defined, respectively, as level 1, level 2 and level 3 units. The kth observation (failure/censoring time) for the jth patient in the ith hospital is denoted by T_{ijk}. Motivated by Yau (2001), we extend the proportional-hazards model of Cox (1972) by including a random effect or frailty for the patient effect and the hospital effect, respectively. The conditional hazard function of T_{ijk} given the unobserved frailty random variable (random effect) U_{ij} of the jth patient in the ith hospital and the fixed observed covariate vector $\mathbf{x}_{ijk}$ corresponding to the kth observation for the jth patient of the ith hospital is

$$h_{ijk}(t \mid U_{ij}, x_{ijk}) = h_0(t) \exp(\eta_{ijk}), \quad \eta_{ijk} = \mathbf{x}'_{ijk}\boldsymbol{\beta} + U_{ij} \tag{1}$$

where $i = 1, 2, \cdots, b$, $j = 1, 2, \cdots, m_i$, $k = 1, 2, \cdots, n_{ij}$, $b =$ number of hospitals, $m_i =$ number of patients in the ith hospital, $n_{ij} =$ random number of multiple failure time observations for the jth patient in the ith hospital, $\sum_{i=1}^{b} m_i = M =$ total number of patients, $\sum_{j=1}^{m_i} n_{ij} = n_i =$ total number of observations in the ith hospital, $\sum_{i=1}^{b} n_i = N =$ total number of observations, β is a vector of fixed effect parameters and $h_0(t)$ is an unknown baseline hazard function common to every subject. The $\mathbf{x}_{ijk}$ may be time-dependent. This model is called a conditional proportional-hazards model by Shih and Louis (1995).

The random effect U_{ij} can be further decomposed as $U_{ij} = E_i + F_{ij}$, where E_i is the random effect of the ith hospital and F_{ij} is the random effect of the jth patient within the ith hospital. Let $\mathbf{e} = (E_1 \ E_2 \ \cdots \ E_b)'$, $\mathbf{f}_i = (F_{i1} \ F_{i2} \ \cdots \ F_{in_i})'$ and $\mathbf{f} = (\mathbf{f}'_1 \ \mathbf{f}'_2 \ \cdots \ \mathbf{f}'_b)'$. The random effects E_i and F_{ij} are assumed to be independent and identically distributed for the ith hospital and the jth patient within the ith hospital having some parametric distribution, respectively. Examples on parametric distribution of the random effects are given in Clayton and Cuzick (1985) and Oakes (1986, 1989). In this article the distribution of the random effect is assumed as follows.

First, the distributions of $\mathbf{e}$ and $\mathbf{f}$ are taken, respectively, to be multivariate normals $N(0, \theta_2 I_b)$ and $N(0, \theta_1 I_M)$. Second, we develop a flexible class of log-skew-t frailty distribution for modelling the dependence. The class includes the log-normal distribution along with other heavy tailed distributions such as the log-t distribution as special cases. As is well known, inference based on models with the use of heavy tailed frailty distributions is more robust in incorporating outliers. That is, we assume multivariate $St_{\nu_e}(0, \theta_2 I_b)$ and $St_{\nu_f}(0, \theta_1 I_M)$ for $\mathbf{e}$ and $\mathbf{f}$, respectively, where $St_\nu(\mu, \Sigma)$ is multivariate skew-t distribution with mean μ, variance-covariance matrix Σ and degrees of freedom ν and I_b and I_M are identity matrices with dimensions being specified by the subscripts. A class of skew elliptical distributions in the context of regression problem is presented in Sahu, Dey and Branco (2003).

2.2 Baseline Hazard Function

Although the Weibull distribution has been extensively used for modelling the baseline hazard, recent advances in Bayesian nonparametric survival analysis permit a wide range of choices for smooth or correlated prior processes for this purpose. In practice, often there might not be sufficient prior information on the baseline hazard except for some knowledge on how smooth we expect the function to be. The piecewise exponential models with correlated prior processes provide a very flexible framework since the smoothness of the baseline hazard rather than the actual baseline hazard is available as prior information. See for example, Leonard (1978), Gamerman (1991), Arjas and Gasberra (1994), Sinha and Dey (1997) and Sahu and Dey (2005) for a review. Information from a previous study, when available, could provide further prior information. Alternatively, one might be able to obtain such prior information from hypothetical data simulated using the prior prediction of the observable survival times and covariates. The setup is as follows.

Suppose that time is divided into g pre-specified intervals $I_l = (\tau_{l-1}, \tau_l]$ for $l = 1, 2, \cdots, g$ where $0 = \tau_0 < \tau_1 < \cdots < \tau_g < \infty$, τ_g being the last survival or censored time. Assume that the baseline hazard is constant within each interval, that is,

$$h(t_{ijk}) = h_l, \quad \text{for } t_{ijk} \in I_l \tag{2}$$

A discrete-time martingale process (see Arijas and Gasbarra, 1994; Aslanidou et al., 1998) is used to correlate the h_l's in adjacent interval, thus introducing some smoothness. Following Sahu and Dey (2005) we assume a martingale process prior for $\lambda_l = \log(h_l)$. We assume that

$$\lambda_l = \lambda_{l-1} + \epsilon_l, \quad \epsilon_l \sim N(0, \sigma^2) \tag{3}$$

with $\lambda_0 = 0$. Let $\boldsymbol{\lambda} = (\lambda_1, \cdots, \lambda_g)'$.

Several authors have provided different choices for the number of grid points g. Prentice (1973) recommended that these jump times be selected independently of the data. More recent solutions to this problem suggest leaving g unspecified. In view of that, we assume that the jump times $\tau_1, \tau_2, \cdots$ form a time-homogeneous Poisson process with rate a. This has several advantages over a fixed value of g. One such advantage is that the number and positions of the grid points need not be fixed in advance. In practical situations, g is determined based on the design. A random choice of g will lead to a posterior distribution with variable dimensions, in this case, sampling techniques other than the Gibbs sampler can be used to complete the posterior distribution, e.g., Green (1995) generalized the classical Metropolis-Hastings algorithm by preserving reversibility of the Markov chain when moving between subspaces of different dimensions.

3. Model Specification

3.1 Likelihood Specification

For the kth failure time observation of the jth patient in the ith hospital the triplet $(t_{ijk}, \nu_{ijk}, \mathbf{x}_{ijk})$ is observed in the lth time interval. Here, ν_{ijk} denotes the indicator variable taking value 1 if the observation corresponding to the kth failure time of the jth patient in the ith hospital falls in that interval and value 0 otherwise. Also, t_{ijk} is a failure time if $\nu_{ijk} = 1$ and a censoring time otherwise. Finally, $\mathbf{x}_{ijk}$ be the covariate for each failure time observation. Now, all such triplets $(t_{ijk}, \nu_{ijk}, \mathbf{x}_{ijk})$ for each subject in each interval are denoted as a whole by $(\mathbf{Y}, \mathbf{X})$. The vectors of unobserved E_i's and F_{ij}'s (the frailty random variables or random effects), denoted by $\mathbf{e}$ and $\mathbf{f}$, respectively, are called the augmented data, and the $(\mathbf{e}, \mathbf{f}, \mathbf{Y}, \mathbf{X})$ called the complete data. We only consider right-censored survival data and assume that the censoring is non-informative.

The likelihood is derived as follows. The kth failure time observation of the jth patient in the ith hospital has a constant hazard of $h_{ijk} = h_l \exp(\mathbf{x}'_{ijk}\boldsymbol{\beta} + E_i + F_{ij})$ in the lth interval $(l = 1, \cdots, g)$ given the unobserved frailty E_i and F_{ij}. If the kth failure time observation has not occurred beyond the lth interval, that is, $t_{ijk} > \tau_l$, the likelihood contribution is $\exp[-h_l\Delta_l \exp(\mathbf{x}'_{ijk}\boldsymbol{\beta} + E_i + F_{ij})]$, where $\Delta_l = \tau_l - \tau_{l-1}$. If the kth failure time observation of the jth patient in the ith hospital has occurred or was censored in the lth interval, i.e., $\tau_{l-1} < t_{ijk} \leq \tau_l$, then the likelihood contribution is as follows:

$$[h_l \exp(\mathbf{x}'_{ijk}\boldsymbol{\beta} + E_i + F_{ij})]^{\nu_{ijk}} \exp[-h_l(t_{ijk} - \tau_{l-1}) \exp(\mathbf{x}'_{ijk}\boldsymbol{\beta} + E_i + F_{ij})].$$

Therefore, the complete data likelihood based on $(\mathbf{e}, \mathbf{f}, \mathbf{Y}, \mathbf{X})$ is given as

$$L(\boldsymbol{\beta}, \boldsymbol{\lambda}, \mathbf{e}, \mathbf{f}, g \mid \mathbf{Y}, \mathbf{X}) = \prod_{i=1}^{b}\prod_{j=1}^{m_i}\prod_{k=1}^{n_{ij}}\left[\prod_{l=1}^{g_{ijk}} \exp\left\{-h_l\Delta_l \exp\ (\mathbf{x}'_{ijk}\boldsymbol{\beta} + E_i + F_{ij})\right\}\right]$$

$$\times \left[h_{g_{ijk}+1} \exp(\mathbf{x}'_{ijk}\boldsymbol{\beta} + E_i + F_{ij})\right]^{\nu_{ijk}} \exp\left\{-h_{g_{ijk}+1}(t_{ijk} - \tau_{l-1}) \exp(\mathbf{x}'_{ijk}\boldsymbol{\beta} + E_i + F_{ij})\right\}, \quad (4)$$

where g_{ijk} is such that $t_{ijk} \in (\tau_{g_{ijk}}, \tau_{g_{ijk}+1}] = I_{g_{ijk}+1}$. The representation in (4) will ease computation by using the MCMC algorithm.

3.2 Prior Specification

The joint prior distribution of all the parameters is given by

$$\prod_{i=1}^{b}\left\{\pi(E_i \mid \cdots)\prod_{j=1}^{m_i}\pi(F_{ij} \mid \cdots)\right\}\pi(\cdots)\pi(g)\pi(\boldsymbol{\lambda} \mid g)\pi(\boldsymbol{\beta}) \qquad (5)$$

where $\cdots$ denote the hyperparameters of the frailty distribution. We assume vague prior distributions for the components of $\boldsymbol{\beta}$. Thus, each component of $\boldsymbol{\beta}$ is assumed to follow the normal distribution with mean zero and a large variance (say 10^4) independently. The frailty distribution can be any one of the multivariate normal, multivariate skew-t distributions given by Sahu, Dey and Branco (2003). In the log-normal frailty case the hyper-parameters are θ_1 and θ_2. Also we assume that θ_i follows an Inverse Gamma distribution with parameters $(\frac{a}{2}, \frac{a}{2})$ and a log-normal distribution with mean 0 and variance σ^2, respectively, $i = 1, 2$.

In the log-skew-t frailty distribution, we consider a skew elliptical class of distribution by using the transformation given by Sahu, Dey and Branco (2003) as follows:

$$E_i = \delta_e z_{ei} + \epsilon_{ei}, \quad z_{ei} \sim G_e, \quad \epsilon_{ei} \sim F_e \tag{6}$$

$$F_{ij} = \delta_f z_{fij} + \epsilon_{fij}, \quad z_{fij} \sim G_f, \quad \epsilon_{fij} \sim F_f \tag{7}$$

where z_{ei}, ϵ_{ei}, z_{fij} and ϵ_{fij} are independent, F_e and F_f are the Standard t distribution functions with d.f. ν_e and ν_f, respectively, and G_e and G_f are the cumulative distribution functions of the half-standard t distribution with d.f. ν_{ze} and ν_{zf}, respectively. These are so-called skew-t-link model. The skewness of the marginal distributions of E_i and F_{ij} are characterized by δ_e and δ_f and the distributions of z_{ei} and z_{fij}. That is, for negative values of δ_e and δ_f, the marginal distributions of E_i and F_{ij} are skewed to the negatively (left), and for positive values, they are skewed to the positively (right). Since the direct work with Student t distribution causes the computational difficulty, we use the fact that Student t distribution can be represented as a scale mixtures of normal distribution. That is, define the auxiliary variables for $i = 1, \ldots, b$ and $j = 1, \ldots, m_i$, as

$$\lambda_{ei} \sim IG(\nu_e, \nu_e), \alpha_{ei} \sim IG(\nu_{ze}, \nu_{ze})$$

$$\lambda_{fij} \sim IG(\nu_f, \nu_f), \alpha_{fij} \sim IG(\nu_{zf}, \nu_{zf}) \tag{8}$$

where $IG(u, v)$ denotes the inverse gamma distribution with shape parameter u and scale parameter v. Then, log-skew-t frailty distribution is formulated as, for $i = 1, \cdots, b$ and $j = 1, \cdots, m_i$,

$$E_i = \delta_e z_{ei} + \epsilon_{ei}, \quad F_{ij} = \delta_f z_{fij} + \epsilon_{fij}$$

$$\epsilon_{ei} \mid \lambda_{ei} \sim N(0, \lambda_{ei}\theta_2), \quad \lambda_{ei} \sim IG(\nu_e, \nu_e)$$

$$z_{ei} \mid \alpha_{ei} \sim N^+(0, \alpha_{ei}), \quad \alpha_{ei} \sim IG(\nu_{ze}, \nu_{ze})$$

$$\epsilon_{fij} \mid \lambda_{fij} \sim N(0, \lambda_{fij}\theta_1), \quad \lambda_{fij} \sim IG(\nu_f, \nu_f)$$

$$z_{fij} \mid \alpha_{fij} \sim N^+(0, \alpha_{fij}), \quad \text{and} \quad \alpha_{fij} \sim IG(\nu_{zf}, \nu_{zf}). \tag{9}$$

We assume that θ_1 and θ_2 follow the Inverse Gamma distribution with $(\frac{a}{2}, \frac{a}{2})$ and a log-normal distribution with mean 0 and variance σ^2, respectively, $i = 1, 2$. Also Assume that δ_e and δ_f, the respective skewness parameters of E_i and F_{ij}, follow the uniform distribution in an interval $[-q, q]$ with pre-specified q. We restrict δ_e and δ_f to a bounded interval in order to avoid the conflict between fat tail and skewness.

We now assume a suitable prior distribution for ν_e, ν_f, ν_{ze} and ν_{zf}, the degrees of freedom parameter. Observe that the log-skew-t frailty distribution is well defined for any positive values of d.f. Thus, we treat ν_e, ν_f, ν_{ze} and ν_{zf} as continuous random variables taking positive values. We assume the exponential distributions each with mean κ as the prior distributions for ν_e, ν_f, ν_{ze} and ν_{zf}. Finally, we assume a moderate value of κ (between 5 and 15) so that the heavy tailed log-skew-t distributions are put forward as prior distributions.

The time-homogeneous Poisson process assumption on the jump times implies that $g \sim \text{Poisson}(a\tau_{\max})$, where $\tau_{\max}$ is the maximum observed survival time. Given g, the martingale specification implies that

$$\boldsymbol{\lambda} \sim N_g(0, \sigma_g^2 C^{-1}) \tag{10}$$

where the $g \times g$ matrix C has all elements zero except for $c_{ll} = 2, l = 1, \cdots, g - 1$ and $c_{gg} = 1$ and $c_{l,l+1} = -1, l = 1, \cdots, g - 1$ and $c_{l-1,l} = -1, l = 2, \cdots, g$.

Let ζ denote the collection of parameters, for which the prior distribution is given by (5). The joint posterior density of ζ is simply proportional to the likelihood (3.1) times the prior (5), i.e.,

$$\pi(\zeta \,|\, \mathbf{Y}, \mathbf{X}) \propto L(\boldsymbol{\beta}, \boldsymbol{\lambda}, \mathbf{e}, \mathbf{f}, g \,|\, \mathbf{Y}, \mathbf{X}) \prod_{i=1}^{b} \left\{ \pi(E_i \,|\, \cdots) \prod_{j=1}^{m_i} \pi(F_{ij} \,|\, \cdots) \right\} \pi(\cdots) \pi(g) \pi(\boldsymbol{\lambda} \,|\, g) \pi(\boldsymbol{\beta}). \quad (11)$$

3.3 Hyper-parameter Values and Prior Sensitivity

As Sahu and Dey (2005) mentioned, several hyper-parameters and simulation constants need to be specified in order to successfully adopt the Bayesian approaches. The aim is to keep the prior distributions as vague as possible so that the data, rather than the prior, drives the inference. Also sensitivity of the assumed values to statistical inference need to be checked. The adopted hyper-parameter values for our examples in Section 5 are discussed below.

For Inverse Gamma distribution prior on θ_1 and θ_2, assume that a is taken to be 0.001. For the log-normal distribution prior, the components θ_1 and θ_2 are assumed to follow the log-normal distributions independently, each with mean zero and a large variance (say 10^4) independently. The skewness parameters, δ_e and δ_f, were given uniform priors in $[-10, 10]$. The resulting interval is large enough to capture the skewness in the frailty distribution. The degrees of freedom parameters ν_e, ν_f, ν_{ze} and ν_{zf} are each given an exponential prior distribution with mean 10. This is to specify a moderate to heavy tail of the assumed t distribution.

The number of jump times, g is assumed to be a Poisson distribution with mean 10, and truncated in the interval 1 to 20. This automatically specifies a to be $10/\tau_{\max}$. The prior variance σ_g^2 of the log baseline hazard levels is assumed to be 1. Bayesian inference is largely insensitive to changes in the values of these hyper-parameters. In particular, we have always obtained a robust posterior distribution of g with about 6-8% acceptance in the reversible jump steps. The values $g = 6, \cdots, 12$ always accounted for more than 90% of the probability mass. Observe that the dimension of ζ changes as the number of jump times g associated with the baseline hazard function changes. Hence, we fit the full Bayesian model using the reversible jump Markov chain Monte Carlo method, see Green (1995); and McKeague and Tighiouart (2000). The algorithm in detail is described in McKeague and Tighiouart (2000); and Sahu and Dey (2005).

We have adopted the deviance information criterion (DIC) for model comparison. The DIC introduced by Spiegelhalter et al. (2002), directly inspired by linear and generalized linear models, but it is not so naturally defined for missing (latent) data models. Celeux et al. (2003) reassessed the criterion for such models, testing the behavior of various extensions in the cases of mixture and random effect models. As our model has random effects, we applied the complete DIC of Celeux et al. (2003). This complete DIC is given as follows. Let z, which is non-observed denote missing (latent) variables. Then

$$\mathrm{DIC} = -4E_{\theta,z}[\,\log P(y, z \,|\, \theta) \,|\, y] + 2E_z[\,\log P(y, z \,|\, E_\theta[\theta \,|\, y, z]) \,|\, y] \quad (12)$$

where the *effective dimension* P_D is given by

$$P_D = -2E_{\theta,z}[\,\log P(y, z \,|\, \theta) \,|\, y] + 2E_z[\,\log P(y, z \,|\, E_\theta[\theta \,|\, y, z]) \,|\, y]. \quad (13)$$

4. Examples

The CGD data set in Fleming and Harrington (1991) consists of a placebo-controlled randomized trial of gamma interferon (γ-IFN) in chronic granulomatous disease. The study was conducted by the International CGD Cooperative Study Group in the late 1980s. In this study, 128 patients from 13 hospitals were followed for about 1 year. The number of patients in hospital ranges from 4 to 26. Out of the 63 patients in the treatment group, 14 patients experienced at least one infection and a total of 20 infection observations were recorded. In the placebo group, 30 patients experienced at least one infection and a total of 56 infection observations were recorded. The aim of the trial was to investigate the effectiveness of γ-IFN in reducing the rate of serious infections in CGD patients. A detailed description of this study and the CGD data set can be found in Fleming and Harrington (1991). That is, applying the three-level survival model by considering that hospitals and patients are both random, the treatment effect (γ-IFN) is estimated. We consider the following four models for this data set.

Model I: log-normal frailty model with Inverse Gamma prior distribution for θ_i's.

Model II: log-normal frailty model with log-normal prior distribution for θ_i's.

Model III: log-skew-t frailty model with Inverse gamma prior distribution for θ_i's.

Model IV: log-skew-t frailty model with log-normal prior distribution for θ_i's.

Using the above prior information in Section 3.3, not all the posterior distributions turned out to be of standard form. Therefore, Metropolis algorithm was used to overcome the problem. To ensure that the samples drawn were capable of correctly identifying the posteriors, we performed the Raftery and Lewis convergence diagnostic (Raftery and Lewis, 1992). We used 2,000 iterations to burn in the Gibbs sampler, and then generated 50,000 iterations to compute all desired posterior quantities including posterior mean, posterior standard deviation, 95% credible interval and complete DIC of Celeux et al. (2003).

Table 1. Estimates of parameters and complete DIC criterion

Models	Parameter	Posterior mean	S.D.	95% credible interval	P_D	DIC
Model I	β	−1.1622	0.2740	(−1.7138, −0.6292)	325.4866	2455.6860
	θ_1	0.6073	0.1064	(0.4228, 0.8187)		
	θ_2	0.1328	0.1939	(0.0005, 0.6542)		
Model II	β	−1.1696	0.2637	(−1.7049, −0.6758)	154.6999	1695.5456
	θ_1	0.3356	0.0852	(0.1379, 0.5138)		
	θ_2	2.1e-6	1.6e-5	(3.e-17, 1.0e-5)		
Model III	β	−1.4634	0.4385	(−2.3389, −0.6421)	9.4679	1003.4760
	θ_1	0.0385	0.0700	(0.0003, 0.2670)		
	θ_2	0.0841	0.2427	(0.0004, 0.8308)		
	δ_e	−1.9898	0.7992	(−3.7003, −0.3438)·		
	δ_f	−4.2204	0.4328	(−4.9039, −3.7175)		
	ν_e	10.3117	9.7624	(1.2265, 36.1340)		
	ν_f	10.4736	11.6136	(1.5329, 44.8653)		
	ν_{ze}	12.6387	10.3463	(1.6142, 41.1060)		
	ν_{zf}	12.4651	9.7233	(2.7389, 37.6595)		
Model IV	β	−1.2396	0.2660	(−1.7796, −0.7225)	22.2153	1087.6042
	θ_1	6.6e-15	2.9e-14	(1.8e-18, 8.0e-14)		
	θ_2	7.0e-10	4.5e-09	(8.4e-18, 6.0e-09)		
	δ_e	−0.5426	0.0001	(−0.5494, −0.5424)		
	δ_f	−1.1528	8.9e-08	(−1.1529, −1.1527)		
	ν_e	9.6882	9.0589	(1.2267, 35.3374)		
	ν_f	14.1022	11.9995	(1.6229, 44.5951)		
	ν_{ze}	12.3931	9.9344	(1.5842, 38.7894)		
	ν_{zf}	14.1372	8.2586	(3.9712, 35.6385)		

Table 1 shows the posterior mean, standard deviation, 95% credible intervals, the effective dimension P_D and complete DIC values for all the parameters. The parameter β is farthest from zero under all four models. This shows that the γ-IFN significantly reduces the rate of serious infection for CGD patients. In the log-normal frailty model, θ_1 is significant and θ_2 is not. This estimate is similar with result of Yau (2001). But in the log-skew-t frailty model, θ_1 and θ_2 are closest to zero. The skewness parameters δ_e and δ_f are also significant, farthest from zero and have negative values. The estimate of the degrees of freedom parameters ν_e, ν_f, ν_{ze} and ν_{zf} suggests that a log-t model is better than a log-normal model. It is of interest to check whether the form of the assumed frailty distribution affects the posterior distribution of the number of jump times, g. To investigate this, the posterior distribution of g for each of the four models is plotted in Fig. 1. The estimated distributions seem not to differ too much and the distribution under log-normal model tends to increase dispersion than log-skew-t model. Also to compare the models, we have calculated the DIC under proposed models. Clearly, from Table 1, the DIC value of Model III is smallest which suggests strong evidence in favor of log-skew-t frailty model, that is, log-skew-t frailty model with Inverse Gamma prior distribution is much better than the other competing models.

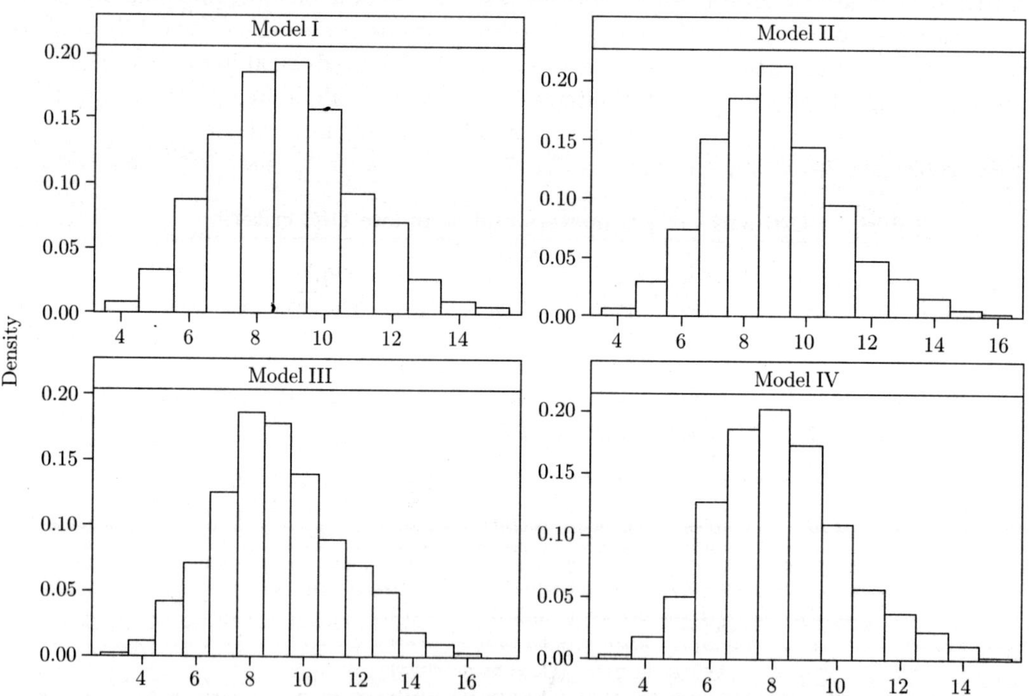

Fig. 1. The posterior distribution of g under the four models.

5. Conclusion

This article has extended the multivariate survival models in two directions by incorporating multilevel structure and the use of log-skew-t distribution for modeling frailty distributions. The new development is shown to provide better model fit and comparison using the complete deviance information criterion. The model selection criterion justifies the consideration of log-skew-t frailty model.

The new frailty distribution cannot incorporate multi-modal shape, however. If it is desired to obtain multi-modal frailty distribution then mixture distributions or a non-parametric specification in infinite dimensional parameter space (Walker and Mallick, 1997) should be considered.

The baseline hazard function conditional on the frailty distribution is modeled using a flexible martingale process. It imposes smoothing using its neighbors. The choice of jump times of the baseline hazard function is also made quite flexible using a time-homogeneous Poisson process. This removes the ad-hoc assumptions often made when specifying the number and position of the jump times.

Our methods can be extended to perform data analysis in many other scenarios including the models with time-dependent covariate effects.

Various aspects of the new models have been quantified using output of our MCMC implementation. The complete DIC values show an order of magnitude improvement provided by the new methods. These improvements in model fit have been illustrated and explained using various diagnostic plots. The proposed models are shown to be viable alternatives to the gamma models. Thus the contribution of this article can be seen in the following comment made by Hougaard (2000):

"Finding the right tools for a given problem is more exciting than using a single tool for all problems."

References

Arjas, E. and Gasbarra, D. (1994). Nonparameteric Bayesian inference for right-censored survival data, using the Gibbs sampler. *Statistica Sinica*, **4**, 505-524.

Aslanidou, H., Dey, D.K. and Sinha, D. (1998). Bayesian analysis of multivariate survival data using Monte Carlo methods. *Canadian Journal of Statistics*, **26**, 33-48.

Celeux, G., Forbes, F., Robert, C.P. and Titterington, D.M. (2003). Deviance Information Criteria for Missing Data Models. *Cahiers du Ceremade*, 0325.

Clayton, D.G. (1991). A Monte Carlo method for Bayesian inference in frailty models. *Biometrics*, **47**, 467-485.

Clayton, D.G. and Cuzick, J. (1985). Multivariate generalizations of the proportional hazards model(with discussion) Semiparametric Bayesian analysis of multiple event time data. *Journal of the Royal Statistical Society, A*, **148**, 82-117.

Cox, D.R. (1972). Regression models and life tables (with discussion). *Journal of the Royal Statistical Society, Series B*, **34**, 187-220.

Fleming, T.R. and Harrington, D.P. (1991). *Counting Processes and Survival Analysis*. New York: Wiley.

Gamerman, D. (1991). Dynamic Bayesian models for survival data. *Applied Statistics*, **40**, 63-79.

Goldstein, R.J. (1995). *Multilevel Statistical Models*. London: Arnold.

Gelfand, A.E. and Smith, A.F.M. (1990). Sampling based approaches to calculating marginal densities. *Journal of the American Statistical Association*, **85**, 398-409.

Gray, R.J. (1994). A Bayesian analysis of institutional effects in a multicenter cancer clinical trial. *Boimetrics*, **50**, 244-253.

Green, P.J. (1995). Reversible jump Markov chain Monte Carlo computation and Bayesian model determination. *Biometrika*, **82**, 711-732.

Gustafson, P. (1997). Large hierarchical Bayesian analysis if multivariate survival data. *Boimetrics*, **53**, 230-242.

Hougaard, P. (1984). Life table methods for heterogeneous populations: Distributions describing the heterogeneity. *Biometrika*, **71**, 75-83.

Hougaard, P. (1986a). A class of multivariate failure time distributions. *Biometrika*, **73**, 671-678.

Hougaard, P. (1986b). Survival models for heterogeneous populations derived from stable distribution. *Biometrika*, **73**, 387-396.

Ibrahim, J.G., Chen, M. H. and Sinha, D. (2001). *Bayesian Survival Analysis*. New York: Springer.

Kim, S.D. and Dey, D.K. (2004). Multivariate Survival Models with Random Effects. *Technical Report.*

Leonard, T. (1978). Density estimation, stochastic processes and prior information. *Journal of the Royal Statistical Society, Series B*, **40**, 113-146.

Mckeague, I. W. and Tighiouart (2000). Bayesian Estimators for Conditional Hazard Functions. *Biometrics*, **56**, 1007-1015.

Oakes, D. (1986). Semiparametric inference in a model for association in bivariate survival data. *Biometrika*, **73**, 353-361.

Oakes, D. (1989). Bivariate survival models induced by frailties. *Journal of the American Statistical Association*, **84**, 487-493.

Prentice, R.L. (1973). Exponential survivals with censoring and explanatory variables. *Boimetrika*, **60**, 279-288.

Raftery, A.E. and Lewis, S.M. (1992). How many iterations in the Gibbs sampler ? In *Bayesian Statistics 4*, eds. J. M. Bernardo, J.O. Berger, A.P. Dawid and A.F.M. Smith, Oxford: Oxford University Press.

Sahu, S.K. and Dey, D.K. (2005). On multivariate survival models with a skewed frailty. Skew-elliptical distribution and their applications: A Journey Beyond Normality, Eds. Marc G. Genton, 321-336, Chapman and Hall/CRC, 2004.

Sahu, S.K., Dey, D.K., Aslanidou, H. and Sinha, D. (1997). A Weibull Regression Model with Gamma Frailties for Multivariate Survival Data. *Lifetime Data Analysis*, **3**, 123-137.

Sahu, S.K., Dey, D.K. and Branco, M.D. (2003). A New Class of Multivariate Skew Distributions with Applications to Bayesian Regression Models. *Canadian Journal of Statistics*, **31**, 129-150.

Sinha, D. (1993). Semiparametric Bayesian analysis of multiple event time data. *Journal of the American Statistical Association*, **88**, 979-983.

Sinha, D. and Dey, D.K. (1997). Semiparametric Bayesian Analysis of Survival Data. *Journal of the American Statistical Association*, **92**, 1192-1212.

Shih, J.A. and Louis, T.A. (1995). Assessing gamma frailty models for clustered failure time data. *Lifetime Data Analysis*, **1**, 205-220.

Spiegelhalter, D.J., Best, N.G., Carlin, B.P. and van der Linde, A. (2002). Bayesian measures of model complexity and fit. *Journal of the Royal Statistical Society, Series B*, **64**, 583-640.

Stangl, D.K. (1995). Prediction and decision making using Bayesian hierarchical models. *Statistics in Medicine*, **14**, 2173-2190.

Stangl, D.K. and Greenhouse, J.B. (1995). Assessing placebo response using Bayesian hierarchical survival models. *Discussion paper 95-01*. Institute of Statistics and Decision Sciences, Duke University.

Yau, K.K.W. (2001). Multilevel Models for Survival Analysis with Random Effects. *Biometrics*, **57**, 96-102.

Bayesian Statistics and Its Applications
Edited by S.K. Upadhyay, U. Singh and D.K. Dey
Anamaya Publishers, New Delhi, India

Modeling Rainfall Data Using a Bayesian Kriged-Kalman Model

Giovanna Jona Lasinio[1], Sujit K. Sahu[2] and Kanti V. Mardia[3]

[1]University of Rome, "La Sapienza", Roma, Italy
[2]University of Southampton, Southampton, UK
[3]Department of Statistics, School of Mathematics, University of Leeds, Leeds LS2 9JT, UK

Abstract

A suitable model for analyzing rainfall data needs to take into account variation in both space and time. The method of kriging is a popular approach in spatial statistics which makes predictions for spatial data. Kalman filtering using dynamic models is often used to analyze temporal data. These approaches have been combined in a classical framework termed kriged Kalman filter (KKF) model. In the combined model, the kriging predictions dictate the optimal regression surface for incorporating spatial structure and the dynamic linear model framework is used to learn about temporal factors such as trends, autoregressive components and cyclical variations. In this article we consider a full Bayesian KKF (BKKF) model for rainfall data and its MCMC implementation. The MCMC techniques provide unified estimation of spatio-temporal effects and allow optimal predictions in time and space. The methods are illustrated with two real data examples. Using many well known validation methods we highlight the advantages of the BKKF model.

1. Introduction

Rainfall data observed at many locations over a number of time points typically covary in space and time. These spatio-temporal data types arise in many other contexts such as environmental pollution monitoring and surveillance, disease mapping, and economic monitoring of real estate prices to name a few. As a result the spatio-temporal models and data handling techniques used for data from other applications can be used to analyze rainfall data. Often the primary interests in analyzing spatio-temporal data are to smooth and predict time evolution of the response variable over a certain spatial domain. In recent years there has been a tremendous growth in the statistical models and techniques to solve such problems.

Typical rainfall data show strong spatial and seasonal effects and some other covariates, such as the elevation of the observation locations, may also significantly influence the rainfall amounts. Weather radar data are a useful covariate for short-term, e.g. hourly or daily predictions, see e.g., Brown et al. (2001) and Sahu et al. (2005). In this article we model monthly rainfall data observed over several years. Using the models we aim to produce annual maps for comparing different years and thereby detect long-term trend in rainfall.

Several statistical models exist for modeling precipitation; the literature is quite extensive. We only discuss a few papers that inspired the current work, the interested reader is referred to the references therein. In Rodriguez-Iturbe et al. (1987, 1988) stochastic point processes based models in space and time are used. A different approach is considered in Smith (1994) where, as in Stern and Coe (1984), they distinguish between processes for wet and dry periods and introduce a positively skewed distribution for the amount of rainfall, conditionally on a wet period.

The amount of rainfall is a continuous random variable, however, the measuring process often implies the rounding of observed values and the occurrences of zero rainfall with positive probability. Many authors have developed methods for handling zero rainfalls using censoring mechanisms, see for example, Dunn (2003), Allcroft and Glasbey (2003), and Sansó and Guenni (1999, 2000). Sahu et al. (2005) extend the modeling to account for several rounded discrete rainfall values occurring with non-zero probabilities by incorporating a latent continuous random variable. In their approach a part of the support of the latent variable is categorized to account for the discrete values and the remaining part is left for modeling the continuous rainfall values. The same approach is adopted here.

Geostatistical approaches are usually considered quite sensible when treating rainfall data and radar rainfall data. The use of co-kriging and kriging with external drift in rainfall fields reconstruction started in the early 1990's (Seo et al., 1990a, 1990b and Raspa et al., 1997) and since then geostatistical models have been used very often in both classical and Bayesian framework (see, e.g., Cassiraga et al., 2004 and Orasi et al., 2005).

The work in the general area of space time modeling has a long history (see Sahu and Mardia (2005b) for a recent review). In a discussion paper Mardia et al. (1998) have introduced a combined approach on kriged-Kalman filter (KKF) modeling. Recent papers within this broad framework include Kent and Mardia (2002), Kyriakidis and Journel (1999), Sahu and Mardia (2005a), and Wikle and Cressie (1999).

Kent and Mardia (2002) provide a unified approach to spatio-temporal modeling through the use of drift and/or correlation in space and/or time to accommodate spatial continuity. For drift functions, they have emphasized the use of so-called principal kriging functions, and for correlations they have discussed the use of a first order Markov structure in time combined with spatial blurring. Here we adopt one of their strategies but in a full Bayesian framework.

We work here with a process which is continuous in space and discrete in time. The underlying spatial drift is modeled by the principal kriging functions and the time component in observed sites is modeled by a vector random-walk process. The dynamic random-walk process models stochastic trend and the resulting Bayesian analysis essentially leads to Kalman filtering which is a computational method to analyze dynamic time series data, see e.g., Mardia et al. (1998) and Sahu and Mardia (2005a). In addition, the proposed models are presented in a hierarchical framework. This allows the inclusion of a 'nugget' term in the spatial part of the model. Furthermore, we incorporate model components for many interesting spatio-temporal effects, e.g., seasonal effects which influence the rainfall and the effect of additional covariates including elevation. We introduce model terms which describe and incorporate these interesting spatio-temporal phenomena.

Sahu et al. (2005) compare two approaches in modeling radar-rainfall data obtained from a controlled cloud seeding operation. There we successfully use a Gaussian random effect (GRE) model with a separable space-time covariance structure, that performs better than BKKF. In the present work we do not adopt the GRE model as the two monthly rainfall data sets are non-stationary in both space and time and a non-separable model must be used. That is why BKKF is a more sensible choice as it can easily handle non-separable and non-stationary phenomena.

The full Bayesian model is hierarchical, non-linear and incorporates a huge number of parameters. The model is fitted and used for forecasting in a unified computational framework using Markov chain Monte Carlo (MCMC) methods. The MCMC methods allow the estimation of principal kriging functions of space and replace the task of Kalman filtering using random-walk model in time. In addition, all the parameters are estimated using the implemented Markov chain. Optimal spatial predictions and temporal forecasts using predictive Bayesian techniques are also obtained as a byproduct of the coded MCMC methods.

Section 2 describes the Italian rainfall data. Section 3 extends the hierarchical BKKF model to include seasonality and covariates. Section 4 illustrates the methods with two examples while a few summary remarks are given in Section 5.

2. Italian Rainfall Data

This section describes a large dataset of monthly rainfall (mm) recorded by 226 raingauges located in south-west Italy (Fig. 1) from 1972 to 1980. The rainguage locations are given by their latitudes and longitudes. Moreover, the elevation information in meters above the sea level is also available. In what follows we shall denote by $Z(\mathbf{s}_i, t)$, $i = 1, \ldots, 226$ and $t = 1, \ldots, 108$ (12 months in 9 years), the amount of rainfall at site $\mathbf{s}_i$ and time t. The dataset comes from the historical data warehouse of the APAT (Italian Environmental Protection Agency). During this period the network was managed by the *Ufficio Compartimentale di Catanzaro* and it covered an area spanning more then 350 km in the North-South direction and about 100 km in the East-West direction. This network covers an area with strong morphological differences: on the southern part there is the Mediterranean on the two sides and a chain of mountains (Sila) in the middle; the northern part is a relatively dry region with few hills and large plains in between (Fig. 2). This figure has been constructed by linearly interpolating elevation information from 226 rainguage locations using the routines `interp` in the software packages `S-Plus` and `R`.

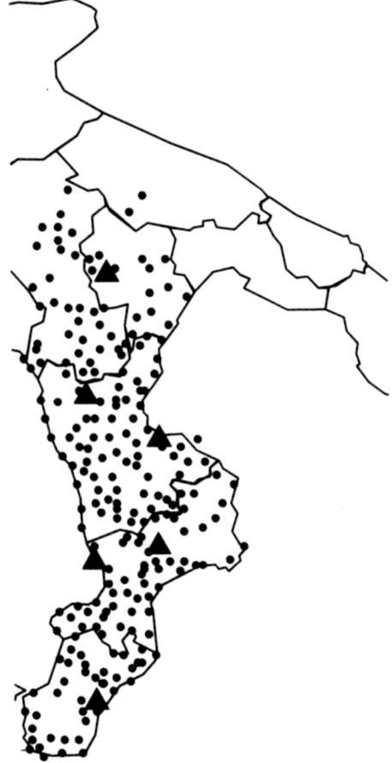

**Fig. 1.   The Catanzaro raingauges network in South Italy. Dots are modeling
sites and solid triangles are validation locations.**

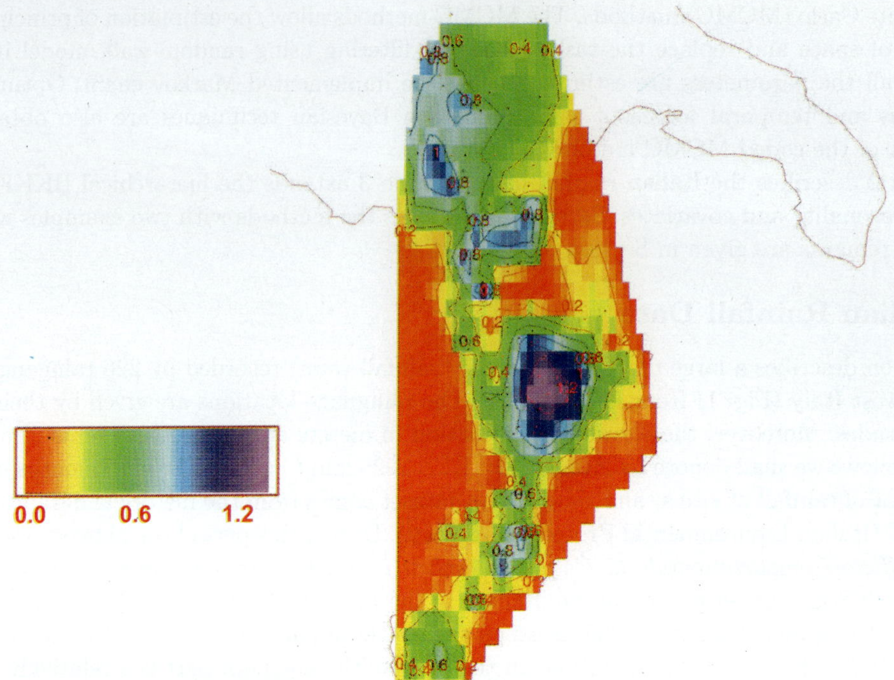

Fig. 2.   Elevation surface (km).

Out of the 226 available rainguages we set aside six sites for validating our model. These validation sites are carefully chosen using the spatial optimal design utility[1] of the package `fields` in the software R. This function finds the set of points on a discrete grid (the coordinate set, i.e., the set of 226 rainguage sites) which minimize a geometric space-filling criterion based, by our choice, on the great circle distance, see the R help manual for further details.

There are 23,760 monthly data points from the 220 modeling sites over the 9 year period. There are no missing data and simple summary statistics are reported in Table 1. In the data set about 7% observations are zero values and the rainfall values under one millimeter have been rounded to the 10th of a millimeter. Hence, we shall adopt the latent variable approach detailed in Section 3.1. See also Sahu et al. (2005) for more details.

Table 1.   Summary statistics of monthly and annual rainfall amount (in mm)

Time	Min.	1st Q.	Median	Mean	3rd Q.	Max.
Monthly data	0.00	19.80	58.00	85.36	120.60	1528.00
Annual data	77	698.2	1024.2689	943.5	1272.15	3049

There is considerable variability and non-stationarity in the data due to variation in space and time. Monthly data on the raw scale show a clear mean-variance relationship (see Fig. 3 (a)) which is almost eliminated by log-transforming observed values (see Fig. 3 (b)). We shall model on the log scale since it allows us to model the mean and variance independently.

---

[1] `cover.design` in the software R, version 2.01, http://www.R-project.org

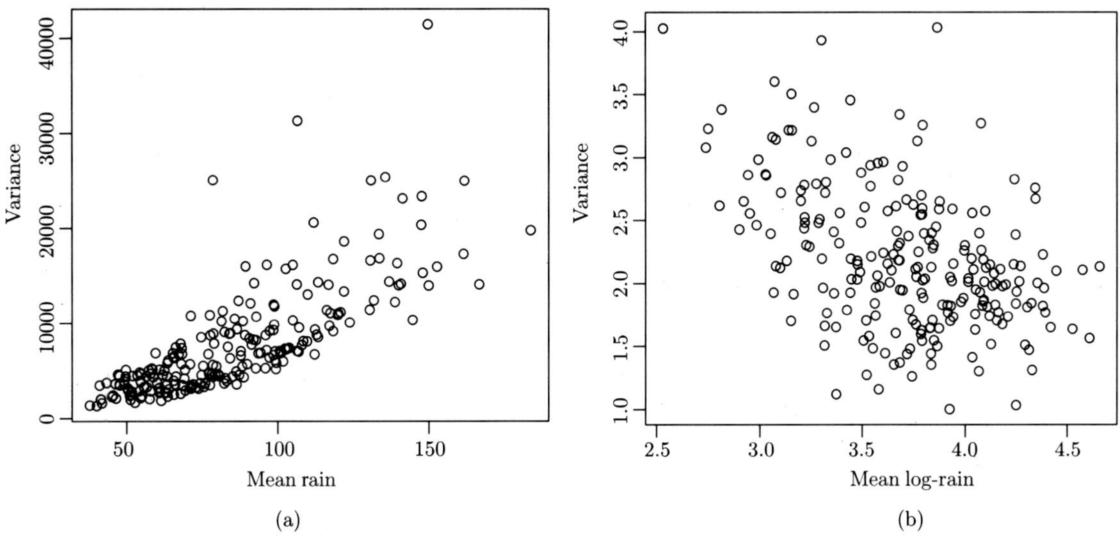

(a) (b)

Fig. 3. Mean versus variance plots on: (a) original scale, and (b) log-scale.

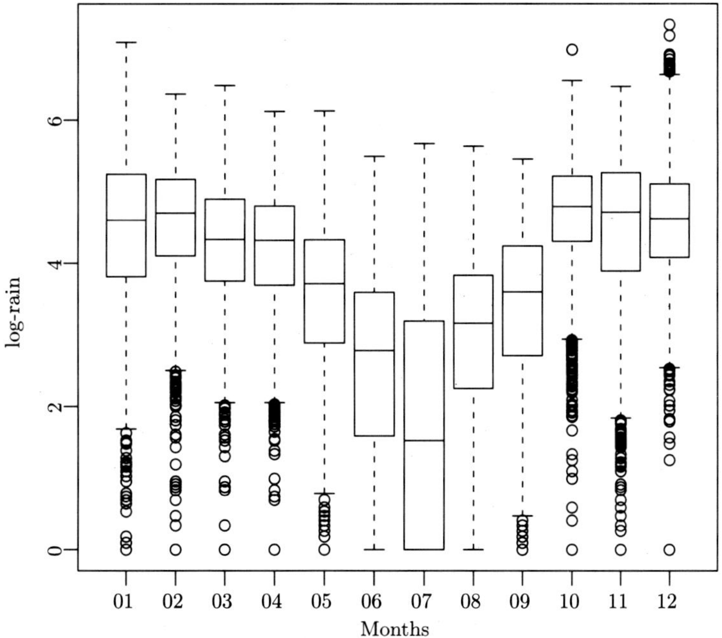

Fig. 4. Boxplot of log-rain measurements against months.

To see monthly seasonal variations, a boxplot of log-rainfall values is given in Fig. 4. July is the driest while December and January are the wettest months on an average. Fig. 5 (a-c) provides scatter-plots (with linear regression lines superimposed) of log-rainfall and three possible covariates: elevation, latitude and longitude. The plots indicate that latitude is not going to be a significant covariate while longitude can be a worthwhile covariate to include in the model. The strongest covariate information is provided by elevation and this is in accord with the well-known hydrologists' knowledge that elevation is one of the most discriminating factors for raingauge classification. We model elevation in kilometers rather than in meters simply to avoid large values.

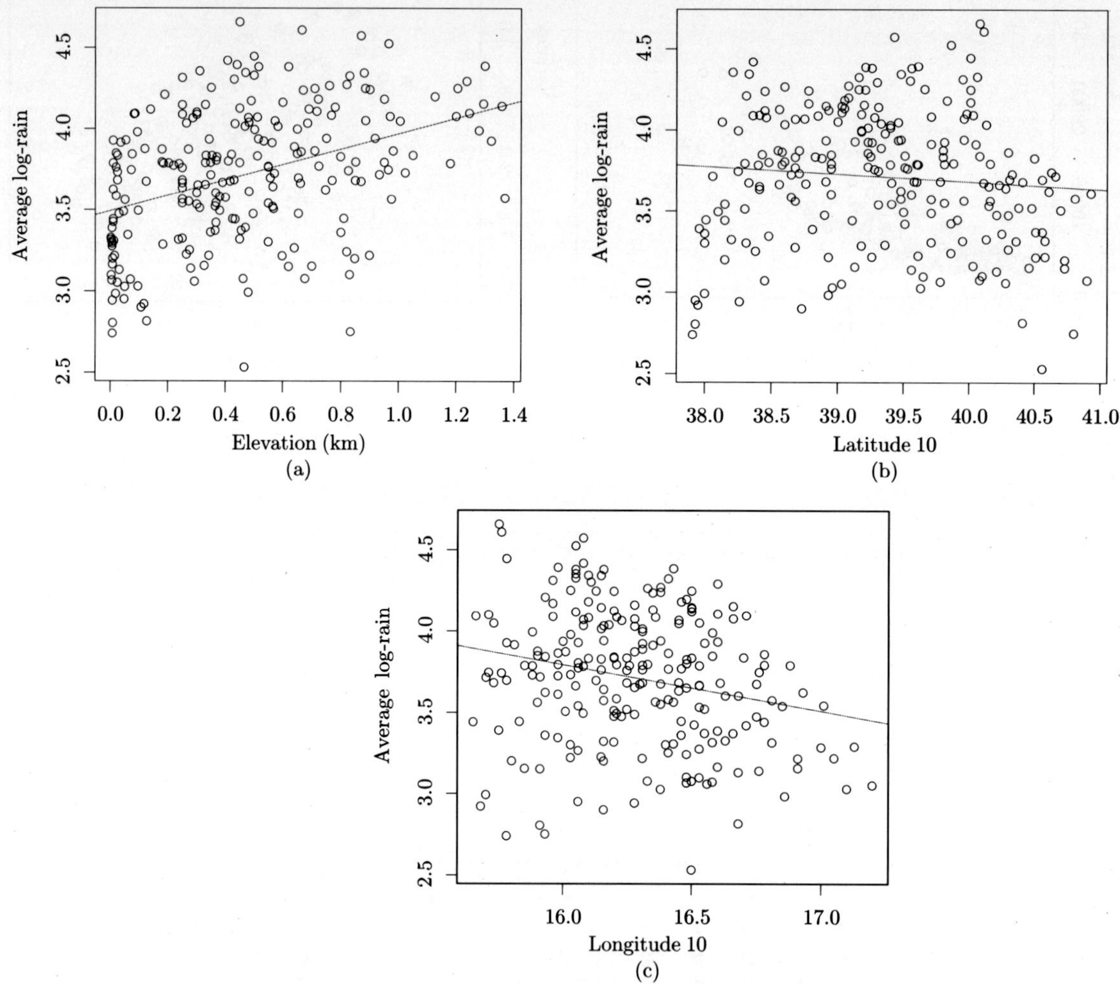

Fig. 5. Plots of log-rainfall averages of 226 sites and regression lines against: (a) elevation above sea level, (b) latitude, and (c) longitude.

The site means show evidence of spatial variation and non-stationarity (see Fig. 6) where averages of log-rainfall, classified according to their quantiles are shown in the map.

We obtain an empirical variogram of the data to investigate spatial variation. We first obtain the residuals after fitting a linear model with month as a factor variable and elevation and longitude as continuous covariates. Let $W(\mathbf{s}_i, t)$ denote the residuals. We suppose that $W(\mathbf{s}_i, t), t = 1, \ldots, T$, are independent replications at location $\mathbf{s}_i, i = 1, \ldots, n$, since we have de-trended the data. We now consider the average variogram defined by

$$\gamma(d_{ij}) = \frac{1}{2T} \sum_{t=1}^{T} E[\{W(\mathbf{s}_i, t) - W(\mathbf{s}_j, t)\}^2]$$

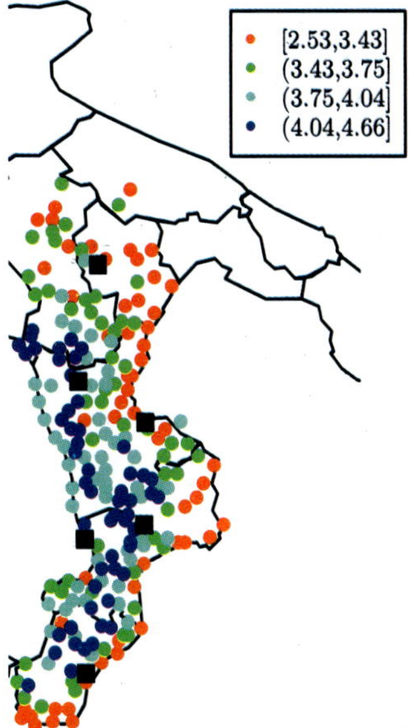

Fig. 6. **Map of log-rainfall site averages (1972-1980) classified according to their quantiles. Black squares denote validation sites.**

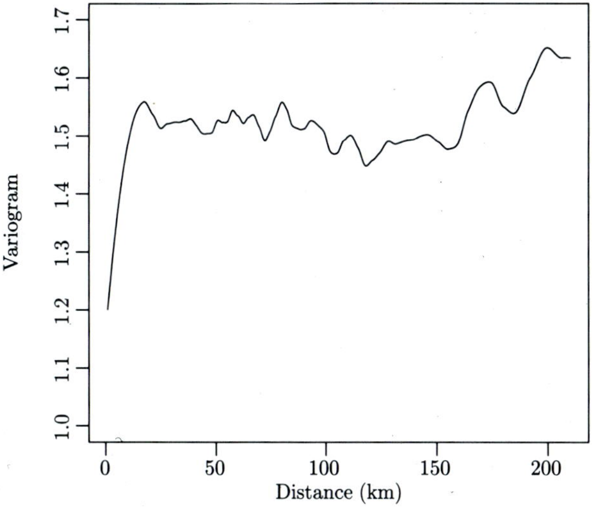

Fig. 7. **A smoothed variogram of residuals.**

where d_{ij} is the distance between the spatial locations $\mathbf{s}_i$ and $\mathbf{s}_j$. The quantity $\gamma(d_{ij})$ is estimated by

$$\hat{\gamma}(d_{ij}) = \frac{1}{2T} \sum_{t=1}^{T} \{w(\mathbf{s}_i, t) - w(\mathbf{s}_j, t)\}^2.$$

The empirical variogram cloud is obtained by plotting $\hat{\gamma}(d_{ij})$ against d_{ij} for the $n(n-1)/2$ possible pairs of locations. Fig. 7 provides a smooth loess curve[2] obtained from the variogram cloud which reveals a clear nugget effect. It also shows oscillations at higher distances which may point to possible non-stationarity. The underlying shape of the variogram can be approximated by an exponential covariogram which we assume in Section 3.

Figs. 6 and 7 and the time series plots in the validation plot in Fig. 9, all point to non-stationary variation in time and space. That is why we describe a non-stationary model with non-separable covariance structure in the next section for these data.

3.   The BKKF Model

The general model discussed here is for rainfall data recorded at n sites $\mathbf{s}_i$, $i = 1, \ldots, n$, over a period of T equally spaced time points. Let $\mathbf{Z}_t = (Z(\mathbf{s}_1, t), \ldots, Z(\mathbf{s}_n, t))^T$ denote the n-dimensional observation vector at time point t; $t = 1, \ldots, T$. The data may also include covariate information which can be time varying (see Section 3.3 for modeling details).

3.1   Latent Variables to Model Discrete Values

As mentioned in Section 2, often rainfall data are rounded and zero rainfall corresponding to dry periods occurs with positive probability. We follow Sahu et al. (2005) to model these discrete values with a continuous latent variable, denoted by $\mathbf{X}_t = (X(\mathbf{s}_1, t), \ldots, X(\mathbf{s}_n, t))'$ at time t; $t = 1, \ldots, T$.

Let there be k particular values of log rainfall $\lambda_1, \lambda_2, \ldots, \lambda_k$ which may occur with positive probabilities. Let $c_1, \ldots, c_k$ be constants such that $\lambda_i < c_i$, $i = 1, \ldots, k$. We suppose that the observed log-rainfall value at a site $\mathbf{s}$ at time t is given by

$$
\log Z(\mathbf{s}, t) = \begin{cases}
\lambda_1 & \text{if } X(\mathbf{s}, t) < c_1 \\
\lambda_2 & \text{if } c_1 \leq X(\mathbf{s}, t) < c_2 \\
\vdots & \qquad \vdots \\
\lambda_k & \text{if } c_{k-1} \leq X(\mathbf{s}, t) < c_k \\
X(\mathbf{s}, t) & \text{otherwise}
\end{cases}
$$

where set $k = 11$ and the values of $\lambda_1, \ldots, \lambda_k$ are $\lambda_1 = \log(0.02)$, $\lambda_2 = \log(0.1)$, $\ldots$, $\lambda_{11} = \log(1)$. We choose the constants $c_1, \ldots, c_k$ to be the logarithms of the numbers 0.05, 0.15, $\ldots$, 1.05, which are the mid-points of the successive intervals formed of the values 0, 0.1, 0.2, and so on.

The latent random variable $X(\mathbf{s}, t)$ for any observed rainfall bigger than 1 millimeter on the original scale is the actual log amount of rainfall. The values of $X(\mathbf{s}, t)$ corresponding to the k discrete values of $Z(\mathbf{s}, t)$ are simulated from the hierarchical model (1) given below, and are restricted to lie in the intervals implied by the above relationships. That is, for example, if $\log Z(\mathbf{s}, t) = \lambda_1$ then the corresponding $X(\mathbf{s}, t)$ will be simulated in the interval $(-\infty, c_1)$. The inverse relationship between $X(\mathbf{s}, t)$ and $Z(\mathbf{s}, t)$ is needed for predictive purposes and is straightforward to obtain from the above discussion.

3.2   Hierarchical Models

We follow Sahu and Mardia (2005a) to construct a hierarchical BKKF model for the latent variable.

---

[2] The curve is obtained using the R2.01 function `loess` with smoothing parameter equal to 0.1.

We assume the hierarchical model

$$X(\mathbf{s}_i, t) = Y(\mathbf{s}_i, t) + \epsilon(\mathbf{s}_i, t) \tag{1}$$

where $\mathbf{Y}_t = (Y(\mathbf{s}_1, t), \ldots, Y(\mathbf{s}_n, t))'$ is an unobserved but scientifically meaningful process (signal) and $\boldsymbol{\epsilon}_t = (\epsilon(\mathbf{s}_1, t), \ldots, \epsilon(\mathbf{s}_n, t))$ is a white noise process. Thus, we assume $\epsilon(\mathbf{s}_i, t)$ are i.i.d. normal random variables with mean zero and unknown variance σ_ϵ^2.

The space-time process $\mathbf{Y}_t$ is modeled as the sum of a mean process and a spatially colored process

$$Y(\mathbf{s}_i, t) = \mu(\mathbf{s}_i, t) + \gamma(\mathbf{s}_i, t) \tag{2}$$

where the mean process $\mu(\mathbf{s}_i, t)$ is described below and $\boldsymbol{\gamma}_t = (\gamma(\mathbf{s}_1, t), \ldots, \gamma(\mathbf{s}_n, t))$ is assumed to be zero mean Gaussian with covariance matrix Σ_γ which has elements

$$\sigma(\mathbf{s}_i, \mathbf{s}_j) = \mathrm{Cov}\left(\gamma(\mathbf{s}_i, t),\ \gamma(\mathbf{s}_j, t)\right) \tag{3}$$

for $i, j = 1, \ldots, n$. We assume exponential covariance structure, i.e., $\sigma(\mathbf{s}_i, \mathbf{s}_j) = \sigma_\gamma^2 \exp(-\phi\ d_{ij})$, where d_{ij} is the distance between sites $\mathbf{s}_i$ and $\mathbf{s}_j$.

3.3 Models for the Mean Process

Now we turn to modeling the mean process, $\boldsymbol{\mu}_t = (\mu(\mathbf{s}_1, t), \ldots, \mu(\mathbf{s}_n, t))'$ comprised of three terms: (i) a kriged-Kalman filter term as described in Sahu and Mardia (2005a), (ii) a term for modeling the seasonal effects, and (iii) a term to adjust covariate effects. We first describe the last two terms.

As a first attempt the seasonal effects can be modeled by the monthly indicators. However, in many practical situations, as in our second example on Venezuelan rainfall data, this may not be sufficient and a more complex set of seasonal harmonics are required. In these situations the seasonal effects are modeled by Fourier representations since those provide adequate flexible models, see e.g., West and Harrison (1997, Chapter 8). Let m be the known periodicity of the data and define $K = m/2$ if m is even and $(m-1)/2$ otherwise. In this article we are going to use two different seasonal terms, one for each example. For the Italian data analyzed in Section 4.1 we apply the following simple model:

$$S_t(\mathbf{s}) = \sum_{j=1}^{12} \rho_j(\mathbf{s})\delta_j(t)$$

where $\delta_j(t)$ are monthly indicators and the unknown coefficients $\rho_j(\mathbf{s})$ may depend on the location $\mathbf{s}$. For the Venezuelan example in Section 4.2, the seasonal term is given by

$$S_t(\mathbf{s}) = \sum_{r=1}^{K} [c_r(\mathbf{s}) \cos(2\pi tr/m) + d_r(\mathbf{s}) \sin(2\pi tr/m)]$$

where the unknown coefficients $c_r(\mathbf{s})$ and $d_r(\mathbf{s})$ may depend on the site $\mathbf{s}$. When m is even we let $d_r(\mathbf{s}) = 0$ so that $m - 1$ free seasonal parameters are kept in the model; the remaining parameter is obtained through the requirement that the seasonal effects cancel each other, i.e., they sum to zero. If there are no justifications for having spatially varying seasonal effects, we can work with the simpler model $\rho_j(\mathbf{s}) = \rho_j$, $c_r(\mathbf{s}) = c_r$ and $d_r(\mathbf{s}) = d_r$, as we do in our examples.

If there are J covariates with values $w(\mathbf{s}, t, j)$ for the jth covariate ($j = 1, \ldots, J$) at site $\mathbf{s}$ and at time t, the regression model

$$\sum_{j=1}^{J} \beta_{jt} w(\mathbf{s}, t, j)$$

can be used, where β_{jt} denotes the time-varying regression coefficient. This allows for the possibility of including time varying covariate effects, which can be significant at certain times and not significant at others. Letting $\beta_{jt} = \beta_j$ for all t in the above, we obtain a simpler model where a constant (over time) regression effect is present.

The mean process is now assumed to be

$$\mu(\mathbf{s}_i, t) = \sum_{j=1}^{p} h_{\mathbf{s}_i j} \alpha_{tj} + S_t(\mathbf{s}) + \sum_{j=1}^{J} \beta_{jt} w(\mathbf{s}_i, t, j) \tag{4}$$

where the quantities $h_{\mathbf{s}_i, j}$ are defined below, $\boldsymbol{\alpha}_t = (\alpha_{t1}, \ldots, \alpha_{tp})'$ is the state vector of dimension p. We assume that $\boldsymbol{\alpha}_t = \boldsymbol{\alpha}_{t-1} + \boldsymbol{\eta}_t$, and $\boldsymbol{\eta}_t \sim N(\mathbf{0}, \Sigma_\eta)$. For the initial value we assume that $\boldsymbol{\alpha}_0 \sim N(\mathbf{0}, C_\alpha I)$ with a large value of C_α where I is the identity matrix.

Let H be the matrix of order $n \times p$ with elements $h_{\mathbf{s}_i, j}$. The matrix H is constructed by using what are known as principal kriging functions (see Mardia et al. (1998) and Sahu and Mardia (2005a) for full details). In this implementation we take the first column of H to be the unit vector $\mathbf{1}$. The other columns are obtained as follows.

First obtain

$$B = \Sigma_\gamma^{-1} - \frac{1}{\mathbf{1}' \Sigma_\gamma^{-1} \mathbf{1}} \Sigma_\gamma^{-1} \mathbf{1} \mathbf{1}' \Sigma_\gamma^{-1}.$$

Now perform the spectral decomposition of B, $B = UEU'$, $B\mathbf{u}_i = e_i \mathbf{u}_i$, where $U = (\mathbf{u}_1, \ldots, \mathbf{u}_n)$ and $E = \mathrm{diag}(e_1, \ldots, e_n)$, and assume without loss of generality, that the eigenvalues are in non-decreasing order, $e_1 = 0 < e_{2+1} \leq \cdots \leq e_n$. Finally, the matrix H is taken as

$$H = (\mathbf{1}, e_2 \Sigma_\gamma \mathbf{u}_2, \ldots, e_p \Sigma_\gamma \mathbf{u}_p). \tag{5}$$

3.4 Prior Distributions

The full Bayesian BKKF model is completed by assuming suitable prior distributions for the parameters. The prior distributions for σ_ϵ^2 and σ_γ^2 are assumed to be the inverse gamma distribution with parameters a and b, $IG(a, b)$. We take $a = 2$ and $b = 1$ to have a proper but diffuse prior distribution with mean 1 and infinite variance. Let $\boldsymbol{\theta}$ denote the regression coefficients $\boldsymbol{\beta}$ and the seasonal parameters defining $S_t(\mathbf{s})$. We assume that $\boldsymbol{\theta} \sim N(\mathbf{0}, C_\theta I)$ for a large value of C_θ, say 10^4, to have a flat normal prior distribution for the regression coefficients.

For $Q_\eta = \Sigma_\eta^{-1}$, we assume the conjugate Wishart prior distribution

$$Q_\eta \sim W_p(2a_\eta, 2b_\eta)$$

where $2a_\eta$ is the prior degrees of freedom ($\geq p$) and b_η is a known positive definite matrix.

We say that $\mathbf{X}$ has the Wishart distribution $W_p(m, R)$ if its density is proportional to

$$|R|^{m/2} |x|^{\frac{1}{2}(m-p-1)} e^{-\frac{1}{2} \mathrm{tr}(Rx)}$$

if x is a $p \times p$ positive definite matrix, see e.g., Mardia et al. (1979, 85). (Here $\mathrm{tr}(A)$ is the trace of a matrix A.) To obtain diffuse but proper prior distributions we choose $a_\eta = p/2$ and following Sahu and Mardia (2005a), we take b_η to be 0.01 times the identity matrix.

3.5 Joint Posterior Distribution

The above model has the following set of parameters: the error variance σ_ϵ^2, the latent process $\mathbf{Y}_t$, the spatial variance σ_γ^2, the dynamic parameters $\boldsymbol{\alpha}_t$ and their precision matrix Q_η, and the seasonal and the regression coefficients in the last two terms in (4), $\boldsymbol{\theta}$. Let $\boldsymbol{\xi}$ denote these parameters. The log of the joint posterior density of $\boldsymbol{\xi}$, denoted by $\pi(\boldsymbol{\xi}|\mathbf{z}_1,\dots,\mathbf{z}_T)$, is given by (upto a constant of proportionality),

$$-\tfrac{Tn}{2}\log(\sigma_\epsilon^2) - \tfrac{1}{2\sigma_\epsilon^2}\sum_{t=1}^{T}(\mathbf{x}_t - \mathbf{y}_t)'(\mathbf{x}_t - \mathbf{y}_t) - \tfrac{T}{2}\log|\Sigma_\gamma| - \tfrac{1}{2}\sum_{t=1}^{T}(\mathbf{y}_t - \boldsymbol{\mu}_t)'\Sigma_\gamma^{-1}(\mathbf{y}_t - \boldsymbol{\mu}_t)$$

$$-\tfrac{T}{2}\log|\Sigma_\eta| - \tfrac{1}{2}\sum_{t=1}^{T}(\boldsymbol{\alpha}_t - \boldsymbol{\alpha}_{t-1})'\Sigma_\eta^{-1}(\boldsymbol{\alpha}_t - \boldsymbol{\alpha}_{t-1}) + \tfrac{1}{2}(2a_\eta - p - 1)\log|Q_\eta| - \tfrac{1}{2}\mathrm{tr}(2b_\eta Q_\eta)$$

$$-\tfrac{1}{b\sigma_\epsilon^2} - (a+1)\log(\sigma_\epsilon^2) - \tfrac{1}{b\sigma_\gamma^2} - (a+1)\log(\sigma_\gamma^2) - \tfrac{1}{2C_\alpha}\boldsymbol{\alpha}_0'\boldsymbol{\alpha}_0 - \tfrac{1}{2C_\theta}\boldsymbol{\theta}'\boldsymbol{\theta}.$$

All complete conditional distributions are standard and are straightforward to obtain, see Sahu and Mardia (2005a) for more details. The posterior predictive distributions are used to make Bayesian predictions in both space and time.

4. Examples

4.1 Italian Rainfall Data

We now return to modeling the Italian rainfall data described in Section 2. The computation problem here is huge since there are 23,760 monthly data points from 220 sites over 9 years. We have implemented the full BKKF model using MCMC methods and performed the usual MCMC diagnostic analyses, the Bayesian sensitivity analyses and the model choice methods for choosing and tuning various simulation parameters and the spatial smoothing parameter ϕ in the BKKF model. Those are omitted for brevity, the interested reader can see Sahu and Mardia (2005a) and Sahu et al. (2005) for more details.

Using a grid search we choose the value 0.05 for ϕ. The value 0.05 provides an effective range of 60 km which is a reasonable choice given that the study region is roughly 300 km $\times$ 100 km. Lastly, we work with the state vector of dimension $p = 10$ in Equation (5). This choice is again based on predictive performance of the model for different values of p. In the following we discuss the main results and

- show the residual plots for assessing model adequacy,
- validate the model by performing out of sample predictions,
- present the parameter estimates,
- illustrate monthly prediction maps,
- obtain annual prediction maps.

We have checked all the residual plots (not shown) for our model for all the 220 modeling sites. Fig. 8 provides the residuals for four randomly chosen sites out of the 220 modeling ones. Residuals do not show any cause for concern and we can state that the model seems to be adequate for the data. This discussion is continued below where we compare observed data surfaces and the fitted surfaces for a typical month later in this section.

We now return to the validation data for six sites which we have set aside. We consider validation at all the 108 time points. Fig. 9 provides the validations plots. Overall few points seem to be poorly predicted. Indeed if we set an error threshold of $\pm$ 150 mm for the monthly data less than 10% of the 648 predicted points are poorly estimated with a prevalence of over estimation. However, we

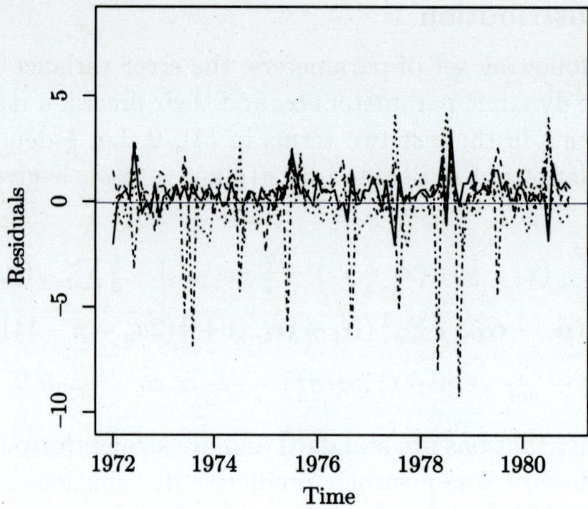

Fig. 8.   **Time series plot of residuals for 4 randomly chosen sites, out of the 220 modeling ones.**

emphasize that all the observed values are within the 95% prediction intervals. We do not show the prediction intervals because all the lower limits are zero (rainfall cannot be negative) and the upper limits are high due to large variability on the original rainfall measurement scale.

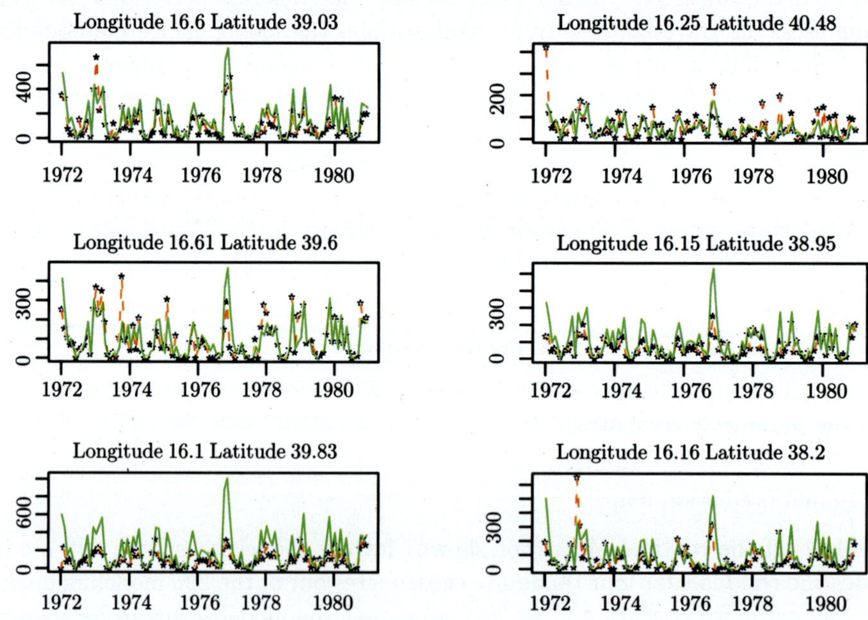

Fig. 9.   **The observed and predicted values at 6 validation sites. Observed values are shown in red (points and dotted lines) and predicted values are shown in green (solid lines).**

Table 2 gives the parameter estimates for our model. The regression coefficient for elevation is seen to be positively significant as expected (see Fig. 5(a)). The coefficient for longitude is negatively significant showing that western regions are wetter than the eastern sites on an average. The error variance σ_ϵ^2 is roughly equal to the nugget effect (see Fig. 7(b)). Monthly indicators turn out to be insignificant and that is why we have not included their estimates in Table 2. This, in our opinion, is an indication that the random walk part of the model is enough to account for most of the temporal variation. Furthermore, we have computed the variance of the residuals categorized by month. Subsequently we have found the ratio of the maximum and minimum variances to be 2.72 which shows that the residuals do not vary by month, i.e., the residuals are homoscedastic.

Table 2.   Parameter estimates

Parameter	Mean	sd	95% interval
Elevation	0.475	0.035	(0.407, 0.542)
Longitude	-0.013	0.005	$(-0.022, -0.003)$
σ_γ^2	0.925	0.036	(0.855, 0.997)
σ_ϵ^2	1.389	0.024	(1.341, 1.436)

In Fig. 10 three spatial surfaces[3] illustrate the model for monthly data. In Fig. 10 (a) a rough linear interpolation of observed data shows small scale variability (roughness) of the data. Fig. 10 (b) reports the same interpolation performed on the fitted values. As expected the fitted surface is smoother than the observed one, but it reproduces the observed rainfall field fairly well. This behavior is common to all months, even when very little rain is recorded. Fig. 10 (c) shows the standard error surface for the fitted values. All values are on the original scale (mm).

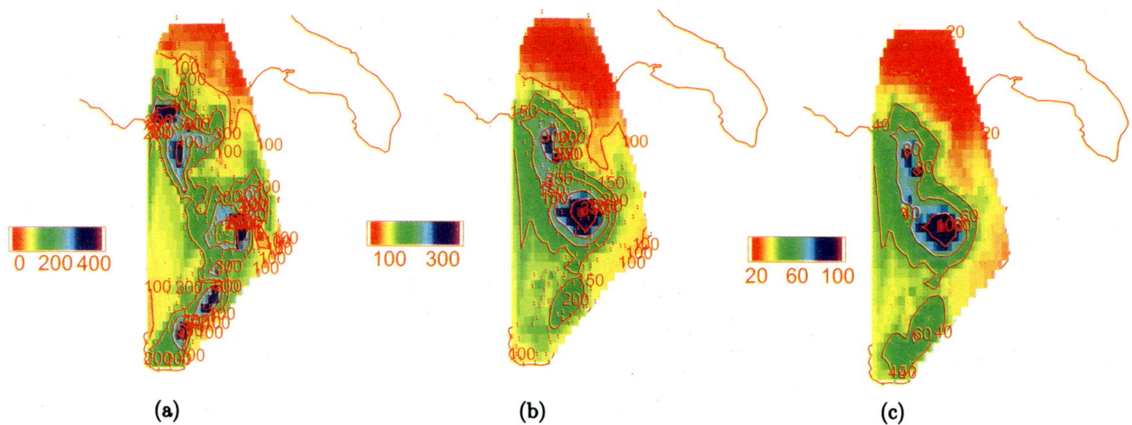

Fig. 10.   Fitted maps for December 1976: (a) observed values (mm), (b) fitted values (mm) and (c) standard errors.

From the fitted monthly values on the original scale we build annual maps of total rainfall amount simply by summation. The maps are presented in Figs. 11 and 12. In panel (a) of Fig. 11, observed values for 1976 (the wettest year) are compared to fitted values (b) for the same year. The same comparison is performed in Fig. 12 for 1977 which happens to be the driest year. Again, the fitted

---

[3] Surfaces are obtained using the `interp` function and the library `map` in Splus

surfaces are smoother than the observed ones and there is general agreement between the two. There are many similarities between the two fitted surfaces for the two years. This is generally expected since rainfall patterns in space do not change rapidly, that is, wet areas remain wet in the following year and so on. However, there are dissimilarities between the two fitted maps, for example, the south-east corner is considerably drier in 1977 (driest year) than 1976 (wettest year). The observed data maps agree with these findings.

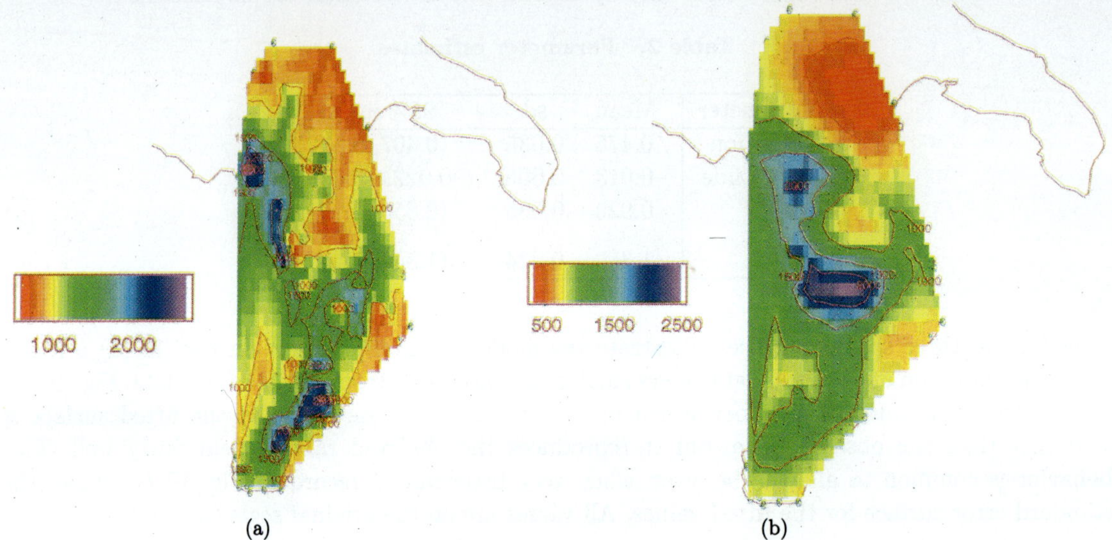

Fig. 11. The fitted annual maps in 1976. The wettest year: (a) observed total rainfall amount (mm) and (b) fitted values (mm).

4.2 Venezuelan Rainfall Data

The previous example provides many detailed analyses of the BKKF model fitted to the Italian rainfall data. The objective of the second example on Venezuelan rainfall data is to compare our methods with some currently available techniques. We consider a data set analyzed by Sansó and Guenni (1999, 2000), and Stroud et al. (2001). The data set consists of monthly rainfall (mm) for 16 years from 80 stations in the Venezuelan state of Guarcio. See Sansó and Guenni (1999) for some preliminary analyses of the data.

We use data for first 15 years from 40 randomly chosen stations to fit the proposed model and use the observations from the remaining one year and 40 stations to validate the model. Of course, better estimation and prediction can be achieved by using all the data (for 16 years and from 80 stations) to fit the models. But here our purpose is to demonstrate the effectiveness of the methods using cross-validation in both space and time.

By considering an annual periodicity ($m = 12$), Sansó and Guenni (2000) have modeled the data so that the mean and variance for a particular month remain the same over all the years at a particular site. We do not impose this restriction in our model. We, however, adopt their time varying transformation approach. Let $z(\mathbf{s}, t)$ denotes the observed rainfall at time t in the site $\mathbf{s}$. We

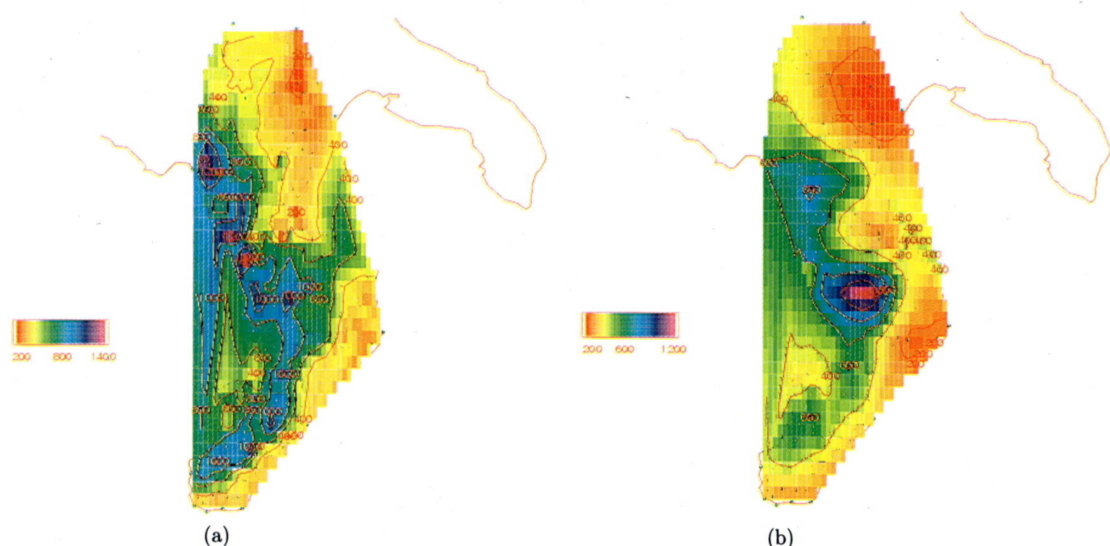

(a)   (b)

Fig. 12.   The fitted annual maps in 1977. The driest year: (a) observed total rainfall amount (mm) and (b) fitted values (mm).

suppose that

$$z(\mathbf{s}, t) = \begin{cases} x(\mathbf{s}, t)^{\beta_t} & \text{if } x(\mathbf{s}, t) > 0 \\ 0 & \text{otherwise} \end{cases}$$

where $\beta_t = \beta_l$ whenever $t = l$ modulo m. The vector of transformed variables $\mathbf{X}_t$ for $n = 40$ stations is assumed to follow the hierarchical model (1). Sansó and Guenni used a time varying variance $\sigma^2_{\epsilon,t}$ with $\sigma^2_{\epsilon,t} = \sigma^2_{\epsilon,l}$ whenever $t = l$ modulo m. However, their analysis shows that β_t and $\sigma^2_{\gamma,t}$ are highly correlated and it is not satisfactory to estimate both simultaneously. This is why we do not assume a time varying variance parameter.

The seasonal harmonics described in Section 3 are significant here since the data show strong periodicity (Fig. 13). This figure shows the fitted means and forecasts for a randomly chosen site, which was used in fitting. The model based estimates generally agree with the observed data points, validating the forecasting abilities of the proposed methods. The error intervals are not superimposed since those were not very informative, though all the observations fell within the two limits. The lower limits were mostly zeros giving a straight line for the lower interval.

The fitted means and forecast plot from the Sansó and Guenni model are shown in Fig. 14. The forecasts from the Sansó and Guenni model are slightly better than those from the proposed model while the fitted means are better under the proposed model. This is expected since the proposed model is dynamic and incorporates more number of parameters than the Sansó and Guenni model. A better model fit is achieved using more number of parameters while forecasting using a smaller number of parameters is seen to be better.

The main difference between the proposed model and the Sansó and Guenni model is in estimation of spatial characteristics. Our model is expected to achieve better results in predicting in the spatial domain as it includes the important principal kriging functions. In order to see this we calculate the weighted distance criterion D^2 proposed in Sahu and Mardia (2005a) to compare the models both spatially and temporally. The distance value for the proposed model is 3547, which should be

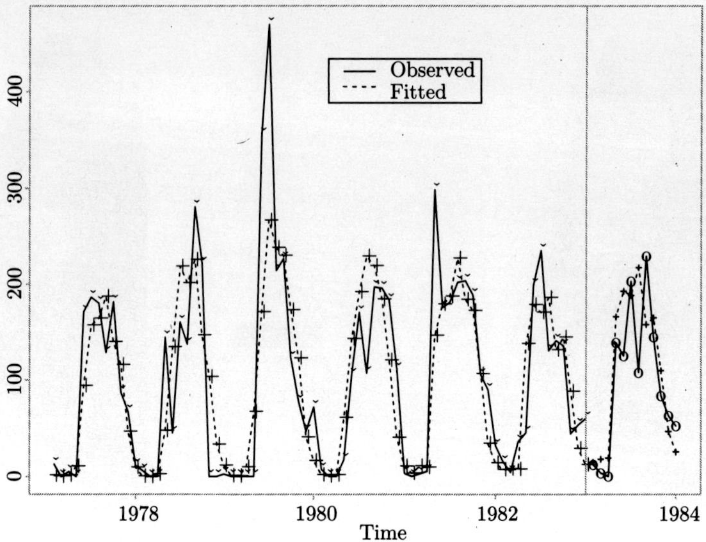

Fig. 13.   The fitted means and forecasts for a randomly chosen site using the proposed method.

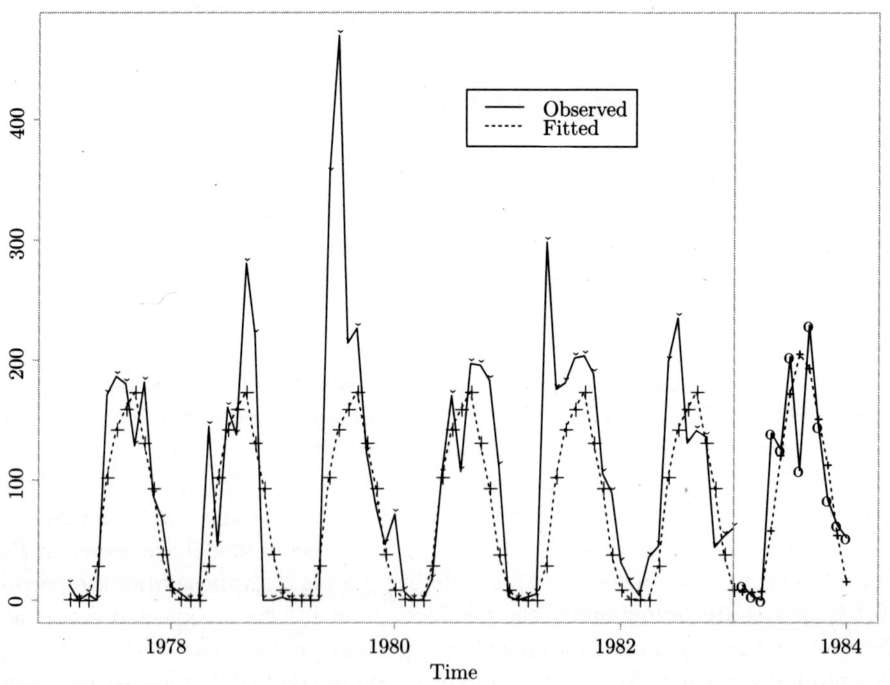

Fig. 14.   The fitted means and forecasts using the Sansó and Guenni method for the same site as in Fig. 13.

compared to a theoretical cut-off point from the χ^2 distribution with 7680 ($= 192 \times 40$) degrees of freedom. Since the observed value is quite small compared to the degrees of freedom we conclude that the model is performing a good job in prediction. The weighted distance for the model proposed by Sansó and Guenni is estimated to be 4996, a value much higher than the distance under the proposed model. Thus the new model performs much better prediction at new locations. This is expected since the proposed model uses the optimal ('kriging') spatial prediction surface while the earlier model does not.

5. Discussion

In this article we have proposed the BKKF model for analyzing rainfall data. Our approach solves in a unified framework the problems of accounting for discreteness in the data, spatial variation, temporal variation and joint space-time non-separable variation. It easily allows us to incorporate different types of seasonal effects which depend on climatic and morphological conditions typical of each region. The model has been fully implemented through MCMC on two different data sets, and it has shown advantages over some current methods. The MCMC implementation of the model allows us to predict in space and forecast in time using Bayesian methods.

Acknowledgements

We would like to thank Dr. Attilio Colagrossi of the *Servizio Raccolta e Gestione Dati* (APAT) for kindly giving us the data and for many useful discussions. This work has been partially supported by the MIUR COFIN project 2004-2006 Statistical Methods for Forecasting Improvement in Environmental Sciences.

References

Allcroft, D. J. and Glasbey, C. A. (2003). A latent Gaussian Markov random-field model for spatiotemporal rainfall disaggregation. *Journal of the Royal Statistical Society, C, Applied Statistics*, **52**, 487–498.

Brown, P. E., Diggle, P. J., Lord, M. E. and Young, P. C. (2001). Space-time calibration of radar rainfall data. *Applied Statistics*, **50**, 221–241.

Cassiraga, E. F., Guardiola-Albert, C. and Gomez-Hernandez, J. J. (2004). Automatic modeling of cross-covariances for rainfall estimation using reengauge and radar data. GEOENV IV - Geostatistics for environmental applications: Proceedings quantitative geology and geostatistics 13, 391–399, Kluwer Academic Publishing, Dordrecht.

Dunn, P. (2003). Precipitation occurrence and amount can be modelled simultaneously. Working Paper, Series SC-MC-0305, Faculty of Sciences, USQ.

Kent, J. T. K. and Mardia, K. V. (2002). Modelling Strategies for Spatial-Temporal Data. In *Spatial Cluster Modelling*. (Eds A. Lawson and D. Denison), London: Chapman and Hall, 214–226.

Kyriakidis, P. C. and Journel, A. G. (1999). Geostatistical space-time models: A review. *Mathematical Geology*, **31**, 651–684.

Mardia, K. V., Kent, J. T. and Bibby, J. M. (1979). *Multivariate Analysis*. London: Academic Press.

Mardia K.V., Goodall C., Redfern E.J., and Alonso F.J. (1998). The Kriged Kalman filter (with discussion). *Test*, **7**, 217–252.

Orasi, A., Jona Lasinio, G. and Ferrari, C. (2005). Comparison of calibration methods for the reconstruction of space-time rainfall fields in Southern Italy. Submitted. Available from http://w3.uniroma1.it/dspsa/Rapporti_tecnici/orasijonaferrari.pdf

Raspa, G., M. Tucci, and R. Bruno (1997). Reconstruction of rainfall fields by combining ground raingauges data with radar maps using external drift method (Kluwer Academic Publishers ed.), Volume 2 of Geostatistics Wollongong 96, 1306–1315. E.Y. Baafi and N.A. Schofield eds.

Rodriguez-Iturbe, I., Cox, D.R. and Isham, V. (1987). Some models for rainfall based on stochastic point processes. *Proceedings of the Royal Society of London*, A, **410**, 269–288.

Rodriguez-Iturbe, I., Cox, D.R. and Isham, V. (1988). A point process model for rainfall: further developments. *Proceedings of the Royal Society of London*, A, **417**, 283–298.

Sahu, K.S., Jona Lasinio G., Orasi, A. and Mardia, K.V. (2005). A comparison of spatio-temporal Bayesian models for reconstruction of rainfall fields in a cloud seeding experiment. *Journal of Mathematics and Statistics*, **1** (4), 273-281.

Sahu, S. K. and Mardia, K. V. (2005a). A Bayesian Kriged-Kalman model for short-term forecasting of air pollution levels. *Journal of the Royal Statistical Society*, C, **54**, 223–244.

Sahu, S. K. and Mardia, K. V. (2005b). Recent Trends in Modeling Spatio-Temporal Data. In: Proceedings of the Special Meeting on Statistics and Environment, Società Italiana, oli Statistica, Messina September 21-23, 2005, Invited Papers, 69-83, CLEUP.

Sansó, B. and Guenni, L. (1999). Venezuelan rainfall data analysed by using a Bayesian space-time model. *Applied Statistics*, **48**, 345–362.

Sansó, B. and Guenni, L. (2000). A nonstationary multisite model for rainfall. *Journal of the American Statistical Association*, **95**, 1089–1100.

Smith, R. (1994). Spatial modelling of rainfall data. In *Statistics for the environment 2: Water Related Issues*, eds. V. Barnett, and K. Feridum Turkman, Wiley: New York, 19–42.

Stern, R. and R. Coe (1984). A model fitting analysis of rainfall data. *Journal of the Royal Statistical Society*, A, **147**, 1–34.

Stroud, J. R., Müller, P. and Sansó, B. (2001). Dynamic models for Spatio-temporal data. *Journal of the Royal Statistical Society*, B, **63**, 673–689.

West, M. and Harrison, J. (1997). *Bayesian Forecasting and Dynamic Models*. New York: Springer.

Wikle, C. K. and Cressie, N.A.C. (1999). A dimension-reduced approach to space-time Kalman filtering. *Biometrika*, **86**, 815–829.

Bayesian Statistics and Its Applications
Edited by S.K. Upadhyay, U. Singh and D.K. Dey
Anamaya Publishers, New Delhi, India

A Line Finding Assignment Problem and Rock Fracture Modelling

K.V. Mardia[1], A.N. Walder[1], C. Xu[2], P.A. Dowd[3], R.J. Fowell[2], V.B. Nyirongo[1] and J.T. Kent[1]

[1]Department of Statistics, School of Mathematics, University of Leeds, Leeds LS2 9JT, UK
[2]School of Process, Environmental & Materials Engineering, University of Leeds, Leeds LS2 9JT, UK
[3]Faculty of Engineering, Computer & Mathematical Sciences, University of Adelaide, Adelaide, Australia

Abstract

This article deals with a stochastic stereologic problem of estimating fracture lines, given only the data at boreholes. We formulate an appropriate model. The problem is challenging since neither the lines (slope, intercept) are known, nor their number. We give an MCMC implementation where all the parameters are allowed to vary. We examine sensitivity to priors. All this is for an important case study arising in the risk assessment for storage of nuclear waste. Some extensions and open problems are also discussed.

1. Introduction

The work in this paper has been motivated by a problem of considerable importance in risk assessment for the safe storage of hazardous wastes in underground repositories; in particular, the need for accurate prediction of pollutant trajectories in the subsurface of the earth (and their likely consequences). As a first step it is necessary to know the location and extent of any fractures in the vicinity of the proposed repositary. However, the only information available to us will be in the form of boreholes, so we have a challenging stereological problem to estimate the fracture lines.

This article considers the simplest rock fracture modelling problem of a plane section containing line segment fractures, shown in Fig. 1, which also shows vertical boreholes, represented by lines parallel to the y-axis. On each of these we would have been given the depths at which the boreholes intersect the fractures, see Fig. 2. (Thus, Fig. 1 shows the ground truth of some real fractures; superimposed are some virtual boreholes, intersecting the fractures, whereas Fig. 2 shows the observed data, leading to the observations at the open circles.) The unknowns are the number of fractures, and for each fracture its intercept, slope and extent. Having established (in Section 2) a plausible model for the likelihood of such a data set, we estimate the parameters for the fractures using an implementation of Markov chain Monte Carlo (in Section 3). Each intersection point is given a label corresponding to the current fracture it has been associated with; part of our problem involves switching these labels to improve the assignment of fractures.

In Markov chain Monte Carlo, the label switching problem is a challenging one. In the context of mixtures, see a recent article by Jasra et al. (2005). The nature of the problem in label switching

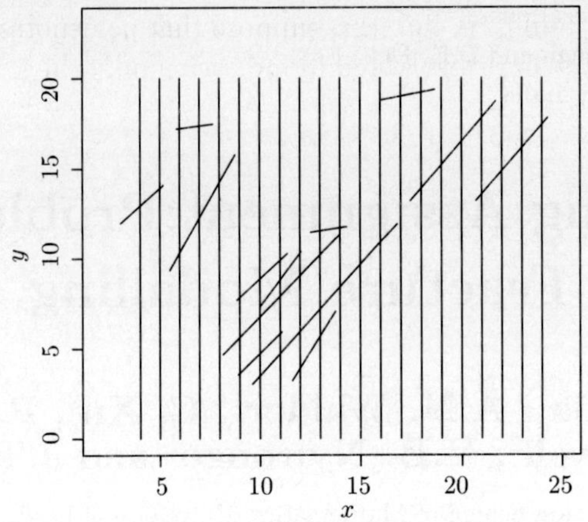

Fig. 1. True fractures, intersecting boreholes.

for assignment here is somewhat more intricate as there is only partial information available, since fracture lines are not observed, only the intersections where they meet the boreholes. A similar problem is matching two unlabelled configurations when the underlying matching transformation is unknown and when only a subset provides a solution. In this case, reversible jump can be avoided (Green and Mardia, 2006). Here, we are looking for correspondence of points with unknown lines, and so reversible jump is essential.

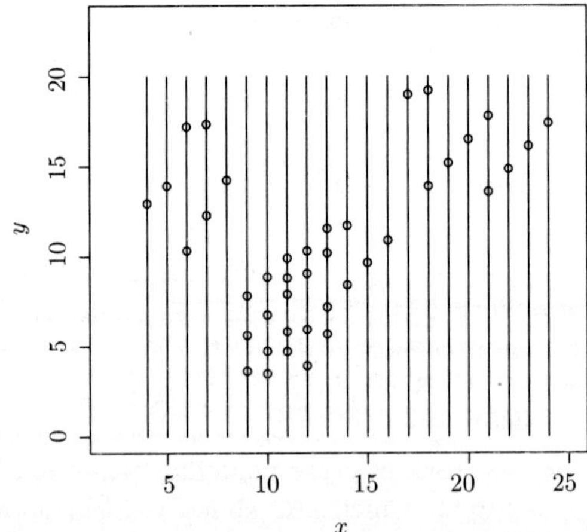

Fig. 2. Observed data as the intersection points on boreholes.

Section 4 presents a case study based on some real data, from the Yucca mountains (Barton and Larson, 1985) and the details of the implementation used, and the results obtained, together with sensitivity studies. Section 5 discusses some extensions and other open problems.

2. The Model

We first introduce some notation. Let us label the boreholes by $i = 1, ..., I$ and the intersection points on each borehole by $j = 1, ..., n_i$, where n_i is the number of points on the ith borehole. Let x_i be the horizontal location of the ith borehole. Further, suppose that y_{ij} denotes the vertical position of the jth intersection point on the ith borehole, listed in increasing order in j, for each i, and let $Y = \{y_{ij}\}$ denote this dataset. Also, let $k = 1,\ldots,K$ denote the (unknown) fracture lines. When lines are fitted to the data then the kth line will intersect a consecutive set of boreholes near $\alpha_k + \beta_k x_i$.

A key parameter is a "matching function" $\pi(i,j) \in \{1,\ldots,K\}$ that associates each intersection point with one of the fracture lines. Denote the *shortest* orthogonal squared distances between the intersection points and the kth fracture line by $d^2_{i,j;k}$ (Fig. 3). If d is normally distributed, with zero mean, then the likelihood for Y, π, K and the $\{\alpha_k, \beta_k\}$ is given by

$$L(Y;\pi,K,\{\alpha_k,\beta_k\}) = \left(\frac{1}{\sqrt{2\pi}\sigma}\right)^n \prod_{i=1}^{I}\prod_{j=1}^{n_i} \exp\left(-\frac{d^2_{ij,\pi(i,j)}}{2\sigma^2}\right)$$

where $n = \sum n_i$. Note that $d^2_{ij;\pi(i,j)}$ can be written as

$$\frac{(\alpha_k + \beta_k x_i - y_{ij})^2}{1 + \beta_k^2}$$

and we shall also write σ_k^2 for $\sigma^2(1 + \beta_k^2)$.

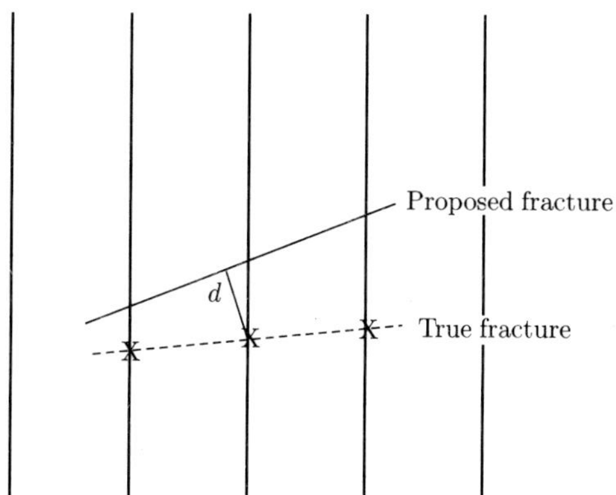

Fig. 3.   Distance to proposed fracture.

We introduce the following priors:

- For the intercept α, assume that the uniform distribution is a suitable prior, over the length of the borehole.

- For the slope, let $\beta = \tan\theta$, and have a prior for 2θ that it is distributed as von Mises

$$f(\theta) = \frac{1}{\pi I_0(k)}\exp\{k\cos(2\theta - 2\mu)\}, (-\pi/2 \leq \theta \leq \pi/2),$$

where $I_0(\kappa)$ is the Bessel function of order 0 and κ is the concentration parameter. The Jacobian of the transformation is $\frac{d\beta}{d\theta} = 1 + \beta^2$. It can be shown that

$$f(\beta) = \frac{1}{\pi I_0(k)(1 + \beta^2)}\exp\left\{\frac{\kappa}{(1 + \beta^2)}\left[(1 - \beta^2)\cos 2\mu + 2\beta\sin 2\mu\right]\right\}, (-\infty \leq \beta \leq \infty).$$

Letting $\gamma = \tan \mu$, then this can be written as

$$f(\beta) = \frac{1}{\pi I_0(\kappa)(1 + \beta^2)} \exp\left\{ \frac{\kappa}{(1 + \beta^2)(1 + \gamma^2)} \left[(1 - \beta^2)(1 - \gamma^2) + 4\beta\gamma \right] \right\}. \qquad (1)$$

Note that when $\kappa = 0$, this is simply the Cauchy distribution.

- For the number of fractures K, use the Poisson distribution.

Combining the likelihood above with prior densities for K, $\{\alpha_k, \beta_k\}$ gives us the posterior density $p(K, \{\alpha_k, \beta_k\}|Y)$. One way to calculate the posterior mean is by a simulation method which does not depend on the complicated normalising constant: a popular technique is Markov chain Monte Carlo. This procedure generates a Markov chain whose equilibrium distribution is the posterior density $p(K, \{\alpha_k, \beta_k\}|Y)$. For simplicity write $\Theta = (K, \{\alpha_k, \beta_k\})$.

3. MCMC Implementation

Recall that, for the MCMC implementation, at each iteration, generate Θ_{new} (usually from a proposal distribution g) and calculate the Hastings ratio

$$p = \frac{p[\Theta_{\text{new}}|Y]g(\Theta_{\text{old}}|\Theta_{\text{new}})}{p[\Theta_{\text{old}}|Y]g(\Theta_{\text{new}}|\Theta_{\text{old}})}.$$

In some cases the proposal density is symmetric, $(g\Theta_{\text{old}}|\Theta_{\text{new}}) = g(\Theta_{\text{new}}|\Theta_{\text{old}})$, in which case $p = p[\Theta_{\text{new}}|Y]/p[\Theta_{\text{old}}|Y]$. We accept $\Theta = \Theta_{\text{new}}$ with probability $\min(1, p)$, otherwise we keep $\Theta = \Theta_{\text{old}}$, i.e. if $p > 1$, we take $\Theta = \Theta_{\text{new}}$, whereas if $p < 1$, we perform a further randomisation by drawing a random sample from uniform $(0, 1)$, and accept $\Theta = \Theta_{\text{new}}$ with probability p. Typically, a burn-in period would be allowed for the initial simulations and then an average is taken of a subset of the remaining simulations, although this is not quite appropriate here.

The question of how to construct a Markov chain whose stationary distribution is our target distribution has been addressed by Metropolis et al. (1953), with the method generalised by Hastings (1970). At each iteration we first propose a type of change, chosen at random, from the following proposals:

- *Slope*: Choose a line at random, and vary the slope, $\beta = \tan \theta$, of a line as follows. Generate a new value of θ from a von Mises distribution, with a fairly large concentration parameter, say $\kappa = 2$. This will be a symmetric proposal, as $g(\theta_{\text{new}}|\theta_{\text{old}}) = g(\theta_{\text{old}}|\theta_{\text{new}})$, where

$$g(\theta_{\text{new}}|\theta_{\text{old}}) = \frac{1}{\pi I_0(\kappa)} \exp\{\kappa \cos(2\theta_{\text{new}} - 2\theta_{\text{old}})\}. \qquad (2)$$

The likelihood will only change for this single line. Assume it is the kth line that has been moved, then we can write the Hastings ratio p as

$$\frac{f(\beta_{\text{new}})}{f(\beta_{\text{old}})} = \prod_{\pi(i,j)=k} \exp\left\{ -\frac{1}{2} \left(\frac{(y_{ij} - \alpha_{k(i,j)} - \beta_{\text{new},k(i,j)}x_i)^2}{\sigma_{\text{new},k}^2} - \frac{(y_{ij} - \alpha_{k(i,j)} - \beta_{\text{old},k(i,j)}x_i)^2}{\sigma_{\text{old},k}^2} \right) \right\}.$$

The product is over the intersection points y_{ij} lying on the kth fracture where $f(.)$ is likelihood.

- *Intercept:* Similarly, choose a line at random, and change the intercept, α. Allow a move up/down that is no more than 5 or 10% of the length of the borehole (proposed from a uniform distribution). The ratio p will be similar to that given above, but the prior terms for

α_{new} and α_{old} cancel out

$$\prod_{\pi(i,j)=k} \exp\left\{-\frac{1}{2\sigma_k^2}\left((y_{ij} - \alpha_{\text{new},k(i,j)} - \beta_{k(i,j)}x_i)^2 - (y_{ij} - \alpha_{\text{old},k(i,j)} - \beta_{k(i,j)}x_i)^2\right)\right\}.$$

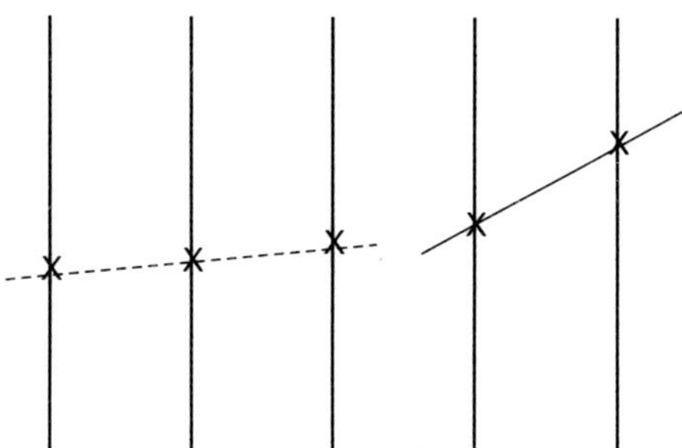

Fig. 4. Two adjacent lines with boreholes at $i-2$, $i-1$, i, $i+1$, $i+2$.

- *Endpoints:* Choose at random two lines which have their endpoints at adjacent boreholes. This requires a list of all adjacent pairs of lines, the direction of the swap to be chosen at random. Assuming it has at least two points, move the last point of the left hand line to be the first point of the right hand line (or vice versa) (Figs. 4 and 5; the lines originally cover boreholes at $\{x_{i-2}, x_{i-1}, x_i\}$ and $\{x_{i+1}, x_{i+2}\}$; the swap changes them to $\{x_{i-2}, x_{i-1}\}$ and $\{x_i, x_{i+1}, x_{i+2}\}$). The likelihood will only change at a single point. With two lines involved k_1 (left hand one), and k_2 (right hand one), the ratio becomes

$$\exp\left\{-\frac{1}{2}\left(\frac{(y_{ij} - \alpha_{k_2} - \beta_{k_2}x_i)^2}{\sigma_{k_2}^2} - \frac{(y_{ij} - \alpha_{k_1} - \beta_{k_1}x_i)^2}{\sigma_{k_1}^2}\right)\right\}.$$

- *Labels:* Choose a borehole at random, and from it choose two adjacent intersection points. Swap the fracture labels associated with each intersection point, so each is now identified with a different fracture line. The likelihood will only change at two points on borehole i, at j_1 and j_2 say. Assume these are originally associated with lines k_1 and k_2, respectively. In this case the ratio can be written

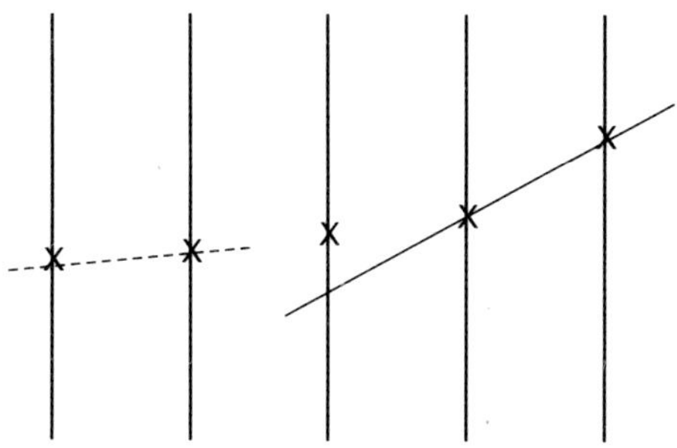

Fig. 5. Move endpoint of one line to the other line.

$$\exp\left\{-\frac{1}{2}\left(\frac{(y_{ij_1}-\alpha_{k_2}-\beta_{k_2}x_i)^2}{\sigma_{k_2}^2}-\frac{(y_{ij_1}-\alpha_{k_1}-\beta_{k_1}x_i)^2}{\sigma_{k_1}^2}-\frac{(y_{ij_2}-\alpha_{k_2}-\beta_{k_2}x_i)^2}{\sigma_{k_2}^2}+\frac{(y_{ij_2}-\alpha_{k_1}-\beta_{k_1}x_i)^2}{\sigma_{k_1}^2}\right)\right\}.$$

- *Split:* Choose a line at random, and split it into two. In this case the ratio can be written as follows (assume that it is the kth line that is being split into k_1 and k_2):

$$\frac{p(K+1)\text{prior}(\beta_k)}{p(K)\text{prior}(\beta_{k_1})\text{prior}(\beta_{k_2})}\times\prod_{\pi(i,j)=k_1}\exp\left\{-\frac{1}{2}\left(\frac{(y_{ij}-\alpha_k-\beta_kx_i)^2}{\sigma_k^2}-\frac{(y_{ij}-\alpha_{k_1}-\beta_{k_1}x_i)^2}{\sigma_{k_1}^2}\right)\right\}$$

$$\times\prod_{\pi(i,j)=k_2}\exp\left\{-\frac{1}{2}\left(\frac{(y_{ij}-\alpha_k-\beta_kx_i)^2}{\sigma_k^2}-\frac{(y_{ij}-\alpha_{k_2}-\beta_{k_2}x_i)^2}{\sigma_{k_2}^2}\right)\right\}$$

$$\times\frac{\phi(\alpha_k)\phi(\beta_k)}{\phi(\alpha_{k_1})\phi(\beta_{k_1})\phi(\alpha_{k_2})\phi(\beta_{k_2})}$$

where $\phi(.)$ is proposal density and Gaussian is used, i.e.,

$$\phi(\alpha_k)\equiv\text{N}\left(\frac{\alpha_{k_1}+\alpha_{k_2}}{2},\sigma_{p\alpha}^2\right);\phi(\alpha_{k_2})\equiv\phi(\alpha_{k_1})\equiv\text{N}(\alpha_k,\sigma_{p\alpha}^2)$$

$$\phi(\beta_k)\equiv\text{N}\left(\frac{\beta_{k_1}+\beta_{k_2}}{2},\sigma_{p\beta}^2\right);\phi(\beta_{k_1})\equiv(\beta_{k_1})\equiv\text{N}(\beta_k,\sigma_{p\beta}^2).$$

Obviously this is slightly more involved than previous proposals; this is due to the number of lines changing and proposals involved in achieving this.

- *Join:* Choose at random two lines where endpoints are at adjacent boreholes, and join them together. Again the number of lines is changing, which makes for a slightly more complicated ratio (assume this time that it is lines k_1 and k_2 that are to be joined, denote the new, longer line by k):

$$\frac{p(K-1)\text{prior}(\beta_{k_1})\text{prior}(\beta_{k_2})}{p(K)\text{prior}(\beta_k)}$$

$$\times\prod_{\pi(i,j)=k_1}\exp\left\{-\frac{1}{2}\left(\frac{(y_{ij}-\alpha_k-\beta_kx_i)^2}{\sigma_k^2}-\frac{(y_{ij}-\alpha_{k_1}-\beta_{k_1}x_i)^2}{\sigma_{k_1}^2}\right)\right\}$$

$$\times\prod_{\pi(i,j)=k_2}\exp\left\{-\frac{1}{2}\left(\frac{(y_{ij}-\alpha_k-\beta_kx_i)^2}{\sigma_k^2}-\frac{(y_{ij}-\alpha_{k_2}-\beta_{k_2}x_i)^2}{\sigma_{k_2}^2}\right)\right\}$$

$$\times\frac{\phi(\alpha_{k_1})\phi(\beta_{k_1})\phi(\alpha_{k_2})\phi(\beta_{k_2})}{\phi(\alpha_k)\phi(\beta_k)}.$$

- *Standard deviation σ:* This is also allowed to vary, using a Gaussian random walk. That is, set σ to $\sigma_{\mathrm{new}} = |\sigma_{\mathrm{old}} + N(0, 0.25)|$. The ratio becomes

$$p = \left(\frac{\sigma_{\mathrm{old}}}{\sigma_{\mathrm{new}}} \right)^n \exp \left\{ -\frac{1}{2} \left[\frac{1}{\sigma_{\mathrm{new}}^2} - \frac{1}{\sigma_{\mathrm{old}}^2} \right] \sum_{\pi(i,j)=k} \frac{(y_{ij} - \alpha_{k(ij)} - \beta_{k(i,j)} x_i)^2}{1 + \beta_{k(i,j)}^2} \right\}.$$

Note that the last two proposals either increase or decrease by one the number of fractures we are fitting, and so require the reversible jump MCMC algorithm; see Green (1995). For each of these, calculate the ratio, and accept/reject the change.

4. A Case Study

Fig. 6 shows a subset of a real 2D fracture dataset, which is a fracture mapping at Yucca Mountain collected by Barton and Larson (1985). The dataset was analysed by Lee et al. (1990) using a hierarchical model, which subdivided the dataset into two characteristic sub-sets. Set one (Fig. 6) includes fractures with nearly parallel orientations and with very few intersections. Set two consists of the remaining fractures. We test our MCMC implementation with the data described in Section 1, and displayed in Fig. 2.

Fig. 7(a) shows intersection of the boreholes with actual fractures. We start with each point assigned its own line with slope 0.75 and intercept initialised to pass through the intersections. Parameters used for the prior distribution for β are $\gamma = 1.3$ and $\kappa = 8$ (Eq. (1)), whilst $\kappa = 2$ was used for the proposal distribution of θ (Eq. (2)). A Gaussian random walk with a standard deviation of 0.5 was used to update σ from an initial value $\sigma_0 = 1.2$.

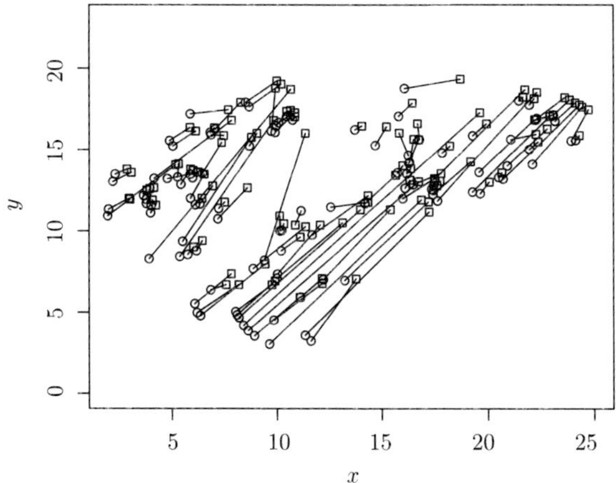

Fig. 6. Yucca mountain data.

We used $\lambda = 4$ for the Poisson prior distribution for the number of fractures. Gaussian proposal distributions with variance 0.1 were used for α and β. In this run acceptance ratios for α and β were 5376/14438 and 901/14326. Acceptance ratios for fracture endpoint change, join, split and intersection label change were 1918/14190, 76/14294, 46/14121 and 442/14306, respectively.

Fig. 7(b) is a snapshot of fitted fractures after a total of 1 million updates, and we took 500,000 iterations for the burn-in period. Only fractures with more than one intersection are shown (so there are 6 not shown). Thus the posterior solution is reasonable. Further, all fractures are recovered except two for which their intersections are not joined. Comparing Fig. 7(b) with Fig. 7(a), we note that taking the average for intercept and slope for fractures most present after the burn-in period gives a similar picture. This shows that the solution is quite stable.

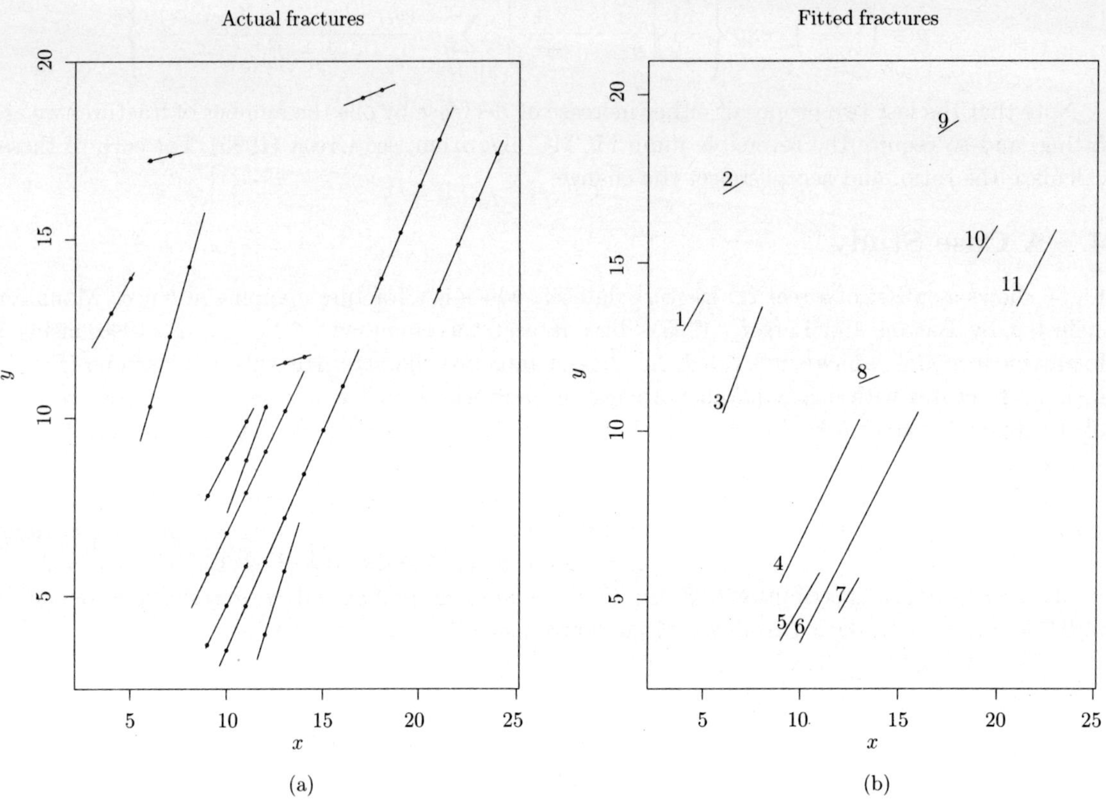

Fig. 7.   **(a) Actual fractures, borehole intersections and (b) fitted fractures.**

Fig. 8 shows posterior distributions of intercepts and slopes for all fractures. The distribution for the intercept is trimodal while that of slope is bimodal. Three modes in the intercept distribution reflect groups consisting of lines: (i) 1 and 3, (ii) 2, 8 and 9 and (iii) 4, 5, 6, 7, 10 and 11. The distribution for slope reflects grouping of lines: (i) 2, 8 and 9 and (ii) 1, 3, 4, 5, 6, 7, 10 and 11. Table 1 gives modal values for intercepts and slopes.

Table 1.   **Summary statistics for intercept and slopes for all lines**

Parameter	Modes		
α	11.00	$-$ 7.00	1.00
β	0.45	0.85	

Fig. 9 shows the trace and posterior distribution for standard deviation on a log scale. Mean and variance for σ after burn-in are given in Table 2. A Gaussian random walk Metropolis-Hastings algorithm was used to explore the state space for this parameter. Using Gibbs sampling from the conditional posterior distribution for updating σ gives similar results. In this case, the conditional posterior distribution for σ^2 is an inverse Gamma distribution. Table 2 gives mean and variance for

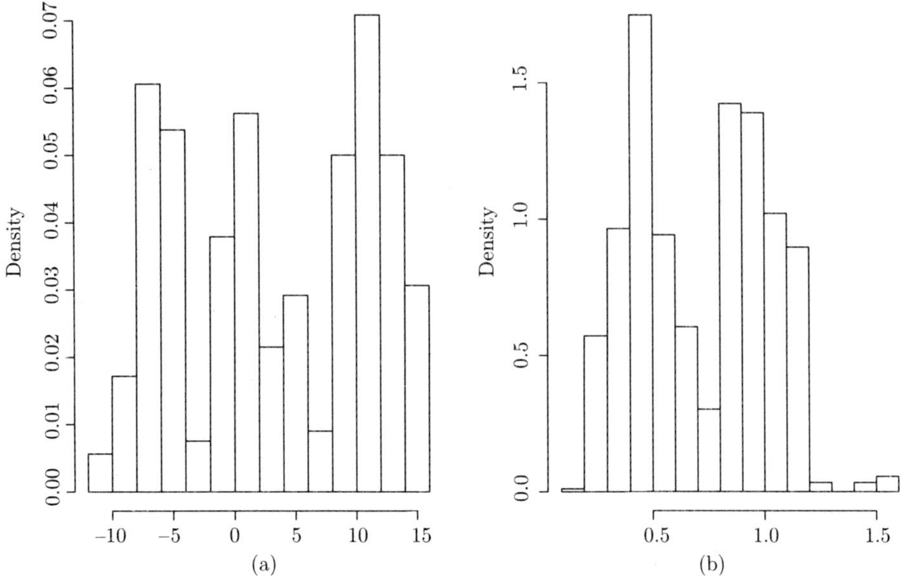

Fig. 8.   Posterior distributions for (a) intercepts and (b) slopes.

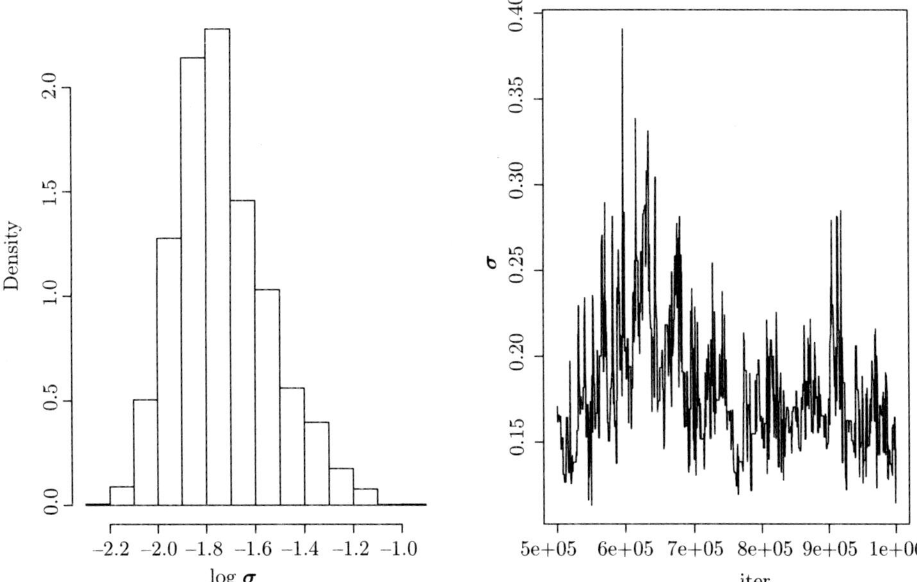

Fig. 9.   Posterior distribution and trace for standard deviation (on log scale). The trace is based on a thinned sample of 2000.

line parameters (after burn-in) for each of the fitted 11 lines. Trace plots for parameters for each line were inspected and showed good convergence. Parameter histograms were mostly unimodal and peaked, showing convergence and stability of the solution as well.

Fig. 10 shows trace plots for the number of fractures and the log posterior after burn-in. The number of fractures hovers around 16 and 19. A few of these fractures may consist of a single

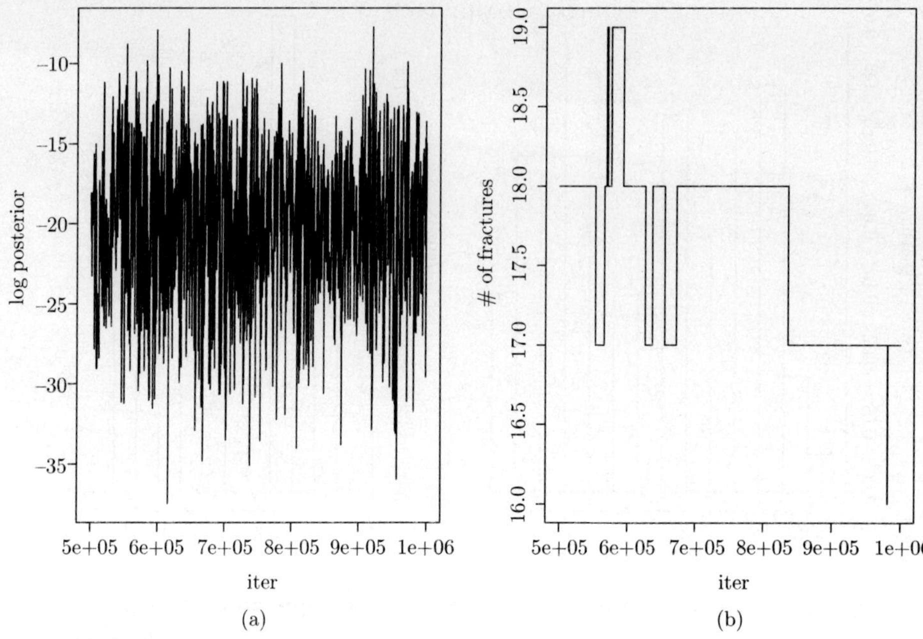

Fig. 10.   (a) The log likelihood, and (b) the trace for number of fracture (the log likelihood plot is based on a thinned sample of 2000).

intersection. The log posterior is also quite stable after burn-in. Mean and variance for the number of fractures K after burn-in are given in Table 2.

Table 2.   Summary statistics for variance, number of fractures (K), intercepts (α) and slopes (β) for each line

Parameter	Mean	Variance	Parameter	Mean	Variance
α	0.18	0.0013	α_6	-7.54	0.0909
K	17.63	0.3120	β_6	1.13	0.0006
α_1	9.80	0.8064	α_7	-7.10	1.0851
β_1	0.75	0.0426	β_7	0.95	0.0059
α_2	14.67	0.2363	α_8	7.46	0.6241
β_2	0.40	0.0050	β_8	0.28	0.0020
α_3	1.79	1.3852	α_9	10.87	1.1650
β_3	1.40	0.05655	β_9	0.47	0.0039
α_4	-4.86	0.1405	α_{10}	-1.13	0.3423
β_4	1.16	0.001	β_{10}	0.86	0.0009
α_5	-4.63	0.5722	α_{11}	-8.49	1.9354
β_5	0.95	0.0052	β_{11}	1.06	0.0034

5.   Discussion

We now discuss some extensions, alternate approaches and open problems.

5.1 Partial MCMC: An Alternative Approach

The 'endpoints' and 'labels' proposals effectively change the matching function $\pi(i,j) \in \{1, \ldots, K\}$. They do not alter directly the configurations of the fractures such as the number of fractures and their slopes and intercepts. To speed up the converging process, we also tried a partial MCMC approach. In this approach, $\pi(i,j)$ changes at each iteration step. The matching is done in such a way that maximises the likelihood, i.e.,

$$\pi(i,j) = \operatorname{argmax}(L(Y;\pi,K,\{\alpha_k,\beta_k\})),$$

which ensures the optimal matching function is always used throughout the simulation. 'Endpoints' and 'labels' proposals are not used in this approach.

The optimal matching function is derived based on the assignment algorithm widely used in operational research (Cohen, 1985, pp 76-83). The algorithm gives answers to problems such as assigning n number of variable A to m number of variable B in such a way that an objective function is maximised (minimised), subjected to the condition that each variable is used not more than once. A variant of the algorithm is used in the partial MCMC approach as all intersection points (variable A) must be used once and only once during the matching assignment but the proposed fractures (variable B) can be used repeatedly.

5.2 Extension to 3D

In our work so far we have obtained satisfactory results from both simulated and actual 2D data. Work is now proceeding to implement a 3D version, with line fractures replaced by planes. Specific problems this causes for our proposals include (i) extending the idea of adjacency for swapping "edgepoints" (formerly endpoints), and (ii) splitting a plane, instead of a line. In 2D we have lines for fractures, and make particular use of their beginning and endpoints on boreholes (it is also important to note that each fracture is 'connected'–all boreholes in between its endpoints must have one intersection point associated with the fracture–there can be no gaps). Extending this to 3D requires knowing the edge/outline of the fitted plane (and also requires that there are no gaps within the plane!) This will be achieved by noting either the first or second order neighbourhood graph structure. "Endpoints" could only be swapped if they were adjacent, in the sense of being (first/second order) neighbours, and the same is true for the joining of planes (we note here that care will be required to check that the new plane is also connected). This will be reasonably straightforward for a regular grid of vertical boreholes. The proposals for intercept/slope/labels will not cause any problems, although splitting a plane will be interesting to implement.

At Leeds we have recently acquired for the first time a real 3D data set. Originally this was a metre cubed block of granite, which was cut into slices, and all the fractures visible were carefully mapped. More information on this 'Leeds rock data set' can be found in Martin et al. (2005). This data will be used as the basis for our future 3D work.

5.3 Other Comments

We have assumed that fractures spread linearly as an idealisation of real fractures. Besides, it may be assumed that many fractures can be represented as concatenation of straight line segments. Indeed, in the case of Yucca mountain data, many fractures extend linearly (see also, Lee et al., 1990).

With the form of the data available, we can only measure the extent of a fracture from the leftmost to the rightmost boreholes it intersects. The true fracture length is obviously marginally longer. With a suitable choice of distribution, the extent of fractures can be better estimated.

Other factors involved are the "width/apertures" of the fractures, which have not been incorporated here. Also we might observe the direction of the fracture line. These can be easily incorporated in our model. The MCMC implementation should then be faster.

The spacing between boreholes is obviously very important. If the spacing could be narrowed, we would expect the fit of our fractures to be better, and the distributions for the parameters to be less variable. Conversely, wider spacing will be used to test the robustness of our methods.

Finally, we note that there is no need for regular spacing between boreholes in 2D. All of the above is immediately applicable. However, if boreholes are not regular (but still vertical) in 3D, then a Dirichlet tessellation would first be required to obtain some kind of neighbourhood structure.

Acknowledgements

The work reported in this paper was funded by EPSRC (Engineering and Physical Sciences Research Council) Research Grant number GR/R94602/01. We are grateful to Wally Gilks, Charles Taylor and Robert Ackroyd for their helpful comments.

References

Barton, C.C. and Larson, E. (1985). Fractal geometry of two-dimensional fractures networks at Yucca Mountain, Southwest Nevada. In *Proceedings of the International Symposium on Fundamentals of Rock Joints*, 78-84, Stephanson, O. (ed.), Centek Publishers.

Cohen, S.S. (1985). *Operational Research.* Edward Arnold, London.

Green, P.J. (1995). Reversible jump Markov chain Monte Carlo computation and Bayesian model determination. *Biometrika*, **82**, 711-732.

Green, P.J. and Mardia, K.V. (2006). Bayesian alignment using hierarchical models, with applications in protein bioinformatics. *Biometrika*, **93**, 235-254.

Hastings, W.K. (1970). Monte Carlo sampling methods using Markov chains and their applications. *Biometrika*, **57**, 97-109.

Jasra, A., Holmes, C.C. and Stephens, D.A. (2005). Markov chain Monte Carlo methods and the label switching problem in Bayesian mixture modelling. *Statistical Science*, **20**, 50-67.

Lee, J.S., Veneziano, D. and Einstein, H.H. (1990). Hierarchical fracture trace model. *Rock mechanics contributions and challenges, Proceedings of the 31th US Rock Mechanics Symposium*, 261-268, Hustrulid, W.A. & Johnson, G. A. (eds.), Balkema, Rotterdam.

Martin, J.A., Dowd, P.A., Mardia, K.V., Fowell, R.J. and Xu, C. (2005). A three dimensional data set of the fracture network of a granite. Submitted to *International Journal of Rock Mechanics & Mining Sciences*.

Metropolis, N., Rosenbluth, A.W., Rosenbluth, M.N., Teller, A.H. and Teller, E. (1953). Equations of state calculations by fast computing machines. *J. Chem. Phys.*, **21**, 1087-1091.

Bayesian Statistics and Its Applications
Edited by S.K. Upadhyay, U. Singh and D.K. Dey
Anamaya Publishers, New Delhi, India

On Bayesian Models Incorporating Covariates in Reliability Analysis of Repairable Systems

Loretta Masini, Antonio Pievatolo, Fabrizio Ruggeri and Emanuela Saccuman

CNR–IMATI, Via Bassini 15, I-20133 Milano, Italy

Abstract

Stemming from a consulting project about a gas distribution network, new Bayesian models are proposed to describe failures in repairable systems when covariates (e.g. diameter, depth and lay location for gas pipes) are recorded along with failure data. We consider covariates both in parameter models and prior distributions on them. We use the proposed models to illustrate some relevant features in the Bayesian analysis of repairable systems.

1. Introduction

In the last decade CNR–IMATI has been very active in analysing failures in complex systems, e.g., gas distribution networks and underground trains. We proposed various models, taking in account different features of data, and we used them in statistical analyses, as illustrated in the review paper by Ruggeri (2005) and the references therein. The analysed systems shared the property that they could be repaired by replacing a component and returned to regular operation; they are known as *repairable systems*, in contrast with the *nonrepairable systems*.

Sometimes, minimal repairs bring the system reliability back to the same level it had before the failure, whereas "perfect" repairs bring the reliability back to the state of the system at the start of the operation. Reliability in the former case is usually told to be *bad as old* whereas *good as new* is used for the latter case, although some authors prefer to call them *same as old* and *same as new*, respectively. Failures of the former repairable systems are often described by means of non-homogeneous Poisson processes (NHPP), whereas renewal processes usually describe the latter systems. The reliability could be constant over time, as it happens in the homogeneous Poisson process (HPP), the only NHPP which is a renewal process as well. In this paper we focus on HPPs and, more in general, on NHPPs, when failure data are collected along with covariates (e.g., diameter, depth and lay location for gas pipes). We illustrate some properties of the proposed models, along with a quick review of some problems faced in the statistical analysis of the reliability of repairable systems. A general review of NHPP's and their applications to the reliability of repairable systems can be found in Rigdon and Basu (2000).

The main contribution of the paper is the systematic introduction of covariates in NHPPs used in reliability problems. Although covariates are widely used in survival analysis (e.g., Cox model), we are unaware of any work about covariates in the analysis of the reliability of repairable systems, at least in the way we propose in the current paper. Two approaches are presented: in the former the parameters are expressed as function of the covariates, whereas the latter introduces the covariates in the prior distributions on the parameters. The former approach shares some aspects with the work by Lawless (1987) and is quite similar to the one on which Cox models are based (except for the absence of a baseline hazard here). The latter relies on hierarchical models, quite used in Bayesian statistics, and extends to NHPPs the approaches by Albert (1985) and George, Makov and Smith (1993) about HPPs.

The paper improves with respect to previous works by Cagno et al. (1998, 2000a, 2000b) who considered covariates in HPPs but only to define independent subsystems for all possible combinations of the values of the covariates. Here the hierarchical structure and the sharing of parameters allow to combine the information provided by all subsystems.

Section 2 provides an illustration of repairable systems along with the most common models for them and the related quantities of interest. The motivating example about gas escapes is illustrated in Section 3. Covariates in the parameters of a widely used NHPP, the power law process (PLP), are considered in Section 4, whereas they are introduced in the hyperparameters of the priors on the parameters of NHPPs in Section 5. The paper concludes with some remarks in Section 6.

2.   NHPPs for Repairable Systems

Distributions of first failure times, and survival functions, are often used to describe the reliability of nonrepairable systems. Failure times are modelled by renewal processes, given, e.g., by sequences of i.i.d. random variables. Point processes (see, e.g., Thompson, 1988) are the natural tools used to describe failures in the repairable systems where the main interest is about their behaviour over time, e.g., the expected number of failures or the probability of no failures in given intervals. As mentioned in Section 1, we focus only on the repairable systems described by NHPPs, arising when repairs are instantaneous as soon as a failure occurs and they are *minimal* ones since they act only on a small component of the system and bring the reliability back to the same state as before the failure.

2.1   Definitions

We denote NHPP by $N_t, t \geq 0$, or $N(t, s), 0 \leq t < s$, i.e., by the number of failures either in $(0, t]$ or in $(t, s]$, respectively. Its intensity function is given by

$$\lambda(t) = \lim_{\Delta t \to 0} \frac{P(N(t, t + \Delta t] \geq 1)}{\Delta t}$$

The mean value function (m.v.f.) of the NHPP is given by $M(t) = \mathcal{E} N_t$. Note that, assuming no simultaneous failures, $\lambda(t)$ and $M(t)$ are related by $\lambda(t) = M'(t)$ and, consequently, by $M(t) = \int_0^t \lambda(u) du$. More properties of Poisson process can be found in, e.g., Rigdon and Basu (2000).

In this paper we concentrate on the power law process (PLP), whose intensity function and m.v.f. are given, respectively, by $\lambda(t) = M \beta t^{\beta-1}$ and $M(t) = M t^{\beta}$, with $M, \beta, t > 0$.

2.2   Likelihood

Suppose the system is observed in interval $(0, y]$ and failures are recorded at $0 < t_1 < \ldots < t_n \leq y$. The upper limit y could be either a predetermined endpoint or the time of the n-th (predetermined) failure. In the former, we talk about a *time truncated* experiment, whereas the latter is a *failure truncated*. Under both experiments and the intensity function $\lambda(t) = \lambda(t; \theta)$, the likelihood function is

$$L(\theta; \underline{t}) = \prod_i^n \lambda(t_i; \theta) \exp\{-\int_0^y \lambda(t; \theta) dt\} \tag{1}$$

2.3 Quantities of Interest

First, we estimate parameters, specially when they have a clear *physical* meaning, like the parameter β in the PLP. Estimation of β is very important since its value determines if (and how) reliability decays or grows over time ($\beta > 1$ and $\beta < 1$, respectively). As shown in Fig. 1, the intensity function is decreasing for $\beta < 1$, constant for $\beta = 1$ and increasing for $\beta > 1$. The three behaviours correspond, respectively, to growth, steadiness and decay in reliability. Such richness of behaviours, the simple mathematical form and the Weibull distribution of the first failure time are, probably, the main reasons for the relevant role played by the PLP in reliability.

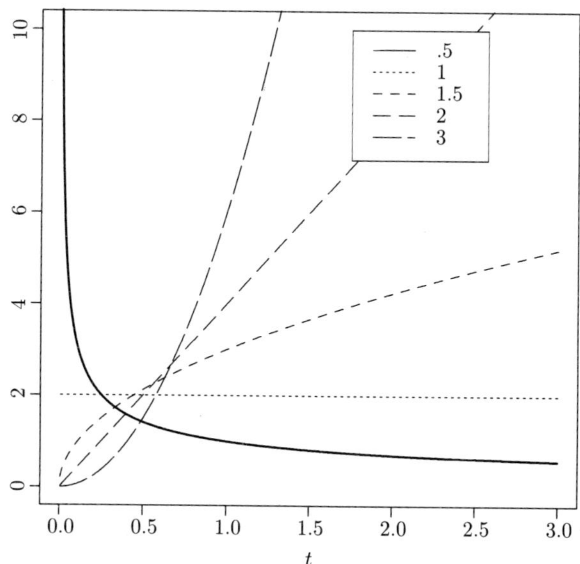

Fig. 1.   **Intensity function $\lambda(t)$ of the PLP for some β's and $M = 2$.**

Some typical quantities of interest are the *reliability $R(t,s) = P$(no failures in $(t,s]$)*, the *expected number of failures $\mathcal{E}N(t,s)$* and the *intensity function $\lambda(t)$ at t*. The first two quantities are relevant in forecasting system reliability; typically t is supposed to be either 0 (when interested in a similar system from its start) or y (when interested in the behaviour of the system under consideration right after the end of the observed period). For the PLP, their values are $R(t,s) = \exp\{-M(s^\beta - t^\beta)\}$ and $\mathcal{E}N(t,s) = M(s^\beta - t^\beta)$. The third quantity is useful when releasing a product (e.g., software) after a testing phase, beyond a time period t, and its future reliability will be determined by its intensity function at time t.

3.   Gas Escapes

Cagno et al. (1998, 2000a, 2000b) considered escapes in a low-pressure gas distribution network (20 mbar over the atmospheric pressure), developed in a large urban area during the last century. One of their aims was to suggest a replacement policy, identifying materials and environmental features more keen to produce gas escapes ("failures"). Classical statistical methods lead to identify the traditional cast iron (*old cast*) as the less reliable material with a failure rate even ten times greater than treated cast iron, spheroidal graphite cast iron, steel and polyethylene pipes. Therefore, they considered 150 escapes observed in 6 years in the 320 km long network (roughly, one fourth of the total network) of *old cast* pipes.

Budget constraints did not allow an immediate, complete replacement of cast-iron pipelines with a different material, e.g., polyethylene. Therefore, further analysis was needed to identify different characteristics (e.g., thickness, diameter, age) and environmental features (e.g., ground properties, traffic, external temperature and moisture) of different sections of the *old cast* network. Classical statistical methods lead to identify diameter, lay depth and location as the most significant features and two levels were chosen for all of them, as shown in Table 1, where the notations "high" and "low" are qualitative rather than quantitative, thus distinguishing more easily two levels of the same factor and simplifying the reading of conclusions. A more detailed illustration of the preliminary statistical analysis and the engineering aspects is available in Cagno et al (2000a).

Table 1.   Significant levels of factors related to failure

Factors	Low level	High level
Lay location	Street	Sidewalk
Diameter	≤ 125 mm	> 125 mm
Lay depth	< 0.9 m	≥ 0.9 m

The three factors with two levels each lead to consider 8 subnetworks and failures in each of them are modelled using a HPP since the small portion of the network to be repaired in a very short period (compared with the operating time) is compatible with the assumption of instantaneous, minimal repair behind the NHPP. Furthermore, the HPP is chosen since the *old cast* pipes were not subject to aging (corrosion) and the reliability of the system was considered constant over time. Aging was considered later, e.g., in Pievatolo and Ruggeri (2004), when gas escapes from steel pipes, subject to corrosion, were analysed.

Therefore, 8 HPPs were considered and Cagno et al. (1998, 2000a, 2000b) collected experts' opinions on the failure rates of each of them using an AHP (analytic hierarchy process) approach and performed a thorough Bayesian analysis, including sensitivity checks, considering mainly Gamma and lognormal priors on the failure rates. Posterior means of failure rates were considered to compare the 8 HPPs and the largest ones lead to identify the conditions under which gas escapes were the most likely (their findings will be illustrated in Section 5 in comparisons with the result of this paper).

The annual failure rates for each subnetwork is presented in Table 2.

Table 2.   Classes of pipelines, according to low/high (L/H) levels of factors

Factors	1	2	3	4	5	6	7	8
Lay location	L	L	L	L	H	H	H	H
Lay depth	L	L	H	H	L	L	H	H
Diameter	L	H	L	H	L	H	L	H
Annual failure rate	.177	.115	.131	.178	.072	.094	.066	.060

A major drawback of this approach is that it considers the 8 classes independently and such assumption could be too simplistic. As an alternative approach, covariates can be directly introduced. Sections 4 and 5 consider covariates in the parameters of the NHPP and in the hyperparameters of the prior distribution on the parameters, respectively.

4. Covariates in the Parameters

We start from the intensity of a PLP, given by $\lambda(t) = M\beta t^{\beta-1}$, and a case, similar to the one discussed in Section 3, where failures depend upon n covariates which have, for simplicity, two possible levels, 0 and 1. Now the total number of possible combinations is 2^n which we denote by the vectors $X_j = (x_{j,1}, \ldots, x_{j,n})$, $j = 0, \ldots, 2^n - 1$, with the base-10 number j corresponding to the binary number $(x_{j,n} \ldots x_{j,1})$.

Consider both parameters of the PLP as functions of the covariates:

$$\beta(X_j) = \beta_0 \prod_{i=1}^{n} \gamma_i^{x_{j,i}} \text{ and } M(X_j) = M_0 \prod_{i=1}^{n} \delta_i^{x_{j,i}}$$

The likelihood function (1), for a PLP observed up to time T (*time truncated experiment*), becomes

$$L(M, \beta; \underline{t}) = M^n \beta^n \left(\prod_{i=1}^{n} t_i \right)^{\beta-1} \exp\{-MT^\beta\}.$$

Now introduce the covariates in the PLP model and suppose that failure data are given by $\underline{d} = (\underline{t}, \mathbf{X})$, where $\underline{t} = (t_1, \ldots, t_n)$ are the failure times and $\mathbf{X}$ is a family of vectors with the corresponding values of the covariates. Simple computations lead to the following representation of the likelihood:

$$L(M(\mathbf{X}), \beta(\mathbf{X}); \underline{t}) = \prod_{j=0}^{2^n-1} \left(M(X_j)^{n_j} \beta(X_j)^{n_j} \left(\prod_{i=1}^{n_j} t_{j,i} \right)^{\beta(X_j)-1} \exp\{-M(X_j)T^{\beta(X_j)}\} \right) \quad (2)$$

where n_j denotes the number of failure times $\{t_{j,i}\}_{i=1}^{n_j}$ recorded under the covariates combination X_j, $j = 0, \ldots, 2^n - 1$.

We now rescale, without loss of generality, the time so that $T = 1$ and we get a simpler version of the likelihood function (2), in which the argument of the exponential function becomes $-M(X_j)$. It is worth mentioning that in this case the likelihood function (2) becomes the product of two functions:

$$f(\beta(\mathbf{X})) \propto \prod_{j=0}^{2^n-1} (\beta(X_j)^{n_j} \exp\{-\beta(X_j)T_j\})$$

$$f(M(\mathbf{X})) = \prod_{j=0}^{2^n-1} (M(X_j)^{n_j} \exp\{-M(X_j)\})$$

where $T_j = -\sum_{i=1}^{n_j} \log t_{j,i}$, $j = 0, \ldots, 2^n - 1$.

We might consider either M or β constant (specially the case $\beta = 1$, corresponding to HPPs), i.e., not dependent upon the covariates but we will concentrate only on the general case. It is worth mentioning that the case of constant β is very similar to a model described in Lawless (1987). More details can be found in Saccuman (1998).

When considering the likelihood function (2), we can choose a conjugate prior for the parameters, as illustrated by the case of a model with two covariates. In this case, (2) is a function of $\underline{\theta} =$

$(\beta_0, \gamma_1, \gamma_2, M_0, \delta_1, \delta_2)$ and is proportional to

$$(\beta_0 M_0)^{n_0} \exp\{-\beta_0 T_0 - M_0\} \cdot (\beta_0 \gamma_1 M_0 \delta_1)^{n_1} \exp\{-\beta_0 \gamma_1 T_1 - M_0 \delta_1\}$$

$$\times (\beta_0 \gamma_2 M_0 \delta_2)^{n_2} \exp\{-\beta_0 \gamma_2 T_2 - M_0 \delta_2\} \cdot (\beta_0 \gamma_1 \gamma_2 M_0 \delta_1 \delta_2)^{n_3} \exp\{-\beta_0 \gamma_1 \gamma_2 T_3 - M_0 \delta_1 \delta_2\}$$

Simple computations lead to the likelihood

$$L(\underline{\theta}|\underline{d}) \propto (\beta_0 M_0)^{\sum_{j=0}^{3} n_j} (\gamma_1 \delta_1)^{n_1 + n_3} (\gamma_2 \delta_2)^{n_2 + n_3}$$

$$\times \exp\{-\beta_0 (T_0 + \gamma_1 T_1 + \gamma_2 T_2 + \gamma_1 \gamma_2 T_3) - M_0 (1 + \delta_1 + \delta_2 + \delta_1 \delta_2)\}. \tag{3}$$

We now consider the joint conjugate distribution given by

$$\pi(\underline{\theta}) \propto \beta_0^A \gamma_1^B \gamma_2^C \exp\{-\beta_0 (D + \gamma_1 E + \gamma_2 F + \gamma_1 \gamma_2 G)\}$$

$$\times M_0^H \delta_1^I \delta_2^L \cdot \exp\{-M_0 (N + P\delta_1 + Q\delta_2 + R\delta_1 \delta_2)\} \tag{4}$$

which corresponds to the conditional Gamma distributions of $\beta_0|\gamma_1, \gamma_2,$ $\gamma_1|\beta_0, \gamma_2,$ $\gamma_2|\beta_0, \gamma_1,$ $M_0|\delta_1, \delta_2,$ $\delta_1|M_0, \delta_2,$ $\delta_2|M_0, \delta_1$.

Combining (3) and (4), a conjugate posterior is found in the class (4). MCMC methods based on Gibbs sampler have been implemented and the *usual* Bayesian quantities of interest are computed. Cases with more covariates can be analysed very easily in a similar way. An example with simulated data was considered in Saccuman (1998).

5.   Covariates in the Hyperparameters

The failure data in Section 3 lead to consider 8 classes (or HPPs) independently. Here we propose models in which the classes influence each other and we pursue such goal using three hierarchical models. More details can be found in Masini (1998).

5.1   Hierarchical Model for HPP

Suppose we consider k HPPs in the intervals $[0, t_i], i = 1, \ldots, k$, respectively. Instead of considering different parameters λ_i for each HPP, as in Section 3, we consider a hierarchical model where all the λ_i's come from the same distribution. Therefore, we consider the following hierarchical model:

$$X_i|\lambda_i \sim \mathcal{P}(\lambda_i t_i), i = 1, \ldots, k \tag{5}$$

$$\lambda_i|\alpha, \beta \sim \mathcal{G}(\alpha, \beta)$$

$$\pi(\alpha) \propto \Gamma(\alpha + 1)^k / \Gamma(k\alpha + a), \ k \text{ integer } \geq 2, a > 0$$

$$\beta \sim \mathcal{G}(a, b)$$

where $\mathcal{P}$ and $\mathcal{G}$ are the Poisson and Gamma distributions, respectively. It can be shown that the prior distribution on α is a proper one and its shape depends on the parameter a, which appears in the prior on β as well. If $a > 1$, then the density is decreasing for all positive α, whereas, for $a \leq 1$, the density is increasing up to its mode $(1 - a)/(k - 1)$, and then it decreases. The choice of the prior on α is motivated not only by its flexibility but also by mathematical considerations. In fact, both Gamma functions in the prior cancel out when computing the posterior distribution of $\underline{\lambda} = (\lambda_1, \ldots, \lambda_k)$, given by

$$\pi(\underline{\lambda}|\underline{d}) \propto \frac{\prod_{i=1}^{k} \lambda_i^{x_i - 1} \exp\{-\lambda_i t_i\}}{(\sum_{i=1}^{k} \lambda_i + b)^a (-\log H(\underline{\lambda}))^{k+1}},$$

where $H(\underline{\lambda}) = \prod_{i=1}^{k} \lambda_i (\sum_{i=1}^{k} \lambda_i + b)^{-k}$. It is to be noted that the computation of posterior quantities is possible using a Metropolis-Hastings algorithm.

The same model was considered by Albert (1985) and George, Makov and Smith (1993), among others. Albert reparametrised the model, considering $\mu = \alpha/\beta$ and $\gamma_i = \beta/(t_i + \beta), i = 1, \ldots, k$. Then he took the noninformative priors $\pi(\mu) = 1/\mu$ and $\pi(\gamma) = \gamma^{-1}(1 - \gamma)^{-1}$ and, using a cycle of approximations, he was able to estimate mean and variance of each λ_i. George et al. (1993) considered exponential priors $\mathcal{E}(1)$ or $\mathcal{E}(0.01)$ for α and Gamma priors $\mathcal{G}(0.1, 1)$ and $\mathcal{G}(0.1, 0.01)$ for β.

Example 1. George, Makov and Smith (1993) considered the number x_i of failures of ten power plants, considered over their operating times t_i, as described in Table 3.

Table 3. Failure data for 10 power plants

Plant i	1	2	3	4	5	6	7	8	9	10
t_i	9.43	1.57	6.29	12.6	.524	3.14	.105	.105	.210	1.05
x_i	5	1	5	14	3	19	1	1	4	22

We consider the hierarchical model (5) with two different choices of parameters: in case (a) we take $a = 0.1$, $b = 1$ and $k = 10$, whereas we set $a = 2$, $b = 1$ and $k = 10$ in case (b). The choice of a leads to a unimodal density in case (a) and a decreasing density in case (b). Estimated failure rates, computed in both cases (a) and (b), are compared in Table 4 with the MLE and the estimates obtained by George, Makov and Smith (1993).

Table 4. Comparison of estimated failure rates λ_i

Plant i	1	2	3	4	5	6	7	8	9	10
MLE	0.53	0.64	0.80	1.11	5.73	6.04	9.54	9.54	19.08	20.99
George et al.	0.60	1.02	0.89	1.16	6.02	6.09	8.91	8.94	15.88	19.94
Case (a)	0.57	0.84	0.85	1.13	5.94	6.08	8.81	9.29	16.35	20.22
Case (b)	0.56	0.83	0.84	1.13	5.37	6.00	6.95	7.16	15.39	19.59

We find our estimates are consistent with the previous ones, falling, in general, between them. The estimates are closer to George's ones than to MLEs, specially in case (a) where we used the same prior on β as George et al. (1993).

5.2 Hierarchical Model for HPP with Covariates

We now extend the hierarchical model (5) to the gas failure data described in Section 3, introducing covariates describing the two levels of each one of the three influential factors (diameter, depth and lay location). In this case we have to consider also the length k_j of each subnetwork determined by the 8 different combinations of levels of the covariates. Therefore, we obtain HPPs with parameters $\lambda_j t_j k_j$, $j = 0, \ldots, 7$ (in general $j = 0, \ldots, 2^n - 1$, for n covariates with two levels as in Section 4). Furthermore, we introduce the covariates in the hyperparameters of the priors on λ_j, considering Gamma distributions $\mathcal{G}(\alpha \exp\{\underline{X}_j^T \underline{\beta}\}, \alpha)$, with the vector of covariates X_j defined as in Section 4.

Therefore, we consider the following hierarchical model:

$$Y_j|\lambda_j \sim \mathcal{P}(\lambda_j t_j k_j), \ j = 0,\dots,2^n - 1$$

$$\lambda_j|\beta \sim \mathcal{G}(\alpha\exp\{\underline{X}_j^T\underline{\beta}\},\alpha)$$

$$\beta \sim \pi(\beta)$$

With this choice of priors on λ_j, we have that their prior mean is $\exp\{\underline{X}_j^T\underline{\beta}\}$, i.e., dependent upon the covariates. We have considered different choices of priors for β; at the end, we have decided to follow an empirical Bayes approach. Given the sample $\underline{Y} = (Y_0,\dots,Y_{2^n-1})$, we look for the value of β maximising $f(\underline{Y}|\underline{\beta}) = \int f(\underline{Y}|\underline{\lambda})\pi(\underline{\lambda}|\underline{\beta})d\underline{\lambda}$, i.e.,

$$f(\underline{Y}|\underline{\beta}) = \prod_{j=0}^{2^n-1} \frac{\alpha^{\alpha\exp\{\underline{X}_j^T\underline{\beta}\}}}{(t_jk_j+\alpha)^{\alpha\exp\{\underline{X}_j^T\underline{\beta}\}+Y_j}} \cdot \prod_{l=0}^{Y_j-1} \frac{\alpha\exp\{\underline{X}_j^T\underline{\beta}\}+l}{l+1} \cdot (t_jk_j)^{Y_j}. \tag{6}$$

We compute the posterior distribution $\pi(\underline{\lambda}|\underline{Y},\underline{\beta})$, which turns out to be the product of independent Gamma distributions for each λ_j, and plug in the value $\widehat{\underline{\beta}}(\underline{Y})$ maximising (6), so that it follows

$$\lambda_j|\underline{Y},\widehat{\underline{\beta}}(\underline{Y}) \sim \mathcal{G}(\alpha\exp\{\underline{X}_j^T\widehat{\underline{\beta}}\} + Y_j, t_jk_j + \alpha).$$

We have set $\alpha = 10$ and have obtained the estimates λ_js and we compare them in Table 5 with MLEs and the findings of Cagno et al. (1998, 2000a, 2000b) who used both a Gamma and a lognormal prior for the failure rates λ_j. The classes are determined by the different levels of the covariates, i.e., we use **W** (under walkway) and **T** (under traffic) for location, **S** (small, <125 mm) and **L** (large, ≤ 125 mm) for diameter, **N** (not deep, <0.9 m) and **D** (deep, ≤ 0.9 m) for depth.

Table 5. Estimates of failure rates for gas data

Class	MLE	Bayes ($\mathcal{LN}$)	Bayes ($\mathcal{G}$)	Hierarchical
TSN	*.177*	**.217**	**.231**	**.170**
TSD	*.115*	*.102*	*.104*	*.160*
TLN	*.131*	*.158*	*.143*	*.136*
TLD	**.178**	*.092*	*.094*	*.142*
WSN	.072	.074	.075	.074
WSD	.094	.082	.081	.085
WLN	.066	.069	.066	.066
WLD	.060	.049	.051	.064

The value in boldface denotes the highest value, whereas those in italics denote the 2nd to 4th highest values. First of all, it is worth observing that location is the most relevant factor, since for all the models the pipes laid under the traffic are the most likely to fail. MLE method gives a different highest value (TLD instead of TSN) but there is an easy explanation: there were 3 failures along 2.8 km (out of a 320 km long network) but pipes in the TLD class are quite unlikely to fail according to the experts. Furthermore, the hierarchical approach *borrows strength* from the other classes with the same levels of covariates.

We performed some sensitivity analysis with respect to the hyperparameter α, considering values between 0 and 50. The changes in the estimates of the failure rates were not very significant, denoting a strong robustness with respect to changes in α.

5.3 Hierarchical Model for NHPP with Covariates

Now use the indexes $1, \ldots, m$ instead of $0, \ldots, 2^n - 1$ to simplify the notation. Here we consider covariates in the parameters of a PLP. The case of a PLP without covariates has been thoroughly analysed from a Bayesian viewpoint in Bar-Lev, Lavi and Reiser (1992). Here we suppose that different levels of covariates lead to m PLPs with parameters (M_i, β_i), with n_i failures $\underline{t}_i = (t_{i,1}, \ldots, t_{i,n_i})$ observed in the interval $(0, y_i]$, $i = 1, \ldots, m$. For each PLP the likelihood is given by

$$L(M_i, \beta_i) = M_i^{n_i} \beta^{n_i} \left(\prod_{j=1}^{n_i} t_{i,j} \right)^{\beta_i - 1} \exp\{-M_i y_i^{\beta_i}\}.$$

Here we consider a *time truncated experiment* but we might consider y_i as the time of the n_i^{th} failure, so that we obtain a *failure truncated experiment*.

We choose the following priors, $i = 1, \ldots, m$:

$$\beta_i | \underline{\delta} \sim \mathcal{G}(\rho \exp\{\underline{X}_i^T \underline{\delta}\}, \rho)$$

$$M_i | \beta_i, \underline{\sigma} \sim \mathcal{G}(\exp\{\underline{X}_i^T \underline{\sigma}\}, y_i^{\beta_i})$$

The choice of the prior is partly motivated by the shape of the likelihood and other considerations, thoroughly discussed in Masini (1998).

As in the previous section, we consider an empirical Bayes approach. Here we need to find $\underline{\delta}$ and $\underline{\sigma}$ maximising

$$f(\underline{t}|\underline{\delta}, \underline{\gamma}) = \int \prod_{i=1}^{m} L(M_i, \beta_i|\underline{t}_i)\pi(\underline{\beta}|\rho, \underline{\delta})\pi(\underline{M}|\beta, \underline{\sigma}) d\underline{M} d\underline{\beta}$$

i.e.,

$$f(\underline{t}|\underline{\delta}, \underline{\gamma}) = \prod_{i=1}^{m} \left(\prod_{j=1}^{n_i} t_{i,j} \right)^{-1} \frac{\Gamma(n_i + \rho \exp\{\underline{X}_i^T \underline{\delta}\})}{\Gamma(\rho \exp\{\underline{X}_i^T \underline{\delta}\})} \frac{\Gamma(n_i + \exp\{\underline{X}_i^T \underline{\sigma}\})}{\Gamma(\exp\{\underline{X}_i^T \underline{\gamma}\})} \tag{7}$$

$$\times \frac{2^{-(n_i + \exp\{\underline{X}_i^T \underline{\sigma}\})} \rho^{\rho \exp\{\underline{X}_i^T \underline{\delta}\}}}{\left(\rho - \sum_{j=1}^{n_i} \log(t_{i,j}/y_i) \right)^{n_i + \rho \exp\{\underline{X}_i^T \underline{\delta}\}}}.$$

We compute the posterior distribution $\pi(\underline{M}, \underline{\beta}|\underline{t}, \underline{\delta}, \underline{\sigma})$, which turns out to be the product of Gamma distributions for each λ_j, and plug in the values $\widehat{\underline{\delta}}(\underline{t})$ and $\widehat{\underline{\sigma}}(\underline{t})$ maximising (7), so that we get the following distributions, $i = 1, \ldots, m$:

$$\beta_i | \underline{t}, \widehat{\underline{\delta}}(\underline{t}) \sim \mathcal{G}(n_i + \rho \exp\{\underline{X}_i^T \widehat{\underline{\delta}}(\underline{t})\}, \rho - \sum_{j=1}^{n_i} \log(t_{i,j}/y_i)),$$

$$M_i | \beta_i, \underline{t}, \widehat{\underline{\sigma}}(\underline{t}) \sim \mathcal{G}(n_i + \exp\{\underline{X}_i^T \widehat{\underline{\sigma}}(\underline{t})\}, 2y_i^{\beta_i}).$$

Analysis with simulated data was performed in Masini (1998).

6. Discussion

Motivated by a consulting project on the reliability of gas distribution networks, we have introduced new models which incorporate covariates in the parameters of NHPPs. There are possible extensions of our works.

Section 4 considered a time truncated experiment, which allowed us to change time scale so that failures occurred in the unit interval; it should be worth considering the case of a general time interval, which leads to a factor $\exp\{-MT^\beta\}$ in the likelihood instead of $\exp\{-M\}$.

Section 5.2 considered a hierarchical model for HPPs, which incorporated covariates in the prior distributions of the failure rates of the HPPs; such distributions were dependent upon a hyperparameter β and no prior was actually specified on it since an empirical Bayes approach was taken. In the future we plan to specify a proper prior on β and perform a new analysis. We performed a sensitivity analysis with respect to the hyperparameter α observing that the estimates were not very influenced by its value. We plan to introduce a prior distribution on α as well. Similar considerations apply to the model on NHPPs, discussed in Section 5.3.

We applied the model in Section 5.2 to the gas failure data in Section 3, to show how our proposed models could be used in the analysis of the reliability of repairable systems. Similarly, it is possible to analyse the same data with the model proposed in Section 4. Other data are needed to apply the model in Section 5.3. We have used NHPPs to describe failure data in steel pipelines in the gas distribution network and in underground train; see Pievatolo and Ruggeri (2004) and Pievatolo, Ruggeri and Argiento (2003), respectively. Unfortunately, the data we collected have very poor information on possible covariates and cannot be used in this context. Finally, in this paper we focussed on parameter estimation of NHPP models. It is straightforward to estimate the reliability measures, described in Section 2.3 as well.

References

Albert, J.H. (1985). Simultaneous estimation of Poisson means under exchangeable and independence models. *Journal of Statistical Computation and Simulation*, **23**, 1-14.

Bar-Lev, S., Lavi, I. and Reiser, B. (1992). Bayesian inference for the Power Law process. *Annals of the Institute of Statistical Mathematics*, **44**, 623-639.

Cagno, E., Caron, F., Mancini, M. and Ruggeri F. (1998). On the use of a robust methodology for the assessment of the probability of failure in an urban gas pipe network. In *Safety and Reliability*, **2**, (S. Lydersen, G.K. Hansen and H.A. Sandtorv eds.), 913-919, Rotterdam: Balkema.

Cagno, E., Caron, F., Mancini, M. and Ruggeri F. (2000a). Using AHP in determining prior distributions on gas pipeline failures in a robust Bayesian approach. *Reliability Engineering and System Safety*, **67**, 275-284.

Cagno, E., Caron, F., Mancini, M., and Ruggeri F. (2000b). Sensitivity of replacement priorities for gas pipeline maintenance. In *Robust Bayesian Analysis* (D. Rios Insua and F. Ruggeri eds.), New York: Springer.

George, E.I., Makov, U.E. and Smith, A.F.M. (1993). Conjugate likelihood distributions. *Scandinavian Journal of Statistics*, **20**, 147-156.

Lawless, J.F. (1987). Regression method for Poisson process data. *Journal of the American Statistical Society*, **82**, 808-815.

Masini, L. (1998). Un approccio bayesiano gerarchico nell'analisi dell'affidabilità dei sistemi riparabili. *B.Sc. Dissertation*, Dipartimento di Matematica, Università degli Studi di Milano.

Pievatolo, A. and Ruggeri, F. (2004). Bayesian reliability analysis of complex repairable systems. *Applied Stochastic Models in Business and Industry*, **20**, 253-264.

Pievatolo, A., Ruggeri, F. and Argiento, R. (2003). Bayesian analysis and prediction of failures in underground trains. *Quality Reliability Engineering International*, **19**, 327-336.

Rigdon, S.E. and Basu, A.P. (2000). *Statistical methods for the reliability of repairable systems.* Wiley: New York.

Ruggeri, F. (2005). On the reliability of repairable systems: methods and applications. *ECMI 2004 Proceedings*, Berlin: Springer.

Saccuman, E. (1998). Modelli che incorporano covariate nell'analisi bayesiana dell'affidabilità di sistemi riparabili. *B.Sc. Dissertation*, Dipartimento di Matematica, Università degli Studi di Milano.

Thompson, W.A. (1988). *Point process models with applications to safety and reliability.* London: Chapman and Hall.

Bayesian Statistics and Its Applications
Edited by S.K. Upadhyay, U. Singh and D.K. Dey
Anamaya Publishers, New Delhi, India

Semi-Parametric Bayesian Models for Population Pharmacokinetics and Pharmacodynamics

Peter Müller and Gary Rosner

Department of Biostatistics, University of Texas M.D. Anderson Cancer Center

1515 Holcombe Baulevard, Box 447, Houston, TX 77030-4009

Abstract

This article describes non- and semi-parametric Bayesian approaches to inference for population pharmacokinetics (PK) and pharmacodynamics (PD). The typical use of non-parametric Bayesian models occurs in the specification of random effects distributions. A traditional population PK/PD model includes a parametric model as sampling model for the observed data, indexed by a vector of patient-specific parameters, i.e., random effects. In many models, these parameters allow a physiological interpretation, for example, traditional PK parameters such as volume parameters and elimination rates. The model is completed with a prior probability model for the distribution of random effects across patients, the random effects distribution. While the random effects distribution is not usually of interest in itself, modeling choices have critical implications on the inference of interest. For example, appropriate description of within- and between-patient variability is critical to report honest uncertainties in posterior predictive inference.

1. Introduction

This article reviews recent research in semi-parametric Bayesian inference for population pharmacokinetic (PK) and pharmacodynamic (PD) data. Population PK and PD studies seek to identify within a large sample of patients those characteristics that significantly alter a drug's disposition and a patient's response to the drug. Population PK studies often presuppose some sort of model relating drug concentration to time. PD studies of treated patients examine relationships of clinical outcomes, such as toxicity or response, to characteristics of the drug's pharmacokinetic profile in each patient. Ratain et al. (1990) review clinical pharmacology of anticancer drugs. Davidian and Giltinan (1995) review statistical methods appropriate for carrying out population PK and PD analyses.

From the perspective of statistical modeling, the analysis of population PK/PD data is longitudinal data analysis. Let y_{ij} denotes the j-th measurement on the i-th patient, θ_i denotes a random effects vector for patient i, and x_i denotes patient-specific covariates. The common structure of population PK/PD models is

$$p(y_{ij}|\theta_i), \quad p(\theta_i|x_i, \phi), \quad p(\phi) \tag{1}$$

where $p(y_{ij}|\theta_i)$ is typically a parametric non-linear regression for response over time, such as a compartmental PK model for drug concentrations. The second level of the model specifies the prior for the random effects vectors θ_i, possibly including a regression on patient-specific covariates x_i, and ϕ denoting the hyperparameters. Bayesian models similar to (1) have been considered in Zeger and Karim (1991) for generalized linear mixed models and Wakefield, Smith, Racine-Poon and Gelfand (1994) for the general population model assuming a multivariate normal population distribution. Wakefield (1996) discusses the use of such models specifically in population PK. Wakefield et al. (1999) review current approaches.

Heterogeneity in the patient population, outliers and over-dispersion make a strict parametric model for the population distribution $p(\theta_i|x_i, \phi)$ unrealistic, leading us to non-parametric extensions. In maximum likelihood-based inference, popular approaches to non-parametric extensions are NPML (Mallet 1986), with no restrictions on the distribution of the random-effects in the model but yielding a discrete estimate of this distribution; and SNP (Davidian and Gallant, 1992, 1993), a method that assumes the model's random effects have a "smooth" density and produces estimates from such a class. Bayesian approaches to non-parametric extensions are described in Walker and Wakefield (1994), with Dirichlet process priors, and in Rosner and Müller (1997), Müller and Rosner (1997), Kleinman and Ibrahim (1998a,b), all with Dirichlet process mixture priors.

One of the inference challenges that arise in the analysis of population PK/PD data is joint inference across multiple studies. During the stages of drug development, patients treated in early phase studies are monitored more closely for safety reasons. Aside from evaluating safety, the investigators also often collect information on the pharmacokinetics of the agents under study. In later phase studies, especially large randomized phase III studies, there is usually less close monitoring of patients, often because of the difficult logistics or cost. Thus, early phase studies typically collect more data per patient but treat relatively few patients, compared to large randomized phase III studies. Methods for combining the fuller data collected on patients enrolled in earlier phase studies with sparse data collected as part of a phase III study help us learn more about PK and PD variability in the population. This requires the extension of the basic model (1) to encompassing hierarchical models to allow joint analysis of several related studies. Such meta-analysis is a common theme in statistical inference (Morris and Normand, 1992; Stangl, 1995), including hierarchical models like (1). There is, however, little work on meta-analysis over related non-parametric models of the type described in Section 2. Mallick and Walker (1997) discuss an approach based on partitioning the covariate space. The method is limited to few levels of covariates, and does not include borrowing strength across the submodels.

One possible approach is to link submodels of the form (1) with an additional level of the hierarchy. Let ϕ_k denotes the hyperparameter in the k-th study or submodel. The submodels are linked by assuming a common prior, $\phi_k \sim p(\phi_k|\eta)$, an approach taken in Müller and Rosner (1998) and Walker and Wakefield (1994). Fig. 1 is a graphical representation of this approach (panel (b)) and the opposite extreme of assuming exchangeable subjects across studies (panel (a)). For many applications, the first case allows too little borrowing of strength across studies, and the latter enforces too much borrowing by assuming essentially one population.

In the context of fully parametric models, it is straightforward to consider approaches between the two extremes of independent submodels (no borrowing strength) and one common population (too much borrowing strength). For example, in an ANOVA-like fashion, ϕ_k could be decomposed into common effects and study-specific effects. In our non-parametric setting, ϕ_k are random distributions, i.e., infinite dimensional random quantities, complicating the use of similar structure.

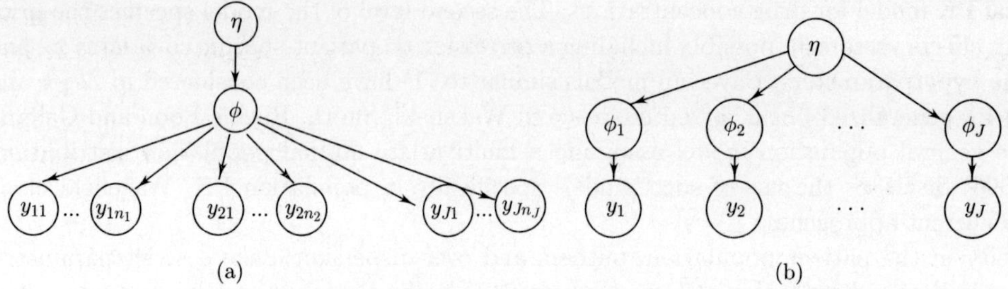

Fig. 1.   (a) **Combining data from related studies assuming exchangeable subjects across studies and (b) independent submodels. The desired level of borrowing strength across the submodels is in-between these two extremes.**

Another interesting class of inference problems are optimal design questions in models for population PK/PD studies. Investigators wishing to carry out large population studies need to reduce the potential logistic problems presented by requiring multiple samples from patients drawn at precise times after initiating chemotherapy. One of several traditional approaches to this problem is a decision theoretic solution.

Formally, decision making under uncertainty is choosing an action d to maximize expected utility

$$U(d) = \int u(d,\theta,y)\, p_d(\theta,y)\, d\theta\, dy \tag{2}$$

where $u(d,\theta,y)$ is a utility function modeling preferences over consequences and $p_d(\theta,y)$ is a probability distribution of parameter θ and observation y, possibly influenced by the chosen action d.

Chaloner and Verdinelli (1995) review Bayesian approaches to decision problems traditionally known as optimal design. Berry and Stangl (1996) discuss general issues related to the use of Bayesian optimal design methods in medical decision problems.

In the context of population PK/PD models, the relevant probability model $p_d(\theta,y)$ is a hierarchical model like (1), and the decision d might be the choice of sampling times for the next observation, the next batch of patients, or the next study. The choice of utility function formalizes the objective of the study. For example, $u(d,\theta,y)$ might be related to some summary of posterior predictive inference for a future patient, like predicted area under the time-concentration curve (AUC). Other important questions that take the form of a decision problem include the choice of the number of patients, number of biopsies, number of sections, etc.

Section 2 reviews semi-parametric models to formalize the general structure in (1). Section 3 describes hierarchical modeling across related studies. Section 4 discusses optimal design issues. and Section 5 concludes with a final discussion.

2. Semi-parametric Population PK/PD Models

In Müller and Rosner (1997) and Rosner and Müller (1997), we characterized the prior distribution $p(\theta_i|\phi)$ for the model parameters in model (1) as a mixture of multivariate normals. Using a mixture of simpler distributions, such as multivariate normals, provides great flexibility in terms of characterizing almost any probability distribution regardless of shape. In addition, we included patient covariates (x_i) with the model parameters (θ_i), by extending the mixture of multivariate normals to a joint distribution $p(\theta_i, x_i | \phi)$ on the pairs (θ_i, x_i). Being a mixture of normals, the implied regression $p(\theta_i | x_i, \phi)$ becomes a locally-weighted mixture of linear regressions, as illustrated in Fig. 2. Mallet et al. (1988) and Mentré and Mallet (1994) describe a similar approach for covariate analysis.

Thus, at the second level of the hierarchy in the general population model (1), we assume

$$(\theta_i, x_i) \sim \underbrace{\sum_h w_h N(\mu_h, \Sigma)}_{= \int N(\mu, \Sigma) dG(\mu)}. \quad G \sim p(G). \tag{3}$$

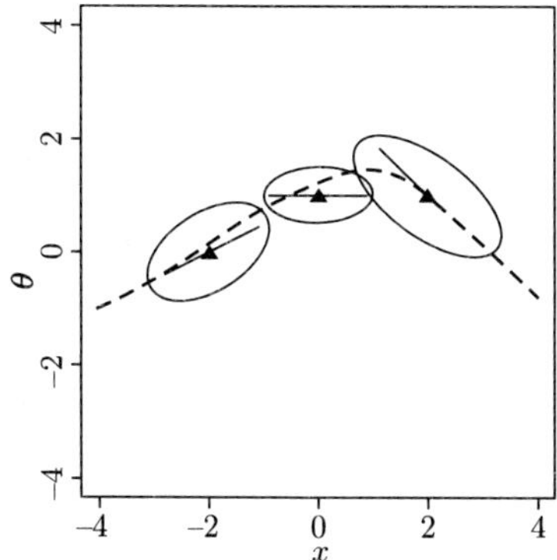

Fig. 2. Non-parametric regression of θ_i on covariate x_i arising from including the covariate in the mixture of normals model. Regression $p(\theta_i | x_i, \phi)$ takes the form of a locally weighted mixture of linear functions.

The mixture is parameterized in terms of the implied mixing measure G with a multivariate normal kernel in the mixture. Here (G, Σ) play the role of ϕ in the general model (1). Let $\delta_x(\cdot)$ denotes a point mass at x. The mixing measure G is the discrete measure $G(\mu) = \sum w_h \delta_{\mu_h}(\mu)$. We complete the model with a Dirichlet process (DP) prior for the unknown mixing measure G

$$G \sim \mathrm{DP}(G_0, \alpha) \tag{4}$$

Ferguson (1973) and Antoniak (1974) discuss Dirichlet processes and mixtures of DPs. A DP defines a probability model for random distributions and requires two parameters: the base measure G_0 and the total mass parameter α. The base measure defines the prior mean, i.e., $E[G(A)] = G_0(A)$ for any (measurable) set A. The total mass parameter is a concentration parameter. The larger the α, the closer G will be to G_0. A critical property of the DP model is the almost-sure discreteness of the random measure G (we can write G as a mixture of point masses, $G = \sum w_h \, \delta_{\mu_h}$). We implicitly used this property when writing the random mixture of normals in (3) as a summation.

We fit this model to data collected as part of a phase I study conducted by the Cancer and Leukemia Group B (CALGB). The goal of study CALGB 8881 (Lichtman et al, 1993) was to find the highest dose of cyclophosphamide that could be given intravenously over one hour every two weeks in the outpatient setting. Patients also received GM-CSF subcutaneously on days 3 through 10 to support hematologic recovery. The study enrolled 52 patients at different doses of cyclophosphamide and GM-CSF. During the first cycle of chemotherapy, there are 684 white blood cell counts (WBC) measurements. The WBC profiles follow a sort of bathtub shape as shown by the data in Fig. 3.

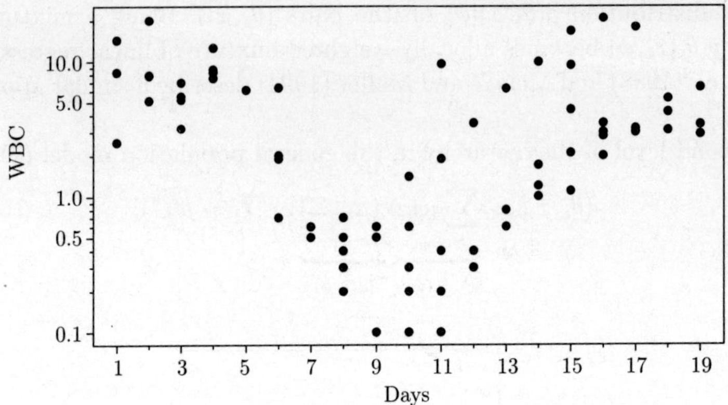

Fig. 3.   **Semilog plot of WBC measured during the first cycle of chemotherapy for patients getting 3 g/m^2 cyclophosphamide and 5 μg/kg of GM-CSF.**

The key features of these profiles are that they exhibit a decline a few days after the patient received the chemotherapy, followed by an S-shaped recovery. We modeled these features on the logarithmic scale by combining a horizontal baseline value (day 0 through τ_{1i}), a logistic function for the recovery period (after day τ_{2i}), and a straight line connecting the two random changepoints τ_{1i} and τ_{2i}. Fig. 4 shows the piecewise linear and logistic regression function $f(\cdot)$. We use this regression to define the top level likelihood $p(y|\theta_i)$

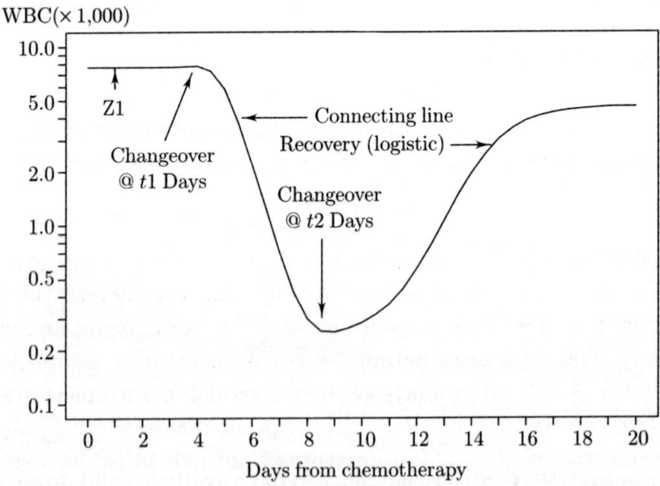

Fig. 4.   **Changepoint model for the logarithm of WBC measured over time: a horizontal line at the initial baseline for $t < \tau_{1i}$, a linear decline for $\tau_{1i} \leq t < \tau_{2i}$, and a logistic recovery for $t \geq \tau_{2i}$. See (5) for a formal definition.**

$$y_{ij} = f(\theta_i, t_{ij}) + \epsilon_{ij}, \quad \epsilon_{ij} \overset{iid}{\sim} N(0, \sigma^2)$$

with

$$f(\theta_i, t_{ij}) = \begin{cases} z_{1i} & \text{if } t_{ij} < \tau_{1i}, \\ r \cdot z_{1i} + (1-r)g(\theta_i, \tau_{2i}) & \text{if } \tau_{1i} \le t_{ij} < \tau_{2i}, \\ g(\theta_i, t_{ij}) & \text{if } t_{ij} \ge \tau_{2i} \end{cases}$$

$$g(\theta_i, t) = z_{2i} + z_{3i} \cdot / \left[1 + \exp\{-(\beta_{0i} + \beta_{1i}(t - \tau_{2i}))\} \right], \tag{5}$$

and $r = (\tau_{2i} - t_{ij})/(\tau_{2i} - \tau_{1i})$. Each profile is characterized by the parameter vector $\theta_i = (z_{1i}, z_{2i}, z_{3i}, \tau_{1i}, \tau_{2i}, \beta_{0i}, \beta_{1i})$. The second level of the model is specified by a mixture model (3), completed with the DP prior (4). Using Markov chain Monte Carlo methods (Gilks et al., 1996), we fit the model (3) to (5) to the study data. Fig. 5 shows observed WBC and fitted profiles for selected patients treated with different combinations of the two study drugs in CALGB 8881. From our Bayesian model, we could estimate predicted profiles for a future patient treated with different doses of cyclophosphamide, say, with GM-CSF at 5 μg/kg (see Fig. 6a), as well as predict various summary measurements of hematologic toxicity (Fig. 6b, c).

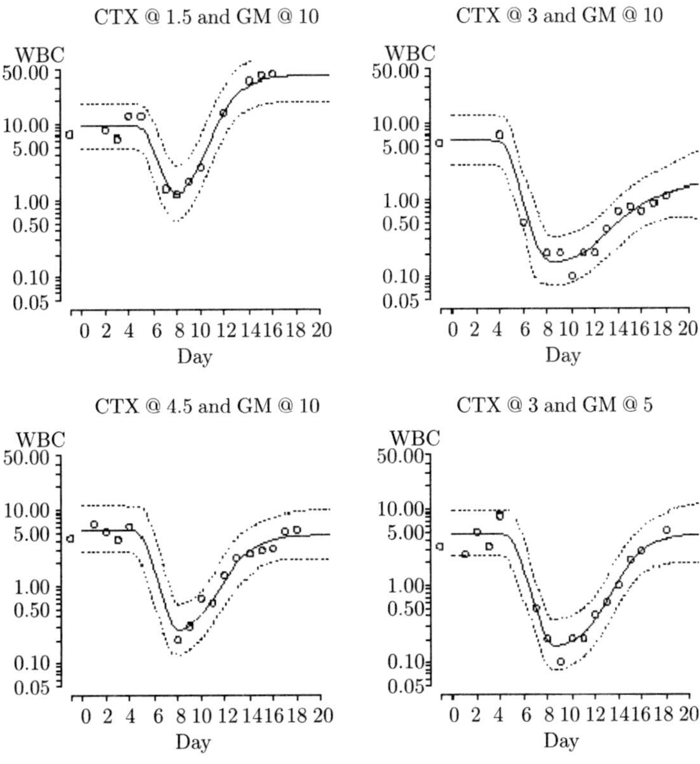

Fig. 5.   **Observed WBCs (circles) and fitted profiles (solid line) during the first course of chemotherapy for selected patients in CALGB 8881 by drug dose. Dashed lines show two posterior standard deviation margins.**

Freitas Lopes et al. (2003) developed an alternative approach based on finite mixtures of normals. Unrelated with applications in population models, Green and Richardson (1998) and Neal (2000)

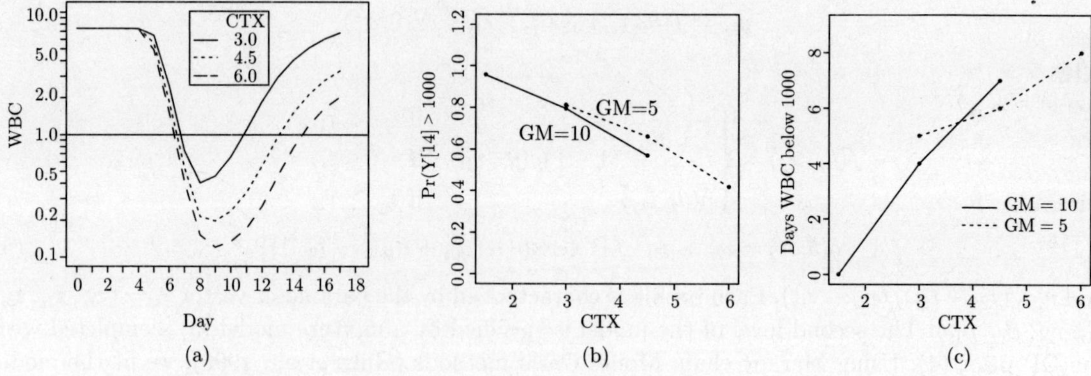

Fig. 6.   **Predictive inference for a future patient's course of chemotherapy by dose of cyclophosphamide (g/m^2). Panel (a) shows the predicted mean daily WBC by CTX dose with GM-CSF at 5 μg/kg. (All three predictive curves are conditioned on the same baseline WBC.) Panel (b) shows the probability of recovery beyond WBC = 1000/μL by day 14. Panel (c) plots the expected number of days below the critical level WBC = 1000. Panels (b) and (c) show regressions as a function of dose of cyclophosphamide, with two separate curves for GM-CSF at 5 μg/kg and at 10 μg/kg.**

argue for the use of finite mixture of normal models in place of Dirichlet process mixtures as in (3), citing issues of computational efficiency, flexibility of prior specifications, and interpretability. Specifically, we use in place of (3) a finite mixture of normals

$$p(\theta_i|\phi) = \sum_{m=1}^{M} \pi_m N(\mu + d_m, S),\qquad(6)$$

where M is the (random) number of terms in the mixture of normals, μ is an overall mean, and d_m are offsets for each of the terms, i.e., $\phi = (\mu, S, M, d_1, \ldots, d_M)$ are the hyperparameters in the general model (1).

3.   A Hierarchical Model for Multiple Studies

The CALGB followed study 8881 with another study using the best 8881 doses. CALGB 9160 (Budman et al., 1998) was a randomized study to evaluate if the drug Amifostine would reduce hematologic toxicity when combined with 3 g/m^2 of cyclophosphamide and 5 μg/kg GM-CSF. The study enrolled 46 patients and collected blood counts three times a week.

We developed a model that combined data from the first study, CALGB 8881, with blood count data collected as part of CALGB 9160.

A hierarchical model that includes a submodel of the form (1) for each study adds an extra level:

$$p(y_{kij}|\theta_{ki}), \quad p(\theta_{ki}|x_{ki}, \phi_k), \quad p(\phi_k|\eta), \quad p(\eta)\qquad(7)$$

for studies $k = 1, \ldots, K$, patients $i = 1, \ldots, n_k$, and occasions $j = 1, \ldots, n_{ki}$. Here $p(\theta_{ki}|x_{ki}, \phi_k)$ is the random effects distribution in study k, and $p(\phi_k|\eta)$ is a common prior on the study-specific parameters ϕ_k. The challenge was including the extra level in the hierarchy for multiple studies while keeping the mixture of normals model.

A straightforward mechanism to combine information from the first study with the second one uses posterior information from the analysis of CALGB 8881 as prior information for CALGB 9160. This was implemented in Müller and Rosner (1998). The approach essentially amounts to the right panel in Fig. 1. Given the hyperparameter η, the submodels are independent. Such linkage between the two submodels typically provides too little borrowing strength. For example, in the context of mixture model (3), only information about the distribution of the weights w_h. and the marginal distribution of the locations μ_h would be shared across studies.

In Freitas Lopes et al. (2003) we achieved the desired level of borrowing strength by extending the finite mixture of normal model (6). The basic idea is borrowed from the analysis of variance (ANOVA) model. In the one-way ANOVA model, we partition the expected value of an observation into a sum of a (population) grand mean and group-specific deviations from the population mean. Similarly, we decompose the distribution $p(\theta_{ki}|\phi_k)$ into a population measure p_0 and a study-specific measure p_k. The hyperparameters are split analogously into a subvector ϕ_0, the population measure's parameters, and subvectors ϕ_k, parameters of the study-specific measures.

$$p(\theta_{ki}|\phi_0, \phi_k) = \varepsilon \underbrace{p_0(\theta_{ki}|\phi_0)}_{\substack{\text{Population} \\ \text{measure}}} + (1 - \varepsilon) \underbrace{p_k(\theta_{ki}|\phi_k)}_{\substack{\text{Study-specific} \\ \text{measure}}}$$

In the application to modeling hematologic profiles in different studies, the model formally describes the variability across different studies regarding the toxicity over time.

We consider each component measure (population and study specific) as a finite mixture of multivariate normal densities as in (6). That is, we write the population measure p_0 as a mixture of M multivariate normals $p_0(\theta_{ki}|\phi_0) = \sum_{m=1}^{M} \pi_m N(\mu + d_m, S)$, where π_m are the mixing weights, μ is

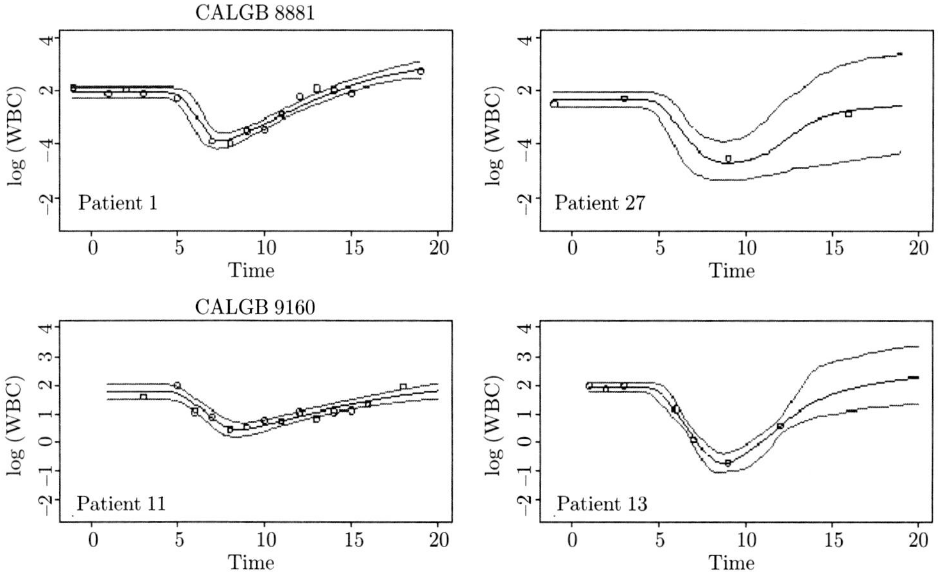

Fig. 7. Observed and fitted ln(WBC) profiles for four patients' first course of chemotherapy in CALGB studies 8881 and 9160. The figure also shows 95% pointwise (posterior) confidence bands for the fits. All the patients received 3 g/m^2 cyclophosphamide and 5 μg/kg GM-CSF.

the overall mean (location), and d_m are component-wise deviations or offsets from μ. We set $d_1 \equiv 0$ to ensure identifiability. (Mengersen and Robert, 1995, discuss this parameterization for finite mixtures of normals.) Similarly, the study-specific measures p_k are considered finite mixtures of multivariate normal densities, each with M' components. The full mixture model, including covariates, is given as follows:

$$p(\theta_{ki}, x_{ki}|\phi_k, \phi_0) = \varepsilon \sum_{m=1}^{M} \pi_m N(\mu + d_m, S) + (1 - \varepsilon) \sum_{m'=1}^{M'} \pi_{km'} N(\mu + d_{km'}, S), \qquad (8)$$

where $\phi_k = (d_{k1}, \pi_{k1}, \dots, d_{kM}, \pi_{kM})$ are study-specific parameters and $\phi_0 = (\mu, d_1, \pi_i, \dots, d_M, \pi_M, \varepsilon, S)$ parameterize the population measure. We fit this mixture of normal meta-model to all cycle 1 WBC data from both studies.

We have also worked out the details for allowing the number of mixture components to be random. The fitting algorithm requires reversible jump Markov Chain Monte Carlo methods (Green 1995). Fig. 7 shows the data and fitted profiles for selected patients from both CALGB studies. Fig. 8 shows the magnitude of the mixing weights in the population and study-specific measures.

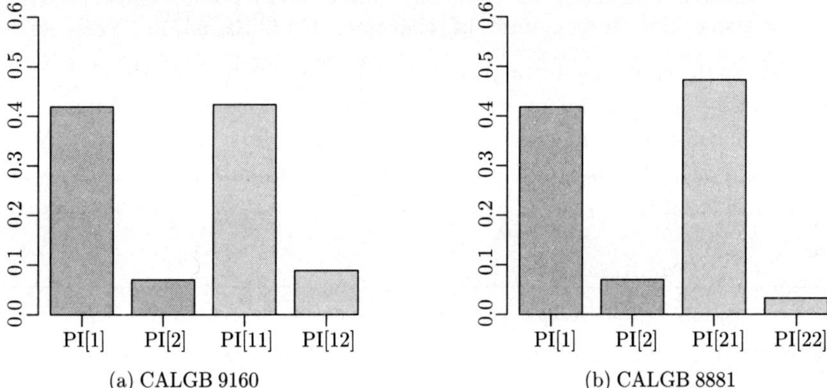

(a) CALGB 9160   (b) CALGB 8881

Fig. 8.   **Posterior mean mixing weights for CALGB 8881 (left panel) and CALGB 9160 (right panel) from the two-component model (8). The barplots show the posterior means for π_1, π_2, π_{11}, π_{12} (left panel) and π_1, π_2, π_{21}, π_{22} (right panel).**

The next task we tackled was finding a way to let these two studies, with relatively frequent blood count monitoring, make more precise our inference in a large phase III study in adjuvant breast cancer. CALGB 8541 (Wood et al., 1994) enrolled over 1500 women with stage II, node positive breast cancer and randomized them to three treatment regimens differing in dose intensity (i.e., higher total doses over some period of time). The chemotherapeutic agents were cyclophosphamide, doxorubicin, and 5-fluorouracil at different doses and schedules, shown in Table 1. The study's protocol called for monitoring blood counts once a week. Fig. 9 shows the relative frequency of blood count measurements during the first cycle of therapy. Patients treated with the most dose-intense regimen (treatment arm I in Table 1) experienced the most myelosuppression during cycle 1, and we limit attention to this treatment regimen for the remainder of this discussion. Our goal is to use this large data set to model relationships between patient characteristics and the risk of severe myelosuppression. In Müller et al. (2005) we have developed the methodology to incorporate the

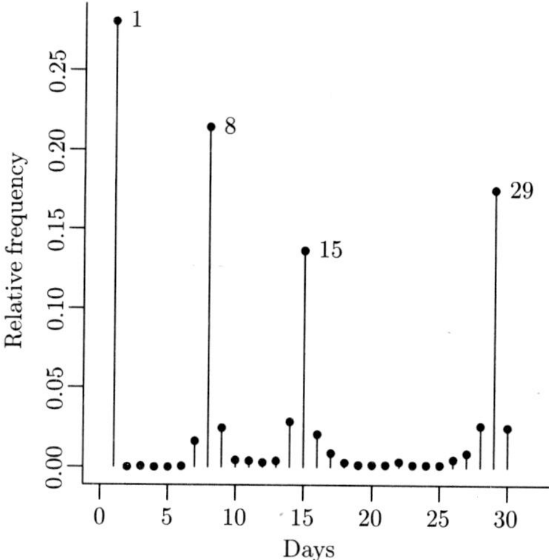

Fig. 9. **Relative frequency of days on which blood counts were measured during the first course of therapy (CALGB 8541). Patients are those randomized to the most dose-intense treatment arm (arm I in Table 1). Peaks at 1, 8, 15 and 29 days are due to the nearly universal weekly measurements.**

information from the two CALGB studies (8881 and 9160) to make more precise inference about CALGB 8541 patients' hematologic profiles. The third study is qualitatively different from the earlier two studies. Study number 8541 enrolled a different patient population and evaluated different treatment regimens than CALGB studies 8881 or 9160. It is, therefore, not *a priori* exchangeable with the other two studies. This is also evident in plots of the data (not shown). We achieve the desired generalization by replacing the normal means μ_h in (3) by a linear model $A_h d_i$. Here d_i is a design vector for the i-th patient. For the sake of argument, ignore for a moment all other covariates, and consider only effects for the three studies. Then $d_i = (1, ST_1, ST_2)$, with ST_1 and ST_2 being indicators for patient i being in study 1 (CALGB 8881) or 2 (CALGB 9160). The matrices A_h have three columns, with the first column representing main effects common to patients from all studies, and columns two and three containing offsets for study 1 and study 2, respectively. Other effects, including effects for different doses of the chemotherapy agent, can be added in the usual

Table 1.   Doses and schedules for CALGB 8541

	Treatment Arm		
	I	II	III
Number of 28-day cycles	4	6	4
Cyclophosphamide (day 1)	600 mg/m^2	400 mg/m^2	300 mg/m^2
Doxorubicin (day 1)	60 mg/m^2	40 mg/m^2	30 mg/m^2
5-Fluorouracil (days 1 and 8)	600 mg/m^2	400 mg/m^2	300 mg/m^2

ANOVA-like fashion. The non-parametric mixture in (3) is now over the matrix A_h, i.e.,

$$p(\theta_i \mid x_i, \Sigma, G) \sim \underbrace{\sum_h w_h N(A_h d_i, \Sigma)}_{= \int N(A d_i, \Sigma) dG(A)}, \quad G \sim p(G). \tag{9}$$

Note that the regression on covariates x_i is now included in the defintion of the design vector d_i. Deteails of this approach are discussed in Müller et al. (2005).

4. Optimal Design for Sampling Times

In Stroud, Müller and Rosner (2001), we considered decision theoretic solutions to the problem of choosing optimal sampling times in a population PK model. We find Bayesian optimal designs for the anticancer agent paclitaxel (Taxol) given by 3-hour infusion. This drug exhibits fairly complex pharmacokinetics. We used a three-compartment model that includes Michaelis-Menten kinetics as likelihood $p(y_{ij}|\theta_i)$. Fig. 10 shows posterior fitted curves for three patients who received 3-hour infusions of paclitaxel with constant dosage of 50, 75 and 100 mg/m^2. As seen in the plots, the concentration-versus-time profiles vary greatly in shape from patient to patient, showing the existence of patient-to-patient PK variability.

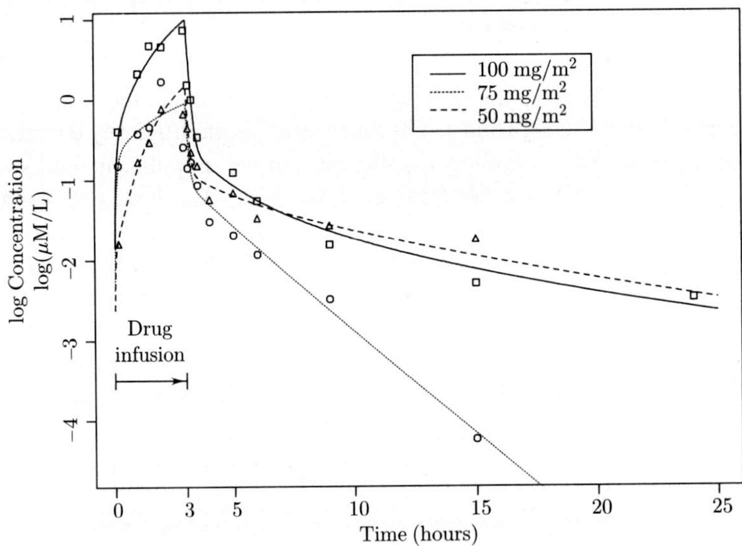

Fig. 10.   **Observed log-concentrations and fitted curves for selected patients receiving 3-hour infusions of Taxol.**

We pursued two complimentary strategies to find optimal sampling times. Both approaches evaluate expected utilities by Monte Carlo integration. Essentially, the expected utility integral (2) is approximated by a sample average $\hat{U}(d) = 1/M \sum u(d, \theta_i, y_i)$, where $(\theta_i, y_i) \sim p_d(\theta, y)$, $i = 1, \ldots, M$, are simulated experiments generated from the relevant probability model. Exploiting continuity of the expected utility function $U(d)$, the simple sample average $\hat{U}(d)$ is replaced by a smooth surface fit through a scatterplot of designs d_j and observed utilities $u(\theta_j, y_j)$ in simulated experiments $(\theta_j, y_j) \sim p_{d_j}(\theta, y)$, simulating only few experiments for each design d_j. The experiments d_j are chosen on a regular grid over the design space.

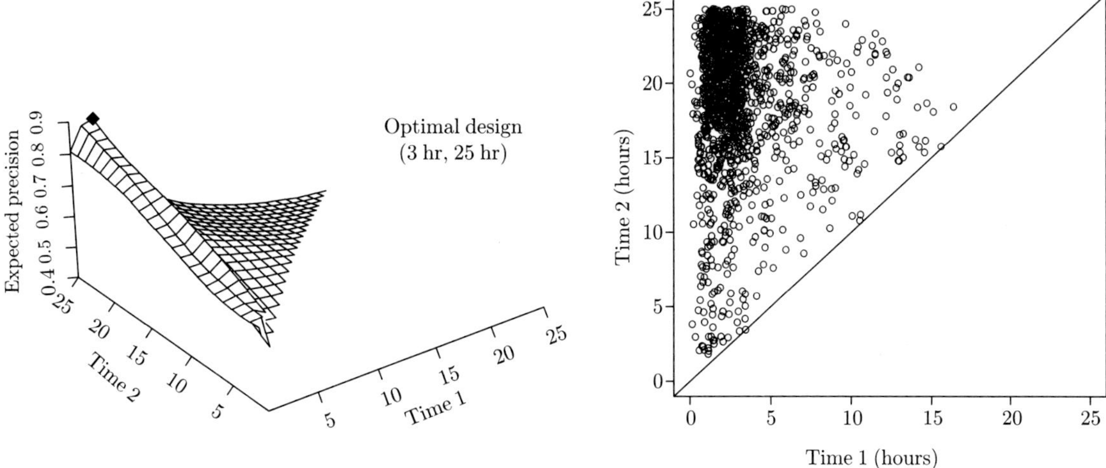

Fig. 11. **Optimal design (d_1, d_2) for two sampling times (criterion: Taxol AUC). The left panel shows the fitted surface of expected utilities. The right panel shows the design choices for which experiments were simulated. The chosen designs cluster around choices with high expected utility, i.e., the simulation effort is concentrated in promising parts of the design space.**

A variation of this first strategy focuses the simulation effort around promising designs, rather than on a regular grid over the design space. This is achieved by generating designs d_j as realizations of a Markov chain defined such that the implied stationary distribution in d is proportional to expected utility. Simulating from this Markov chain generates designs that cluster in areas of high expected utility, i.e., exactly where increased simulation effort is desirable. We defined such Markov chains and develop simulation strategies to implement efficiently the simulation. Fig. 11 shows the estimated expected utility surface as a function of two sampling times.

5. Conclusion

We discussed semi-parametric Bayesian models for inference in population PK/PD studies. All models were based on variations of mixed effects models with non-parametric mixtures of normal random effects distributions. The strength of such models is the ease of computation, and straightforward use even in higher dimensions. Naturally, there are several limitations. For example, the inclusion of binary or multinomial responses is not immediately possible. Another common format for repeated measurement responses are fractions of cells staining for some attribute of interest, leading to data restricted between 0 and 1, including positive probabilities at 0 and 1.

Most discussed models included as underlying non-parametric Bayesian model some variation of DP mixture models. Several arguments can be made about the limitations of such models. A straightforward extension is the use of species sampling models (Pitman, 1996) and stick-breaking priors (Ishwaran and James, 2001 and 2003).

Acknowledgement

Research was partially supported by NIH under grant R01CA75981.

References

Antoniak, C.E., (1974). Mixtures of Dirichlet Processes with Applications to Bayesian Nonparametric Problems. *Annals of Statistics*, **2**, 1152–1174.

Berry, D.A., and Stangl, D.K. (1996). *Bayesian Biostatistics*, New York: Marcel Dekker.

Budman, D.R., Rosner, G.L., Lichtman, S.M., Miller, A., Ratain, M.J. and Schilsky, R.L. (1998). A randomized trial of WR-2721 (Amifostine) as a chemoprotective agent in combination with high-dose cyclophosphamide and molgramostim (GM-CSF). *Cancer Therapeutics*, **1**, 164-167.

Chaloner, K. and Verdinelli, I. (1995). Bayesian experimental design: a review. *Statistical Science*, **10**, 273-304.

Davidian, M. and Gallant, A.R. (1992). Smooth nonparametric maximum likelihood estimation for population pharmacokinetics, with application to quinidine. *Journal of Pharmacokinetics and Biopharmaceutics*, **20**, 529-556.

Davidian, M. and Gallant, A.R. (1993). The nonlinear mixed effects model with a smooth random effects density. *Biometrika*, **80**, 475-488.

Davidian, M. and Giltinan, D.M. (1995). *Nonlinear Models for Repeated Measurement Data*. Chapman and Hall, London.

Ferguson, T.S. (1973). A Bayesian analysis of some nonparametric problems. *Annals of Statistics*, **1**, 209–230.

Freitas Lopes, H., Müller, P. and Rosner, G. (2003). Meta-Analysis for Longitudinal Data Models using Multivariate Mixture Priors. *Biometrics*, **59**, 66-75.

Gilks, W.R., Richardson, S. and Spiegelhalter, D.J. (1996). *Markov Chain Monte Carlo in Practice*. London, U.K.: Chapman and Hall.

Green, P.J. (1995). Reversible jump Markov chain Monte Carlo computation and Bayesian model determination. *Biometrika*, **82**, 711-732.

Green, P.J. and Richardson, S. (1998). Modelling heterogeneity with and without the Dirichlet process. Technical report, University of Bristol.

Ishwaran, H. and James, L.J. (2003), Generalized weighted Chinese restaurant processes for species sampling mixture models. *Statistica Sinica*, **13**, 1211–1235.

Ishwaran, H. and James, L.J. (2001). Gibbs Sampling Methods for Stick-Breaking Priors. *Journal of the American Statistical Association*, **96**, 161–173.

Kleinman, K.P. and Ibrahim, J.G. (1998a). A Semi-parametric Bayesian Approach to the Random Effects Model. *Biometrics*, **54**, 921–938.

Kleinman K.P. and Ibrahim, J.G. (1998b). A Semi-parametric Bayesian Approach to Generalized Linear Mixed Models. *Statistics in Medicine*, **17**, 2579–2596.

Lichtman, S.M., Ratain, M.J., Van Echo, D.A. Rosner, G.L., Egorin, M.J., Budman, D.R., Vogelzang, N.J., Norton, L. and Schilsky, R.L. (1993). Phase I trial of granulocyte-macrophage colony stimulating factor plus high-dose cyclophosphamide given every 2 weeks: A Cancer and Leukemia Group B study. *Journal of the National Cancer Institute*, **85**, 1319-1326.

Mallet, A. (1986). A maximum likelihood estimation method for random coefficient regression models. *Biometrika*, **73**, 645-656.

Mallet A., Mentré, F., Gilles, J., Kelman, A.W., Thomson, A.H., Bryson, S.M. and Whiting, B. (1988). Handling covariates in population pharmacokinetics with an application to gentamicin. *Biomed Meas Inf Contr*, **2**, 138-146.

Mallick, B.K. and Walker, S.G., (1997). Combining Information from Several Experiments with Nonparametric Priors. *Biometrika*, **84**, 697–706.

Mengersen, K. and Robert, C.P. (1995). Testing for mixtures: a Bayesian entropic approach. In *Bayesian Statistics 5*, J.O. Berger, J.M. Bernardo, A.P. Dawid, D.V. Lindley, A.F.M. Smith, eds., London: Oxford University Press.

Mentré, F. and Mallet, A. (1994). Handling covariates in population pharmacokinetics. *International Journal of Bio-Medical Computing*, **36**, 25–33.

Morris, C. and Normand, S.L. (1992). Hierarchical models for combining information and for meta-analyses. In *Bayesian Statistics 4*, J.M. Bernardo, J.O. Berger, A.P. Dawid and Smith, A.F.M., eds., 321-335.

Müller, P. and Rosner, G.L. (1997). A Bayesian population model with hierarchical mixture priors applied to blood count data. *Journal of the American Statistical Association*, **92**, 1279–1292.

Müller, P. and Rosner, G.L. (1998). Semiparametric PK/PD models. In *Practical Nonparameteric and Semiparametric Bayesian Statsitics*. Dey, D., Müller, P., and Sinha, D., eds., New York: Springer-Verlag.

Müller, P., Rosner, G.L., De Iorio, M. and MacEachern, S. (2005). A Nonparametric Bayesian Model for Inference in Related Studies. *Applied Statistics*, **54 (3)**, 611-626.

Neal, R.M. (2000). Markov chain sampling methods for Dirichlet process mixture models, *Journal of Computational and Graphical Statistics*, **9**, 249–265.

Pitman, J. (1996). Some Developments of the Blackwell-MacQueen Urn Scheme. In *Statistics, Probability and Game Theory. Papers in Honor of David Blackwell*, eds. T.S. Ferguson, L.S. Shapeley, and J.B. MacQueen, IMS Lecture Notes, 245–268, IMS.

Ratain, M.J., Schilsky, R.L., Conley, B.A. and Egorin, M.J. (1990). Pharmacodynamics in cancer therapy. *Journal of Clinical Oncology*, **8**, 1739-1753.

Rosner, G.L. and Müller, P. (1997). Bayesian population pharmacokinetic and pharmacodynamic analyses using mixture models. *Journal of Pharmacokinetics and Biopharmaceutics*, **25**, 209–233.

Stangl, D.K. (1995). Prediction and decision making using Bayesian hierarchical models. *Statistics in Medicine*, **14**, 2173-2190.

Stroud, J.R., Müller, P. and Rosner G.L. (2001). Optimal Sampling Times in Population Pharmacokinetic Studies. *Applied Statistics*, **50**, 345-359.

Wakefield, J.C (1996). The Bayesian analysis of population pharmacokinetic models. *Journal of the American Statistical Association*, **91**, 62-75.

Wakefield, J.C., Smith, A.F.M., Racine-Poon, A. and Gelfand, A. (1994). Bayesian analysis of linear and nonlinear population models using the Gibbs sample. *Applied Statistics*, **43**, 201-221.

Wakefield, J.C., Aarons, L. and Racine-Poon, A. (1999). The Bayesian approach to population pharmacokinetic/pharmacodynamic modelling. In *Case Studies in Bayesian Statistics* (eds B.P. Carlin, A.L. Carriquiry, C. Gatsonis, A. Gelman, R.E. Kass, I. Verdinelli and M. West). Springer-Verlag.

Walker, S.G. and Wakefield, J. (1994). Population Models with a Nonparametric Random Coefficient Distribution. Technical Report, Imperial College London.

Wood, W.C., Budman, D.R., Korzun, A.H., Cooper, M.R., Younger, J., Hart, R.D., Moore, A., Ellerton, J.A., Norton, L., Ferree, C.R., Ballow, A.C., Frei, E. and Henderson, I.C. (1994). Dose and dose intensity of adjuvant chemotherapy for stage II, node positive breast cancer. *New England Journal of Medicine*, **330**, 1253-1259.

Zeger, S.L. and Karim, M.R. (1991). Generalized linear models with random effects. *Journal of the American Statistical Association*, **86**, 79-86.

Bayesian Statistics and Its Applications
Edited by S.K. Upadhyay, U. Singh and D.K. Dey
Anamaya Publishers, New Delhi, India

Bayesian Predictive Inference Under Informative Sampling via Surrogate Samples

Balgobin Nandram

Department of Mathematical Sciences, Worcester Polytechnic Institute,
100 Institute Road, Worcester, MA 01609

Abstract

A sample is drawn from a finite population, but because of selection bias, the sample is not a random sample from the original finite population. In fact, the original sample is a random sample from a weighted distribution, and one can convert this sample to a surrogate sample from the original distribution. This surrogate sample can now be used to make inference about the original finite population without any further consideration about the biased sample. In a simple example on the finite population proportion, we have the first-order inclusion probabilities for the sampled units but not those for the nonsampled units. The bias comes from a relation between the selection probabilities and the characteristic as in probability proportional to size sampling. Using the joint posterior density as a prior in the census model, a beta-binomial model, the prior predictive distribution of the sample is then obtained. A surrogate sample is obtained from the prior predictive distribution, thereby transforming the biased sample into a "random" sample. Then, the census model is used to make a posteriori inference about the finite population. Our simulation study and illustrative example show that one should consider using our method.

1. Introduction

We consider the problem in which a random (representative) sample is taken from a finite population, and this sample is not observed. Instead, a sample, significantly perturbed by some mechanism, is observed. For example, in many complex sample surveys, sample units are drawn with probability proportional to some measure of size. Thus, the sample is not a random (representative) sample from the finite population. Here, the model holding for the sample could be different from the model holding for the rest of the population (i.e., there is selection bias). Models for this type of sample design can be formed using weighted distributions (Patil and Rao, 1978). We use a Bayesian method to infer about a finite population proportion under selection bias.

If a biased sample is drawn from a finite population, one cannot make inference about the nonsampled values unless the nature of the bias is clearly understood. One example is when a sample is drawn from a finite population with probability proportional to size (PPS). In this case the sampled values are large and the nonsampled values tend to be small. It is possible to construct an appropriate selection model for the sample data, and this will turn out to be a weighted distribution. Predictive inference from the weighted distribution is not straightforward.

It is standard practice to view the finite population as a random sample from a superpopulation. While there is no bias in "drawing" the finite population from the superpopulation, there is selection bias in drawing the sample from the finite population. The parameters of both the finite population and the superpopulation are of interest.

Therefore, the main task is to estimate the distribution of all the values in the finite population. One simple approach is to use the Horvitz-Thompson estimator (Cochran, 1977). This may be a bad idea for two reasons. First, the Horvitz-Thompson estimator may not be efficient because although the selection probabilities may be linearly related to the measurements, the relation is not necessarily through the origin. Second, only the first-order selection probabilities are assumed known; the second-order inclusion probabilities are not known so that a sensible measure of the standard error does not exist (Cochran, 1977).

For many establishment surveys, the selection probabilities of the establishments are proportional to the volume (in dollars) of business in the previous year. However, only the sampled data for the current year are available, and for confidentiality or convenience only the selection probabilities for the sampled units are presented to secondary data analysts. Clearly, in PPS sampling these selection probabilities contain substantial information about the nonsampled values, and therefore, the mean of the finite population.

In PPS sampling the selection probabilities are proportional to some measure of size, and this measure of size, in turn, is proportional to the characteristic of interest. Henceforth, we let $y_i, i = 1, \ldots, N$, denote the finite population values and $\pi_i, i = 1, \ldots, N$, denote the selection probabilities of the entire population of size N. That is, $\pi_i = \beta_0 + \beta_1 y_i, i = 1, \ldots, N$, albeit with some error. In a survey of hospitals, the selection probabilities are typically proportional to a measure of size (e.g., number of beds). Larger hospitals have more beds, and they tend to have better quality of care. So that with interest on the quality of care, as a first approximation, one can take the selection probabilities to be linearly related to the quality of care.

In concrete terms we briefly discuss the weighted distribution. Let $p(y \mid \underset{\sim}{\theta}_1)$ denote the probability distribution that describes the finite population (i.e., the census). When a random sample is taken from this finite population, it is perturbed by the weight function $w(y; \underset{\sim}{\theta}_1, \underset{\sim}{\theta}_2)$ to produce a sample from the new probability distribution $q(y \mid \underset{\sim}{\theta}_1, \underset{\sim}{\theta}_2)$. That is, a representative sample is observed from

$$q(y \mid \underset{\sim}{\theta}_1, \underset{\sim}{\theta}_2) = w(y; \underset{\sim}{\theta}_1, \underset{\sim}{\theta}_2) p(y \mid \underset{\sim}{\theta}_1).$$

One example is when a sample of fish is drawn from a pond, and the probability a fish is taken is proportional to its length, creating a length bias (Patil and Rao, 1978). See Silliman (1997) for a recent discussion of such a weighted distribution. The question that arises is "Can we recreate the sample from $p(y \mid \underset{\sim}{\theta}_1)$?" That is, can we create a surrogate (representative) sample from the original finite population? The answer is "yes", and inference can be made about the original finite population using this surrogate sample via $p(y \mid \underset{\sim}{\theta}_1)$. In model-based analysis, PPS sampling is a special case within this framework, and it deserves special attention.

There are several approaches to the superpopulation problem. Pfeffermann, Krieger and Rinott (1998) use weighted distributions in the spirit of Patil and Rao, (1978); see also Pfeffermann and Sverchkov (1999). Chambers, Dorfman and Wang (1998) also consider the case in which the π_i are related to the y_i and other covariates as well; see also Burgos and Nandram (2003) for a Bayesian extension. In the work of Chambers, Dorfman and Wang (1998) there is improved precision over alternative estimators, but there is a correction only for small selection bias.

For the estimation of a finite population quantity, the problem is more complex than for a superpopulation parameter because if there is a bias which tends to make the sampled values large, the nonsampled values would tend to be small. Such an adjustment is difficult to carry out. Generally, it has been assumed that the sample size is much smaller than the population size, and this eliminates the finite population estimation problem. However, we note that Sverchkov and Pfeffermann (2004) use a method to estimate the finite population total. They define the sample and sample-complement distributions for developing design consistent predictors of the finite population total. Essentially they define the distributions of the sampled values and the nonsampled values as two separate weighted distributions of the census distribution (see Patil and Rao, 1978).

It is pertinent to describe the work of Malec, Davis and Cao (1999), where a sample selection derivation is given in an Appendix of their paper. A difficulty in including the selection probabilities directly in the model forces them to make an ad hoc adjustment to the likelihood function, and to use a Bayes empirical Bayes approach. Let $y_i|p \overset{\text{i.i.d.}}{\sim}$ Bernoulli (p), $i = 1, \ldots, N$, and, again, let $\pi_i, i = 1, \ldots, N$, be the corresponding selection probabilities. Malec, Davis and Cao (1999) obtain the profile likelihood function corresponding to the vector of sampled values y_s

$$\text{Lik}\,(p|y_s, \underset{\sim}{\pi}) = \frac{1}{(pW_1^{-1} + (1-p)W_0^{-1})^n} p^t(1-p)^{n-t}, 0 < p < 1, \tag{1}$$

where $W_1 = \sum_{i=1}^{n} \pi_i^{-1} y_i / t$, $W_0 = \sum_{i=1}^{n} \pi_i^{-1}(1 - y_i)/(n - t)$, and $t = \sum_{i=1}^{n} y_i$. Then, assuming a priori that $p \sim$ Uniform $(0, 1)$, the posterior density of p given the vector of sampled values y_s is

$$\pi(p|y_s) = A^{-1} \frac{B(t+1, n-t+1)}{(pW_1^{-1} + (1-p)W_0^{-1})^n} \frac{p^t(1-p)^{n-t}}{B(t+1, n-t+1)},$$

where $A = \int_0^1 \frac{B(t+1, n-t+1)}{(pW_1^{-1} + (1-p)W_0^{-1})^n} \frac{p^t(1-p)^{n-t}}{B(t+1, n-t+1)} dp$ and $B(.,.)$ is the beta function. Without any selection bias, inference is based on

$$\pi(p|y_s) = \frac{p^t(1-p)^{n-t}}{B(t+1, n-t+1)},$$

the standard beta probability density function. Malec, Davis and Cao (1999) would have obtained the nonsampled values y_{ns} using the composition method,

$$p(y_{ns}|y_s) = \int_0^1 p(y_{ns}|p)\pi(p|y_s)dp$$

after samples are drawn from $\pi(p|y_s)$; they did not state this explicitly. It appears that this procedure will need a relatively large sample for W_1 and W_0 to have reasonable properties.

There are two approaches to the problem. The first approach incorporates the nonsampled selection probabilities in a model (e.g., Nandram and Sedransk, 2005). This is a computational problem because the nonsampled part of the population is much larger than the sample. The second approach proposes two models, one for the sample, called the *survey model*, and the other for the population, called the *census model*. We will call this second approach the *surrogate sampling approach*.

The surrogate sampling approach is somewhat different from all others. It permits us to obtain a surrogate random sample from the census model, and then prediction is done via the census model. We use a fully Bayesian method; we do not use a Bayes empirical Bayes method nor do we use a non-Bayesian method. However, our method is related to that of Chambers, Dorfman and Wang (1998) and Nandram and Sedransk (2005), but somewhat different because we work with only the sampled data to construct the surrogate sample from the census model. Our work is related to that of Malec, Davis and Cao (1999) in that we are estimating the finite population proportion under selection bias. We believe that the work of Malec, Davis and Cao (1999) is the only one which treats the estimation of the finite population *proportion under selection bias*; others work on continuous data. The computation time for our method is small in comparison with that of an analogous method of Nandram and Sedransk (2005) because we do not have to generate latent variables for the nonsampled values, an enormous improvement.

In this paper, we illustrate the use of surrogate sampling to estimate a finite population proportion under selection bias. Incorporating the information on how the sample is selected can undo the selection bias, and a surrogate representative sample can be generated from the finite population. Then, model-based inference can be made about the finite population using the surrogate sample.

This paper has six more sections. In Section 2 we describe the problem in the case of the finite population proportion. In Section 3, as part of the procedure, we show how to make inference about the superpopulation proportion and, in Section 4, we describe the requisite computation. In Section 5 we describe how to make inference about the finite population proportion. In Section 6 we describe a simulation study to assess changes in our estimators as sample size and variability increase. In Section 7 we describe an illustrative example from the National Health Interview Survey. Finally, in Section 8 we present the conclusion.

2. Description for a Finite Population Proportion

Suppose a simple random sample of size n is taken from a finite population of size N, and suppose each individual does, $y = 1$, or does not have a characteristic, $y = 0$ (i.e., binary data). Thus, we assume that

$$y_1 \ldots , y_N | p \overset{\text{i.i.d.}}{\sim} \text{Bernoulli } (p)$$

and the first n values are the sample. That is, $\underset{\sim}{y}_s = (y_1, \ldots , y_n)'$ and $\underset{\sim}{y}_{ns} = (y_{n+1}, \ldots , y_n)'$ are the sampled and nonsampled values. From a Bayesian point of view, to make inference about p, we use the "noninformative" prior

$$p \sim \text{uniform } (0, 1),$$

a natural choice when there is no information about p. It follows that a posteriori

$$p | \underset{\sim}{y}_s \sim \text{Beta } (t + 1, n - t + 1), \tag{2}$$

where $t = \sum_{i=1}^{n} y_i$, and inference about p is straightforward.

There is general interest about the finite population proportion. Letting $T = \sum_{i=1}^{N} y_i$ denotes the finite population total, the finite population proportion is $P = T/N = f\overline{y}_s + (1 - f)\overline{y}_{ns}$, where $\overline{y}_s = t/n$, $\overline{y}_{ns} = \sum_{i=n+1}^{N} y_i/(N - n)$ and $f = \dfrac{n}{N}$ are, respectively, the sample proportion, the non-sample proportion and the sample fraction. Inference about P is first obtained by making inference about $T_{ns} = \sum_{i=n+1}^{N} y_i$ as follows:

$$g(T_{ns}|y_s) = \int_0^1 g(T_{ns}|p)\pi(p|y_s)dp, \qquad (3)$$

where $\pi(p|y_s)$ is the posterior density in (2) and $T_{ns}|p \sim$ Binomial $(N-n, p)$. Thus, inference about P can be obtained by drawing random samples from (3). This can be done easily using the composition method.

In a biased sample (2) and (3) no longer hold. One situation is when the sampled values are taken with unequal probabilities of selection, and these probabilities are related to the characteristic y. For example, in the 1996 National Health Interview Survey when we considered activity limitation for adults older than 30 years, we found significant selection bias in two domains formed by crossing education, race and sex; the selection probabilities for adults, who have activity limitation, tend to be smaller than for adults, who do not have activity limitation. Thus, there is a relation between the π_i and the y_i. In addition, the selection probabilities for the nonsampled individuals are unknown or unreported.

In our surrogate sampling approach, the posterior distribution of p is obtained using the sample data and the selection probabilities only. Then this posterior density is used as a prior distribution in the census model to generate a surrogate random sample from the finite population, ignoring the sampled values already obtained. The Bayesian analysis is used to "convert" the biased sample into a random sample from the finite population. This sample is then used as in (2) and (3).

Recalling that $\pi_1, \ldots, \pi_n$ are the selection probabilities for the sample, the sample weights are $\omega_i = \pi_i^{-1}/\sum_{i=1}^{n} \pi_i^{-1}, i = 1, \ldots, n$. Note that $\omega_i \propto 1/\pi_i, i = 1, \ldots, n$, and $\sum_{i=1}^{n} \omega_i = 1$. Our survey model will include $\omega_i, i = 1, \ldots, n$, which are fixed and known. Note that $\sum_{i=1}^{n} \omega_i = 1$. When there is a selection bias, we may want to link the ω_i to the y_i (i.e. nonignorable selection). One possibility is to take

$$\omega_i \propto \beta_0 + \beta_i y_i + e_i, \quad e_i \stackrel{\text{i.i.d.}}{\sim} \text{Normal}(0, \sigma_e^2), \quad i = 1, \ldots, n.$$

That is, $\omega_i = (\beta_0 + \beta_i y_i + \sigma_e z_i)/n(\beta_0 + \beta_1 \overline{y}_s + \sigma_e \overline{z}_s)$, $z_i \stackrel{\text{i.i.d.}}{\sim} \text{Normal}(0, 1)$ with $\overline{z}_s = \sum_{i=1}^{n} z_i/n$. Because the ratio is invariant to scale, the parameters β_0, β_1 and σ_e^2 are not identifiable. Setting β_1 or σ_e^2 to one is a standard choice; we prefer $\beta_1 = 1$ because this is more convenient. Thus,

$$\omega_i = \frac{\beta_0 + y_i + \sigma_e z_i}{n(\beta_0 + \overline{y}_s + \sigma_e \overline{z}_s)}, \quad z_i \stackrel{\text{i.i.d.}}{\sim} \text{Normal}(0,1), i = 1, \ldots, n. \qquad (4)$$

In this formulation σ_e^2 must be nonnegative, and we will argue that β_0/σ_e should be bounded from below.

Note that in (4), if $\sigma_e^2 \to \infty, \omega_i \to \dfrac{z_i}{n\overline{z}_s}$, and if $\sigma_e^2 \to 0$, $\omega_i \to \dfrac{\beta_0 + y_i}{n(\beta_0 + \overline{y}_s)}$. That is, when σ_e^2 is large, the ω_i do not depend on y_i. Thus, if σ_e^2 is large, there will be little selection bias, and if σ_e^2 is small, there will be large selection bias.

The survey model is completed by putting prior densities on β_0 and σ_e^2. We take

$$p(\beta_0) = 1, \quad p(\sigma_e^2) \propto \frac{1}{\sigma_e^2}, \quad \beta_0/\sigma_e > \mu_0, \quad \sigma_e^2 > 0, \tag{5}$$

where μ_0 is to be specified. Our census model is

$$y_1, \ldots, y_N | p \overset{\text{i.i.d.}}{\sim} \text{ Bernoulli } (p). \tag{6}$$

Note that the survey model has (6) for $y_1, \ldots, y_n$ as well.

We will show how to obtain the joint posterior density of $p, \beta_0, \sigma_e^2 | y_s$, and how to obtain a sample from $p, \beta_0, \sigma_e^2 | y_s$, and in particular, a sample form $\pi(p|y_s)$ under selection bias in a procedure comparable to (2). Then, in our surrogate sampling procedure $\pi(p|y_s)$ will be used as a prior for p, together with the census model in (6) as in (2) and (3).

Note that if the sampled values tend to be too large, the nonsampled values will tend to be too small, because of the PPS sampling. Thus, our method corrects for this bias in the finite population proportion. Inference for p follows from $\pi(p|y_s)$ using the biased data, corrected through the model.

3. Inference about the Superpopulation Proportion

Inference about p is obtained using samples from the joint posterior density. Then, inference about the finite population proportion can be obtained using $\pi(p|y_s)$. Thus, we show how to obtain samples from the joint posterior density of $p, \beta_0, \sigma_e^2 | y_s$.

We simplify the analysis enormously by introducing the latent variables $\nu_1, \ldots, \nu_n$, where

$$\nu_i = \beta_0 + y_i + e_i, \quad e_i | \sigma_e^2 \overset{\text{i.i.d.}}{\sim} \text{ Normal } (0, \sigma_e^2), \quad i = 1, \ldots, n.$$

That is, $\nu_i | \beta_0, y_i, \sigma_e^2 \overset{\text{i.i.d.}}{\sim} \text{ Normal } (\beta_0 + y_i, \sigma_e^2), i = 1, \ldots, n$. Thus, $\omega_i = \nu_i / \sum_{i'=1}^{n} \nu_{i'}, i = 1, \ldots, n$. Here ν_i can all be nonnegative or nonpositive; so without loss of generality, we take them to be nonnegative.

Note that

$$f(\nu_i | \beta_0, y_i = 0, \sigma_e^2) = \frac{\frac{1}{\sigma_e} \phi\left(\frac{\nu_i - \beta_0}{\sigma_e}\right)}{\Phi\left(\frac{\beta_0}{\sigma_e}\right)}, \nu_i \geq 0, \tag{7}$$

where $\phi(\cdot)$ is the standard normal density function, $\Phi(t) = \int_{-\infty}^{t} \phi(z)dz$, and (7) is true for all β_0 and σ_e^2. Then, using (7) it is easy to show that

$$E(\nu_i - \beta_0 | \beta_0, y_i = 0, \sigma_e^2) = \sigma_e \frac{\phi\left(\frac{\beta_0}{\sigma_e}\right)}{\Phi\left(\frac{\beta_0}{\sigma_e}\right)}.$$

Thus, because $E(\nu_i | \beta_0, y_i = \sigma_e^2) \geq 0$, we have

$$\frac{\beta_0}{\sigma_e} \Phi\left(\frac{\beta_0}{\sigma_e}\right) + \phi\left(\frac{\beta_0}{\sigma_e}\right) \geq 0 \quad \text{for all} \quad \beta_0, \sigma_e^2.$$

Now, consider the function $g(x) = x\Phi(x) + \phi(x)$. It is true that $g(x) \geq 0$ for all x in $(-\infty, \infty)$, $g(x)$ is strictly increasing, and as x approaches $-\infty$, $g(x)$ approaches 0. Thus, there is no restriction on $\beta_0 | \sigma_e$. However, for all practical purposes, consistent with a standard normal random variable, we take $\beta_0 / \sigma_e \geq \mu_0$, and we can take μ_0 to be -3 or -4. This condition hardly imposes any restriction, and it is a sufficient condition for propriety of the joint posterior density, as we will show.

It is easy to show that

$$f(\nu_i, y_i | p, \beta_0, \sigma_e)^2 \propto \frac{1}{\sqrt{2\pi\sigma_e^2}} e^{-\frac{1}{2\sigma_e^2}(\nu_i - (\beta_0 + y_i))^2} p^{y_i}(1-p)^{1-y_i}, \nu_i > 0, y_i = 0, 1.$$

So that

$$f(\nu_i, y_i | p, \beta_0, \sigma_e^2) = \frac{\frac{1}{\sqrt{2\pi\sigma_e^2}} e^{-\frac{1}{2\sigma_e^2}(\nu_i - (\beta_0 + y_i))^2}}{p\Phi\left(\frac{\beta_0 + 1}{\sigma_e}\right) + (1-p)\Phi\left(\frac{\beta_0}{\sigma_e}\right)} p^{y_i}(1-p)^{1-y_i}, \nu_i > 0, y_i = 0, 1. \tag{8}$$

Observe that ν_i and y_i are correlated, but note that the pairs (ν_i, y_i) given p, β_0, σ_e^2 are independent over i. Note that in Sverchkov and Pfeffermann (2004) the y_i are also independent.

Next, we must incorporate the condition that $\sum_{i=1}^{n} \omega_i = 1$ and the ω_is themselves into (8). Note that the joint density of $\{(\nu_i, y_i), i = 1, \ldots, n\}$ is

$$f(\underset{\sim}{\nu}, \underset{\sim}{y} | p, \beta_0, \sigma_e^2) = \frac{\left(\frac{1}{2\pi\sigma_e^2}\right)^{n/2} e^{-\frac{1}{2\sigma_e^2}\sum_{i=1}^{n}(\nu_i - (\beta_0 + y_i))^2}}{\left[p\Phi\left(\frac{\beta_0 + 1}{\sigma_e}\right) + (1-p)\Phi\left(\frac{\beta_0}{\sigma_e}\right)\right]^n} p^t(1-p)^{n-t}. \tag{9}$$

We incorporate the condition $\sum_{i=1}^{n} \omega_i = 1$ into our analysis using $\omega_i = \nu_i / \sum_{i'=1}^{n} \nu_{i'}, i = 1, \ldots, n$. Observe that this latter condition is the same as $\nu_i = \omega_i \sum_{i'=1}^{n} \nu_{i'}, i = 1, \ldots, n$. Thus, to use the condition $\nu_i = \omega_i \sum_{i'=1}^{n} \nu_{i'}, i = 1, \ldots, n$, we make the transformation $\gamma_i = \nu_i - \omega_i \sum_{i'=1}^{n} \nu_{i'}, i = 1, \ldots, n-1$, and we will take $\gamma_i = 0, i = 1, \ldots, n-1$. First, we pretend that the γ_i are unrestricted, and we find the joint density of $\underset{\sim}{\gamma}, y_s | p, \beta_0, \sigma_e^2$. Second, we set $\gamma_i = 0, i = 1, \ldots, n-1$, in this joint density. Note that in this transformation we only need $n-1$ new variables, not n, because $\sum_{i=1}^{n} \omega_i = 1$ and that one ω_i is redundant. Thus,

$$\underset{\sim}{\gamma}_{n-1} = B'_{n-1,n} \underset{\sim}{\nu} \quad \text{and} \quad B'_{n-1,n} = (I_{n-1} - \underset{\sim}{\omega}_{(n)} \underset{\sim}{J}'_{n-1} : -\underset{\sim}{\omega}_{(n)}),$$

where I_{n-1}, $\underset{\sim}{J}_{n-1}$ and $\underset{\sim}{\omega}_{(n)}$ are respectively the identity matrix, a vector of ones, and a vector of $\omega_1, \ldots, \omega_{n-1}$ (i.e., ω_n is excluded).

Then, making the transportation $\underset{\sim}{\gamma} = B'\underset{\sim}{\nu}$ in (9) we have

$$f(\underset{\sim}{\gamma}, y_s | p, \beta_0, \sigma_e^2) = \frac{\frac{1}{|2\pi\sigma_e^2 B'B|^{1/2}} e^{-\frac{1}{2}[\underset{\sim}{\gamma} - B'(\beta_0 \underset{\sim}{1} + y_s)]'(B'B\sigma_e^2)^{-1}[\underset{\sim}{\gamma} - B'(\beta_0 \underset{\sim}{1} + y_s)]}}{\left[p\Phi\left(\frac{\beta_0 + 1}{\sigma_e}\right) + (1-p)\Phi\left(\frac{\beta_0}{\sigma_e}\right)\right]^n} p^t(1-p)^{n-t}, \tag{10}$$

where $\underset{\sim}{1}$ is a vector of ones. Note that $\nu_i = \omega_i \sum_{i'=1}^{n} \nu_{i'}, i = 1, \ldots, n-1$ is the same as $\underset{\sim}{\gamma} = \underset{\sim}{0}$. Also,

note that $\sum_{i=1}^{n-1} \nu_i = \left(\sum_{i=1}^{n-1} \omega_i\right) \sum_{i=1}^{n} \nu_i = (1 - \omega_n) \sum_{i=1}^{n} \nu_i = \sum_{i=1}^{n} \nu_i - \omega_n \sum_{i=1}^{n} \nu_i$; so that $\nu_n = \omega_n \sum_{i=1}^{n} \nu_i$ as well.

Thus, the probability function of interest is obtained by taking $\underset{\sim}{\gamma} = \underset{\sim}{0}$ in (10), and

$$f(\underset{\sim}{\gamma} = 0, \underset{\sim}{y}_s \,|\, p, \beta_0, \sigma_e^2) = \frac{\frac{1}{|2\pi\sigma_s^2 B'B|^{1/2}} e^{-\frac{1}{2\sigma_e^2}(\beta_0 \underset{\sim}{1} + \underset{\sim}{y}_s)' B(B'B)^{-1} B'(\beta_0 \underset{\sim}{1} + \underset{\sim}{y}_s)}}{\left[p\Phi\left(\frac{\beta_0+1}{\sigma_e}\right) + (1-p)\Phi\left(\frac{\beta_0}{\sigma_e}\right)\right]^n} p^t (1-p)^{n-t}. \qquad (11)$$

Note that the first term in (11) is the adjustment for selection bias, and in particular, it depends on $\underset{\sim}{y}_s$ as well as p, β_0, σ_e. One might think of (11) as the likelihood function corrected for selection bias. It is interesting to note that if β_0 is much bigger than unity, the ratio (i.e., the weight function) in front of $p^t(1-p)^{n-t}$ in (11) will not depend on p, and in a Bayesian analysis p cannot be corrected for selection bias.

There is a striking resemblance between (1) and (11). In Malec, Davis and Cao (1999) the adjustment for the selection bias is $1/(pW_1^{-1} + (1-p)W_0^{-1})^n$ and in (11) the adjustment is

$\frac{1}{|2\pi\sigma_e^2 B'B|^{1/2}} e^{-\frac{1}{2\sigma_e^2}(\beta_0 \underset{\sim}{1} + \underset{\sim}{y}_s)' B(B'B)^{-1} B'(\beta_0 \underset{\sim}{1} + \underset{\sim}{y}_s)} \Big/ \left[p\Phi\left(\frac{\beta_0+1}{\sigma_e}\right) + (1-p)\Phi\left(\frac{\beta_0}{\sigma_e}\right)\right]^n$. So that $\Phi\left(\frac{\beta_0+1}{\sigma_e}\right)$ and $\Phi\left(\frac{\beta_0}{\sigma_e}\right)$ replace W_0^{-1} and W_1^{-1}, respectively. However, while there is a single parameter p in (1), there are two nuisance parameters (β_0, σ_e^2) in (11) as well.

By Bayes' theorem, the joint posterior density of p, β_0, σ_e^2, denoted by $\pi(p, \beta_0, \sigma_e^2 | \underset{\sim}{y}_s)$ under selection bias, is

$$\pi(p, \beta_0, \sigma_e^2 | \underset{\sim}{y}_s) \propto \frac{(\sigma_e^2)^{-\frac{n+1}{2}} e^{-\frac{1}{2\sigma_e^2}(\beta_0 \underset{\sim}{1} + \underset{\sim}{y}_s)' B(B'B)^{-1} B'(\beta_0 \underset{\sim}{1} + \underset{\sim}{y}_s)}}{\left[p\Phi\left(\frac{\beta_0+1}{\sigma_e}\right) + (1-p)\Phi\left(\frac{\beta_0}{\sigma_e}\right)\right]^n} p^t (1-p)^{n-t}, \qquad (12)$$

$0 < p < 1, \beta_0/\sigma_e \geq \mu_0, \sigma_e^2 > 0$. We choose to make inference about p by drawing samples from the joint posterior density $\pi(p, \beta_0, \sigma_e^2 | \underset{\sim}{y}_s)$. These samples are also used to make inference about the finite population proportion via surrogate sampling.

In Appendix A we provide a simplification of (12), which we state explicitly here for convenience. We show in (A.1) of Appendix A that the joint posterior density is

$$\pi(p, \beta_0, \sigma_e^2 | \underset{\sim}{y}_s) \propto (\sigma_e^2)^{-\frac{n+1}{2}} \exp\left\{-\frac{1}{2\sigma_e^2}\left[\left(n - 1\Big/\sum_{i=1}^{n} \omega_i^2\right)(\beta_0 + \overline{y}_\omega)^2 + S_\omega^2\right]\right\}$$

$$\times \frac{p^t(1-p)^{n-t}}{\left[p\Phi\left(\frac{\beta_0+1}{\sigma_e}\right) + (1-p)\Phi\left(\frac{\beta_0}{\sigma_e}\right)\right]^n}, \quad 0 < p < 1,\ \beta_0/\sigma_e \geq \mu_0,\ \sigma_e^2 > 0; \qquad (13)$$

$\overline{y}_\omega$ and S_ω^2 are defined in Appendix A. In Appendix B we also show that the posterior density, $\pi(p, \beta_0, \sigma_e^2 | \underset{\sim}{y}_s)$ in (13) is proper.

4. Requisite Computation for the Superpopulation Proportion

To perform the computations, one can use numerical integration or a sampling-based method. Because we want to draw a sample from the posterior density of p, we use a sampling based method. However, to carry out a sampling based method, one would need some kind of rejection sampling. The Metropolis-Hastings sampler can be used, but this will need a Metropolis step for each conditional posterior density, and each step has to be tuned; this is somewhat inefficient. Thus, we perform a sampling-based method without using Markov chains. This is attractive because we will simply subsample a random sample from an approximate joint posterior density.

First, it is convenient to write (13) as

$$\pi(p, \beta_0, \sigma_e^2 | y_s) = \pi_1(\beta_0, \sigma_e^2 | y_s)\pi_2(p | \beta_0, \sigma_e^2, y_s), \tag{14}$$

where

$$\pi_1(\beta_0, \sigma_e^2 | y_s) = \pi_{1a}(\beta_0, \sigma_e^2 | y_s)R(\beta_0, \sigma_e^2; y_s),$$

$$\pi_{1a}(\beta_0, \sigma_e^2 | y_s) \propto (\sigma_e^2)^{-\frac{n+1}{2}} \exp\left\{ -\frac{1}{2\sigma_e^2}\left[\left(n - 1/\sum_{i=1}^{n}\omega_i^2\right)(\beta_0 + y_\omega)^2 + S_w^2 \right] \right\}, \ \beta_0/\sigma_e > \mu_0, \ \sigma_e^2 > 0,$$

$$R(\beta_0, \sigma_e^2; y_s) = \int_0^1 \frac{p^t(1-p)^{n-t}}{\left[p\Phi\left(\frac{\beta_0+1}{\sigma_e}\right) + (1-p)\Phi\left(\frac{\beta_0}{\sigma_e}\right) \right]^n} dp,$$

$$\pi_2(p | \beta_0, \sigma_e^2, y_s) \propto \frac{p^t(1-p)^{n-t}}{\left[p\Phi\left(\frac{\beta_0+1}{\sigma_e}\right) + (1-p)\Phi\left(\frac{\beta_0}{\sigma_e}\right) \right]^n}, 0 < p < 1.$$

In Appendix C we show how to draw a sample $\left\{ \left(\beta_0^{(h)}, \sigma_e^{2(h)}\right), h = 1, \ldots, M \right\}$ of size M from $\pi_{1a}(\beta_0, \sigma_e^2 | y_s)$. Then, we use the sampling importance resampling (SIR) algorithm to draw a sample with probability proportional to weights (i.e. weighted bootstrap); see Smith and Gelfand (1992) and Nandram and Erhardt (2005).

We now describe how to compute $R(\beta_0, \sigma_e^2; y_s)$. First, it is easy to show that

$$R(\beta_0, \sigma_e^2; y_s) = \frac{B(t, n-t)}{\{\Phi((\beta_0+1)/\sigma_e)\}^t \{\Phi(\beta_0/\sigma_e)\}^{n-t}} I(\beta_0, \sigma_e^2; t),$$

where

$$I(\beta_0, \sigma_e^2; t) = \int_0^1 A(q; \beta_0, \sigma_e^2)dF(q),$$

$$A(q; \beta_0, \sigma_e^2) = q\Phi(\beta_0/\sigma_e)(1-q)\Phi((\beta_0+1)/\sigma_e) / \{q\Phi(\beta_0/\sigma_e) + (1-q)\Phi((\beta_0+1)/\sigma_e)\}^2,$$

$$F(q) = \int_0^q \frac{x^{t-1}(1-x)^{n-t-1}}{B(t, n-t)} dx.$$

It follows that a simple efficient estimator of $I(\beta_0, \sigma_e^2; t)$ is

$$\hat{I}(\beta_0, \sigma_e^2; t) = \sum_{k=1}^{G} A\left\{(q_{k-1} + q_k)/2; \beta_0, \sigma_e^2\right\} \left\{F(q_k) - F(q_{k-1})\right\}$$

with $0 \equiv q_0 < q_1 < \ldots < q_G \equiv 1$ for $G - 1$ points in $(0, 1)$.

We subsample $\left\{(\beta_0^{(h)}, \sigma_e^{2(h)}), h = 1, \ldots, M\right\}$ with replacement; we found similar results using sampling without replacement. The subsampling probabilities, $\tilde{\omega}_h$, are proportional to $R(\beta_0, \sigma_e^2; y_s)$,

$$\tilde{\omega}_h \propto R\left(\beta_0^{(h)}, \sigma_e^{2(h)}; y_s\right), \quad h = 1, \ldots, M.$$

This procedure is efficient if $\pi_{1a}\left(\beta_0, \sigma_e^2 | y_s\right)$ is a good approximation to $\pi_1\left(\beta_0, \sigma_e^2 | y_s\right)$ (i.e., the $\tilde{\omega}_h$ are very close to M^{-1}). Our procedure satisfies this restriction; see Nandram and Erhardt (2005) on how to tune the algorithm.

We can draw a sample from $\pi_2(p | \beta_0, \sigma_e^2, y_s)$ using a rejection procedure. We can consider

$$\pi_2(p | \beta_0, \sigma_e^2, y_s) \propto \left[\frac{\Phi\left(\frac{\beta_0}{\sigma_e}\right)}{p\Phi\left(\frac{\beta_0+1}{\sigma_e}\right) + (1 - p)\Phi\left(\frac{\beta_0}{\sigma_e}\right)}\right]^n p^t (1 - p)^{n-t}, 0 < p < 1, \tag{15}$$

a weighted distribution. We can draw $p | \beta_0, \sigma_e^2, y_s \sim \text{Beta}\,(t + 1, n - t + 1)$ and accept it with probability $\left[\Phi\left(\frac{\beta_0}{\sigma_e}\right) / \left\{p\Phi\left(\frac{\beta_0+1}{\sigma_e}\right) + (1 - p)\Phi\left(\frac{\beta_0}{\sigma_e}\right)\right\}\right]^n$. But this probability can be very small making this rejection sampling procedure inefficient. However, in Appendix C we show how to improve its efficiency.

A random sample has now been obtained from $\pi(p, \beta_0, \sigma_e^2 | y_s)$. Our posterior density of primary interest is $\pi(p | y_s)$, and so we have a sample $p^{(1)}, \ldots, p^{(M)}$ from $\pi(p | y_s)$. Thus, inference about p is made in the usual empirical manner using $p^{(1)}, \ldots, p^{(M)}$.

5. Inference about the Finite Population Proportion

It is possible to make inference about the finite population proportion P using the census model. But, note again that the biased sample originally obtained is not a random sample from the census model. Thus, our objective is to construct surrogate random samples from the census model, and then conditional on these samples, inference can be made about the finite population proportion.

We now consider the finite population values to be $z_1, \ldots, z_n, z_{n+1}, \ldots, z_N$; the first n values being the random sample, and the last $N - n$ values are to be predicted.

We have shown how to obtain the posterior distribution $\pi(p | y_s)$, which we now use as a prior distribution in the census model to obtain a prior predictive distribution of $z_1, \ldots, z_n$. Here,

$$p(\underset{\sim}{z_s}) \int_0^1 p(\underset{\sim}{z_s} | p) \pi(p | y_s) dp. \tag{16}$$

Thus, it is easy to draw samples from $p(\underset{\sim}{z_s})$, and in fact, because $p(\underset{\sim}{z_s})$ is discrete, we have at our convenience the entire distribution of $\underset{\sim}{z_s}$. Any random sample from $p(\underset{\sim}{z_s})$ will replace the biased sample. This sample is a surrogate sample from the finite population, and is representative of the finite population as long as the model fits (the model is almost nonparametric).

Our sampling-based algorithm provides a sample, $p^{(h)}, h = 1, \ldots, M$, from the "prior" $\pi(p|y_s)$.

Thus, it is easy to construct a sample of the vector $\underset{\sim}{z}_s$ say $\underset{\sim}{z}_s^{(h)}, h = 1, \ldots, M$, with $\underset{\sim}{z}_s^{(h)}$ corresponding to $p^{(h)}$. That is, we can obtain M surrogate samples $\underset{\sim}{z}_s^{(h)}$ from $p(\underset{\sim}{z}_s)$. We now set $\overline{z}_s^{(h)} = n^{-1} \sum_{i=1}^{n} z_i^{(h)}, h = 1, \ldots, M$. Finally, we construct

$$\overline{z}_s = M^{-1} \sum_{h=1}^{M} \overline{z}_s^{(h)}$$

for $M = 1000$ iterates from the sampling-based method. Now, letting $n' = [n\overline{z}_s]$, where $[t]$ is the smallest integer larger than t, our surrogate sample $\underset{\sim}{z}_s$ has n' ones and $n - n'$ zeros for its components.

In fact, we only need $\sum_{i=1}^{n} z_i$ because this is a sufficient statistic in the census model. Thus, we can now fit the model

$$z_1, \ldots, z_N | p \overset{\text{i.i.d.}}{\sim} \text{Bernoulli } (p), p \sim \text{Uniform } (0, 1),$$

where $z_1, \ldots, z_n$, or equivalently $\sum_{i=1}^{n} z_i$, are observed data.

Inference can now be made about P as in the case without selection bias. The procedure can be performed for several random samples, and an average position can be taken.

6.   Simulation Study

We have performed a simulation study to assess our estimators of the finite population proportion under ignorability and nonignorability with respect to sample size and variance.

We kept the population size fixed at $N = 100, p = .25, \beta_0 = 1$ and $\beta_1 = 1$. Then we generated 1000 datasets at $n = 10, 20, 30, 40$ and $\sigma_e^2 = 1, 10, 25$. Thus, there are twelve design points. Our population consists of $N = 100$ independent values $y_1, \ldots, y_N$ from the Bernoulli distribution with parameter $p = .25$. The selection probabilities $\pi_i, i = 1, \ldots, N$, were obtained by drawing N independent normal random variables $\nu_1, \ldots, \nu_N$ with mean $\beta_0 + \beta_1 y_i, i = 1, \ldots, N$, truncated to be nonnegative, and then $\pi_i = n\nu_i / \sum_{i=1}^{N} \nu_i$. Then, finally a PPS sample of size n is taken. This produces the biased sample.

For each of the 1000 datasets we calculated the finite population proportion $P = N^{-1} \sum_{i=1}^{N} y_i$ (i.e., P is known). Based on each of the 1000 samples we have calculated the Horvitz-Thompson estimator $\text{HTE} = N^{-1} \sum_{i=1}^{n} y_i / \pi_i$. The HTE should perform well under the conditions of our simulations (e.g., see Little and Rubin 2002), but note that β_0 is not necessarily 0.

We then fit the ignorable (ig) and nonignorable (nig) selection models to each of the 1000 datasets. Note that the weights in the SIR algorithm are approximately .001, showing good performance. We calculated HTE, the posterior (PM), posterior standard deviation (PSD) and the 95% credible interval for P. We studied the ratios

$$R_1 = \text{HTE}/P, \ R_2 = \text{PM}_{\text{ig}}/P, \ R_3 = \text{PM}_{\text{nig}}/P.$$

We also studied C_{ig} and C_{nig}, the proportion of the 1000 95% credible intervals containing P, and calculated the widths of these intervals W_{ig} and W_{nig}. In Table 1 we have reported the averages of these quantities over the 1000 datasets.

Table 1. **Simulation study: Comparisons of the Horvitz-Thompson estimator (HTE), the ignorable and nonignorable selection models using the posterior distributions of the finite population proportion (P) by sample size (n) and variance (σ_e^2)**

		Point Estimator			Interval Estimator			
n	σ_e^2	R_1	R_2	R_3	C_{ig}	C_{nig}	W_{ig}	W_{nig}
10	1	0.973	1.416	1.400	0.916	0.941	0.458	0.471
	10	1.029	1.266	1.297	0.961	0.971	0.446	0.462
	25	1.369	1.250	1.286	0.970	0.978	0.444	0.462
20	1	0.994	1.423	1.283	0.822	0.966	0.342	0.350
	10	0.952	1.173	1.098	0.939	0.987	0.325	0.328
	25	0.986	1.127	1.061	0.945	0.978	0.323	0.324
30	1	0.991	1.414	1.255	0.719	0.894	0.278	0.260
	10	0.991	1.147	1.045	0.934	0.969	0.263	0.246
	25	1.003	1.100	1.005	0.950	0.971	0.260	0.243
40	1	0.990	1.371	1.193	0.652	0.905	0.223	0.214
	10	0.975	1.118	0.993	0.950	0.974	0.210	0.199
	25	1.012	1.099	0.975	0.954	0.974	0.209	0.198

NOTE: $R_1 = \text{HTE}/P$, $R_2 = P_{\text{ig}}/P$, $R_3 = P_{\text{nig}}/P$ where P_{ig} and P_{nig} are the posterior means of P under the ignorable and nonignorable selection models, respectively; C_{ig} and C_{nig} are, respectively, the probability contents of the 95% credible intervals for P under the ignorable and the nonignorable selection models; W_{ig} and W_{nig} are, respectively, the widths of the 95% credible intervals for P under the ignorable and the nonignorable selection models; P_{ig}, P_{nig}, W_{ig} and W_{nig} are averages based on 1000 simulated experiments at each design point, and C_{ig} and C_{nig} are the proportions of the 1000 95% credible intervals containing P.

Except for one design point $(n = 10, \sigma_e^2 = 25)$ HTE works as it should. R_2 and R_3 go down with n and σ_e^2. (Recall that smaller σ_e^2 means larger selection bias.) As expected, except at $n = 10$ the nonignorable selection model corrects for the selection bias, and is to be preferred over the ignorable selection model; even at $n = 20$ the nonignorable selection model does a better job. Next, we consider the intervals. Except for a few design points, the 95% credible intervals under the nonignorable selection model are conservative, and those under the ignorable selection model can be much smaller than the nominal value of 95%. The widths of the intervals are comparable.

Thus, the nonignorable selection model is potentially useful to correct for the bias. Of course, HTE works well under these situations. But one cannot calculate the 95% confidence interval for P because we need the second order inclusion probabilities. These are not available in our framework. However, one possibility is $s/\sqrt{n}$, where $s^2 = \sum_{i=1}^{n} (y_i/\pi_i - a^2)/(n - 1)$ with $a = n^{-1}\sum_{i=1}^{n} y_i/\pi$, but we prefer not to do so.

7. Illustrative Example on the National Health Interview Survey

The National Health Interview Survey (NHIS) has been conducted every year by the National Center for Health Statistics (NCHS) to measure an aspect of health status of the U.S. population (Adams and Marano, 1995). Through this sample survey, NCHS conducts surveys on chronic and acute conditions, doctor visits, hospital episodes, disability, house-hold and personal information, and other special aspects of health of the U.S. population. We use data on individuals with chronic conditions from the 1996 NHIS to illustrate our methodology.

The NHIS frame is basically a two-stage stratified sample survey design of probability proportional to population size. The first stage is the selection of primary sampling units within strata, and the second stage is the selection of segments. On the average each segment includes about 4-12 households, and all the sample households in each segment are interviewed. While the primary sampling units (psu's) in the NHIS design are drawn with unequal probabilities (large psu's are drawn with certainty), within the psu's the segments are drawn with equal probabilities. We incorporate this design feature into our model because the selection probabilities may be related to the characteristic measured.

One of the variables we use in the NHIS is activity limitation status (ALS) which is categorized as "unable to perform major activity" $(y = 1)$ versus not so severely limited $(y = 0)$ for each member of a household. We are interested in the proportion of people in the entire finite population with severe limitation. There are twelve domains formed by crossing education (1-8 years, 9-11 years, college), sex (male, female) and race (white, non-white); we have considered only individuals who are younger than 54 years. We have the sample weights for each individual; so that we have the size of each domain (i.e., the sum of the sample weights). The selection probabilities are the reciprocal of the sample weights.

In Table 2 we present the data by domain. Specifically, we present the population size (N), sample size (n), a measure of selection bias (SB) and the Horvitz-Thompson estimator (HTE). Here, $\text{SB} = (\overline{\pi}_0 - \overline{\pi}_1)/\overline{\pi}_1$ where $\overline{\pi}_0$ and $\overline{\pi}_1$ are the averages of the π_i corresponding to $y_i = 0$ and $y_i = 1$ respectively. Larger values of SB show larger selection bias; Domains 1 and 3 have relatively large selection bias. Note that the domains are very large making the superpopulation proportion virtually the same as the finite population proportion. However, we have studied the finite population proportion P.

Table 2.   **1996 NHIS data on the population size (N), sample size (n), a measure of selection bias (SB) and the Horvitz-Thompson estimator (HTE) by domain**

Domain	N	n	SB	HTE
1	2.532	923	.333	.149
2	0.465	116	.046	.296
3	2.475	993	.333	.123
4	0.478	129	−.024	.172
5	23.878	5552	−.009	.059
6	4.903	1179	−.053	.072
7	23.914	5853	−.001	.042
8	5.507	1497	−.035	.056
9	27.645	5914	−.109	.017
10	4.613	1059	−.043	.025
11	28.271	6245	−.040	.023
12	5.719	1459	−.054	.024

NOTE: The domains are formed by crossing education (elementary, high, at least college), sex (male, female) and race (white, non-white). $N \times 10^6$ is the finite population size and n is the sample size. $\text{SB} = (\overline{\pi}_0 - \overline{\pi}_1)/\overline{\pi}_1$ where $\overline{\pi}_0$ and $\overline{\pi}_1$ are the averages of the π_i corresponding to $y_i = 0$ and $y_i = 1$, respectively; HTE is the Horvitz-Thompson estimator of P.

In Table 3 we present the characteristics of the finite population mean by domain. We note that the weights in the SIR algorithm for all twelve domains are between .0004 and .004, showing good performance. The Horvitz-Thompson estimator (HTE) differs from the estimates of the ignorable and the nonignorable selection models in these domains; see Table 2. HTE is to the right of the 95% credible interval of P in both domains. The PM's for the ignorable selection model are larger than those for the nonignorable selection model, but the PSD's from the ignorable selection model are similar to those from the nonignorable selection model. Thus, the 95% credible intervals for the nonignorable selection model overlap on the left of those for the ignorable selection model.

Table 3. **1996 NHIS Example: Comparisons of the ignorable and nonignorable selection models using the posterior distributions of the finite population proportion (P) by domain**

Domain	Ignorable Model			Nonignorable Model		
	PM	PSD	I	PM	PSD	I
1	.112	.010	(.092, .133)	.097	.009	(.079, .117)
2	.314	.042	(.233, .395)	.292	.041	(.210, .374)
3	.090	.009	(.073, .109)	.081	.009	(.065, .099)
4	.175	.032	(.116, .242)	.158	.031	(.097, .223)
5	.060	.003	(.055, .066)	.054	.003	(.048, .060)
6	.095	.008	(.080, .112)	.085	.008	(.070, .101)
7	.044	.003	(.039, .050)	.037	.002	(.033, .042)
8	.064	.006	(.052, .077)	.055	.006	(.044, .067)
9	.019	.002	(.016, .023)	.016	.002	(.013, .019)
10	.035	.006	(.025, .046)	.028	.005	(.019, .039)
11	.024	.002	(.021, .028)	.019	.002	(.016, .023)
12	.029	.005	(.021, .039)	.023	.004	(.017, .032)

NOTE: See note to Table 2. Here, PM, PSD and I are, respectively, the posterior mean, posterior standard deviation and 95% credible interval of P.

8. Conclusion

We have described a Bayesian method to study the finite population proportion when the sample is biased. The PPS sampling scheme is considered a simple random sample perturbed in which the perturbation process is understood. Our method eliminates the perturbation to produce a representative surrogate sample from the finite population.

Our method is very simple, and can produce many samples from the finite population quickly. The computation is sampling-based, but it uses independent samples rather than Markov chains. It can perform better than the Horvitz-Thompson estimator, a randomization-based estimator. There are substantial differences between the ignorable selection model, which assumes that the original sample is random, and our nonignorable selection model.

Our illustrative example on the NHIS and the simulation study show that our method is to be preferred.

Acknowledgement

The author is grateful to Dr. Myron Katzoff, National Center for Health Statistics (CDC, USA), for a very careful reading of an earlier version of the manuscript.

References

Adams, P.F. and Marano, M.A. (1995). Current Estimate from the National Health Interview Survey, 1994. Series 10, Report No. 193, National Center for Health Statistics, CDC.

Burgos, C.G.B. and Nandram, B. (2003). Bayesian Predictive Inference and Selection Bias. *Journal of Research in Science and Engineering*, **1**, 1-10.

Chambers, R., Dorfman, A. and Wang, S. (1998). Limited Information Likelihood Analysis of Survey Data. *Journal of the Royal Statistical Society*, Series B, 60, 397-411.

Devroye, L. (1986). *Non-Uniform Random Variate Generation*, New York. Springer-Verlag.

Little, R.J.A. and Rubin, D.B. (2002), *Statistical Analysis with Missing* Data. New York: John Wiley & Sons.

Nandram, B. and Erhardt, E. (2005). Fitting Bayesian Two-Stage Generalized Linear Models Using Random Samples via the SIR Algorithm. *Sankhya*, B, **66**, 733-755.

Nandram, B. and Sedransk, J. (2005). Bayesian Predictive Inference for Finite Population Quantities Under Informative Sampling. *Technical Report*, Worcester Polytechnic Institute.

Malec, D., Davis, W. and Cao, X. (1999). Model-based Small Area Estimates of Overweight Prevalence Using Sample Selection Adjustment. *Statistics in Medicine*, **18**, 3189-3200.

Patil, G.P. and Rao, C.R. (1978). Weighted Distributions and Size-Biased Sampling With Applications to Wildlife Populations and Human Families. *Biometrics*, **34**, 179-189.

Pfeffermann, D., Krieger, A.M. and Rinott, Y. (1998). Parametric Distributions of Complex Survey Data under Informative Probability Sampling. *Statistica Sinica*, **8**, 1087-1114.

Pfeffermann, D. and Sverchkov, M. (1999). Parametric and Semi-parametric Estimation of Regression Models Fitted to Survey Data. *Sankhya*, B, **61**, 1-21.

Silliman, N.P. (1997). Hierarchical Selection Models With Applications in Meta-Analysis. *Journal of the American Statistical Association*, **92**, 926-936.

Smith, A.F.M. and Gelfand, A.E. (1992). Bayesian Statistics Without Tears: A Sampling-Resampling Perspective. *The American Statistician*, **46**, 84-88.

Sverchkov, M. and Pfeffermann, D. (2004). Prediction of Finite Population Totals Based on the Sample Distribution. *Survey Methodology*, **30**, 79-92.

Appendix A: Simplification of the Joint Posterior Density

We show that

$$Q = (\beta_0 \underset{\sim}{1} + y_s)'B(B'B)^{-1}B'(\beta_0 \underset{\sim}{1} + y_s) = \left(n - 1/\sum_{i=1}^{n}\omega_i^2\right)(\beta_0 + \overline{y}_\omega)^2 + S_\omega^2, \qquad (A.1)$$

$$\overline{y}_\omega = \sum_{i=1}^{n}\left(1 - \omega_i/\sum_{i=1}^{n}\omega_i^2\right)y_i/\sum_{i=1}^{n}\left(1 - \omega_i/\sum_{i=1}^{n}\omega_i^2\right),$$

$$S_\omega^2 = t - \left\{\left(\sum_{i=1}^{n}\omega_i y_i\right)^2 + \left(n\sum_{i=1}^{n}\omega_i^2 - 1\right)\overline{y}_\omega^2\right\}/\sum_{i=1}^{n}\omega_i^2,$$

where $t = \sum_{i=1}^{n}y_i$. Let $\underset{\sim}{\omega} = (\omega_1,\ldots,\omega_n)$, $\underset{\sim}{\omega}_{(n)} = (\omega_1,\ldots,\omega_{n-1})$, I_{n-1} be an $(n-1)\times(n-1)$ identity matrix, $\underset{\sim}{1}_{n-1}$ be a $(n-1)$ vector of ones and $J_{n-1} = \underset{\sim}{1}_{n-1}\underset{\sim}{1}'_{n-1}$.

Because $B' = (I_{n-1} - \underset{\sim}{\omega}_{(n)} \underset{\sim}{1}'_{n-1} : -\underset{\sim}{\omega}_{(n)})$, it is easy to show that

$$B'B = I_{n-1} - \frac{J_{n-1}}{n} + \frac{(\underset{\sim}{1}_{n-1} - n\omega_{(n)})(\underset{\sim}{1}_{n-1} - n\omega_{(n)})'}{n}.$$

Now setting $A^{-1} = I_{n-1} - \frac{J_{n-1}}{n}$ and $\underset{\sim}{u} = \frac{\underset{\sim}{1}_{n-1} - n\underset{\sim}{\omega}_{(n)}}{\sqrt{n}}$ we can use the identity $(A^{-1} + \underset{\sim}{u}\,\underset{\sim}{u}')^{-1} = A - A\underset{\sim}{u}\,\underset{\sim}{u}'A/(1 + \underset{\sim}{u}'A\underset{\sim}{u})$ twice to get $A = I_{n-1} + J_{n-1}$ and

$$(B'B)^{-1} = I_{n-1} + J_{n-1} - \frac{(\omega_n \underset{\sim}{1}_{n-1} - \underset{\sim}{\omega}_{(n)})(\omega_n \underset{\sim}{1}_{n-1} - \underset{\sim}{\omega}_{(n)})'}{\underset{\sim}{\omega}'\underset{\sim}{\omega}}.$$

Next, we compute $B(B'B)^{-1}B'$. First, note that $B' = (D : -\underset{\sim}{\omega}_{(n)})$ and $(B'B)^{-1} = A - \underset{\sim}{a}\,\underset{\sim}{a}'$ where $D = I_{n-1} - \underset{\sim}{\omega}_{(n)} \underset{\sim}{1}'_{n-1}$ and $\underset{\sim}{a} = (\omega_n \underset{\sim}{1}_{n-1} - \underset{\sim}{\omega}_{(n)})/\sqrt{\underset{\sim}{\omega}'\underset{\sim}{\omega}}$. It follows that

$$B(B'B)^{-1}B' = \begin{bmatrix} D'AD - D'\underset{\sim}{a}\,\underset{\sim}{a}'D, & -(D'A\underset{\sim}{\omega}_{(n)} - D'\underset{\sim}{a}\,\underset{\sim}{a}'\underset{\sim}{\omega}_{(n)}) \\ -(\underset{\sim}{\omega}'_{(n)}AD - \underset{\sim}{\omega}'_{(n)}\underset{\sim}{a}\,\underset{\sim}{a}'D), & \underset{\sim}{\omega}'_{(n)}A\underset{\sim}{\omega}_{(n)} - \underset{\sim}{\omega}'_{(n)}\underset{\sim}{a}\,\underset{\sim}{a}'\underset{\sim}{\omega}_{(n)} \end{bmatrix}.$$

It is worth noting that the columns of B are orthogonal to $\underset{\sim}{\omega}$ because $B'\underset{\sim}{\omega} = \underset{\sim}{\omega}_{(n)} - \underset{\sim}{\omega}_{(n)}(1 - \omega_n) - \omega_n \underset{\sim}{\omega}_{(n)} = 0$. Then, it is easy to show that

$$D'AD = I_{n-1} - 2\underset{\sim}{\omega}_{(n)} \underset{\sim}{1}'_{n-1} + \underset{\sim}{\omega}'\underset{\sim}{\omega} J_{n-1},$$

$$D'\underset{\sim}{a}\,\underset{\sim}{a}D = S_\omega J_{n-1} - 2\underset{\sim}{\omega}_{(n)} \underset{\sim}{1}'_{n-1} + \frac{\underset{\sim}{\omega}_{(n)} \underset{\sim}{\omega}'_{(n)}}{\underset{\sim}{\omega}'\underset{\sim}{\omega}},$$

$$\underset{\sim}{\omega}'_{(n)}A\underset{\sim}{\omega}_{(n)} = \underset{\sim}{\omega}'_{(n)} \underset{\sim}{\omega}_{(n)} + (1 - \omega_n)^2,$$

$$\underset{\sim}{\omega}'_{(n)}\underset{\sim}{a}\,\underset{\sim}{a}'\underset{\sim}{\omega}_{(n)} = \frac{\omega_n^2}{\underset{\sim}{\omega}'\underset{\sim}{\omega}} - 2\omega_n + \underset{\sim}{\omega}'\underset{\sim}{\omega},$$

$$D'A\underset{\sim}{\omega}_{(n)} = \underset{\sim}{\omega}_{(n)} - \underset{\sim}{\omega}'\underset{\sim}{\omega} \underset{\sim}{1}_{n-1} + \omega_n \underset{\sim}{1}_{n-1},$$

$$D'\underset{\sim}{a}\,\underset{\sim}{a}'\underset{\sim}{\omega}_{(n)} = \omega_n \underset{\sim}{1}_{n-1} - \frac{\omega_n \underset{\sim}{\omega}_{(n)}}{\underset{\sim}{\omega}'\underset{\sim}{\omega}} - \underset{\sim}{\omega}'\underset{\sim}{\omega} \underset{\sim}{1}_{n-1} + \underset{\sim}{\omega}_{(n)}. \tag{A.2}$$

Using the results in (A.2), it follows that

$$B(B'B)^{-1}B' = I_n - \underset{\sim}{\omega}\,\underset{\sim}{\omega}'/\underset{\sim}{\omega}'\underset{\sim}{\omega}. \tag{A.3}$$

Then, letting $C = B(B'B)^{-1}B'$,

$$Q = (\beta_0 \underset{\sim}{1} + \underset{\sim}{y}_s)'C(\beta_0 \underset{\sim}{1} + \underset{\sim}{y}_s).$$

First, note that

$$Q = Q_1 + Q_2,$$

where $Q_1 = \underset{\sim}{1}'C\underset{\sim}{1}\left[\beta_0^2 + 2\left(\dfrac{\underset{\sim}{1}'C\underset{\sim}{y}_s}{\underset{\sim}{1}'C\underset{\sim}{1}}\right)\beta_0 + \left[\dfrac{\underset{\sim}{1}'C\underset{\sim}{y}_s}{\underset{\sim}{1}'C\underset{\sim}{1}}\right]^2\right]$ and $Q_2 = \underset{\sim}{y}_s'C\underset{\sim}{y}_s - (\underset{\sim}{1}'C\underset{\sim}{y}_s)^2/\underset{\sim}{1}'C\underset{\sim}{1}$. Note

also that $Q_1 \geq 0$, and by the Cauchy-Schwarz inequality $Q_2 \geq 0$ with equality if for $\underset{\sim}{y}_s$ all the components are ones or all the components are zeros. It is easy to show that

$$\underset{\sim}{1}'C = \underset{\sim}{1}' - \underset{\sim}{\omega}'/\underset{\sim}{\omega}'\underset{\sim}{\omega}'. \tag{A.4}$$

It follows from (A.4) that

$$\underset{\sim}{1}'C\underset{\sim}{1} = \sum_{i=1}^{n}(1 - \omega_i/\underset{\sim}{\omega}'\underset{\sim}{\omega}) = n - 1/\underset{\sim}{\omega}'\underset{\sim}{\omega},$$

$$\underset{\sim}{1}'C\underset{\sim}{y}_s = \sum_{i=1}^{n}(1 - \omega_i/\underset{\sim}{\omega}'\underset{\sim}{\omega})y_i \text{ and } \underset{\sim}{y}_s'C\underset{\sim}{y}_s = \sum_{i=1}^{n}y_i^2 - \left(\sum_{i=1}^{n}\omega_iy_i\right)^2/\underset{\sim}{\omega}'\underset{\sim}{\omega}. \tag{A.5}$$

Then,

$$Q_1 = (n - 1/\underset{\sim}{\omega}'\underset{\sim}{\omega})\left\{\beta_0 + \overline{y}_\omega\right\}^2,$$

and using (A.4), (A.5) and $y_i^2 = y_i$, $i = 1,\ldots,n$ ($y_i = 0$ or 1), it is easy to show that

$$Q_2 = t - \left\{\left(\sum_{i=1}^{n}\omega_iy_i\right)^2 + (n\underset{\sim}{\omega}'\underset{\sim}{\omega} - 1)\overline{y}_w^2\right\}/\underset{\sim}{\omega}'\underset{\sim}{\omega} = S_\omega^2.$$

Appendix B: Propriety of the Joint Posterior Density

We need to show that

$$J = \int_0^\infty \int_{\mu_0\sigma_e}^\infty \int_0^1 \left(\frac{1}{\sigma_e^2}\right)^{\frac{n}{2}} \exp\left\{-\frac{1}{2\sigma_e^2}\left[\left(n - 1/\sum_{i=1}^{n}\omega_i^2\right)(\beta_0 + \overline{y}_\omega)^2 + S_\omega^2\right]\right\}$$

$$\times \frac{p^t(1-p)^{n-t}}{\left[p\Phi\left(\dfrac{\beta_0+1}{\sigma_e}\right) + (1-p)\Phi\left(\dfrac{\beta_0}{\sigma_e}\right)\right]^n}\,dp\,d\beta_0\,d\sigma_e^2 < \infty. \tag{B.1}$$

First, consider

$$J_1 = \int_0^1 \frac{p^t(1-p)^{n-t}}{\left[p\Phi\left(\dfrac{\beta_0+1}{\sigma_e}\right) + (1-p)\Phi\left(\dfrac{\beta_0}{\sigma_e}\right)\right]^n}\,dp$$

Because $\Phi\left(\dfrac{\beta_0+1}{\sigma_e}\right) \geq \Phi\left(\dfrac{\beta_0}{\sigma_e}\right)$ and

$$\Phi\left(\frac{\beta_0}{\sigma_e}\right) \leq p\Phi\left(\frac{\beta_0+1}{\sigma_e}\right) + (1-p)\Phi\left(\frac{\beta_0}{\sigma_e}\right) \leq \Phi\left(\frac{\beta_0+1}{\sigma_e}\right)$$

for all β_0, σ_e and p, we have

$$J_1 \leq \int_0^1 p^t(1-p)^{n-t}dp \bigg/ \left[\Phi\left(\frac{\beta_0}{\sigma_e}\right)\right]^n = \frac{1}{(n+1)\dbinom{n}{s}} \frac{1}{\left[\Phi\left(\dfrac{\beta_0}{\sigma_e}\right)\right]^n}, \quad s \geq 0. \tag{B.2}$$

Now $\beta_0/\sigma_e \geq \mu_0$ implies $\Phi\left(\dfrac{\beta_0}{\sigma_e}\right) \geq \Phi(\mu_0)$. Thus,

$$J_1 \leq \frac{1}{(n+1)\dbinom{n}{s}[\Phi(\mu_0)]^n} < \infty \tag{B.3}$$

and we are left to show that

$$J_2 = \int_0^\infty \int_{\mu_0}^\infty \left(\frac{1}{\sigma_e^2}\right)^{\frac{n}{2}} \exp\left\{-\frac{1}{2\sigma_e^2}\left[\left(n-1\bigg/\sum_{i=1}^n \omega_i^2\right)(\beta_0 + \overline{y}_\omega)^2 + S_\omega^2\right]\right\} d\beta_0 d\sigma_e^2 < \infty.$$

Now, making the transformation $\beta_0 = \mu\sigma_e$ with σ_e^2 untransformed, we have

$$J_2 = \int_0^\infty \left(\frac{1}{\sigma_e^2}\right)^{\frac{(n-1)}{2}} e^{-\frac{1}{2\sigma_e^2}t(1-\overline{y}_\omega)} \left\{\int_{\mu_0\sigma_e}^\infty e^{-\frac{1}{2\sigma_e^2}\left(n-1/\sum_{i=1}^n \omega_i^2\right)(\mu\sigma_e+\overline{y}_\omega)^2} d\mu\right\} d\sigma_e^2.$$

Excluding the case $\omega_i = 1/n, i = 1,\ldots,n$, and using the standard variance inequality (i.e., $1/\sum_{i=1}^n \omega_i^2 < n$, strictly),

$$\int_{\mu_0\sigma_e}^\infty e^{-\frac{1}{2\sigma_e^2}(n-1/\sum_{i=1}^n \omega_i^2)(\mu\sigma_e+\overline{y}_\omega)^2} d\mu$$

$$= \frac{\sqrt{2\pi}}{\sqrt{n-1/\sum_{i=1}^n \omega_i^2}}\left[1 - \Phi\left\{(\mu_0 + \sigma_e^{-1}\overline{y}_\omega)\sqrt{n-1\bigg/\sum_{i=1}^n \omega_i^2}\right\}\right] < \infty \text{ for all } \sigma_e^2 > 0.$$

Finally, we need to show that $\displaystyle\int_0^\infty \left(\frac{1}{\sigma_e^2}\right)^{(n-1)/2} e^{-\frac{S_\omega^2}{2\sigma_e^2}} d\sigma_e^2 < \infty$. Excluding the cases $y_i = 1$, $i = 1,\ldots,n$ and $y_i = 0, i = 1,\ldots,n$, we have

$$\int_0^\infty \left(\frac{1}{\sigma_e^2}\right)^{\frac{n-3}{2}+1} e^{-\frac{1}{2\sigma_e^2}S_\omega^2} d\sigma_e^2 = \Gamma((n-3)/2)/\{S_\omega^2\}^{n/2-1} < \infty.$$

Thus, the joint posterior density is proper provided $n \geq 4, \omega_i \neq 1/n$ for some $i, i = 1,\ldots,n$ and $t \neq 0$ or n.

Appendix C: Sampling the Joint Posterior Density

Here we show how to sample $\pi_{1a}(\beta_0, \sigma_e^2|y_s)$ and $\pi_2(p|\beta_0, \sigma_e^2, y_s)$ in (14).

First, consider $\pi_{1a}(p, \beta_0, \sigma_e^2 | y_s)$. Making the transformation $\beta_0 = \mu \sigma_e$, with σ_e^2 untransformed, in $\pi_{1a}(\beta_0, \sigma_e^2 | y_s)$, we have

$$\pi_{1a}(\beta_0, \sigma_e^2 | y_s) \propto \exp\left\{ -(n - 1/\sum_{i=1}^{n} \omega_i^2)(\mu + \overline{y}_\omega/\sigma_e)^2 \right\} (\sigma_e^2)^{-((n-2)/2+1)} \exp\left\{ -S_\omega^2/2\sigma_e^2 \right\},$$

$\mu > \mu_0, \sigma_e^2 > 0$. Thus, under $\pi_{1a}(\beta_0, \sigma_e^2 | y_s)$, we have

$$\mu | \sigma_e^2, y_s \sim \text{Normal}\ \{ -\overline{y}_\omega/\sigma_e, 1/(n - 1/\sum_{i=1}^{n} \omega_i^2) \}, \mu > \mu_0, \tag{C.1}$$

$$\sigma_e^{-2} | \beta_0, y_s \sim \text{Gamma}\left(\frac{n-1}{2}, \frac{S_\omega^2}{2} \right). \tag{C.2}$$

We use one-one sampling to draw from (C.1); see Devroye (1986).

Thus, it is easy to draw a random sample from $\pi_{1a}(\beta_0, \sigma_e^2 | y_s)$. We simply use the composition method to draw a sample from (C.1) and (C.2) by first drawing σ_e^2 from (C.2) and then μ from (C.1); β_0 is obtained using $\beta_0 = \mu \sigma_e$.

Once β_0 and σ_e^2 are obtained, a sample of p is drawn from $\pi_2(p | \beta_0, \sigma_e^2, y_s)$ as follows. Making the transformation $q = p\Phi((\beta_0 + 1)/2)/\{p\Phi((\beta_0 + 1)/2) + (1 - p)\Phi(\beta_0/2)\}$, we have

$$\pi_2(q | \beta_0, \sigma_e^2, y_s) \propto \left[\frac{\Phi(\beta_0/\sigma_e)}{q\Phi(\beta_0/\sigma_e) + (1 - q)\Phi((\beta_0 + 1)/\sigma_e)} \right]^2 q^t (1 - q)^{n-t}, \ \ 0 < q < 1.$$

Thus, we draw $q | t \sim \text{Beta}\ (t + 1, n - t + 1)$. Then, with probability, $[\Phi(\beta_0/\sigma_e)/\{q\Phi\ (\beta_0/\sigma_e) + (1 - q)\Phi((\beta_0 + 1)/\sigma_e)\}]^2$, we accept it. We get p by computing

$$p = q\Phi(\beta_0/\sigma_e)/\{q\Phi(\beta_0/\sigma_e) + (1 - q)\Phi((\beta_0 + 1)/\sigma_e)\}.$$

This is more efficient than the original procedure for drawing p because it has a higher acceptance rate; see (15).

Bayesian Statistics and Its Applications
Edited by S.K. Upadhyay, U. Singh and D.K. Dey
Anamaya Publishers, New Delhi, India

Research in Elicitation

Anthony O'Hagan

Department of Probability and Statistics, University of Sheffield, Sheffield S3 7RH, UK

Abstract

This article reviews some areas of research in the probabilistic elicitation of expert judgements. In addition, the importance of this research is emphasised, with a view to encouraging Bayesian researchers to work in elicitation.

---

1. Introduction

Elicitation is the process of formulating the knowledge of an expert regarding one or more uncertain quantities as a probability distribution for those quantities. It is an important area of research because elicitation is widely used. In Bayesian statistics, elicitation should be the basis for formulating the prior distribution, which will then be combined with the likelihood via Bayes' theorem to yield the posterior distribution. In this context, the 'expert' is the person for whom the analysis is being performed. This may be the scientist who collected the data and who wishes to draw conclusions from them, or it may be a decision-maker who wishes to make inferences on the basis of the available data and her prior knowledge.

However, elicitation is also used extensively in contexts where the elicited distribution will not be combined with evidence from data, because the expert opinion is essentially all the available knowledge. Expert knowledge is elicited in situations where it is not possible to make direct observations, such as when studying the safety of nuclear installations or assessing the risks of terrorist attacks. Indeed, it is such applications, rather than eliciting prior distributions for Bayesian statistics, which have driven much of the research in elicitation.

1.1 Eliciting a Univariate Distribution

The basic approach to elicitation has two stages:

1. Elicit from the expert a (usually small) number of specific judgements, such as quantiles, credible intervals or mode(s). These judgements are summaries of the expert's underlying probability distribution.

2. Fit a particular distribution to the elicited summaries. This will usually be some standard distribution, such as a normal or beta distribution.

For example, to elicit an expert's probability distribution for the concentration X of a particular chemical in a river such that 50% of fish will be killed, we might ask for the expert's median m and upper quartile q. We could then fit a normal distribution with mean m and standard deviation 1.48 $(q - m)$ and declare this to be the elicited distribution to represent the expert's knowledge.

Before rejecting this as a hopelessly crude technique, it is worth noting a few points.

First, there is basically no alternative. Expert knowledge lies in the expert's head, and cannot be measured neatly like we can measure the length of a piece of wood. It is not practical to elicit more than a few summaries from the expert, but if those summaries capture the most important features of the expert's knowledge then the resulting fitted distribution should be useful. Simply to elicit a measure of location (such as the median) and a measure of spread (such as the distance between median and upper quartile) captures in a crude way the range of values that the expert thinks most likely for X, and an indication of the strength of her knowledge.

We can often supplement these simple summaries with some additional probabilities or quantiles, to refine the distribution or to check that the fitted distribution is appropriate. Nevertheless, there will always be many distributions that might have fitted the elicited summaries equally well. It is important to recognise this when using the elicited distribution. It may be that the choice of fitted distribution does not matter, in the sense that any other plausible distribution that fits the summaries would lead to essentially the same conclusions. For instance, given reasonable data, the prior distribution in a Bayesian analysis does not need to be elicited very accurately because quite large variations in the prior distribution will affect the posterior distribution only trivially. In other situations, though, the choice of fitted distribution may materially affect the inference or decision to be derived.

In planning an elicitation for a single unknown quantity X, there is good guidance in the literature on what kinds of summaries to elicit. Experts may be expert in their field but are rarely experts in probability or statistics. It is generally unwise to ask for moments, like the mean or standard deviation, because these are not well understood and are not assessed accurately in practice (even by statisticians!). Asking for particular quantiles, such as the median or quartiles, is much more useful because these are easier to explain to experts. Similarly, credible intervals are generally well understood and fairly accurately assessed provided we do not ask for intervals with high coverage probabilities (e.g., 90% or above).

1.2 Multivariate Elicitation

In practice, we often wish to obtain a joint distribution from the expert for two or more uncertain quantities, and here there is much less research to guide us in planning such an elicitation. It obviously helps enormously if the variables are independent, since that means we can simply elicit their marginal distributions separately. But remember that independence here is a property of the expert's knowledge, the property that obtaining more evidence about one variable will not alter the expert's beliefs about the other, and we cannot dictate or assume independence without confirming that this does indeed reflect her beliefs.

If the variables are not independent, then we may be able to identify a transformation to other variables which are independent. For instance, if we wish to elicit an expert clinician's joint distribution for the efficacies X_1 and X_2 of two different drugs, these are unlikely to be independent because learning the true value of one will affect the expert's judgements of the other. However, if the first drug is a standard treatment and the second is a new drug to be tested against the standard, then it may be that the expert would judge X_1 and the relative efficacy X_2/X_1 to be independent. Appropriately structuring the elicitation becomes important in multivariate contexts.

If we cannot use independence, then it is less clear what kinds of summary we should elicit. Based on experience with a single variable, we should again ask about probabilities in some form.

Some marginal probabilities will be useful, but we obviously need to elicit some joint or conditional probabilities in order to identify the association between the variables. Even with two variables, these are more complex things for the expert to think about, and with three or more variables it will be very difficult for the expert to assess suitable values for joint or conditional probabilities. There is little research on how well experts perform on these tasks.

1.3 Outline

There are several good reviews of the field of elicitation, most recently Garthwaite, Kadane and O'Hagan (2005). A very substantial review is in preparation as part of the BEEP (Bayesian Elicitation of Experts' Probabilities) project; further information may be obtained from the website http://www.shef.ac.uk/beep/. The reader is referred to these for a general introduction to the subject and to the literature. The present article simply highlights some areas where research is needed. These are organised into empirical topics, covered in Section 2, and more technical topics in Section 3.

2. Psychology, Biases and Empirical Research

2.1 How Experts Construct Responses

It is tempting to think of elicitation in terms of the expert retrieving the values of summaries from her memory, but this is not the case. The probabilities that we ask for are not sitting, pre-formed, in the expert's brain waiting to be expressed. Instead, the expert constructs a judgement in response to the request. In doing so, she accesses whatever information seems relevant, from her memory, and processes it to produce an answer. What information is retrieved, and how it is processed, is affected by context and in particular by how the question is formed. There is a substantial literature in the field of psychology about how people form judgements of probability.

Psychologists refer to various 'heuristics' that people use and which may lead to biases in their judgements. We give two examples here, but more detail can be found in Garthwaite et al. (2005).

Experts will tend to access information that is most recent or has strongest associations with the question. This can lead to biases in their judgement by excluding older information or an accumulation of weaker items of information. An example is when people are asked about risks of being involved in air accidents or terrorist attacks. The high degree of media attention surrounding such incidents leads to them having high *availability*, and so to the risks being over-estimated. On a more mundane level, we *always* seem to pick the wrong queue at the supermarket, so that we are standing waiting our turn while people in neighbouring queues are processed more quickly. This perception is partly because the times when we are in the wrong queue are more irritating, go on longer and are more memorable than the times when we choose the right queue and are quickly out of the supermarket. To combat availability bias, experts should be asked to review all the relevant evidence explicitly and fully before being asked to make specific judgements.

Another source of bias is known as *anchoring and adjustment*. When experts are asked a series of related questions, they tend to answer later questions by adjusting their answers to previous ones. This leads to them being 'anchored' to the earlier answer and not adjusting far enough. For instance, consider the example of eliciting the 50% lethal concentration of a chemical. When giving the value of the upper quartile q, the expert may be anchored by the previously given median m, and this may lead to q not being assessed sufficiently far from q. The result will be under-statement of the standard deviation. Indeed, a common (but not universal) problem with expert judgements is over-confidence, which manifests itself as under-statement of uncertainty. Anchoring and adjustment may be one explanation for over-confidence, but there are probably others. O'Hagan (1998) presents an elicitation scheme with various devices to combat anchoring and adjustment bias, but there is no empirical evidence that this or other similar approaches work in practice.

A more philosophical question that does not seem to have been considered in the literature is the relationship between elicited judgements and the definition of subjective probability. The pioneering work of Savage (1972) and DeGroot (1970) defined subjective probabilities in terms of coherent preferences between gambles. According to these theories, the expert's true subjective probability distribution would be revealed implicitly by choices that she makes between gambles with uncertain consequences. However, elicitation rarely offers the expert such choices, and it is rare for her to receive any kind of reward or consequence based on the judgements that she makes. What, then, is the relationship between her elicited summaries and her 'true' underlying probability distribution? In practice, we assume that her answers are (admittedly faulty and potentially biased) summaries of that distribution, but it seems to me at least possible that they are summaries of a different distribution, one that represents some other description of her knowledge than the probability distribution implied by Savage's or DeGroot's axioms.

2.2 Evaluating Experts

How can we determine whether an elicitation has been successful? Ultimately, the criterion for success should be whether the elicited distribution accurately represents the expert's knowledge, and this is impossible to decide. It is usual to compare elicited distributions for quantities with their true values. For instance, in experiments experts are often asked about quantities or events whose truth is known to the experimenter but not to the expert. Then the expert's performance can be assessed by comparing her judgements with the truth. However, this confounds two separate things, the accuracy of the elicited distributions and the expert's expertise. The expert could be very inaccurate in her beliefs, so that the truth is well outside the range of values that she believes it will probably lie in. This will lead to poor performance, even if those inaccurate beliefs are accurately reflected in the elicited distribution.

In some respects, it is unimportant whether an expert performs badly because her expertise is faulty or because her knowledge is poorly represented in her elicited distributions. Cooke (1991) describes the elicitation of distributions for the same uncertain quantity from several experts, with a view to deriving a distribution that somehow synthesises their expertise. He first judges their performance in expressing distributions for several quantities whose true values are known, and then uses this to derive weights to attach to the experts' judgements on the unknown quantities that are the objective of the elicitation. Poorly performing experts will have low weights.

One measure of an expert's performance over a long series of elicited judgements is whether she is *calibrated*. That is, if she assigns probability p to each of a large number of events, these judgements are calibrated if a long-run proportion p of them actually occurs. If her assessments accord in this way with empirical frequencies over the whole range of values of p from 0 to 1, then she is said to be calibrated. Calibration is widely used as a measure of performance, but it needs to be interpreted with care. Consider the case $p = 0.5$. For every event E to which she gives probability 0.5, the expert implicitly also gives probability 0.5 to the complement of E. One of these two is sure to occur, so in this sense she is automatically calibrated on all her judgements of probability 0.5.

The argument can be taken even further: as soon as she has specified a probability distribution for a single continuous quantity X, then there are an infinite number of sets A to which this distribution assigns probability 0.5, and exactly half of them will occur. The argument applies to any rational

p, and so effectively (to any required degree of approximation) to all p. So after specifying a single continuous probability distribution the expert is automatically calibrated.

The idea of calibration therefore should be applied only to sets of events that have some common basis for judgement. Suppose, for example, that we make a long series of probability statements about the outcomes of horse races. By the above argument, in all the cases that we give probability 0.5 to a particular horse winning we also give probability 0.5 to it losing, and half of these events will occur. But it is entirely possible for the proportion of winners in this sequence to be different from 0.5. It might be q, and the proportion of losers would then be $1 - q$. We would not be calibrated on the sequence of winners alone, even if we are calibrated (trivially) overall. This would suggest a deficiency in my performance in judging the probabilities that specified horses win. Wherever a set of probability assessments are made for a group of related events, particularly where the expert might be expected to use similar evidence and reasoning in making those assessments, then it is meaningful to consider calibration on that set, and failure of calibration would suggest deficiency in the use of the evidence.

2.3 Empirical Research Challenges

A number of areas for research have been identified in the above discussion.

- Psychologists have described various sources of potential bias, due to the heuristic reasoning processes that people employ, and statisticians (and others who practice elicitation) have developed schemes that they hope will avoid these biases. But there is no consensus on which approaches work, and no relevant empirical research. These elicitation protocols have generally been developed to work with genuine experts, whereas much psychological research has been conducted on ordinary people (such as psychology students). This is understandable since the time of genuine experts is seriously limited and may be expensive, but it is important to understand how elicitation methods work with actual experts.

- There has been very little study of methods for eliciting multivariate distributions, and still less into how to structure a multivariate elicitation problem to facilitate accurate elicitation.

- There appears to be no way to evaluate the accuracy of elicitation, separately from the expertise of the expert.

- There is a fundamental question about the relationship between the elicited distribution and the expert's true underlying probability distribution (as defined by subjective probability theory). Is it profitable to speculate on any difference?

3. Imprecision and Uncertainty

3.1 Expert Imprecision and Inference Sensitivity

The preceding discussion has identified two sources of imprecision or uncertainty in the fitted distribution. First, we have remarked that there are many distributions that would fit the expert's stated summaries. There is therefore uncertainty regarding how well the chosen distribution approximates the expert's true distribution. Second, it is also clear that the expert cannot in general specify summaries precisely. In addition to biases introduced by the various heuristics discussed in Section 2.1, the expert's statements will almost inevitably be rounded. So there is further uncertainty induced by the numerical imprecision in the expert's stated summaries.

The key questions here are whether we can in some meaningful way quantify the uncertainty in the elicited distribution, and whether that uncertainty matters. Whether it matters is a question of how sensitive the inferences or decisions that will rest on the elicited distribution are to errors in its specification.

The usual approach to both is informal. Uncertainty in the fitted distribution is characterised by identifying a selection of alternative distributions that are more or less equally consistent with the expert's statements, often in an ad hoc way, and then recomputing the desired inferences/decisions using these new distributions. If the variation in the conclusions, as a result of varying the elicited distribution, is immaterial in the context of the relevant inference or decision process, then the uncertainty surrounding the expert's true distribution is unimportant.

This informal kind of sensitivity analysis may often be seen where elicitation is used to specify the prior distribution for a Bayesian inference problem. Then explicit recognition of the uncertainty in the fitted prior distribution is usually confined to checking whether a few alternative prior distributions lead to appreciably different inferences. A more formal and analytical approach to quantifying sensitivity to the prior distribution was developed in the 1980s and 1990s under the title of *robust Bayes* methods (Berger, 1994). However, in practice the range of alternative prior distributions was still a matter of ad hoc choice and the theory was very difficult to apply except in the simplest kinds of situations, with the result that this body of research is no longer used today.

3.2 Bayesian Nonparametric Elicitation

A recently developed approach has the potential to quantify realistically at least some aspects of uncertainty in the elicited distribution and the sensitivity of consequent inferences or decisions. Oakley and O'Hagan (2005) propose a Bayesian formulation in which the expert's elicited summaries are treated as data and used to update the prior distribution of an external agent called the *facilitator*. This framework is not new, having been described in the 1970s by Lindley, Tversky and Brown (1979). They considered two different perspectives for the facilitator. In one, the facilitator is a decision-maker, and it is the facilitator's own beliefs about the uncertain quantity X that are updated in the light of the expert's statements. This version of the analysis was extended by several other authors, but our interest here is in Lindley et al's second formulation, in which the expert's role is simply to assist in the elicitation of the expert's distribution $\pi_X(.)$ for X. In this case, the facilitator has a prior distribution for $\pi_X(.)$, and it is this that is updated by the elicited summaries.

This aspect of Lindley et al. (1979) is developed by Oakley and O'Hagan (2005) using a Gaussian process formulation to model the facilitator's prior distribution for $\pi_X(.)$. The facilitator's posterior distribution then provides not only a 'best' fitted distribution (for instance, the facilitator's posterior mean) but also a description of uncertainty about $\pi_X(.)$. The facilitator's distribution can be used to compute an implied distribution for any inference or decision that might be consequent on the elicitation. The approach therefore quantifies both the uncertainty in the elicited distribution and the sensitivity of the resulting inferences. Oakley and O'Hagan (2005) show an example in which there is very little uncertainty about $\pi_X(.)$, after the expert has specified just a few probabilities.

There is ongoing research at Sheffield to extend this work, some of it forming a component of the BEEP project. First, as described above, the method handles the uncertainty that is implicit in fitting a distribution to the elicited summaries, but it does not represent the additional imprecision in the expert's statements themselves. Oakley and O'Hagan (2005) indicate how this can be addressed, but it requires judgements about how accurately the expert can specify any given summary. Their work suggests that adding in this expert imprecision component can have a big impact on the posterior uncertainty regarding $\pi_X(.)$. There is some ongoing research into this question.

Another current area of research is into what would be a reasonable prior distribution for the facilitator. It is clearly important that the facilitator's prior distribution does not impose extra knowledge or beliefs that might distort the elicitation of the expert's distribution. But Oakley and O'Hagan's prior distribution certainly, and deliberately, adds some information, such as that the expert's distribution has a smooth density function, and that it is more likely to approximate to a simple unimodal and symmetric density than to any more complex shape. The nature of the facilitator's prior distribution is another obvious area for research.

Finally, Oakley and O'Hagan dealt only with the case of eliciting a univariate distribution, and there is ongoing work to extend it to multivariate elicitation.

3.3 Technical Research Challenges

Analytical work in elicitation seems to be confined at present to the Oakley and O'Hagan (2005) approach, but there are a variety of questions raised by this research.

- Is this the way to go, and if not what is? Are there other effective ways of quantifying the uncertainty implicit in selecting a single representative fitted distribution from a set of (imprecise) expert judgements? Or is it better just to address this uncertainty in an informal, unquantified way?

- What accuracy is achievable in elicitation? Asking the expert for more summaries would in principle eventually identify her true distribution precisely, if her judgements are perfectly accurate. But they are not, and it seems reasonable to suppose that there is a limit to the accuracy that can be achieved, after even an unlimited number of summaries, because once a few simple summaries have been elicited subsequent ones might be answered with even greater imprecision. Can the Oakley and O'Hagan approach shed light on this?

- The areas of ongoing research outlined above–modelling imprecision, the facilitator's prior and multivariate elicitation–are examples of a range of technical challenges in this methodology.

4. Concluding Remarks

This is an important area for research. In particular, Bayesian statisticians have for too long failed to address the use of genuine subjective prior information. It is true that prior information is often irrelevant when the data are substantial, but this situation is far less common than it might seem. Statisticians, and particularly Bayesian statisticians, are building increasingly complex models for real applications, and these models have large numbers of parameters, many of which are only weakly informed by the model. Prior information matters in such situations. The essence of the Bayesian paradigm is to use **all** the available evidence, including prior information, yet one almost never finds genuine prior information in published articles.

What I see all the time is the use of improper, so-called noninformative priors. There is a huge industry in generating these, with so many different names—weak priors, ignorance priors, default priors, reference priors, objective priors. They are useful in practice, and of course I have used them myself frequently, but they must **never** be used without first thinking whether genuine and informative prior knowledge exists. The convenience and illusory objectivity of these priors should never be an excuse to throw away information.

I would particularly encourage young colleagues to engage in research in elicitation. There are many exciting research questions to be addressed. Progress in elicitation will be enormously useful, to those who already recognise and use elicitation as an important tool, and to our Bayesian colleagues who currently fail to use it out of laziness or a misguided aversion to honest subjective judgements. Elicitation is a far more respectable area for research than the dangerous heresy that today goes by the name of 'objective Bayes'!

References

Berger, J.O. (1994). An overview of robust Bayesian analysis (with discussion). *Test*, **3**, 5–124.

Cooke, R.M. (1991). *Experts in Uncertainty: Opinion and Subjective Probability in Science.* Oxford: Oxford University Press.

DeGroot, M.H. (1970). *Optimal Statistical Decisions.* McGraw-Hill: New York.

Garthwaite, P.H., Kadane, J.B. and O'Hagan, A. (2005). Statistical methods for eliciting probability distributions. *Journal of the American Statistical Association*, **100**, 680-701.

Lindley, D.V., Tversky, A. and Brown, R.V. (1979). On the reconciliation of probability assessments (with discussion). *J. Roy. Statist. Soc.*, **A 142**, 146–180.

Oakley, J.E. and O'Hagan, A. (2005). Uncertainty in prior elicitations: a nonparametric approach. *Research Report No. 521/02* Department of Probability and Statistics, University of Sheffield.

O'Hagan, A. (1998). Eliciting expert beliefs in substantial practical applications. *The Statistician*, **47**, 21–35 (with discussion, 55–68).

Savage, L.J. (1972). *The Foundations of Statistics*, 2nd edition. Dover: New York.

Bayesian Statistics and Its Applications
Edited by S.K. Upadhyay, U. Singh and D.K. Dey
Anamaya Publishers, New Delhi, India

On a Characterization of Dirichlet Distribution

R.V. Ramamoorthi[1] and Laura M. Sangalli[2]

[1]Department of Statistics and Probability, Michigan State University, East Lansing, Michigan 48824, USA
[2]Departimento di Matematica, Università degli Studi di Pavia, Via Ferrata 1, 27100 Pavia, Italy

Abstract

Let $Z = (X, Y)$ be a random variable taking values in a finite set $I \times J$. For any distribution P on $I \times J$, let $P_{(\cdot)}$ be the marginal distribution of X and let $P_{\cdot|i}$ be the conditional distribution of Y given $X = i$. Define $Q_{(\cdot)}, Q_{\cdot|j}$ be the marginal and conditional distribution of Y and X given $Y = j$. Geiger and Heckerman [The Annals of Statistics, Vol. 25, No. 3 (Jun., 1997), pp. 1344-1369] showed that if a prior Π for the distribution of Z, satisfies, under Π,

(1) $P_{(\cdot)}$ is independent of $\{P_{\cdot|i} : i \in I\}$ and

(2) $\{P_{\cdot|j} : j \in J\}$ are independent among themselves

(3) a similar statement holds for $Q_{(\cdot)}, Q_{\cdot|j}$

then Π is a Dirichlet distribution.

We provide a simpler proof of the above characterization. Then we indicate an extension of the result to general Dirichlet processes and also examine the distribution of the conditional distribution $P_{|X}$ for a general Dirichlet process.

1. Introduction

Let $\mathbf{X}$, $\mathbf{Y}$ be a pair of random variables taking values in some finite sets I and J, respectively. If P is a joint distribution for $(\mathbf{X}, \mathbf{Y})$, denote by $P^{\mathbf{X}}$ the marginal distribution of $\mathbf{X}$, and by $P^{\mathbf{Y}}(\cdot|i)$ the conditional distribution of $\mathbf{Y}$ given $\mathbf{X} = i$.

Let $\mathbf{P}$ be a random probability measure on $I \times J$. If $D_{c\alpha}$ is a Dirichlet prior distribution for $\mathbf{P}$, i.e., $(\mathbf{P}(i, j); i \in I, j \in J)$ has Dirichlet distribution with parameters $(c\alpha(i, j); i \in I, j \in J)$, where $c > 0$ and α is a probability measure on $I \times J$ (with $\alpha(i, j) > 0$, for all (i, j)), then it is well known that

(a) $\mathbf{P}^{\mathbf{X}}$ and $(\mathbf{P}^{\mathbf{Y}}(\cdot|i); i \in I)$ are independent
(b) $\mathbf{P}^{\mathbf{Y}}(\cdot|i), i \in I$, are mutually independent
(c) $\mathbf{P}^{\mathbf{X}}$ is distributed as $D_{c\alpha^{\mathbf{X}}}$, and
(d) $\mathbf{P}^{\mathbf{Y}}(\cdot|i)$ is distributed as $D_{c\alpha^{\mathbf{X}}(i)\alpha^{\mathbf{Y}}(\cdot|i)}$

where $\alpha^{\mathbf{X}}(i) = \sum_{j \in J} \alpha(i, j)$ and $\alpha^{\mathbf{Y}}(j|i) = \alpha(i, j)/\alpha^{\mathbf{X}}(i)$.

Geiger and Heckerman (1997) show that if Π is a prior for $\mathbf{P}$, with positive density, such that (a) and (b) above hold and similar conditions hold starting with $\mathbf{Y}$, i.e.,

(a') $\mathbf{P}^{\mathbf{Y}}$ and $(\mathbf{P}^{\mathbf{X}}(\cdot|j); j \in J)$ are independent (b′) $\mathbf{P^X}\,(\cdot|j), j \in J$, are mutually independent

with $\mathbf{P^Y}$ and $\mathbf{P^X}(\cdot|j)$ denoting respectively the marginal distribution of $\mathbf{Y}$ and the conditional distribution of $\mathbf{X}$ given $\mathbf{Y} = j$, then Π is a Dirichlet distribution. They then explore the consequences of this characterization for priors in the context of Bayesian networks.

This note is motivated by the result of Geiger and Heckerman and has three parts. In the first part we give a simpler proof of the Geiger-Heckerman characterization. Their proof uses sophisticated analysis involving functional equations. We, on the other hand, use the following result of Doksum (1974), which in turn is based on a result of Darroch and Ratcliff (1971).

Theorem 1.1 (Doksum (1974), Corollary 2.1) *Let C_1 denote the class of all random probabilities* $\mathbf{P}$ *such that either*

 (i) $\mathbf{P}$ *is degenerate at a given point P_0*

 (ii) $\mathbf{P}$ *concentrates at a random point, or*

 (iii) $\mathbf{P}$ *concentrates on two non-random points.*

The Dirichlet process is the only process not in C_1 such that for each Borel set A, the posterior distribution of $\mathbf{P}(A)$, given a sample $\mathbf{X}_1$, $\mathbf{X}_2$, ..., $\mathbf{X}_n$ from $\mathbf{P}$, depends only on the number n_A of $\mathbf{X}$'s that fall in A.

In the second part we extend the characterization to Dirichlet processes. This result naturally raises the question whether (a)-(d) hold for a Dirichlet process, and this is the objective of the third part. That the analog of (a), (b) and (c) hold is not hard to see. The analog of (d) is the one that is intriguing. A bit informally, let $\mathcal{X}$ and $\mathcal{Y}$ be two complete separable metric spaces, $\mathbf{P}$ a discrete random probability measure on $\mathcal{X} \times \mathcal{Y}$, and $D_{c\alpha}$ a Dirichlet process prior for $\mathbf{P}$, with α continuous. Fix x and consider the conditional distribution $\mathbf{P^Y}(\cdot|\mathbf{X} = x)$. We show that the random measure, $\mathbf{P^Y}(\cdot|\mathbf{X} = x)$ is supported by the set of degenerate measures, i.e.,

$$D_{c\alpha}\{P^Y(\cdot|\mathbf{X} = x) = \delta_y \text{ for some } y\} = 1.$$

This result seems to be known but not widely so. In any case we have not seen it in print.

2. Finite Case

Let $\mathcal{Z}$ be a finite set and Π be a prior distribution on the set of probability measures on $\mathcal{Z}$. The set up that we use is: $\mathbf{P}$ is a random probability on $\mathcal{Z}$ distributed as Π, and given $\mathbf{P} = P$, $\mathbf{Z}_1$, $\mathbf{Z}_2$, ..., $\mathbf{Z}_n$ are i.i.d. P. Throughout this section we assume that the prior distribution is in Γ, where Γ is the set of all Π such that

 (i) $\Pi(C_1) = 0$, where C_1 is the class described in the Introduction,

 (ii) $\Pi\{P : P(\omega) > 0 \text{ for all } \omega\} = 1$.

Let $\Pi(\cdot|z_1, z_2, \ldots, z_n)$ denote the posterior distribution of $\mathbf{P}$ given $\mathbf{Z}_1 = z_1$, $\mathbf{Z}_2 = z_2$, ..., $\mathbf{Z}_n = z_n$. When we want to think of $\Pi(\cdot|z_1, z_2, \ldots, z_n)$ as a prior we will write it as $\Pi_{z_1, z_2, \ldots, z_n}$. Thus $\Pi_{z_1, z_2, \ldots, z_n}(\cdot|z_{n+1})$ will stand for the posterior distribution of $\mathbf{P}$ given $\mathbf{Z} = z_{n+1}$, when $\mathbf{P} \sim \Pi_{z_1, z_2, \ldots, z_n}$, and given $\mathbf{P} = P, \mathbf{Z} \sim P$. Note that condition (ii) above ensures that the posterior distribution is uniquely defined and also that $\Pi_{z_1, z_2, \ldots, z_n} \in \Gamma$ for all $z_1, z_2, \ldots, z_n$. For any set A, we write $\Pi^A_{z_1, z_2, \ldots, z_n}$ for the posterior distribution of $\mathbf{P}(A)$ under $\Pi_{z_1, z_2, \ldots, z_n}$.

Let $\mathcal{A} = \{A_1, A_2, \ldots, A_N\}$, where $N \geq 2$, be a partition of $\mathcal{Z}$, and P a probability measure on $\mathcal{Z}$. Let $\underline{P}^{\mathcal{A}} = (P(A_1), P(A_2), \ldots, P(A_N))$ stand for the marginal distribution on $\mathcal{A}$, and $P(\cdot|A_i)$ stand for the conditional distribution given A_i, for $i = 1, 2, \ldots, N$. Set $\underline{P}(\cdot|\mathcal{A}) = (P(\cdot|A_1), \ldots, P(\cdot|A_N))$.

Definition 2.1 $\Pi \in \Gamma$ is said to satisfy Global Independence (GI) and Local Independence (LI) with respect to $\mathcal{A}$ if, for $\mathbf{P} \sim \Pi$,

(GI) $\underline{\mathbf{P}}^{\mathcal{A}}$ and $\underline{\mathbf{P}}(\cdot|\mathcal{A})$ are independent,

(LI) $\mathbf{P}(\cdot|A_1), \ldots, \mathbf{P}(\cdot|A_N)$ are mutually independent.

Let

$$\Gamma(\mathcal{A}) = \{\Pi \in \Gamma : \Pi \text{ satisfies GI and LI with respect to } \mathcal{A}\}.$$

Suppose $\Pi \in \Gamma(\mathcal{A})$. For ease of exposition we write Π as a product of densities $g(\underline{P}^{\mathcal{A}}) \prod_1^N g_i(P(\cdot|A_i))$ with respect to some product measure $\mu \times \prod_1^N \mu_i$. Then the posterior density of $\mathbf{P}$ given an observation $\mathbf{Z} = z, z \in A_j$, is given by

$$\frac{P(A_j)g(\underline{P}^{\mathcal{A}})P(z|A_j)g_j(P(\cdot|A_j)) \prod_{i \neq j} g_i(P(\cdot|A_i))}{\int P(A_j)g(\underline{P}^{\mathcal{A}}) \int P(z|A_j)g_j(P(\cdot|A_j)) \prod_{i \neq j} \int g_i(P(\cdot|A_i))} \tag{2.1}$$

It is immediate from the product structure of the posterior density that $\Pi(\cdot|z) \in \Gamma(\mathcal{A})$. Repeatedly applying this, we get the following lemma.

Lemma 2.2 *Suppose* $\mathbf{P} \sim \Pi$, *with* $\Pi \in \Gamma(\mathcal{A})$, *and given* $\mathbf{P} = P, \mathbf{Z}_1, \mathbf{Z}_2 \ldots, \mathbf{Z}_n$ *are i.i.d.* P. *Then for all* $z_1, z_2, \ldots, z_n, \Pi_{z_1,z_2,\ldots,z_n} \in \Gamma(\mathcal{A})$.
Here is another nice consequence of GI and LI.

Proposition 2.3 *Suppose* $\Pi \in \Gamma(\mathcal{A})$. *If* $z_0 \in A_{i_0}$, *then for each* $j \neq i_0$, $\Pi^{\{z_0\}}(\cdot|z)$, *as a function of* z, *is constant in* A_j, *i.e., if* $z_1, z_2 \in A_j$, *then* $\Pi^{\{z_0\}}(\cdot|z_1) = \Pi^{\{z_0\}}(\cdot|z_2)$.

Proof After integrating out $\{P(\cdot|A_i) : i \neq i_0\}$ in (2.1) the joint posterior density of $(\underline{\mathbf{P}}^{\mathcal{A}}, \mathbf{P}(\cdot|A_{i_0}))$ is

$$\frac{P(A_j)g(\underline{P}^{\mathcal{A}})g_{i_0}(P(\cdot|A_{i_0}))}{\int P(A_j)g(\underline{P}^{\mathcal{A}}) \int g_{i_0}(P(\cdot|A_{i_0}))}. \tag{2.2}$$

This expression depends on z only through j. Consequently the posterior density of $(\underline{\mathbf{P}}^{\mathcal{A}}, \mathbf{P}(\cdot|A_{i_0}))$ and hence that of $\mathbf{P}(z_0)$ is the same for all z in A_j. $\qquad\square$

Next, for any $k \geq 2$ partitions $\mathcal{A}^1, \ldots, \mathcal{A}^k$ of $\mathcal{Z}$, where $\mathcal{A}^t = \{A_1^t, A_2^t, \ldots, A_{N_t}^t\}$ with $N_t \geq 2$, let $\Gamma(\mathcal{A}^1, \ldots, \mathcal{A}^k) = \Gamma(\mathcal{A}^1) \cap \ldots \cap \Gamma(\mathcal{A}^k)$, i.e.,

$$\Gamma(\mathcal{A}^1, \ldots, \mathcal{A}^k) = \{\Pi \in \Gamma : \Pi \text{ satisfies GI and LI with respect to each } \mathcal{A}^t, t = 1, 2, \ldots, k\}.$$

We now consider $k \geq 2$ partitions $\mathcal{A}^1, \ldots, \mathcal{A}^k$ of $\mathcal{Z}$, such that

(C) $A_{i_1}^1 \cap A_{i_2}^2 \cap \ldots \cap A_{i_k}^k$ is nonempty and is a singleton for any choice of $i_1, \ldots, i_k$.

Theorem 2.4 (Geiger and Heckerman) *If* $\Pi \in \Gamma(\mathcal{A}^1, \ldots, \mathcal{A}^k)$, *with* $\mathcal{A}^1, \ldots, \mathcal{A}^k$ *such that* (C) *holds* $(k \geq 2)$, *then* Π *is a Dirichlet distribution.*
Geiger and Heckerman formulation of the theorem can be recovered with $\mathcal{Z} = I \times J$, $\mathcal{A}^1 = \{i \times J : i \in I\}$ and $\mathcal{A}^2 = \{I \times j : j \in J\}$.

Proof By Theorem 1.1, it suffices to show that for each $z_1, z_2, \ldots, z_n, \Pi^A_{z_1, z_2, \ldots, z_n}$ depends only on $n_A = \sum_1^n I_A(z_i)$. In Proposition 2.5 we show that, if $\Pi \in \Gamma_0$, then $E_\Pi(\mathbf{P}(A)|z)$ is a function of $I_A(z)$. We complete the proof using Proposition 2.6, where we show that, if $E_\Pi(\mathbf{P}(A)|z)$ depends only on $I_A(z)$, then $\Pi^A(\cdot|z_1, z_2, \ldots, z_n)$ depends only on n_A. $\qquad\square$

Proposition 2.5 *If* $\Pi \in \Gamma(\mathcal{A}^1, \ldots, \mathcal{A}^k)$, *with* $\mathcal{A}^1, \ldots, \mathcal{A}^k$ *such that* (C) *holds* $(k \geq 2)$, *then*

(i) *for any* $z_0 \subset \mathcal{Z}$, $\Pi^{\{z_0\}}(\cdot|z)$ *is constant in* $\{z_0\}^c$,

(ii) *for any* $A \in \mathcal{Z}, E_\Pi(\mathbf{P}(A)|z)$ *depends only on* $I_A(z)$.

Proof (i) We prove (i) by invoking Proposition 2.3 to the partitions $\{\mathcal{A}^t; t = 1, 2, \ldots, k\}$.

Let $z_1 \neq z_2$ and both be distinct from z_0. We need to show that

$$\Pi^{\{z_0\}}(\cdot|z_1) = \Pi^{\{z_0\}}(\cdot|z_2).$$

Since $z_1 \neq z_0$ and $z_2 \neq z_0$, there exist t_1 and t_2 such that z_0 and z_1 belong to different elements in the partition $\mathcal{A}^{t_1}$, and z_0 and z_2 belong to different elements in the partition $\mathcal{A}^{t_2}$. Let

$$z_0 \in A^{t_1}_{i_0} \cap A^{t_2}_{j_0}, \qquad z_1 \in A^{t_1}_{i_1} \cap A^{t_2}_{j_1}, \qquad z_2 \in A^{t_1}_{i_2} \cap A^{t_2}_{j_2}$$

with $i_1 \neq i_0$ and $j_2 \neq j_0$. Pick $y \in A^{t_1}_{i_1} \cap A^{t_2}_{j_2}$ (by assumption, this intersection is not void). Since z_1 and y belong to the same element $A^{t_1}_{i_1}$ of the partition $\mathcal{A}^{t_1}$, and z_0 belongs to a different element of this partition, then, by Proposition 2.3,

$$\Pi^{\{z_0\}}(\cdot|z_1) = \Pi^{\{z_0\}}(\cdot|y).$$

On the other hand, since z_2 and y belong to the same element $A^{t_2}_{j_2}$ of the partition $\mathcal{A}^{t_2}$, and z_0 belongs to a different element of this partition, then

$$\Pi^{\{z_0\}}(\cdot|y) = \Pi^{\{z_0\}}(\cdot|z_2).$$

Thus,
$$\Pi^{\{z_0\}}(\cdot|z_1) = \Pi^{\{z_0\}}(\cdot|z_2)$$

and this proves (i).

One would then like to prove that, for any event A, $\Pi^A(\cdot|z)$ is constant in A^c. In (i) we showed that this holds when A consists of one point, but the proof cannot be directly extended to the case in which A consists of more than one point, since this would involve the joint posterior distribution. So we work with conditional expectations first.

(ii) For any $A \subset \mathcal{Z}$ and for any $z \in \mathcal{Z}$,

$$E(\mathbf{P}(A)|z) = \sum_{z' \in A} E(\mathbf{P}(z')|z). \tag{2.3}$$

By (i), if $\{z_1, z_2\} \subset A^c$, then $\Pi^{\{z'\}}(\cdot|z_1) = \Pi^{\{z'\}}(\cdot|z_2)$ for any $z' \in A$. Thus, in particular, $E(\mathbf{P}(z')|z_1) = E(\mathbf{P}(z')|z_2)$ for every $z' \in A$. By (2.3) this implies that $E(\mathbf{P}(A)|z)$ is constant in A^c. Next, since

$$E(\mathbf{P}(A)|z) = 1 - E(\mathbf{P}(A^c)|z)$$

and the right hand side is constant in A, so is $E(\mathbf{P}(A)|z)$. This proves that $E(\mathbf{P}(A)|z)$ depends only on $I_A(z)$. $\qquad\square$

In the next proposition (i) is Doksum's condition and (iv) has been shown to hold when $\mathcal{A}^1, \ldots, \mathcal{A}^k$ are such that (C) holds $(k \geq 2)$.

Proposition 2.6 *Let $\mathcal{A}^1, \ldots, \mathcal{A}^k$ be any k partitions of $\mathcal{Z}$. The following are equivalent:*

(i) *for every $\Pi \in \Gamma(\mathcal{A}^1, \ldots, \mathcal{A}^k)$, for any $A \subset \mathcal{Z}$, for any $z_1, z_2, \ldots z_n, \Pi^A_{z_1, z_2, \ldots, z_n}$ depends only on $n_A = \sum_i^n I_A(z_i)$, the number of z's that fall in A;*

(ii) *for every $\Pi \in \Gamma(\mathcal{A}^1, \ldots, \mathcal{A}^k)$, for any $A \subset \mathcal{Z}, \Pi^A(\cdot|z)$ depends only on $I_A(z)$;*

(iii) *for every $\Pi \in \Gamma(\mathcal{A}^1, \ldots, \mathcal{A}^k)$, for any $A \subset \mathcal{Z}$, for any $z_1, z_2, \ldots, z_n, E_\Pi(\mathbf{P}(A)|z_1, z_2, \ldots, z_n)$ depends only on n_A;*

(iv) *for every $\Pi \in \Gamma(\mathcal{A}^1, \ldots, \mathcal{A}^k)$, for any $A \subset \mathcal{Z}, E_\Pi(\mathbf{P}(A)|z)$ depends only on $I_A(z)$.*

Proof We will show that (i)$\Leftrightarrow$(ii), (iii)$\Leftrightarrow$(iv) and (ii)$\Leftrightarrow$(iii).

(i) $\Leftrightarrow$ (ii). The implication (i) $\Rightarrow$ (ii) is immediate. For the reverse implication let $z_1, z_2, \ldots, z_n$ and $z_1^*, z_2^*, \ldots, z_n^*$ be such that $\sum I_A(z_i) = \sum I_A(z_i^*)$. Since $\prod^A_{z_1, z_2, \ldots, z_n}$ is exchangeable, we assume, without loss of generality, that $I_A(z_i) = I_A(z_i^*)$, for $i = 1, 2, \ldots, n$. Note that $\Pi^A_{z_1, z_2, \ldots, z_{n-1}, z_n} = \Pi^A_{z_1, z_2, \ldots, z_{n-1}}(\cdot|z_n)$, and, by Lemma 2.2, $\Pi_{z_1, z_2, \ldots, z_{n-1}}$ is in $\Gamma(\mathcal{A}^1, \ldots, \mathcal{A}^k)$. Hence, by (ii), since $I_A(z_n) = I_A(z_n^*)$,

$$\Pi^A_{z_1, z_2 \ldots, z_n} = \Pi^A_{z_1, z_2, \ldots, z_{n-1}}(\cdot|z_n) = \Pi^A_{z_1, z_2, \ldots, z_{n-1}}(\cdot|z_n^*) = \Pi^A_{z_1, z_2, \ldots, z_{n-1}, z_n^*}.$$

Next consider $\Pi^A_{z_1, z_2, \ldots, z_{n-1}, z_n^*} = \Pi^A_{z_1, z_2, \ldots, z_{n-2}, z_n^*}(\cdot|z_{n-1})$. Since $I_A(z_{n-1}) = I_A(z_{n-1}^*)$,

$$\Pi^A_{z_1, z_2 \ldots, z_{n-1}, z_n^*} = \Pi^A_{z_1, z_2, \ldots, z_{n-2}, z_n^*}(\cdot|z_{n-1}) = \Pi^A_{z_1, z_2, \ldots, z_{n-2}, z_n^*}(\cdot|z_{n-1}^*) = \Pi^A_{z_1, z_2, \ldots, z_{n-1}^*, z_n^*}.$$

Continuing this argument, (i) follows.

(ii)$\Leftrightarrow$(iii). Since (i) $\Leftrightarrow$(ii) and (i) $\Rightarrow$ (iii), we have (ii)$\Rightarrow$(iii). For the reverse implication, let z, z^* be in A. We need to show that $\Pi^A(\cdot|z) = \Pi^A(\cdot|z^*)$. Equivalently, for all k,

$$\int [P(A)]^k \Pi(dP|z) = \int [P(A)]^k \Pi(dP|z^*). \tag{2.4}$$

For $k = 1$ this is just (iii), when $n = 1$. Assume that we have proved (2.4) for $k = m - 1$, i.e.,

$$\int [P(A)]^{m-1} \Pi(dP|z) = \int [P(A)]^{m-1} \Pi(dP|z^*). \tag{2.5}$$

To avoid notational mess, denote $\Pi(\cdot|z)$ by Π_1 and $\Pi(\cdot|z^*)$ by Π_2. It is convenient to think of $\mathbf{P} \sim \Pi_i$ and given $\mathbf{P} = P, \mathbf{Z}_1, \mathbf{Z}_2, \ldots, \mathbf{Z}_k$ i.i.d. P (for $i = 1, 2,$ and $k = 1, 2, \ldots$). Let λ_i denotes the corresponding marginal distribution,

$$\lambda_i^{[k]}(z_1, \ldots, z_k) = \int \prod_{j=1}^k P(z_j) \Pi_i(dP).$$

Setting $D = \{(z_1, \ldots, z_{m-1}) : z_j \in A, j = 1, \ldots, m-1\}$, (2.5) can be rewritten as

$$\lambda_1^{[m-1]}(D) = \lambda_2^{[m-1]}(D).$$

We want to show that

$$\lambda_1^{[m]}(D \times A) = \lambda_2^{[m]}(D \times A).$$

Now,

$$\lambda_1^{[m]}(D \times A) = \sum_D E(\mathbf{P}(A)|z_1, \ldots, z_{m-1}, z) \lambda_1^{[m-1]}(z_1, \ldots, z_{m-1})$$

and

$$\lambda_2^{[m]}(D \times A) = \sum_D E(\mathbf{P}(A)|z_1, \ldots, z_{m-1}, z^*)\lambda_2^{[m-1]}(z_1, \ldots, z_{m-1}).$$

Moreover, on $D, I_A(z) + \sum_{j=1}^{m-1} I_A(z_j) = I_A(z^*) + \sum_{j=1}^{m-1} I_A(z_j)$, and hence, by (iii), $E(\mathbf{P}(A)|z_1, \ldots, z_{m-1}, z)$ and $E(\mathbf{P}(A)|z_1, \ldots, z_{m-1}, z^*)$ are both equal to some constant c.
Hence

$$\lambda_1(D \times A)^{[m]} = c\lambda_1^{[m-1]}(D) = c\lambda_2^{[m-1]}(D) = \lambda_2^{[m]}(D \times A)$$

where the middle equality follows from the induction hypothesis $\lambda_1^{[m-1]}(D) = \lambda_2^{[m-1]}(D)$.

The proof of (iii)$\Leftrightarrow$ (iv) is along the same lines as the proof of (i) $\Leftrightarrow$ (ii) and we omit the details. $\square$

We illustrate applications of Theorem 2.4 with an example.

Example 2.7 Let $\mathbf{X}_1, \mathbf{X}_2, \mathbf{X}_3, \mathbf{X}_4$ be random variables taking values in some finite sets $\mathcal{X}_1, \mathcal{X}_2, \mathcal{X}_3, \mathcal{X}_4$. For any joint distribution P of $(\mathbf{X}_1, \mathbf{X}_2, \mathbf{X}_3, \mathbf{X}_4)$, let P^1 denotes the marginal distribution of $\mathbf{X}_1$, and $P(\cdot|\mathbf{X}_1)$ denotes the conditional distribution of $\mathbf{X}_2$, $\mathbf{X}_3$, $\mathbf{X}_4$ given $\mathbf{X}_1$; let P^2 denotes the marginal distribution of $\mathbf{X}_2$, and $P(\cdot|\mathbf{X}_2)$ denote the conditional distribution of $\mathbf{X}_1$, $\mathbf{X}_3$, $\mathbf{X}_4$ given $\mathbf{X}_2$. Define P^j and $P(\cdot|\mathbf{X}_j)$, for $j = 3, 4$, similarly. Let Π be a prior for $\mathbf{P}$ with full support. If under Π, for each i,

(a) $\mathbf{P}^i$ and $\mathbf{P}(\cdot|\mathbf{X}_i)$ are independent

(b) $\mathbf{P}(\cdot|\mathbf{X}_i = x), x \in \mathcal{X}_i$, are mutually independent

then Π is a Dirichlet distribution. This result follows by setting for $\mathcal{A}^1 = \{x \times \mathcal{X}_2 \times \mathcal{X}_3 \times \mathcal{X}_4 : x \in \mathcal{X}_1\}$, $\mathcal{A}^2 = \{\mathcal{X}_1 \times x \times \mathcal{X}_3 \times \mathcal{X}_4 : x \in \mathcal{X}_2\}$ and so on. Note that this example is different from the multivariate example discussed in Geiger and Heckerman (1997).

3. General Case

In this section we study the case when $\mathcal{Z} = \mathcal{X} \times \mathcal{Y}$, where $\mathcal{X}$ and $\mathcal{Y}$ are complete separable metric spaces.

Let $\mathbf{Z} = (\mathbf{X}, \mathbf{Y})$ be a random vector, with $\mathbf{X}$ taking values in $\mathcal{X}$ and $\mathbf{Y}$ in $\mathcal{Y}$. Let $M(\mathcal{Z}) = M(\mathcal{X} \times \mathcal{Y})$ be all joint distributions of $(\mathbf{X}, \mathbf{Y})$, $M(\mathcal{X})$ and $M(\mathcal{Y})$ be all distributions of $\mathbf{X}$ and $\mathbf{Y}$.

An easy generalization of Theorem 2.4 is the following:

Theorem 3.1 *For any partition $\mathcal{A} = \{A_1, A_2, \ldots, A_n\}$ of $\mathcal{X}$ and any partition $\mathcal{B} = \{B_1, B_2, \ldots, B_m\}$ of $\mathcal{Y}$ by Borel sets, let $\Omega = \{A_i \times B_j : A_i \in \mathcal{A}, B_j \in \mathcal{B}\}$. If Π is prior on $M(\mathcal{Z})$ such that it satisfies LI and GI with respect to $\Omega = \mathcal{A} \times \mathcal{B}$, for every pair of partitions $\mathcal{A}$ and $\mathcal{B}$, then Π is a Dirichlet process distribution.*

Proof If Π is indeed a Dirichlet process distribution then it will have parameter $c\alpha$, where α is a measure on $\mathcal{Z}$ defined by $\alpha(C) = E_\Pi(\mathbf{P}(C))$, for any Borel set $C \subset \mathcal{Z}$, and the constant c is equal to $(E_\Pi(\mathbf{P}(C))(1 - E_\Pi(\mathbf{P}(C)))/Var_\Pi(\mathbf{P}(C))) - 1$, where C is some Borel subset of $\mathcal{Z}$ such that $Var_\Pi(\mathbf{P}(C))$ is positive. Let $D_{c\alpha}$ be the Dirichlet process with parameter $c\alpha$. We need to show that $\Pi = D_{c\alpha}$. By Proposition 2.5.1 in Ghosh and Ramamoorthi (2003) it is enough to show that the two measures agree on the σ-algebra generated by functions of the form $P \mapsto P(A \times B)$, for Borel subsets A of $\mathcal{X}$ and B of $\mathcal{Y}$, which in turn would follow if the moments, i.e. the expectations of $\Pi_{i,j}\mathbf{P}(A_i \times B_j)^{n_{i,j}}$, are the same under both Π and $D_{c\alpha}$. Now $\Pi_{i,j}\mathbf{P}(A_i \times B_j)^{n_{i,j}}$ can be written as a sum of terms $\Pi_{k,l}P(C_k \times D_l)^{n_{k,l}}$ where the C_ks are disjoint and the D_ls are disjoint ($k = 1, \ldots, m_k, l = 1, \ldots, m_l$). Arguing as in the finite case, since the C_ks are disjoint and the D_ls are disjoint, and Π satisfies LI and GI with respect to $\Omega = \mathcal{A} \times \mathcal{B}$, for every pair of partitions $\mathcal{A}$ and $\mathcal{B}$, one has that $(\mathbf{P}(C_k \times D_l); k = 1, \ldots, m_k, l = 1, \ldots, m_l)$ has Dirichlet distribution. Thus $E_\Pi(\Pi_{k,l}\mathbf{P}(C_k \times D_l)^{n_{k,l}}) = E_{D_{c\alpha}}(\Pi_{k,l}\mathbf{P}(C_k \times D_l)^{n_{k,l}})$, and this completes the proof. $\qquad\square$

We next move on to a formulation more in line with that in the last section, and leave the language of partitions.

Let $M_D(\mathcal{X} \times \mathcal{Y})$ be the set of *discrete* measures on $\mathcal{X} \times \mathcal{Y}$. For any $P \in M_D(\mathcal{X} \times \mathcal{Y})$, let $P^\mathbf{X}$ denotes the marginal distribution of $\mathbf{X}$ and $Q_P(\cdot|\mathbf{X})$ denote a version of the conditional distribution of $\mathbf{Y}$ given $\mathbf{X}$. We consider versions that are measurable in the sense that for any Borel set $B \subset \mathcal{Y}, (P,x) \mapsto Q_P(B|x)$ is jointly measurable in (P,x). We will sometimes consider the entire conditional distribution as an element of $M_D(\mathcal{Y})^\mathcal{X}$, in which case we will denote it by $\tilde{Q}_P$.

Let Γ_0 be the set of all priors Π on $M_D(\mathcal{X} \times \mathcal{Y})$ such that $\Pi(C_1) = 0$, where C_1 is the class described in the Introduction, and, for $\mathbf{P} \sim \Pi$,

(i) $\mathbf{P}^\mathbf{X}$ and $\tilde{Q}_\mathbf{P}$ are independent;

(ii) $Q_\mathbf{P}(\cdot|x), x \in \mathcal{X}$, are mutually independent;

(iii) $\mathbf{P}^\mathbf{Y}$ and $\tilde{Q}'_\mathbf{P}$ are independent, where $\mathbf{P}^\mathbf{Y}$ is the marginal distribution of $\mathbf{Y}$, and $\tilde{Q}'_\mathbf{P}$ is the conditional distribution of $\mathbf{Y}$ given $\mathbf{X}$, as an element of $M_D(\mathcal{X})^\mathcal{Y}$;

(iv) $Q'_\mathbf{P}(\cdot|y), y \in \mathcal{Y}$, are mutually independent.

For any prior Π for $\mathbf{P}$, let $\Pi^\mathbf{X}$ denote the distribution of $\mathbf{P} \mapsto \mathbf{P}^\mathbf{X}$ and by $\tilde{\Pi}$ the distribution of $\mathbf{P} \mapsto \tilde{Q}_\mathbf{P}$. Note that if $\Pi \in \Gamma_0$ then $\Pi = \Pi^\mathbf{X} \times \tilde{\Pi}$. For each x, let Π_x denote the distribution of $\mathbf{P} \mapsto Q_\mathbf{P}(\cdot|x)$. Further, set

$$\lambda(\cdot) = \int P(\cdot)\Pi(dP), \quad \lambda^\mathbf{X}(\cdot) = \int P^\mathbf{X}(\cdot)\Pi^\mathbf{X}(dP^\mathbf{X}), \quad \lambda_x(\cdot) = \int Q(\cdot|x)\Pi_x(dQ).$$

Let $\Pi \in \Gamma_0$. Then, Kolomogorov's consistency theorem gives rise to the product measure $\Pi^\mathbf{X} \times \Pi_{x \in \mathcal{X}}\Pi_x$ in the space $M_D(\mathcal{X}) \times M_D(\mathcal{Y})^\mathcal{X}$. Here, on $M_D(\mathcal{Y})$ we consider the Borel σ-algebra and on $M_D(\mathcal{Y})^\mathcal{X}$ the product σ-algebra. Let $\Delta_D \subset M_D(\mathcal{Y})^\mathcal{X}$ be

$$\Delta_D = \{\tilde{Q} \in M_D(\mathcal{Y})^\mathcal{X} : \text{ for each Borel set } B \subset \mathcal{Y}, x \mapsto Q(B|x) \text{ is measurable}\}.$$

We will show in the appendix that under $\prod_{x \in \mathcal{X}} \Pi_x$ the outer measure of Δ_D is 1. We will also show that, on $M_D(\mathcal{X}) \times \Delta$, the function $(P^\mathbf{X}, \tilde{Q}) \mapsto \int_A Q(B|x)P^\mathbf{X}(dx)$ is measurable. All this will show that to specify a $\Pi \in \Gamma_0$, it is enough to specify $\Pi^\mathbf{X}$ and $\{\Pi_x : x \in \mathcal{X}\}$. In the appendix we also discuss the difficulty that arises if we consider all of $M(\mathcal{X})$ rather than $M_D(\mathcal{X})$.

If $\Pi \in \Gamma_0$, it is not hard to see that

$$\lambda(C \times D) = \int P(C \times D)\Pi(dP) = \int \left[\int_C Q(D|x)P^\mathbf{X}(dx)\right]\Pi_x(dQ)\Pi^\mathbf{X}(dP^\mathbf{X}) = \int_C \lambda_x(D)\lambda^\mathbf{X}(dx).$$

Here is the analog of Lemma 2.2.

Lemma 3.2 *Let* $\Pi \in \Gamma_0, \mathbf{P} \sim \Pi$ *and given* $\mathbf{P} = P, (\mathbf{X}^*, \mathbf{Y}^*) \sim P$. *The following is a version of the posterior* $\Pi(\cdot|(\mathbf{X}^*, \mathbf{Y}^*))$

(1) *for all* $(x^*, y^*), \Pi(\cdot|(x^*, y^*)) \in \Gamma_0$;

(2) $\Pi^\mathbf{X}(\cdot|(x^*, y^*)) = \Pi^\mathbf{X}(\cdot|x^*),$

where $\Pi^{\mathbf{X}}(\cdot|x^*)$ is the posterior arising from the model: $\mathbf{P}^{\mathbf{X}} \sim \Pi^{\mathbf{X}}$, and given $\mathbf{P}^{\mathbf{X}} = P^{\mathbf{X}}$, $\mathbf{X}^* \sim P^{\mathbf{X}}$;

(3) *if $x \neq x^*$, $\Pi_x(\cdot|(x^*, y^*)) = \Pi_x(\cdot)$,*
if $x = x^*$, $\Pi_x(\cdot|(x^*, y^*)) = \Pi_{x^*}(\cdot|y^*)$,
where $\Pi_{x^*}(\cdot|y^*)$ is the posterior given y^*, when Π_{x^*} is the prior on $M_D(\mathcal{Y})$, and given Q in $M_D(\mathcal{Y})$, $\mathbf{Y}^* \sim Q$.

Proof We need to show that for Borel sets $C \times D \subset \mathcal{X} \times \mathcal{Y}$, as E varies over Borel sets of $M_D(\mathcal{X} \times \mathcal{Y})$, the following holds:

$$\int_E P(C \times D)\Pi(dP) = \int_{C \times D} \Pi(E|(x^*, y^*))\lambda_{x^*}(dy^*)\lambda^{\mathbf{X}}(dx^*).$$

It is enough to show that for any Borel sets $A_1 \times B_1, A_2 \times B_2, \ldots, A_k \times B_k$, and any integers $n_1, n_2, \ldots n_k$,

$$\int P(C \times D) \left\{ \prod_1^k [P(A_i \times B_i)]^{n_i} \right\} \Pi(dP)$$

$$= \int_{C \times D} \left(\int \left\{ \prod_1^k [P(A_i \times B_i)]^{n_i} \right\} \Pi(dP|(x^*, y^*)) \right) \lambda_{x^*}(dy^*)\lambda^{\mathbf{X}}(dx^*). \tag{3.1}$$

Setting $n_i' = \sum_{j=1}^i n_j$ (for $i = 1, \ldots, k$), $n_0' = 0, n = n_k'$, and setting $\underline{x} = (x_1, \ldots, x_{n_1}, x_{n_1+1}, \ldots, x_{n_2'}, x_{n_2'+1}, \ldots, x_{n_k'})$, since

$$[P(A_i \times B_i)]^{n_i} = \int_{A_i^{n_i}} \prod_{j=n_{i-1}'+1}^{n_i'} Q_P(B_i|x_j)P^{\mathbf{X}}(dx_j)$$

it turns out that

$$\int P(C \times D) \left\{ \prod_{i=1}^k [P(A_i \times B_i)]^{n_i} \right\} \Pi(dP)$$

$$= \int \left[\int_{C \times F} Q(D|x^*) \left\{ \prod_{i=1}^k \prod_{j=n_{i-1}'+1}^{n_i'} Q(B_i|x_j) \right\} P^{\mathbf{X}}(dx^*) \prod_{j=1}^n P^{\mathbf{X}}(dx_j) \right] \tilde{\Pi}(dQ)\Pi(dP^{\mathbf{X}}),$$

where $F = \prod_{i=1}^k A_i^{n_i}$. Setting $C_i(\underline{x}, x^*) = \{j : n_{i-1}' + 1 \leq j \leq n_i', x_j = x^*\}, m_i(\underline{x}, x^*) = \#C_i(\underline{x}, x^*)$, and $D_i(\underline{x}, x^*) = \{j : n_{i-1}' + 1 \leq j \leq n_i', x_j \neq x^*\}$,

$$\int P(C \times D) \left\{ \prod_{i=1}^k [P(A_i \times B_i)]^{n_i} \right\} \Pi(dP) = \int \left[\int_{C \times F} Q(D|x^*) \left\{ \prod_{i=1}^k [Q(B_i|x^*)]^{m_i(\underline{x}, x^*)} \right\} \right.$$

$$\left. \times \left\{ \prod_{i=1}^k \prod_{j \in D_i(\underline{x}, x^*)} Q(B_i|x_j) \right\} P^{\mathbf{X}}(dx^*) \prod_{j=1}^n P^{\mathbf{X}}(dx_j) \right] \tilde{\Pi}(dQ)\Pi(dP^{\mathbf{X}}).$$

Let

$$I_i(x^*, Q) = Q(D|x^*) \prod_{i=1}^{k} [Q(B_i|x^*)]^{m_i(\underline{x},x^*)} \quad \text{and} \quad I_2(\underline{x}, Q) = \prod_{i=1}^{k} \prod_{j \in D_i(\underline{x},x^*)} Q(B_i|x_j).$$

Denote by $\prod_{\underline{x},x^*}$ the joint distribution of $\{Q_\mathbf{P}(\cdot|x_j) : j \in D_i(\underline{x}, x^*), i = 1, \ldots, k\}$. $\prod_{\underline{x},x^*}$ can be described more explicitly as a product of $\prod_x$ over distinct elements of $\underline{x}$, however we do not need that here. Using the independence of $\{Q_\mathbf{P}(\cdot|x_j) : j \in D_i(\underline{x}, x^*), i = 1, \ldots, k\}$ and $Q_\mathbf{P}(\cdot|x^*)$ and changing the order of integration

$$\int P(C \times D) \left\{ \prod_1^k [P(A_i \times B_i)]^{n_i} \right\} \Pi(dP)$$

$$= \int \int_{C \times F} \left[\int I_1(x^*, Q)\Pi_{x^*}(dQ) \right] \left[\int I_2(\underline{x}, Q)\Pi_{x,x^*}(dQ) \right] P^\mathbf{X}(dx^*) \prod_{j=1}^{n} P^\mathbf{X}(dx_j)\Pi(dP^\mathbf{X}). \qquad (3.2)$$

We now move to the right hand side of (3.1) which by similar calculations is equal to

$$\int_C \int_D \int \left(\int_F \left[\int \left\{ \prod_{j=1}^{k} [Q(B_i|x^*)]^{m_i(\underline{x},x^*)} \right\} \Pi_{x^*}(dQ|(x^*, y^*)) \right] \right.$$

$$\times \left. \left[\int I_2(\underline{x}, Q)\Pi_{\underline{x},x^*}(dQ|(x^*, y^*)) \right] \prod_{j=1}^{n} P^\mathbf{X}(dx_j) \right) \Pi^\mathbf{X}(dP^\mathbf{X}|(x^*, y^*))\lambda_{x^*}(dy^*)\lambda(dx^*)$$

where $\Pi_{\underline{x},x^*}(\cdot|(x^*, y^*))$ denotes the joint posterior distribution of $\{Q_\mathbf{P}(\cdot|x_j) : j \in D_i(\underline{x}, x^*), i, \ldots, k\}$, given (x^*, y^*). Now, for fixed (x^*, y^*), $\Pi_{\underline{x},x^*}(\cdot|(x^*, y^*)) = \Pi_{\underline{x},x^*}(\cdot)$, which gives

$$\int I_2(\underline{x}, Q)\Pi_{\underline{x},x^*}(dQ|(x^*, y^*)) = \int I_2(\underline{x}, P)\Pi_{\underline{x},x^*}(dQ).$$

Further, since $\Pi_{x^*}(dQ|(x^*, y^*)) = \Pi_{x^*}(dQ|y^*)$,

$$\int_D \int \left\{ \prod_{j=1}^{k} [Q(B_i|x^*)]^{m_i(\underline{x},x^*)} \right\} \Pi_{x^*}(dQ|(x^*, y^*))\lambda_{x^*}(dy^*) = \int I_1(x^*, Q)\Pi_{x^*}(dQ).$$

Hence, since $\Pi^\mathbf{X}(\cdot|(x^*, y^*)) = \Pi^\mathbf{X}(\cdot|x^*)$, the right hand side of (3.1) is

$$\int_C \int \left(\int_F \left[\int I_1(x^*, Q)\Pi_{x^*}(dQ) \int I_2(\underline{x}, Q)\Pi_{\underline{x},x^*}(dQ) \right] \prod_1^{n} P^\mathbf{X}(dx_i) \right) \Pi^\mathbf{X}(dP^\mathbf{X}|x^*)\lambda(dx^*).$$

The term inside square brackets is a function of P, and the integrals amount to taking expectation of the conditional expectation given $\mathbf{X}^*$, and it is easy to see that it is same as the right hand side of (3.2). $\qquad \square$

Theorem 3.3 *If $\Pi \in \Gamma_0$ then Π is a Dirichlet process distribution.*

Proof Consider a set of the form $A \times B$. Then

$$E(\mathbf{P}(A \times B)|(x^*, y^*)) = \int P(A \times B)\Pi(dP|(x^*, y^*))$$

$$= \int I_A(x)Q(B|x)\Pi_x(dQ|(x^*, y^*))P^\mathbf{X}(dx)\Pi(dP^\mathbf{X}|x^*). \qquad (3.3)$$

Since for $x^* \notin A$ and $x \in A, \Pi_x(\cdot|(x^*, y^*)) = \Pi_x, E(\mathbf{P}(A \times B)|(x^*, y^*))$ is constant in y^*, for each $x^* \notin A$. Invoking properties (iii) and (iv) in the definition of Γ_0 we get $E(\mathbf{P}(A \times B)|(x^*, y^*))$ is constant in $(A \times B)^c$. This property easily extends to sets E which are finite disjoint union of sets of the form $A_i \times B_i$. Since this class is closed under complements $E(\mathbf{P}(E)|(x^*, y^*))$ depends only on $I_E(x^*, y^*)$. Now consider the class of all sets E for which $E(\mathbf{P}(E)|(x^*, y^*))$ depends only on $I_E(x^*, y^*)$. It is easily seen that this class is all Borel sets in $\mathcal{X} \times \mathcal{Y}$. Thus, Theorem 1.1 yields the result.

4. Distribution of the Conditional Distribution

As before, let $\mathbf{Z} = (\mathbf{X}, \mathbf{Y})$ be a random vector, where $\mathbf{X}$ and $\mathbf{Y}$ take values on the complete separable metric spaces $\mathcal{X}$ and $\mathcal{Y}$ respectively. For any discrete distribution P of $\mathbf{Z}$, let $P^{\mathbf{X}}$ denote the marginal distribution of $\mathbf{X}$ and $Q_P(\cdot|x)$ the conditional distribution of $\mathbf{Y}$ given $\mathbf{X} = x$. Let Π be a prior for the distribution $\mathbf{P}$ of $\mathbf{Z}$. Denote by $\Pi^{\mathbf{X}}$ the distribution of $\mathbf{P}^{\mathbf{X}}$. For each x, let Π_x be the distribution of the conditional distribution $Q_{\mathbf{P}}(\cdot|x)$—more formally, Π_x is the probability measure on $M_D(\mathcal{Y})$ arising from the map $\mathbf{P} \mapsto Q_{\mathbf{P}}(\cdot|x)$, where $\mathbf{P} \sim \Pi$.

Let α be a continuous probability measure on $\mathcal{X} \times \mathcal{Y}, \alpha^{\mathbf{X}}(\cdot) = \alpha(\cdot \times \mathcal{Y})$, and $\alpha(\cdot|x)$ such that, for any Borel $B \subset \mathcal{Y}, \alpha(B|x)$ is Borel measurable in x and $\int_A \alpha(B|x)\alpha^{\mathbf{X}}(dx) = \alpha(A \times B)$ for all Borel $A \subset \mathcal{X}$. Let $D_{c\alpha}$ be a Dirichlet process distribution.

Proposition 4.1   *If* $\mathbf{P} \sim \Pi = D_{c\alpha}$*, then*

(i) $\mathbf{P}^{\mathbf{X}} \sim D_{c\alpha^{\mathbf{X}}}$

(ii) $\mathbf{P}^{\mathbf{X}}$ *and* $(Q_{\mathbf{P}}(\cdot|x); x \in \mathcal{X})$ *are independent;*

(iii) $Q_{\mathbf{P}}(\cdot|x), x \in \mathcal{X}$, *are independent among themselves;*

(iv)   *for* $\alpha^{\mathbf{X}}$ *almost all* x,

$$\Pi_x\{\delta_y : y \in \mathcal{Y}\} = 1$$

and, for any Borel subset B *of* $\mathcal{Y}$,

$$\Pi_x\{\delta_y : y \in B\} = \alpha(B|x).$$

Proof   (i) is essentially part of the definition of a Dirichlet process. (ii) and (iii) can be obtained by writing

$$Q_P(\cdot|x) = \lim_{k \to \infty} \sum_{i=1}^{m_k} \frac{P(A_{ki} \times \cdot)}{P^{\mathbf{X}}(A_{ki})} I_{A_{ki}}(x) \tag{4.1}$$

where A_k is sequence of algebras, with atoms $\{A_{k1}, A_{k2}, \ldots, A_{km_k}\}$, which increases and generates the Borel σ-algebra on $\mathcal{X}$, and using an appropriate martingale convergence theorem. The details are technical but routine.

To see why one would heuristically expect (iv) to hold, suppose that $A_k, k = 1, 2, \ldots,$ is a decreasing sequence of sets such that $\cap_k A_k = \{x\}$ and $P(\cdot|A_k) \to P(\cdot|x)$ and $\alpha(\cdot|A_k) \to \alpha(\cdot|x)$, as $k \to \infty$. Now, $\mathbf{P}(\cdot|A_k)$ is Dirichlet with parameter $c\alpha(A_k)\alpha(\cdot|A_k)$, and $\alpha(A_k) \to 0$ as $k \to \infty$. Hence we have a situation where we have a sequence of $D_{c_k\mu_k}$, with $\mu_k \to \alpha(\cdot|x)$ and $c_k \to 0$, and (ii) of Theorem 3.2.6 in Ghosh and Ramamoorthi (2003) suggests the conclusion.

To turn to a more formal argument, fix a Borel subset B of $\mathcal{Y}$. We will denote by Q a generic distribution of $\mathbf{Y}$. What we will show is

(a) $\int Q(B)\Pi_x(dQ) = \alpha(B|x)$ a.e. $\alpha^{\mathbf{X}}$

(b) $\int Q^2(B)\Pi_x(dQ) = \int Q(B)\Pi_x(dQ)$ a.e. $\alpha^{\mathbf{X}}$.

Note that (b) would imply that a.e. Π_x, $\mathbf{Q}^2(B)$ is 0 or 1. Running B through a countable algebra one gets that a.e. Π_x, $\mathbf{Q}$ is a 0-1 measure and any 0-1 measure is of course degenerate.

Consider the model $\mathbf{P} \sim D_{c\alpha}$, i.e., choose $\mathbf{P}^{\mathbf{X}}$ according to $D_{c\alpha^{\mathbf{X}}}$ and $\tilde{Q}_{\mathbf{P}}$ according to the distribution arising from the $D_{c\alpha}$ prior (independent of the distribution of $\mathbf{P}^{\mathbf{X}}$). Denote by Π_x the distribution of $Q_{\mathbf{P}}(\cdot|x)$, and let $\Pi^{\mathbf{X}} = D_{c\alpha^{\mathbf{X}}}$. Given $\mathbf{P} = P, (\mathbf{X}_1, \mathbf{Y}_1), (\mathbf{X}_2, \mathbf{Y}_2)$ are i.i.d. P.

For any Borel subset A of $\mathcal{X}$,

$$\int I_A(x)Q(B)\Pi_x(dQ)P^{\mathbf{X}}(dx)\Pi^{\mathbf{X}}(dP^{\mathbf{X}}) = \Pr(\mathbf{X} \in A, \mathbf{Y} \in B) = \int I_A(x)\alpha(B|x)\alpha(dx).$$

Since the above holds for all A, we have (a).

Now for the Dirichlet

$$\Pr((\mathbf{X}_2, \mathbf{Y}_2) = (x, y)|(\mathbf{X}_1, \mathbf{Y}_1) = (x, y)) = \frac{1}{c+1}.$$

Since

$$\Pr(\mathbf{X}_2 = x|\mathbf{X}_1 = x) = \frac{1}{c+1}$$

we have

$$\Pr(\mathbf{Y}_1 = \mathbf{Y}_2|X_1 = x, \mathbf{X}_2 = x) = 1$$

from which it follows that for any Borel subset B of $\mathcal{Y}$,

$$\Pr(\mathbf{Y}_1 \in B, \mathbf{Y}_2 \in B|\mathbf{X}_1 = x, \mathbf{X}_2 = x) = \Pr(\mathbf{Y}_1 \in B|\mathbf{X}_1 = x, \mathbf{X}_2 = x). \tag{4.2}$$

Since $\mathbf{Y}_1$ and $\mathbf{X}_2$ are conditionally independent given $\mathbf{X}_1$, the right hand side is equal to $\Pr(\mathbf{Y}_1 \in B|\mathbf{X}_1 = x) = \alpha(B|x)$. Since the term in the left hand side of (4.2) is $\int Q^2(B)\Pi_x(dQ)$, we have the result. $\square$

Acknowledgements

We like to thank Prof. Antony O'Hagan for his remarks and his suggestion to cast the first part in terms of partitions. This work was done when the first author was visiting University of Pavia. We both like to express our gratitude to Prof. E. Regazzini for his kindness and encouragement.

One of the authors was partially supported by the Ministero dell'Istruzione dell'Università e della Ricerca (Italian Minister of University and Research), research projects *"Bayesian networks and causal inference: methods and applications"* and *"Bayesian nonparametric methods and their applications"*.

References

Darroch, J.N. and Ratcliff, D. (1971). A characterization of the Dirichlet distribution, *J. Amer. Statist. Assoc.* **66**, 641–643.

Doksum, K. (1974). Tailfree and neutral random probabilities and their posterior distributions, *Ann. Probability*, **2**, 183–201.

Geiger, D. and Heckerman, D. (1997). A characterization of the Dirichlet distribution through global and local parameter independence. *Ann. Statist.*, **25**, 3, 1344–1369.

Ghosh, J.K., Hjort, N.L., Messan, C. and Ramamoorthi, R.V. (2006). Bayesian bivariate survival estimation. *J. Statist. Plann. Infer.*, **136**, 2297-2308.

Ghosh, J.K. and Ramamoorthi, R.V. (2003). *Bayesian nonparametrics*. Springer Series in Statistics, Springer-Verlag, New York.

Srivastava, S.M. (1998). *A course on Borel sets*, Graduate Texts in Mathematics, vol. 180, Springer-Verlag, New York.

Appendix

1. The first measure theoretic issue is the existence of $Q_P(\cdot|x)$ that is jointly measurable in (P, x). The construction is standard and we indicate it when $\mathcal{Y} = \mathbb{R}$. Let $\mathbb{Q}$ be the set of rationals, and $B_q = (-\infty, q)$, for $q \in \mathbb{Q}$. Fix a sequence of algebras $\mathcal{A}_k$ with atoms $\{A_{k1}, A_{k2}, \ldots, A_{km_k}\}$ which increase and generate the Borel σ-algebra on $\mathcal{X}$. Define

$$Q_P(B_q|x) = \lim_{k \to \infty} \sum_{i=1}^{m_k} \frac{P(A_{ki} \times B_q)}{P^{\mathbf{X}}(A_{ki})} I_{A_{ki}}(x), \text{ when the limit exists.}$$

The above limit exists with $P^{\mathbf{X}}$ probability 1. Since $\mathbb{Q}$ is countable it is easy to see that, outside a null set, the above limits would be monotone in q and right continuous at each q. This can then be extended to be right continuous at all real numbers. Thus for each x outside a null set we have a distribution function, which can then be extended to a measure on all of the Borel σ-algebra. If x is in the null set we fix $Q_P(\cdot|x)$ to be a fixed probability measure Q_0. This is easily checked to be a measurable version.

2. The second issue is to show that the measure $\tilde{\Pi} = \Pi_x \, \Pi_x$ on $M(\mathcal{Y})^{\mathcal{X}}$, given by the Kolmogorov consistency theorem, gives outer measure 1 to $\Delta \subset M(\mathcal{Y})^{\mathcal{X}}$, where

$$\Delta = \{\tilde{Q} \in M(\mathcal{Y})^{\mathcal{X}} : \text{ for each Borel set } B \subset \mathcal{Y}, x \mapsto Q(B|x) \text{ is measurable}\}.$$

The proof is similar to that appearing in Ghosh et al. (2006). Let E be any measurable set in $M(\mathcal{Y})^{\mathcal{X}}$ containing Δ. We will show that E is the whole space. Since E is measurable there is a countable set $\{x_1, x_2, \ldots\}$ such that E is in the σ-algebra generated by the projections to these coordinates. Hence for any $\tilde{Q}_1, \tilde{Q}_2$, if $Q_1(\cdot|x_i) = Q_2(\cdot|x_i), i = 1, 2, \ldots$ then either both $\tilde{Q}_1$ and $\tilde{Q}_2$ belong to E or neither would be in E. Suppose $\tilde{Q} \in M(\mathcal{Y})^{\mathcal{X}}$, and pick a $\tilde{Q}_0$ from Δ. Define $\tilde{Q}^*$ by

$$\begin{cases} Q^*(\cdot|x_i) = Q(\cdot|x_i), & \text{for } i = 1, 2, \ldots \\ Q^*(\cdot|x) = Q_0(\cdot|x), & \text{otherwise.} \end{cases}$$

Since $Q^*(\cdot|x) = Q_0(\cdot|x)$ but at countably many points x, measurability still holds and hence $\tilde{Q}^* \in \Delta$. Now $Q^*(\cdot|x_i) = Q(\cdot|x_i)$, for $i = 1, 2, \ldots$, hence $\tilde{Q}$ is also in E.

3. Let $M_D(\mathcal{X})$ be all discrete measures on $\mathcal{X}$. We show that on $M_D(\mathcal{X}) \times \Delta$, for Borel sets $A \subset \mathcal{X}$ and $B \subset \mathcal{Y}$, the function $(P^{\mathbf{X}}, \tilde{Q}) \mapsto \int_A Q(B|x) P^{\mathbf{X}}(dx)$ is jointly measurable in $\left(P^X, \tilde{Q}\right)$.

Let

$$G = \{(P^{\mathbf{X}}, x) : P^{\mathbf{X}}\{x\} > 0\}.$$

By Lusin's theorem [Theorem 5.10.3, Srivastava (1998)] there exist measurable functions $\phi_1(P^{\mathbf{X}}), \phi_2(P^{\mathbf{X}}), \ldots$, such that the graphs of $\phi_i(P^{\mathbf{X}})$, for $i = 1, 2, \ldots$, are disjoint and the union of graphs is G.

For any $A \times B$, our interest is in the function

$$(P^{\mathbf{X}}, \tilde{Q}) \mapsto \sum_{i:\phi_i(P^{\mathbf{X}}) \in A} Q(B|\phi_i(P^{\mathbf{X}}))P^{\mathbf{X}}(\phi_i(P^{\mathbf{X}})).$$

We want to argue that this is jointly measurable in $(P^{\mathbf{X}}, \tilde{Q})$. It is enough to show that $Q(B|\phi_i(P^{\mathbf{X}}))$ is jointly measurable in $(P^{\mathbf{X}}, \tilde{Q})$. This is immediate if ϕ_i is a simple function for then $\tilde{Q}$ depends only on finitely many coordinates. The general case follows since ϕ_i is a limit of simple functions.

4. In the general case $M(\mathcal{X}) \times \Delta$, for Borel sets $A \subset \mathcal{X}$ and $B \subset \mathcal{Y}$, the function $(P^{\mathbf{X}}, \tilde{Q}) \mapsto \int_A Q(B|x)P^{\mathbf{X}}(dx)$ is NOT jointly measurable in $(P^{\mathbf{X}}, \tilde{Q})$.

   Suppose it is, then $E = \left\{(P^{\mathbf{X}}, \tilde{Q}) : \int_A Q(B|x)P^{\mathbf{X}}(dx) > a\right\}$ is measurable set in $M(\mathcal{X}) \times M(\mathcal{Y})^{\mathcal{X}}$. Hence for each $P^{\mathbf{X}}$, the $P^{\mathbf{X}}$-section, $E_{P^{\mathbf{X}}}$, of E is measurable. So, for a fixed $P^{\mathbf{X}}$, there exists a countable set $\{x_1, x_2, \ldots\}$ such that for any $\tilde{Q}_1, \tilde{Q}_2$, if $Q_1(\cdot|x_i) = Q_2(\cdot|x_i)$, for $i = 1, 2, \ldots$, then either both $\tilde{Q}_1$ and $\tilde{Q}_2$ belong to $E_{P^{\mathbf{X}}}$ or neither is in $E_{P^{\mathbf{X}}}$. But if $P^{\mathbf{X}}$ is continuous this fails, for the value on a countable set has no effect on $\int_A Q(B|x)P^{\mathbf{X}}(dx)$.

5. In the general case, when we do not confine to discrete measures, a way is to define the σ-algebra on $M(\mathcal{X}) \times \Delta$ to be the one that makes the functions $(P^{\mathbf{X}}, \tilde{Q}) \mapsto \int_A Q(B|x)P^{\mathbf{X}}(dx)$ measurable. This is larger than the σ-algebra obtained by product of the Borel σ-algebra on $M(\mathcal{X})$, and the product σ-algebra on $M(\mathcal{Y})^{\mathcal{X}}$, where $M(\mathcal{Y})$ is endowed with its Borel σ-algebra. Note that the σ-algebra we are here taking on $M(\mathcal{Y})^{\mathcal{X}}$, when restricted to finitely many coordinates, is the same as the product field. Suppose we are given the marginal distributions Π_x on each coordinate and want to extend the product measure on $M(\mathcal{X}) \times M(\mathcal{Y})^{\mathcal{X}}$ to the σ-algebra generated by the maps $(P^{\mathbf{X}}, \tilde{Q}) \mapsto \int_A Q(B|x)P^{\mathbf{X}}(dx)$. One way would be to define for any Borel sets $A_1 \times B_1, A_2 \times B_2, \ldots, A_k \times B_k$, on $\mathcal{X} \times \mathcal{Y}$, and integers $n_1, n_2, \ldots, n_k$,

$$\int \left\{\prod_{i=1}^{k}[P(A_i \times B_i)]^{n_i}\right\} \Pi(dP) = \int \left[\int_F \left\{\prod_{i=1}^{k} \prod_{j=n'_{i-1}+1}^{n'_i} Q(B_i|x_j)\right\} \prod_{j=1}^{n} P^{\mathbf{X}}(dx_j)\right] \tilde{\Pi}(dQ)\Pi(dP^{\mathbf{X}})$$

where $F = \Pi_{i=1}^{k} A_i^{n_i}$, $n'_i = \sum_{j=1}^{i} n_j$ (for $i = 1, \ldots, k$), $n'_0 = 0, n = n'_k$. These define moments, from which one can assert the existence of a measure on the σ-algebra generated by the functions $(P^{\mathbf{X}}, \tilde{Q}) \mapsto \int_{A_i} Q(B_i|x)P^{\mathbf{X}}(dx), i = 1, 2, \ldots, k$. Then, we would have to check consistency conditions to ensure that the measure exists on the whole σ-algebra. These would take us on a detour and we do not pursue it here.

Bayesian Statistics and Its Applications
Edited by S.K. Upadhyay, U. Singh and D.K. Dey
Anamaya Publishers, New Delhi, India

Simpler Calculation of Posterior Distributions of the Parameters in Structural Equation Model

Kazuo Shigemasu and Takahiro Hoshino

Department of Cognitive and Behavioral Science, The University of Tokyo, Tokyo, Japan

Abstract

The obvious advantages of Bayesian approach lie in that the more precise inference for all unknown quantities in SEM is possible whereas the traditional methods provide only point estimates of the parameters and the asymptotic approximation of their variances, but the required calculation to derive all relevant posterior distributions can be very tedious. In order to circumvent this problem, we propose the numerical procedure which replaces latent variables in the conditional distributions by their expected values. The simulation study was conducted using 1,000 data sets. It was shown that the estimation performance by the proposed method is better than the ML estimates given by AMOS. Although the posterior variance of the parameters by the proposed method is slightly smaller than the fully Bayesian method, the EAP estimates are as precise as those obtained by the fully Bayesian method. Finally, the proposed method was applied to real data.

1. Introduction

Although Structural Equation Model (SEM) is widely used in social sciences as a very influential method, some methodological problems in SEM have not been solved completely. For example, there has still been no clear practical guideline for identifiable model specification. Also, our feeling is that a large number of criteria (goodness of fit indices, information criteria, RMSEA, etc.) for model evaluation have not been integrated properly and sound and exact statistical inference for estimation and prediction has not yet been practical.

This article focuses on the statistical inference of the parameters in SEM. In this respect, the state-of-the-art techniques available are far from complete. Many techniques heavily depend on the asymptotic approximation and they often provide inconsistent results.

For the very complex models such as SEM, the Bayesian approach is most suitable in that it enables exact and coherent statistical inferences about parameters and unobserved variables. More concretely, the Bayesian approach for SEM is advantageous because of:

(1) SEM can be regarded as a hierarchical model from the Bayesian viewpoint. Then it is easier to understand the meaning of each model in SEM setting and to build a new model to be appropriate to solve real problems.

(2) Exact statistical inference can be made possible based on the exact posterior distributions of parameters, which are derived numerically.

(3) Predictive distribution for future unobserved variables is available. Also, the marginal posterior distribution for latent variables (factor scores) is available for prediction. Of course, the Bayesian approach for SEM is not a new idea, and a number of papers have been published (e.g., Arminger and Muthén, 1998; Scheines, Hoijtink and Boomsma, 1999; Shi and Lee, 1997, 1998).

Shigemasu, Ohmori and Hoshino (2002) briefly reviewed these papers, and they proposed a fully Bayesian method, which used the generalized natural conjugate priors (Press, 1982). In their development of numerical procedure, latent variables were not integrated out, and the full conditional distributions turned out to belong to standard distributions, and hence the Gibbs Sampler was used to obtain posterior distributions of the parameters. Gibbs Sampler is appropriate in that it is easier to program and it provides the posterior distributions for all relevant unknown quantities including factor scores. But a plausible problem is that it can be computationally burdensome. The purpose of the present paper is to make the Bayesian procedure developed in Shigemasu and Nakamura (1993), and Shigemasu, Ohmori and Hoshino (2002) more practical and feasible to analyze the data with many observations. There have been added more recent Bayesian work for SEM. For example, Ansari, Jedidi and Dube (2002) specified various multilevel models of mean and co-variance heterogeneity in confirmatory factor analysis and developed Bayesian estimation method. Hogan and Tchernis (2004) proposed a Bayesian hierarchical model for factor analysis of spatially correlated multivariate data. Lopes and West (2004) explored reversible jump MCMC methods that uses parallel Gibbs sampling based analyses to determine the number of latent factors.

2.   Model and Distributional Assumptions

SEM is a combination of confirmatory factor analysis model and simultaneous equation model. The observation matrix $\boldsymbol{X}$ ($n \times p$; n: number of persons, p: number of variables) is expressed by the linear model of latent variables $\boldsymbol{F}$ ($n \times m$; m: number of factors) as

$$\boldsymbol{X} = \boldsymbol{F}\boldsymbol{\Lambda}^t + \boldsymbol{E}_1,$$

where $\Lambda(p \times m)$ is a weight matrix, and $\boldsymbol{E}$ is a matrix of residual terms. Latent variables or factor scores $\boldsymbol{F}$ are divided into two parts, i.e., dependent latent variables $\boldsymbol{F}_1(n \times m_1)$ and independent latent variables $\boldsymbol{F}_2$ ($n \times m_2, m = m_1 + m_2$). The weight matrix or factor loading matrix $\boldsymbol{\Lambda}$ is divided accordingly into two parts $\boldsymbol{\Lambda}_1$, and $\boldsymbol{\Lambda}_2$, i.e., $\boldsymbol{\Lambda} = (\boldsymbol{\Lambda}_1, \boldsymbol{\Lambda}_2)$. Thus

$$\boldsymbol{X} = \boldsymbol{F}_1\boldsymbol{\Lambda}_1^t + \boldsymbol{F}_2\boldsymbol{\Lambda}_2^t + \boldsymbol{E}_1.$$

Two kinds of factor scores are connected linearly as

$$\boldsymbol{F}_1 = \boldsymbol{F}_1\boldsymbol{B}_0^t + \boldsymbol{F}_2\boldsymbol{\Gamma}^t + \boldsymbol{E}_2,$$

where $\boldsymbol{B}_0(m_1 \times m_1)$ and $\boldsymbol{\Gamma}(m_1 \times m_2)$ are appropriate weight matrices. That is, the dependent latent variables $\boldsymbol{F}_1$ are explained simultaneously as the linear combination of $\boldsymbol{F}_1$ and independent latent variables $\boldsymbol{F}_2$.

Now we specify the distributions of random variables. Each row of $\boldsymbol{F}_2$($\boldsymbol{f}_{i2}$), each row of $\boldsymbol{E}_1$ ($\boldsymbol{\varepsilon}_{1i}$), and each row of $\boldsymbol{E}_2$ ($\boldsymbol{\varepsilon}_{2i}$) are assumed to be distributed as the multivariate normal distributions; $\boldsymbol{f}_{i2} \sim N(0, \boldsymbol{\Phi})$, $\boldsymbol{\varepsilon}_{1i} \sim N(0, \boldsymbol{\Psi})$, $\boldsymbol{\varepsilon}_{2i} \sim N(0, \boldsymbol{\Omega})$. Defining $\boldsymbol{B}_0^t = (\boldsymbol{I} - \boldsymbol{B}_0)^t$, the model distribution of

X and F given the structural parameters is

$$p(X, F | \Lambda, B, \Gamma, \Phi, \Psi, \Omega)$$

$$\propto |\Omega|^{-\frac{n}{2}} |\Psi|^{-\frac{n}{2}} |\Phi|^{-\frac{n}{2}} \exp\left\{ -\frac{1}{2}\mathrm{tr}[F_2 \Phi^{-1} F_2^t + (X - F\Lambda^t)\Psi^{-1}(X - F\Lambda^t)^t \right.$$

$$\left. + (F_1 B^t - F_2 \Gamma^t)\Omega^{-1}(F_1 B^t - F_2 \Gamma^t)^t] \right\}.$$

2.1   Prior Distributions

Prior distributions for variance matrices are specified as

$$p(\Phi) \propto |\Phi|^{-\nu_\Phi/2} \exp\left\{ -\frac{1}{2}\mathrm{tr}\Phi^{-1}G_\Phi \right\},$$

$$p(\Psi) \propto |\Psi|^{-\nu_\Psi/2} \exp\left\{ -\frac{1}{2}\mathrm{tr}\Psi^{-1}G_\Psi \right\},$$

and

$$p(\Omega) \propto |\Omega|^{-\nu_\Omega/2} \exp\left\{ -\frac{1}{2}\mathrm{tr}\Omega^{-1}G_\Omega \right\}.$$

These prior distributions are abbreviated as $W^{-1}(\nu_\Phi, G_\Phi)$ and likewise. As for the other parameters Λ, B and Γ, we assume the (locally) uniform distributions for convenience. Then the joint posterior distributions are obtained as the product of the model distribution and the prior distribution.

2.2   Full Conditional Distributions for Parameters

In this article Gibbs Sampler is used to derive the posterior distributions. To make use of Gibbs Sampler, we need full conditional distributions.

The conditional distributions for Φ, Ψ, Ω, given X and all the other parameters, are

$$W^{-1}(\nu_\Phi + n, G_\Phi + F_2^t F_2),$$

$$W^{-1}(\nu_\Psi + n, G_\Psi + (X - F\Lambda^t)^t(X - F\Lambda^t)),$$

and

$$W^{-1}(\nu_\Omega + n, (F_1 B^t - F_2 \Gamma^t)^t(F_1 B^t - F_2 \Gamma^t)),$$

respectively. Also, the conditional distributions for F_1 and F_2 are the following matrix normal distributions.

$$F_1 \sim N\left\{ [(X - F_2\Lambda_2^t)\Psi^{-1}\Lambda_1 + F_2\Gamma^t\Omega^{-1}B]C_1^{-1}, \ I_n \otimes C_1^{-1} \right\},$$

and

$$F_2 \sim N\left\{ [(X - F_1\Lambda_1^t)\Psi^{-1}\Lambda_2 + F_1 B^t\Omega^{-1}\Gamma]C_2^{-1}, \ I_n \otimes C_2^{-1} \right\},$$

where

$$C_1 = \Lambda_1^t\Psi^{-1}\Lambda_1 + B^t\Omega^{-1}B$$

and

$$C_2 = \Phi + \Lambda_2^t\Psi^{-1}\Lambda_2 + \Gamma^t\Omega^{-1}\Gamma.$$

The density functions of the full conditional distributions for Λ, B and Γ can be expressed as

$$p(\Lambda|X \text{ and other parameters}) \propto \exp\left\{-\frac{1}{2}\mathrm{tr}\Psi^{-1}(\Lambda - XF(F^tF)^{-1})(F^tF)(\Lambda - X(F^tF)^{-1})^t\right\},$$

$$p(B|X \text{ and other parameters}) \propto \exp\left\{-\frac{1}{2}\mathrm{tr}\Omega^{-1}(B-\Gamma F_2^t F_1(F_1^t F_1)^{-1})(F_1^t F_1)(B-\Gamma F_2^t F_1(F_1^t F_1)^{-1})^t\right\},$$

$$p(\Gamma|X \text{ and other parameters}) \propto \exp\left\{-\frac{1}{2}\mathrm{tr}\Omega^{-1}(\Gamma-BF_1^t F_2(F_2^t F_2)^{-1})(F_2^t F_2)(\Gamma-BF_1^t F_2(F_2^t F_2)^{-1})^t\right\}.$$

By vectorizing the weight matrices Λ, B, and Γ, these density functions appear to be multivariate normal densities. Let $\lambda = vec(\Lambda^t)$, $\beta = vec(B^t)$ and $\gamma = vec(\Gamma^t)$, then

$$p(\lambda|X \text{ and other parameters}) \propto \exp\left\{-\frac{1}{2}(\lambda - \bar{\lambda})^t(\Psi \otimes (F^tF)^{-1})^{-1}(\lambda - \bar{\lambda})\right\},$$

$$p(\beta|X \text{ and other parameters}) \propto \exp\left\{-\frac{1}{2}(\beta - \bar{\beta})^t(\Omega \otimes (F_1^t F_1)^{-1})^{-1}(\beta - \bar{\beta})\right\},$$

and
$$p(\gamma|X \text{ and other parameters}) \propto \exp\left\{-\frac{1}{2}(\gamma - \bar{\gamma})^t(\Omega \otimes (F_2^t F_2)^{-1})^{-1}(\gamma - \bar{\gamma})\right\},$$

where $\bar{\lambda} = vec[(X^t F(F^tF)^{-1})^t]$, $\bar{\beta} = vec[(\Gamma F_2^t F_1(F_1^t F_1)^{-1})^t]$ and $\bar{\gamma} = vec[(BF_1^t F_2(F_2^t F_2)^{-1})^t]$.

It should be reminded that the model employed is the confirmatory factor analysis model, and some of the elements of the parameters are fixed in advance. Without loss of generality, this condition is expressed by using the permutation matrices.

Now let the permutation matrix K_λ changes the location of the elements of λ so that the non-zero elements of the vector λ^* are put at the top and the zero elements at the bottom, that is,

$$\begin{pmatrix}\lambda^* \\ 0\end{pmatrix} = K_\lambda\lambda.$$

Similarly,

$$\begin{pmatrix}\beta^* \\ 0\end{pmatrix} = K_\beta\beta,$$

and
$$\begin{pmatrix}\gamma^* \\ 0\end{pmatrix} = K_\gamma\gamma.$$

Also, let us now write $\bar{\lambda}^* = K_\lambda\bar{\lambda}, \bar{\beta}^* = K_\beta\bar{\beta}, \bar{\gamma}^* = K_\gamma\bar{\gamma}, \Sigma_\lambda^{-1} = K_\lambda(\Psi \otimes (F^tF)^{-1})^{-1}K_\lambda^t, \Sigma_\beta^{-1} = K_\beta(\Omega \otimes (F_1^t F_1)^{-1})^{-1}K_\beta^t$, and $\Sigma_\gamma^{-1} = K_\gamma(\Omega \otimes (F_2^t F_2)^{-1})^{-1}K_\gamma^t$.

The conditional distribution for λ^* with fixed zero elements is given as

$$p(\lambda^*|X \text{ and other parameters}) \propto \exp\left\{-\frac{1}{2}\left[(\lambda^* - \bar{\lambda}_1^*)^t\Sigma_{11\lambda}(\lambda^* - \bar{\lambda}_1^*) - \bar{\lambda}_2^{*t}\Sigma_{21\lambda}(\lambda^* - \bar{\lambda}_1^*)\right]\right\}$$

$$\propto \exp\left\{-\frac{1}{2}(\lambda^* - \bar{\lambda}_c^*)^t\Sigma_{\lambda_c}^{-1}(\lambda^* - \bar{\lambda}_c^*)\right\},$$

where $\overline{\boldsymbol{\lambda}}_c^* = \overline{\boldsymbol{\lambda}}_1^* - \boldsymbol{\Sigma}_{12\lambda}^{-1}\boldsymbol{\Sigma}_{22\lambda}^{-1}\overline{\boldsymbol{\lambda}}_2^*, \quad \boldsymbol{\Sigma}_{\lambda c} = \boldsymbol{\Sigma}_{11\lambda} - \boldsymbol{\Sigma}_{12\lambda}\boldsymbol{\Sigma}_{22\lambda}^{-1}\boldsymbol{\Sigma}_{21\lambda}, \quad \boldsymbol{\Sigma}_\lambda = \begin{bmatrix} \boldsymbol{\Sigma}_{11\lambda} & \boldsymbol{\Sigma}_{12\lambda} \\ \boldsymbol{\Sigma}_{21\lambda} & \boldsymbol{\Sigma}_{22\lambda} \end{bmatrix}, \quad \overline{\boldsymbol{\lambda}}^* = \begin{pmatrix} \overline{\boldsymbol{\lambda}}_1^* \\ \overline{\boldsymbol{\lambda}}_2^* \end{pmatrix}.$

The conditional distribution for $\boldsymbol{\lambda}^*$ is $N(\overline{\boldsymbol{\lambda}}_c^*, \boldsymbol{\Sigma}_{\lambda c})$. Similarly,

$$\boldsymbol{\beta}^* \sim N(\overline{\boldsymbol{\beta}}_c^*, \ \boldsymbol{\Sigma}_{\beta,c}), \quad \overline{\boldsymbol{\beta}}_c^* = \overline{\boldsymbol{\beta}}_1^* - \boldsymbol{\Sigma}_{12\beta}\boldsymbol{\Sigma}_{22\beta}^{-1}\overline{\boldsymbol{\beta}}_2^*, \quad \boldsymbol{\Sigma}_{\beta c} = \boldsymbol{\Sigma}_{11\beta} - \boldsymbol{\Sigma}_{12\beta}\boldsymbol{\Sigma}_{22\beta}^{-1}\boldsymbol{\Sigma}_{21\beta},$$

$$\boldsymbol{\gamma}^* \sim N(\overline{\boldsymbol{\gamma}}_c^*, \ \boldsymbol{\Sigma}_{\gamma,c}), \quad \overline{\boldsymbol{\gamma}}_c^* = \overline{\boldsymbol{\gamma}}_1^* - \boldsymbol{\Sigma}_{12\gamma}\boldsymbol{\Sigma}_{22\gamma}^{-1}\overline{\boldsymbol{\gamma}}_2^*, \quad \boldsymbol{\Sigma}_{\gamma c} = \boldsymbol{\Sigma}_{11\gamma} - \boldsymbol{\Sigma}_{12\gamma}\boldsymbol{\Sigma}_{22\gamma}^{-1}\boldsymbol{\Sigma}_{21\gamma}.$$

3. Gibbs Sampler and Its Improvement

The numerical procedure used in Shigemasu, Ohmori and Hoshino (2002) is as follows. The hyper parameters were fixed at

$$\boldsymbol{G}_\Psi = \boldsymbol{G}_\Omega = \boldsymbol{G}_\Phi = 0.2\nu I, \text{ and } \nu = 2p + 3.$$

Let $\boldsymbol{\theta}_1 = vec(\Lambda, \boldsymbol{B}, \boldsymbol{\Gamma}), \boldsymbol{\theta}_2 = vec(\Psi, \Phi, \Omega), \boldsymbol{\theta}_3 = vec\ \boldsymbol{F}$, their algorithm considered repetition of following steps until convergence:

Step 0: Initialize $\boldsymbol{\theta}_2$ and $\boldsymbol{\theta}_3$.

Step 1: Generate random observation $\boldsymbol{\theta}_1$ from $p(\boldsymbol{\theta}_1|\boldsymbol{\theta}_2, \boldsymbol{\theta}_3)$.

Step 2: Generate random observation $\boldsymbol{\theta}_2$ from $p(\boldsymbol{\theta}_2|\boldsymbol{\theta}_1, \boldsymbol{\theta}_3)$.

Step 3: Generate random observation $\boldsymbol{\theta}_3$ from $p(\boldsymbol{\theta}_3|\boldsymbol{\theta}_1, \boldsymbol{\theta}_2)$.

As mentioned in Section 1, when the parameter vector $\boldsymbol{\theta}_3$ (factor scores) has too many elements, the above procedure can be too time-consuming and as such, may not be practical for real data analysis. In fact, $\dim(\boldsymbol{\theta}_3)$ is equal to the number of persons times the number of variables, which can typically be more than 1,000. It is often burdensome to get random draws for all parameters.

Therefore, approximation is needed to reduce the computational burden to a feasible level. It is sometimes good enough to make inferences about other parameters with $\boldsymbol{\theta}_3$ fixed to its expected value, i.e., $\boldsymbol{\tau} = \boldsymbol{E}(\boldsymbol{\theta}_3|\boldsymbol{\theta}_1, \boldsymbol{\theta}_2)$, which is obtained analytically when $\boldsymbol{\theta}_1$ and $\boldsymbol{\theta}_2$ are given. So we propose a simple version for the Gibbs sampler algorithm. This algorithm consists of two parts. One part is calculation of expected values of $\boldsymbol{\theta}_3$ when $\boldsymbol{\theta}_1$ and $\boldsymbol{\theta}_2$ are given and the other part is generating random numbers with $\boldsymbol{\theta}_3$ fixed to its expected values. The steps are

Step 0: Set up the initial values of $\boldsymbol{\theta}_1$ and $\boldsymbol{\theta}_2$ and calculate the expected value of $\boldsymbol{\theta}_3$,

Step 1: Generate random observation $\boldsymbol{\theta}_1$ from $p(\boldsymbol{\theta}_1|\boldsymbol{\theta}_2, \boldsymbol{\tau})$.

Step 2: Generate random observation $\boldsymbol{\theta}_2$ from $p(\boldsymbol{\theta}_2|\boldsymbol{\theta}_1, \boldsymbol{\tau})$.

Step 3: Calculate the expected value $\boldsymbol{\tau} = \boldsymbol{E}(\boldsymbol{\theta}_3|\boldsymbol{\theta}_1, \boldsymbol{\theta}_2)$ with the updated values of $\boldsymbol{\theta}_1$ and $\boldsymbol{\theta}_2$.

Step 4: Repeat Steps 1 and 3 until the convergence is attained.

To obtain conditional posterior mean of $\boldsymbol{F}$ (Step 3), let us write, $\boldsymbol{D} = (\boldsymbol{B}, -\boldsymbol{\Gamma})$ and $I_{m_2}^* = (0^t, I_{m_2})^t$. We then have

$$\boldsymbol{F} \sim N(\boldsymbol{X}\boldsymbol{\Psi}^{-1}\boldsymbol{\Lambda}\boldsymbol{\Sigma}_F^{-1}, I_n \otimes \boldsymbol{\Sigma}_F^{-1}),$$

where
$$\boldsymbol{\Sigma}_F = \boldsymbol{\Lambda}^t\boldsymbol{\Psi}^{-1}\boldsymbol{\Lambda} + I_{m_2}^*\boldsymbol{\Phi}^{-1} + \boldsymbol{D}^t\boldsymbol{\Omega}^{-1}\boldsymbol{D}.$$

Hence, the expected value of $\boldsymbol{F}$ is given by

$$\boldsymbol{\tau} = vec(\boldsymbol{X}\boldsymbol{\Psi}^{-1}\boldsymbol{\Lambda}\boldsymbol{\Sigma}_F^{-1}).$$

In the following way, we can justify the use of expected value for random draws in the above procedure. Instead of generating random numbers from $\boldsymbol{\theta}_1 \sim p(\boldsymbol{\theta}_1|\boldsymbol{\theta}_2,\boldsymbol{\theta}_3)$ and $\boldsymbol{\theta}_2 \sim p(\theta_2|\boldsymbol{\theta}_1,\boldsymbol{\theta}_3)$, if the marginalization is feasible, we can generate

$$\boldsymbol{\theta}_1 \sim p(\boldsymbol{\theta}_1|\boldsymbol{\theta}_2) = \int p(\boldsymbol{\theta}_1|\boldsymbol{\theta}_2,\boldsymbol{\theta}_3)p(\boldsymbol{\theta}_3|\boldsymbol{\theta}_2)d\boldsymbol{\theta}_3.$$

$$\boldsymbol{\theta}_2 \sim p(\boldsymbol{\theta}_2|\boldsymbol{\theta}_1) = \int p(\boldsymbol{\theta}_2|\boldsymbol{\theta}_1,\boldsymbol{\theta}_3)p(\boldsymbol{\theta}_3|\boldsymbol{\theta}_1)d\boldsymbol{\theta}_3.$$

Using the result of Plug-in posterior predictive distribution of Dunsmore (1976)

$$p(\boldsymbol{\theta}_1|\boldsymbol{\theta}_2) = \int p(\boldsymbol{\theta}_1|\boldsymbol{\theta}_2,\boldsymbol{\theta}_3)p(\boldsymbol{\theta}_3|\boldsymbol{\theta}_2)d\boldsymbol{\theta}_3$$

can be approximated by

$$p(\boldsymbol{\theta}_1|\boldsymbol{\theta}_2) \approx p(\boldsymbol{\theta}_1|\boldsymbol{\theta}_2,\boldsymbol{\tau})$$

where

$$\boldsymbol{\tau} = \boldsymbol{E}(\boldsymbol{\theta}_3|\boldsymbol{\theta}_1,\boldsymbol{\theta}_2).$$

We apply the above result by letting $\boldsymbol{\theta}_3 = vec(\boldsymbol{F})$.

4. Simulation Study

The model used for simulation study is as follows (see Fig. 1):

$$\boldsymbol{\Lambda}_1 = \begin{pmatrix} 1^* & 0^* \\ \lambda_{21} & \lambda_{22} \\ \lambda_{31} & \lambda_{32} \\ 0^* & 1^* \end{pmatrix},$$

$$\boldsymbol{\Lambda}_2 = \begin{pmatrix} \lambda_{53} & 0^* & 0^* \\ \lambda_{63} & \lambda_{64} & 0^* \\ \lambda_{73} & \lambda_{74} & 0^* \\ 0^* & \lambda_{84} & \lambda_{85} \\ 0^* & \lambda_{94} & \lambda_{95} \\ 0^* & 0^* & \lambda_{105} \end{pmatrix},$$

$$\boldsymbol{B} = \begin{pmatrix} 1^* & \beta_{21} \\ 0^* & 1^* \end{pmatrix},$$

$$\boldsymbol{\Gamma} = \begin{pmatrix} \gamma_{11} & 0^* \\ \gamma_{21} & 0^* \\ 0^* & \gamma_{32} \end{pmatrix}.$$

The assigned "True" values and their estimates are given in Tables 1 and 2. Assuming that the number of persons (n) is 100 and 400, we conducted the simulation study using 1,000 sets of randomly generated data taking the same structure for both $n = 100$ and $n = 400$. The mean squared errors (MSE) for estimation are also given in Tables 1 and 2.

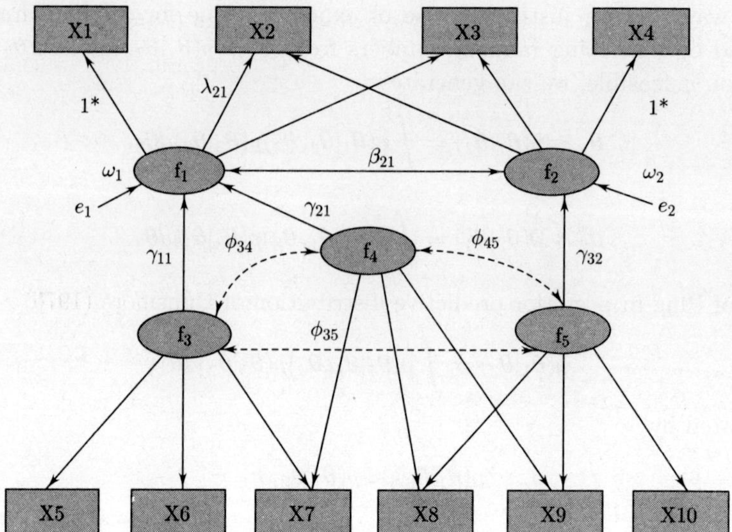

Fig. 1.   True model for simulation.

Table 1.   Estimates and standard errors ($N = 100$)

	TRUE	Estimates Gibbs	MLE	Prop	Standard Errors Gibbs	MLE	Prop		TRUE	Estimates Gibbs	MLE	Prop	Standard Errors Gibbs	MLE	Prop
λ_{21}	0.70	0.684	0.716	0.678	0.092	0.090	0.088	ψ_1	0.20	0.213	0.215	0.212	0.059	0.054	0.061
λ_{22}	0.30	0.311	0.306	0.318	0.098	0.092	0.096	ψ_2	0.20	0.215	0.212	0.224	0.068	0.062	0.066
λ_{31}	0.30	0.308	0.297	0.307	0.093	0.086	0.093	ψ_3	0.20	0.192	0.183	0.194	0.062	0.055	0.058
λ_{32}	0.70	0.710	0.777	0.690	0.088	0.088	0.090	ψ_4	0.20	0.206	0.208	0.210	0.064	0.069	0.062
λ_{53}	1.50	1.474	1.423	1.542	0.102	0.096	0.090	ψ_5	0.30	0.274	0.276	0.273	0.060	0.066	0.064
λ_{63}	1.70	1.640	1.643	1.598	0.089	0.084	0.087	ψ_6	0.30	0.301	0.316	0.310	0.071	0.030	0.064
λ_{73}	1.00	1.000	0.976	0.942	0.093	0.077	0.086	ψ_7	0.30	0.288	0.313	0.260	0.048	0.040	0.050
λ_{74}	0.70	0.645	0.672	0.639	0.082	0.088	0.089	ψ_8	0.30	0.901	0.309	0.306	0.069	0.060	0.060
λ_{84}	0.50	0.554	0.512	0.572	0.100	0.091	0.101	ψ_9	0.30	0.301	0.310	0.313	0.067	0.064	0.063
λ_{85}	1.20	1.335	1.367	1.244	0.080	0.076	0.090	ψ_{10}	0.30	0.287	0.200	0.291	0.063	0.067	0.058
λ_{94}	0.80	0.804	0.775	0.776	0.087	0.081	0.088	γ_{11}	0.50	0.476	0.495	0.480	0.062	0.069	0.057
λ_{95}	0.70	0.672	0.670	0.665	0.092	0.087	0.089	γ_{21}	0.60	0.630	0.508	0.633	0.061	0.061	0.062
λ_{105}	1.50	1.513	1.498	1.456	0.097	0.081	0.088	γ_{32}	0.70	0.661	0.721	0.676	0.064	0.064	0.058
ϕ_{12}	0.50	0.503	0.488	0.461	0.046	0.038	0.090	ω_1	0.20	0.206	0.211	0.187	0.062	0.048	0.056
ϕ_{13}	0.30	0.300	0.300	0.327	0.077	0.072	0.060	ω_2	0.30	0.285	0.288	0.292	0.064	0.060	0.090
ϕ_{23}	0.50	0.517	0.496	0.553	0.073	0.066	0.067	β_{21}	−0.40	−0.397	−0.419	−0.396	0.060	0.068	0.054
								MSE		263.25	322.98	281.72			

Table 2.   Estimates and standard errors ($N = 400$)

		Estimates			Standard Errors		
	TRUE	Gibbs	MLE	Prop	Gibbs	MLE	Prop
λ_{21}	0.70	0.722	0.721	0.718	0.060	0.055	0.060
λ_{22}	0.30	0.302	0.287	0.308	0.061	0.053	0.059
λ_{31}	0.30	0.284	0.285	0.283	0.061	0.056	0.056
λ_{32}	0.70	0.714	0.725	0.716	0.061	0.067	0.060
λ_{53}	1.50	1.424	1.404	1.417	0.060	0.055	0.056
λ_{63}	1.70	1.656	1.636	1.654	0.061	0.056	0.059
λ_{73}	1.00	1.012	1.002	1.019	0.062	0.058	0.060
λ_{74}	0.70	0.684	0.701	0.682	0.061	0.057	0.056
λ_{84}	0.50	0.484	0.478	0.486	0.061	0.056	0.058
λ_{85}	1.20	1.278	1.258	1.285	0.061	0.056	0.063
λ_{94}	0.80	0.786	0.801	0.785	0.061	0.057	0.058
λ_{95}	0.70	0.700	0.686	0.702	0.061	0.055	0.061
λ_{105}	1.50	1.442	1.496	1.445	0.061	0.038	0.067
ϕ_{12}	0.50	0.497	0.508	0.488	0.040	0.037	0.096
ϕ_{13}	0.30	0.318	0.314	0.317	0.040	0.034	0.041
ϕ_{23}	0.50	0.475	0.514	0.476	0.041	0.039	0.035

		Estimates			Standard Errors		
	TRUE	Gibbs	MLE	Prop	Gibbs	MLE	Prop
ψ_1	0.20	0.220	0.224	0.220	0.040	0.006	0.043
ψ_2	0.20	0.189	0.175	0.190	0.041	0.069	0.044
ψ_3	0.20	0.196	0.210	0.194	0.039	0.038	0.036
ψ_4	0.20	0.184	0.184	0.185	0.040	0.035	0.035
ψ_5	0.30	0.312	0.308	0.311	0.040	0.034	0.040
ψ_6	0.30	0.311	0.329	0.312	0.041	0.039	0.040
ψ_7	0.30	0.209	0.296	0.290	0.041	0.037	0.037
ψ_8	0.30	0.288	0.288	0.288	0.041	0.036	0.042
ψ_9	0.30	0.307	0.272	0.308	0.040	0.033	0.039
ψ_{10}	0.30	0.277	0.270	0.277	0.040	0.034	0.034
γ_{11}	0.50	0.477	0.480	0.477	0.041	0.034	0.036
γ_{21}	0.60	0.557	0.571	0.568	0.041	0.036	0.033
γ_{32}	0.70	0.632	0.675	0.701	0.042	0.036	0.039
ω_1	0.20	0.103	0.190	0.102	0.040	0.036	0.036
ω_2	0.30	0.204	0.296	0.264	0.040	0.037	0.029
β_{21}	−0.40	−0.416	−0.426	−0.414	0.040	0.035	0.040
MSE		154.21	173.08	162.94			

The simulation study indicates that in terms of the precision of estimation, the proposed method performs better than the MLE (calculated by AMOS, Arbuckle, 1995), but somewhat worse than the fully Bayesian method.

Tables 1 and 2 show that the standard deviations of the estimates (or standard errors) were smaller for MLEs than Bayesian methods. But this does not imply that MLE is more precise. Note that the asymptotic approximation based on the inverse of observed Fisher Information matrix is only the lower bound of variances of MLEs. As we anticipated, the calculation time required for the proposed model is remarkably reduced (see Fig. 2). This performance was compared using the personal computer with Pentium 4 CPU with 2.8 GHz.

5.   Extension of Model and Analysis of Ordinal Data

To demonstrate the utility of the proposed method, we applied the model to a real data. The data analysis used for demonstration aims to predict some typical behavior by the basic personality structure. The personality structure is often believed to have five basic factors. Thus the independent variables are five factor scores (F1: Conscientiousness, F2: Extraversion, F3: Openness, F4: Neuroticism, F5: Agreeableness). The dependent variable is "How long do you use the Internet in a day? (more than 4 hr/day, 2-4 hr, 1-2 hr, 4-7 hr/week, 1-4 hr/week, none)" (see Fig. 3). The five factors are latent, and the observed are the responses to 15 items. These responses are categorical on 5 point scale. Then we need a data generation model which relates the categorical responses to

N	100	400	2000
Full Gibbs	101.32	392.55	1892.37
Proposed	65.39	88.22	119.35

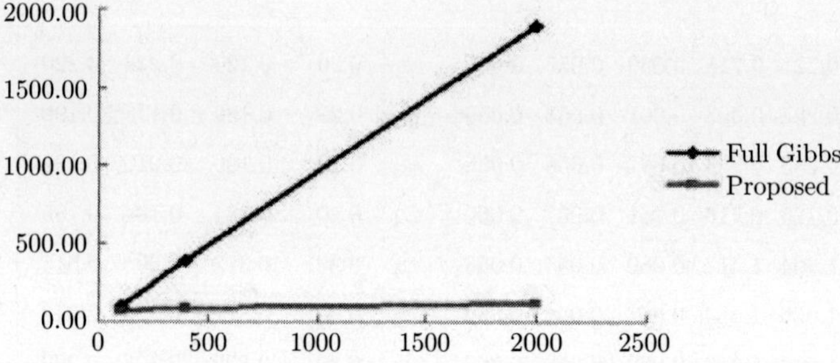

Fig. 2.   **Average of calculation time.**

the continuous latent variables. We introduce the additional latent variable (u), which is expressed by the linear model, that is

$$u_i \sim N(\boldsymbol{f}_i^t \boldsymbol{\omega}, 1)$$

and we assume that the person i responds to the category k, when

$$c_{k-1} < u < c_k$$

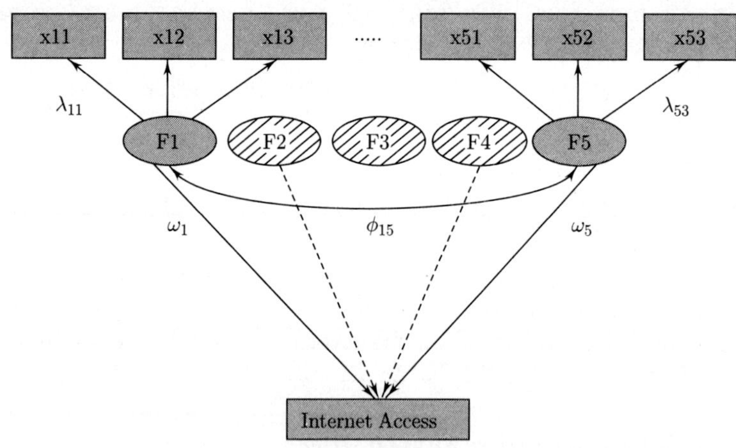

Fig. 3.   **Model for real data analysis.**

Further we employ the non-informative prior distributions for c's and $\boldsymbol{\omega}$. The posterior distribtion for $\boldsymbol{\omega}$ is,

$$\boldsymbol{\omega}|\boldsymbol{Fu} \sim N((F^t F)^{-1} F^t \boldsymbol{u}, (F^t F)^{-1})$$

The conditional posterior distribution of c_k is uniform in the interval

$$\text{(lowerbound)} \max\{\max\{\max\{u : x = k\}, c_{k-1}\}\}$$
$$\text{(upperbound)}\ \ \min\{\min\{u : x = k + 1\}, c_{k+1}\}\}$$

(see Albert and Chib, 1993).

The estimates are given in Tables 3, 4 and 5. The total number of iterations was 7000. The number of burn-in iterations was 3000. Convergence of this sampling was confirmed using shrink factor of Gelman and Rubin (Gelman and Rubin, 1992).

Table 3.   Estimates of parameters λ

| | | | | Upper estimate | | | |
				Lower standard deviation			
λ_{11}	λ_{12}	λ_{13}	λ_{14}	λ_{21}	λ_{21}	λ_{23}	λ_{24}
1.3487	0.7855	0.7425	1.0747	0.8429	1.2773	0.9312	0.6273
0.0180	0.0250	0.0247	0.0203	0.0210	0.0193	0.0219	0.0174
λ_{31}	λ_{32}	λ_{33}	λ_{34}	λ_{41}	λ_{42}	λ_{43}	λ_{44}
0.8602	0.8569	0.9640	1.0418	0.9057	0.9063	0.7830	0.09324
0.0180	0.0253	0.0197	0.0298	0.0200	0.0251	0.0192	0.0202
λ_{51}	λ_{52}	λ_{53}	λ_{54}				
1.0788	1.0516	1.8985	0.8598				
0.0199	0.0062	0.0216	0.0130				

Table 4.   Estimates of ϕ and ω

| | | | | Upper: estimate | | | | | |
				Lower: standard deviation					
ϕ_{12}	ϕ_{13}	ϕ_{14}	ϕ_{15}	ϕ_{23}	ϕ_{24}	ϕ_{25}	ϕ_{34}	ϕ_{35}	ϕ_{45}
0.2130	0.1895	0.0904	1.940	0.2053	0.1749	0.1403	0.1791	0.0855	0.1693
0.0228	0.0167	0.0217	0.0288	0.0613	0.0136	0.0225	0.0115	0.0133	0.0184

	ω_1	ω_2	ω_3	ω_4	ω_5
	0.1362	−0.2352	0.0428	0.1715	0.0383
	0.0417	0.0147	0.0230	0.0189	0.0148

Table 5.   Estimates of ψ

| | | | | Upper: estimate | | | | | |
				Lower: standard deviation					
ψ_1	ψ_2	ψ_3	ψ_4	ψ_5	ψ_6	ψ_7	ψ_8	ψ_9	ψ_{10}
1.4582	1.8302	2.0904	1.7490	1.8043	2.8034	2.2168	1.8524	2.0683	1.8470
0.0283	0.0233	0.0268	0.0162	0.0182	0.0271	0.0123	0.0191	0.0142	0.0148
ψ_{11}	ψ_{12}	ψ_{13}	ψ_{14}	ψ_{15}	ψ_{16}	ψ_{17}	ψ_{18}	ψ_{19}	ψ_{20}
3.1074	2.6631	2.3793	1.9471	2.3352	2.5928	1.8701	2.3124	2.6141	2.0247
0.0088	0.0127	0.0293	0.0199	0.0230	0.0150	0.0192	0.0216	0.0161	0.0228

The result implies that more conscientious, less extroverted and more neurotic people use the internet more frequently. This corresponds to the common sense, and can be an evidence that the proposed model should work effectively for real data analysis.

References

Albert, J.H. and Chib, S. (1993). Bayesian analysis of binary and polytomous response data. *Journal of the American Statistical Association,* **88**, 669–679.

Ansari, A., Jedidi, L. and Dube, L. (2002). Heterogeneous factor analysis models: A Bayesian approach. *Psychometrika,* **67**, 49–77.

Arbuckle, J.L. (1995). *Amos for Windows. Analysis of moment structures (Version 3.5).* Chicago, IL: Small Waters.

Arminger, G. and Muthén, B.O. (1998). A Bayesian approach to nonlinear latent variable models using the Gibbs sampler and the Metropolis-Hasting algorithm. *Psychometrika,* **63**, 271–300.

Dunsmore, I.R. (1976). Asymptotic prediction analysis. *Biometrika,* **63**, 627–630.

Gelman, A. and Rubin, D.B. (1992). Inference from iterative simulation using multiple sequences. *Statistical Science,* **7**, 503–511.

Hogan, J.W. and Tchernis, R. (2004). Bayesian factor analysis for spatially correlated data, with applicaton to summarizing area-level material deprivation from census data. *Journal of the American Statistical Association,* **99**, 314–324.

Lopes, H.F. and West, M. (2004). Bayesian model assessment in factor analysis. *Statistica Sinica,* **14**, 41–67.

Press, S.J. (1982). *Applied multivariate analysis.* Florida: Krieger Publishing.

Scheines, R., Hoijtink, H. and Boomsma, A. (1999). Bayesian estimation and testing of structural equation models. *Psychometrika,* **64**, 37–52.

Shi, J-Q. and Lee, S.Y. (1998). Bayesian sampling-based approach for factor analysis models with continuous and polytomous data. *British Journal of Mathematical and Statistical Psychology,* **51**, 233–252.

Shi, J.Q. and Lee, S-Y., (1997). A Bayesian estimation of factor score in confirmatory factor model with polytomous, censored or truncated data. *Psychometrika,* **62**, 29-50.

Shigemasu, K. and Nakamura, T. (1993). A Bayesian Numerical Estimation Procedure in Factor Analysis Model. *E.S.T. Research Report,* **93-6,** Tokyo Institute of Technology.

Shigemasu, K., Ohmori, T., and Hoshino, T. (2002). Bayesian Analysis of Structural Equation Modeling. *In Measurement and Multivariate Analysis* (S. Nishisato, Y. Baba, H. Bozdogan, and K. Kanefuji, eds.). Tokyo: Springer-Verlag, 207–216.

Bayesian Statistics and Its Applications
Edited by S.K. Upadhyay, U. Singh and D.K. Dey
Anamaya Publishers, New Delhi, India

Nonparametric Empirical Bayes Two-Tail Tests for Scale Exponential Family when Random Variables in the Sequence are Negatively Associated

R.S. Singh[1] and L.S. Wei[2]

[1]Department of Mathematics and Statistics, University of Guelph,
Guelph, Ontario, N1G2W1, Canada

[2]Department of Mathematics and Statistics, University of Science & Technology of China,
Heifel, Anhui, China

Abstract

Bayes decision procedures are derived for two-tail tests under weighted loss function for scale exponential family with unknown prior distribution. For the case of negatively associated (NA) samples, nonparametric empirical Bayes (EB) decision procedures using kernel method of estimation are constructed under no assumption whatsoever on the parametric form of the prior distribution. The asymptotic optimality and convergence rates of the proposed EB decision procedures are obtained under suitable conditions. The paper is concluded with a few remarks indicating how the methods here can be extended to some other types of tests.

---

1. Introduction

Since Robbins (1955, 1964) introduced the empirical Bayes (EB) approach to statistical problems, EB test and estimation problems have been studied extensively in the literature. Johns and Van Ryzin (1971, 1972) considered EB two-action problems for discrete and continuous one-parameter exponential family. Van Houwelingen (1976), Liang (1988) and Karunamuni and Yang (1995) discussed the monotone EB test problems for above two families. Singh (1976, 1979) examined EB estimation in regular exponential family, while Singh and Wei (2000) studied the EB two-tail test problem for scale exponential family. However, the EB test procedures in these and other existing literature mostly involve independent random variables, whereas in many problems, including reliability theory, percolation model and multivariable analysis, the random variables are not necessarily independent but are negatively associated (NA). Because of its wide applications, the notions of NA sequences have received more and more attention in recent years. One may refer to

---

AMS (1991) Subject Classification: 62C12. 62F05

Joag-Dev and Proschan (1983) and Block, Savits and Shaked (1982) for some fundamental properties and applications of NA sequence. For the limit theorems of NA sequence, see Newman (1984), Matula (1992) and Su et al. (1996), among others. For further references of research on NA sequence and for examples of some well known distributions where the r.v.'s are NA, see Hu (2000). Wei (2001) studied the consistency properties for kernel-type density estimates in the case of NA samples. The following definition of NA random variables is first given by Joag-Dev and Proschan (1983).

Random Variables X_1, X_2, X_3, ..., X_n $(n \geq 2)$ are said to be negatively associated, if for every pair of disjoint subsets A_1, A_2 of $\{1, 2, \ldots, n\}$,

$$\mathrm{cov}\,\{f_1(X_i), i \in A_1,\ f_2(X_j), j \in A_2\} \leq 0$$

whenever f_1 and f_2 are non-increasing (or non-decreasing) componentwise, provided the covariance exists.

As noted in Joag-Dev and Proschan (1983), an infinite family $\{X_j,\ j \in N\}$ of random variables is said to be NA if and only if every finite subfamily is negatively associated. Also, a set of independent random variables is NA, and union of independent sets of NA random variables is NA. For further sufficient conditions for NA r.v.'s, see Hu and Hu (1999).

In an EB context, we consider a sequence of identical Bayes decision problems with unknown prior distribution of the parameter in question and at each stage in the sequence, we construct a decision rule based on the current and past observations so that its Bayes risk is close to the (minimum) Bayes risk of a Bayes decision rule which we would be able to construct and use only if the prior distribution was known. A typical problem in the sequence is called a component problem, and a typical decision rule so constructed is called an EB decision rule.

The article aims to develop and study the asymptotic properties of EB tests for scale exponential family when the random variables in the sequence are NA. Note that this includes the case of independent random variables. Section 2 describes the model and the component problem, and then derives Bayes decision rule. Section 3 constructs EB test procedure at the nth $(n > 1)$ stage using the current and the past observations and using the kernel method of density estimation. Section 4 gives some useful Lemmas and then establishes asymptotic optimality of our EB test procedure and the minimum speed of convergence with which the risk of EB procedure approaches the minimum Bayes risk. Finally, Section 5 concludes with some remarks as to how this work can be extended to some other types of test problems.

2. The Model, Two-Tail Test Problem and the Bayes Decision Rule

The model we consider in this article is the scale exponential family of probability densities given by

$$f(x|\theta) = u(x)c(\theta)\exp\left(-x/\theta\right) \tag{2.1}$$

where $u(x) > 0$, $x \in X = (0, \infty)$ is the sample space and

$$\Theta = \left\{\theta > 0:\ 0 < \int u(x)\exp\left(-x/\theta\right)dx < \infty\right\}$$

is the parameter space and the prior distribution G on Θ is unknown and unspecified. Here, and throughout the paper, the range of integration is over $(0, \infty)$, unless stated otherwise.

The component problem that we examine here is the two-tail test problem for testing

$$H_0 : \theta_1 \leq \theta \leq \theta_2 \quad \text{against } H_1 : \theta < \theta_1 \text{ or } \theta > \theta_2 \tag{2.2}$$

based on a random observation X (or a sufficient statistic X for θ) having the probability density given by (2.1), where $0 < \theta_1 < \theta_2 < \infty$ are known constants. The loss function we consider in this paper is the most widely used in the literature for such a two-tail test involving a scale parameter, which is given by

$$L_0(\theta, \, d_0) = \begin{cases} 0 & \text{if } \theta_1 \leq \theta \leq \theta_2 \\ a\left((\theta - \theta_1)/\theta\right)\left((\theta - \theta_2)/\theta\right) & \text{if } \theta < \theta_1 \text{ or } \theta > \theta_2 \end{cases}$$

and

$$L_1(\theta, \, d_1) = \begin{cases} 0 & \text{if } \theta < \theta_1 \text{ or } \theta > \theta_2 \\ a\left((\theta - \theta_1)/\theta\right)\left((\theta_2 - \theta)/\theta\right) & \text{if } \theta_1 \leq \theta \leq \theta_2 \end{cases}$$

where $a > 0$ is a constant of proportionality, and d_0 and d_1 are, respectively, decisions of accepting H_0 and H_1. What the loss function really says that if we take action d_0 and accept H_0 while H_1 happens to be true (i.e. Type II error occurs), then the loss is proportional to the product of the scaled distances of θ being away from θ_1 and θ_2. A similar interpretation can be seen when we take action d_1 and accept H_1 while H_0 happens to be true, i.e. when we incur a type I error.

If we introduce the notation

$$\theta_0 = (\theta_1 + \theta_2)/2 \quad \text{and} \quad b = (\theta_2 - \theta_1)/2$$

then the test problem (2.2) is equivalent to testing

$$H_0 : \quad |\theta - \theta_0| \leq b \quad \text{against } H_1 : \quad |\theta - \theta_0| > b \tag{2.3}$$

and the loss functions can be re-expressed as

$$L_0(\theta, \, d_0) = \begin{cases} 0 & \text{if } |\theta - \theta_0| \leq b \\ a((\theta - \theta_0)^2 - b^2)/\theta^2 & \text{if } |\theta - \theta_0| > b \end{cases}$$

and

$$L_1(\theta, \, d_1) = \begin{cases} a(b^2 - (\theta - \theta_0)^2)/\theta^2 & \text{if } |\theta - \theta_0| \leq b \\ 0 & \text{if } |\theta - \theta_0| > b \end{cases} \tag{2.4}$$

which in turn, for $i = 0, 1$, can be written as

$$L_i(\theta, d_i) = (1 - i)a\theta^{-2}\left\{(\theta - \theta_0)^2 - b^2\right\} I\left[|\theta - \theta_0| > b\right] + ia\theta^{-2}\left\{b^2 - (\theta - \theta_0)^2\right\} I\left[|\theta - \theta_0| \leq b\right]$$

where $I(A)$ is the indicator function of set A.

If we define a randomized decision function δ on the sample space X onto the interval $[0, 1]$ by

$$\delta(x) = P(\text{accepting } H_0 \mid X = x) \tag{2.5}$$

that is $\delta(x)$ and $1 - \delta(x)$ are respectively the probabilities of accepting H_0 and H_1 when $X = x$ is observed, then the Bayes risk of this randomized rule δ with respect to the prior distribution G on Θ is given by

$$R(\delta, \, G) = \int \int \left[L_0(\theta, d_0)f(x|\theta)\delta(x) + L_1(\theta, d_1)f(x|\theta)(1 - \delta(x))\right] dx \, dG(\theta)$$

where, here and throughout the paper, all the integrals are over the interval $(0, \infty)$, unless stated otherwise. After some simplification, it can be seen that

$$R(\delta, G) = a \int \alpha(x)\delta(x)dx + C_G \tag{2.6}$$

with

$$C_G = \int L_1(\theta, d_1)dG(\theta) \tag{2.7}$$

$$\begin{aligned}
\alpha(x) &= \int \left(\left((\theta - \theta_0)^2 - b^2 \right)/\theta^2 \right) f(x|\theta)dG(\theta) \\
&= f(x) + 2\theta_0 u(x)p^{(1)}(x) + (\theta_0^2 - b^2)u(x)p^{(2)}(x)
\end{aligned} \tag{2.8}$$

where

$$f(x) = \int f(x|\theta)dG(\theta) = u(x)p(x), \tag{2.9}$$

$$p(x) = \int c(\theta)\exp\left(-x/\theta\right)dG(\theta), \tag{2.10}$$

$$p^{(i)}(x) = \int (-\theta)^{-i}c(\theta)\exp\left(-x/\theta\right)dG(\theta), \qquad i = 1, 2, \ldots . \tag{2.11}$$

Note that $f(x)$ is the marginal density of X and $p^{(1)}(x)$ and $p^{(2)}(x)$ are, respectively, the first and the second derivatives of $p(x)$.

Thus, from (2.6) it follows that the Bayes test rule that minimizes the Bayes risk $R(\delta, G)$ with respect to G over the class of all randomized decision procedures δ, is given by

$$\delta_G(x) = \begin{cases} 1 & \text{if } \alpha(x) \leq 0 \\ 0 & \text{if } \alpha(x) > 0, \end{cases} \tag{2.12}$$

i.e., when $X = x$ is observed and $\alpha(x) \leq 0$, then accept H_0 with probability one, otherwise accept H_1 with probability one. The minimum Bayes risk attained by $\delta_G(x)$ is

$$R(G) = \inf_\delta R(\delta, G) = R(\delta_G, G) = a \int \alpha(x)\delta_G(x)dx + C_G, \tag{2.13}$$

Note that the minimum Bayes risk test procedure δ_G is available for use only if the prior distribution G is completely known, which is rare in practice. We take the practical point of view, and as stated earlier assume that G is completely unknown and unspecified, and hence δ_G is not directly available for use. This leads us to the use of Robbins' EB approach to exhibit sequence of decision rules whose risks are close to $R(G)$ as the sequence gets large.

Remark. Note that the loss function taken in Singh and Wei (2000) is the usual product of linear distances of true θ being away from θ_1 and θ_2, and the r.v.'s in the sequence are independent. In contrast, our loss function in the current paper is the product of scaled distances, which is more appropriate when dealing with a scale parameter, and the r.v.'s are negatively associated, which includes the case of independence.

3.   Proposed Empirical Bayes Decision Rules

Suppose that the component problem described in Section 2 occurs sequentially, and our present problem is, in fact, the $(n + 1)^{\text{th}}$ problem in the sequence. Thus, for the present test problem we have acquired data $X_1, X_2, \ldots, X_n$ from the past n occurrences of the test problem and $X_{n+1} = X$ from the present problem. We assume that $\{X_n, n \geq 2\}$ is a weakly stationary and identically distributed NA sequence with each X_i having the marginal probability density $f(x)$ given in (2.9). Let C_i be the class of all real valued nonnegative functions on $(0, \infty)$ whose i^{th} derivative exist, are continuous and bounded above. For our asymptotic optimality results, we assume that $f \in C_1$ and $p \in C_i$ for $i = 1,\ 2$, and for the speed of convergence results we assume that

$$f \in C_s \text{ and } p \in C_s \text{ for some } s \geq 3. \tag{3.1}$$

To exhibit EB test procedure in our context, we proceed as follows. Following Singh (1979), for $i = 0,\ 1,\ 2$ and an integer $s > 2$, let $\mathcal{K}_i$ be the class of all Borel-measurable real valued bounded functions K_i vanishing outside $(0,1)$ satisfying the following Condition A.
   Condition A

$$\frac{1}{j!} \int x^j K_i(x) dx = \begin{cases} 1 \text{ if } j = i \\ 0 \text{ if } j \neq i, \end{cases} \quad j = 0, 1, 2, \ldots, s - 1,$$

K_i is differentiable and $\sup_{x} |K_i'(x)| < \infty$.

Further, for the main results we assume that the NA sequence $\{X_n\}$ satisfies the following covariance structure, which we refer to as Condition B.
   Condition B

$$\sum_{j=1}^{\infty} |\text{cov}(X_1, X_j)| < \infty$$

Let $0 \leq h_n \to 0$ as $n \to \infty$. We abbreviate h_n by h throughout this paper. Then, using the observations $X_1, X_2, \ldots, X_n$ from the past n occurrences of the component problem, we estimate, as in Singh and Wei (2000), the probability density function $f(x)$ by

$$f_n(x) = (nh)^{-1} \sum_{i=1}^{n} K_0((X_i - x)/h) \tag{3.2}$$

and the functions $p^{(i)}(x)$ appeared in (2.11) for $i=1,\ 2$, by

$$p_n^{(1)}(x) = (nh^2)^{-1} \sum_{j=1}^{n} K_1((X_i - x)/h)/u(X_j) \tag{3.3}$$

and

$$p_n^{(2)}(x) = (nh^3)^{-1} \sum_{j=1}^{n} K_2((X_i - x)/h)/u(X_j), \tag{3.4}$$

respectively. Finally, examining the structure of $\alpha(x)$ that appeared in (2.8) we estimate $\alpha(x)$ by

$$\alpha_n(x) = f_n(x) + 2\theta_0 u(x) p_n^{(1)}(x) + (\theta^2 - b^2) u(x) p_n^{(2)}(x). \tag{3.5}$$

An EB procedure at the $(n + 1)^{\text{th}}$ stage, with X_{n+1} denoted as X, is a $(X_1, \ldots, X_n\ ;\ X)$-measurable function $\delta_n(X) = \delta_n(X_1, \ldots, X_n; X)$ with values in $[0, 1]$, and in our context, it is the probability of accepting H_0 at the $(n+1)^{\text{th}}$ stage when $X_1, \ldots, X_n$ and X are observed. Following (2.18) and (3.5) our proposed EB test procedure at the $(n+1)^{\text{th}}$ stage is given by

$$\delta_n(X) = \begin{cases} 1 & \text{if } \alpha_n(X) \leq 0 \\ 0 & \text{if } \alpha_n(X) > 0, \end{cases} \tag{3.6}$$

i.e. at the $(n+1)^{\text{th}}$ stage we accept H_0 w.p. 1 if $\alpha_n(X) \leq 0$, otherwise we accept H_1 w.p. 1. Following the arguments to arrive at (2.6) for any rule δ, it can be seen that the overall Bayes risk due to $\delta_n(X)$ is

$$R_n = R(\delta_n, G) = a \int \alpha(x) E(\delta_n(x)) dx + C_G \tag{3.7}$$

where E is the expectation operator w.r.t. all the random variables involved.

If $\lim_{n \to \infty} R_n = R(G)$, then we say the sequence of test procedures $\{\delta_n\}$ is asymptotically optimal (a.o.). If $(R_n - R(G)) = O(n^{-q})$ for some $q > 0$, then the convergence rates are said to be of the order $O(n^{-q})$.

We assume throughout the remainder of the paper that kernels K_0, K_1 and K_2 satisfy Condition A and the sequence $\{X_n\}$ satisfy co-variance structure Condition B.

4.   Main Results and Some Useful Lemmas

This section presents some lemmas followed by our main results in Theorems 4.1 and 4.2. Theorem 4.1 establishes the asymptotic optimality of our EB procedures δ_n and Theorem 4.2 establishes the speed of convergence. The lemmas, which are important in their own contexts, are used to prove the main results. In what follows, $c, c_0, c_1, c_2, \ldots$ are simply positive constants, although they may not represent the same value at different places.

Lemma 4.1. *Let $R(G)$ and R_n be given by (2.13) and (3.7) respectively. Then*

$$0 \leq R_n - R(G) \leq a \int |\alpha(x)|\, P\left(|\alpha_n(x) - \alpha(x)| > |\alpha(x)|\right) dx.$$

Proof. See Lemma 1 of Johns and Van Ryzin (1972).

Lemma 4.2. *Let X, Y be NA random variables with finite variances. Then for any two differentiable functions g_1 and g_2*

$$|\text{cov}(g_1(X), g_2(Y))| \leq \sup_x |f'(x)| \sup_y |g'(y)| \left[-\text{cov}(X, Y)\right]$$

where $g_1{}'$ and $g_2{}'$ denote the derivatives of g_1 and g_2, respectively.

Proof. See Lemma 1 of Pan (1997).

Lemma 4.3. *Let $f_n(x)$ be defined by (3.2) and let $\{X_n, n \geq 2\}$ be a weakly stationary and identically distributed NA sequence satisfying Condition B. Further, let $u(x)$ be a non-decreasing function. If $f(x)$ is continuous and $nh^4 \to \infty$ and $h \to 0$ as $n \to \infty$, then we have*

(i) $$\lim_{n \to \infty} E|f_n(x) - f(x)|^2 = 0 \text{ for } x > 0$$

and if for some integer $s \geq 2$, $f(x) \in C_s$ and $h \propto n^{-1/(4+2s)}$, then we have

(ii) $$E|f_n(x) - f(x)|^r \leq cn^{-rs/(2s+4)} \text{ for } x > 0 \text{ and } 0 < r \leq 2.$$

Proof. Note that

$$E\left|f_n\left(x\right) - f(x)\right|^2 = \left|Ef_n(x) - f(x)\right|^2 + \mathrm{var}(f_n(x)). \tag{4.1}$$

Since

$$E\left(f_n\left(x\right)\right) = h^{-1}E(K_0((X_1 - x)/h)) = \int K_0(u)f(x + hu)du$$

by the properties of K_0 and the continuity of f, we have from Singh (1979),

$$\lim_{n\to\infty}\left|Ef_n(x) - f(x)\right| \leq \lim_{n\to\infty}\int |K_0(u)|\,|f(x + hu) - f(x)|du = 0. \tag{4.2}$$

Next we will evaluate $\mathrm{var}(f_n(x))$. Note that

$$\mathrm{var}(f_n(x)) = (n^2h^2)^{-1}\mathrm{var}\sum_{j=1}^{n} K_0\left((X_j - x)/h\right) = I_{21} + I_{22} \tag{4.3}$$

where

$$I_{21} = (n^2h^2)^{-1}\sum_{j=1}^{n} \mathrm{var}\left(K_0\left((X_j - x)/h\right)\right),$$

and

$$I_{22} = 2(n^2h^2)^{-1}\sum_{1\leq j<k\leq n} \mathrm{cov}\left(K_0\left((X_j - x)/h\right), K_0\left((X_k - x)/h\right)\right).$$

Note that

$$\begin{aligned}
I_{21} &\leq \left(nh^2\right)^{-1} E\left(K_0^2\left((X_1 - x)/h\right)\right)\\
&= (nh)^{-1}\int K_0^2\left(u\right)f\left(x + hu\right)du \leq c\left(nh\right)^{-1}
\end{aligned} \tag{4.4}$$

for some positive constant c since f is continuous. Denote $g_n(x, y) = h^{-1}K_0\left((y - x)/h\right)$. Then by the Conditions A and B, Lemma 4.2, and the weakly stationary property of NA sequence $\{X_n, n \geq 2\}$, we have

$$I_{22} \leq 2n^{-2}\sum_{1\leq j<k\leq n} \left|\mathrm{cov}\left(g_n(x, X_j), g_n(x, X_k)\right)\right|$$

$$\leq 2n^{-2}\sum_{1\leq j<k\leq n}\sup_{y} \left|g_n'(x, y)\right|^2\left[-\mathrm{cov}(X_j, X_k)\right]$$

$$\leq c_0(nh^4)^{-1}\sum_{k=1}^{\infty}\left|\mathrm{cov}(X_1, X_k)\right| \leq c(nh^4)^{-1} \tag{4.5}$$

for some positive constant c. Now substituting (4.4) and (4.5) into (4.3), we obtain

$$\mathrm{var}(f_n(x)) \leq c(nh^4)^{-1} \tag{4.6}$$

Hence, as $nh^4 \to \infty$ and $h \to 0$ as $n \to \infty$, we have

$$\lim_{n\to\infty}\mathrm{var}(f_n(x)) = 0. \tag{4.7}$$

By (4.2) and (4.7), the proof of (i) of the lemma is now proved.

To prove (ii) of the lemma, we proceed as follows.
By c_r-inequality, we have for $0 < r < 2$,

$$E\,|f_n(x) - f(x)|^r \le 2\,|Ef_n - f(x)|^r + 2\,[\mathrm{var} f_n(x)]^{r/2}\,. \tag{4.8}$$

From the proof of (i), we have

$$(Ef_n(x) - f(x)) = \int K_0(u)\,(f(x + hu) - f(u))du.$$

Since by Taylor expansion

$$(f(x + hu) - f(x)) = \frac{f^{(1)}(x)}{1!}(hu) + \frac{f^{(2)}(x)}{2!}(hu)^2 + \ldots + \frac{f^{(s-1)}(x)}{(s-1)!}(hu)^{s-1}$$

$$+ \frac{f^{(s)}(x + \xi hu)}{s!}(hu)^s \text{ for some } 0 < \xi < 1$$

applying Condition A and the fact that $f \in C_s$, we have from Singh (1979),

$$|Ef_n(x) - f(x)| \le \int |K_0(u)|\,\left|f^{(s)}(x + \xi hu)\right|(hu)^s\,du \le ch^s.$$

Hence

$$|Ef_n(x) - f(x)|^r \le ch^{rs}. \tag{4.9}$$

Now taking $h \propto n^{-1/(4+2s)}$ and substituting (4.9) and (4.6) into (4.8), we obtain

$$E\,|f_n(x) - f(x)|^r = cn^{-rs/(2s+4)}. \tag{4.10}$$

This completes (ii) of the lemma.

The proof of the lemma is now complete.

Lemma 4.4. *For $i=1$, 2, let $p_n^{(i)}(t)$ be given by (3.3) and (3.4), respectively, and let $\{X_n, n \ge 2\}$ be a weakly stationary and identically distributed NA sequence satisfying Condition B. Further, let $u(x)$ be a non-decreasing function. Then we have*

(i) *If $p \in C_2$, then with $nh^4 \to \infty$ and $h \to 0$ as $n \to \infty$, we have for $i=1$, 2*

$$\lim_{n \to \infty} E\,\left|p_n^{(i)}(x) - p^{(i)}(x)\right|^2 = 0 \text{ for } x > 0, \tag{4.11}$$

    and

(ii) *if $p \in C_s$ for some $s > 2$, then with $h \propto n^{-1/(2s+4)}$, we have*

$$E\,\left|p_n^{(i)}(x) - p^{(i)}(x)\right|^r \le cn^{-r(s-2)/(2s+4)}\left(\left|p^{(s)}(x)\right|^r + (p(x)/u(x))^{r/2} + (u(x))^{-r}\right),$$

for $x > 0$, $0 < r \le 2$ and $s \ge 3$.

Proof. Note that for $i=1$, 2

$$E\,\left|p_n^{(i)}(x) - p^{(i)}(x)\right|^2 = \left|Ep_n^{(i)}(x) - p^{(i)}(x)\right|^2 + \mathrm{var}(p_n^{(i)}(x)). \tag{4.12}$$

By Taylor expansion and moment conditions on K_i, we obtain

$$\left(Ep_n^{(i)}(x) - p^{(i)}(x)\right) = h^{-i}\int K_i(v)p(x+hv)dv - p^{(i)}(x)$$

$$= (i!)^{-1}\int v^i K_i(v)(p^{(i)}(x+\xi hv) - p^{(i)}(x))dv \qquad (4.13)$$

for some $0 < \xi < 1$. Since $p(x) \in C_2$, we have for $i=1,\ 2$, from Singh (1979),

$$\lim_{n\to\infty}\left|Ep_n^{(i)}(x) - p^{(i)}(x)\right| \le \lim_{n\to\infty}\int v^i|K_i(v)|\left|p^{(i)}(x+\xi hv) - p^{(i)}(x)\right|dv = 0. \qquad (4.14)$$

Now for $x > 0$, we have

$$\mathrm{var}(p_n^{(i)}(x)) = (n^2 h^{2+2i})^{-1}\mathrm{var}(\sum_{j=1}^{n}K_i((X_j - x)/h)/u(X_j)) = I_1 + I_2,\ \text{say}, \qquad (4.15)$$

where

$$I_1 = (n^2 h^{2+2i})^{-1}\sum_{j=1}^{n}\mathrm{var}(K_i((X_j - x)/h)/u(X_j)),$$

and

$$I_2 = 2(n^2 h^{2+2i})^{-1}\sum_{1\le j<k\le n}\mathrm{cov}\left(K_i((X_j - x)/h)/u(X_j), K_i((X_k - x)/h)/u(X_k)\right).$$

We now obtain bounds on I_1 and I_2. By the monotonicity of $u(x)$, we have

$$I_1 = (nh^{2+2i})^{-1}E(K_i((X_1 - x)/h)/u(X_1))^2$$

$$= (nh^{1+2i})^{-1}\int\frac{K_i^2(v)}{u(x+hv)}p(x+hv)dv \le c(nh^{1+2i})^{-1}\left(p(x)/u(x)\right). \qquad (4.16)$$

To bound I_2, first we note that $u(X_j)$ and $u(X_k)$ are bounded below by $u(x)$ for $x < X_j \le x+h$ and $x < X_k \le x+h$. Now, let $g_n(x,y) = h^{-1}K((y-x)/h)$. By Condition A on kernels K_i, covariance structure Condition B of $\{X_n\}$, Lemma 4.2, and from the fact that the NA sequence $\{X_n, n \ge 2\}$ is weakly stationary, we have

$$I_2 \le 2\left(n^2 h^{2i}u^2(x)\right)^{-1}\sum_{1\le j<k\le n}|\mathrm{cov}\left(g_n(x, X_j), g_n(x, X_k)\right)|$$

$$\le 2\left(n^2 h^{2i}u^2(x)\right)^{-1}\sum_{1\le j<k\le n}(\sup_y|g_n'(x,y)|^2\left(-cov(X_j, X_k)\right)$$

$$\le 2\left(nh^{4+2i}u^2(x)\right)^{-1}\sum_{k=1}^{\infty}|\mathrm{cov}(X_j, X_k)| \le c\left(nh^{4+2i}u^2(x)\right)^{-1}. \qquad (4.17)$$

Thus, by (4.15), (4.16) and (4.17), we have

$$\mathrm{var}(p_n^{(i)}(x)) \le c(nh^{4+2i})^{-1}((p(x)/u(x)) + u^{-2}(x)). \qquad (4.18)$$

Hence with $nh^8 \to \infty$ and $h \to 0$ as $n \to \infty$, we have for $i = 1,\ 2$

$$\lim_{n\to\infty}\mathrm{var}(p_n^{(i)}(x)) = 0 \text{ for } x > 0. \qquad (4.19)$$

By (4.12) and (4.19), part (i) of the lemma follows.

Now, we prove part (ii) of the Lemma. For $0 < r < 2$

$$E\left|p_n^{(i)}(x) - p^{(i)}(x)\right|^r \leq 2\left|Ep_n^{(i)}(x) - p^{(i)}(x)\right|^r + 2(\mathrm{var}(p_n^{(i)}(x)))^{r/2}. \tag{4.20}$$

Since $E(p_n^{(i)}(x)) = h^{-i} \int K_i(v)p(x+hv)dv$, replacing $p(x+hv)$ by its Taylor series expansion,

$$p(x+hv) = p(x) + \frac{p'(x)}{1!}(hv) + \ldots + \frac{p^{(s-1)}(x)}{(s-1)!}(hv)^{s-1} + \frac{p^{(s)}(x+\xi hv)}{s!}(hv)^s \text{ for } 0 < \xi < 1$$

and applying the moment conditions of K_i and the fact that $p(x) \in C_s$, we have

$$\left|Ep_n^{(i)}(x) - p^{(i)}(x)\right| \leq h^{s-i} \int |K_i(v)|\, v^s \left|p^{(s)}(x+\xi hv)\right| dv \leq ch^{s-i}\left|p^{(s)}(x)\right|. \tag{4.21}$$

Hence
$$\left|Ep_n^{(i)}(x) - p^{(i)}(x)\right|^r \leq ch^{r(s-i)}\left|p^{(s)}(x)\right|^r. \tag{4.22}$$

Also, from (4.18), we have

$$\left(\mathrm{var}(p^{(i)}(x))\right)^{r/2} \leq c\left(nh^{4+2i}\right)^{-r/2}\left((p(x)/u(x))^{-r/2} + u^{-r}(x)\right). \tag{4.23}$$

Now taking $h \propto n^{-1/(2s+4)}$ and substituting (4.22) and (4.23) into (4.20), we obtain the conclusion of the part (ii) of the Lemma.

Lemma 4.5. *Let $\alpha(x)$ be given by (2.8). If $E\left(\theta^{-2}\right) < \infty$, then $\int |\alpha(x)|\, dx < \infty$.*
Proof. By (2.8) we know that

$$\int |\alpha(x)|\, dx \leq c_0 \int f(x)\, dx + c_1 \int u(x)\left|p^{(1)}(x)\right| dx + c_2 \int u(x)\left|p^{(2)}(x)\right| dx$$

$$\leq c_0 + c_1 \int \theta^{-1} \int f(x|\theta)dxdG(\theta) + c_2 \int \theta^{-2} \int f(x|\theta)dxdG(\theta)$$

$$\leq c_0 + c_1 E\left(\theta^{-1}\right) + c_2 E\left(\theta^{-2}\right) < \infty$$

Lemma 4.6. *Let $\alpha(x)$ be given by (2.8). Suppose that $u(x)$ is non-decreasing function, $p(x) \in C_2$, and there exist a $M > 0$ and a $\gamma > 0$ such that $u(x) \leq Mx^\gamma$ for large enough x. If for some $s > 2$, $E\left(\theta^{-s}\right) < \infty$ and for some $\varepsilon > 0$ and $0 < \lambda < 1$, $E(\theta^\delta) < \infty$, $E\left(\theta^{\tau+\gamma}C(\theta)\right) < \infty$ and $E(\theta^{\tau+\gamma-1}C(\theta)) < \infty$ with $\delta = \max\{\tau, 1\}$, $\tau = (1+\varepsilon)\lambda(1-\lambda)^{-1}$, then we have*

(1) $\int |\alpha(x)|^{1-\lambda} f^{\lambda/2}(x)\, dx < \infty$,

(2) $\int |\alpha(x)|^{1-\lambda} \left|u(x)p^{(s)}(x)\right|^\lambda dx < \infty$,

(3) $\int |\alpha(x)|^{1-\lambda} dx < \infty$.

Proof. To prove (1), we note that by Holder-inequality,

$$\int |\alpha(x)|^{1-\lambda} f^{\lambda/2}(x)\, dx < \left(\int |\alpha(x)|\, dx\right)^{1-\lambda}\left(\int f^{1/2}(x)dx\right)^\lambda = I_1^{1-\lambda}I_2^\lambda.$$

But I_1 is finite from Lemma 4.5, and I_2 is finite from Lemma A.6 of Wei (1998). This completes the proof of (1).

To prove (2), again by Holder-inequality, we have

$$\int |\alpha(x)|^{1-\lambda} \left| u(x) p^{(s)}(x) \right|^{\lambda} dx < \left(\int |\alpha(x)| dx \right)^{1-\lambda} \left(\int u(x) \left| p^{(s)}(x) \right| dx \right)^{\lambda} = I_1^{1-\lambda} I_3^{\lambda},$$

where I_1 is finite from Lemma 4.5, and

$$I_3 = \int u(x) \left| p^{(s)}(x) \right| dx = \int \int \theta^{-s} f(x|\theta) dx dG(\theta) = E\left(\theta^{-s}\right) < \infty$$

To prove (3), we note from the definition of $\alpha(x)$ in (2.8) that

$$\int |\alpha(x)|^{1-\lambda} dx \le c_0 \int f^{1-\lambda}(x) dx + c_1 \int \left(u(x) p^{(1)}(x) \right)^{1-\lambda} dx + c_2 \int \left(u(x) p^{(2)}(x) \right)^{1-\lambda} dx$$

$$= c_0 Q_0 + c_1 Q_1 + c_2 Q_2, \text{ say.}$$

From Lemma A.6 of Wei (1998), it follows that if $E\left(\theta^{\delta}\right) < \infty$ with $\delta = \max\left\{(1+\varepsilon)\lambda(1-\lambda)^{-1}, 1\right\}$, for some $\varepsilon > 0$, then $Q_0 < \infty$. Now

$$Q_1 = \int \left(u(x) p^{(1)}(x) \right)^{1-\lambda} dx = \int_0^1 \left(u(x) p^{(1)}(x) \right)^{1-\lambda} dx + \int_1^{\infty} \left(u(x) p^{(1)}(x) \right)^{1-\lambda} dx$$

$$= Q_{11} + Q_{12}, \text{ say.}$$

By the fact that $u(x)$ is non-decreasing function and $p(x) \in C_2$, we see that $Q_{11} \le c < \infty$. Also by Holder-inequality we obtain

$$Q_{12} = \int_1^{\infty} x^{-(1+\varepsilon)\lambda} x^{(1+\varepsilon)\lambda} (u(x) p^{(1)}(x))^{1-\lambda} dx$$

$$\le \left(\int_1^{\infty} x^{-(1+\varepsilon)} dx \right)^{\lambda} \left(\int_1^{\infty} x^{\tau} u(x) p^{(1)}(x) dx \right)^{1-\lambda}. \tag{4.24}$$

The first factor on the right hand side of (4.24) is finite. Since $u(x)$ is non-decreasing function and $u(x) \le M x^{\gamma}$ for large enough x, for large enough $T > 0$, the integral in the second factor is, with $\tau = (1+\varepsilon)\lambda(1-\lambda)^{-1}$,

$$\int_1^T x^{\tau} u(x) p^{(1)}(x) dx + \int_T^{\infty} x^{\tau} u(x) p^{(1)}(x) dx$$

$$\le c + M \int \theta^{\tau+\gamma} C(\theta) \int_0^{\infty} \theta^{-(\tau+\gamma+1)} x^{\tau+\gamma} \exp\{-x/\theta\} dx dG(\theta)$$

$$\le c + c_1 \int \theta^{\tau+\gamma} C(\theta) dG(\theta) = c + c_1 E(\theta^{\tau+\gamma} C(\theta)) < \infty. \tag{4.25}$$

Hence $Q_{12} < \infty$ and thus we conclude that Q_1 is finite.

Similarly, since $p(x) \in C_2$, it can be shown that if $E(\theta^{\tau+\gamma-1} C(\theta)) < \infty$, then $Q_2 < \infty$. This completes the proof of (3), which finally completes the proof of the lemma.

The following theorem is about the asymptotic optimality property of our proposed EB decision rules.

Theorem 4.1. *Let $\delta_n(x)$ be defined by (3.6). Suppose that $\{X_n, n \ge 2\}$ is a weakly stationary and identically distributed NA sequence satisfying Condition B. If $u(x)$ is non-decreasing function, $f(x)$ is continuous, $p(x) \in C_2$, $nh^4 \to \infty$ and $h \to 0$ as $n \to \infty$ and*

$$E\left(\theta^{-2}\right) < \infty, \tag{4.26}$$

then
$$\lim_{n\to\infty} R_n = R\,(G).$$

Proof. By Lemma 4.1, we have

$$0 \leq R_n - R\,(G) \leq a \int B_n(x)dx \tag{4.27}$$

where $B_n\,(x) = |\alpha\,(x)|\,P\,(|\alpha_n\,(x) - \alpha\,(x)| \geq |\alpha\,(x)|)$. Since $B_n\,(x)$ is bounded by $|\alpha\,(x)|$, which is integrable by Lemma 4.5, by dominant convergence theorem, all we need to prove that for almost all $x \in X$,

$$\lim_{n\to\infty} B_n\,(x) = 0 \tag{4.28}$$

Now, since by Markov inequality $B_n\,(x) \leq E\,|\alpha_n\,(x) - \alpha\,(x)|$, we have from the definitions of $\alpha_n(x)$ and $\alpha(x)$,

$$B_n(x) \leq c_0 \left\{ E\,|f_n\,(x) - f\,(x)| + u\,(x)\,E\left|p_n^{(1)}\,(x) - p^{(1)}\,(x)\right| + u\,(x)\,E\left|p_n^{(2)}\,(x) - p^{(2)}\,(x)\right| \right\}$$

$$\leq c_0\{(E\,|f_n\,(x) - f\,(x)|^2)^{\frac{1}{2}} + u\,(x)\,(E\left|p_n^{(1)}\,(x) - p^{(1)}\,(x)\right|^2)^{1/2} + u(x)(E\left|p_n^{(2)}\,(x) - p^{(2)}\,(x)\right|^2)^{1/2}\}$$

The proof of (4.27), and hence proof of the theorem is now complete from parts (i) of Lemmas 4.3 and 4.4.

Theorem 4.2. *Let (3.1) hold, and $\delta_n\,(x)$ be given by (3.6) with $h \propto n^{-1/(2s+4)}$ for some integer $s \geq 3$. Suppose that $\{X_n, n \geq 2\}$ is a weakly stationary and identically distributed sequence satisfying Condition B. Let for some $\varepsilon > 0$, $0 < \lambda < 1$ and for the integer s in the definition of h, all the conditions of Lemma 4.6 be satisfied and $f(x)$ and $p(x)$ be in C_s, then*

$$(R_n - R\,(G)) = O\left(n^{-\lambda(s-2)/2(s+2)}\right). \tag{4.29}$$

Proof. By Lemma 4.1 and Markov-inequality, we have for $0 < \lambda < 1$

$$0 \leq R_n - R\,(G) \leq a \int |\alpha\,(x)|P\,(|\alpha_n\,(x) - \alpha\,(x)| \geq |\alpha\,(x)|)\,dx$$
$$\leq a \int |\alpha\,(x)|^{1-\lambda}\,E\,|\alpha_n\,(x) - \alpha\,(x)|^\lambda\,dx. \tag{4.30}$$

Now from the definitions of α_n and α, we have

$$E\,|\alpha_n\,(x) - \alpha\,(x)|^\lambda \leq c_0\{E\,|f_n\,(x) - f\,(x)|^\lambda + u^\lambda\,(x)\,E_n\left|p_n^{(1)}\,(x) - p^{(1)}\,(x)\right|^\lambda$$
$$+ u^\lambda\,(x)\,E_n\left|p_n^{(2)}\,(x) - p^{(2)}\,(x)\right|^\lambda\}$$
$$\leq c_0 n^{-\lambda s/2(s+2)} + n^{-\lambda(s-2)/(2s+4)}(c_1\left|u\,(x)\,p^{(s)}\,(x)\right|^\lambda + c_2 f^{\lambda/2}\,(x) + c_3)$$

from parts (ii) of Lemmas 4.3 and 4.4. Thus from (4.30),

$$0 \leq R_n - R\,(G) \leq c_0 n^{-\lambda(s-2)/(2s+4)}\left(\int |\alpha\,(x)|^{1-\lambda}\left|u\,(x)\,p^{(s)}\,(x)\right|^\lambda\,dx\right.$$
$$\left. + \int |\alpha\,(x)|^{1-\lambda}f^{\lambda/2}\,(x)\,dx + \int |\alpha\,(x)|^{1-\lambda}\,dx\right).$$

The proof of the theorem is now complete from Lemma 4.6.

Remark and Examples. Theorem 4.1 says that under appropriate conditions the risk of our test procedures approaches the minimum risk attained by the Bayes procedure as n gets large.

Theorem 4.2 establishes the speed of this convergence. It is easy to see that if s is large enough and λ is arbitrarily close to 1, then the convergence rates in Theorem 4.2 can be arbitrarily close to $O(n^{-1/2})$.

As far as some examples are concerned, it can be verified that for the exponential densities $f(x|\theta) = \theta^{-1}\exp(-x/\theta)I(x > 0)$ and for gamma densities $f(x|\theta) = x^{\gamma-1}(\Gamma(\gamma)\theta^{\gamma})^{-1}\exp(-x/\theta)$, all the conditions of Theorems 4.1 and 4.2 are satisfied for any integer $s > 2$ and for any $0 < \lambda < 1$ if the support of the prior distribution G is $\Theta = (a,\ b)$ for some $0 < a < b < \infty$.

5. Concluding Remarks

This article has dealt with the two-tail test problem with the scaled distance product loss function when nothing is known about the parametric form of the prior distribution and the random variables in the sequence are negatively associated. The method can be as well applied to one tail test, i.e. to test

$$H_0' : \theta \leq \theta_0 \quad \text{against } H_1' : \theta > \theta_0 \tag{5.1}$$

for some known $\theta_0 > 0$. If we take the scaled distance product loss function for one-tail test problem for a scale parameter, given by

$$\begin{aligned}
L_0'(\theta,\ d_0) &= (a(\theta - \theta_0)/\theta)I(\theta > \theta_0) \\
L_1'(\theta,\ d_1) &= (a(\theta_0 - \theta)/\theta)I(\theta \leq \theta_0)
\end{aligned} \tag{5.2}$$

and if $\delta'(x) = P[\text{accepting } H_0|X = x]$ is a randomized decision rule, then its Bayes risk, can be expressed as, after some simplification,

$$R\left(\delta',G\right) = a \int \alpha'\left(x\right)\delta'\left(x\right)dx + C_G \tag{5.3}$$

where

$$\alpha'\left(x\right) = f\left(x\right) + \theta_0 u\left(x\right)p^{(1)}\left(x\right) \tag{5.4}$$

and C_G, $f\left(x\right)$ and $p^{(1)}\left(x\right)$ are given by (2.7), (2.9) and (2.11), respectively. Thus the Bayes procedure that minimizes the risk (5.3) is $\delta_G(x) = I(\alpha'(x) \leq 0)$. Now nonparametric empirical Bayes procedures can be constructed and their asymptotic optimality and rates of convergence can be established along the lines of this paper.

For the test problem

$$H_0'' : \theta = \theta_0 \quad \text{against} \quad H_1'' : \theta \neq \theta_0 \tag{5.5}$$

it is difficult to define an appropriate loss function. However, this test problem can be closely approximated by the two-tail test

$$H_0^* : \theta_0 - \varepsilon \leq \theta \leq \theta_0 + \varepsilon \quad \text{against} \quad H_1^{**} : \theta < \theta_0 - \varepsilon \text{ or } \theta > \theta_0 + \varepsilon \tag{5.6}$$

for some arbitrarily small $\varepsilon > 0$, which can now be handled by the method developed here.

Acknowledgements

Project was supported partly by the National Natural Science Foundation of China (No: 19971085) and the Natural Sciences and Engineering Research Council of Canada (Grant No. 4631).

References

Block, H.W., Savits, T.H. and Shaked, M. (1982). Some concepts of negative dependence, *Ann. Probab.* **10**, 765-772.

Hu, Taizhong and Hu Jinjin (1999). Sufficient conditions for negative association of random variables. *Statist. Prob. Letters*, **45**, 167-173.

Hu, Taizhong (2000). Negatively super additive dependence of random variables with applications. *China J. Appl. Prob. and Statist*, **16**, 133-144.

Joag-Dev, K. and Proschan, F. (1983). Negative Association of Random Variables with applications. *Ann. Statist.*, **11**, 286-295.

Johns, M.V. Jr. and Van Ryzin, J. (1971). Convergence Rates in Empirical Bayes Two-action Problem II. Discrete Case. *Ann. Math. Statist.*, **42**, 1521-1539.

Johns, M.V. Jr. and Van Ryzin, J. (1972). Convergence Rates in Empirical Bayes Two-action Problem II. Continuous Case. *Ann. Math. Statist.*, **43**, 934-947.

Karunamuni, R. J. and Yang, H. (1995). On convergence rates of monotone empirical Bayes test for the continuous one-parameter exponential family. *Statist. Decision*, **13**, 181-192.

Liang, T. (1988). On the convergence rates of empirical Bayes rules for two-action problems: Discrete case. *Ann. Statist.*, **16**, 1635-1642.

Matula, P. (1992). A note on the almost sure convergence of sums of negatively dependent random variables, *Statist, Prob. Lett.*, **15**, 209-213.

Newman, C.M. (1984). Asymptotic independence and limit theorems for positively and negatively dependent random variables. *Inequalities in Statistics Probability* (ed. Y.L. Tong), 127-140, IMS, Hayward, California.

Pan, J.M. (1997). On the convergence rates in the central limit theorems for negatively associated sequences, *Chinese Journal of Appl. Prob. and Stat.*, **13**, 183-192.

Robbins, H. (1955). An empirical Bayes approach to Statistics. *Proc. Third Berkeley Sump. Math. Statist. Prob.*, **1**, Univ. California Press, 157-163.

Robbins, H. (1964). The empirical Bayes approach to statistical decision problems. *Ann. Math. Statsit.*, **35**, 1-20.

Singh, R.S. (1976). Empirical Bayes estimation with convergence rates in non-continuous Lebesque-exponential families, *Ann. Statist.*, **4**, 431-439.

Singh, R.S. (1979). Empirical Bayes estimation in Lebesgue-exponential family with convergence rates near the best possible rate, *Ann. Statist.* **7**, 890-902.

Singh, R.S. and Wei, L.S. (2000). Nonparametric empirical Bayes procedures, asymptotic optimality and rates of convergence for two-tail tests in exponential family, *Nonparametric Statistics*, **12**, 475-501.

Su, C., Zhao, L.C. and Wang, Y.B. (1996). Moment inequalities and weak convergence for NA sequence. *Science in China Ser.* A, **26**, 1091-1099.

Van Houwelingen, J.C. (1976). Monotone empirical Bayes test for the continuous one-parameter exponential family. *Ann. Statisti.*, **4**, 981-989.

Wei, L.S. (2001). The consistencies for the kernel-type density estimation in the case of NA samples. *J. Sys. Sci. & Math. Scis.*, **21**, 79-87.

Wei, L.S. (1998). Convergence rates in empirical Bayes estimation in a class of linear models. *Statistica Sinica*, **8**(2), 589-605.

Bayesian Statistics and Its Applications
Edited by S.K. Upadhyay, U. Singh and D.K. Dey
Anamaya Publishers, New Delhi, India

Bayesian Methods in Educational Measurement

Sandip Sinharay

Center for Statistical Theory and Practice, Educational Testing Service, Princeton, NJ 08541

Abstract

This article describes how Bayesian statistical analysis has solved important problems in educational measurement, often when frequentist methods have been found inadequate. First, a brief overview of the key statistical models in educational measurement, and a brief history of application of Bayesian methods in the field are provided. Then the article describes three recent applications of Bayesian analysis to educational measurement.

---

1. Introduction

Educational measurement provides test administrators all around the globe with tools to ensure that scores from a test are fair to all examinees. Educational measurement primarily deals with analysis of data from educational tests; naturally, statistical methods form an integral part of educational measurement. Even though statistical methods have been used in education since at least Cattell (1890), Bayesian methods have not played a major role in the field except probably for a brief period in the 1970's when Mel Novick and his colleagues employed Bayesian methods to a number of problems. However, there has been a recent surge in the use of Bayesian methods in educational measurement due to the increasing complexity of the problems faced in the area and the advent of new, more efficient and relatively easy-to-use Markov chain Monte Carlo (MCMC) methods (that can be readily used to fit complicated models), and personal computers and software that support this work. The types of educational applications that lend themselves well to Bayesian modeling include sequential testing, hierarchical structures, considerable bodies of experience and data that might be exploited and multiple sources of evidence that need to be integrated.

Section 2 provides an overview of the key statistical models used in educational measurement and a history of the use of Bayesian methods in the field. Section 3 describes three applications of Bayesian methods to educational measurement and Section 4 gives a brief conclusion. The intended audience of this article consists of those who have some knowledge of Bayesian methods, but little or no knowledge of the field of educational measurement.

2. Educational Measurement and the Role of Bayesian Statistics

2.1 Classical Test Theory

The set of statistical methods that have been mostly used over the years in educational measurement fall under the general umbrella of classical test theory (CTT). While a number of books like Gulliksen

(1950), Lord and Novick (1968), and Thissen and Wainer (2001) provide the details of CTT, the following discussion provides a very brief description of its key ideas. The basis of CTT is the set of assumptions

$$x_i = \tau_i + e_i, \quad V(\tau_i) = \sigma_\tau^2, \quad E(e_i) = 0, \quad Var(e_i) = \sigma_e^2, \quad i = 1, 2, \ldots n, \tag{1}$$

where, for examinee i, x_i is the observed score (usually, a number-correct score or a "formula score" obtained using the number of correct, incorrect, and omitted answers produced by the examinee) on an assessment, τ_i is the corresponding true score, and e_i is the error of measurement, τ_i being independent of e_i. The variance component σ_τ^2 is called the *true score variance* and σ_e^2 is called the *error variance*. The *total variance* σ_x^2 of an examinee's observed score is then the sum of σ_τ^2 and σ_e^2.

Two tests are defined to be *parallel* to each other if an examinee could fairly be given either one of these two tests and we would expect the same result. A key concept in CTT is the *reliability* $\rho_{xx'}$ of a test, which is defined as the correlation coefficient of the scores in a test with those in another test parallel to it. The reliability can be expressed as

$$\rho_{xx'} = \frac{\sigma_\tau^2}{\sigma_x^2}. \tag{2}$$

For a test with dichotomous items, reliability is usually estimated using the Kuder-Richardson formula (Kuder & Richardson, 1937) given by

$$\hat{\rho}_{xx'} = \frac{k}{k-1}\left(1 - \frac{\sum_i p_i(1 - p_i)}{\sigma_x^2}\right), \tag{3}$$

where k is the number of items and p_i is the proportion of examinees who get item i correct.

An estimate of the true score is provided by Kelley's equation (Kelly, 1927)

$$\widehat{\tau} = \hat{\rho}_{xx'}\, x + (1 - \hat{\rho}_{xx'})\hat{\mu}, \tag{4}$$

where x is the observed score, $\hat{\rho}_{xx'}$ is the estimated reliability of the test, and $\hat{\mu}$ is the average observed score. Kelley's derivation involved a direct estimation of the parameters of the linear regression equation of true score on observed score.

Validity is another basic concept in CTT. Validity of a test score is the extent to which that score measures the attribute of the respondents that the test is employed to measure, in the population(s) for which the test is used. There are different notions of validity, e.g., *predictive validity, content validity, construct validity* etc.; the *predictive validity* is the most common form of validity and is measured by calculating the degree of association of the test scores with a *criterion variable* (like the degree of success of the examinees at a later point in time).

Equating is a statistical process that is used to adjust scores on test forms so that scores on the forms can be used interchangeably; the technique adjusts for differences in difficulty among forms that are built to be similar in difficulty and content. Mathematically, equating finds $y(x)$ so that a score x on one test form corresponds to a score $y(x)$ in another test form.

It is clear from the above discussion that CTT mostly deals with linear models.

2.2 Item Response Theory

Item response theory (IRT; Lord and Novick, 1968; Hambleton and Swaminathan, 1985) has a number of features that are difficult or impossible to obtain using CTT. Most of these are outcomes of the fact that IRT defines a scale for the underlying *latent variable* (called *ability/proficiency*) that is measured by the test items. This aspect of IRT means that comparable scores may be computed for examinees who did not answer the same questions, without intermediate equating steps (Thissen and Wainer, 2001, p. 73). As an outcome, a large number of alternate forms of a test can be used in theory. Computerized adaptive testing (CAT) based on IRT makes use of this feature, resulting in different test forms for virtually every examinee, yet allowing fair scoring of the examinees. Even in large-scale paper-and-pencil testing programs, the need to develop alternate forms scored on a common scale may provide sufficient motivation to use IRT; for example, National Assessment of Educational Progress (NAEP; more details in Section 3.2) uses IRT models as it employs a number of alternate forms to ensure wide content coverage.

Consider an educational assessment, consisting of J items, that is given to I individuals. Unidimensional IRT models, the simplest of the IRT models, assume that individual i has a univariate proficiency/ability θ_i that the assessment attempts to measure. Item j has a corresponding item parameter vector $\boldsymbol{\eta}_j$, measuring aspects of the item like its difficulty, discrimination power, etc. Let Y_{ij} denotes the score of the i-th individual to the j-th item.

The next step expresses the probability distribution of Y_{ij} in terms of θ_i and $\boldsymbol{\eta}_j$. For example, the *three-parameter logistic model* (3PL; Lord, 1980), dealing with dichotomous items, has three item parameters for each item, i.e., $\boldsymbol{\eta}_j = (a_j, b_j, c_j)$, with a_j, b_j, and c_j respectively being the slope, difficulty and guessing parameters of item j; the model expresses the probability of a correct response as

$$P(Y_{ij} = 1 \mid \theta_i, \boldsymbol{\eta}_j) = c_j + (1 - c_j) \operatorname{logit}^{-1}\left(a_j \left(\theta_i - b_j\right)\right), \tag{5}$$

where $\operatorname{logit}^{-1}(x) = \frac{\exp(x)}{1+\exp(x)}$. In the special case of $c_j = 0$ for all j, the 3PL model becomes the *2PL model* (Lord and Novick, 1968), while in the special case of $c_j = 0$, and $a_j = a$ for all j, the 3PL model becomes the *Rasch/1PL model* (e.g., Lord & Novick, 1968).

IRT models make the additional assumption that given the proficiency variable, the observed outcomes for different items are independent. This is called the assumption of *local independence.*

For polytomous items (those scored in more than two categories, e.g., essay-type questions), one can use the generalized partial credit model (GPCM) (Muraki, 1992)

$$P(Y = y|\theta_i, \boldsymbol{\eta}_j) = \frac{\exp(a_j(y\theta_i - b_{jy}))}{1 + \sum_k \exp(a_j(k\theta_i - b_{jk}))}, \tag{6}$$

where y may take a value in $\{0, 1, .., n_j\}$, a_j is the slope parameter, b_{jx} is the cumulative step parameter with $b_{jx} \equiv \sum_{m=1}^{x} \beta_{jm}$ (β_{jm} being the step parameter for score category m) for item j.

The above models are the most popular models in psychometrics and the testing companies around the world use these models operationally. A standard normal population distribution, $\theta_i \overset{\text{i.i.d}}{\sim} \mathcal{N}(0, 1)$, $i = 1, 2, \ldots I$, is usually assumed to fix the scale of the problem.

2.3   Bayesian Methods in Educational Measurement

Although the work of Calandra (1941), who suggested a Bayesian approach to score examinees in a test with dichotomous items, is probably the first known application of the Bayesian methodology in educational measurement, the work of Edwards, Lindman and Savage (1963) most likely represents the first real effort to encourage practitioners of psychological and educational measurement to think from a Bayesian point of view. Novick (1969a, p. 9) commented:

The transition from classical to Bayesian methods of inference has proceeded more slowly than many of us had hoped following the publication of the seminal article "Bayesian statistical inference for psychological research" by Edwards, Lindman and Savage (1963).

Works like Novick (1964), Novick and Hall (1964), and Meyer (1966) followed. Edwards, Lindman, and Savage (1963) assert that the classical mode of inference "has failed, inevitably" although Novick (1964, p. 25) disagrees with the assertion, recommending the use of both classical and Bayesian methods.

Lord (1969), Lindley (1969a, 1969b, 1969c, 1970), Novick (1969a, 1969b), Novick and Thayer (1969), Novick, Thayer, and Jackson (1970), Novick, Jackson, and Thayer (1971), and Jackson, Novick, and Thayer (1971), Novick, Jackson, Thayer, and Cole (1971, 1972), and Novick and Jackson (1975), in a series of works produced mostly at Educational Testing Service (ETS), applied Bayesian analysis to a variety of problems in educational measurement. Novick (1969a) and Novick and Jackson (1975, pp. 139-140) showed that Kelley's (1927) true score formula (4) can be expressed as the Bayesian estimate of the true score given the observed score under the assumptions of normally distributed observed score and true score. Novick (1969a) also suggested a Bayesian approach for adaptive testing under an IRT framework.

During his tenure at Educational Testing Service, Donald Rubin, along with his colleagues, applied Bayesian methods to a number of problems in educational measurement, which resulted in publications like Rubin (1980, 1981, 1983, 1984) and Braun, Jones, Rubin, and Thayer (1983).

Most of the above applications were in the context of CTT and normal-theory linear models (and often exploited the advantage of the empirical Bayes method that it pools information from other similar observations). Item response theory (IRT) involves considerable statistical modeling, more so than CTT, and provides more flexibility. Naturally, Bayesian methods found plenty of applications in IRT. While Birnbaum (1967) applied Bayesian analysis to estimate the ability parameters in IRT, Swaminathan and Gifford (1986, and their two earlier papers referenced therein), Mislevy (1986), etc. suggested Bayesian methods to estimate the item parameters.

Albert (1992) and Patz and Junker (1999a, 1999b) showed how a Bayesian approach combined with the Markov Chain Monte Carlo (MCMC) algorithms could be applied to IRT models; what has followed has been a surge in the use of the algorithm in IRT and other areas in educational measurement. As the test users are becoming more demanding, and computing power is increasing exponentially, there has been an advent of more and more complicated IRT models. Because it is often difficult to find the maximum likelihood estimates (MLE) for these models, the researchers typically have adopted a Bayesian approach and have fitted the models using the MCMC algorithm.

Examples of complicated IRT models fitted using the MCMC algorithms include multidimensional IRT models (Beguin and Glas, 2001), multilevel IRT models (Fox and Glas, 2001, 2003), a hierarchical model for analyzing data where there is dependence among item parameters even after conditioning on the ability (Glas and van der Linden, 2003; Sinharay, Johnson, and Williamson, 2003; Johnson and Sinharay, 2005), a testlet model that incorporates dependence among an examinee's responses to items sharing a common stimulus (Bradlow, Wainer, and Wang, 1999), a hierarchical model incorporating rater effects (Patz, Junker, Johnson, and Mariano, 2002), and a non-parametric IRT model (Karabatsos, and Sheu, 2004). Rupp, Dey and Zumbo (2004) discussed a number of these papers.

Recently, there has been an increasing level of interest in the testing community in *formative/diagnostic assessment* (which provides a profile of the state of acquisition of a variety of knowledge, skills and proficiencies for each examinee), rather than in *summative assessment* (which reports only an overall/total score for each examinee). As a result, there has been a surge of *cognitively diagnostic statistical models*. Most of these are rather complicated and it is difficult to find MLEs for the model parameters. Therefore, they are fitted using an MCMC algorithm. Some examples of such models are Bayesian networks (e.g., Mislevy, 1995; Sinharay, 2006) and fusion model (Hartz, Roussos and Stout, 2002). These models assume that solving a test item requires the presence of a number of skills (e.g., solving a mathematical item may require the presence of certain levels of proficiency in algebra and geometry), and include a discrete proficiency variable for each examinee for each skill. After fitting these models, researchers make inferences on the parameters related to the distribution of the proficiency variables.

Other important miscellaneous applications of Bayesian statistics to educational measurement include work on the use of covariate information about items in estimation of item parameters (Mislevy, 1988), Bayesian item selection criteria for computer adaptive tests (CAT) (e.g. Owen, 1975; van der Linden, 1998), fitting of a multidimensional latent class model (Hoijtink and Molenaar, 1997), application of Bayesian decision theory in a pass-fail test (Lewis and Sheehan, 1990) and mastery decisions (Huynh, 1982), factor analysis from a Bayesian viewpoint (Lee, 1981), and multidimensional scaling from a Bayesian view (Ramsay, 1978).

3. Three Applications

This section describes three ongoing research projects, all funded by Educational Testing Service, where Bayesian analysis was found to be the method of choice. The first example shows how a Bayesian model checking procedure reveals problems with an IRT model where it is not clear how frequentist methods could be applied. The second example describes how a Bayesian hierarchical model (combined with an MCMC algorithm to fit the model) combines the three steps of estimation of a complicated model used in National Assessment of Educational Progress (NAEP). The third example describes how use of past information in a prior distribution promises to improve estimation in assessing differential item functioning (DIF) with small samples.

3.1 Application of a Bayesian Model Checking Method

3.1.1 Model Checking in Educational Measurement

Model checking remains a major hurdle to overcome for effective implementation of IRT (e.g., Sinharay, 2005). The possible number of responses (2^I for a test with I binary items) is sufficiently large for even a moderate number of items so that the standard χ^2 test of goodness of fit does not directly apply. Therefore, investigators assess the fit of IRT models by examining if the model can explain various summaries (or collapsed versions) of the original data, although the null distribution of those are rarely easy to establish. Tests have been suggested to assess unidimensionality/local independence (e.g., Stout, 1987), item fit (e.g. Sinharay, 2003, and the references therein), differential item functioning (e.g., Lord, 1980; Holland, 1985), and person fit (e.g. Glas and Meijer, 2003). One can also analyze residuals. Still, there is no unanimously agreed-upon measure in any of the above areas and the null distributions of a number of IRT fit statistics suggested in the literature are not well-established. Standard IRT software packages lack reliable model-fit indices (e.g., Sinharay, 2005).

The posterior predictive model checking (PPMC) method (Guttman, 1967; Rubin, 1981, 1984; Gelman, Meng and Stern, 1996) is a popular Bayesian model checking tool because of its simplicity, strong theoretical basis, and obvious intuitive appeal. The method primarily consists of comparing the observed data with *replicated data* (those predicted by the model) using a number of test statistics. The aspect of the method that makes it especially attractive for application to educational measurement is that the method is relatively straightforward once an MCMC algorithm has been used to fit a model—there is no need to determine the theoretical null distribution of the test statistic used. Therefore, one can check several aspects of IRT model fit using this method without too much difficulty.

Sinharay and Johnson (2003) and Sinharay (2003, 2005) used the PPMC method to assess several aspects of fit of IRT models using different *discrepancy measures* (which are like classical test statistics) that are natural and observed quantities in the context of test data. These papers used appropriately created graphical displays to demonstrate the ability of the PPMC method to provide graphical evidence of misfit. The issue of evaluating practical consequences of model misfit has been given little attention in the model checking literature in educational measurement. Sinharay (2005) evaluated practical consequences of model misfit with a number of real data applications. Following is a hitherto unpublished example of the application of the PPMC method to educational measurement.

3.1.2 *Application of the PPMC Method to an Admission Test Data*

Let us consider the responses of a random subset of 10,000 examinees to 78 multiple choice items constituting the verbal measure of a widely used admission test. We treat the omitted and not reached responses as wrong answers. The verbal measure consists of eight parts. Each part has either (a) stand-alone/discrete items or (b) *testlets* (e.g., Bradlow, Wainer and Wang, 1999; a set of items that share a common stimulus, like a reading passage); Bradlow et al. argue that there is often a *testlet effect*, an extra association among these items, that an unidimensional IRT model cannot explain.

The eight parts in the verbal measure, in the order in which they appear in the test booklet received by examinees, are as follows:

1. Items 1-10: 10 discrete items requiring an examinee to complete a sentence ("completion items" henceforth)
2. Items 11-23: 13 discrete items involving word relationships ("WR items" for future reference)
3. Items 24-36: 13 testlet items (based on one shared stimulus)
4. Items 37-45: 9 discrete completion items
5. Items 46-51: 6 WR items
6. Items 52-57: 6 testlet items
7. Items 58-66: 9 testlet items
8. Items 67-78: 12 testlet items

The first three parts form the first separately-timed section. Parts 4-7 form the second section and the part 8 forms the third section. There is a strict time-limit for each of the three sections and evidence suggests that the test may be slightly speeded.

We fit the 3PL model to the data set using an MCMC algorithm, employing non-informative prior distributions used in Sinharay and Johnson (2003). The next step is to assess the fit of the 3PL model using the PPMC method.

Fig. 1 shows the observed standard deviation (SD), using a dotted vertical line, and the posterior predictive distribution, using a histogram, of the point biserial correlation coefficients (e.g., Lord, 1980, p. 33), which are correlation coefficients of an individual item score and number-correct score in the 78-item test, and natural quantities of interest in the context of educational test data. The plots show that the 3PL model significantly under-predicts the variation of the point biserial correlation coefficients. Because the null distribution of the point biserial correlations or their SD under the 3PL model is unknown, it would be impossible to apply a frequentist approach here; the Bayesian approach provides a simple solution.

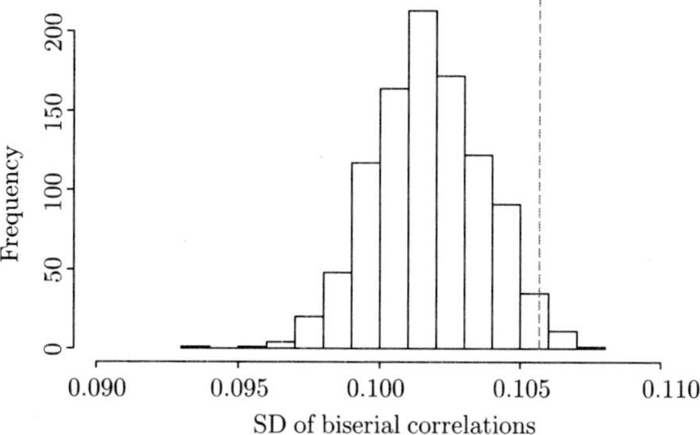

Fig. 1. The observed and predicted SD of the point biserial correlations for 3PL model fit to the admission test data.

Fig. 2 summarizes the posterior predictive p-values (PPP-value) for the odds ratios (OR; e.g., Agresti, 2002) for item pairs predicted by the 3PL model, using symbols described in the legend of the figure. Examining model prediction of ORs should be useful in detecting any extra dependence among the items that the common IRT models cannot explain; such dependence may occur, e.g., in tests that measure multiple skills. Chen and Thissen (1997) attempted to use ORs as frequentist test statistics for assessing local independence of IRT models, found that the null distribution of standardized ORs did not follow the hypothesized distribution, and hence ended up not finding the ORs useful. The PPMC method overcomes this difficulty as the method provides a natural simulation-based null distribution of the test statistics. A solid triangle (see Fig. 2) indicates a

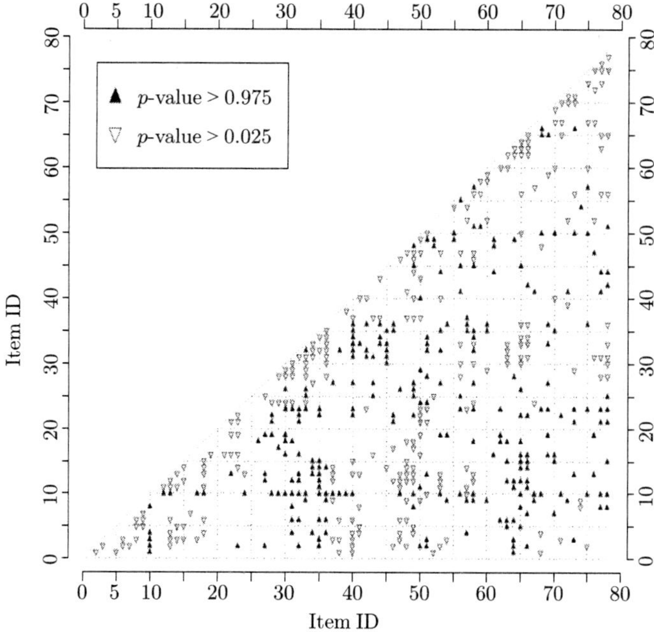

Fig. 2. PPP-values for odds ratios for 3PL model fit to the admission test data.

significant over-prediction by the model while an inverted hollow triangle indicates a significant under-prediction. Horizontal and vertical grid-lines at multiples of 5, axis-labels on the right side and top of the figure, and a diagonal line are there for convenience.

The plot has too many triangles and inverted triangles, indicating that the 3PL model fails to predict the associations among the items. The plot shows that there exists some association among the items that the model cannot predict adequately. The percentage of pairs for which a PPP-value is significant at the 5% level is 17.5, substantially more than what can be attributed to chance. So the 3PL model appears to be inadequate for the data set when predicting the association among the items is the concern.

Further careful examination of the plot reveals a number of interesting patterns regarding association among the items. Stout (1987) and many other researchers argued that for tests where local independence does not hold, association among items is more than what a unidimensional model can predict for within-cluster items (those that are influenced by a trait T_b that is different from the one trait T_a that the test intends to measure; the trait T_a is common to each item in the test) and less than what a unidimensional model can predict for pairs of between-cluster items (those that are influenced by different traits). Thus patterns of under-prediction and over-prediction might reveal useful information on clustering of items.

Fig. 2 shows a few clusters of items with inverted triangles (which means under-prediction by the model) for most of the pairs within the clusters. Most visible of these clusters are those with Items 28-36 and Items 62-66. Interestingly, both these clusters consist of testlet items and appear toward the end of their respective sections. Hence, probably a combination of a testlet effect and a speededness effect (examinees getting rushed because of time pressure) cause the failure of the 3PL model to explain the association among items within these clusters. Note that there is significant overall under-prediction for pairs of items consisting of one item from Items 30-36 (all testlet items) and another from Items 63-66 (all testlet items). There exists moderate overall under-prediction within the clusters with Items 56-62 and 67-78, all testlet items. These phenomena indicate that the testlet items define a dimension by themselves, i.e., measure a trait that is different from the overall trait that the test attempts to measure. Other clusters with a high or a moderately high number of inverted triangles are the clusters involving Items 11-23 and 46-50—all WR items; under-prediction also occurs for item pairs with one item from the set 12-18 (all WR items) and another from 46-49 (all WR items). Under-prediction is also prevalent for completion items, although to a lesser extent. For example, there are few inverted triangles for the set 37-45 and for the pairs with one from 1-10 and another from 37-45.

Many of the significant over-predictions in Fig. 2 (denoted by triangles) occur for item pairs with one testlet item and one discrete item. For example, there are a significant number of over-predictions involving item pairs consisting of one item from Items 1-23 and another item from Items 24-36 or Items 58-78. This is probably because the testlet items capture a different skill from the discrete items (a hypothesis that seems reasonable from a substantive point of view). The 3PL model also significantly over-predicts the associations involving very difficult items (e.g., items 10, 32, 35, 45, 51).

Scoring (and ranking) of examinees using the type of test considered in this example does not take into account the dependence among the item responses exhibited by Fig. 2. A number of papers, such as, Bradlow, Wainer, and Wang (1999) and the references therein, find that if there exists a dependence structure in the data, an analysis ignoring the structure will result in an overstatement of the precision of proficiency estimates. Bradlow, Wainer, and Wang (1999) comment that:

The primary goal of exam scores is to rank individuals. In particular the rankings at the upper end of the ability scale are often used to designate award winners, admit applicants to highly competitive programs, etc.

The same authors find through a simulation study that their testlet model predicts better the ranks of the individuals than does a simple IRT model, specifically for the high-scoring examinees. Therefore, Fig. 2, along with the findings in Bradlow, Wainer, and Wang (1999), indicate that a scoring technique that can take into account the association prevalent in the data will result in a fairer ranking of the examinees. Also, works like Beguin, Hanson, and Glas (2000) showed that if an investigator was to perform IRT equating with these data, a model taking the lack of local independence (observed in these data) into account will perform better than a univariate IRT model.

3.2   Application of a Bayesian Hierarchical Model to Data from NAEP

3.2.1   A Brief Description of NAEP

The National Assessment of Educational Progress (NAEP) is the only regularly administered and Congressionally mandated national assessment program (Beaton and Zwick, 1992). NAEP is an ongoing survey of the academic achievement of the school students in the U.S. in a number of subject areas like reading, writing, mathematics, etc. It is administered by the National Center for Education Statistics (NCES), a part of the U.S. Department of Education, to a sample of students in Grades 4, 8, and 12. To ensure a wide content coverage while not burdening a student with too many questions, NAEP employs *matrix sampling,* i.e., an examinee receives only a subset of all questions for that test.

A document called the Nation's Report Card[1] reports the results of NAEP for a number of academic subjects on a regular basis. Comparisons of results are provided with the results from previous assessments in each subject area. Academic achievement is represented by the average level of proficiency of all students in the U.S. and by the percentage of students attaining predefined levels of proficiency (indicating what students should know at each grade level) in different subjects. In addition to producing these numbers for the nation as a whole, NAEP reports the same results for states, urban districts, and other jurisdictions as well as for different subpopulations (based on gender, race, school-type, etc.) of the student population.

3.2.2   The Statistical Model Used in NAEP

The reporting method used in NAEP (and other educational survey assessments such as IALS (International Adult Literacy Study) and TIMSS (Trends in Mathematics and Science Study)) relies on a model that consists of two components.

The first component of the model (an IRT part) defines the responses of examinee i to a set of cognitive items to be dependent on a p-dimensional latent *trait /proficiency* vector $\boldsymbol{\theta}_i = (\theta_{i1}, \ldots \theta_{ip})$. The possible values of p are 2, 3, or 5, depending on the subject tested.

The dichotomous items and polytomous items are modeled using the 3PL model given by (5) and the GPCM given by (6), respectively.

In the second component (a linear regression part), the examinee proficiency vector $\boldsymbol{\theta}_i$ is assumed to follow a multivariate normal prior distribution, i.e.,

$$\boldsymbol{\theta}_i | \boldsymbol{x}_i \sim N(\Gamma' \boldsymbol{x}_i, \Sigma),$$

---

[1] Available in the website: http://nces.ed.gov/nationsreportcard/

conditional on $\boldsymbol{x}_i$, where $\boldsymbol{x}_i = (x_{i1}, x_{i2}, \ldots x_{im})$ represents m demographic and educational characteristics for the examinee. The mean parameter matrix Γ and the variance matrix Σ are assumed to be the same for all examinees.

3.2.3 NAEP Estimation Technique

Let us denote the response vector to the examination items for examinee i as $\boldsymbol{y}_i = (\boldsymbol{y}_{i1}, \boldsymbol{y}_{i2}, \ldots, \boldsymbol{y}_{ip})$, where, $\boldsymbol{y}_{ik}$, a vector of responses, contributes information about θ_{ik}. The likelihood for an examinee is given by

$$f(\boldsymbol{y}_i|\boldsymbol{\theta}_i) = \prod_{q=1}^{p} f_1(\boldsymbol{y}_{iq}|\boldsymbol{\theta}_{iq}) \equiv L(\boldsymbol{\theta}_i; \boldsymbol{y}_i). \tag{7}$$

The terms $f_1(\boldsymbol{y}_{iq}|\boldsymbol{\theta}_{iq})$ above are those from a univariate IRT model, usually 3PL (for binary items) or GPCM (for items scored in more than 2 categories). To simplify notation, the dependence of $f(\boldsymbol{y}_i|\boldsymbol{\theta}_i)$ on item parameters are suppressed. Under this model, $L(\Gamma, \Sigma|\boldsymbol{Y}, \boldsymbol{X})$, the (marginal) likelihood function for (Γ, Σ) based on the data $(\boldsymbol{X}, \boldsymbol{Y})$, is given by

$$\prod_{i=1}^{n} \int f_1(\boldsymbol{y}_{i1}|\boldsymbol{\theta}_{i1}) \ldots f_1(\boldsymbol{y}_{ip}|\boldsymbol{\theta}_{ip})\phi(\boldsymbol{\theta}_i|\Gamma'\boldsymbol{x}_i, \Sigma)d\boldsymbol{\theta}_i, \tag{8}$$

where n is the number of examinees, and $\phi(.|.,.)$ is the multivariate normal density function. Mislevy (1985) showed that it is not straightforward to find MLEs for the NAEP model, mainly because of the presence of the multivariate integral in (8).

To fit the above model operationally, NAEP uses the following three-stage estimation process that requires several different software programs and makes a number of approximations that have been questioned recently (see, e.g., von Davier and Sinharay, 2004).

- The first stage, *scaling*, fits the IRT part (consisting of 3PL and GPCM terms) of the model to the examinee response data, and estimates the item parameters. This stage ignores the linear regression part of the model.

- The second stage, *conditioning*, assumes that the item parameters are fixed at the estimates found in *scaling*, and fits the model (8) to the data, i.e., estimates Γ and Σ, using an EM algorithm (Mislevy, 1985) whose E-step involves a Laplace approximation of a multiple integral.

- The third stage of the NAEP estimation process, *variance estimation*, estimates the variances corresponding to the examinee subgroup averages using multiple imputation procedure (e.g. Rubin, 1987) in combination with a jackknife approach (see, e.g. Johnson and Jenkins, 2004, and the references therein).

For more details about the NAEP model and the current estimation technique, see, e.g., von Davier and Sinharay (2004) or Johnson and Jenkins (2004).

3.2.4 A Bayesian Hierarchical Model for NAEP

Johnson and Jenkins (2004) and Johnson (2002) suggested a Bayesian approach that combines the three steps (scaling, conditioning, and variance estimation) of NAEP estimation process. They employed a Bayesian hierarchical model that includes the item response theory component and the linear regression component of the NAEP model discussed earlier; in addition, their model accommodates clustering of the examinees within schools and primary sampling units (PSU; school district, state, etc.), using a linear mixed effects (LME) model of the form:

$$\boldsymbol{\theta}_i | \nu_{s(i)}, \Gamma, \boldsymbol{x}_i, \Sigma \sim N(\nu_{s(i)} + \Gamma' \boldsymbol{x}_i, \Sigma),$$
$$\nu_{s(i)} | \eta_{p(s)}, T \sim N(\eta_{p(s)}, T),$$
$$\eta_{p(s)} | \Omega \sim N(0, \Omega),$$

where $s(i)$ is the school examinee i attends, $p(s)$ is the PSU that school s belongs to.

Suitable prior distributions are assumed where required. Gibbs sampler samples from the joint posterior distribution of the item and ability parameters, Γ and Σ. Johnson and Jenkins (2004) deal with unidimensional proficiency while Johnson (2002) handles data with 3-dimensional proficiency. Some results from the Bayesian hierarchical model are different from those obtained under the frequentist approach. Unlike the current (frequentist) estimation process, the Bayesian approach requires neither questionable approximations nor multiple software programs.

3.3 A Bayesian Approach to Assess Differential Item Functioning

Differential item functioning (DIF) is the difference in test item performances between two comparable group of examinees, that is, groups that are matched with respect to the construct measured by the test. For example, research on Scholastic Aptitude Test (SAT) reading passages has shown that content related to technical aspects of science (as opposed to the history or philosophy of science) appears to be more difficult for women than for a matched group of men (Lawrence, Curley, and McHale, 1988). From an IRT point of view, no DIF for an item means that the item has the same item response function in every group of examinees. Detecting DIF and finding a remedy is an important part in ensuring the fairness of a test.

Holland (1985) suggested the use of the Mantel-Haenszel (MH) common odds ratio estimate to detect whether an item shows DIF. In an application of the Mantel-Haenszel test, the examinees are divided into K matching groups, typically based on their raw scores. For an item, the data from the k-th matched group of focal (the examinee group of interest; mostly, it is the female or black or Hispanic examinees) and reference (those other than the focal group) group members are arranged as a 2×2 table. Then the MH odds ratio estimate (Mantel & Haenszel, 1959), which pools information from the K 2×2 tables, is computed. A function of the statistic is used to determine the DIF status, which could be "C" (representing moderate to large DIF), "B" (representing slight to moderate DIF) or "A" (negligible DIF). It is also possible to compute the Mantel-Haenszel χ^2 test statistic, which pools information from the K 2×2 tables and asymptotically follows a χ_1^2 distribution.

However, test administrators often face the dilemma of detecting DIF with small samples. Because the application of the Mantel-Haenszel test requires large samples, questions arise as to what to do with small samples. Often, the practice is not to perform any DIF analysis at all for samples of size less than a prespecified value, even though some researchers feel differently. Lewis and Thayer (2000), for example, recommended that to produce a licensing test that is as fair as possible for all test takers, a DIF analysis should be performed even for small sample sizes.

One advantage of the Bayesian methods over frequentist methods is that the former can incorporate, in the form of a prior distribution, existing information on the inference problem at hand. Further, for any operational test, a huge volume of past data is available and for any item appearing in the current test, there is a high chance that a number of similar items have appeared in past operational forms of the test; therefore, ideally, it will be possible to use this past information as a prior distribution in a Bayesian DIF analysis. In an ongoing research project, Sinharay and

Dorans (2004) are exploring the use of past information to improve assessment of DIF for small samples.

The basic statistical model to be used is that suggested by Zwick, Thayer and Lewis (1999), where the Mantel-Haenszel DIF statistic MH_i for item i (which is computed in the operational DIF analysis) is assumed to follow a $\mathcal{N}(\theta_i, \sigma_i^2)$ distribution. Zwick et al. (1999) assign a $\mathcal{N}(\mu, \tau^2)$ prior distribution to the θ_i's and adopt an empirical Bayes approach to estimate μ and τ^2 from available data.

The approach of Sinharay and Dorans (in press) assumes the same prior distribution $\mathcal{N}(\mu_T, \tau_T^2)$ for the θ_i's whose corresponding items are of a specific type T, e.g., on reading passages with content related to technical aspects of science. An analysis of past data on similar items should provide good estimates of μ_T's and τ_T^2's. It is also possible to allow μ_T's and τ_T^2's to be random variables and assume hyper-priors on them. Preliminary analyses show that this approach has potential to provide more accurate results (than an empirical Bayes method or a frequentist method) in DIF analyses with small sample sizes.

4. Conclusions

The examples in this article demonstrate how Bayesian statistics can be useful in educational measurement, especially in situations where frequentist methods fall short. As computers become faster and access to data becomes easier, Bayesian statistics and MCMC algorithms are sure to find more applications in educational measurement. However, this situation also suggests the use of caution. An application of an MCMC algorithm often does not reveal identifiability problems with the model and unless the investigator is careful, the resulting Bayesian inference may be inappropriate. Therefore, before using a very complicated over-parameterized model, a researcher should always make sure that the model performs considerably better than a simpler alternative, both substantively and statistically, and that the model is well-identified. It is important to remember the following comment in Rubin (1983):

William Cochran once told me that he was relatively unimpressed with statistical work that produced methods for solving non-existent problems or produced complicated methods which were at best only imperceptibly superior to simple methods already available. Cochran went on to say that he wanted to see statistical methods developed to help solve existing problems which were without currently acceptable solutions.

Acknowledgements

The author thanks Charles Lewis, Robert Mislevy, Russell Almond, Shelby Haberman, Daniel Eignor, the anonymous reviewer, and Satyanshu Upadhyay for useful advice. Any opinions expressed in this article are that of the author and not of Educational Testing Service.

References

Agresti, A. (2002). *Categorical data analysis*. New York: Wiley.

Albert, J.H (1992). Bayesian estimation of normal ogive item response curves using Gibbs sampling. *Journal of Educational Statistics*, **17**, 251-269.

Beaton, A. and Zwick, R. (1992). Overview of the National Assessment of Educational Progress. *Journal of Educational and Behavioral Statistics*, **17**, 95-109.

Beguin, A.A., and Glas, C.A.W. (2001). MCMC estimation and some model-fit analysis of multidimensional IRT models. *Psychometrika,* **66**, 541-561.

Beguin, A.A. Hanson, B.A. and Glas, C.A.W. (2000). Effect of multidimensionality on separate and concurrent estimation in IRT equating. Paper presented at the annual meeting of the National Council on Measurement in Education, New Orleans, LA.

Birnbaum, A. (1967). *Statistical theory for logistic mental test models with a prior distribution of ability.* (ETS RB-67-12). Princeton, NJ: ETS.

Bradlow, E.T., Wainer, H. and Wang, X. (1999). A Bayesian random effects model for testlets. *Psychometrika,* **64**, 153-168.

Braun, H.I., Jones, D.H., Rubin, D.B. and Thayer, D.T. (1983). Empirical Bayes Estimation of Coefficients in the general linear model from data of deficient rank. *Psychometrika,* **48**, 171-181.

Calandra, A. (1941). Scoring formulas and probability considerations. *Psychometrika,* **6**, 1-9.

Cattell, J. McK. (1890). Mental tests and measurements. *Mind,* **15**, 373-381.

Chen, W. and Thissen, D. (1997). Local dependence indexes for item pairs using item response theory. *Journal of Educational and Behavioral Statistics,* **22**, 265-289.

Edwards, W., Lindman, H. and Savage, L.J. (1963). Bayesian statistical inference for psychological research. *Psychological Review,* **70**, 193-242.

Fox, J.P. and Glas, C.A.W. (2001). Bayesian estimation of a multi-level IRT model using Gibbs sampling. *Psychometrika,* **66**, 271-288.

Fox, J.P. and Glas, C.A.W. (2003). Bayesian modeling of measurement error in predictor variables using item response theory. *Psychometrika,* **68**, 169-191.

Gelman, A., Meng, X.L. and Stern, H. (1996). Posterior predictive assessment of model fitness via realized discrepancies. *Statistica Sinica,* **6**, 733-807.

Glas, C.A.W. and Meijer, R.R. (2003). A Bayesian approach to person fit analysis in item response theory models. *Applied Psychological Measurement,* **27**, 217-233.

Glas, C.A.W. and van der Linden, W.J. (2003). Computerized adaptive testing with item cloning. *Applied Psychological Measurement,* **27**, 247-261.

Gulliksen, H. (1950). *Theory of mental tests.* New York: Wiley.

Guttman, I. (1967). The use of the concept of a future observation in goodness-of-fit problems. *Journal of the Royal Statistical Society B,* **29**, 83-100.

Hambleton, R.K. and Swaminathan, H. (1985). *Item response theory: Principles and applications.* Boston: Kluwer Nijhoff Publishing.

Hartz, S., Roussos, L. and Stout, W. (2002). *Skill diagnosis: Theory and practice.* User Manual for Arpeggio software. ETS.

Hoijtink, H. and Molenaar, I.W. (1997). A multidimensional item response model: Constrained latent class analysis using the Gibbs sampler and posterior predictive checks. *Psychometrika,* **62**, 171-189.

Holland, P.W. (1985). On the study of differential item performance without IRT. *Proceedings of the 27th annual conference of the Military Testing Association Vol. I,* 282-287, San Diego, CA: Navy Personnel Research and Development Center.

Huynh, H. (1982). A Bayesian Procedure for Mastery Decisions Based on Multivariate Normal Test Data. *Psychometrika,* **47**, 309-319.

Jackson, P.H., Novick, M.R. and Thayer, D.T. (1971). Estimating regressions in m groups. *British Journal of Mathematical and Statistical Psychology,* **24**, 129-153.

Johnson, M.S. (2002). *A Bayesian Hierarchical Model for Multidimensional Performance Assessments.* Paper presented at the Annual Meeting of the National Council on Measurement in Education, New Orleans, LA, USA.

Johnson, M.S. and Jenkins. F. (2004). *A Bayesian Hierarchical Model for Large-scale Educational Surveys: An Application to the National Assessment of Educational Progress.* (ETS RR-04-38). Princeton, NJ: ETS.

Johnson, M.S. and Sinharay, S. (2005). Calibration of polytomous item families using Bayesian hierarchical modeling. *Applied Psychological Measurement,* **29**, 364-400.

Karabatsos, G. and Sheu, C.F. (2004). Order-constrained Bayes inference for dichotomous models of unidimensional non-parametric IRT. *Applied Psychological Measurement, 28*(2), 110-125.

Kelley, T.L. (1927). *The Interpretation of Educational Measurements.* New York: World Book.

Kuder, G.F. and Richardson, M.W. (1937). The theory of estimation of test reliability. *Psychometrika,* **2**, 151-160.

Lawrence, I.M., Curley, W.E. and McHale, F.J. (1988). *Differential item functioning for males and females on SAT-Verbal reading sub-score items* (RR-88-04). Princeton, NJ: ETS.

Lee, S.Y. (1981). A Bayesian approach to confirmatory factor analysis. *Psychometrika,* **46**, 237-263.

Lewis, C. and Sheehan, K.M. (1990). Using Bayesian decision theory to design a computerized mastery test. *Applied Psychological Measurement,* **14**, 367-386.

Lewis, C. and Thayer, D.T. (2000). *An investigation of the effectiveness of Mantel-Haenszel procedures to detect DIF using pretest data from an adaptive licensure test.* Unpublished technical report. Princeton, NJ: ETS.

Lindley, D.V. (1969a). *A Bayesian solution for some educational prediction problems.* (ETS RB-69-57). Princeton, NJ: ETS.

Lindley, D.V. (1969b). *A Bayesian estimate of true scores that incorporate prior information.* (ETS RB-69-75). Princeton, NJ: ETS.

Lindley, D.V. (1969c). *A Bayesian solution for some educational prediction problems, II.* (ETS RB-69-91). Princeton, NJ: ETS.

Lindley, D.V. (1970). *A Bayesian solution for some educational prediction problems, III.* (ETS RB-70-33). Princeton, NJ: ETS.

Lord, F.M. (1969). Estimating true-score distributions in psychological testing (an empirical Bayes estimation problem). *Psychometrika,* **34**, 259-299.

Lord, F.M. (1980). *Applications of item response theory to practical testing problems.* Hillsdale, NJ: Lawrence Erlbaum Associates.

Lord F.M. and Novick M.R. (1968). *Statistical theories of mental test scores.* Reading, MA: Addison-Wesley.

Mantel, N. and Haenszel, W. (1959). Statistical aspects of the analysis of data from retrospective studies of disease. *Journal of the National Cancer Institute,* **22**, 719-748.

Meyer, D.L. (1966). Bayesian statistics. *Review of Educational Research,* **36**, 503-516.

Mislevy, R.J. (1985). Estimation of latent group effects. *Journal of the American Statistical Association,* **80**, 993-997.

Mislevy, R.J. (1986). Bayes modal estimation in item response models. *Psychometrika,* **51**, 177-195.

Mislevy, R.J. (1988). Exploiting Auxiliary Information About Items in the Estimation of Rasch Item Difficulty Parameters, *Applied Psychological Measurement,* **12**, 281-296.

Mislevy, R.J. (1995). Probability-based inference in cognitive diagnosis. In P. Nichols, S. Chipman and R. Brennan (Eds.), *Cognitively diagnostic assessment,* 43-71, Hillsdale, NJ: Erlbaum.

Muraki, E. (1992). A generalized partial credit model: Application of an EM algorithm. *Applied Psychological Measurement,* **16**, 159-177.

Novick, M.R. (1964). *On Bayesian logical probability.* (ETS RB-64-22). Princeton, NJ: ETS.

Novick, M.R. (1969a). *Bayesian methods for psychological testing.* (ETS RB-69-31). Princeton, NJ: ETS.

Novick, M.R. (1969b). *A Bayesian approach to the selection of predictor variables.* (ETS RB-69-58). Princeton, NJ: ETS.

Novick, M.R. and Hall, W.J. (1964). *On the choice of prior distribution.* (ETS RB-64-23). Princeton, NJ: ETS.

Novick, M.R. and Jackson, P.H. (1975). *Statistical Methods for Educational and Psychological Research.* New York: McGraw-Hill.

Novick, M.R., Jackson, P.H. and Thayer, D.T. (1971). Bayesian Inference and the Classical Test Theory Model: Reliability and True Scores. *Psychometrika*, **36**, 261-288.

Novick, M.R. Jackson, P.H. Thayer, D.T. and Cole, N. (1971). *Applications of Bayesian Methods to the Prediction of Educational Performance*. (ETS RB-71-18). Princeton, NJ: ETS.

Novick, M.R., Jackson, P.H., Thayer, D.T. and Cole, N. (1972). Estimating multiple regression in m groups: a cross-validation study. *British Journal of Mathematical and Statistical Psychology*, **25**, 33–50.

Novick, M.R., and Thayer, D.T. (1969). *A Comparison of Bayesian Estimates Of True Score*. (ETS RB-69-74). Princeton, NJ: ETS.

Novick, M.R., Thayer, D.T. and Jackson, P. H. (1970). *Bayesian Inference And The Classical Test Theory Model II Validity And Prediction*. (ETS RB-70-32). Princeton, NJ: ETS.

Owen, R.J. (1975). A Bayesian sequential procedure for quantal response in the context of adaptive mental testing. *Journal of the American Statistical Association*, **70**, 351-356.

Patz, R. and Junker, B. (1999a). A straightforward approach to Markov chain Monte Carlo methods for item response models. *Journal of Educational and Behavioral Statistics*, **24**, 146-178.

Patz, R. and Junker, B. (1999b). Applications and extensions of MCMC in IRT: Multiple item types, missing data, and rated responses. *Journal of Educational and Behavioral Statistics*, **24**, 342-366.

Patz, R. Junker, B., Johnson, M. S. and Mariano, L. (2002). The hierarchical rater model for rated test items and its application to large-scale educational assessment data. *Journal of Educational and Behavioral Statistics*, **27**, 341-384.

Ramsay, J.O. (1978). Confidence regions for multidimensional scaling analysis. *Psychometrika*, **43**, 145-160.

Rubin, D.B. (1980). Using Empirical Bayes Techniques in the Law School Validity Studies. *Journal of the American Statistical Association*, **75**, 801-816.

Rubin, D.B. (1981). Estimation in parallel randomized experiments. *Journal of Educational Statistics*, **6**, 377-401.

Rubin, D.B. (1983). Some Applications of Bayesian Statistics to Educational Data. *The Statistician*, **32**, 55-68.

Rubin, D.B. (1984). Bayesianly justifiable and relevant frequency calculations for the applied statistician. *Annals of Statistics*, **12**, 1151-1172.

Rubin, D.B. (1987). *Multiple imputation for nonresponse in surveys*. John Wiley & Sons Inc.

Rupp, A.A., Dey, D.K. and Zumbo, B.D. (2004). To Bayes or not to Bayes, from whether to when: Applications of Bayesian methodology to item response modeling. *Structural Equation Modeling*, **11**, 424-451.

Sinharay, S. (2003). *Bayesian item fit analysis for dichotomous item response theory models*. (ETS RR-03-34). Princeton, NJ: ETS. Retrieved November 1, 2004, from http://www.ets.org/research/newpubs.html.

Sinharay, S. (2005). Assessing fit of unidimensional item response theory models using a Bayesian approach. *Journal of Educational Measurement*, **42**, 375-394.

Sinharay, S. (2006). Model diagnostics for Bayesian networks. To appear in *Journal of Educational and Behavioral Statistics*, **31**, 1, 1-34.

Sinharay, S. and Dorans, N. (in press). Small-sample DIF estimation from a Bayesian point of view. Manuscript in preparation.

Sinharay, S. and Johnson, M.S. (2003). *Simulation studies applying posterior predictive model checking for assessing fit of the common item response theory models*. (ETS RR-03-28). Princeton, NJ: ETS.

Sinharay, S., Johnson, M.S. and Williamson, D.M. (2003). Calibrating item families and summarizing the results using family expected response functions. *Journal of Educational and Behavioral Statistics*, **28**, 295-313.

Swaminathan, H. and Gifford, J.A. (1986). Bayesian estimation in the three-parameter logistic model. *Psychometrika*, **51**, 581-601.

Stout, W.F. (1987). A non-parametric approach for assessing latent trait dimensionality. *Psychometrika*, **52**, 589-617.

Thissen, D. and Wainer, H. (Eds) (2001). *Test scoring*. Hillsdale, NJ: Lawrence Erlbaum Associates.

van der Linden, W.J. (1998). Bayesian item-selection criteria for adaptive testing. *Psychometrika,* **63**, 201-216.

von Davier, M. and Sinharay, S. (2004). *An Importance Sampling EM Algorithm for Latent Regression Models.* To appear in *Journal of Educational and Behavioral Statistics.*

Zwick, R., Thayer, D.T. and Lewis, C. (1999). An empirical Bayes approach to Mantel-Haenszel DIF analysis. *Journal of Educational Measurement,* **36**, 1-28.

Bayesian Statistics and Its Applications
Edited by S.K. Upadhyay, U. Singh and D.K. Dey
Anamaya Publishers, New Delhi, India

A Bayes Analysis of the Birnbaum-Saunders Distribution Using the Gibbs Sampler Approach

S.K. Upadhyay and M. Peshwani

Department of Statistics, Banaras Hindu University, Varanasi-221 005, India

Abstract

In the recent past Gibbs sampler has become an important Markov chain Monte Carlo procedure for exploring complicated posterior surfaces, which are difficult, or often even impossible to analyze by alternative non-sample based approaches. In this article we have proposed a new form of three-parameter Birnbaum-Saunders distribution and presented its full posterior analysis in addition to the posterior analysis of usual two-parameter form. The posterior analysis in each case is carried out using the Gibbs sampler algorithm. It is to be mentioned here that the study ultimately aims to provide a comparison of the two-parameter versus three-parameter forms of the models and, therefore, the Bayesian tools used in the article always concentrate on our ultimate goal, that is, model comparison/validation. Numerical illustrations are provided on both real and simulated data sets from the models.

1. Introduction

It is well known that fatigue data obtained under a process can be modelled by a number of families. In fact, in the region of central tendency most of the common families such as the Weibull, gamma, lognormal, etc. can all be fitted by parametric estimation and because of the availability of small sample sizes, hardly any one can be rejected by the usual goodness of fit tests criteria. However, when the question of predicting safe life arises, there is a wide discrepancy among these commonly arising models. For this and several other reasons (see, for example, Birnbaum and Saunders (1969a)), a family of distributions obtained from the consideration of basic characteristics of fatigue process should be more persuasive than any ad hoc family chosen for certain extraneous reasons. One such distribution was derived by Birnbaum and Saunders (1969a) to model time to failure for metals subject to fatigue from assumptions concerning the behavior of fatigue crack growth under cyclic loading. The life length obtained was the mathematical representation of the number of cycles needed to force the fatigue crack to exceed a critical value (see also Mann et al. (1974)).

The probability density function (pdf) of the resulting distribution, known as two parameter Birnbaum-Saunders (B-S) distribution, is given by

$$f(x; \alpha, \beta) = \frac{x + \beta}{2\sqrt{2\pi}\alpha\beta^{1/2}x^{3/2}} \exp\left\{ -\frac{1}{2a^2} \left(\frac{x}{\beta} + \frac{\beta}{x} - 2 \right) \right\} ; x > 0 \tag{1}$$

where positive parameters α and β are the shape and scale parameters, respectively.

As in the cases of Weibull, gamma and lognormal fatigue models, B-S distribution can also be extended by introducing a threshold parameter μ such that a simple displacement of x (say, $x - \mu$) follows a two parameter B-S distribution (1). Thus, the range of x is $\mu < x < \infty$. In other words, x follows three parameter B-S distribution with shape, scale and threshold parameters α, β and μ, respectively. The two parameter B-S distribution is then the special case for which $\mu = 0$. The pdf for the three-parameter model can be written as

$$f(x; \alpha, \beta, \mu) = \frac{x - \mu + \beta}{2\sqrt{2\pi}\alpha\beta^{1/2}\left(x - \mu\right)^{3/2}} \exp\left\{-\frac{1}{2\alpha^2}\left(\frac{x - \mu}{\beta} + \frac{\beta}{x - \mu} - 2\right)\right\}; x > \mu, \alpha, \beta, \mu > 0. \quad (2)$$

Thus, the expected lifetime (EL) for model (2) is

$$\text{EL} = \mu + \beta\left(1 + \frac{\alpha^2}{2}\right) \quad (3)$$

whereas the reliability and the hazard functions at time t are

$$R\left(t\right) = 1 - \phi\left[\frac{1}{\alpha}\left\{\left(\frac{t - \mu}{\beta}\right)^{1/2} - \left(\frac{\beta}{t - \mu}\right)^{1/2}\right\}\right], \quad (4)$$

$$h(t) = \frac{f(t; \alpha, \beta, \mu)}{R(t)}. \quad (5)$$

respectively. The quantity $\phi\left(z\right)$ is the standard normal cumulative distribution function. It can be seen that the hazard rate for the model initially increases, attains a maximum and then asymptotically approaches a positive constant value. Similar expressions for expected lifetime, reliability and hazard rate can be obtained for model (1) by substituting $\mu = 0$ in (3)-(5).

Classical inferences for model (1) have been considered by a number of authors in the literature. Mann et al. (1974) compiles most of the classical results for the distribution up to that time. The classical maximum likelihood estimators (MLEs) for the model were considered by Birnbaum and Saunders (1969b). These authors also obtained the "mean mean" estimator as an alternative to MLE for the scale parameter β. Ahmad (1988) obtained a jackknife estimator for β based on its "mean mean" estimator. Engelhardt et al. (1981) used Monte Carlo techniques and asymptotic results based on MLEs to derive certain test statistics and confidence intervals concerning the parameters. Other significant work on the model includes Rieck and Nedelman (1991), Chang and Tang (1994), Owen and Padgett (1999, 2000), etc. (see also Johnson et al. (1994)).

Bayesian inferences for the model have proved to be meager only perhaps because of the dependency of concerned posteriors on integrals that are difficult to solve analytically. Achcar (1993) considered Bayes inferences for the model (1) using Laplace approximation. He developed the procedures based on two non-informative prior distributions for α and β. If, however, β is known, he suggested using a gamma prior for the other parameter but his resulting posterior appears to be inappropriate with one of his priors. Padgett (1982) is another important reference, which deals with the Bayesian estimation of the reliability function corresponding to the model (1).

The form of the model given in (2) is an extension of the original B-S distribution in the sense that it includes an additional parameter μ. This family appears to be new and practically nothing has appeared in the literature with regard to either classical or Bayesian inferential procedures. As mentioned earlier the Bayesian inferences are conceptually straightforward except that they are hard to work with by the complexities of analytically intractable integrals and, therefore, we advocate the use of an important Markov chain Monte Carlo procedure, namely the Gibbs sampler algorithm. The reason behind choosing the Gibbs sampler simply lies in its routine implementation although other algorithm can also be used sometimes efficiently.

Once the Gibbs sampler output is successfully obtained from the models (1) and (2), a question often arises as to which of the concerned models is most compatible with the given set of data. The problem is equivalent to checking whether for a given data set, one can safely take μ to be zero and go for an easy two-parameter form. Model comparison or validation, which sometimes leads to the choice of an appropriate model, is often a difficult task. It has become a common practice to recommend a model according to the parsimony principle, that is, a simple model gets a priority over a complicated one. The recommendation according to this principle may be profitable because it appears that the developments of inferential procedures based on simple model are comparatively easier than those based on difficult one. However, the choice according to this principle may be, sometimes, disingenuous particularly when the resulting inferences from the individual models are compared and found different. In such situations, the experimenter may consider a model not just because of its simplicity but also because of its intended use (see, for example, Upadhyay et al. (2001)). Thus it looks as if the practitioner must pay little attention on some of these inferences in order to help him decide the appropriate model.

Bayesian methods provide a number of tool-kits to study the compatibility of the models with the given data set and hence for recommending the appropriate model(s) in the concerned situation. One such important strategy is based on predictive simulation ideas where Gibbs sampler output may be used to draw predictive or future data from each of the concerned models. Now using the predictive data along with the given data, one can draw the empirical distribution function (edf) plots under each model to have an informal impression of the actual compatibility. Similarly, one can calculate the Bayesian p-value for an assumed discrepancy measure and acclaim a model for which the evaluated p-value is large (see, for example, Gelman et al. (1996), Upadhyay et al. (2001), etc. for details).

Once the model compatibility is established, Bayes factor can be used for comparing and recommending a model. The Bayes factor, which provides the odds in favour of one model against the other given by the data, can be defined in a straightforward manner but its evaluation may be often difficult especially when dealing with non-informative priors or non-regular families. Recently some useful versions of the same such as the posterior Bayes factor, intrinsic Bayes factor and fractional Bayes factor, etc. have been defined in the literature (see, for example, Aitkin (1991), Berger and Perrichhi (1996) and O'Hagan (1995), etc.), which have been used quite successfully in a variety of awkward situations.

The outline of the present article is as follows. In the next section, we provide a brief sketch of the Gibbs sampler algorithm and summarize the full conditional forms corresponding to B-S distribution for the implementation of the algorithm. Both model forms given in (1) and (2) are considered for the latter developments. Throughout, our discussion is based on the assumption of independent vague priors for the parameters. Section 3 provides the complete unrestricted posterior analysis of the models based on both real and simulated data sets. It is important to mention here that the section is not just aimed to analyze the models but also to use the results for comparing and recommending the models and, therefore, besides the model parameters some other commonly used reliability characteristics are also inferred under each concerned model. In section 4 some other Bayesian tool-kits, as mentioned in the previous paragraphs, are discussed to examine the issue of model comparison or validation. Numerical results supported with discussion are given in section 5. Finally, a brief conclusion is given in section 6.

2. The Gibbs Sampler and Its Implementation

The Gibbs sampler is a Markovian updating scheme for extracting samples from posterior, typically available up to proportionality, obtained as a product of the likelihood function and a prior. The scheme proceeds via iterated sampling from the various full conditional forms again specified up to proportionality from the joint posterior, treating, for each unknown in turn, every other quantity as fixed known constant. At first, some arbitrarily chosen values are assigned to the variates and then the generation proceeds in a cyclic order; that is, every time a value is generated, it is used immediately in the next step of the cycle. It can be shown that after a large number of iterations generated samples converge in distribution to a random sample from the true posterior distribution.

Sometimes, besides obtaining the samples from the original posterior, we might be interested in generating samples from the posterior of a non-linear function of the original variates. The Gibbs sampler algorithm provides a straightforward answer to such a problem as well. To clarify, let us consider that we have samples from a posterior $p(\theta)$ and we wish to obtain the samples from a posterior $p(\varphi)$ where φ is a non-linear (or any) function of θ. An easy way of doing that is to substitute the samples of θ in the expression of φ to get the same from the posterior of φ.

It is therefore obvious that for the implementation of the Gibbs sampler scheme we only need to ensure the availability of various full conditionals for the concerned variates where the term 'availability' means that samples can be straightforwardly and efficiently generated from the concerned full conditionals. Therefore, in this section, we emphasize the various full conditional forms and concerned random variate generations assuming that once this task is satisfactorily accomplished, the Gibbs sampler can be easily implemented. However, for details, readers are referred to Upadhyay et al. (2001) and the references cited therein.

In order to provide the various full conditional forms for the implementation of the Gibbs sampler, let us begin with the assumption that n items be subjected to testing and let $x_1, x_2, .., x_n$ be the observed (ordered) failure times. The likelihood function corresponding to model (2) may be written as

$$L\left(x; \alpha, \beta, \mu\right) = \left(\frac{1}{2\sqrt{2\pi}}\right)^n \left(\frac{1}{\alpha\beta^{1/2}}\right)^n \prod_{i=1}^{n} \left\{\frac{x_i - \mu + \beta}{(x_i - \mu)^{3/2}}\right\} \exp\left\{-\frac{1}{2\alpha^2}\sum_{i=1}^{n}\left(\frac{x_i - \mu}{\beta} + \frac{\beta}{x_i - \mu} - 2\right)\right\}.$$

$$(6)$$

Considering independent (vague) priors for α, β and μ as

$$g_1\left(\beta\right) \propto \frac{1}{\beta},$$

$$g_2\left(\alpha\right) \propto \frac{1}{K}, \quad 0 < \alpha < K$$

$$g_3\left(\mu\right) \propto \text{a constant},$$

$$(7)$$

where K is some known constant, the joint posterior up to proportionality can be specified as

$$p\left(\alpha, \beta, \mu \,|\, x\right) \propto \left(\frac{1}{\alpha}\right)^n \left(\frac{1}{\beta}\right)^{\frac{n}{2}+1} \prod_{i=1}^{n}\left\{\frac{x_i - \mu + \beta}{(x_i - \mu)^{3/2}}\right\} \exp\left\{-\frac{1}{2\alpha^2}\sum_{i=1}^{n}\left(\frac{x_i - \mu}{\beta} + \frac{\beta}{x_i - \mu} - 2\right)\right\}. \quad (8)$$

From (8) the various full conditional forms for α, β and μ are

$$p\left(\alpha \,|\, \beta, \mu, x\right) \propto \left(\frac{1}{\alpha}\right)^n \exp\left\{-\frac{1}{2\alpha^2}\sum_{i=1}^{n}\left(\frac{x_i - \mu}{\beta} + \frac{\beta}{x_i - \mu} - 2\right)\right\}, \qquad (9)$$

$$p\left(\beta \mid \alpha, \mu, x\right) \propto \left(\frac{1}{\beta}\right)^{\frac{n}{2}+1} \left\{\prod_{i=1}^{n}(x_i - \mu + \beta)\right\} \exp\left\{-\frac{1}{2\alpha^2}\sum_{i=1}^{n}\left(\frac{x_i - \mu}{\beta} + \frac{\beta}{x_i - \mu}\right)\right\}, \qquad (10)$$

$$p\left(\mu \mid \alpha, \beta, x\right) \propto \left[\prod_{i=1}^{n}\left\{\frac{x_i - \mu + \beta}{(x_i - \mu)^{3/2}}\right\}\right] \exp\left\{-\frac{1}{2\alpha^2}\sum_{i=1}^{n}\left(\frac{x_i - \mu}{\beta} + \frac{\beta}{x_i - \mu}\right)\right\}. \qquad (11)$$

It is clear from (9) that $\lambda = 1/\alpha^2$ follows gamma distribution with shape parameter $(n-1)/2$ and scale parameter $2\Big/\left\{\sum_{i=1}^{n}\left(\frac{x_i-\mu}{\beta} + \frac{\beta}{x_i-\mu} - 2\right)\right\}$ and thus α can be generated using any of the standard gamma generating routines (see Devroye (1986)). β and μ can be generated through specifically designed rejection algorithms by choosing appropriate envelope densities as

$$f_e\left(\beta\right) \propto \frac{\beta + Q}{\alpha \beta^{3/2} Q^{1/2}} \exp\left\{-\frac{1}{2\alpha^2}\left(\frac{\beta}{Q} + \frac{Q}{\beta}\right)\right\}, \qquad (12)$$

$$f_e(\mu) \propto \frac{\overline{x} - \mu + \beta}{(\overline{x} - \mu)^{3/2}} \exp\left\{-\frac{1}{2\alpha^2}\left(\frac{\overline{x} - \mu}{\beta} + \frac{\beta}{\overline{x} - \mu}\right)\right\}, \qquad (13)$$

for β and μ, respectively, where $Q = \frac{1}{n}\sum_{i=1}^{n}(x_i - \mu)$ and $\overline{x} = \frac{1}{n}\sum_{i=1}^{n}x_i$.

It is to be noted that $f_e\left(\beta\right)$ is a two parameter B-S distribution with shape and scale parameters α and Q respectively, and, therefore, β can be generated from (12) using the routine for generating the B-S distribution. Generation of μ from (13) is given in appendix. The algorithms for the proposed rejection method were tested on a variety of data sets and it was found that the algorithms perform quite satisfactorily in each case. It may be further noted that the expressions given above can be easily converted to those for two-parameter model given in (1) if one replaces μ by zero everywhere. In the following sections, we use (8a) to denote the posterior corresponding to two-parameter B-S distribution which can be obtained by taking $\mu = 0$ in (8).

3.   Posterior Results and Concerned Discussions

This section considers the complete analysis of the posteriors arising from models (1) and (2) on the basis of both real and simulated data sets. By complete analysis we mean extracting samples from the concerned posteriors and then inferring on some important model parameters or characteristics. The ultimate aim of the article is to provide routine Bayes procedures for analyzing the concerned models using a straightforward Gibbs algorithm and then to see how the two models behave for the given data sets. This latter study may often motivate the experimenter in choosing and finalizing an appropriate model in a given situation.

To begin let us consider extensive data sets on fatigue lives of 6061-T6 aluminum coupons cut parallel to the direction of rolling and oscillated at 18 cycles per second. Three groups of more than a hundred life lengths each, with maximum stress per cycle are presented in Tables 1 to 3. The data sets (henceforth referred to as the 1st data, 2nd data and 3rd data, respectively, from Tables 1, 2 and 3) were initially considered by Birnbaum and Saunders (1969b) but later on reanalyzed by a number of authors (see also Birnbaum and Saunders (1958)).

The Gibbs sampler algorithm was applied separately on the posteriors (8) and (8a) using each of the above data sets and convergence monitoring was confirmed on the basis of ergodic

Table 1. Lifetimes in cycles $\times 10^{-3}$ (101 observations, maximum stress per cycle 31,000 psi)

70	90	96	97	99	100	103	104	104	105	107	108	108	108	109
109	112	112	113	114	114	114	116	119	120	120	120	121	121	123
124	124	124	124	124	128	128	129	129	130	130	130	131	131	131
131	131	132	132	132	133	134	134	134	134	134	136	136	137	138
138	138	139	139	141	141	142	142	142	142	142	142	144	144	145
146	148	148	149	151	151	152	155	156	157	157	157	157	158	159
162	163	163	164	166	166	168	170	174	196	212	-	-	-	-

Table 2. Lifetimes in cycles $\times 10^{-3}$ (102 observations, maximum stress per cycle 26,000 psi)

233	258	268	276	290	310	312	315	318	321	321	329	335	336	338
338	342	342	342	344	349	350	350	351	351	352	352	356	358	358
360	362	363	366	367	370	370	372	372	374	375	376	379	379	380
382	389	389	395	396	400	400	400	403	404	406	408	408	410	412
414	416	416	416	420	422	423	426	428	432	432	433	433	437	438
439	439	443	445	445	452	456	456	460	464	466	468	470	470	473
474	476	476	486	488	489	490	491	503	517	540	560	-	-	-

Table 3. Lifetimes in cycles $\times 10^{-3}$ (101 observations, maximum stress per cycle 21,000 psi)

370	706	716	746	785	797	844	855	858	886	886	930	960
988	990	1000	1010	1016	1018	1020	1055	1085	1102	1102	1108	1115
1120	1134	1140	1199	1200	1200	1203	1222	1235	1238	1252	1258	1262
1269	1270	1290	1293	1300	1310	1313	1315	1330	1355	1390	1416	1419
1420	1420	1450	1452	1475	1478	1481	1485	1502	1505	1513	1522	1522
1530	1540	1560	1567	1578	1594	1602	1604	1608	1630	1642	1674	1730
1750	1750	1763	1768	1781	1782	1792	1820	1868	1881	1890	1893	1895
1910	1923	1940	1945	2023	2100	2130	2215	2268	2440	-	-	-

averages obtained through a single long run of the Gibbs chain. Large number of iterations (around 60,000) was needed for achieving convergence with model (2) perhaps because of the high posterior correlations among various pairs of concerned variates. For model (1), however, the number of iterations required was comparatively low. For the starting values of the parameters α and β, the classical estimators obtained by Birnbaum and Saunders (1969b) were used for both (8) and (8a). For μ, however, one can either take x_1 as the starting value or can start iterating from the full conditional of μ itself in the Gibbs cycle and, therefore, may not require the initial estimate of μ. The value of prior hyper parameter K was taken to be 10.0 throughout, although it was observed that it has no considerable affect on the final results.

Finally, samples of size 1000 from each of the two posteriors ((8) and (8a)) were obtained by picking up equally spaced (every 10th) outcomes. The equal spacing was merely done to avoid serial correlation among the variates in the generating chain. In case of (8), therefore, the samples can be regarded as those of (α, β, μ) whereas in case of (8a) the same corresponds to those of (α, β) with individual components from the corresponding marginals. We also obtained the samples of size 1000 each from EL, $R(t)$ and $h(t)$ merely by substitution (see, for example, Upadhyay and Smith (1994)). The mission time t was arbitrarily fixed at 120, 400 and 1000 cycles ($\times 10^{-3}$), respectively, for the 1st, 2nd and 3rd data set.

Some of the important sample based posterior characteristics are shown in Tables 4 and 5 for each of the two models. The second column in each case corresponds to the values for two-parameter model. The marginal density estimates for the concerned posteriors are presented in Figs. 1 to 3 in the form of histograms. The pictures are shown for α, β and μ only although one can similarly draw for EL, $R(t)$ and $h(t)$ as well. One conclusion that immediately arises is the ease of implementation of the Gibbs sampler algorithm in such low dimensional but complicated situations. By extracting the samples from the concerned posteriors one can not only provide the global overview of the entire posterior but simultaneously draw any inferences he or she intends. It is obvious that most of the estimated posteriors are skewed (except one or two pictures of α and β which are close to symmetry) and, therefore, the usual practice of taking posterior means may not be advisable rather they may be misleading in some of the situations (see Figs. 1 to 3). The bivariate density estimates are not shown although they can be easily drawn for any pair of variates. The estimated posterior correlations were, however, found to be quite large in case of model (2) for all the three data sets.

Table 4.   Sample based posterior characteristics for α, β and μ

Data set	Characteristics	α		β		μ	
	Minimum	0.144	0.139	56.412	124.185	9.35 e-5	0.000
	Maximum	0.570	0.231	136.300	138.314	67.264	0.000
1st data	Mean	0.200	0.173	117.439	131.782	14.053	0.000
	Mode	0.169	0.169	125.877	131.386	3.889	0.000
	Minimum	0.135	0.135	175.847	364.350	0.186	0.000
	Maximum	0.486	0.207	405.093	412.500	211.497	0.000
2nd data	Mean	0.196	0.165	338.654	392.609	52.993	0.000
	Mode	0.180	0.162	373.812	394.277	11.698	0.000
	Minimum	0.272	0.258	1.03 e3	1.20 e3	0.006	0.000
	Maximum	0.525	0.400	1.45 e3	1.50 e3	179.210	0.000
3rd data	Mean	0.333	0.315	1.29 e3	1.34 e3	42.462	0.000
	Mode	0.320	0.316	1.30 e3	1.34 e3	8.170	0.000

Table 5.   Sample based posterior characteristics for EL, $R(t)$ and $h(t)$

Data set	Characteristics	EL		$R(t)$		$h(t)$	
	Minimum	126.866	126.267	0.547	0.574	0.012	0.017
	Maximum	150.014	140.778	0.816	0.799	0.034	0.032
1st data	Mean	133.818	133.773	0.699	0.705	0.024	0.024
	Mode	133.149	133.270	0.704	0.705	0.024	0.024
	Minimum	379.345	369.135	0.339	0.282	0.65 e-2	0.94e-2
	Maximum	421.507	418.513	0.580	0.574	1.79 e-2	1.85e-2
2nd data	Mean	398.080	397.987	0.449	0.455	1.33 e-2	1.33e-2
	Mode	398.544	399.052	0.457	0.461	1.36 e-2	1.34e-2
	Minimum	1261.364	1263.695	0.669	0.710	6.77 e-4	6.21e-4
	Maximum	1550.674	1572.112	0.908	0.905	1.51 e-3	1.48e-3
3rd data	Mean	1405.542	1403.636	0.816	0.822	1.03 e-3	1.00e-3
	Mode	1389.667	1398.277	0.818	0.829	1.02 e-3	0.99e-3

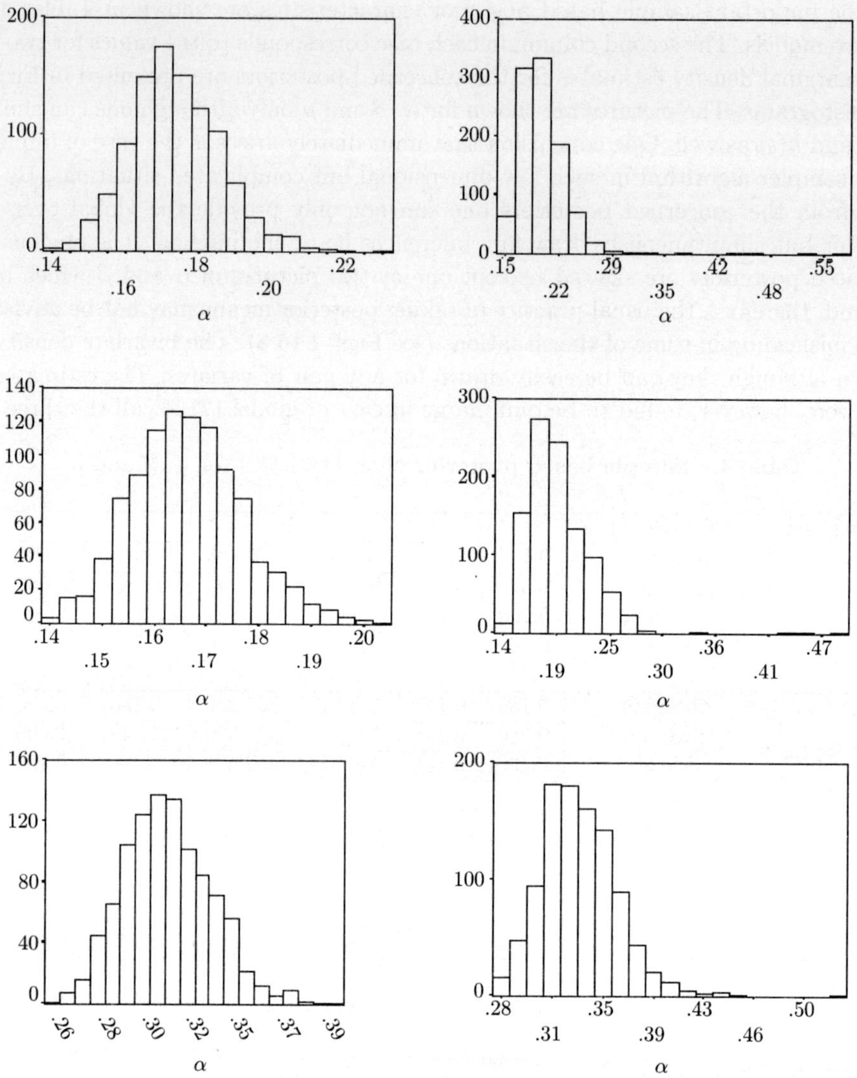

Fig. 1. Histograms showing the density estimates of α (right hand pictures correspond to three-parameter model).

A number of other conclusions can likewise be drawn on the basis of the results obtained from the Gibbs output but without going into the details of most of these usual conclusions, let us concentrate on comparing the two models with each other for the considered data sets. It is to be noted here that the previous works concentrated only on model (1) for any kind of study based on these data sets. Owen and Padgett (1999, 2000), however, considered an accelerated Birnbaum-Saunders distribution which is a three-parameter family, but the third parameter considered by them is not a shift or threshold parameter as in the present case. Thus, an obvious question arises that, do we gain anything extra if we consider the present three-parameter form. The answer is not so easy but one thing that immediately comes from the comparison of the results is differences in the behaviour if one considers (2) instead of (1). Say, for example, if we concentrate on posterior mean or mode, etc., the two models provide quite close results but the two-parameter form mostly differs in its behaviour from the three-parameter one in extreme regions. This fact can be verified by comparing the maximum and the minimum values in the tables. It may be further seen that these differences are quite often significant in such regions (see Tables 4 and 5 and Figs. 1, 2 and 3) and, therefore, the three-parameter model should not just be discarded over the two-parameter form simply because of its inherent complexity. We rather feel that the parameter μ is an added advantage with the three-parameter form which provides additional information in the form of minimum guarantee. No doubt, there are some additional complications by adding an extra parameter in the model (1) but this is not a deterrent issue with the availability of straightforward computational procedures, say, for example, the Gibbs sampler. It is, therefore, obvious from the results (see Table 4 and Fig. 3) that μ can never be taken straightforwardly as zero and so the three-parameter model, if there is no other prejudice, is always justified.

To emphasize our last conclusion we considered the estimated posterior probabilities based on the final Gibbs output of μ from the three-parameter model. The values are shown in Table 6 for arbitrarily chosen small ranges of μ. It is obvious that the estimated posterior probabilities, for all the three data sets, in no way guarantee for the suitability of the two-parameter model rather the values shown in Table 6 quite logically advocate for the use of three-parameter form. There may be some variation in the values of estimated posterior probabilities from one data set to another but these are always close to zero for ranges of μ close to zero.

Table 6. Estimated posterior probabilities based on the final Gibbs output from the marginal posterior of μ

Ranges of μ	Estimated posterior probabilities for		
	1st data	2nd data	3rd data
$\mu \leq 10.0$	0.405	0.121	0.180
$\mu \leq 5.0$	0.219	0.055	0.086
$\mu \leq 1.0$	0.038	0.008	0.018
$\mu \leq 0.5$	0.016	0.005	0.009

We also considered a number of simulated data sets from the models. Two of which are reported here as 4th data and 5th data generated using $\alpha = 0.5$, $\beta = 50$, $\mu = 100$, $n = 20$ and $\alpha = 0.5$, $\beta = 150$, $\mu = 200$, $n = 100$. We do not provide here the complete set of generated observations except the following characteristics given in Table 7 where $X_{\min}$ and $X_{\max}$ denote the minimum and the maximum of the generated observations.

Table 7. Some of the reported characteristics for the two simulated data sets

Characteristics	Arithmetic average	Geometric average	Variance	$X_{\min}$	$X_{\max}$
4th data	160.33	158.54	603.00	126.92	210.73
5th data	375.86	368.29	6076.76	250.48	586.30

The posterior analyses were done exactly as in the case of real data sets assuming both (1) and (2) to be the true models. The prior hyper parameter K was found to have no considerable effect; however, for the purpose of illustration it was assigned a value as 10.0 in each case. The results for some of the posterior characteristics are reported in Table 8. The second column in each case corresponds to the values for the two-parameter model. Here t was arbitrarily taken to be 150 and 300, respectively, for 4th and 5th data sets.

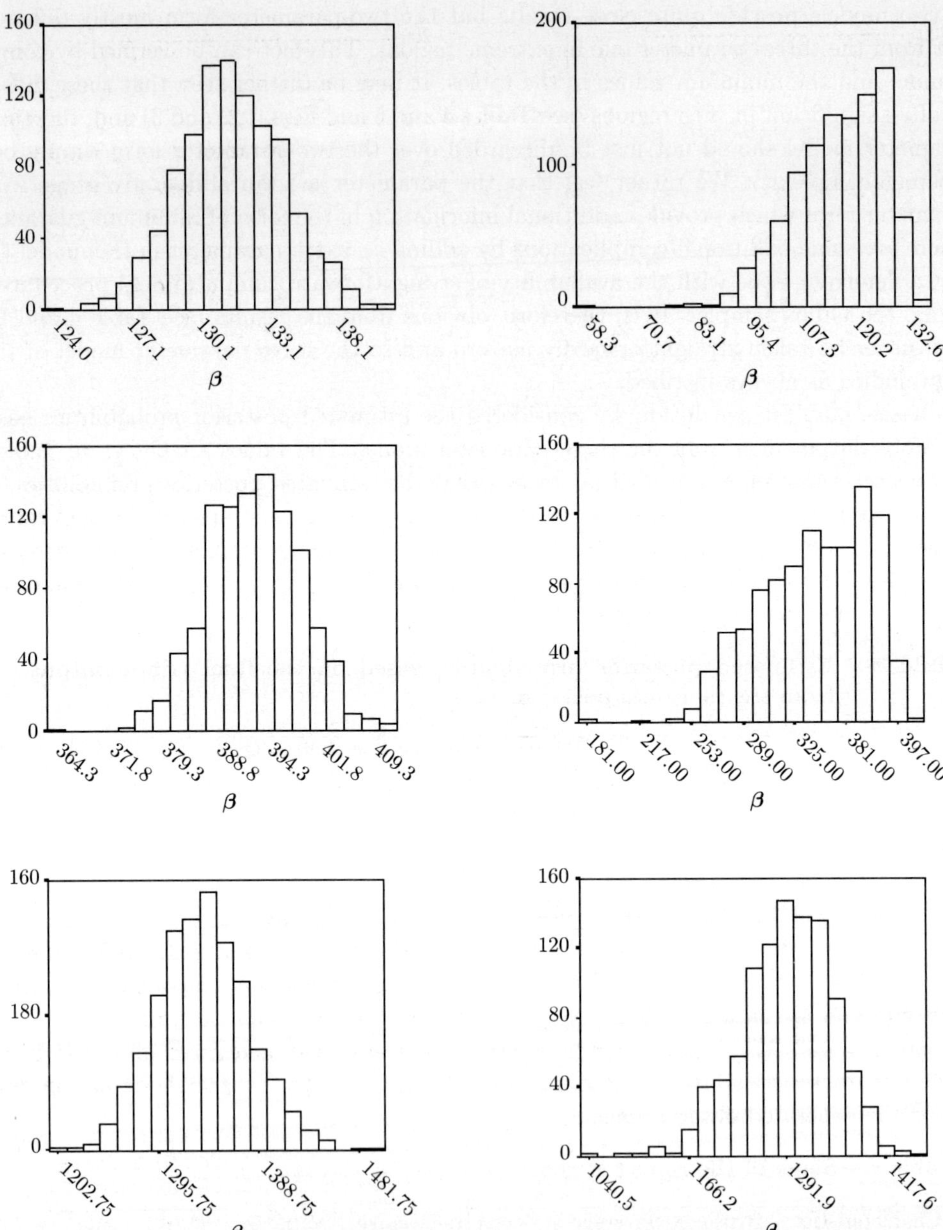

Fig. 2. **Histograms showing the density estimates of β (right hand pictures correspond to three-parameter model).**

We notice that the results are, in general, quite supportive to the three-parameter model but much surprising in a few cases. To focus on what we claim, let us observe the values shown in Table 8 little more closely. It can be seen that the two models behave quite differently with regard to the model parameters α, β (and μ) but very much similar with regard to EL, $h(t)$, etc. An experimenter not able to guess the actual parameters might even go for model (1) and can claim, at least on the

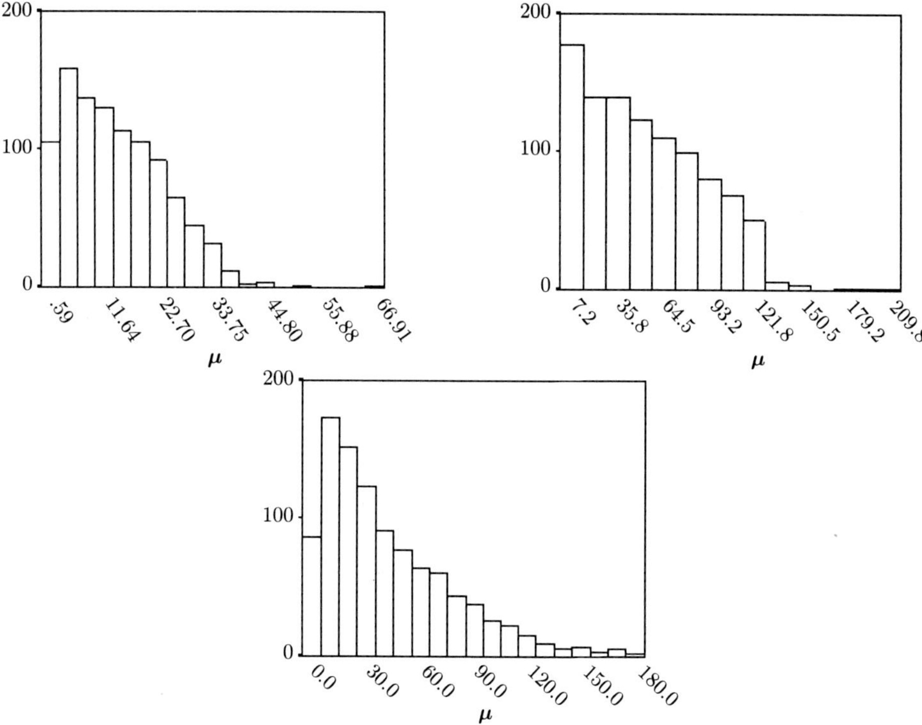

Fig. 3. Histograms showing the density estimates of μ corresponding to three-parameter model.

basis of EL, $h(t)$, etc., that there is no problem in choosing the same but may be in real trouble if the estimates of α and β are also seen. The estimates of α and β under the two models differ significantly with each other and those under the two parameter model are quite far apart from the actual parametric values that generated the two data sets. But none of these estimates can help us to know which of the models actually resulted in simulated data sets unless one goes to examine the estimates of μ as well. Perhaps only the parameter μ appears to provide an answer to the fact that which model was actually used to get the simulated data sets and, therefore, the importance of the estimates of μ plays a significant role in choosing an appropriate model. Moreover, the experimenter may be quite safe in going for the model (2) as it provides the estimates of α, β and μ at least closer to the actual parameter values and no real difference in the estimates of EL, $h(t)$, etc. over the corresponding values obtained under the assumption of model (1). It should be noticed that a similar behaviour in favour of model (1) was observed when data were simulated from the same but due to paucity of space these results are not presented. Before we finish the section, let us emphasize the need for some other tool-kits on model comparison or validation in the forthcoming sections in order to attempt to reach the final conclusion.

Table 8. Sample based posterior characteristics for α, β, μ, **EL**, $R(t)$ and $h(t)$

Data set	Characteristics	α		β		μ	
4th data	Minimum	0.123	0.096	13.191	140.829	0.260	0.000
	Maximum	1.715	0.292	172.653	187.336	125.243	0.000
	Mean	0.576	0.164	54.441	158.612	100.105	0.000
	Mode	0.475	0.159	37.969	158.398	119.086	0.000
5th data	Minimum	0.201	0.170	103.907	346.981	11.171	0.000
	Maximum	0.801	0.269	353.309	393.870	240.394	0.000
	Mean	0.430	0.205	183.995	368.149	176.313	0.000
	Mode	0.470	0.202	155.737	365.997	205.586	0.000

Data set	Characteristics	EL		$R(t)$		$h(t)$	
4th data	Minimum	143.542	143.564	0.270	0.354	1.38e-2	0.88e-2
	Maximum	207.984	192.948	0.841	0.892	6.33e-2	5.04e-2
	Mean	162.269	160.810	0.570	0.631	3.01e-2	2.49e-2
	Mode	160.316	160.272	0.597	0.623	2.64e-2	2.21e-2
5th data	Minimum	346.738	354.552	0.694	0.748	3.25e-3	3.00e-3
	Maximum	409.430	403.916	0.926	0.915	9.82e-3	6.61e-3
	Mean	376.438	375.932	0.840	0.841	6.09e-3	4.71e-3
	Mode	375.508	376.893	0.842	0.855	5.87e-3	4.84e-3

4. Other Bayesian Tool-Kits for Model Comparison/Validation

4.1 Predictive Simulation Techniques

Predictive simulation techniques are basically meant to offer model checking criteria rather than model comparison (or choice) criteria although the results obtained on the basis of the study of a number of models independently can be used to decide which of the various available models provide better compatibility with the given set of data. Model checking through predictive simulation techniques can be conducted in a variety of ways (see, e.g. Gelman et al. (1996) and Upadhyay et al. (2001), etc.). This article focusses on some of these available techniques and applies them for studying the two models (1) and (2) for a given set of data.

The predictive simulation technique is based on the comparison of predictive or future data with the observed or given data in order to see if the two data sets appear to have come from the same model distribution and have no otherwise differences in their representations. It is to be noted here that the predictive data can be easily obtained from the parent sampling distribution once the concerned model parameters are specified through the Gibbs output. The comparison can be carried out in a variety of ways. For example, the empirical distribution function (edf) plots of the two data sets can tell us (at least informally) that the two data sets might have arisen from the same model. Similarly, one can motivate Bayesian versions of the usual goodness of fit criterion which is based on evaluating the tail area probability for a particular measure of discrepancy assuming that the model under consideration is true. Bayesian versions of various classical discrepancy measures have already been defined in the literature. This tail area probability, better known as the p-value, can be used to study the compatibility of the model with the given data set. For instance, the large p-values may be used to provide enough evidence to favour a particular model. We definitely do not suggest that the p-values form the basis of any model choice criterion rather they should be used to draw an informal conclusion that the given data set can be assumed to have come from the model under consideration. It is usually recommended that one should use complete Bayesian analysis to come across any final conclusion on the choice of the model. This article attempts to provide a little flavor of what we claim.

Tail area probability or Bayesian p-value is some sort of averaging of the classical p-value with respect to a specified distribution of θ where θ may be a vector valued parameter of, say, k components. A number of choices have been suggested in the literature for the distribution of θ giving rise to different versions of Bayesian p-values. Some of the well-known versions are the prior, posterior, conditional and partial predictive p-values (see, e.g., Bayarri and Berger (1998) for details) but each suffers from its own limitations. For example, the conditional and partial predictive p-values are conceptually very much encouraging though they are computationally quite awkward. Similarly, the posterior predictive p-values are computationally not that difficult but they are conceptually not much appealing from a number of common viewpoints (see, for example, Bayarri and Berger (1998)). Prior predictive p-values are definitely the most appealing measure but they have their own limitations in the sense that they require a proper prior. As a matter of fact, we decided to use the posterior p-values based on Bayesian versions of certain classical discrepancy measures (as advocated by Gelman et al. (1996) and Upadhyay et al. (2001)) in spite of being aware of the fact that they have some serious drawbacks (see Bayarri and Berger (1998) and Dey et al. (1997) for details). It is important to mention here that getting conceptually appealing and computationally straightforward version of Bayesian p-value is still an open problem for the future researchers.

To provide the details of the criterion used here let us assume that the random variable X has the distribution $f(x|\theta)$ and let $T(.)$ be the discrepancy measure between the observed sample and the model quantities. Then the corresponding posterior p-value can be defined as

$$p = \Pr[(T(y) \geq T(x)|f, x)] = \int \Pr(T(y) \geq T(x)|f, \theta).p(\theta|f, x)d\theta. \tag{14}$$

In (14) x is the observed data, $p(\theta|f, x)$ the posterior distribution under the model f and the prior, say, g and y is the predicted data, that is, the data that would have been observed in the future with the same experiment and the model that produced x today. In Bayesian analysis one may often feel that rejecting a model may be because of the ill-chosen prior and not because of the likelihood. Well, that can be rightly pointed out but in our analysis we have considered weak (vague) priors for the parameters under the two models and, therefore, no such possibility is feasible. Moreover, the priors chosen for the common parameters are same giving rise to almost the same comparative effects on the two models.

To extend the ideas further let the discrepancy measure be chi-square (see Upadhyay et al. (2001)), then we have

$$T = \sum_{i=1}^{n} \frac{(t_i - E(t_i|\theta))^2}{V(t_i|\theta)}, \tag{15}$$

where t_i may be x_i or y_i $(i = 1, ..., n)$ as the case may be and n is the sample size. On the other hand, if the discrepancy measure is the one used by Kolmogorov-Smirnov, T may be written as

$$T = \sup_{t} |\hat{F}_n(t) - F_0(t|\theta)|. \tag{16}$$

where $F_0(t|\theta)$ is the cumulative distribution function (cdf) of the posited model and it is completely specified. $\hat{F}_n(t)$ is the empirical distribution function (edf), given by

$$\hat{F}_n(t) = \frac{\text{Number of } t_i\text{'s} \leq t}{n} \tag{17}$$

In order to calculate the posterior p-values corresponding to the above discrepancy measures, we first generate θ from the concerned posterior either directly or by using one of the available MCMC techniques. We then simulate predictive data from the model distribution f using θ generated in the previous step and calculate $T(y)$ and $T(x)$ as given in either (15) or (16). We repeat these two steps and estimate the p-value as the proportion of times $T(y)$ exceeds $T(x)$ (see, for example, Gelman et al. (1996)).

4.2 Posterior and Fractional Bayes Factors

It is generally accepted that p-values, based on tail area probabilities, offer us an informal basis for the choice of a model based on the given set of data. If one is really concerned with some formal or authentic summary, one should go for complete Bayesian analysis where two or more competitive models are specified and derive the conclusion on the basis of likelihoods or odds ratio rather than on the basis of tail areas. To be specific, let us consider the two models or the hypotheses being specified as H_j: data X belongs to model M_j, $j = 1, 2$; where M_1, M_2 are the models under consideration. Under M_j, $X \sim f(x|\theta_j)$ and the parameter vector θ_j are unknown. Hence the Bayes factor for M_2 against M_1 may be defined as

$$B_{21} = \frac{\int L(x; \theta_2)\pi(\theta_2)d\theta_2}{\int L(x; \theta_1)\pi(\theta_1)d\theta_1}, \tag{18}$$

where likelihood function $L(x; \theta_j) = \prod_{i=1}^{n} f(x_i \mid \theta_j)$ and $\pi(\theta_j)$ is the prior distribution of θ_j.

The Bayes factor given in (18) is conceptually quite straightforward except when the prior distributions $\pi(\theta_j)$ $(j = 1, 2)$ are non-informative. It is to be noted that the non-informative priors $\pi(\theta_j)$ are often improper and defined only up to arbitrary constants C_j. Hence the Bayes factor in (18) is defined only up to C_2/C_1, which is itself arbitrary. In such a situation one can either use conventional proper priors or go for some crude approximations to the Bayes factor.

In previous decade a number of alternative forms of Bayes factors were suggested in the literature for dealing with the situations involving non-informative or improper priors. Some of these are pseudo Bayes factor (see Geisser and Eddy (1979)), posterior Bayes factor, PBF (see Aitkin (1991)), intrinsic Bayes factor, IBF (Berger and Pericchi (1996)) and fractional Bayes factor, FBF (O'Hagan (1995)), etc. Most of these forms have their own merits and demerits and, therefore, none can be considered to provide overall satisfactory performance in any given situation. The IBF and the FBF are definitely advantageous and have some very nice properties in comparison to others but the former is computationally quite awkward especially for non-regular families. Readers are referred to Berger and Pericchi (1996) for a detailed discussion and a complete set of references. In this article we shall be confined to PBF and FBF only perhaps because of their ease of implementation in spite of being aware with some severe shortcomings of PBF (see Dey et al. (1997)).

Posterior Bayes Factor In (18), replacing the prior distribution $\pi(\theta_j)$ with the posterior $p(\theta|x)$, one can get

$$B_{21}^p = \frac{\int L(x;\theta_2)\, p\,(\theta_2\,|\,x)\, d\theta_2}{\int L(x;\theta_1)\, p\,(\theta_1\,|\,x)\, d\theta_1},\qquad(19)$$

which is the PBF for M_2 against M_1 defined by Aitkin (1991).

It can be seen that PBF is the ratio of the posterior mean of likelihood functions under the two models. Aitkin (1991) suggested that the value of B_{21}^p less than $1/20$, $1/100$ or $1/1000$ constitute strong, very strong and overwhelming evidence against M_2 in favour of M_1.

Fractional Bayes Factor O'Hagan (1995) proposed an alternative form of Bayes factor known as FBF to tackle the problem of improper priors. Here, in the first step, the fraction $b(= m/n)$ of the likelihood function is taken to obtain the proper posterior using the improper prior. In the second step, the Bayes factor is calculated using this posterior and the remaining fraction $(1 - b)$ of the likelihood function. The quantity m is called the minimal training sample size (see, e.g., Berger and Pericchi (1996), O'Hagan (1995), etc.). Thus

$$B_{21}^f = \frac{\int L^{1-b}(x;\theta_2)\, p'(\theta_2\,|\,x)\, d\theta_2}{\int L^{1-b}(x;\theta_1)\, p'(\theta_1\,|\,x)\, d\theta_1},\qquad(20)$$

where $p'(\theta_j\,|\,x) = \dfrac{L^b(x;\theta_j)\,\pi(\theta_j)}{\int L^b(x;\theta_j)\,\pi(\theta_j)\, d\theta_j}$ for $j = 1,\,2$. O'Hagan (1995) has also suggested for certain choices of the fraction b in evaluating the FBF. According to him b can be taken as m/n when the robustness is no concern or $(1/n)$ Max $(m, \sqrt{n})$ when the robustness is a serious concern. He also suggested a value of b as Max $\{m, \log n\}/n$ that can be taken as an intermediate option.

5. Numerical Results Continued

We continue with the numerical results once again on the basis of same data sets reported in section 3 but this time with an aim to provide the computations as detailed in section 4. First of all let us consider the edf plots for both the observed and predictive data sets on the same scale assuming that a particular model, either (1) or (2), is true. Using the Gibbs sampler algorithm we generated samples of size 50 from each of $(\alpha,\ \beta,\ \mu)$ (equation (8)) or $(\alpha,\ \beta)$ (equation (8a)) and, in each case, obtained predictive samples of size equal to the size of observed data from the corresponding model forms. Thus we have 50 predictive samples from each of the two models considering either 1st, 2nd or 3rd data set. Fig. 4 shows the edf plots for the 2nd data set. The pictures for the other two data sets can be similarly drawn. In each figure the continuous line represents the cdf, dotted lines represent the edf based on predictive data and the step line represents the edf based on observed data corresponding to the considered model (Fig. 4). At this stage, it often becomes almost clear that a particular model is not compatible with the observed data. However, for this particular data set it is obvious from the figure that none of the models can be immediately rejected. We, however, find slightly better compatibility for the three-parameter model particularly if one looks on the right tail where one finds a little scope to discriminate between the competitive models. Therefore, we stick to our earlier conclusion that if there is a possibility of choosing the appropriate model, one should definitely go for the three-parameter form as there can be an added gain in the form of threshold and no substantial cost to pay in its analysis with the availability of routine algorithm in

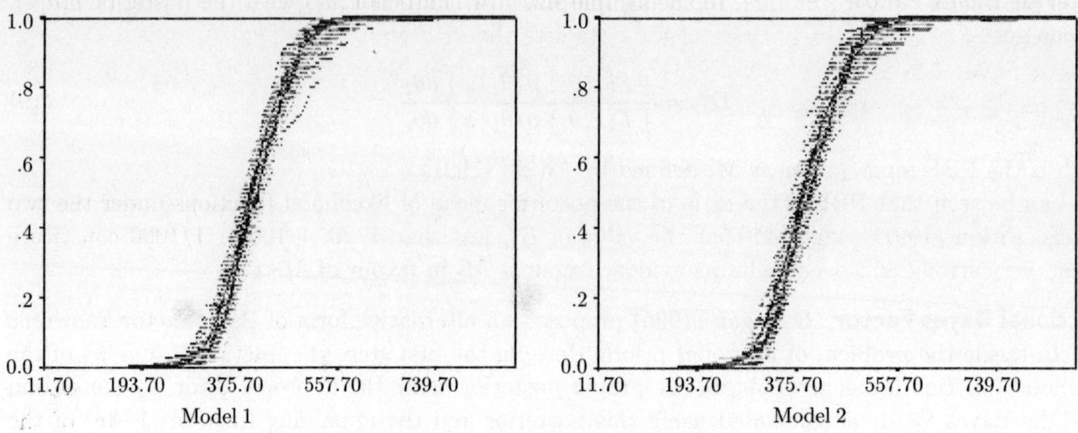

Fig. 4.   Edf plots for 2nd data.

the form of Gibbs sampler. Long run of the Gibbs sampler chain is perhaps not a stringent issue from the point of view of practitioners.

In order to evaluate the Bayesian posterior p-value for the discrepancy measures defined in subsection 4.1, we next generated 10^3 posterior samples and corresponding predictive samples of size equivalent to the size of the observed data from each of the two models. The evaluated p-values shown in Table 9 are quite large and, therefore, we come across a similar conclusion that none of the models can be rejected on the basis of evaluated p-values. So, the question arises that why not to go for model (2) which, as pointed out earlier, has an added advantage. The values in Table 9 corresponding to 4th and 5th data are meant to demonstrate the correctness of our procedure as these values definitely favour model (2) from which the data were actually drawn.

Table 9.   Posterior p-values based on chi-square and Kolmogorov-Smirnov discrepancy measures

Considered data set	Posterior p-values based on			
	Chi-square discrepancy measure		Kolmogorov-Smirnov discrepancy measure	
	Model 1	Model 2	Model 1	Model 2
1st data	0.654	0.614	0.379	0.383
2nd data	0.628	0.709	0.587	0.556
3rd data	0.817	0.895	0.357	0.331
4th data	0.402	0.612	0.419	0.605
5th data	0.349	0.614	0.332	0.625

Finally, we evaluated the posterior and the fractional Bayes factors defined in (19) and (20) for all the five data sets considered in section 3. The results are shown in Table 10. For the value of m in FBF, we considered $\sqrt{n}$ as the only choice because the other small values (or even the minimal training size) led to very poor convergence in the Gibbs sampler algorithm corresponding to the fractional likelihood in the first step (see section 4.2). It is obvious that none of the values provide strong evidence against any of the models except those for the 5th data where we have modest support for model (2) against model (1). This support is obvious as well because the data set was generated from the model (2) itself. The results for 4th data may, however, be surprising where we do not find enough evidence for model (2) although the data set was actually generated from it. The reason may perhaps be attributed to the sample size n which is not too big in case of 4th data but quite large for 5th data set. For rest of the data sets the evaluated Bayes factors are not sufficient either to advocate strongly for model (1) or to do so for model (2) except slightly in case of 1st data where model (1) can be suggested to some extent.

Table 10.   Posterior and fractional Bayes factors for all the five data sets

Data set	B_{21}^{p}	B_{21}^{f}
1st data	0.247	0.516
2nd data	0.314	0.598
3rd data	0.305	0.552
4th data	2.208	1.392
5th data	11.889	5.350

Another important result may be once again based on the threshold parameter μ. It is to be noted that both B_{21}^{p} and B_{21}^{f} require μ and also n to be large enough (see, for example, the values used for generating the 5th data) to support model (2); otherwise, they fail to do so even when the data actually comes from it (for example, the 4th data). For situations where either μ and/or n may be small, these measures are not capable of providing the real choice in favour of model (2); rather they advocate for the use of either of the two models. One can therefore say that if the two models are really distinct, the Bayesian measures suggested here will tell something formally about the true model otherwise they may not be appropriate measures to provide any kind of discrimination unless compounded by some severe restrictions either in the form of priors or otherwise.

6.   Conclusion

This article provides the complete posterior analysis of two different forms of Birnbaum-Saunders distribution, which has never been considered earlier in the literature. The analysis shows that model (2) that has been proposed here perhaps for the first time appears to be quite logical as it has an additional threshold parameter offering the minimum guarantee. It seems as if the complicated posterior form (8) and/or the likelihood form (6) may be important reasons for the denial of model (2) for quite a long time which is now quite straightforward with the availability of the Gibbs sampler algorithm.

In addition, this article also examines the comparison of model (1) with the generalized form (2) using some of the available Bayesian tool-kits. The developments suggest that the usual Bayesian measures for model comparison or validation may fail to offer clear cut conclusions in situations where the competitive models are quite close to each other and, therefore, the different posterior based characteristics must also be taken into account to provide the final recommendations. Since the generalized model may always be beneficial, one should simply not discard it because of its inherent complexities.

References

Achcar, J.A. (1993). Inferences for the Birnbaum-Saunders fatigue life model using Bayesian methods. *Computational Statistics & Data Analysis*, **15**, 367-380.

Ahmad, I.A. (1988). Jackknife estimation for a family of life distributions. *J. Statist. Comp. Simul.*, **29**, 211-223.

Aitkin, M. (1991). Posterior Bayes Factors. *J. Roy. Statist. Soc. B*, **53**, 111-142.

Bayarri, M.J. and Berger, J.O. (1998). Quantifying surprise in the data and model verification, Bayesian statistics 6, eds. J.M. Bérnardo, J.O. Berger, A.P. Dawid and A.F.M. Smith, Oxford University Press, 53-82.

Berger, J.O. and Pericchi, L.R. (1996). The intrinsic Bayes factor for model selection and prediction. *J. Amer. Stat. Assoc.*, **91**, 109-121.

Birnbaum, Z.W. and Saunders, S.C. (1958). A statistical model for life length of material. *J. Amer. Statist. Assoc.*, **53**, 151-160.

Birnbaum, Z.W. and Saunders, S.C. (1969a). A new family of life distribution. *J. Appl. Prob.*, **6**, 319-327.

Birnbaum, Z.W. and Saunders, S.C. (1969b). Estimation for a family of life distributions with applications to fatigue. *J. Appl. Prob.* , **6**, 328-347.

Chang, D.S. and Tang, L.C. (1994). Percentile bounds and tolerance limits for the Birnbaum-Saunders distribution. *Comm. Statist.-Theory and Methods*, **23**, 2853-2863.

Dey, D.K., Ghosh, S.K. and Chang, H. (1997). Measuring the effect of observations using the posterior and the intrinsic Bayes factors with vague prior information. *Sankhya*, **59**, 376-391.

Devroye, L. (1986). *Non-Uniform Random Variate Generation.* Springer-Verlag: New York.

Engelhardt, M., Bain, L.J. and Wright, F.T. (1981). Inferences on parameters of the Birnbaum-Saunders fatigue life distribution based on maximum likelihood estimation. *Technometrics,* **23**, 251-256.

Geisser, S. and Eddy, W.F. (1979). A predictive approach to model selection. *J. Amer. Statist. Assoc.*, **74**, 153-160.

Gelman, A., Meng, X.L. and Stern, H.S. (1996). Posterior predictive assessment of model fitness via realized discrepancies. *Statistica. Sinica*, **6**, 733-807.

Johnson, N.L., Kotz, S. and Balakrishnan, N. (1994). *Continuous univariate distribution.* Vol. 2, 2nd ed. Chapter 33.

Mann, N.R., Schafer, R.E. and Singpurwalla, N.D. (1974). *Methods for Statistical analysis of reliability and lifetime data.* Wiely, New York.

O'Hagan, A. (1995). Fractional Bayes factors for model comparisons. *J. Roy. Statist. Soc.*, B, **57**, 99-138.

Owen, W.J. and Padgett, W.J. (1999). Acceleration models for system strength based on Birnbaum-Saunders distributions. *Lifetime Data Analysis*, **5**, 133-147.

Owen, W.J. and Padgett, W.J. (2000). Birnbaum-Saunders accelerated life model. *IEEE Trans. Reliab.*, **49**, 224-229.

Padgett, W.J. (1982). On the Bayes estimation of reliability for the Birnbaum-Saunders fatigue model. *IEEE Trans. on Reliab.*, **31**, 436-438.

Rieck, J.R. and Nedelman, J.R. (1991). A loglinear model for the Birnbaum-Saunders distribution. *Technometrics*, **33**, 51-60.

Upadhyay, S.K. and Smith, A.F.M. (1994). Modelling complexities in reliability, and role of Bayes simulation. *International Jr. Cont. Eng. Educ.: Sp. Issue on Applied Probability Mod.*, **4**, 93-104.

Upadhyay, S.K., Vasishta, N. and Smith, A.F.M. (2001). Bayes inference in life testing and reliability via Markov chain Monte Carlo simulation. *Sankhya, A,* **63** (1), 15-40.

Appendix: Algorithm for Generation of μ from $f_e(\mu)$

1. Generate z from normal $N(0, 1)$.
2. Set $\mu = \overline{x} - \beta \left[1 + \frac{1}{2}\alpha^2 z^2 + \left(1 + \frac{1}{4}\alpha^2 z^2 \right)^{1/2} \alpha z \right]$.
3. If $0 < \mu < x_1$, then accept μ otherwise go to step 1.

Bayesian Statistics and Its Applications
Edited by S.K. Upadhyay, U. Singh and D.K. Dey
Anamaya Publishers, New Delhi, India

Bayesian Approaches to Content-based Image Retrieval

Simon P. Wilson and Georgios Stefanou

Department of Statistics, Trinity College, Dublin 2, Ireland

Abstract

This article describes the problem of content-based image retrieval (CBIR) and applies Bayesian and decision theoretic methodologies to develop a CBIR system. This is evaluated using the idea of target testing. Comments on the performance of the system, and how it might be extended, are given. Then we propose a decision theoretic solution to the problem of identifying sets of images to show the user of the system during the search process.

1. Introduction

Searching for images in a computer database is a more difficult task than searching for text because of the considerably more complex semantics that are possible. It is certainly unrealistic to expect a successful searching for images from a text-based query, such as is done in current internet search engines for example. Furthermore, textual annotation of images to aid retrieval is an expensive process that is unrealistic for many large databases, and is itself subject to the subjective interpretation of the annotator.

Content-based image retrieval (CBIR) systems attempt to make image searches more successful by questioning the user as to the type of image required; see Smeulders et al. (2000) for a comprehensive review of the early literature on the problem. In such a system, a search starts by the computer displaying a subset of images. These may be randomly selected from the whole database, or from a subset that can be defined a priori, perhaps using meta-data supplied with the database. For example, in searching for a painting of a rural scene from a database of paintings, we can eliminate all paintings that have been recorded in the meta-data as being portraits. The user then selects an image or images from the displayed subset that best match those that are required. The system then returns another set based on the user's selection. This process is known in the literature as relevance feedback and is repeated until the user encounters an image that meets his or her requirements.

CBIR systems are becoming relevant to the management of many large image databases that now exist and are growing quickly within media organisations, medical records departments, museums and research groups of many kinds. They are also relevant to making more efficient searches within that ultimate contemporary image database, the internet.

One vital aspect of a CBIR system is the relevance feedback algorithm. This is just a learning process, and if we view the images displayed and selected as data then we can contemplate using a Bayesian method. This was first realised in Cox et al. (2000), who developed a Bayesian CBIR system called PicHunter.

In this article we discuss how the PicHunter algorithm may be extended to other aspects of the relevance feedback process. In particular, we consider other aspects of the searching process that we can learn about, and consider the problem of which images to display at each iteration. This latter question is a decision problem, and as such it is to be solved within the Bayesian paradigm by the methods of decision theory.

We start with a description of how distances between images, important in the definition of the model, can be defined. Section 3 describes a model for the process of relevance feedback and the Bayesian learning algorithm for CBIR, which is then evaluated in Section 4. Section 5 continues with a description of a decision theoretic solution to the display strategy problem, illustrated in Section 6 by some examples.

2. Defining a Distance Measure on Images

An important component of the model for relevance feedback is a distance between images. A digital image $T = \{T_{ij} \,|\, i = 1, \dots, n_1; j = 1, \dots, n_2\}$ is a matrix of pixels, each pixel having one colour T_{ij}. Colour is defined by a vector of 3 values $T_{ij} = (T_{ij1}, T_{ij2}, T_{ij3})$, typically defining the red, green and blue components of the pixel colour, although there are many other parameterisations that are possible and useful.

An image feature is defined to be any real-valued function of T. Many statistical features—mean, variance, autocorrelations, histograms—can be defined. Letting $f(T) \in \mathbb{R}^d$ be a vector of d features of the image T, then we can define the distance between two images T_1 and T_2 to be the distance in $\mathbb{R}^d$ between their feature vectors:

$$d(T_1, T_2) = \|f(T_1) - f(T_2)\|$$

Euclidean or a Mahalanobis distance is usually chosen. The intention is that images that are close "semantically" are also close in feature space. Of course this can never be true for all circumstances and all interpretations of an image, a fact that is known as the semantic gap.

In what follows we use a set of some 600 features for each image: summary statistics such as means and variances, colour histograms, autocorrelations between neighbouring pixels, colour coherence vectors (a measure of how large are areas of similar colours) and the location and size of objects in the image, obtained from a segmentation algorithm. These are divided into three groups: global colour features (such as summary statistics of the entire image and histograms), texture features (such as autocorrelations) and segmentation features. For each group, a principal components analysis was conducted, reducing the dimension of the feature space to about 100, which were then normalised to lie in $(0, 1)$.

For definitions and more discussion of image features, see Jain (1989) or Lewis (1990).

3. Bayesian Inference for CBIR

Our approach is an extension of Cox et al. (2000). We consider a database of images $\mathcal{I} = \{T_1, \dots, T_N\}$. The objective is to determine the "target" image $T \in \mathcal{I}$ that the user requires. T could be a specific known image in the database or more generally that image in $\mathcal{I}$ which best satisfies the user's subjective search criteria. The determination of T is accomplished by displaying a set of N_D images from $\mathcal{I}$, from which the user picks one that best satisfies what is being looked for. The system uses this information to select another image set, from which the user picks one, and so on. We define $D_k \subseteq \mathcal{I}$ to be the set of displayed images at the kth iteration of this process, and $A_k \in D_k$ to be the image picked. We define $H_t = \{D_1, A_1, D_2, A_2, \ldots, D_t, A_t\}$ to be the history of displayed images and user actions up to the tth iteration.

The learning algorithm is based around a probability model for which image the user will select to be the best match to that required from a set of images D_k when the true target image that the user seeks is T. This defines the likelihood for H_t. Specifically, we assume that the probability that a user picks an image A_k from D_k, given the target image T, is:

$$P(A_k \,|\, D_k, T, \sigma, F) = \frac{\exp\left(-d_F(A_k, T)/\sigma\right)}{\sum_{T_j \in D_k} \exp\left(-d_F(T_j, T)/\sigma\right)}, \; A_k \in D_k, \tag{1}$$

where σ is a precision parameter and d_F is a normalised distance measure in the set of image features F. The parameter σ is a measure of the performance of the feature space in describing the user's query; a small value of σ implies that there is a small region of feature space that contains images that satisfy the user's query, whereas a large value implies that images satisfying the query are not well clustered in feature space. In this case we have 3 sets of features: global colour, texture and segmentation features, so $F \in \{GC, TX, SG\}$. The idea is that users may search according to different feature types, which we should learn about through inferring F.

The unknowns are T, σ and F. Given H_t, our knowledge about these unknowns is given by the posterior distribution:

$$P(T, \sigma, F \,|\, H_t) \propto \left(\prod_{k=1}^{t} P(A_k \,|\, D_k, T, \sigma, F) \right) P(T)\, P(\sigma)\, P(F), \tag{2}$$

where $P(T)$, $P(\sigma)$ and $P(F)$ are prior distributions that we assume are uniform: $P(T = T_i) = N^{-1}, i = 1, \ldots, N$, $P(\sigma) = 1$, $0 \leq \sigma \leq 1$ and $P(F) = 1/3, F \in \{GC, TX, SG\}$.

We discretise the range of σ into 20 values $0.025, 0.075, \ldots, 0.975$, from which we can compute this posterior exactly at all combinations of (T, F, σ) without the need for any approximations. This is important because CBIR systems do not have the luxury of long computation times; the user is waiting at the computer for a response after each iteration. Of interest in this report is the marginal posterior distribution of T:

$$P(T = T_i \,|\, H_t) = \int_0^1 \sum_{F \in \{GC, TX, SG\}} P(T_i, \sigma, F \,|\, H_t)\, d\sigma, \; i = 1, \ldots, N, \tag{3}$$

which can also be computed using the discretisation of σ. This is used to decide which set of images to display at the next iteration. In this system the N_D images with the highest marginal posterior probabilities are selected for D_{t+1}. So the CBIR system works as follows:

1. Initialise by randomly or otherwise selecting a subset $D_1 \subset \mathcal{I}$ of N_D images and display to the user. Let $t = 1$.

2. Repeat until the user finds the target image:
   - User selects $A_t \in D_t$;
   - $P(T \,|\, H_t)$ is computed;
   - D_{t+1}, the N_D highest probability images from $P(T \,|\, H_t)$, is displayed;
   - $t = t + 1$.

Table 1. **Analysis of variance for number of iterations to find images as a function of system, image and person**

Factor	SS	df	p-value for F test
Intercept	56525.32	1	< 0.000
System	79.07	1	0.494
Person	481.39	7	0.896
Image	9038.79	14	< 0.000
Error	30813.02	183	

4.   Evaluation

Evaluation of the system is done by the procedure of target testing. In target testing, an image from the database is displayed and then the user employs the CBIR system to find it. The number of iterations until the image is found is then recorded. If the image is not discovered by 60 iterations, the search is abandoned as being unsuccessful and is recorded as a missing value.

We tested this algorithm against the original system of Cox et al. (2000), which only conducted inference on T. In that system, distance was defined on all features together, rather than by subset, and σ was fixed.

We used a database of $N = 1066$ paintings supplied to us by the Bridgeman Art Library, London, with $N_D = 9$ images displayed per iteration. Eight professors and graduate students in the Statistics Department volunteered to look for 15 images. We recorded the number of iterations taken by each person to find each image, with the original system and with the system described in Section 3. Fig. 1 shows 9 of the 15 images, and is representative of the many different subjects and style of painting that are in the database.

Table 1 shows the results of an ANOVA analysis of the number of iterations taken as a function of the 3 factors: system, image and person. The only significant factor is found to be image. We conclude that there is no significant difference in the number of iterations taken to find an image between the 2 systems, or between individuals, but that different images can take different numbers of iterations to locate.

We also observe that the mean number of iterations to find an image is 17, which is significantly less than the number expected by simply randomly displaying N_D images without replacement from n, in this case $1066/18 = 59.2$.

The conclusion of the ANOVA is that we have not seen any difference in performance between the 2 systems. We have not taken account of how many times each system failed to find an image, which was 7.5% for the original PicHunter and 20% for our system. So our system failed to find more images than the original PicHunter. We concluded that the problem lies with the relatively small amount of information that the data give us. We have defined the simplest possible model; one image is selected from the display set. More complicated selection procedures, involving more than one image, perhaps ordered by preference, and identifying objects in the image, would yield more information and perhaps allow us to exploit our more general model.

Having proposed a CBIR algorithm and evaluated it, we now move on to describe a strategy to decide what images to display at each iteration.

5.   Deciding the Next Display Set D_{t+1}

Based on $P(T = T_i \,|\, H_t)$, which set of images D_{t+1} should be displayed next? This is a decision problem; we must decide which subset of $\mathcal{I}$ of size N_D to display. We define a utility $U(D, T)$ of

Fig. 1.   Images used in the target testing.

picking the set D to display when the target image is T. Since T is unknown, we compute for each possible D the expected utility with respect to $P(T = T_i \mid H_t)$:

$$\mathcal{U}(D) = \sum_{i=1}^{N} U(D, T_i)\, P(T = T_i \mid H_t). \tag{4}$$

The optimal set to display is that D which maximises expected utility

$$D_{t+1} = \arg \max_{\substack{D \subseteq \mathcal{I} \\ |D| = N_D}} \mathcal{U}(D). \tag{5}$$

5.1   The Most Probable Display Scheme

The most obvious display scheme is to display those N_D images with the highest posterior probability. We observe that if we define

$$U_I(D,T) = \begin{cases} 1 & \text{if } T \in D, \\ 0 & \text{otherwise,} \end{cases}$$

then

$$\mathcal{U}_I(D) = \sum_{T_i \in D} P(T = T_i \,|\, H_t),$$

which is clearly maximised by those images with highest probability. We call this the indicator utility. This strategy was used in Section 3.

5.2   Other Display Strategies

A property of the most probable display scheme is that it tends to quickly display images in a small region of the feature space, clustered about the user actions, and ignores all images outside it. While this may be ultimately what is needed during a query, it may be more worthwhile to display images that maximise information to the system, at least in the early stages of the query process. We propose 2 utilities to model this idea.

Variance Utility We display a set of images that are widely dispersed in feature space. We can use the variance of the distances between images in D and T to define a measure of dispersion, thus

$$U_V(D,T) = \frac{1}{N_D - 1} \sum_{T_i \in D} (d(T_i, T) - \bar{d})^2$$

where $d(T_i, T)$ is a normalised distance measure in feature space and

$$\bar{d} = \sum_{T_i \in D} d(T_i, T)/N_D$$

is the mean distance of images in D to T.

Entropy Utility A measure of information is the reduction in entropy in the distribution of T by selecting a particular display set. So we can define a utility based on the negative expected entropy of the posterior of T from picking an image in D, expectation over the images in D:

$$U_E(D,T) = - \sum_{A_j \in D} \mathcal{E}(A_j, D) \, P(A_j \,|\, D, T) \tag{6}$$

where $\mathcal{E}(A_j, D) = - \sum_{i=1}^{N} P(T = T_i | A_j, D) \, \log(P(T = T_i | A_j, D))$ is the entropy of the posterior distribution of T given that A_j is picked from D (following Equations 2 and 3) and

$$P(A_j \,|\, D, T) = \frac{\exp\left(-d(A_j, T)/\bar{\sigma}\right)}{\sum_{T_j \in D} \exp\left(-d(T_j, T)/\bar{\sigma}\right)}$$

is the likelihood term as in Equation 1 but, in order to keep the computation time practical, using the distance measure d over the entire feature space and $\bar{\sigma}$ is the posterior mean of σ.

5.3   Combining Utility Functions

The two new utilities that we have proposed—variance and entropy—are primarily of use in the early stages of a query, when the objective is to learn as much as possible about the user's target.

Ultimately, one will want to resort to a utility that displays images close to the target, such as the indicator. An obvious way to do this is to consider a utility that is a convex weighted combination of the indicator utility with one of the other two, with the weight on the indicator utility increasing to 1 with the iteration, for example at the tth display set:

$$U(D, T) = \alpha_t U_I(D, T) + (1 - \alpha_t) U_E(D, T)$$

with $0 \leq \alpha_t \leq 1$ and $\alpha_t \to 1$, and the entropy utility normalised from that in Equation 6 so that it lies in $[0, 1]$ like $U_I(D, T)$. This is the subject of current work.

5.4 Optimisation Methods

Because $\mathcal{U}(D)$ is separable in each element of D, the optimal D for the indicator utility can be easily computed. This is not the case if one moves to using the variance or entropy utilities. Evaluation of the expected utility for all possible D is not an option as the number is large i.e. for $N = 1000$ and $N_D = 6$ we have about 1.37×10^{15} possible subsets. For these, we have to resort to methods that are not guaranteed to find the optimal. We propose two Monte Carlo optimisation schemes.

Random Generation: We randomly generate without replacement K subsets $D^1, \ldots, D^K$. Then we let

$$D_{t+1} = \arg\max_{k=1}^{K} \mathcal{U}(D^k). \tag{7}$$

Each element of a set D can be simulated from any distribution on $\mathcal{I}$; obvious choices are the uniform and $P(T \mid H_t)$. In this paper we choose the latter.

Simulated Annealing: For simulated annealing, we define a "neighbour" of a subset D to be another subset with one different image. A simulated annealing algorithm then runs as follows:

1. Define an initial temperature T_0, a final temperature $T_{\min}$ and a cooling schedule $T_1, T_2, \ldots$. Randomly generate without replacement a set D^0, using $P(T \mid H_t)$. Let $k = 0$.
2. While $T_k > T_{\min}$
   - $k = k + 1$.
   - Select at random one image in D^{k-1} and replace with another image in $\mathcal{I} - D^{k-1}$, randomly generated according to $P(T \mid H_t)$. Call this new set D^{new}.
   - With probability $\min\{1, \exp((\mathcal{U}(D^{\text{new}}) - \mathcal{U}(D^{k-1}))/T_k)\}$, let $D^k = D^{\text{new}}$ else $D^k = D^{k-1}$.
3. $D_{t+1} = D^k$.

Initial and final temperatures were decided on by using the methods of Kirkpatrick et al. (1983) and Lundy and Mees (1986). We looked at several different cooling schedules, and found that inverse linear ($T_k = a/(1 + bk)$) performed best. The choice of $P(T \mid H_t)$ to generate D^0 and D^{new} can be changed, to for example the uniform, but we found that the method was not particularly sensitive to this choice. Finally, our definition of neighbour can be made more or less strict, by for example allowing two changes for D^{new} or, conversely, only favouring replacement of one image in D^{new} that is close in feature space to that image replaced. However, we found that our choice was a compromise between a too small and too large change that offered a good acceptance rate.

Finally we note that computation time is limited in a live implementation of either optimisation scheme, so typically we can compute only a small number of expected utilities.

Table 2.　Average of the expected utility for D_2 over 1000 runs for the three utility functions and three computation methods. All runs use the example of Fig. 2

Utility	Exact computation over all subsets	Random generation of 100 subsets	Simulated annealing 100 iterations
Indicator	0.2643	0.2624	0.2591
Variance	0.2299	0.2264	0.2246
Entropy	−2.6869	−2.6879	−2.6883

6.　Comparison of Display Strategies and Optimisation Methods

To compare the display strategies and evaluate the performance of the optimisation approaches, we have a simulated database of only $N = 15$ images, each with only 2 features, for which queries are implemented by displaying $N_D = 3$ images. This is clearly an unrealistically small example but it has the advantages of allowing us to display what happens in feature space and, since there are only 455 possible subsets of size 3, to compute the exact optimal subset under all three utilities and compare with the results obtained by random sampling and simulation.

Figure 2 shows the feature space of this very simple database of 15 images. In each plot, the location of each numbered image is shown in the feature space. Assume that one of the three displayed images is selected as relevant. In the top left plot, the 3 images in the initial display set D_1, numbers 2, 3 and 5, are marked by a $\times$; thus $D_1 = \{T_2, T_3, T_5\}$ in our notation. We assume that image 3 is selected as relevant (thus $A_1 = T_3$). In the top right plot, the three images of the subsequent display set D_2 under the indicator utility are marked by a $\times$. The two plots on the bottom also mark by a $\times$ the three images in D_2, but under the variance (to the left) and entropy (to the right) utilities respectively. We see that, under the indicator utility, images close to the selected image 3 appear in D_2. For the variance and entropy utilities, images that are widely separated in feature space are chosen to be in D_2.

To explore the effectiveness of the optimisation methods, we repeated this experiment 1000 times, computing the optimal D_2 according to the random generation method and simulated annealing. For the random subset generation, we simulated $K = 100$ subsets. For the simulation annealing we used an inverse linear cooling schedule $T_k = a/(1 + bk)$ with a and b chosen so that the final temperature was reached in 100 iterations, thus both methods took the same time to compute. The results are compared with the exact calculation in Table 2 and we see that both non-exact methods are sub-optimal, but nevertheless do manage on average to find subsets with expected utility close to the optimal. Random generation appears to do slightly better than simulated annealing.

Finally, we move to the Bridgeman Art Library database used in Section 4. In this case we cannot enumerate all possible subsets and so D_2 under the variance and entropy utilities is computed by the two optimisation schemes only. Table 3 compares the optimisation methods over 10 runs for an example from this database where D_1 consists of 6 images; note that we only have the exact result for the indicator utility. The computation per iteration was about the same for all three utilities, at about 30 seconds, based on MATLAB code and a 1.5 GHz PC. From the results for the indicator utility, it appears that both optimisation methods can be significantly sub-optimal. From all the three utilities it appears that random generation performs better than simulated annealing.

Note that in Table 2, both sampling approaches have sampled 100 out of 455 subsets, a substantial percentage. For the random generation, it is not surprising that one does not discover the optimal subset but the performance of the simulated annealing is disappointing. This poor performance is also reflected in Table 3, where 100 images represent a tiny proportion of possible subsets of the BAL

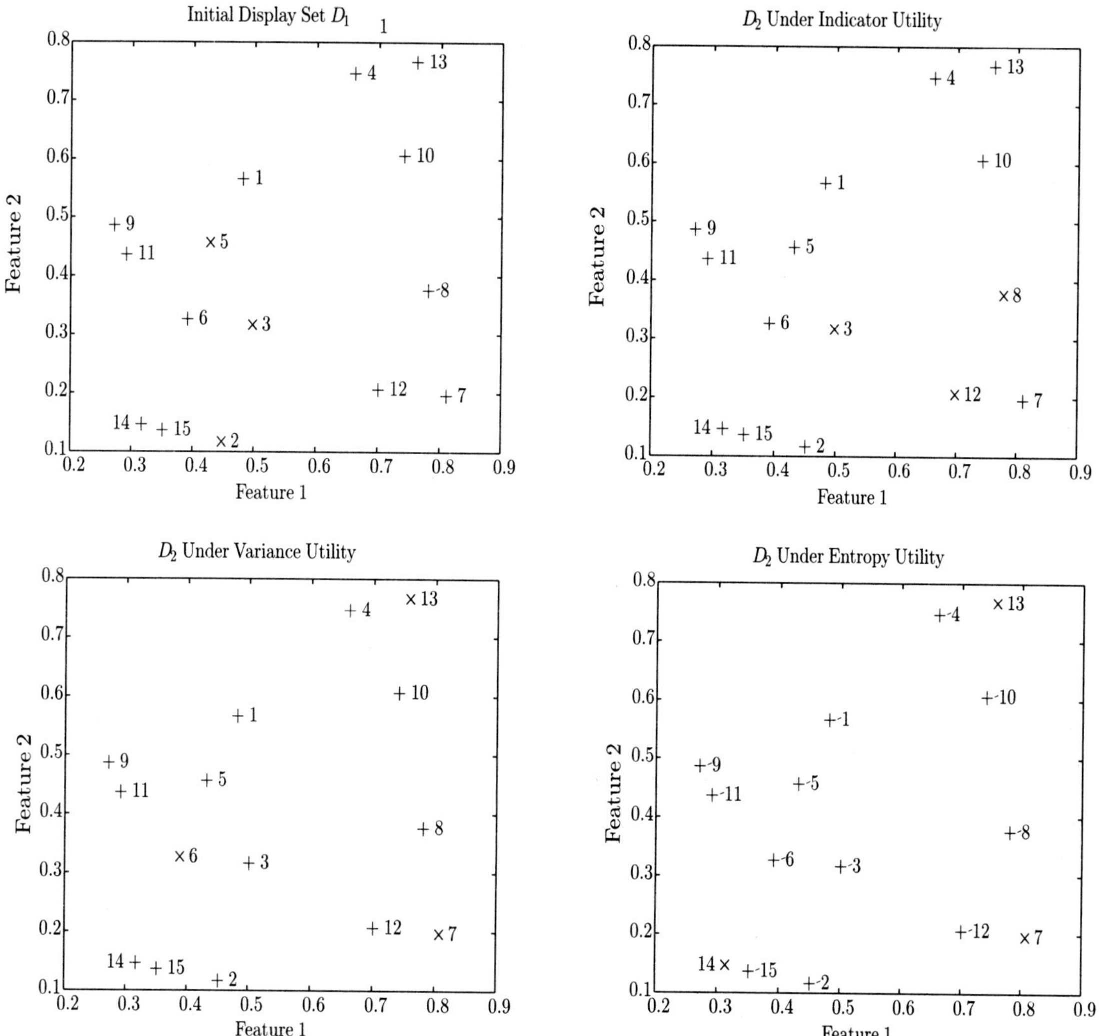

Fig. 2. Feature space plots for the selection of D_2 for a simple database of 15 images given $D_1 = \{T_2, T_3, T_5\}$ and $A_1 = T_3$. Top left shows the initial display set with D_1 highlighted by $\times$. Top right shows D_2 under the indicator utility, bottom left shows D_2 under the variance utility and bottom right shows D_2 under the entropy utility. Images in D_2 are highlighted by $\times$ in each figure.

Table 3. Average of the expected utility for D_2 over 10 runs for the three utility functions and three computation methods using the BAL database

Utility	Exact computation over all subsets	Random generation of 100 subsets	Simulated annealing 100 iterations
Indicator	0.0157	0.0106	0.0100
Variance	—	0.7767	0.7583
Entropy	—	−6.9642	−6.9647

database, yet random generation still outperforms simulated annealing. However, there is potential for improvement with this annealing algorithm by altering the proposal mechanism; for example, proposing to change 2 images, or proposing changes to images close in feature space. Nevertheless, it is clear that a large number of iterations is required to achieve even a moderately successful search.

Fig. 3 is an example from a database of paintings from the Bridgeman Art Library, using sets of 6 images. The upper display set is D_1, then the two lower sets are D_2 under the indicator and entropy utilities. They clearly show the large effect that the choice of utility has on the search process. For the indicator utility, images with very similar colour schemes and locations of objects are returned, several of which are also portraits; this is quite a successful retrieval. For the entropy utility, the goal is to display a set of images that is widely dispersed in feature space, and hopefully in meaning and interpretation.

Fig. 3. An example of display set selection with paintings. Top six images are for D_1. Male portrait (centre, top row) is selected. Bottom left: D_2 under the indicator utility. Bottom right: D_2 under the entropy utility.

7. Conclusion

We have described a Bayesian CBIR system and a decision-theoretic approach to the problem of the display set strategy.

The model that we described is for the simplest possible relevance feedback. One result of this work is the realisation that there is not much information per iteration in such a scheme. Psychological studies have shown that we cannot expect a user to complete more than 30 iterations before becoming frustrated or bored. Hence, the goal is to produce an algorithm to extract more information per iteration by, for example, allowing more than one to be selected, ranking of images, images to be chosen that are to be excluded in future, and identification of objects in images that are of specific relevance. Better specification of a prior for T, by colour and object shapes, for example, could also help. Future work will look towards priors and models for these more complicated processes.

The notion of utility is, we believe, a useful and intuitive way to quantify display strategy objectives. It allows great flexibility, as one is free to define any utility function at all, as long as computational issues can be successfully addressed.

There are many issues raised in this problem. We mention three of them in conclusion. First, the choice of an initial display set D_1 is an interesting problem and often one can do better than a random sample. One can base this set on a sketch from the user, or perhaps choose a representative set, for example, with high feature variance. This is called the "page zero" problem and is an active area of research; see Boujemaa et al. (2004). Second, evaluation is difficult, especially when it concerns retrieval of images having complex semantics, and where no pre-classification is available, like in the BAL data. Another issue is the choice of the display set size N_D. It must be a compromise between being large enough to give a good representation of the database, but not too big so that a user is overwhelmed with data. Studies have shown that $N_D > 20$ is not desirable. The sequence of images retrieved is clearly dependent on N_D, more so with the alternatives to the indicator utility where at least as the display set for smaller N_D is a subset of that for larger N_D; this is not necessarily the case for entropy and variance utilities.

Acknowledgements

The images of paintings are courtesy of the Bridgeman Art Library, London. This work has been made possible by the research and training network MOUMIR (`http://www.moumir.org`), and by the network of excellence MUSCLE (`http://www.muscle-noe.org`), both funded by the European Union.

References

Boujemaa, N., Fauqueur, J. and Gouet, V. (2004). What's beyond query by example? *Trends and Advances in Content-Based Image and Video Retrieval*, L. Shapiro, H. P. Kriegel, and R. Veltkamp, eds., Springer-Verlag.

Cox, I.J., Miller, M.L., Minka, T.P., Papathomas, T.V. and Yianilos, P.N. (2000). The Bayesian image retrieval system, PicHunter: theory, implementation and psychophysical experiments. *IEEE Transactions on Image Processing*, **9**, 20–37.

Jain, A.K. (1989). *Fundamentals of Digital Image Processing*. Prentice-Hall, New Jersey.

Kirkpatrick, S., Gelatt, C.D. and Vecchi, M.P. (1983). Optimization by simulated annealing. *Science*, **220**, 671–680.

Lewis, R. (1990). *Practical Digital Image Processing*. Prentice-Hall, New Jersey.

Lundy, M. and Mees, A. (1986). Convergence of an annealing algorithm. *Mathematical Programming*, **34**, 111–124.

Smeulders, A.W.M., Worring, M., Santini, S., Gupta, A. and Jain, R. (2000). Content-based image retrieval at the end of the early years. *IEEE Transactions on Pattern Analysis and Machine Intelligence*, **22**, 1349-1380.

Bayesian Statistics and Its Applications
Edited by S.K. Upadhyay, U. Singh and D.K. Dey
Anamaya Publishers, New Delhi, India

Theoretical and Applied Bayesian Information Processing

Arnold Zellner

University of Chicago, 5807 S. Woodlawn Ave., Chicago, IL 60637

Abstract

For many years, traditional Bayesian (TB) and information theoretic (IT) procedures for learning from data were viewed as distinctly different approaches. Derivations of the TB and IT learning models are reviewed and compared. Then the 1988 synthesis of the TB and IT learning models and generalizations of them are described along with descriptions of selected applications. Included are learning procedures that do not require use of likelihood functions and/or priors. Works by leading Bayesians and information theorists are cited and related to TB/IT issues.

1. Introduction

This article presents some observations on aspects of Bayesian information processing or learning. As is widely appreciated, learning from data and experience is a fundamental objective of all the sciences. Thus, it is important to have good learning models available for scientific use. It is a fact, that many non-Bayesians do not use a formal learning model in their work and learn informally. Bayesians, who generally use Bayes' Theorem as a fundamental learning model have shown that it works well in analyzing a broad range of problems. An information theoretic procedure for deriving Bayes' Theorem and generalizations of it will be reviewed below. In particular, it will be shown how to perform inverse inference without the use of a prior density and likelihood function, or with just a likelihood function and no prior, or with quality adjusted priors and likelihood functions. Various applications of these information processing approaches will be cited and procedures for formally comparing them with traditional Bayesian approaches will be described. Last, references to works in the literature in which these new approaches have been applied will be provided.

Section 2 reviews derivations of traditional Bayesian and new information processing rules. Section 3 describes the works in the literature using the new information processing rules. In some of these works, the new procedures have been compared to traditional Bayesian procedures using posterior odds, predictive tests, etc. A summary and some thoughts on possible future developments are presented in Section 4.

2. Derivation of Learning Models

2.1 The Traditional Bayesian Learning Model

In Bayesian texts, the Bayesian learning model, Bayes' theorem is usually derived by use of the product rule of probability. That is, the joint probability density function (pdf) for an observation vector y, and a parameter vector θ, is given by:

$$p(y, \theta) = f(y|\theta)\pi(\theta) = g(\theta|y)h(y). \tag{1a}$$

Then, we have

$$g(\theta|y) = \pi(\theta)f(y|\theta)/h(y), \tag{1b}$$

where g is the posterior density, π the prior density, f viewed as function of θ is the likelihood function and h is the marginal density of the observations. To obtain (1b), the product rule of probability (1a) has been employed. Note that Jeffreys (1998, pp. 24-25) in his discussion of the proof of the product rule states that, "The proof has assumed that the alternatives considered are equally probable ... . It has not been found possible to prove the theorem without using this condition ... . But it is necessary to further developments of the theory that we shall have some way of relating probabilities on different data, and Theorem 9 suggests the simplest general rule that they can follow if there is one at all. We therefore take the more general form as an axiom ..". In the proof of the product rule, to which Jeffreys refers, the elements of the sets A, B and of the intersection of A and B are assumed equally likely to be drawn in deriving $\Pr(AB) = \Pr(A)\,\Pr(B$ given $A)=\Pr(B)\Pr(A$ given $B)$. Jeffreys mentioned that he was unsuccessful in his attempts to prove Bayes' theorem without this "equally likely to be drawn" assumption.

It is clear from this discussion of the proof of the product rule of probability, as with all proofs, that assumptions are being made that may not be satisfied in all circumstances. Thus, some years ago, it occurred to me that it would be desirable to have a new way of producing learning models, such as Bayes' theorem. In this effort, I decided to proceed in a pragmatic way, pretty much the way an engineer might, to consider informational inputs and outputs and to derive a relation between them that would result in the output information to be as close as possible to the input information in order not to lose any information. Using the traditional inputs, a prior density, denoted by π above and a likelihood function, f and outputs g and h, the problem was how to obtain an optimal rule relating the outputs to the inputs. Using information theoretic measures of information in probability density functions, namely the negative entropy relative to uniform measure, that is, the expectation of the logarithm of a density function, $E \ln f(x) = \int f(x) \ln f(x)dx$, that I reinterpreted as the expected ln height of f, a very descriptive measure of the information in a density function, I formulated the following criterion functional:

$$\Delta(g) = \int g \ln g d\theta + \int g \ln h d\theta - \int g \ln f d\theta - \int g \ln \pi d\theta \tag{2}$$

$$= \int g \ln[g/(\pi f/h)]d\theta \geq 0.$$

The problem is to minimize the difference $\Delta(g)$ between the output and input information measures in (2) with respect to the output density g subject to it being a proper density. In 1988, I solved this problem using a calculus of variations approach. I also mentioned in my response to the discussants that independently Bruce Hill, Udi Makov and Robert McCulloch pointed out to me that the terms in the first line of (2) could be collected, as shown in the second line of (2), to provide a form of the non-negative Jeffreys-Kullback-Leibler measure of the distance between g and $\pi f/h$. (see, e.g., Kullback (1959, p.14ff) for a proof of the non-negativity of this distance measure).

The solution to the minimization problem using either the calculus of variations or the non-negative distance measure approach is

$$g* = \pi f/h \tag{3}$$

that is precisely Bayes' Theorem. Further, when the optimal solution in (3) is substituted in (2), it is the case that the input information is exactly equal to the output information and thus no information is lost in this information processing procedure. That is, it is 100% efficient.

Before proceeding to discuss variants of the above problem, it is useful to review various reactions to the result in (3). The eminent physicist Edwin T. Jaynes (1988) commented as follows in his discussion of my result, "...entropy has been a recognized part of probability theory since the work of Shannon 40 years ago, and the usefulness of entropy maximization as a tool in generating probability distributions is thoroughly established in numerous new applications including statistical mechanics, spectrum analysis, image reconstruction and biological macromolecular structure determination .... This makes it seem scandalous that the exact relation of entropy to the other principles of probability is still rather obscure and confused. But now we see that there is, after all, a close connection between entropy and Bayes' theorem. Having seen this start, other such connections may be found, leading to a more unified theory of inference in general. Thus in my view, Zellner's work is probably not the end of an old story but the beginning of a new one" (pp. 280-281).

As Jaynes points out, the usefulness of entropy maximization as a tool in generating probability distributions has been recognized by many. In economics and econometrics, Davis (1941) was an early pioneer in using maxent to produce income distributions, firm size distributions and a model of consumer behavior. In addition, see Lisman (1949) for comments on Davis's entropic theory of the household, Maasoumi (1990) and Zellner (1991) for reviews of research on entropy in economics and econometrics by many leading workers, and Golan (2002), Mittelhammer, Judge and Miller (2002) and Soofi (1996, 2000) for descriptions of new developments in information theory as it relates to economics, econometrics and statistics. Note also, that entropic procedures have been used by many to produce prior densities for parameters as well as density functions for observations or likelihood functions in many fields of science. Last, Barnard (1951) provides a discussion of Shannon's and Fisher's measures of information, information theory and statistics in a paper with many invited discussants and his reply. Bartlett commented, " ... he [Barnard] has done the Society a service by discussing the communication engineer's modern use of the word "information" and its implications for mathematicians and statisticians".

Further, the statistician Hill (1988) commented on the need to consider time coherence and referred to some of his and others' related work. See Zellner (2000) for results on dynamic information processing that show that it is optimal to update using Bayes' theorem, a procedure that is a solution to a dynamic programming problem, a dynamic version of the optimization problem described in equation (2). In addition, Hill commented as follows, "Zellner is to be congratulated for clearly formulating the conservation property implicit in Bayes' theorem, and holding it up for our careful scrutiny. ... If one does not wish to conserve this property, as is the case in all strictly non-Bayesian analyses of data, then it should be incumbent upon the statistician to state explicitly from whence the violation arises" (p. 281). He goes on to illustrate this point using the procedure of "size-biased sampling" as an example.

Bernardo (1988) in his interesting discussion of my paper pointed to his and others' earlier work that involved using utility theory to derive optimal or proper scoring rules. (see, e.g., Bernardo's (1979) article, "Expected Information as Expected Utility," and related work by I. J. Good and L.J. Savage that he cites). In this work, the utility of a density function, denoted by g, is employed, say $u(g) = a \ln g + b(\theta)$ and inference is viewed as an expected utility maximizing process that has yielded the Bayesian learning model as a solution. My response to Bernardo's thoughtful comments was as follows: "As regards Savage's and others' proofs that Bayes' theorem or the Bayesian IPR [Information Processing Rule] is an expected utility-maximizing solution, this is a fundamentally different result from that in my article, where no utility considerations enter and there is no assumption that the expected utility hypothesis is in some sense "valid" or "rational." My result deals with information-processing, not utility-maximizing behavior" (p. 284). However, it is interesting to note that the utility function, $u(g)$, shown above is a monotonic function of $\ln g$, the log height of the density function g that is also an argument of the standard Gibbs-Shannon entropy information measure, as shown explicitly above. More general utility functions and/or information measures are available, as is well known, and use of them in formulating and solving the above optimization problem will lead to alternatives to the traditional Bayesian learning model, Bayes' Theorem.

Of course this response to Bernardo's comments does not rule out the use of the concepts of the utility of information or the price per unit of information. Indeed, in some economic models of firms, consumers and investors, information has been viewed as an input and decision-makers have been modeled as Bayesian learners in a number of studies. For some early work, see e.g. Zellner and Chetty (1965), Grossman (1975), Bawa, Brown and Klein (1979), Boyer and Kihlstrom (1984) and Cyert and DeGroot (1987). For more recent work on Bayesian portfolio analysis, see Quintana, Chopra and Putnam (1995). Whether these economic information processing models are useful in modeling the production of scientific research output is of course a difficult issue. In this regard, note that there is still much controversy regarding the axiomatic foundations of utility theory, as noted intuitively by Jeffreys (1998, p. 30ff) who declined to base his axiom system for probability theory on utility considerations or "expectation of benefit" as Bayes, Ramsey and others had done but was not against use of his theory in analyzing problems involving utility considerations. See also Machina and Schmeidler (1992) for work on an axiom system for probability theory that involves "the separation of an individual's preferences from his beliefs." (p. 748). Also, their axiom system is a "choice-theoretic axiomatization of classical subjective probability which neither assumes nor implies the expected utility hypothesis." (p. 748). As I mentioned to Machina some years ago, this separation was just what Jeffreys adopted in his Theory of Probability many years ago. Thus it appears "reasonable" and useful to entertain the production of information as a "technical" process and to characterize and design optimal or good technical information production and processing procedures that do not involve utility considerations. Then, given these technical results, they can be employed, just as economists employ production functions, in dealing with analyses involving the utility of information, e.g., in decision-making contexts, and possibly in analyses of the price of information and how it is determined, say, in a market for information. There is indeed room for much more work to be done in these areas. However, below, I shall just concentrate on the "technical" aspects and not the utility aspects of information processing.

In the last sentence of my 1988 paper, I stated, "Further research to consider extended variants of the criterion functional used in this study as well as alternative measures of information would be valuable." In subsequent work, described below, I have pursued these and other topics to produce a broader range of information processing rules that can be implemented in practice; see also the innovative information processing rules described by a former student in my course, David Just (2001) that he used to explain paradoxical behavior in nine psychological learning experiments.

Before turning to these results, it is relevant to note that a fuller characterization of the information in a density function occurred to me several years ago that I reported in a lecture at the University of Wisconsin and discussed at length with Ehsan Soofi. Above the information in a density function relative to uniform measure was defined to be the negative entropy or $E \ln g$ that I interpreted as the expected log height of the density. It occurred to me that not only the first but also higher order moments of $\ln g$ might be considered, that is, $E(\ln g)^n = \int (\ln g)^n g d\theta, n = 1, 2, \ldots$ Then with these given moments, a maxent density for the log height of the density, $\ln g$ can be produced. This is very operational, as Soofi mentioned to me after he worked a few problems in the evening following my lecture. For example, in the simple case, if $E \ln g = a$ and Var $\ln g = b$, the maxent density for $\ln g$ is $N(a, b)$ and g is log normally distributed.

As another example of the use of higher order moments of $\ln g$, consider Bayes' Theorem given in (3). On logging both sides and taking the expectation with respect to g, we have: $E \ln g = E \ln \pi + E \ln f - \ln h$. Then, $\ln g - E \ln g = \ln \pi - E \ln \pi + \ln f - E \ln f$. On squaring both sides of this last expression and taking expectations of the terms with respect to g, the result is

$$\text{Var}(\ln g) = E(\ln g - E \ln g)^2$$

$$= E(\ln \pi - E \ln \pi)^2 + E(\ln f - E \ln f)^2 + 2E(\ln \pi - E \ln \pi)(\ln f - E \ln f). \tag{4}$$

Thus the variance of $\ln g$ is decomposed into posterior second moments of $\ln \pi$, the log height of the prior, and $\ln f$, the log height of the likelihood function. In addition, this analysis provides a measure of covariance or correlation between $\ln \pi$ and $\ln f$ that quantifies the extent to which there is dependence between information in a prior density and that in a likelihood function, a topic that has been discussed qualitatively in the literature for many years. Note that if the prior is uniform, the covariance between the information in the prior and that in the likelihood function is equal to zero.

Second, other measures of the information in a density may be employed as alternatives to $E \ln g$, the expected ln height. For example, Silver (1991, 1999) suggested using what he calls the Fisher information, namely, $E \left[\partial g / \partial x / g \right]^2 = \int g \left[\partial g / \partial x / g \right]^2 dx$, the expectation of the squared relative slope of the density g as a measure of information in both univariate and multivariate densities. He notes that minimizing this criterion functional subject to certain side conditions results in a solution in the form of a Schrödinger-like partial differential equation. He suggests that such solution densities may be more general and useful than those produced by minimizing the usual $E \ln g$ subject to given side conditions. In addition, some have considered the Rényi information measure that includes the Shannon $E \ln g$ measure as a special case; (see, e.g., the illuminating paper by Jizba and Arimitsu (2002) for comparative analyses of the Shannon and Rényi measures of information). How use of the Fisher and Rényi measures of information affects solutions to the information processing problem in (2) above, namely minimize the difference between the output and input information with respect to the choice of the form of g is a problem at the top of my "to do" list. In this connection, note that Jizba and Arimitsu (2002) remark, "Although Rényi's information measure offers a very natural ... setting for entropy, it has not found so far as much applicability as Shannon's (or Gibbs's) entropy"(p. 1). However, see Golan and Perloff (2002) for an interesting study that extensively uses Rényi's entropy measure.

Last, it is important to note that some researchers seem to be adverse to the use of information theory on grounds that information theoretic or maximum entropy procedures lack "order invariance." In Zellner (1998), included as Appendix A of this paper, it is shown that this argument is fallacious. Indeed, when the same side conditions, e.g., given zero'th, first and second moments, are employed throughout, then maximum entropy procedures are order invariant. That is the same results are obtained when, e.g., sample 1 is analyzed followed by sample 2 or sample 2 is analyzed before sample 1 or the two samples are analyzed simultaneously. With respect to the critical literature, it is pointed out that authors have changed the number and/or nature of the side conditions, e.g., in going from sample 1 to sample 2, etc. Under such conditions, maxent procedures should not be invariant, as pointed out dramatically by Jaynes in his remark that if they were invariant we wouldn't have the laws of physics. See Appendix A for a more detailed analysis of this invariance issue with references to the literature.

2.2   Some Alternative Optimal Information Processing Rules

Clearly there are many variants of the optimization problem described above. For example, there are situations in which one might wish to input just a likelihood function and not a prior density, as R.A. Fisher did in his fiducial approach. If in the above problem, we omit the prior density input and just input a likelihood function, the solution is to take the post data density for the parameters proportional to the likelihood function, that is, $g* \propto f$. This solution is equivalent to employing a uniform prior for the parameters but there is no need to introduce it in obtaining the above solution. Another problem, analyzed in Zellner (2000) involves adjusting the prior and likelihood functions for quality by raising each of them to fractional powers, i.e., π^{w_1} and f^{w_2}, with the w's having values in the closed interval zero to one. Raising densities to fractional powers usually spreads them out, as noted in the literature on "power priors" (see, e.g., Ibrahim, Chen and Sinha (2003), a paper dealing with "power priors" and the use of information processing analysis to rationalize them). Using the "quality adjusted" input likelihood function and prior density in the criterion functional in (2) and minimizing it with respect to the choice of g subject to it being a proper density yields the following solution:

$$g^{**} = c\pi^{w_1} f^{w_2} \tag{5}$$

where c is a normalizing constant, as shown in Zellner (2000). Thus with these quality adjustments introduced, the optimal solution is in a form different from (3), Bayes' Theorem. Also, the solution in (5) satisfies what Hill (1988) called the "conservation principle", namely information in = information out and thus the information processing rule in (5) is 100% efficient.

It is direct to apply the analysis associated with equation (4) to obtain an expression for the variance of the log of the density in (5), $\ln g^{**}$ and to compare it to that for the variance of $\ln g*$ given in (4). From this calculation, it is the case that raising densities to fractional powers does indeed lower their informational content, as measured by expected log height. In addition, it is the case that only one side condition, the condition that the solution density be proper, was used to produce the optimal information processing rules in (3) and (5). Other possible side conditions have been mentioned in the literature namely moment side conditions, e.g., a given mean for the parameter vector, inequality constraints on parameters' values, differential equation side conditions restricting the solution density to belong to a certain family of densities, e.g., the Pearson system of densities, etc. Thus a rich range of side conditions reflecting given input information can be introduced and will modify forms of derived optimal information processing rules. Alternative rules, as well as combinations of rules can be evaluated using data as has been done in past work by van der Merwe et al (2001), Zellner and Tobias (2001) and other papers listed in the annotated bibliography in Appendix A2. It is clearly desirable to use information in data, as well as analytical tools, to evaluate the performance of alternative learning models.

3. Learning Without Likelihood Functions and Prior Densities

It has long been appreciated that in some circumstances dependable likelihood functions and prior densities may not be available. Without these two inputs, the solution to the optimization problem in (2) is to have g be a uniform density, a not very informative result. Is there anything that can be done to introduce additional information that may be available to working scientists? Some years ago, it occurred to me that "stories" have been made up about the sampling properties of error terms in relations, e.g., the errors are iid $N(0, \sigma^2)$ and raised the question, "Why not make up "stories" about the realized error terms?" When measurements are made in science, as with Millikan and his oil drop experiments, each observation is obtained with a lot of background information regarding its quality, error, etc. (see, e.g., Press (2003) for an intriguing description of Millikan's evaluations of individual data points). Thus, I thought it would be good to make assumptions about the realized error terms' properties which, given a mathematical model for the observations, would imply information about properties of the subjectively random parameters of the model. Note that this is a reversal of the process that was employed in Zellner (1975) and Chaloner and Brant (1988) in which traditional Bayesian posterior densities for parameters were employed to calculate posterior densities for realized error terms and functions of realized error terms. For example, in terms of a standard regression equation, the observed, given observation $y_i = x_i'\beta + u_i$, where x_i is a given input vector of independent variables, β is a vector of regression coefficients with a given posterior density, say multivariate Student t, and u_i is a realized error term that is regarded as being subjectively random. While lecturing many years ago, it dawned on me that the realized error term u_i is a linear function of the elements of β and thus has a univariate Student t density. It did not take long to work out the details and to derive or compute the posterior densities of various functions of the realized error terms, as reported in my (1975) paper and work by Chaloner and Brant (1988), Hong (1989), Min (1992), Albert and Chib (1993) and others, that are very valuable in the diagnostic checking of models' assumptions. Note that all of this work went forward in a traditional Bayesian framework with a given likelihood function and a given prior for the parameters.

A question arose in the early 1990s, namely can I reverse the process described in the previous paragraph by making prior assumptions about the *realized* error terms and/or functions of them that would imply post data moments and other properties of the parameters given the observed data? As usual when I have a new idea, I try it on relatively simple examples. One of them involves given time to failure data $y_i, i = 1, 2, \ldots, n,\ 0 < y_i < \infty$, that are assumed to satisfy the following relation:

$$y_i = \theta + u_i, i = 1, 2, \ldots, n, \tag{6}$$

where θ is a parameter with an unknown value. On summing both sides of (6) and dividing by n, the result is that the observed sample mean time to failure, $\overline{y}$ is given by

$$\overline{y} = \theta + \overline{u}, \tag{7}$$

where $\overline{u} = \sum_{i=1}^{n} u_i/n$. Now to make inference about the parameter θ that we view as subjectively random we have to make some assumptions in view of the old adage, nothing in, nothing out. Let us assume that there are no variables left out of equation (6), that its algebraic form is appropriate, that there are no outliers, and no systematic biases in the measurements. With all of these assumptions, that are usually made, but with no sampling assumptions regarding the errors in (6), we may further be willing to assume that the mean of the realized error terms satisfies, $E\,\overline{u}\,|D = 0$, where E is the subjective mean operator and D is the given data and background prior assumptions, stated above. Given that we have made this zero mean assumption, then from (7) we have

$$E\theta\,|D = \overline{y} - E\,\overline{u}\,|D = \overline{y}. \tag{8}$$

Thus, without a likelihood function, without a prior density and without the use of Bayes' Theorem, we have the result that the post data mean of θ is $\overline{y}$, that is, $E\theta\,|D = \overline{y}$. And it is well known that this mean is an optimal estimate of θ relative to a quadratic estimation loss function $L(\theta,\hat{\theta}) = (\theta - \hat{\theta})^2$.

Further, to obtain a post data density for $\theta, 0 < \theta < \infty$, we can easily find the form of the density for the parameter θ, $g(\theta\,|D)$, that minimizes the criterion functional,

$$\min \Delta(g) = \int g \ln g \, d\theta \tag{9}$$

subject to the side condition given in (8) and that g be a proper density, that is, we seek the least informative density for g in terms of expected ln height subject to the two side conditions mentioned above. The solution to this problem, obtained by standard calculus of variations procedures, is well known to be the exponential density

$$g(\theta|D) = (1/\overline{y}) \exp(-\theta/\overline{y}), 0 < \theta < \infty. \tag{10}$$

Using this density, it is possible to compute the probability that θ lies between any two values, say a and b by evaluating $\int_a^b g(\theta\,|D)d\theta = \Pr(a < \theta < b\,|D)$. Since this procedure solves the problem originally posed by Bayes many years ago, it was named the Bayesian Method of Moments (BMOM) in my first, 1994 paper on this topic. In a 1993 University of Chicago workshop talk on this work, I mentioned that the idea for it came to me about Mothers' Day in 1993 and that I called it the BMOM approach in honor of Bayesian MOMs and MOMs of Bayesians.

Further, note that if we consider a future, as yet unobserved value of the time to failure, $y_f = \theta + u_f$, and assume that there are no measurement biases, that the functional form is satisfactory, etc., we can make assumptions regarding the properties of u_f and deduce the implications for possible values of y_f given that we have a post data density for θ. For example, if we assume $Eu_f\,|D = 0$, we have $Ey_f\,|D = E\theta\,|D = \overline{y}$. Then the maxent density for y_f is an exponential density, namely, $f(y_f\,|D) = (1/\overline{y}) \exp[-y_f/\overline{y}], 0 < y_f < \infty$. In addition, note by use of a conceptual sample, $y_c = \theta\iota + u_c$, where $\iota' = (1,1,...,1)$, additional prior information can be introduced (see Zellner (1997b) for details).

In my first paper on the Bayesian method of moments (BMOM), the procedure was applied to location parameter and multiple regression models, using not only first order moment conditions on the error terms but also second order moment conditions. Post data moments and maxent densities for the parameters and future values of variables were derived and reported at the 1994 Valencia meeting in Alicante, Spain. Among those in the audience was Ed Green who mentioned to me that he and Bill Strawderman were working on an analysis of a forestry model with data provided by forestry scientists. Since the forestry scientists did not provide them with any information about error terms' sampling properties, they were having difficulty in formulating a likelihood function, needed for their traditional Bayesian analysis of the data. On hearing about the BMOM approach, Green remarked that it was just what he and Bill needed and would apply it on his return to the U.S. And indeed their 1996 paper, cited in Appendix B and the references, was the first published application of the BMOM approach, including some ingenious extensions of it.

Another early, innovative application of the BMOM approach was the 1999 study by La France, cited in Appendix B and the references, on inferring the nutrient content of food in which he compared the BMOM approach with other available approaches. He opted for the BMOM approach on the basis of some interesting considerations and showed that it produced useful empirical results.

In Zellner (1997b), I extended the BMOM approach to apply to dichotomous variable models and a number of other models. A reanalysis of the Laplace Rule of Succession problem yielded BMOM results that have the estimated probability of a particular outcome rising more rapidly than provided by the Laplace Rule of Succession given a sequence of outcomes all of one type, say successes in tests of a theory. Some, including Jeffreys, have argued that instead of having the estimated probability of success rise in accord with the Laplacian rule, $(n + 1)/(n + 2)$, where n is the number of successful outcomes in n trials, it should rise more rapidly. The BMOM solution provides a more rapidly increasing probability as successes pile up. Then in other work with J. Tobias and H. Ryu, the BMOM approach was successfully extended and applied to new problems in multiple regression, semi-parametric regression and time series estimation, prediction and other problems. The invited discussant of our BMOM time series, forecasting paper, E. de Alba wrote very favorably about our work and proposed extensions of it. Also, Soofi (2000) in his JASA review paper, "Principal Information Theoretic Approaches" commented favorably on the BMOM approach.

On delightful visits to South Africa in 1996 and 1998, it was a pleasure to present talks on the BMOM at various universities, the 1996 ISBA meeting in Cape Town and the 1998 annual South African Statistics Association meeting. In particular, during my visits to the Department of Mathematical Statistics of the University of the Free State in Bloemfontein I had the good fortune to discuss my work with Abrie van der Merwe and his colleagues. Merwe and C. Vilijoen were the first to analyze the multivariate "seemingly unrelated regression" model using the BMOM approach in a paper presented to the meeting of the South African Statistical Association in November, 1998. Also, Merwe, A. Pretorius, J. Hugo and I in a 2001 paper, published in the *Journal of the South African Statistical Association* analyzed the mixed regression model using the BMOM approach and compared the results to those provided by maximum likelihood and traditional Bayesian methods.

The BMOM approach was applied to general "reduced form," multivariate regression and structural econometric models in my 1998 paper presented at a conference in honor of Carl F. Christ and published in the *Journal of Econometrics Annals Issue* in his honor, edited by the Nobel Prize winner, L.R. Klein. It yielded new, exact finite sample minimum expected loss (MELO) estimates for structural parameters that are very operational as well as exact, finite sample post data densities for parameters and future observations. My discussant at the Christ Conference was Adrian Pagan, the eminent Australian econometrician and economist. After studying my paper and analyzing its results, he stated that he saw some good in the Bayesian approach. I remarked in return for this kind remark that we would no longer regard him as A Pagan.

Then too, it was a pleasure to receive a deep, insightful letter from George A. Barnard (1997) in which he commented about the BMOM approach as follows:

"And above all any method is welcome which, unlike nonparametrics, remains fully quantifiable without paying obeisance to a model which one knows to be false. And your proposal to compare BMOM results with a model-based one should achieve the best of both worlds.

The general point seems to me to be that we should express prior knowledge, as far as we can, in a prior. Then our model—likelihood-producing or moment-producing, or whatever—should help us process the observed data. Then we should go back to compare what we thought we knew before with the result of our data-processing. In arriving at our (for the time being) conclusion the weight that we can attach to the three components of our inference will vary from case to case. BMOM will be specially useful when the latter two stages of the three should predominate."

As the above, brief comments indicate, in general the new BMOM approach was given a warm reception by many. However, with anything new, as is to be expected there were critics. In particular, a referee's report on my first BMOM paper was very critical. It took a few minutes in the evening to discover that the elaborate critical analysis of the referee was based on an assumption that I did not make. As I reported to Jack Lee, the co-editor of the volume in which the paper appeared, if one removes the referee's unwarranted assumption, his critique falls apart. Jack saw the point immediately and accepted my paper for publication. This practice of referees introducing unwarranted assumptions to reach negative conclusions happened not just once but three times. In one case, the editor of a journal accepted and published a paper critical of the BMOM without even sending it to me for review. When I discovered that the critical article had been published, I sent the editor my previously written working paper indicating that the critics had made an assumption that I did not make. And when this erroneous assumption was removed from their paper, there was nothing left to their critique. See the citation in Appendix B to the exchange between Geisser and Seidenfeld and myself published in the *Journal of Applied Statistics*. It seems to me that the behavior of editors in such delicate matters should be more constructive and thorough in seeking the responses of those being criticized before rushing critical papers into print.

After the BMOM approach appeared, many wondered about its exact relation to the traditional Bayesian approach based on Bayes' theorem involving use of a prior density and a likelihood function. Barnard's remarks, presented above, do much to help clarify the situation. Further, it is clear from what has been presented, that both approaches can be derived as solutions to well-defined information processing problems. In one problem there are two informational inputs, the information in a prior and in a likelihood function. In the BMOM problem where it is assumed that the likelihood function and prior density are not available, the input is the information in moment side conditions. Also, if a prior density for the parameters is available as an input, along with moment side conditions for the parameters, the solution to the information processing problem is to take the output density for the parameters proportional to the prior times the maxent density for the parameters given the data. This solution appears in the form of Bayes' theorem with the maxent density for the parameters given the data replacing the likelihood function. Thus, the information processing problem can be formulated and solved for the variety of situations that Barnard described in the excerpt from his letter, presented above, in which we may find ourselves in analyzing data. Formulating, solving and testing solutions for a wide range of information processing problems appear to me to be a good way to make progress.

4.   Summary and Conclusions

In this paper some historical issues surrounding the process of information processing, Bayes' Theorem, the Bayesian method of moments and related topics have been considered. From what has been accomplished in the last few decades, it seems clear that the synthesis of traditional Bayesian and information processing procedures is a productive one that has already led to fruitful, new approaches for learning from data, formulation of explanatory and predictive models and enlarging the capabilities of data analysts. With respect to this last point, it is now possible to perform inverse inference and to derive predictive densities when the likelihood function's and/or the prior density's forms are unknown. Also, with predictive densities available, it is possible to use them in predictive testing of alternative models and/or in combining them and their predictions. That such Bayesian information processing procedures have been applied in published studies by a number of researchers worldwide indicates a need for them and the profession's appreciation of their value. Last, past axiom systems relating to optimization problems involving utility considerations and learning will probably have to be generalized to allow for the fact that various optimal learning models are now available.

Acknowledgements

Research financed in part by the National Science Foundation and by income from the H.G.B. Alexander Endowment Fund, Graduate School of Business, University of Chicago. Thanks to Dr. Satyanshu Upadhyay for his constructive suggestions.

References

Albert, J.H and Chib, S. (1993). Bayesian Analysis of Binary and Polychotomous Response Data. *J. of the Am. Stat. Assoc.*, **88**, 669-679.

Barnard, G.A. (1951). The Theory of Information. *J. of the Royal Statistical Society*, B, 46-59 with invited discussion by M.S. Bartlett, P.A. Moran, N. Wiener, D. Gabor, I.J. Good, F. J. Anscombe, R.L. Plackett and C.A.B. Smith and the author's response, 59-64.

—— (1997), Personal Communication.

Bawa, V.S., Brown, S.J. and Klein, R.W. (1979). *Estimation Risk and Optimal Portfolio Analysis*. Amsterdam: North-Holland Publishing Company.

Bernardo, J.M. (1988). Comment. *The American Statistician*, **42**, 4, 158, reprinted in Zellner (1997a).

—— (1979). Expected Information as Expected Utility. *The Annals of Statistics*, **7**, 686-690.

Berry, D.A., Chaloner, K. and Geweke, J. (1996). *Bayesian Analysis in Statistics and Econometrics: Essays in Honor of Arnold Zellner*, New York: John Wiley & Sons, Inc.

Boyer, M. and Kihlstrom, R.E. (1984). Bayesian Models in Economic Theory. Amsterdam: North-Holland Publishing Company.

Chaloner, K. and Brant, R. (1988). A Bayesian Approach to Outlier Detection and Residual Analysis. *Biometrika*, **75**, 651-659.

Cyert, R.M. and DeGroot, M.H. (1987). *Bayesian Analysis and Uncertainty in Economic Analysis*. London: Chapman and Hall Ltd.

Davis, H.T. (1941). *The Theory of Econometrics*. Bloomington, Indiana: Principia Press.

Golan, A., ed. (2002). Information Theory and Entropy Econometrics. Annals Issue of the *J. of Econometrics*, **107**, 1-2, 374 pp.

Golan, A. and Perloff, J.M., (2002). Comparison of Maximum Entropy and Higher-Order Entropy Estimators. In Golan (2002), 195-211.

Green, E. and Strawderman, W. (1996). A Bayesian Growth and Yield Model for Slash Pine Plantation. *J. of Applied Statistics*, **23**, 285-299.

Grossman, S. (1975). Essays on Rational Expectations, the Informational Role of Futures Markets and Equilibrium Bayesian Experimentation. Ph.D. Thesis, Dept. of Economics, University of Chicago.

Hill, B.M. (1988). Comment. *The American Statistician*, **42**, 4, 281-282, reprinted in Zellner (1997a).

Hong, C. (1989). Forecasting Real Output Growth Rates and Cyclical Properties of Models: A Bayesian Approach. Ph.D. Thesis, Dept. of Economics, University of Chicago.

Ibrahim, J.G., Chen, M.H. and Sinha, D. (2003). On Optimality of the Power Prior. *J.of the American Statistical Association*, **98**, 461, 204-213.

Jaynes, E. T. (1988). Comment. *The American Statistician*, **42**, 4, 280-281, reprinted in Zellner (1997a).

Jeffreys, H. (1998). *Theory of Probability*, 3rd revised 1967 edition, reprinted in Oxford Classics Series, Oxford: Oxford U. Press.

Jizba, P. and Arimitsu, T. (2002). The World According to Rényi: Thermodynamics of Multifractal Systems. ms., Institute of Physics, U. of Tsukuba, Japan, 24 pp.

Just, D.R. (2001). Information and Learning, Ph.D. Thesis, Dept. of Agricultural and Resource Economics, U. of California at Berkeley.

Kullback, S.(1959). *Information Theory and Statistics*. New York: John Wiley & Sons, Inc.

La France, J. (1999). Inferring the Nutrient Content of Food with Prior Information. *American J. of Agricultural Economics*, **81**, 728-734.

Lisman, J.H.C. (1949). Economics and Thermodynamics: A Remark on Davis' *Theory of Budgets. Econometrica*, **17**, 59-62.

Maasoumi, E. (1990). *Information Theory in the New Palgrave Econometrics*. Eatwell, J., Millgate, M. and Newman, P. (eds.), New York: W.W. Norton & Co., 101-112.

Machina, M. and Schmeidler, D. (1992). A More Robust Definition of Subjective Probability. *Econometrica*, **60**, 745-780.

Min, C.K. (1992). Economic Analysis and Forecasting of International Growth Rates Using Bayesian Techniques. Ph.D. Thesis, Dept. of Economics, University of Chicago.

Mittelhammer, R.C., Judge, G.G. and Miller, D.J. (2002). *Econometric Foundations*, Cambridge: Cambridge U. Press.

Press, S. J. (2003). *A Note on Modeling Subjectivity*. ms, Dept. of Statistics, U. of California at Riverside, 9 pp.

Quintana, J., Chopra, V. and Putnam, B. (1995). Global Asset Allocation: Stretching Returns by Shrinking Forecasts. Proceedings Volume. of the Section on Bayesian Statistical Science, American Statistical Association.

Silver, R.N. (1991). Quantum Statistical Inference. ms, Los Alamos National Laboratory, NM, 15pp., published in Grandy, J.W.T. and Milonni, P.W.(eds), Physics & Probability: Essays in Honor of Edwin T. Jaynes, Cambridge: Cambridge University Press.

—— (1999). *Quantum Entropy Regularization.* In von der Linden, W., Dose, V., Fischer, R. and Preuss, R.(eds.), Maximum Entropy and Bayesian Methods, Dordrecht/Boston/London: Kluwer Academic Publishers, 91-98.

Soofi, E.S.(1996). *Information Theory and Bayesian Statistics.* In Berry, D.A., Chaloner, K.M. and Geweke, J.K.(eds.), Bayesian Analysis in Statistics and Econometrics: Essays in Honor of Arnold Zellner, New York: John Wiley & Sons, Inc., 179-189.

—— (2000). Principal Information Theoretic Approaches. *J. of the American Statistical Association*, **95**, 1349-1353..

van der Merwe, A.J. and Vilijoen, C. (1998). Bayesian Analysis of the Seemingly Unrelated Regression Model. ms, Dept. of Mathematical Statistics, U. of the Free State, Bloemfontein, S.A., presented to the annual meeting of the S.A. Statistical Association, November, 1998.

van der Merwe, A.J., Pretorius, A.L., Hugo, J. and Zellner, A. (2001). Traditional Bayes and the Bayesian Method of Moment Analysis for the Mixed Linear Model with an Application to Animal Breeding. *South African Statistical Journal*, **35**, 19-68.

Zellner, A. (1975). Bayesian Analysis of Regression Error Terms. *J. of the American Statistical Association*, **70**, 138-144.

—— (1988). Optimal Information Processing and Bayes's Theorem. *The American Statistician*, **42**, 4, 278-294, with invited discussion and the author's reply.

—— (1991). Bayesian Methods and Entropy in Economics and Econometrics. in Grandy, W.T. and Schick, L.H. (eds.), *Maximum Entropy and Bayesian Methods*, Dordrecht/Boston/London: Kluwer Academic Publishers, 17-31.

—— (1997a). *Bayesian Analysis in Econometrics and Statistics: The Zellner View and Papers*. Cheltenham, UK & Lyme, US: Edward Elgar Publishing Ltd.

—— (1997b). The Bayesian Method of Moments (BMOM): Theory and Applications. *Advances in Econometrics*, **12**, 85-105.

—— (1998). The Finite Sample Properties of Simultaneous Equations' Estimates and Estimators: Bayesian and Non-Bayesian Approaches. invited paper presented at research conference in honor of Prof. Carl F. Christ and published in Klein, L.R. (ed.) Annals Issue of the *J. of Econometrics*, **83**, 185-212.

—— (2000). Information Processing and Bayesian Analysis, presented to the American Statistical Association Meeting, Aug., 2001 and published in A. Golan (ed.), Annals Issue of the *J. of Econometrics*, **107** (2002), 41-50.

Zellner, A. and Chetty, V.K. (1965). Prediction and Decision Problems in Regression Models from the Bayesian Point of View. *J. of the American Statistical Association*, **60**, 608-616.

—— and Tobias, J. (2001). Further Results on Bayesian Method of Moments Analysis of the Multiple Regression Model. *International Economic Review*, **42**, 1, 121-140.

——, Tobias, J.L. and Ryu, H. (1997). Bayesian Method of Moments Analysis of Time Series Models with an Application to Forecasting Turning Points in Output Growth Rates. Estadistica, *J. of the Inter-American Statistical Association*, 49-51, 152-157 (1997-1999), 3-63 with invited discussion by Prof. Enrique de Alba.

——, Tobias, J. and Ryu, H. (1999). Bayesian Method of Moments (BMOM) Analysis of Parametric and Semi-Parametric Regression Models. Summary in 1997 Proceedings Volume of the Section on Bayesian Statistical Science, *Am. Stat. Assoc.*, 211-216, and published in *South African Statistical Journal*, **31**, 41-69.

Appendix A1: On Order Invariance of Maximum Entropy Procedures[1]

Herein we demonstrate the order invariance of maximum entropy procedures, an extension of remarks made at seminars at the University of Toronto and the Hebrew University of Jerusalem in 1997 and of Zellner (1997). After demonstrating the order invariance of maximum entropy procedures, an analysis of some examples that have appeared in the literature will be provided in which the hypothesis space has not been preserved. For example, new moment or other restrictions have been added to an original set and a lack of invariance has been shown in such circumstances. Several years ago such a demonstration led E.T. Jaynes to remark at a seminar presentation, "If maxent were invariant in such circumstances, we would not have the laws of physics." For example, the various gas laws are generated by changing the side conditions associated with maxent solutions. If such laws were invariant in the sense of implying logically equivalent results, as Jaynes remarked we would not have the various gas laws which have logically and empirically different implications.

To illustrate the order invariance of maximum entropy densities derived subject to moment side conditions, consider sample observations $D(1) = y(1)$, where $y(1)$ is an $n \times 1$ vector of observations with sample mean $m(1)$ and sample variance $v(1)$. Then the proper maxent density with these given two moments is well known to be a normal density, $N[m(1), v(1)]$. Now if we observe an independent sample of observations $D(2) = y(2)$, we can combine it with the sample $D(1)$ to obtain the combined sample, $D(1, 2)$, with mean $m(1, 2)$ variance $v(1, 2)$ and proper normal maxent density, $N[m(1, 2), v(1, 2)]$. If alternatively, we observed $D(2)$ first and then $D(1)$, the mean and variance for the combined sample, $D(2, 1)$ will be identical to the mean and variance for the combined sample $D(1, 2)$ and thus the associated normal maxent density, $N[m(2, 1), v(2, 1)] = N[m(1, 2), v(1, 2)]$ since $m(2, 1) = m(1, 2)$ and $v(2, 1) = v(1, 2)$ and clearly we have order invariance when the same moments are updated for two or more independent sets of data. While we have demonstrated order invariance using two moments above, the result applies to any given number of moment side conditions. Note that we do not permit the number of moment side conditions to change from sample to sample. Usually a given subject matter theory or law indicates which moments to employ as side conditions for all data sets.

As a second example, consider a Bayesian method of moments (BMOM) analysis, Zellner (1997 a,

---

[1] The material in this Appendix is in a 1998 H.G.B. Alexander Research Foundation Working paper by A. Zellner.

b) of data relating to a mean time to failure parameter μ and a sample of failure times $y_i = \mu + u_i$, $i = 1, 2, ..., n$. If we assume that the y_i's are given and $E\,\overline{u}\,|y = 0$, where $\overline{u} = \sum_i^n u_i/n$ and E is the subjective post-data expectation operator, we have $E\mu|y = m(y) = \sum_1^n y_i/n$.

Then the proper maxent post data density for μ given that its mean is $m(y)$ is the following exponential density, $f\,[\mu|m(y)] = [1/m(y)]\exp\{-\mu/m(y)\}$. Now if another sample of data becomes available, say $w_i = \mu + v_i$, $i = 1, 2, ..., n$ and we compute the mean of the y's and w's, denoted by $m(y, w)$ which is equal to $E\mu|y, w$, the proper maxent post data density for μ is $f\,(\mu\,|y, w) = [1/m(y, w)]\exp\{-\mu/m(y, w)\}$. Further, if w is observed first and then y is observed, $m(w, y) = m(y, w)$, that is sample means are the same for both orderings of the data, and thus the proper maxent exponential densities subject to either of these means will be the same. That is, the procedure is order invariant. It should be emphasized that in this example the form of the likelihood function is assumed to be unknown and thus the traditional Bayesian analysis and updating procedures, using Bayes' Theorem, can not be utilized.

While many more BMOM and other examples can be provided, the above suffice to indicate that as long as the moment side conditions are not changed, the associated proper maxent densities will be invariant to the order in which the data sets are analyzed. In reviewing the literature on this issue, it appears that authors have changed the side conditions, or as Jaynes might say, changed the laws, and are surprised that results are not invariant. To illustrate, Kass and Wasserman (1996, p. 1349ff), citing Seidenfeld (1987), present the following example, "Consider a six-sided die and suppose that we have information that $E(X) = 3.5$, where X is the number of dots on the uppermost face of the die. Following Seidenfeld, it is convenient to list the constraint set $C_0 = \{E(X) = 3.5\}$. The probability that maximizes the entropy subject to this constraint is P_0 with values (1/6, 1/6, 1/6, 1/6, 1/6, 1/6). Let A be the event that the die comes up odd, and suppose we learn that A has occurred. There are two ways to include this information. We can condition P_0 to obtain $P_0((X\,|A)$, which has values (1/3, 0, 1/3, 0, 1/3, 0), or we can regard the occurrence of A as another constraint; i.e., $E(I_A) = 1$, where I_A is the indicator function of the event A. The probability Q maximizes the entropy subject to the constraint set $C_1 = \{E(X) = 3.5, E(I_A) = 1\}$ has values (0.22, 0, .32, 0.47, 0), which conflicts with $P_0(\cdot\,|A)$."

In this example, the conditioning events have been changed, namely, first $E(X) = 3.5$, second $E(X) = 3.5$ given $E(I_A) = 1$ and third $E(X) = 3.5$ and $E(I_A) = 1$. That the maxent probability mass functions for these three cases differ is to be expected since the side conditions have been changed. By changing the side conditions, different maxent probability mass functions are obtained just as in deriving various gas laws by maxent in physics. Note also, the unusual reaction to observing "the event that the die comes up odd ... " To have one observation lead to the side condition or constraint, $E(I_A) = 1$ is mind-boggling. To go from one odd observation to stating that they are all odd is hardly scientific. If one assumes that only odd sides can appear upward, that indeed is quite a different "law" from that obtained just under the constraint that $E(X) = 3.5$. In line with Jaynes' remark above, we should not expect our results to be invariant under such disparate "laws."

The second example that Kass and Wasserman (1996, p. 1350) provide also involves a change in the constraints. "Consider again the die example. After rolling a die, we typically can see two or three visible surfaces, "that is, in addition to the uppermost side of the die, we can see one or two side faces depending on the orientation of the die. Thus we can record not just the value of the upper face, but also whether the sum of all visible spots on the side faces of the die is less than, equal to, or greater than the value showing. There are 14 such possible outcomes. For example, outcome (3, equal) means the top face shows 3 and the sum of visible side faces equals 3. The original sample space can now be viewed as a partition of this larger sample space. Maximum entropy leads to a probability Q that assigns probability $1/14$ to each outcome. The marginal of Q for the six original outcomes is not P_0. The problem is, then, which probability we should use, Q or P_0?"

Again it is clear that the hypothesis space has been changed, as explicitly recognized in the quotation. Under conditions in which we just observe the upward face of the die, we have P_0; under conditions in which we observe more than just the upward side of the die, we have Q. That P_0 and Q are different is similar to saying that Newton's and Einstein's laws are different which is as it should be since they are based on differing assumptions. Over a range of velocities, they provide similar predictions while outside this range, they provide different predictions. However, if one stays within the Newtonian realm, everything is logically consistent and similarly within the Einsteinian realm.

Maxent is a deductive procedure for producing alternative models or hypotheses that may or may not be consistent with observed data in descriptive and predictive senses. As shown by analysis of examples in Tobias and Zellner (1997), it is possible to use data and model selection techniques to choose among predictive models produced by BMOM maxent procedures and by traditional Bayesian assumptions and methods. While both approaches are logically sound, since they make different assumptions they are obviously not logically equivalent. The empirical validity of models produced by alternative approaches is of course of utmost importance.

References

Kass, R.E. and L. Wasserman (1996). The Selection of Priors by Formal Rules. *J. American Statistical Assoc.*, **91**, 435, 1343-1370.

Seidenfeld, T. (1987). Entropy and Uncertainty. In *Foundations of Statistical Inference*, eds. I.B. MacNeil and G.J. Umphrey, Boston: Reidel, 259-287.

Tobias, J. and A. Zellner (1997). Further Results on the Bayesian Method of Moments Analysis of the Multiple Regression Model. H.G.B. Alexander Research Foundation, Graduate School of Business, U. of Chicago. Presented at Econometric Society meeting, June, 1997 and at world meeting of the Int. Soc. For Bayesian Analysis, August, 1997.

Zellner, A. (1997). Remarks on a 'Critique' of the Bayesian Method of Moments (BMOM). H.G.B. Alexander Research Foundation, Graduate School of Business, U. of Chicago

Zellner, A. (1997a). The Bayesian Method of Moments (BMOM): Theory and Applications. in T.B. Fomby and R.C. Carter, eds., *Advances in Econometrics*, **12**, 85-105.

Zellner, A. (1997b). *Bayesian Analysis in Econometrics and Statistics: The Zellner View and Papers*, Edward Elgar Publ. Ltd., UK and US, 291-304 and 308-318.

Appendix A2: Selected References on New Information Processing and Bayesian Method of Moments (BMOM) Methods

1. General Information Processing Results: Producing Models, Priors and Information Processing Rules (including Bayes' Theorem)

See A. Zellner, Bayesian Analysis in Econometrics and Statistics: The Zellner View and Papers, Elgar, 1997, Part III, "Bayesian Priors, Models and Information Processing," pp.97-175, on reserve in Giannini Hall Library and referred to below as AZ (1997):

Here the problem of model formulation is discussed with many examples. In particular, it is shown how information theory can be employed to derive univariate and multivariate regression and many other commonly employed models and prior densities for their parameters. Also, in the 1988 American Statistician article, "Optimal Information Processing and Bayes' Theorem," with discussion by E.T Jaynes, B.M. Hill, S. Kullback and J. Bernardo and the author's response, pp. 154-160 in AZ (1997), it is shown how to derive Bayes' Theorem as a solution to an information theory optimization problem. In a later article, A. Zellner, "Information Processing and Bayesian Analysis," (2000) presented to the Am. Stat. Assoc. in 2001 and published in J. of Econometrics, Vol. 107 (2002), 41-50, Bayes' Theorem and other learning models, including the Bayesian Method of Moments (BMOM) model are derived as solutions to optimization problems. See also the 2001 doctoral dissertation "Information and Learning," by D.R. Just, Dept. of Agricultural and Resource Economics, U. of California, Berkeley for additional information processing rules and their use in explaining anomalous behavior in psychological learning experiments. It should be appreciated that the BMOM model permits investigators to obtain posterior and predictive densities when likelihood functions and prior densities are not available.

2. References for the Theory and Applications of BMOM

1. Zellner, A. (1994). Bayesian method of moments (BMOM) analysis of mean and regression models. In J.C. Lee, W.O. Johnson and A. Zellner (eds.), Prediction and Modeling Honoring Seymour Geisser, New York: Springer-Verlag, 61-74, reprinted in AZ (1997), 291-304.

2. Green, E. and W. Strawderman (1996). A Bayesian Growth and Yield Model for Slash Pine Plantations. *J. of Applied Statistics*, **23**, 285-299. [The authors did not have enough information to specify a likelihood function and thus used the BMOM in the first serious application of the method.]

3. Zellner, A. (1997). The Bayesian Method of Moments (BMOM): Theory and Applications. *Advances in Econometrics*, **12**, 85-105. [The BMOM approach is applied to a wide range of models.]

4. Zellner, A., Tobias, J.L. and Ryu, H. (1997, 1999). Bayesian Method of Moments (BMOM) Analysis of Parametric and Semi-Parametric Regression Models. In 1997 Proceedings of the Section on Bayesian Statistical Science, *Am. Stat. Assoc.*, 211-216 and in *South African Statistical Journal*, **31**, 1999, 41-69.

5. Zellner, A. (1998). The finite sample properties of simultaneous equations' estimates and estimators: Bayesian and non-Bayesian approaches. *J. of Econometrics*, **83**, 185-212. [The BMOM approach is applied to multivariate regression, unrestricted reduced form and structural estimation problems and results are compared to those yielded by traditional Bayesian and non-Bayesian estimation approaches, e.g. ML, 2SLS, etc.]

6. Zellner, A. (1998). On Order Invariance of Maximum Entropy Procedures. ms., 5 pp., H.G.B. Alexander Research Foundation, Grad. School of Business, U. of Chicago. [It is shown that maximum entropy procedures are order invariant. Arguments to the contrary in the literature are shown to be defective.]

7. La France, J. (1999). Inferring the nutrient content of food with prior information. American *J. of Agricultural Economics*, **81**, 728-734. [An impressive analysis of an important problem using the BMOM approach and comparing it to other possible approaches.]

8. Zellner, A., Tobias, J.L. and Ryu, H. (1997). Bayesian Method of Moments Analysis of Time Series Models with an Application to Forecasting Turning Points in Output Growth Rates. published in Estadistica, *J. of the Inter-American Statistical Institute* with discussion by Prof. Enrique de Alba, **49-51**, 152-157, 1997-1999, 3-63.

9. van der Merwe, A. and Vilijoen, C. (1998). Bayesian Analysis of the Seemingly Unrelated Regression Model. ms., Dept. of Mathematical Statistics, U. of the Free State, Bloemfontein, S.A., presented to the annual meeting of the S.A. Statistical Association, November, 1998.

10. Geisser, S. and T. Seidenfeld (1999). Remarks on the 'Bayesian' method of moments. *J. of Applied Statistics*, **26**, 97-101 and Zellner, A. (2001). Remarks on a 'critique' of the Bayesian Method of Moments. *J. of Applied Statistics*, **28**, 6, 775-778, published version of my 1997 working paper. [It is pointed out that Geisser and Seidenfeld introduced an erroneous assumption that led to their negative conclusion.]

11. Soofi, E.S. (2000). Principal information theoretic approaches. *J. of the American Statistical Association*, **95**, 1349-1353.[Comments on information processing derivations of learning models and the BMOM.]

12. Mittelhammer, R.C., Judge, G.G. and Miller, D.J. (2000). *Econometric Foundations*, Cambridge: Cambridge U. Press, 688-693. [A brief introduction to the BMOM analysis of the multiple regression model.]

13. Zellner, A. and Tobias, J.L. (2001). Further Results on Bayesian Method of Moments Analysis of the Multiple Regression Model. *International Economic Review*, **42**, 1, 121-140.

14. van der Merwe, A.J., Pretorius, A.L., Hugo, J. and Zellner, A. (2001). Traditional Bayes and the Bayesian Method of Moment Analysis for the Mixed Linear Model with an Application to Animal Breeding. *South African Statistical Journal*, **35**, 19-68.

15. Zellner, A. and B. Chen (2001). Bayesian Modeling of Economies and Data Requirements. *Macroeconomic Dynamics*, **5**, 673-700. [BMOM estimation and forecasting techniques are employed, along with others, to forecast annual output growth rates for 11 sectors of the U.S. economy. Sector forecasts are aggregated to produce forecasts of aggregate U.S. GDP growth rates and such forecasts are compared with those derived from aggregate data and models. See also, Zellner, A. and Tobias, J.L. (2000). A Note on Disaggregation and Forecasting Performance. *J. of Forecasting*, **19**, 457-469, and Zellner, A. (2003). "Bayesian Shrinkage Estimates and Forecasts of Individual and Total or Aggregate Outcomes," ms, 25 pp., H.G.B. Alexander Research Foundation, Grad. School of Business, U. of Chicago, for additional results on the effects of disaggregation on forecasting accuracy.]

16. Ibrahim, J.G., Chen, M.H. and Sinha, D. (2003). On Optimality of the Power Prior. *J. of the American Statistical Assoc.* **98**, 461 (March), 204-213. [Discusses the properties of power priors and their relation to earlier work on information processing with "quality corrected" prior densities and likelihood functions.]